Welding Technology Fundamentals

Fourth Edition

by

William A. Bowditch
Career and Technical Education Consultant, Portage, Michigan
Life Member of the American Welding Society
Member of the Association for Career and Technical Education

and

Kevin E. Bowditch
Welding Engineer Specialist, Subaru of Indiana Automotive, Inc., Lafayette, Indiana
Member of the American Welding Society
Member of the Association for Career and Technical Education

and

Mark A. Bowditch
Member of the American Welding Society
Member of the Association for Career and Technical Education

Publisher
The Goodheart-Willcox Company, Inc.
Tinley Park, Illinois
www.g-w.com

Manufactured in the United States of America.

Library of Congress Catalog Card Number 2009020997

ISBN: 978-1-60525-256-8

2 3 4 5 6 7 8 9 – 10 – 14 13 12 11 10

The Goodheart-Willcox Company, Inc. Brand Disclaimer: Brand names, company names, and illustrations for products and services included in this text are provided for educational purposes only and do not represent or imply endorsement or recommendation by the author or the publisher.

The Goodheart-Willcox Company, Inc. Safety Notice: The reader is expressly advised to carefully read, understand, and apply all safety precautions and warnings described in this book or that might also be indicated in undertaking the activities and exercises described herein to minimize risk of personal injury or injury to others. Common sense and good judgment should also be exercised and applied to help avoid all potential hazards. The reader should always refer to the appropriate manufacturer's technical information, directions, and recommendations; then proceed with care to follow specific equipment operating instructions. The reader should understand these notices and cautions are not exhaustive.

The publisher makes no warranty or representation whatsoever, either expressed or implied, including but not limited to equipment, procedures, and applications described or referred to herein, their quality, performance, merchantability, or fitness for a particular purpose. The publisher assumes no responsibility for any changes, errors, or omissions in this book. The publisher specifically disclaims any liability whatsoever, including any direct, indirect, incidental, consequential, special, or exemplary damages resulting, in whole or in part, from the reader's use or reliance upon the information, instructions, procedures, warnings, cautions, applications, or other matter contained in this book. The publisher assumes no responsibility for the activities of the reader.

Library of Congress Cataloging-in-Publication Data

Bowditch, William A.
Welding technology fundamentals / by William A. Bowditch, Kevin E. Bowditch, and Mark A. Bowditch. — 4th ed.
p. cm.
Includes index.
ISBN 978-1-60525-256-8
1. Welding. I. Bowditch, Kevin E. II. Bowditch, Mark A. III. Title.
TS227.B69 2009
671.5'2—dc22 2009020997

About the Authors

William A. Bowditch has an extensive teaching and welding background. He has been a teacher, department head, and supervisor of special needs and vocational programs. In addition to his formal college training in preparation for teaching, Bill has taken several specialized courses in industry, such as the Hobart Welding School and American Welding Society courses. He is a member of the Association for Career and Technical Education and a life member of the American Welding Society. As a coauthor of ***Modern Welding*** and ***Welding Technology Fundamentals***, he has guided those Goodheart-Willcox books through many revisions to keep them up-to-date and technically correct, maintaining their value as authoritative welding texts.

Kevin E. Bowditch is a welding engineer specialist for Subaru of Indiana Automotive Inc. His welding experience includes working for two automotive firms, two aerospace firms, a construction company that builds nuclear plants, and a precision sheet metal firm. While working for one aerospace firm, Kevin designed resistance welding and soldering equipment, special equipment for custom applications, and worked to develop correct welding and soldering schedules for customers. He has a bachelor's degree in welding engineering from Ohio State University and has attended specialized conferences and courses sponsored by the American Welding Society, American Society of Mechanical Engineers, and American National Standards Institute. Kevin has been coauthoring ***Modern Welding*** with his father for twenty years, and has been a coauthor of ***Welding Technology Fundamentals*** since its first edition, published in 1991.

Mark A. Bowditch joined the Bowditch authoring team of welding authors in 1998 by coauthoring ***Oxyfuel Gas Welding*** with his brother Kevin. He holds both a bachelor's degree and a master's degree, having majored in General Science with an emphasis on mathematics. With over ten years of teaching experience, he utilizes the expertise and communication skills of a professional educator in his writing. Mark attended Hobart Welding School in 1997 and is a member of both the American Welding Society and the Association for Career and Technical Education.

Introduction

Welding Technology Fundamentals is written for secondary and postsecondary students, apprentices, journeymen, and individuals who wish to learn to weld.

This book covers the equipment and techniques used for the welding and cutting processes most often employed in industry today. These processes are shielded metal arc welding, gas metal arc welding, flux cored arc welding, gas tungsten arc welding, oxyfuel gas cutting and welding, and resistance welding.

Welding Technology Fundamentals contains information about welding careers and the physics of welding. Technical information regarding weld inspection and testing, welder qualification, drawing interpretation, and welding symbols is also included.

General welding safety is covered in Chapter 1. Safety information and cautions are also written into the text wherever they apply. Safety information and cautions are printed in red, so that they will stand out.

The text is organized into nine sections. Each section is composed of one or more chapters that describe processes, explain procedures, or present general information relating to the topic of that section. A section can be studied independently or in sequence with other sections.

The first section provides general information about welding. The topics presented in these chapters include welding safety, an overview of welding and cutting processes, a brief discussion of the physics behind welding, and a close look at the different weld joints and welding positions.

Sections 2–8 present detailed information about welding and cutting processes. This study of processes begins in Section 2 with a close look at shielded metal arc welding. The subsequent sections present detailed information about gas metal arc welding, gas tungsten arc welding, plasma arc cutting, oxyfuel gas processes, resistance welding, and special welding and cutting processes.

The final section discusses technical information that is of practical importance to welders. The topics covered in these chapters includes interpreting welding symbols, inspecting and testing welds, and welder certification. The end of the book contains several useful appendices and an extensive glossary of technical terms.

You may begin your study of welding with any section and progress in any desired sequence from section to section. However, when you use the Laboratory Manual for ***Welding Technology Fundamentals,*** we recommend that Chapter 33, Welding Symbols, be studied early. Welding symbols are used in the Laboratory Manual to describe the sample joints and welds for each job assignment. Before attempting any welding process that uses a pressurized gas, Chapters 20 and 21 should be studied.

Welding Technology Fundamentals is written in an easy-to-read and understandable style. All welding terms used are those approved by the American Welding Society (AWS). In cases where nonstandard terms are used by some people in the trade, such terms are often given in parentheses after the correct AWS term. The book is extensively illustrated with drawings and photographs to show the various processes or welding techniques.

Many tables and charts are provided to help you select the proper variable values required to make a good weld. Photographs of industrial welding applications have been used, along with photographs of practice welds in progress. Equivalent SI metric measurement units are shown in parentheses following US conventional measurements.

You should read the caption accompanying each illustration, since the caption often gives information that is not covered in the text. Review questions are provided at the end of each chapter to test your knowledge of the information covered. In most chapters, practice exercises are provided to test your skills as a student welder in completing various welding tasks.

It is our sincere hope that *Welding Technology Fundamentals* will help you progress in an organized manner toward a mastery of the essential welding skills.

William A. Bowditch
Kevin E. Bowditch
Mark A. Bowditch

Chapter Listing

Contents

Contents

Section 3
Gas Metal Arc Welding

Section 4
Gas Tungsten Arc Welding

Contents

Contents

Contents

Section 1

Introduction to Welding

Fibre-Metal Products Co.

Chapter 1

Safety in the Welding Shop

Learning Objectives

After studying this chapter, you will be able to:

- List at least seven hazards that exist in the welding shop.
- Identify the clothing items that should be worn when welding or cutting.
- Explain the various causes of fire hazards.
- Describe the safety features present in the welding shop that can be used in an emergency.
- Describe the danger of fumes and airborne contaminants to the welder and the safety precautions that provide respiratory protection.
- Cite at least five general rules to follow when storing compressed gas.

arc flash
combustibles
fire watch
negative-pressure air-purifying respirator
positive-pressure respirator

General Shop Safety

Working and moving about in a welding shop or welding environment can present many dangers, such as heat, sparks, fumes, arc and ionizing radiation, high voltage, hot metal, flammable material, moving vehicles, hazardous machinery, moving overhead cranes and their loads, falling objects, compressed gases, and noise. However, through the use of training, engineering controls, personal protective equipment, and proper work procedures, all these safety hazards can be controlled.

Every professional welder should be familiar with the latest safety and health information. The most complete and influential safety document available today is entitled *Safety in Welding, Cutting, and Allied Processes* (ANSI Z49.1). The American Welding Society (AWS) publishes this document, which outlines the procedures and practices that will keep you and others safe from the many potential dangers of welding, cutting, and related processes.

Three of the most important factors in safety on the job are:

- Staying healthy in mind and body.
- Becoming well-trained in the required job or task and its possible hazards.
- Having a good attitude toward safety rules, equipment, and training on the job.

The management of a welding shop is responsible for making sure that all welders and supervisors are thoroughly trained in correct procedures and the safe operation of their equipment. The management must also clearly communicate all hazards and safety precautions to all workers before work is started. Companies should consistently encourage an "if you don't know, ask" attitude. Each welder has the responsibility of knowing and following all safety rules in order to create a safe workplace wherever welding is done.

Personal Safety and Clothing

Welders should wear work clothes or coveralls. The shirt or coveralls should have covered pockets and have buttons up to the neck. Trousers and coveralls should not have cuffs that could catch hot metal spatter. A cap of some kind should be worn to protect the hair from hot metal spatter. Gloves should be worn to protect against flames, sparks, and hot or sharp metal. Leather gloves with gauntlet-type cuffs are recommended for welders. Safety glasses should be worn by everyone working in an eye hazard area.

Steel-toed safety shoes are recommended for welders and other workers who handle heavy articles. Oxyfuel gas welders and cutters should wear approved goggles with the correct shade lens.

Arc welders should wear an arc welding helmet with the correct filter shade lens for the process and amperage being used, as shown in **Figure 1-1**. See the chart in **Figure 1-2** to determine the recommended filter shade number based on the welding or cutting operation being performed. Welding helmets are designed to protect the eyes and face from arc rays, weld sparks, and weld spatter that directly hit the helmet. Welding helmets with filter lenses are not intended to protect against slag chips, grinding fragments, or other hazards such as wire wheel bristles. Many arc welders also wear a pair of safety goggles or glasses with side shields under their arc welding helmet to protect against these hazards.

An ***arc flash*** occurs when an individual looks directly at a welding arc without the proper eye protection. When a person receives a flash, the eyes will feel burnt and tears will typically follow. A cool, damp cloth and sleep may provide relief following an arc flash. To prevent this hazard from occurring, both the welder and all others in the immediate welding area should wear the proper eye protection.

Welders should be aware of the danger of excessive noise. The Occupational Safety and Health Administration (OSHA) sets limits to the amount of permissible noise exposure. If noise exceeds the set limits, hearing protection, such as earmuffs or earplugs, must be worn. Foam earplugs are most comfortable for long-term use. Be sure to position the earmuffs or insert the plugs correctly in order to avoid noise hazards.

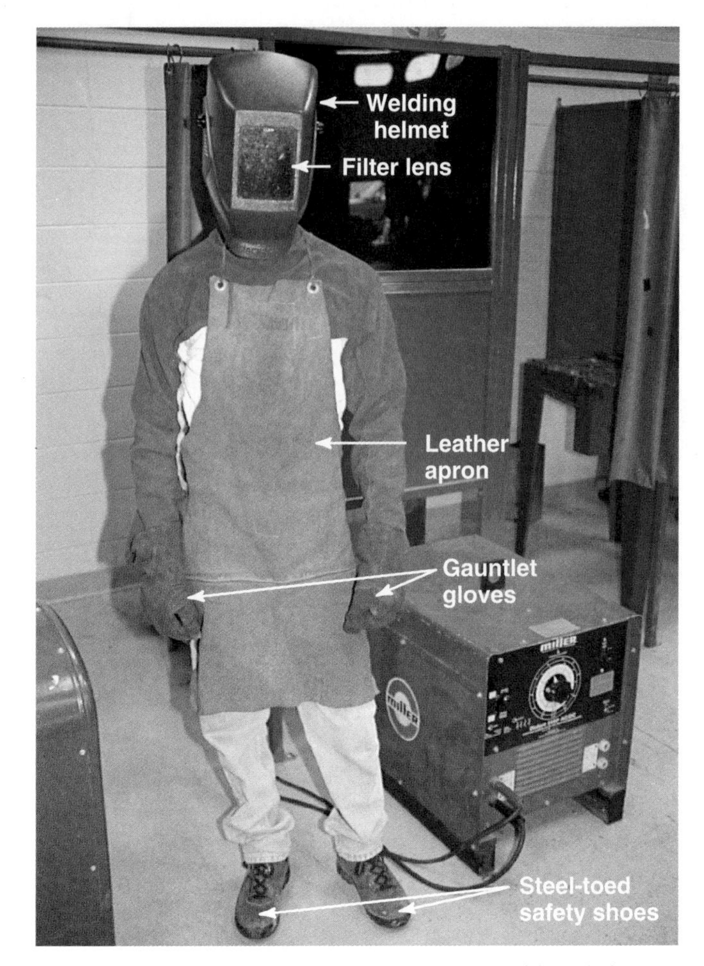

Figure 1-1. *This welder is wearing an arc welding helmet, a cap, leather cape and apron, trousers without cuffs, gauntlet gloves, and high-top steel-toed shoes as protection from hot metal and harmful rays.*

Housekeeping is one of the most important factors in shop safety. Shop housekeeping refers to the process of keeping floors and workbenches clean of dirt, scrap metal, grease, oil, and anything that is not essential to the job at hand. ***Combustibles*** (easily ignited materials, such as wood, paper, rags, and flammable liquids) must be kept clear of all areas where sparks or hot metal may fly. Aisles must be kept clear of hoses and electric cables, which can cause tripping accidents. Hoses and cables could also be run over and possibly damaged.

Fire Hazards

Prior to welding or cutting with any type of equipment, clear the area of all combustible materials. Paint, oil, cleaning chemicals, and other combustibles must be kept in steel cabinets designed for such storage in accordance with local fire codes. Welding and

Filter Shades for Welding and Cutting			
Operations	**Electrode Diameter (Inches)**	**Arc Current (Amperes)**	**Recommended Shade Number**
Shielded Metal Arc Welding	< 3/32 3/32-5/32 5/32-1/4 >1/4	<60 60-160 160-250 250-550	10 10 12 14
Gas Metal Arc Welding		<60 60-160 160-250 250-500	10 11 12 14
Gas Tungsten Arc Welding		<50 50-150 150-500	10 12 14
Plasma Arc Cutting	Light* Medium* Heavy*	<300 300-400 400-800	9 12 14
		Plate Thickness (Inches)	**Recommended Shade Number**
Oxyfuel Gas Cutting	Light Medium Heavy	<1 1 to 6 >6	3-4 4-5 5-6
Oxyfuel Gas Welding	Light Medium Heavy	<1/8 1/8 to 1/2 >1/2	4-5 5-6 6-8
Brazing		any thickness	3-4
Soldering		any thickness	2

* These values apply where the arc is clearly seen. Lighter filters may be used when the arc is hidden by the workpiece.

Figure 1-2. *This chart shows recommended filter shades for welding and cutting operations.*

cutting is considered "hot work" by the National Fire Protection Association (NFPA). When welding or cutting is done outside the shop, a welder may need to obtain a "hot work" permit from the local fire department. More welding environment fires are caused while performing oxyfuel gas or arc cutting than by any other means.

Many fires start when a welder welds or cuts before receiving approval. Supervisors must first check the safety conditions. Weld and cut at least 35 feet from any flammable materials. Any flammable materials that cannot be moved should be covered with flameproof or flame-retardant blankets and also wet down to prevent a fire.

Fire Watch

To help prevent fires, assign at least one fire watch with a fire extinguisher to watch for possible fires. A ***fire watch*** is a person whose job is to carefully watch for fires in the area where welding or cutting is done. When a worker is welding, the darkness of the filter plate may obscure any signs of a fire. A fire watch observes the job site to ensure the safety of workers and property.

Whenever hot work is done, sparks and molten metal droplets can ignite flammable materials. If combustible materials *cannot* be kept at least 35 feet away from any welding, a fire watch is necessary. A fire watch should be assigned for each side of a wall and for each level immediately adjacent to a welding zone. A fire watch must also be aware of heat transfer caused by welding along walls, ceilings, or piping. It is possible that this heat could travel along these surfaces that may be in contact with combustible materials.

The main responsibilities of a fire watch include using a fire extinguisher, notifying workers of a fire, operating a fire alarm system, directing workers through the correct escape routes away from a fire area, and shutting down any equipment. Periodic fire drills should be conducted so that every worker receives proper fire safety instructions and knows where the fire exits are located. Every worker in the shop should be trained in fire extinguisher use. Only attempt to extinguish a fire if your equipment is clearly able to quench the fire.

Fire Extinguishers and Fire Classifications

There are four fire classifications according to the different flammable materials. Class A fires are of solids, such as paper, wood, or cloth. Class B fires are of combustible liquids, such as gasoline, oil, or paint thinner. Class C fires are of electrical equipment, such as switches, fuse boxes, or motors. Class D fires are of combustible metals, such as titanium and magnesium.

Fire extinguishers should be located throughout the shop and clearly marked for easy identification. Each fire extinguisher is marked with a sticker of differing colors and a classification letter in the middle of the sticker. Some fire extinguisher labels also include a picture representing the kind of fire it is designed to quench.

A Class A fire extinguisher, used to extinguish solid material fires, is marked with a green triangle with an *A* in the middle of the triangle. A Class B fire extinguisher, used to extinguish combustible liquid fires, has a red square sticker on it with a *B* in the middle. A Class C fire extinguisher, used to put out electrical fires, has a blue circle sticker with a *C* in the middle. A Class D fire extinguisher, used to extinguish combustible metal fires, has a yellow star sticker with a *D* in the star.

Burn Hazards

Burns are the most common injury encountered in the welding or cutting processes. Burns can occur to the body or eyes. Burns to the body generally are caused by ultraviolet or infrared rays given off when welding. They may also occur from hot metal or molten metal thrown off while welding. As described earlier, wearing appropriate clothing for the work involved is the best way to prevent body burns.

There are three classifications of body burns. A *first degree burn* occurs on the outer surface of the skin. First aid for this type of burn is the application of cold water or cold water compresses to the burn area. With a *second degree burn*, the surface of the skin is severely damaged with small breaks in the skin and blisters appearing on the surface. First aid for such a burn involves applying cold water (not ice) and cold water compresses to the burn area and then covering with sterile bandages or pads. Do *not* apply ointment, butter, or any other home remedies. A doctor should be seen for further treatment.

A *third degree burn* is the most serious burn. The surface of the skin will appear white or charred like burned meat, and the third or deeper layer of the skin may also be severely damaged with nerve and blood vessel damage. For first aid of a third degree burn, place *no* water or ice on the burn area. Also, *no* clothes should be removed from the burned area of the body. Removing clothes may further damage the skin and place the patient in further pain or shock. Do *not* apply ointment, antiseptics, or any home remedies on the burn area. Cover the wound with sterile dressings. Call an ambulance or take the patient to the nearest hospital.

Eye Hazards and Arc Rays

Remember that eye protection must be worn in the shop area at all times to protect the eyes against flying sparks and possible metal chips coming off welding or grinding operations. Damage to the eyes and skin from exposure to ultraviolet or infrared rays is a serious hazard to avoid when welding. Even a short exposure to an arc strike without proper eye protection can cause arc flash. Depending on the length of the exposure, arc flash may temporarily or permanently damage a welder's eyes. A spot will appear in the welder's vision for a short time if the exposure is short. When a longer exposure occurs, vision may be permanently damaged. The welder and those in the welding area must wear approved welding goggles or an arc welding helmet to protect against these hazards to the eyes. Be sure to know what kind of welding is occurring so you can have on an appropriate filter plate.

Electrical Hazards

Electrical devices of various kinds are common in welding shops. All electrical devices are hazardous, but some use extremely high and dangerous voltages. All equipment and areas where 220 volts or more are used must be marked well. The installation and repair of electrical equipment must only be done by well-trained, competent technicians.

Remember that water conducts electricity. Welders must take special precautions when welding in damp conditions in the shop or outside. Standing on wood or rubber is an excellent safeguard when welding in damp or wet conditions. If you are arc welding outside and it begins to rain, stop welding and turn off your machine if your clothing gets wet.

Machinery Hazards

Machinery must be operated only after thorough training. The topics of this training should include how the machine operates, its safety hazards, its safety features, the correct placement of the operator's hands and feet, and the proper sequence of operation.

The most common injuries in the welding shop involve hands, feet, and limbs and are typically caused by machines. For example, during gas metal arc welding, the electrode wire can be fed through the welding gun at high travel speeds. Hands must be kept clear of the gun nozzle since electrode wire can easily pierce welding gloves and skin. The general safety rule is to always keep clear of any moving mechanical parts.

Even the pedestal grinder, a simple piece of equipment, can be one of the most dangerous pieces of machinery in the shop! Failing to adjust the tool rest to its closest safe point and tighten it securely can result in an accident.

All equipment in the welding shop must be used according to the manufacturer's stated operating instructions and the safety guidelines found in ANSI Z49.1. Be sure to also consult the manufacturer's safety information and maintenance information. The welding and cutting equipment in the shop must be inspected on a regular basis to ensure that it remains in safe operating condition. If any equipment does not pass a safety inspection, be sure it is repaired and serviced by qualified personnel before using it again.

Fumes and Ventilation

Dust, fumes, and metal particles can be a hazard to health. Adequate and approved ventilation is required in welding and cutting shops to ensure that toxic fumes, dust, and dirt are removed properly. The air may need to go through filters and cleaners before it is recirculated. Ventilation pickups should be located so that fumes are captured below the level of the

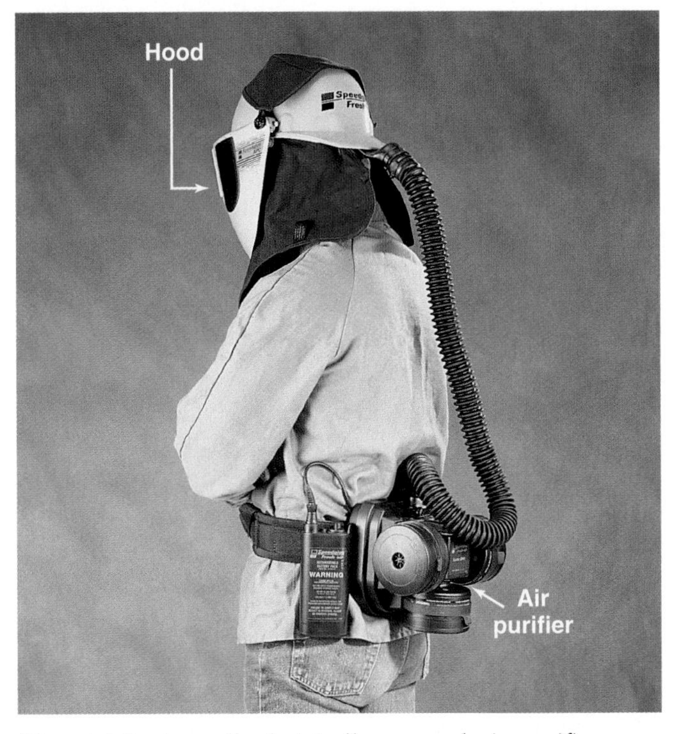

Figure 1-3. *A small, electrically powered air purifier worn around the waist supplies clean air to the welding helmet and hood. (Nederman, Inc.)*

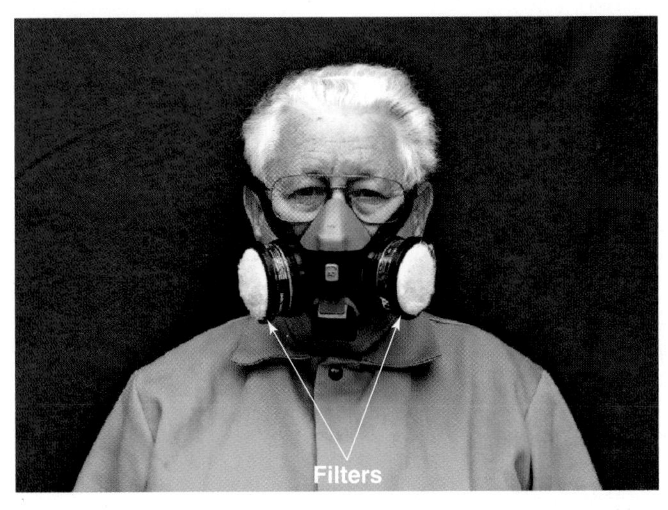

Figure 1-4. *Negative-pressure air-purifying respirators filter dust and other particles from the air but will not protect against toxic fumes.*

welder's nose, before he or she has to inhale them. While installed ventilation systems help greatly in removing dangerous airborne particles, they are frequently not enough to adequately protect the welder.

The need for proper and sufficient respiratory protection is perhaps the most overlooked safety concern in the field of welding. Respiratory protection is most needed when performing arc welding, oxyacetylene cutting, resistance welding, brazing, and soldering. These are the processes that expose welders to the most toxic fumes and gases. Welders have an increased risk of developing respiratory illnesses, such as bronchitis, pneumonia, and even lung cancer. This is especially true for workers who are exposed to dangerous chemical elements, such as nickel, chromium, cadmium, lead, zinc, and beryllium.

There are many safety regulations that companies must adhere to in order to provide protection from harmful fumes. OSHA regulations should be reviewed and enforced in each workplace for the sake of all workers. See the following OSHA regulations: 29 CFR 1910.133, 29 CFR 1910.95, and 29 CFR 1910.251 through 1910.254.

Despite the regulations that exist, each welder should take personal responsibility for wearing the proper respiratory protection. Even though a welder may be working in an area where dangerous fumes and respiratory contaminants fall below the permissible level, the voluntary use of respiratory protection is still a good idea.

Respirators fall into two main groupings: positive-pressure respirators and negative-pressure air-purifying respirators. ***Positive-pressure respirators*** provide a clean flow of breathing air from a compressor or purifier. This clean air travels through a hose into the helmet of the welder. See **Figure 1-3**. ***Negative-pressure air-purifying respirators*** remove dust and other airborn particles by drawing the contaminated air through a filter or cartridge in the respirator, as shown in **Figure 1-4**.

Storing Compressed Gases

Before any individual is involved in the purchasing, handling, or storing of compressed gas, he or she must first receive thorough instruction and training. Every licensed gas producer and distributor will provide reliable safety information about the safe handling, transporting, and general characteristics of each particular gas.

The welding shop must have a safe area set aside for cylinder inventory and storage. This area must be well ventilated and separate from welding areas. Smoking and the use of any flames must be prohibited in the vicinity of the cylinder storage area. Be sure to place hazard warning signs to clearly mark the storage area. All cylinders should be clearly labeled and separated according to the gas type. Oxygen cylinders must be separated from fuel gas cylinders by a minimum of 25 feet (7.6 m) or by a fire-resistant wall.

Full and empty cylinders should also be stored separately, and full cylinders should be stored so that the oldest stock is used first. Stored gas cylinders must have a cylinder cap. The cylinder cap (also called a safety cap) is an important safety feature. It should always be screwed securely in place when the cylinder is in storage. Cylinders should always be secured so they cannot topple or roll. The safety guidelines for the proper handling of cylinders are discussed in Chapter 20.

Lifting

Lifting is always a hazard to the body, but training in how to lift objects safely can reduce the chance of injury. Limits should be set on how much weight a worker may lift. The maximum weight to be lifted may vary from shop to shop and by worker classification. In many shops, a worker whose job involves lifting is required to wear a lifting belt or back brace.

Hazardous Obstacles

Safety hazards or obstacles in the shop should be clearly marked. Signs, fences, or barriers should be erected while temporary hazards are present, so that all workers are fully aware of them. Permanent hazards are often painted with yellow and black stripes to create high visibility.

Hand and Power Tools

All hand and power tools should be examined prior to use for loose parts that may come off and injure the operator or a worker nearby. Power cords should be checked for cut or frayed insulation and exposed wires. Report any defective equipment to the shop supervisor or to the proper maintenance personnel.

Designated Welding and Cutting Areas

All welding and cutting should be done in designated areas of the shop, if possible. These areas should be made safe for welding and cutting operations with concrete floors, filter screens, protective drapes or curtains, and fire extinguishers. No combustibles should be stored nearby.

Suffocation Hazards

Gases that are heavier than air or lighter than air can be extremely dangerous to welders in confined spaces or closed tanks. Argon and carbon dioxide are examples of heavier-than-air gases. Helium is an example of a lighter-than-air gas. These gases are colorless and odorless and will displace the oxygen in a closed space. To minimize the risk of asphyxiation, closed spaces must be well-ventilated when heavier-than-air or lighter-than-air gases are to be used. If proper ventilation cannot be obtained, the welder must go into the space using a positive-pressure respirator.

Welding on Hazardous Containers

Unless you are trained in the proper procedures for doing so, never try to weld or cut a container that has held flammable or hazardous materials. Such a container may explode. Proper procedures are described in the American Welding Society (AWS) Standard F4.1, *Safe Practices for the Preparation of Containers and Piping for Welding and Cutting*. You should also consult with the local fire marshal before welding or cutting such containers.

Additional Safety Publications

The American Welding Society offers many excellent booklets and documents that provide important safety information in addition to the main safety document mentioned at the beginning of the chapter. Many states require the posting of general and specific safety information in the welding shop. Many safety posters and charts are available, **Figure 1-5**. The following is a list of some of the many safety publications that can be obtained from the AWS.

- *Arc Welding and Cutting Noise* (AWN).
- *Effects of Welding on Health* (EWH-1 to EWH-13).
- *Characterization of Arc Welding Fumes* (CAWF).
- *Methods for Sampling and Analyzing Gases from Welding and Allied Processes* (F1.5M).
- *Method for Sampling of Airborne Particulates Generated by Welding and Allied Processes* (F1.41M).
- *A Sampling Strategy Guide for Evaluating Contaminants in the Welding Environment* (F1.3).
- *Welding Handbook* (AWS WHB). Included in this multi-volume work is a chapter on safety and health.
- *Ventilation Guide for Weld Fume* (F3.2M/F3.2).
- *Oxyfuel Gas Welding, Cutting and Heating Safety* (AWS OWS).
- *Safety and Health Fact Sheets.*
- *Braze Safely* (BRS).
- *Fire Safety in Welding and Cutting* (AWS FSW).

Welding Safety Checklist		
Hazard	**Factors to Consider**	**Precautionary Summary**
Electric shock can kill	• Wetness • Welder in or on workpiece • Confined space • Electrode holder and cable insulation	• Insulate welder from workpiece and ground using dry insulation, rubber mat, or dry wood. • Wear dry, hole-free gloves. (Change as necessary to keep dry.) • Do not touch electrically "hot" part or electrode with bare skin or wet clothing. • If wet area and welder cannot be insulated from workpiece with dry insulation, use a semiautomatic, constant-voltage welding machine or stick welding machine with voltage-reducing device. • Keep electrode holder and cable insulation in good condition. Do not use if insulation is damaged or missing.
Fumes and gases can be dangerous	• Confined area • Positioning of welder's head • Lack of general ventilation • Electrode types, i.e., manganese, chromium, etc. • Base metal coatings, galvanizing, paint	• Use ventitilation or exhaust to keep air-breathing zone clear and comfortable. • Use helmet and position of head to minimize fume breathing zone. • Read warnings on electrode container and material safety data sheet for electrode. • Provide additional ventilation/exhaust where special ventilation requirements exist. • Use special care when welding in a confined area. • Do not weld unless ventilation is adequate.
Welding sparks can cause fire or explosion	• Containers that have held combustibles • Flammable materials	• Do not weld on containers that have held combustible materials (unless strict AWS F4.1 explosion procedures are followed). Check before welding. • Remove flammable materials from welding area or shield from sparks, heat. • Keep a fire watch in area during and after welding. • Keep a fire extinguisher in the welding area. • Wear fire-retardant clothing and hat. Use earplugs when welding overhead.
Arc rays can burn eyes and skin	• Process: gas-shielded arc most severe	• Select a filter lens that is comfortable for you while welding. • Always use helmet when welding. • Provide nonflammable shielding to protect others. • Wear clothing that protects skin while welding.
Confined space	• Metal enclosure • Wetness • Restricted entry • Heavier-than-air gas • Welder inside or on workpiece	• Carefully evaluate adequacy of ventilation, especially where electrode requires special ventilation or where gas may displace breathing air. • If basic electric shock precautions cannot be followed to insulate welder from work and electrode, use semiautomatic, constant-voltage equipment with cold electrode or stick welding machine with voltage-reducing device. • Provide welder helper and method of welder retrieval from outside enclosure.
General work area hazards	• Cluttered area	• Keep cables, materials, tools neatly organized.
	• Indirect work (welding ground) connection	• Connect work cable as close as possible to area where welding is being performed. Do not allow alternate circuits through scaffold cables, hoist chains, ground leads.
	• Electrical equipment	• Use only double insulated or properly grounded equipment. • Always disconnect power to equipment before servicing.
	• Engine-driven equipment	• Use only in open, well-ventilated areas. • Keep enclosure complete and guards in place. • Refuel with engine off. • If using auxiliary power, OSHA may require GFCI protection or assured grounding program (or isolated windings if less than 5 KW).
	• Gas cylinders	• Never touch cylinder with the electrode. • Never lift machine with cylinder attached. • Keep cylinder upright and chained to support.

Figure 1-5. *A welding safety checklist similar to those posted on the safety board in shops where welding is performed. (Adapted from The Lincoln Electric Company)*

Summary

- Welders should wear the proper equipment.
- Paint, oil, cleaning chemicals, and other possible combustibles must be kept in steel cabinets designed for such storage in accordance with the local fire codes.
- Welders must take special precautions when welding in damp conditions in the shop or outside.
- Machinery must be operated only after thorough training. Always keep clear of any moving mechanical parts.
- Adequate and approved ventilation is required in welding and cutting shops to ensure that toxic fumes, dust, and dirt are removed properly.
- Before any individual is involved in the purchasing, handling, or storing of compressed gas, he or she must first receive thorough instruction and training.

Review Questions

Write your answers on a separate sheet of paper. Please do *not* write in this book.

1. Name seven hazards that are found in the welding shop.
2. What is the name of the most complete safety document available for the field of welding?
3. What is the recommended filter shade number that should be used in the welding helmet when performing gas tungsten arc welding with an arc current less than 50 amps?
4. Welding helmets protect the eyes and face from arc rays, weld sparks, and weld spatter that directly hit the helmet. What should be worn in addition to a welding helmet to protect against slag chips, grinding fragments, and other hazards, such as wire wheel bristles?
5. What is the name of the potential hazard that results from looking at a welding arc without eye protection? What should be done if a welder experiences this hazard?
6. Name three dangerous chemical elements that pose a respiratory health threat.
7. *True or False?* The negative-pressure air-purifying respirator delivers a constant flow of clean breathing air to the welder through a hose.
8. Oxygen cylinders must be separated from fuel gas cylinders by a minimum distance of _____.
9. When compressed gas cylinders are in storage, what important safety feature must be securely in place on the top of the cylinder?
10. Name three precautions that can be taken to avoid an electric shock.
11. When combustible material cannot be kept at least 35 feet away from a welding area, a _____ is required.

Chapter 2

Welding and Cutting Processes

Learning Objectives

After studying this chapter, you will be able to:

- Cite advantages of welding over other joining processes.
- List the significant developments in the history of welding.
- Identify recent developments in welding and cutting processes.
- Identify several occupations in the welding industry and list the recommended amount of education for each.
- List at least five traits and skills sought by employers.
- List at least five factors that influence a person's ability to obtain and hold a job in the welding industry.

Technical Terms

forge welding (FOW)
fusion welding
occupation
process
prototype parts
thermal spraying
weld
welding
weldment

The Welding Process

A ***process*** is an operation used to produce a product—riveting, forging, casting, cutting, turning, bending, and welding are processes used in industry. Metalworking processes, such as joining metal by riveting and welding, have been used for thousands of years. Blacksmiths have used their skills and knowledge of metalworking processes to work metal into many desired products. Hinges, nails, cooking pots, farm implements, wagon wheels, and swords were produced by forging and forge welding.

Metalworkers are employed today to make prototype parts for cars, airplanes, and other equipment. ***Prototype parts*** are the first models of parts that later may be mass-produced. Prototype parts are generally handcrafted. After a prototype part is approved, machines and equipment are designed to mass-produce it.

Forge welding (FOW) is a process used by blacksmiths to produce welded metal parts. In forge welding, the parts are heated until they are very hot and in a softened state. The heated ends are then placed together on an anvil and struck repeatedly with a hammer. See **Figure 2-1**. This hammering creates a ***weld*** by forcing the hot, softened metal from the pieces to intermix and form a single unit. If the welding process is performed properly, the weld is stronger than the original metal.

Welding is defined as a process of "joining metallic or nonmetallic material in a relatively small area by heating the area to the welding temperature. Pressure may or may not be applied; welds may be made by applying pressure only. Welds may be made with or without the addition of filler material." Not all materials can be welded, but most metals and plastics are weldable.

Most welding processes require the addition of heat to the weld area. ***Fusion welding*** is the process of joining materials (metals or nonmetals) by applying heat until the areas of the weld joint reach their melting points. The materials then become molten (liquid) and flow together, forming a single piece after they cool.

Welds are also made with the metal at or near room temperature. Cold welding and explosion welding are examples of welding processes that are performed at room temperature. Extremely high pressure is applied to the area to be welded. Metal at the surface of one part is driven into the surface of the adjoining part by the pressure. The pieces are thus joined or welded at the surface.

Advantages of Welding and Cutting Processes

Welding and cutting processes offer a number of advantages over other joining and cutting methods. Most welding and cutting equipment is portable. It can be operated inside boilers, furnaces, large containers, or pipes to make repairs. Welding and cutting can be done under water to repair ships or to cut up and remove hazards. Welding and cutting equipment can be transported into the field to repair farm machinery, trucks, and earthmoving equipment. See **Figure 2-2**. Welding is used to construct pipelines, buildings, and ships.

Figure 2-1. A blacksmith is joining a rod to a plate by the forge welding method.

Because of modern welding techniques, engineers are able to design strong parts that are lightweight, complex, and often less expensive than those made using other joining methods. As late as 1920, many steel structures were built using rivets. Automobile and truck chassis also were constructed using rivets. Handles were attached to cooking pots using rivets.

Holes are drilled into parts to be riveted, weakening the parts in that area. Thicker and heavier metal is used to compensate for the weakness. See **Figure 2-3**. Today, many parts are specifically designed to be welded rather than riveted. The welded parts are lighter and generally cheaper to produce than riveted parts. Parts joined by welding are known as ***weldments***.

Many complex shapes are cast or forged and then machined into their final shape. Cast or forged parts are heavy and are produced with large, costly equipment. A large amount of machining time would be required to produce the part in **Figure 2-4A** from 3″ (76 mm) solid stock. The same part can be produced from 3″ (76 mm) and 1 1/2″ (38 mm) stock by cutting the stock to length and welding it at both ends. See **Figure 2-4B**. In general, weldments use less material, take less time, and are cheaper to produce than machined parts. Also, weldments are equal in strength to machined parts. Therefore, welding is a very efficient and widely used industrial process for producing quality parts.

Applications of welding and cutting processes are unlimited. For example, stones can be welded together to build or repair a statue. A layer of weldable metal can be added to the stone surfaces by thermal spraying. The metal can then be welded to join the stones together. In this example, thermal spraying is used to add layers of weldable material to the surface of an unweldable part so it can then be welded. The ***thermal spraying*** process can also be used to increase the surface hardness of soft materials by adding a harder material (metallic or nonmetallic) onto the surface, as shown in **Figure 2-5**.

History of Welding

The use of welding dates back to 2000 BC, but the development of modern forms of welding began in

A

B

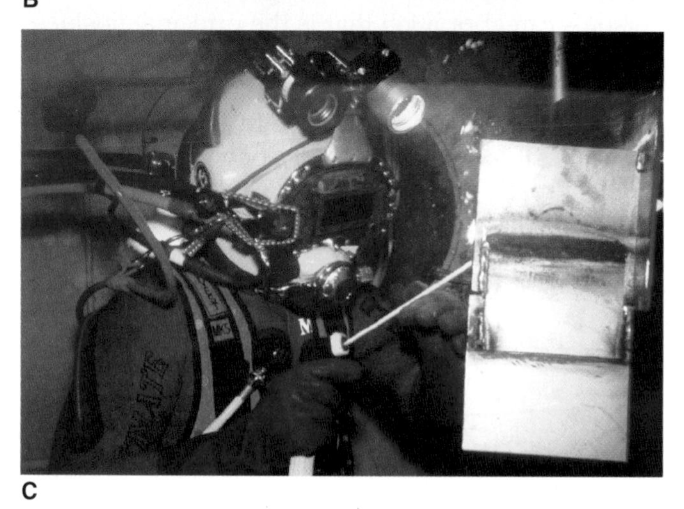

C

Figure 2-2. *Some examples of portable welding and cutting equipment. A—Repairing a blade on a construction site. (Miller Electric Mfg. Co.) B—Building the Grand Canyon Skywalk. (The Lincoln Electric Company) C—Underwater welding a metal structure. (Navy Joining Center)*

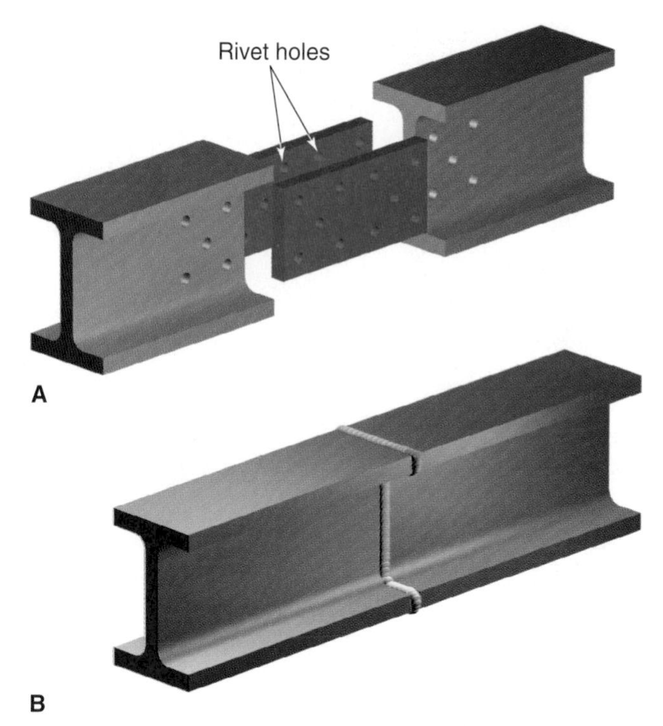

Figure 2-3. *A—On riveted joints, rivet holes weaken the metal at the joint. Thicker metal must be used. B—On welded joints, holes are not required. Thinner metal may be used.*

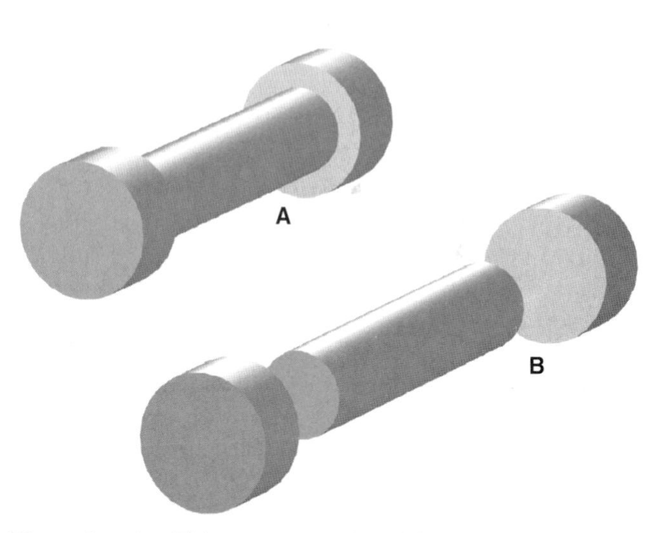

Figure 2-4. *A—This part was machined from one large-diameter rod. Metal removed from the middle section is wasted. B—This part can be produced by welding three individual pieces of stock together. Machining is not required.*

1881. Some of the major dates and developments are as follows:

2000 BC—Forge welding used to join copper and bronze.

1881—Electric arc welding performed by Auguste de Meritens. A carbon electrode was used.

1883–1885—Electric resistance welding developed by Elihu Thomson. It was used to weld the ends of wires together.

Figure 2-5. Thermal spraying is being used to apply a hard surface to a soft metal shaft. A hard surfacing material in rod form is superheated and sprayed onto the surface. (Wall Colmonoy Corporation)

1894—Carbon arc welding used commercially to produce steel barrels.
1902—Arc cutting demonstrated.
1903—Oxyfuel gas welding and cutting torches developed by Fouché and Picard.
1907—Covered steel electrode developed by Kjellborg in Sweden.
1917—Strength of welded joints was first tested by the British and United States governments and by Lloyd's of London (a ship insurance company). Test results proved that welded joints were as strong as riveted joints. Welding became an accepted practice for repair and construction. When the U.S. entered World War I, the Germans severely damaged 105 of their own ships that were in U.S. ports. This was done to prevent their use by the United States. Repairs to boilers and pumps were performed using welding, saving $20,000,000.
1918—First all-welded ship was launched.
1920—First all-welded building was constructed. Built by Electric Welding Company of America, it measured 40′ × 60′.
1940—Submerged arc welding process developed in Russia.
1942—Gas tungsten arc welding (GTAW), or heliarc welding, developed, using two tungsten electrodes and helium as a shielding gas.
1948—Gas metal arc welding (GMAW) developed.
2004—A total of 116 welding and cutting processes are in use with a variety of energy sources and shielding gases. American Welding Society (AWS) approved names and abbreviations are shown in **Figure 2-6**.

Recent Developments in Welding and Cutting Processes

Engineers have developed new welding and cutting processes in recent years. Innovative processes have been developed because new metals and alloys are being used in highly technological products, such as supersonic aircraft, nuclear plants, submarines, and spacecraft. State-of-the-art welding and cutting processes that are now in use include:

- Plasma arc welding and cutting.
- Electron beam welding and cutting.
- Electroslag welding.
- Electrogas welding.
- Laser beam welding and cutting.
- Plasma spraying.
- Electric arc spraying.
- Flame spraying.
- Explosion welding.
- Friction stir welding.
- Exothermic cutting.

Although the future of the welding industry is unknown, it is certain that it will continue to grow. New welding and cutting processes will have to be developed as new metals, alloys, plastics, and ceramics are created.

Obtaining and Holding a Job in the Welding Industry

There are and always will be many occupations available in the welding industry. An ***occupation*** is a person's career or a job that is the principal business of his or her life. Most jobs in the welding industry require at least a high school education or an apprenticeship. Other jobs may require a junior college (associate), four-year (bachelor), or advanced degree.

A number of welding and welding-related jobs, together with the educational level usually required and where this education may be obtained, are listed below.

High School or Technical School

- Gas welder.
- Assembler/brazer.
- Gun welder.
- Arc welder.
- Fitter welder.
- Tack welder.
- Resistance welder.
- Assembler/welder.

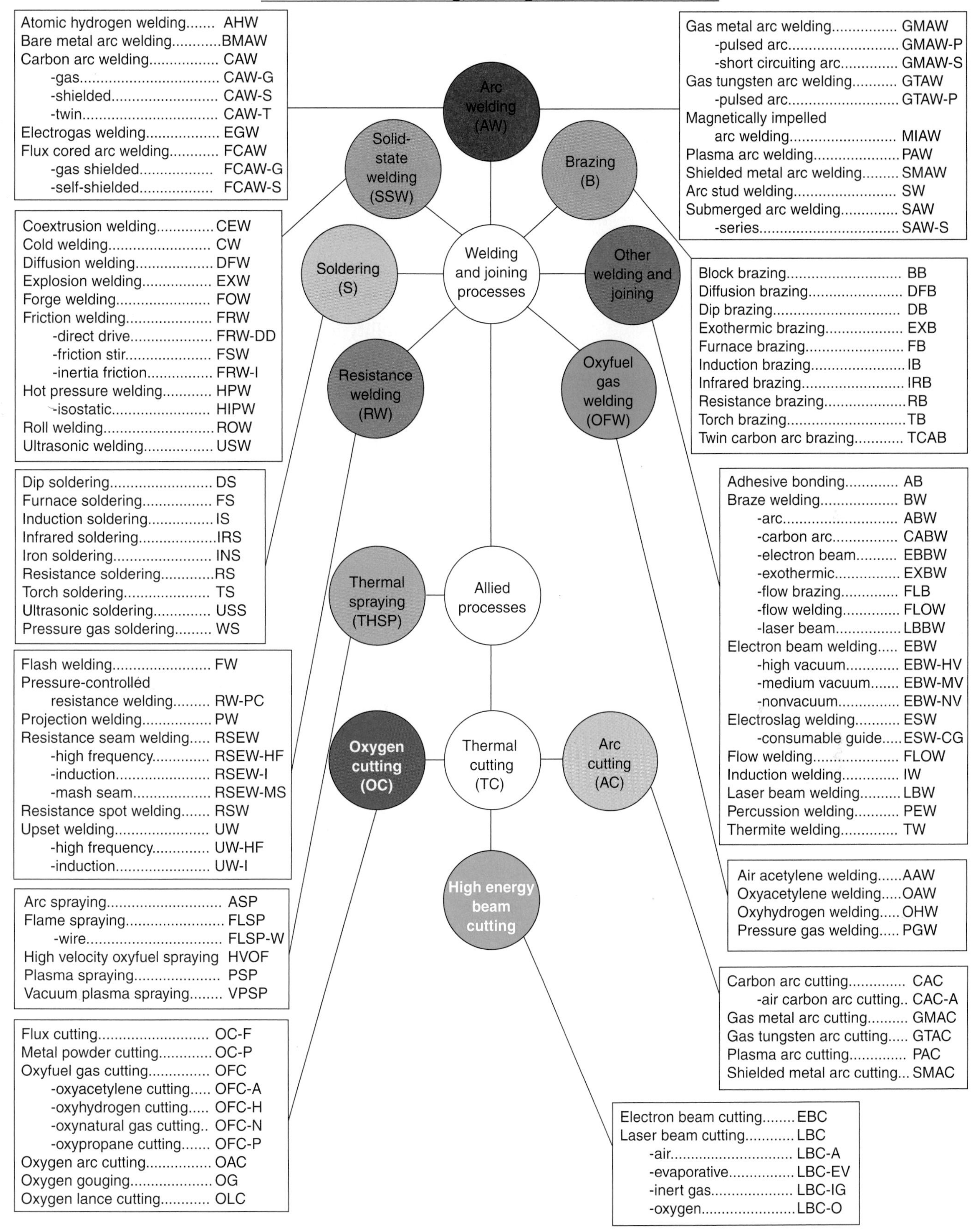

Figure 2-6. *These are the welding and cutting processes as defined by the American Welding Society. (Adapted from AWS A3.0:2001, Figures 54A and B, Master Chart of Welding and Joining Processes and Master Chart of Allied Processes, reproduced with permission from the American Welding Society, Miami, FL)*

- Combination welder.
- Solderer.

Technical School or Community College

- Ultrasonic welding machine operator.
- Welding inspector.
- Laser beam machine operator.
- Welding supervisor.
- Experimental welder.
- Welding machine repair person.
- Robotic welding machine programmer.
- Tool-and-die welder.

Military Training Schools

- Combination welder.
- Diver/welder.
- Repair welder.
- Specialty welder.

Trade Apprenticeship

- Sheet metal worker.
- Pipefitter and steamfitter.
- Structural ironworker.
- Ornamental ironworker.
- Blacksmith for experimental parts.
- Boilermaker.

College or University

- Welding engineer.
- Metallurgist.
- Metallurgical engineer.

Getting a job is sometimes difficult, but often, the job seeker has control over the reasons for that difficulty. To get the job you want, you must take the required courses in school. You must demonstrate the personal traits that employers are looking for in an employee and have the required skills.

Suggested School Subjects

Welding is a technical trade, but to succeed, an employee should know more than how to weld. Some of the school subjects suggested for greater success in finding and holding a job in the welding industry are print reading, mechanical drafting, electricity, electronics, metals, physics, math, algebra, geometry, trigonometry, and language arts classes.

Personal Traits Sought by Employers

Employers want the following traits in an employee:

- Dependability.
- The ability to follow directions.
- The ability to get along with peers.
- Thoroughness.
- Self-confidence.
- Willingness to accept responsibility.
- Initiative.
- The ability to get along with supervisors.
- The ability to communicate written ideas.
- The ability to communicate ideas orally.

Since school is a student's first job, you should learn and demonstrate the traits required by a future employer while you are still in school. All of these traits can be demonstrated in school by participating in sports and clubs and by doing good classroom work in a timely manner.

When applying for a job, fill out the application neatly and completely. Dress appropriately for the job. Arrive on time for all applications and interviews.

Academic Skills Sought by Employers

The academic skills that employers are seeking include the ability to:

- Read and understand written materials.
- Write and understand the technical terms and language of the trade or business in which you desire to work.
- Read and understand graphs and charts.
- Understand basic math.
- Use mathematics to solve problems.
- Research using the Internet, libraries, and various documents.
- Use the tools and equipment involved in the business, including computers.
- Use the scientific method of solving problems.

Factors That Can Lead to Rejection for Employment

Employers have identified a number of personal factors or traits that would cause a person to be rejected for a job. These factors include:

- A poor scholastic record.

- Inadequate personality.
- Lack of goals.
- Lack of enthusiasm.
- Inability to express yourself.
- Unrealistic salary demands.
- Poor personal appearance.
- Lack of maturity.
- Excessive interest in security and benefits.
- A poorly completed job application.

Factors That May Lead to Termination from a Job

Employers have also identified various reasons for a person being overlooked for promotion or even having their job terminated. These reasons include:

- Poor attendance without cause.
- Habitually arriving late.
- Alcoholism.
- Illegal drug use.
- Inability to perform the tasks assigned.
- Inability to work as a team member.
- Fighting and threatening peers.
- Insubordination to directions from a supervisor.
- Talking with others too much and too often.
- Lack of respect for others.
- Lack of respect for others' property.
- Always making excuses.
- Constant complaining.

Summary

- Welding is defined as a process of "joining metallic or nonmetallic material in a relatively small area by heating the area to the welding temperature."
- Fusion welding is the process of joining materials (metals or nonmetals) by applying heat until the areas of the weld joint reach their melting points.
- Welds can also be made with the metal at or near room temperature, using processes such as cold welding and explosion welding.
- Welding and cutting processes offer a number of advantages over other joining and cutting methods, including portability of the equipment, the ability to perform the processes under water, and the cost effectiveness of the processes.
- Most jobs in the welding industry require at least a high school education or an apprenticeship. Other jobs may require a junior college (associate), four-year (bachelor), or advanced degree.

Review Questions

Write your answers on a separate sheet of paper. Please do *not* write in this book.

1. Forge welding has been used since the year _____.
2. *True or False?* In the forge welding process, parts are heated until they are molten.
3. *True or False?* Most welding and cutting equipment is portable.
4. *True or False?* Thicker metal is used for riveted parts to compensate for strength lost due to rivet holes.
5. *True or False?* Welding is performed on both metallic and nonmetallic materials.
6. In what year did welding become an accepted practice for repair and construction?

(continued)

Review Questions *continued*

7. Give three reasons why welding is generally a more efficient process than riveting and machining.
8. Refer to Figure 2-6. Give the correct name for the SMAW process.
9. Refer to Figure 2-6. What AWS abbreviation is used for oxyacetylene welding?
10. Identify one welding occupation that you would like to pursue at each educational level. Discuss your choices with a career counselor. Write a brief summary of each.

Skilled welders are in demand both in small and large industrial companies since manufacturing often depends on the joining of metals. (The Lincoln Electric Company)

Chapter 3

The Physics of Welding

Learning Objectives

After studying this chapter, you will be able to:

- Identify the three general methods by which a weld is achieved.
- Describe the difference between chemical and mechanical properties and give examples of each.
- Explain the effects of welding on metal.
- Identify processes used to heat-treat metal.
- Describe the relationship between voltage and current.
- Give examples of US conventional and SI metric units of measurement.
- Convert US conventional units of measurement to SI metric units.
- Convert SI metric units of measurement to US conventional units.

Technical Terms

amperes
annealing
base metal
brittleness
chemical composition
chemical properties
compressive strength
contraction
corrosion resistance
current
density
ductility
electrode
expansion
grain size
hardness
interpass heating
mechanical properties
normalizing
ohms
open circuit voltage (OCV)
oxidation resistance
physical properties
preheating
quenching
resistance
root opening
shielded metal arc welding (SMAW)
SI metric system
strength
stress-relieving
tempering
tensile strength
toughness
US conventional system
volts
voltage

Welding Theory

Welding is a group of processes used to join metallic or nonmetallic materials. Welding is often done using heat. It can also be done using pressure or a combination of heat and pressure. A filler material may or may not be added to the weld joint.

Welding with Heat

Heat is used to create welds in many welding processes. Filler material is commonly used when welding with heat only. ***Shielded metal arc welding (SMAW)*** is a process that uses heat and filler material. In the SMAW process, heat is created by an arc that is struck between an ***electrode*** and ***base metal*** (metal to be welded). See **Figure 3-1**. The heat causes the end of the electrode and an area of the base metal to melt. When two pieces of base metal are placed together, both are heated by the arc. A portion of each piece melts and the liquid areas flow together. The molten filler material combines with the molten base metal. The molten material cools and becomes solid, creating a weld.

Welding with Heat and Pressure

Some welding processes use both heat and pressure. Filler material is generally not used in these processes. In resistance spot welding, pressure is applied through opposing electrodes. Electrical current flows from one electrode through the base metal to the other electrode. Resistance to the electrical current provides the heat required to join the pieces. A resistance spot welding machine is shown in **Figure 3-2**.

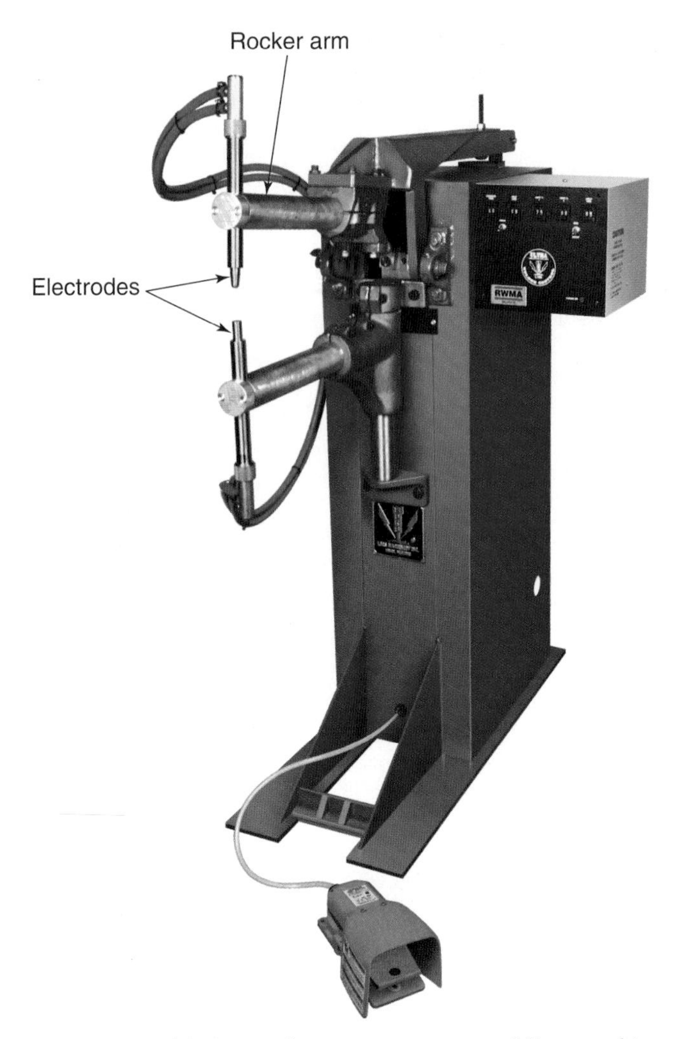

Figure 3-2. This is a rocker-arm-type spot welding machine. Electrodes apply pressure to pieces to be welded. Resistance to electric current between the electrodes results in heat. (LORS Machinery Company)

Welding with Pressure

Metal or other material can be welded together using pressure alone. Heat and filler material are not required. Cold welding is an example of a welding process that does not require the use of heat or filler material. In the cold welding process, very clean pieces of metal are forced together under considerable pressure. The pressure forces the atoms of the materials together to create a weld.

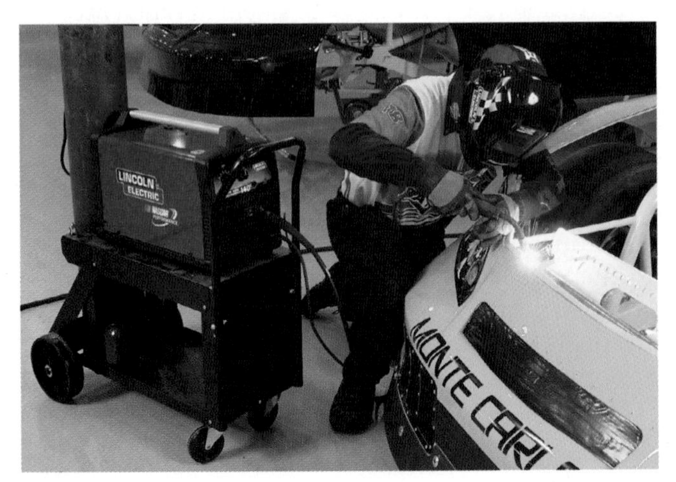

Figure 3-1. This welder is using gas metal arc welding on a NASCAR race car. (The Lincoln Electric Company)

Properties of Metals

The properties of a metal determine how it is used. Properties can be grouped into three categories: physical, chemical, and mechanical. These properties are largely determined by the chemical composition of a metal.

The ***chemical composition*** of a metal alloy consists of the different metals or elements that are combined to produce that alloy. For example, the chemical composition of low carbon steel is a combination of iron and carbon. The chemical composition of stainless steel includes iron, chromium, nickel, manganese, and carbon.

Properties of metals are affected by the method of processing. Processing includes bending, rolling, forming, heat-treating, and welding.

Physical Properties

Physical properties are used to identify or describe a metal. These include color, melting temperature, and density. *Density* is the weight of a particular material per unit volume.

Chemical Properties

Chemical properties determine the way a material reacts in a given environment. Corrosion resistance and oxidation resistance are chemical properties that determine how a material will withstand the effects of the environment. *Corrosion resistance* is the ability of a material to withstand corrosive attack. Acids and saltwater are both corrosive. A material resistant to attack by an acid, however, may not be resistant to the effects of salt water. *Oxidation resistance* is the ability of a material to resist the formation of oxides. Metal oxides occur when oxygen combines with the metal.

Mechanical Properties

Mechanical properties determine how a material reacts under applied loads or forces. Testing procedures for mechanical properties are discussed in Chapter 34.

Strength is the ability of a material to withstand applied loads without failing.

Tensile strength is the ability of a material to resist pulling forces.

Compressive strength is the ability of a material to resist pressing or crushing forces. Tensile strength and compressive strength are exact opposites, with respect to the direction of the applied load. **Figure 3-3** shows the difference between tensile and compressive forces acting on a material.

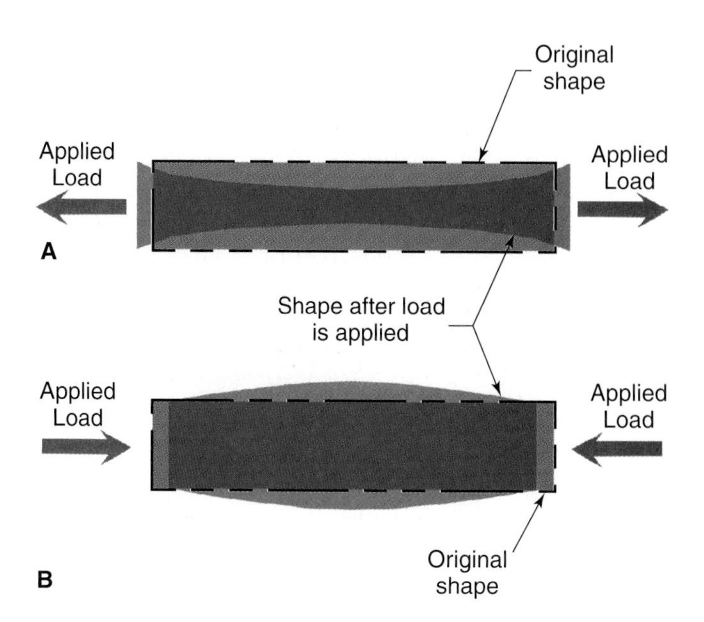

Figure 3-3. *A—When testing tensile strength, the thickness of the material is reduced and length is increased. B—When testing compressive strength, the thickness of material is increased and length is decreased.*

Ductility is the ability of a material to stretch or bend without breaking.

Brittleness is the inability of a material to resist fracturing. Brittleness is the exact opposite of ductility. A material that is brittle has very low ductility and breaks easily.

Toughness is the ability of a material to resist cracking and prevent a crack from progressing. Materials with a high amount of ductility are usually very tough.

Hardness is the ability of a material to resist indentation or scratching. Hardness testing is done by forcing a steel ball or pointed diamond into the surface of a material. Hardness of some steels can be improved through various processes, such as quenching and tempering. Quenching and tempering are discussed later in the chapter.

Grain size is important in determining the mechanical properties of a material. Large or coarse-grained materials have a high amount of brittleness and a low amount of ductility. Fine-grained materials have a low amount of brittleness and a high amount of ductility.

Effects of Welding

The chemical and mechanical properties of a metal are often affected when the metal is welded. Heat produced by welding creates stress in the metal. Heat also affects the ductility and toughness of the metal. If the metal was heat-treated before being welded, the effects of heat-treating are lost in the area around the weld.

A weld that is made correctly is usually stronger than the base metal. An incorrect welding procedure may result in serious problems. For example, a stainless steel pipe that is corrosion-resistant to certain chemicals may lose its resistance at an improperly welded joint. Chemicals within the pipe may attack the weld area, causing the pipe to leak.

Another example where the correct welding procedure is critical is armor plate for the military. Armor plate is very tough. It resists cracks and prevents cracks that do start from getting larger. If an improper welding procedure is used, the metal near the weld area may lose its toughness. If a shell hits the armor, a crack may develop at the weld and continue along the weld joint.

Expansion and Contraction of Metal

When heat is applied to metal, the metal *expands* (increases in size). When heat is removed, the metal cools and *contracts* (reduces in size). ***Expansion*** and ***contraction*** create stress in the metal. The metal may relieve the stress by changing shape (warping). In many welding applications, deformation or movement is not acceptable. Welding jigs or fixtures are commonly used to keep parts from moving. However, while being welded, stress remains in parts that are clamped.

When welding a butt joint, the root opening may be reduced in size toward the end of the weld. The ***root opening*** is the space between the pieces to be welded. This occurs because the weld metal contracts and pulls the pieces together, as in **Figure 3-4A**. To prevent this, the pieces should be tack welded as shown in **Figure 3-4B**. The outer edges of the base metal may also bend toward the weld bead. To prevent this, the pieces may be clamped into a welding fixture or positioned with a reverse angle to compensate for the movement. **Figure 3-5** shows the effects of welding an unrestrained butt joint and a method used to compensate for the movement.

When welding a T-joint, the vertical piece may be pulled toward the weld bead. To prevent this, place the pieces in a welding fixture. Another solution is to tack weld the vertical piece a few degrees from the perpendicular, as shown in **Figure 3-6**.

In general, six techniques may be used to reduce movement and/or stress when welding.

- Tack weld the parts.
- Align the parts to allow for contraction during welding.
- Use welding jigs or fixtures.
- Preheat the parts.
- Heat-treat the welded parts.
- Use the proper welding procedure.

Straightening

The principles of expansion and contraction can be used to straighten parts. If a flat plate or weldment is bent or warped, heat can be used to straighten the piece. An oxyfuel gas torch is used to locally heat the metal, causing it to expand. To be effective, this heating must be done on the correct side of the metal and in the correct location. This metal will not contract to

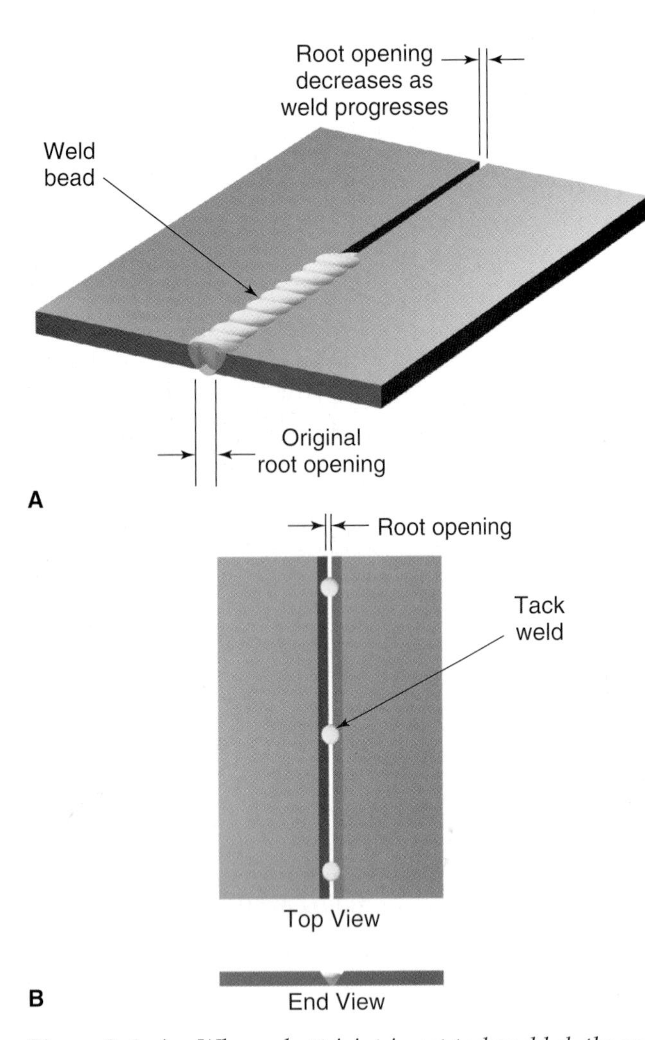

Figure 3-4. A—When a butt joint is not tack welded, the root opening decreases as the weld progresses. In extreme cases, the pieces may overlap. B—Notice the two views of a V-groove butt joint that is tack welded. The root opening is uniform and will be held uniform by the tack welds during the rest of the weld.

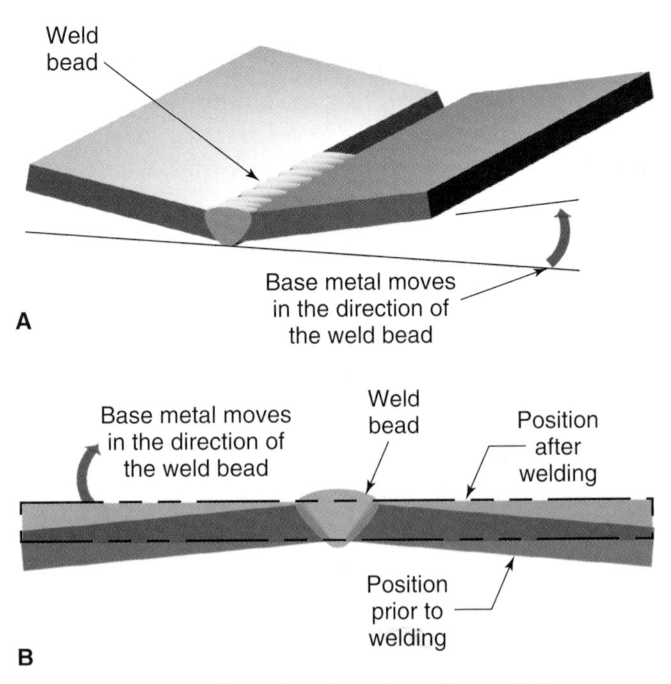

Figure 3-5. A—When the pieces in a butt joint are unrestrained, the outer edges bend toward the weld bead. B—When the pieces in a butt joint are tack welded with a reverse angle, the offset can compensate for the distortion. The amount of offset is determined by experimentation.

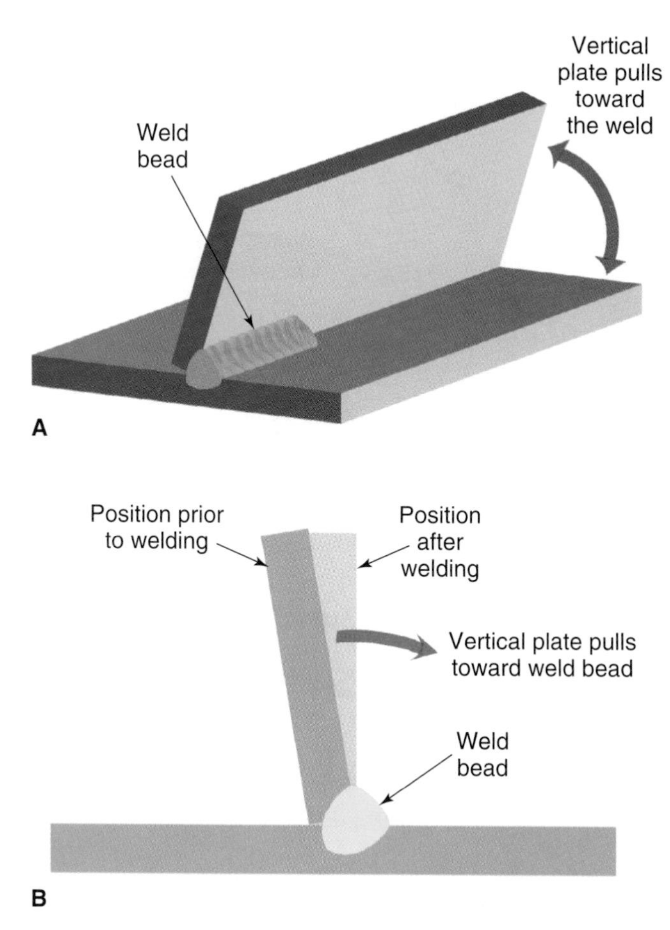

Figure 3-6. A—With unrestrained pieces in a T-joint, the vertical piece is pulled toward the weld bead. B—This T-joint is tack welded a few degrees from vertical. The offset compensates for the movement of the vertical piece.

its original shape or position when it cools. Although straightening a part is not easily performed, it is a skill that can be acquired through practice.

Heat-Treating

Various heat-treating processes are used in industry. Different heat-treating processes are used to accomplish different purposes. Typical processes include annealing, stress relieving, and quenching and tempering. In addition to heat-treating a part after welding, heat may be applied before or during the welding process. Preheating is done just before the welding operation. Interpass heating is done while the weld is being completed.

Preheating

Preheating is used to raise the temperature of the metal before welding. Preheating causes less local expansion of the part during welding. When preheating, the entire part is heated rather than one specific area. When the part cools, less contraction occurs, and less stress is developed.

Interpass Heating

Interpass heating is a method of heating metal while it is being welded or between weld passes. Interpass heating is commonly used on thick plates and pipes. Preheating is used to heat the metal to the desired temperature. Interpass heating is used to maintain an elevated temperature. Preheating and interpass heating are both used to reduce or minimize the amount of expansion, contraction, and stress resulting from welding.

Annealing and Normalizing

Annealing and ***normalizing*** are heat-treating processes in which a metal is heated and allowed to cool slowly. The result is a decrease in the hardness and an increase in the ductility of the metal, making the material easier to bend and machine. The material is also less likely to crack and become distorted.

Normalizing involves heating a metal to a very high temperature (1670°F [910°C] for steel) in an oven or furnace for a required time and then removing the metal and allowing it to cool to room temperature. Although annealing is a process similar to normalizing, it requires a more carefully controlled and much slower cooling process. After the metal has been heated to the proper temperature for the correct amount of time, it is kept in an annealing oven where the temperature is slowly lowered in small increments.

Stress-Relieving

Stress-relieving is similar to annealing, except that lower temperatures are used. For example, to relieve stress, mild steel is heated to only 1200°F (650°C) and is kept at that temperature for a few hours. The metal is then allowed to air cool.

Quenching and Tempering

Quenching and tempering are processes used to harden steel and steel alloys. In the ***quenching*** process, metal is heated to a fairly high temperature. The temperature is maintained for a given period of time. The metal is then cooled quickly by immersing it in a bath of water, oil, or other liquid. This produces a very hard and brittle metal. The metal is then tempered by reheating it to several hundred degrees and cooling it. After ***tempering***, the metal is less hard and no longer brittle. A tempered metal has good toughness.

Electrical Principles

Electricity is produced by the movement of electrons within a circuit. Electricity is measured in terms of voltage and current. ***Voltage*** is the force that causes electrons to flow through a circuit. Voltage is measured in ***volts***. It can be compared to water pressure.

When your kitchen sink faucet is turned off, water pressure is still present. See **Figure 3-7A**. When the faucet is opened, water begins to flow because the pressure is forcing the water out of the faucet. Likewise, voltage is always present in an electrical circuit. Although an arc may not be struck (started) between the base metal and electrode, voltage is still present, **Figure 3-7B**. This is known as ***open circuit voltage (OCV)***. When an arc is struck, voltage forces electrons across the arc.

The air gap between the electrode and base metal offers resistance to the flow of electrons. ***Resistance*** is the opposition to the flow of electrons. Resistance is measured in ***ohms***. A higher voltage setting on electric arc welding equipment allows the arc length to be longer. The arc stops if the arc length is longer than the voltage allows.

Current is the flow of electrons in an electrical circuit. Current is measured in ***amperes***, or amps. Current can be compared to the flow of water from a faucet. When the faucet is turned off, water is unable to flow. When the faucet is opened, water begins to flow, as in **Figure 3-7C**. Likewise, when there is no arc, there is no current flowing. When an arc is struck, current is produced as electrons flow across the arc. See **Figure 3-7D**.

Sometimes, when you turn a faucet all the way on, water flows very fast with a great deal of force. This is due to high water pressure and a high flow rate. If you turn the faucet halfway off, the same pressure is there, but the flow rate is reduced. At other times, when a faucet is turned all the way on, only a small amount of water flows, and it does not have much force. This is due to low water pressure and a low flow rate. If you place your thumb over part of the faucet opening, water is delivered with much greater force. The pressure is higher, although the flow rate has not changed. Similar principles apply to voltage and current in an electrical circuit. In some welding applications, a higher voltage and/or current is required. In other applications, a lower voltage and/or current may be required.

In later chapters, instructions will be given on the setup of different types of welding machines. Most machines require that the welder set the voltage and/or the current that will be used.

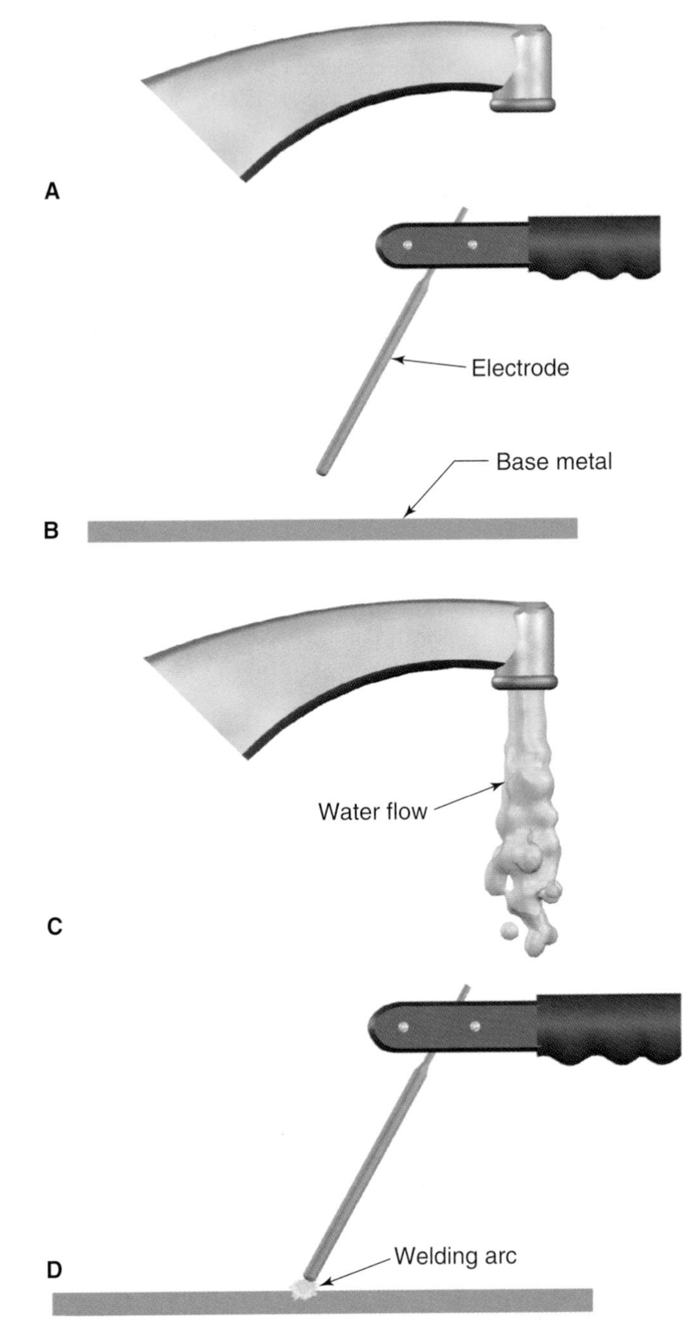

Figure 3-7. A—The faucet is off. Water does not flow, but water pressure exists in the pipe. B—A welding arc has not been struck. Electrons do not flow, but voltage exists in the circuit. C—The faucet is on. Water flows, and the water pressure drops slightly. D—An arc is struck. Electrons flow, and the voltage drops slightly.

Units of Measurement

There are two different measurement systems in common use—the US conventional system and the SI metric system. The ***US conventional system*** is used primarily in the United States. The ***SI metric system*** is the standard in much of the rest of the world and is also used in the United States. Some of the basic units used in the US conventional system are inches, feet, gallons, pounds, and pounds per square inch. Some of the basic units in the SI metric system are millimeters, meters, liters, kilograms, and pascals.

In the US conventional system, the basic unit of measurement is commonly converted and renamed. For example, 12 inches are converted to 1 foot, 3 feet are converted to 1 yard, and 1760 yards are converted to 1 mile. **Figure 3-8** lists common conversions in the US conventional system.

The SI metric system adds a prefix to the basic unit of measurement to increase or decrease the value. For example, there are 1000 millimeters in 1 meter,

Common Conversions: US Conventional System		
Linear Measurement		
12 inches	=	1 foot
3 feet	=	1 yard
36 inches	=	1 yard
1760 yards	=	1 mile
5280 feet	=	1 mile
Liquid and Dry Measurement		
2 pints	=	1 quart
4 quarts	=	1 gallon
31.5 gallons	=	1 barrel
Square Measurement		
144 square inches	=	1 square foot
9 square feet	=	1 square yard
1296 square inches	=	1 square yard
Cubic Measurement		
7.48 gallons	=	1 cubic foot
1728 cubic inches	=	1 cubic foot
27 cubic feet	=	1 cubic yard
202 gallons	=	1 cubic yard

Figure 3-8. *This table displays common conversions in the US conventional system.*

Common SI Metric Prefixes		
micro	=	1/1000000 or .000001
milli	=	1/1 000 or .001
centi	=	1/100 or .01
deci	=	1/10 or .1
kilo	=	1 000
mega	=	1 000 000
The unit of length is the METER		
Examples of prefixes:		
millimeter (mm)	=	.001 or 1/1000 of a meter
centimeter (cm)	=	.01 or 1/100 of a meter
kilometer (km)	=	1000 meters
The unit of weight is the GRAM		
Examples of prefixes:		
milligram (mg)	=	.001 or 1/1 000 or a gram
centimeter (cm)	=	1000 grams
megagram (Mg)	=	1000000 grams
The unit of volume is the LITER		
Examples of prefixes:		
milliliter (mL)	=	.001 or 1/1000 of a liter
kiloliter (kL)	=	1000 liters
The unit of time is the SECOND		
Examples of prefixes:		
millisecond	=	.001 or 1/1 000 sec.
microsecond	=	.000001 or 1/1000000 sec

Figure 3-9. *It is important to be familiar with common prefixes used in the SI metric system. Examples of basic units of measurement and prefixes are also shown.*

and 1000 meters in 1 kilometer. In these examples, *meter* is the basic unit of measurement. The prefixes *milli* and *kilo* are added to it to change the value. See **Figure 3-9** for a list of common prefixes used in the SI metric system.

Conversions between the US conventional system and the SI metric system may need to be made occasionally. Common conversions are listed in **Figure 3-10**.

Property	To Convert From	To	Multiply By
Area	in.^{2}	mm^2	645.16
	in.^{2}	m^2	.00064516
	ft.^{2}	mm^2	92903
	ft.^{2}	m^2	.092903
Current density	A/in.^{2}	A/mm^2	.00155
	A/mm^2	A/in.^{2}	645.16
Deposition rate	lb./h	kg/h	.045
	kg/h	lb./h	2.2
Electrode force	lb. (force)	N	4.4482
	kg (force)	N	9.8067
	N	lb. (force)	.22481
Flow rate	cfh	Lpm	.47195
	gallons/h	Lpm	.06309
	gallons/min.	Lpm	3.7854
	Lpm	cfh	2.1188
Heat input	J/in.	J/m	39.37
	J/m	J/in.	.0254
Length	in.	mm	25.4
	in.	m	.0254
	ft.	mm	304.8
	ft.	m	.3048
	mm	in.	.03937
	mm	ft.	.0032808
Mass	lb.	kg	.45359
Pressure (gas and liquid)	psi	Pa	6894.8
	psi	kPa	6.8948
	N/mm^2	Pa	1,000,000
	kPa	psi	.14504
	kPa	lb/ft^2	20.885
	kPa	N/mm^2	.0001
Tensile strength	psi	kPa	6.8948
	N/mm^2	MPa	1.000
	MPa	psi	145.04
Torque	in.·lb.	N·m	.11298
	in.·lb.	N·m	1.3558
Travel speed	in./min.	mm/s	.42333
	mm/s	in./min.	2.3622
Volume	in.^{3}	mm^3	16387
	in.^{3}	m^3	.000016387
	ft.^{3}	mm^3	28316850
	ft.^{3}	m^3	.028317
	in.^{3}	L	.016387
	ft.^{3}	L	28.317
	gallons	L	3.7854

To convert from °F to °C, use °C = (°F - 32) ÷ 1.8.
To convert from °C to °F, use °F = (1.8 × °C) + 32.

Figure 3-10. *This table shows many common conversions for US conventional and SI metric systems. To convert from units in Column 2 to Column 3, multiply by the value in Column 4. To convert from units in Column 3 to Column 2, divide by the value in Column 4.*

Summary

- Heat is used to create welds in many welding processes. Filler material is commonly used when welding with heat only.
- Some welding processes use both heat and pressure. Filler material is generally not used in these processes.
- Metal or other material can be welded together using pressure alone. Heat and filler material are not required.
- The properties of a metal determine how it is used. Properties can be grouped into three categories: physical, chemical, and mechanical. These properties are largely determined by the chemical composition of a metal.
- The chemical and mechanical properties of a metal are often affected when the metal is welded. A weld that is made correctly is usually stronger than the base metal. Heat treating may be used to obtain desired base metal properties.
- Electricity is produced by the movement of electrons within a circuit. Electricity is measured in terms of voltage and current.
- There are two different measurement systems in common use. In the US conventional system, the basic unit of measurement is commonly converted and renamed. In the SI metric system, a prefix is added to the basic unit of measurement to increase or decrease the value.

Review Questions

Write your answers on a separate sheet of paper. Please do *not* write in this book.

1. A weld can be achieved through what three methods?
2. Briefly define the following terms.
   A. Tensile strength
   B. Ductility
   C. Toughness
   D. Hardness
3. *True or False?* Welding changes the properties of the metal being welded.
4. What happens to the size of base metal when it is heated?
5. What can happen when welding a T-joint that is not tack welded or clamped in a welding jig or fixture?
6. List four ways to reduce stress in a weld.
7. What type of heat-treatment process is used to harden steel and steel alloys?
8. Define the following terms.
   A. Voltage
   B. Current
   C. Resistance
9. What are the two measurement systems in common use?
10. List the unit from the other measurement system that corresponds to the given unit of measurement.
    Length: inch _____
    Mass: pound _____
    Electrode force: pound _____
    Volume: liter _____
    Tensile strength: pounds per square inch (psi) _____

Chapter 4

Weld Joints and Positions

Learning Objectives

After studying this chapter, you will be able to:

- Identify the five basic weld joints.
- Identify the types of welds that can be made on each joint.
- Identify the parts of a fillet weld.
- Identify the parts of a groove weld.
- Describe a stringer bead and a weaving bead.
- List the four welding positions.
- State the conditions for welding in the four welding positions.

base material
base metal
bevel angle
butt joint
corner joint
cover pass
edge joint
edge preparation
effective throat
face reinforcement
filler material
filler pass
fillet weld
fillet weld size
flange joint
flare-groove joint
flat (1G) welding position
groove angle
groove face
groove joint
groove weld
horizontal (2G) welding position
inside corner joint
joint geometry
joint penetration
lap joint
leg
multiple-pass weld
outside corner joint
overhead (4G) welding position
penetration
root face
root pass
root reinforcement
stringer bead
tack weld
T-joint
vertical (3G) welding position
weave bead
weld axis
weld bead
weld face
weld joint
weld pass
weld root
weld size
weld toe
welding positions

Basic Weld Joints

A ***weld joint*** refers to how the parts to be joined are assembled prior to welding. There are five basic types of joints used in welding: butt, lap, corner, T-, and edge. See **Figure 4-1**.

The metal to be joined is called the ***base metal***. If the part to be welded is not metal, it is called ***base material***. It is also known as the workpiece or work.

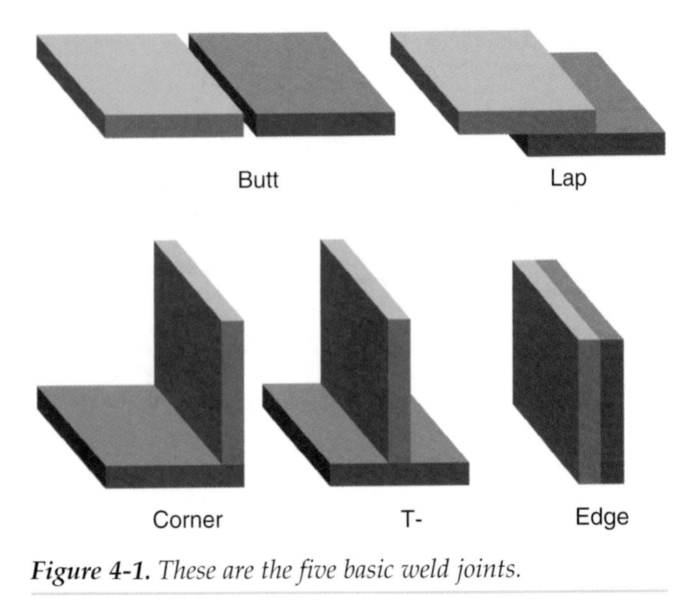

Figure 4-1. These are the five basic weld joints.

The edges of the base metal are often machined, sheared, gouged, flame cut, or bent to prepare them for welding. Weld joint design and metal thickness usually determine how the joint is prepared. Generally, the weld joint design is determined by an engineer.

Butt Joint

Butt joints are used when parts are joined edge-to-edge. Common examples of butt joints are the deck plates on a ship or the pipes of an oil pipeline. Both are assembled end-to-end. There are a variety of butt joint configurations, depending on how the ends of the pieces being joined are prepared. For quality welds to be produced on butt joints, the edges of the base metal often require special preparation before welding.

Edge preparation refers to how the edges of the joint are shaped prior to welding. If the base metal is thin, the edges may just be squared without additional machining or cutting. The edges of thin metal may also be bent to form flare-groove or edge-flange joints, as shown in **Figure 4-2**.

Generally, when base metal over 3/16″ (4.8 mm) thick is used, edges are beveled by machining or flame cutting. Edge preparation is required to allow the weld to penetrate to the required depth. Thick base metal may be machined, gouged, or flame cut along the upper or lower edges of the joint, or both, to form a double-bevel, V-, J-, or U-groove.

A butt joint may be prepared using any of the edge preparations shown in Figure 4-2. A ***groove weld*** is made by fusing molten filler metal into a butt joint that has been set up in a groove formation. These groove formations include a single-square groove; a single or double bevel, V-, J-, or U-groove; or a flared groove, which forms a groove where the edge is bent.

A welder should know the names of the various parts of a ***groove joint***, as shown in **Figure 4-3**. The

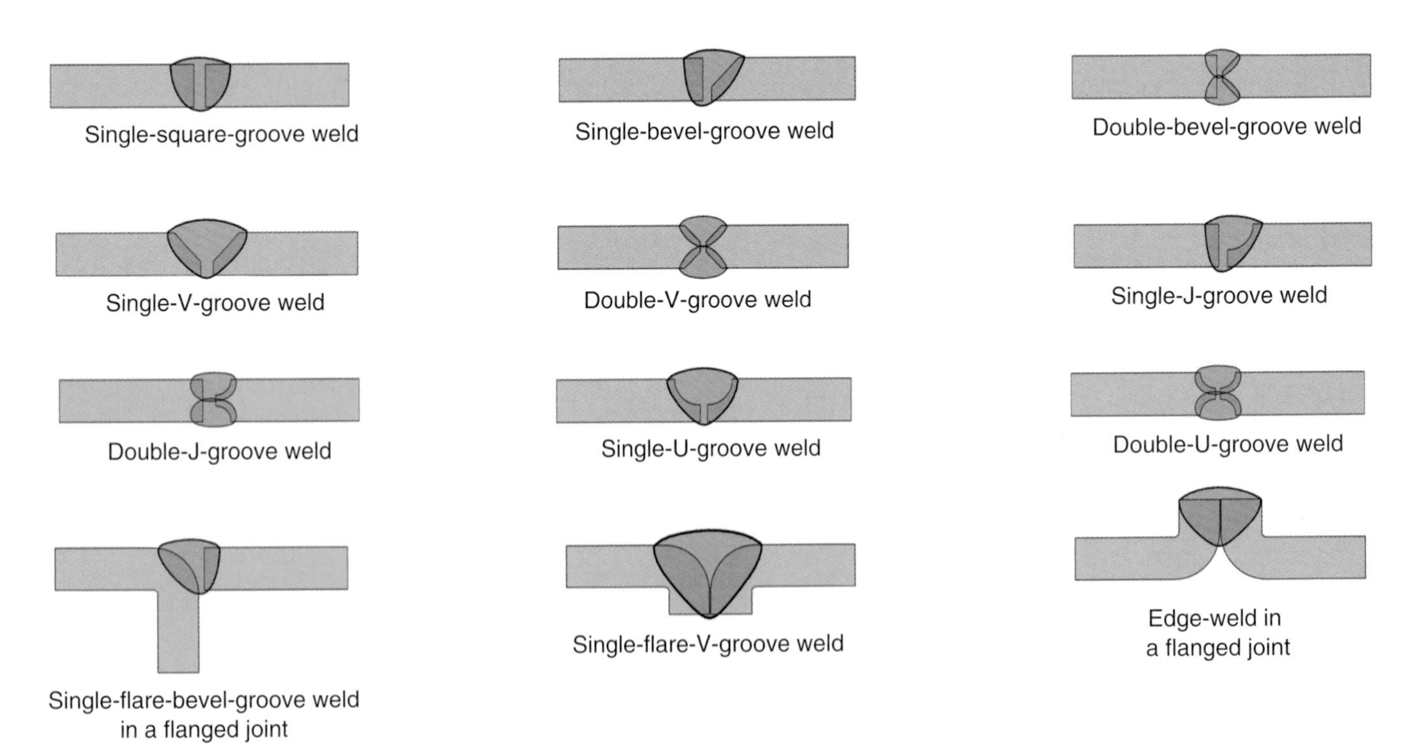

Figure 4-2. Students should become familiar with the various methods of preparing the edges of a butt joint. The completed weld is shown in blue. Double grooves are used on thick metal that is welded from both sides. The base metal is bent to form the bottom three joints.

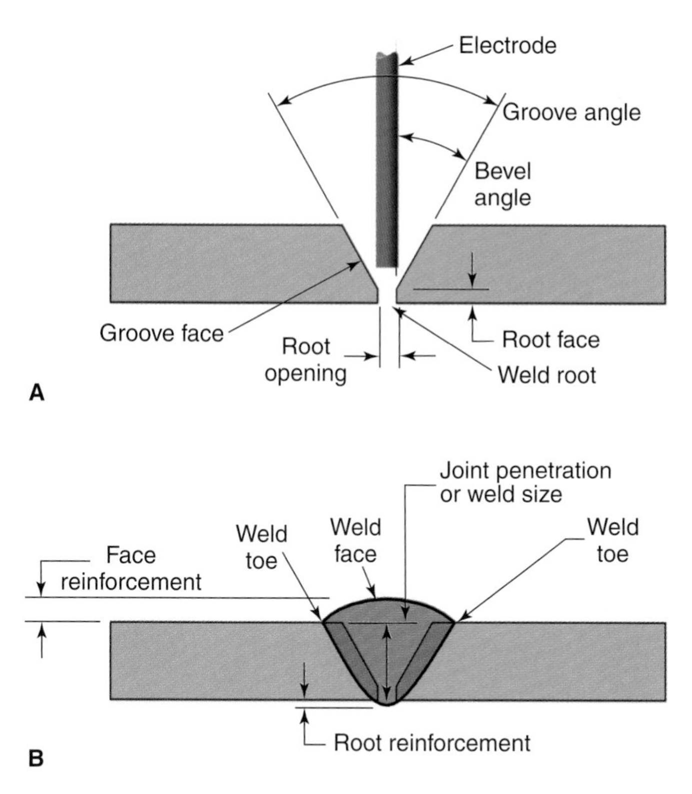

Figure 4-3. A—These are the parts of a grooved butt joint. The groove angle should be just large enough to allow the torch or electrode to reach the root opening. B—The parts of a groove weld.

groove face is the surface formed on the edge of the base metal after it has been machined or flame cut. The total angle formed between the groove face on one piece and the groove face on the other piece is the ***groove angle***. The ***bevel angle*** is the angle between the bevel of the joint and a plane perpendicular to the surface of the base material. The ***weld root*** is the point where the weld intersects the base metal surface near the bottom of the joint. The distance from the weld root to the point where the bevel angle begins is the ***root face***. The root opening is the distance between the two pieces at the root of the weld.

Figure 4-3 shows a cross section of a completed weld. The ***weld face*** is the outer surface of the weld bead on the side the weld was made. ***Face reinforcement*** is the distance from the top of the weld face to the surface of the base metal. The ***weld toe*** is the point where the weld bead contacts the base metal surface. It occurs twice on each weld bead. ***Root reinforcement*** is the distance that the penetration projects from the root side of the joint. ***Joint penetration*** or ***weld size*** is the depth that a weld extends into the joint from the surface.

Lap Joint

A ***lap joint*** is formed by two overlapping pieces of base metal. The top surface of one piece is in contact with the bottom surface of the other, as shown in Figure 4-1. Special edge preparation is not required.

Corner Joint

A ***corner joint*** is formed by placing two pieces of base metal perpendicular or at an angle to one another so that the edge of one piece of base metal intersects the surface of the other piece near its outer edge. At least one edge of the two pieces is exposed, as illustrated in **Figure 4-4**. The pieces may be joined at any angle, but they are commonly welded at a 90° angle. Corner joints may be welded as inside corners, outside corners, or a combination of both. ***Inside corner joints*** are welded along the inside of the intersection of the two pieces. ***Outside corner joints*** are welded along the outside edge of the joint. The edges may be square, beveled, J-grooved, flared, or edge-flanged. See Figure 4-4.

T-Joint

A ***T-joint*** is formed by two pieces of base metal that are at an angle of approximately 90° to one another. The main difference between a corner and T-joint is that a corner joint is formed along the edges of both pieces, while a T-joint is formed at the edge of one piece and away from the edge of the second piece. The edges of the base metal may be prepared as a square, bevel-grooved, J-grooved, or flared-bevel-groove joint, as shown in **Figure 4-5**. Both edges of the base metal may be prepared to form a double-bevel-groove joint.

Edge Joint

An ***edge joint*** is formed when the surfaces of two pieces are in contact and their edges are flush (even). The pieces are joined by welding along at least one of the flush edges. **Figure 4-6** shows the edge preparation for various edge joints.

Flange Joint

A ***flange joint*** is formed when the edge of one or more pieces of the joint is bent to form a flange. The flanged or unflanged edges are aligned and a weld is placed along the specified edges. Figures 4-2, 4-4, 4-5, and 4-6 show flange joints.

Flare-Groove-Joint

Flare-groove joints are formed when the flanged edges of one or both pieces are placed together to form a single-flare-bevel or double-flare-V-groove. The weld is placed in the bevel or V-groove, as shown in Figures 4-2, 4-4, and 4-5.

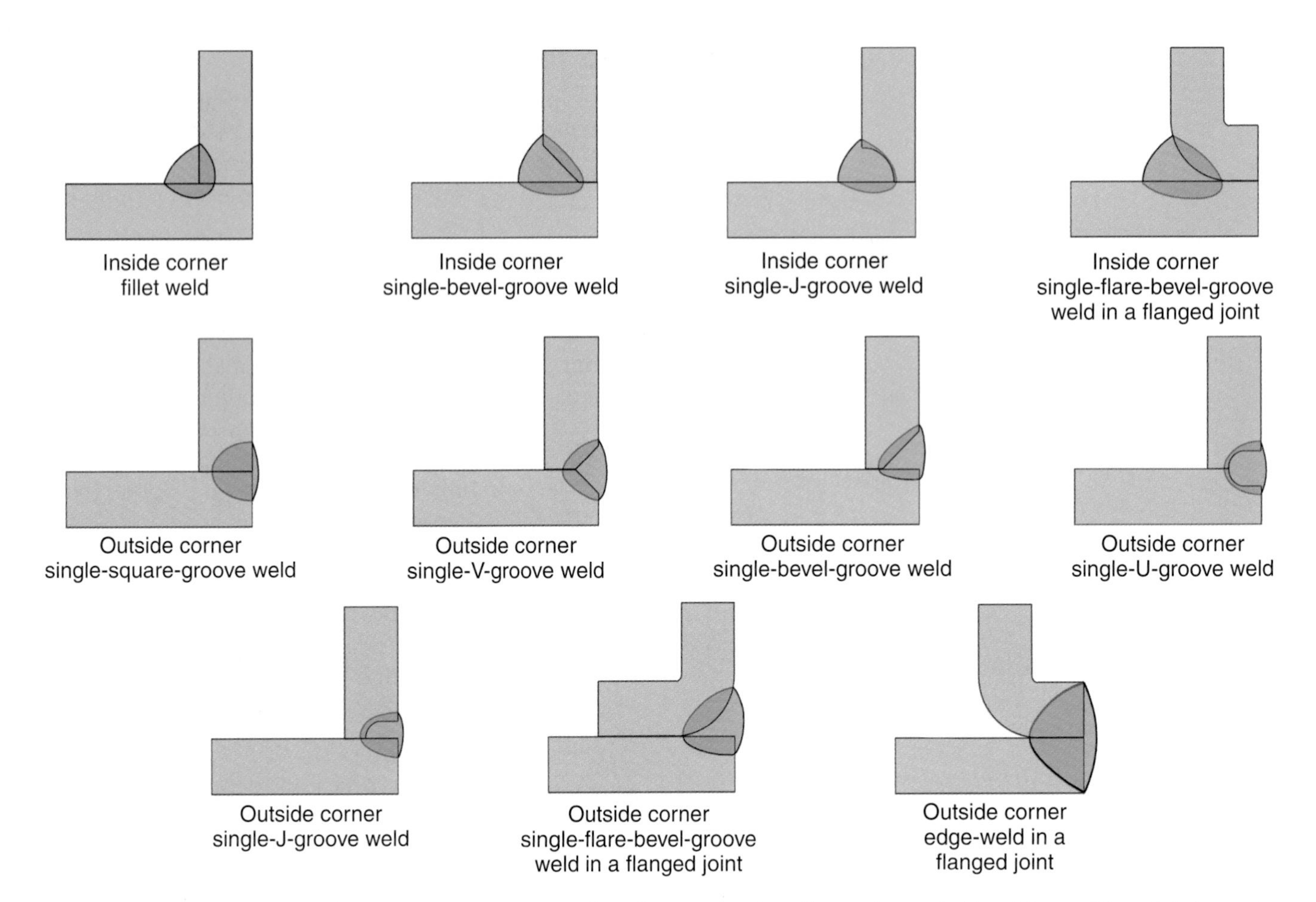

Figure 4-4. *For each method of corner joint edge preparation, the completed welds are shown in blue. Three of these joints require the base metal to be bent for proper formation.*

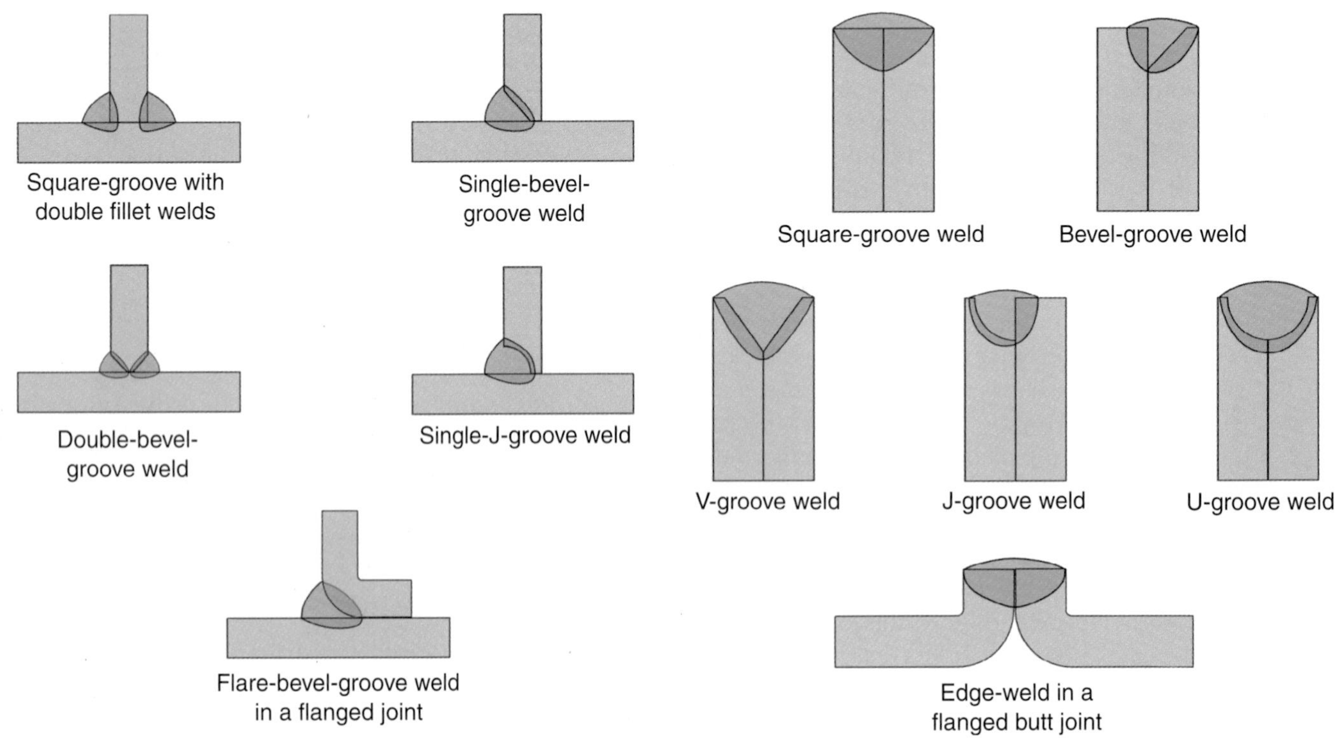

Figure 4-5. *For each method of T-joint edge preparation, the completed welds are shown in blue.*

Figure 4-6. *For each method of edge-type joint edge preparation, the completed welds are shown in blue.*

Types of Welds

A weld is defined as "the blending or mixing of two or more metals or nonmetals by heating them until they are molten and flow together. This may be done with or without the addition of a filler material." See **Figure 4-7**. Welding is the process of making a weld on a joint. ***Fillet welds*** are made at the intersection of a surface and an edge or in a corner where two surfaces meet. Fillet welds are generally triangular in shape, as shown in **Figure 4-8**, and are placed into lap, inside corner, and T-joints. Groove welds can be used on all types of weld joints.

When the edges of thicker metal are machined or flame cut, metal is removed from the pieces. ***Filler material*** must be added to replace the metal that is removed. The addition of filler metal ensures that the completed weld joint is as thick and as strong as the base metal. Edge, flange, or flare-groove joints for thin metal may be welded without the addition of filler material. Figures 4-2, 4-4, 4-5, and 4-6 show edge, flange, and flare-groove joints.

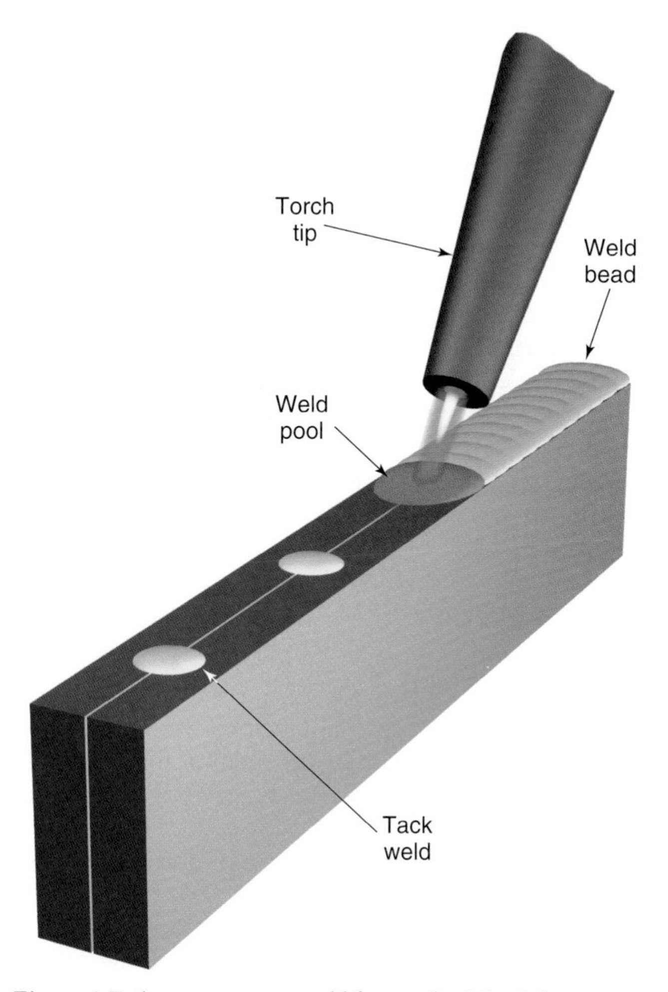

Figure 4-7. *A square-groove weld for an edge joint is in progress. The weld pool extends to the outer edges of the base metal. Filler metal may not be required on thin pieces of base metal.*

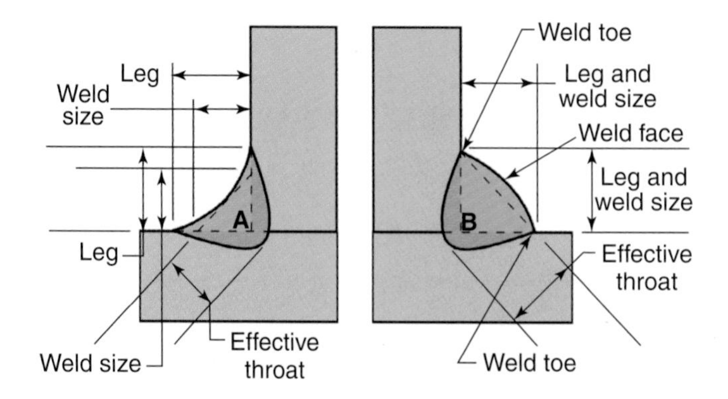

Figure 4-8. *Look at the parts of these fillet welds. Weld A is concave. Its weld size is smaller than Weld B, which is a convex weld. Notice that the leg sizes in Weld A and B are the same, but the weld size is larger with a straight or slightly convex bead.*

The parts and dimensions for fillet welds are the same for lap, inside corner, and T-joints. Refer to Figure 4-8. The weld face is the outer surface of the weld bead. As previously mentioned, the weld toe is the point where the weld face touches the surface of the base metal. A fillet weld is made up of three primary dimensions. The ***fillet weld size*** is the length of one side. The ***leg*** is the shortest distance from the toe to the surface of the other piece of base metal. The ***effective throat*** is the minimum distance from the weld face to the root of the weld without any convexity.

Figure 4-8 shows two fillet welds with the same leg dimensions, but different sizes. The size of the weld with a concave bead, Figure 4-8A, is smaller than the size of the weld with a convex bead, Figure 4-8B. A fillet weld with a convex bead is stronger than one with a concave bead because of the additional filler metal.

Weld Beads and Weld Passes

A ***weld bead*** is one weld pass of filler metal that is added to a weld joint. A ***weld pass*** occurs each time a welder lays one weld bead across a weld joint. Only one weld bead or weld pass is required for fusing thin base metal, **Figure 4-9A**. If the metal is thick, more than one weld pass is required to make a strong joint with complete penetration. See **Figure 4-9B**. A ***multiple-pass weld*** is sometimes required, **Figure 4-10**. The first weld pass is the ***root pass***. Intermediate weld passes are ***filler passes***. The final weld pass is the ***cover pass***. Generally, a weld bead should not be thicker than 1/4" (6.4 mm). A weld bead may be made as a stringer bead or a weave bead.

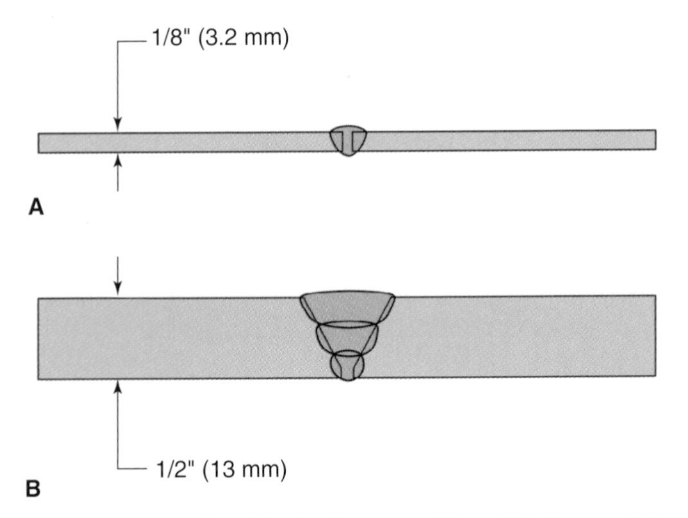

Figure 4-9. A—On this single-pass weld on thin base metal, notice the build-up of weld metal and complete penetration. B—On this multiple-pass weld on thick base metal, the edges have been prepared to form a V-groove joint. Notice the root opening required. Three beads were used, with each bead measuring less than 1/4" (6.4 mm) thick. A weaving bead may be used for the wider, upper bead.

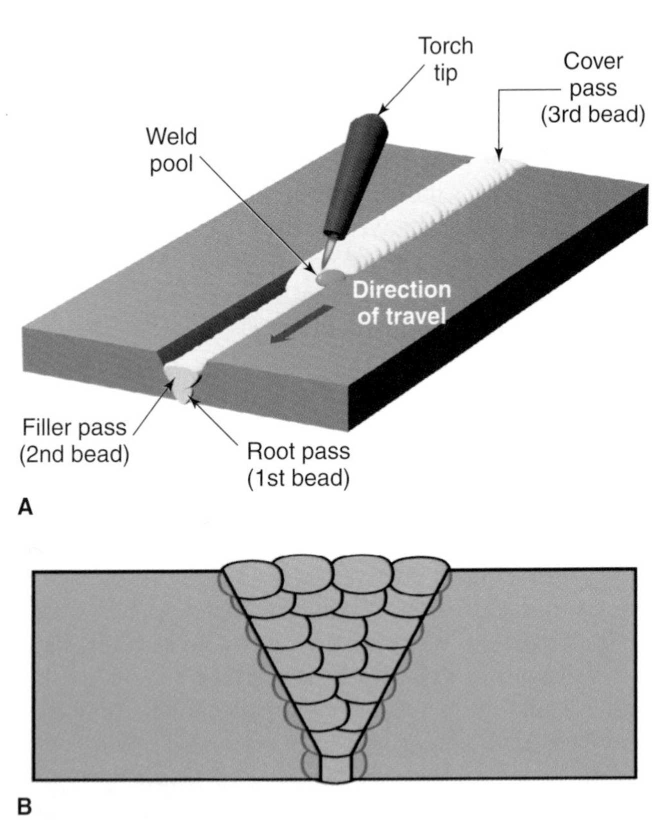

Figure 4-10. This is a multiple-pass weld. A—Three weld passes are used in this example, but any number of passes may be used. B—Twenty passes were made in this weld.

Stringer Bead

A ***stringer bead*** is used when a standard bead width is acceptable. Stringer beads are made by moving the torch or electrode holder along the weld without any side-to-side motion. See **Figure 4-11A**. A deep and wide joint can be filled using several stringer beads.

Weave Bead

A ***weave bead*** is used to create a wider weld pool. A weave bead is formed by moving the torch or electrode holder from side-to-side as the weld pass progresses along the weld joint. See **Figure 4-11B**.

Various torch or electrode movement patterns can be used when making a weave bead. The crescent motion, shown in **Figure 4-11C**, is one of the most popular patterns.

Joint Geometry

The American Welding Society defines ***joint geometry*** as "the shape and dimensions of a (weld) joint, in cross section, prior to welding." Joint geometry is generally determined by a welding engineer or designer. The assembly design and the dimensions of a joint depend on the metal thickness and shape and on the load requirements of the parts. The parts are prepared to ensure that the weld will have adequate penetration. The joint geometry design also provides space for the welder to reach near the bottom of the weld joint with the torch or electrode.

Preparation

The edges of thick metal are prepared for welding by flame cutting, gouging, or machining. Preparation allows the weld to penetrate as deep as required by the engineer or weld designer. A groove joint allows the welder to reach the bottom of the weld joint. The groove angle must be large enough to allow the torch tip or electrode to reach near the bottom of the joint. If the groove angle is too large, filler metal and the welder's time are wasted. This increases the cost of making a weld. See **Figure 4-12A**. A properly designed J-groove or U-groove joint also decreases the groove dimensions while allowing adequate space for welding. See **Figure 4-12B**.

Joint Alignment

The alignment of a joint before welding is very important. In the shop, the alignment of the weld joint is often referred to as "fit-up." A ragged edge or an edge that is not cut straight is hard to weld. See **Figure 4-13**. Edges to be welded must be straight and cut to exact size.

Parts of a weldment should be properly aligned and held in position during the welding operation. Tack welding is usually adequate to hold parts during welding. A ***tack weld*** is a small weld used to hold pieces in alignment. Parts may also be held mechanically during the welding operation because the metal expands,

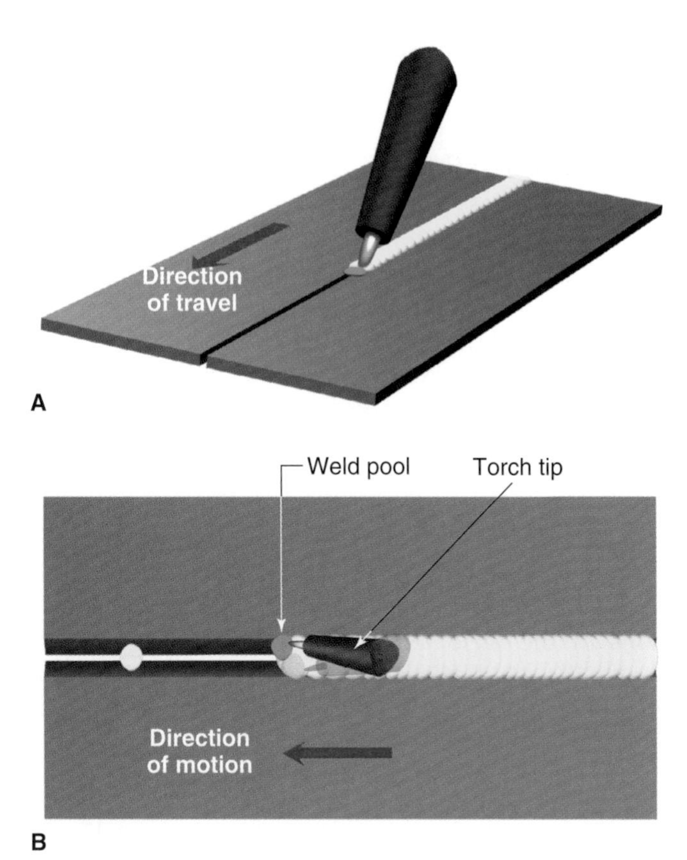

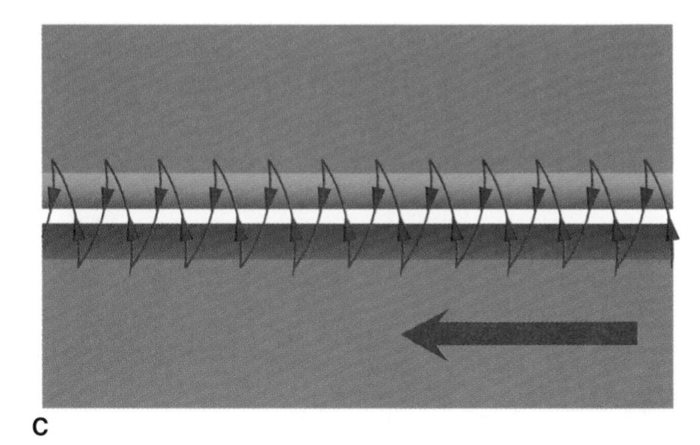

Figure 4-11. A—A stringer bead is in progress on a square-groove butt joint. The bead width is two to three times the metal thickness. B—A weave bead is in progress. The torch tip and weld pool are moved from side to side in the direction of the arrow. C—Here is a suggested motion for a weave bead. The bead width is seldom greater than 3/4″–1″ (19 mm–25 mm).

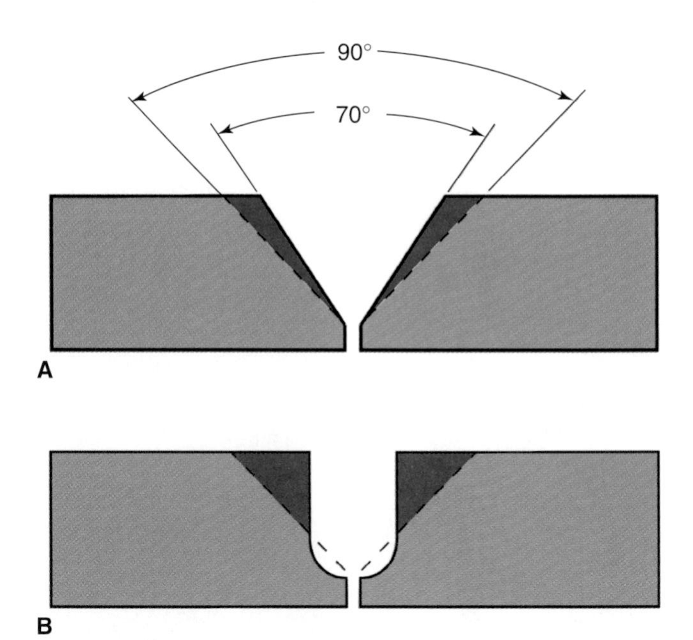

Figure 4-12. A—Compare the 70° and 90° groove angles. A 70° groove angle is cheaper to weld than a 90° groove angle. The shaded area represents an unnecessary cost in filler metal and welder time. B—Look at the U-groove joint. The root of the weld can be reached easily. Little filler metal and welder time are wasted.

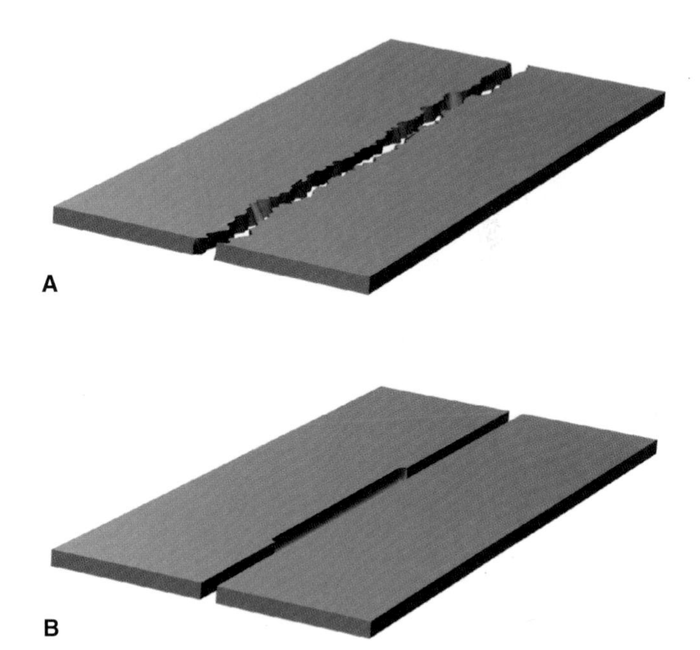

Figure 4-13. Poorly prepared base metal edges. The edges in A are ragged. One edge in B is not cut straight, which changes the width of the joint. Both joints would be difficult to weld.

bends, and changes shape when heated. Clamps or other devices, such as jigs and fixtures, are used to hold weldments when welding. See **Figure 4-14**.

Penetration

A completed weld joint must be at least as strong as the base metal. The weld must penetrate deeply into the base metal to be strong. ***Penetration*** is the depth of fusion of the weld below the surface. Total (100%) penetration occurs when a weld penetrates through the entire thickness of the base metal. Generally, total penetration is required only on a butt joint. The edges of thick metal may need to be machined or flame cut

Figure 4-14. These various clamps are used to position and hold parts to be welded. (James Morton, Inc.)

to achieve 100% penetration. Thick metal also may have to be welded from both sides of the joint.

Welding Positions

Welders often must weld in a variety of positions. Welds may be made in the flat, horizontal, vertical, or overhead welding positions. See **Figure 4-15**. On welding drawings, these positions are often abbreviated in the tail of the welding symbol as F, H, V, and O. The American Welding Society refers to ***welding positions*** with a number and letter combination. Groove joints in the flat, horizontal, vertical and overhead positions are referred to as 1G, 2G, 3G, and 4G, respectively. Fillet joints in the flat, horizontal, vertical, and overhead position are designated as 1F, 2F, 3F, and 4F, respectively.

Welding positions are determined by the positions of the weld axis and weld face. **Figure 4-16** shows the weld axis and weld face. The ***weld axis*** is an imaginary line running lengthwise through the center of a completed weld. The weld face is the exposed surface of a completed weld on the side on which the welding was done.

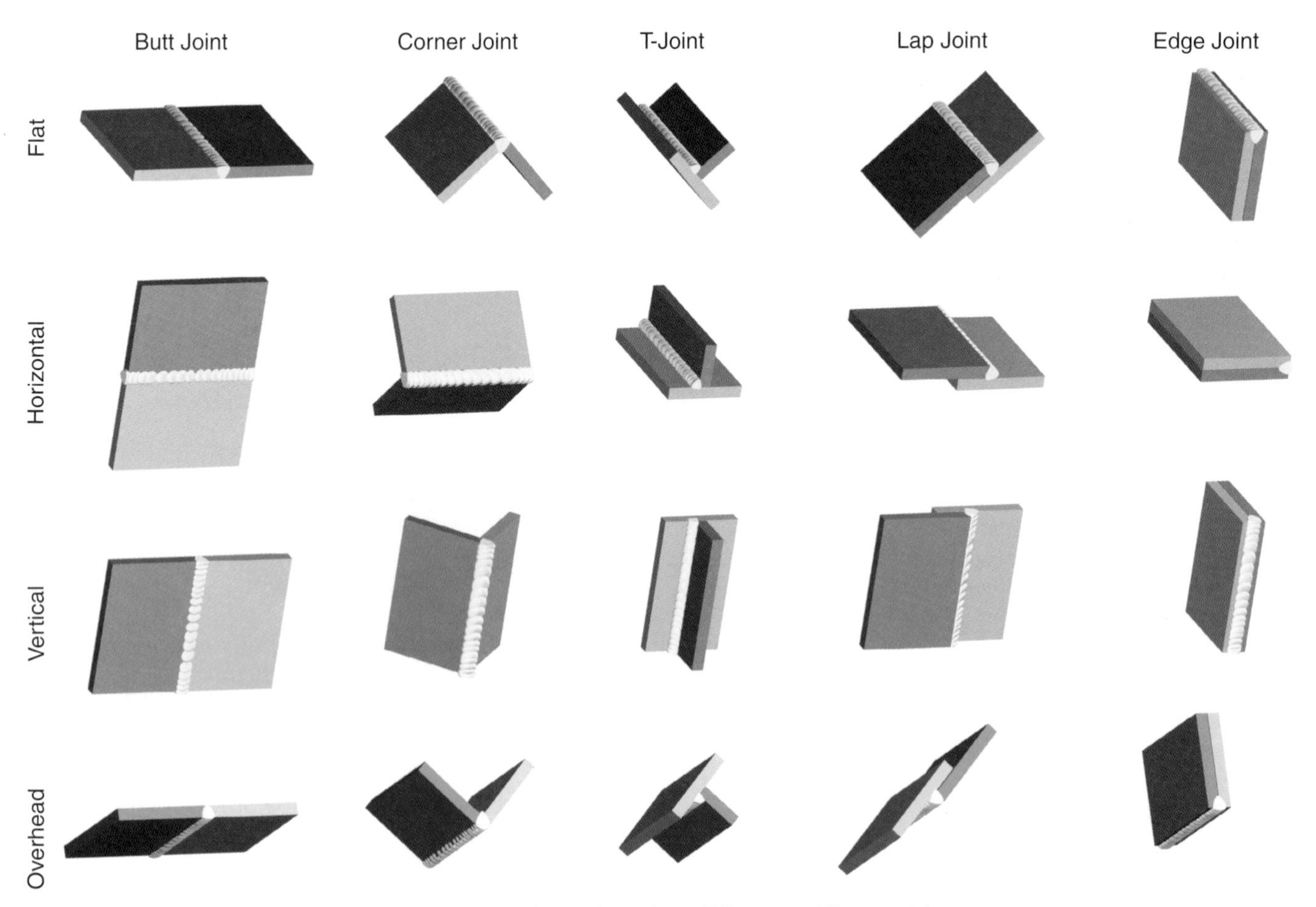

Figure 4-15. Each of the five basic weld joints may be made in four different welding positions.

Flat (1G) Welding Position

Welds made on a groove joint in the ***flat (1G) welding position*** must meet these conditions:

- The weld axis must be within 15° of horizontal. See **Figure 4-17**.
- The weld face is within 30° of horizontal.

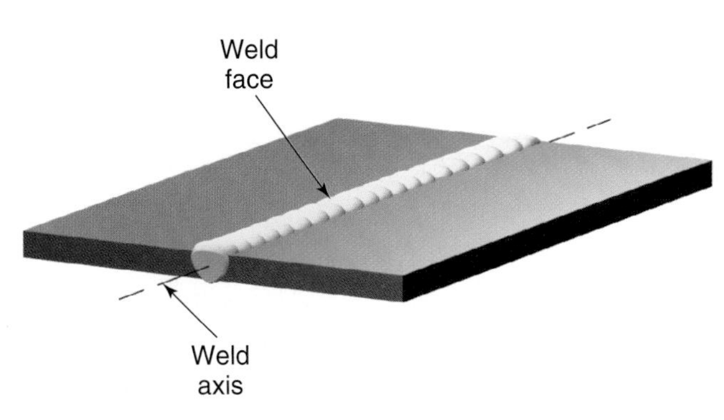

Figure 4-16. The weld axis is an imaginary line running lengthwise through the center of the weld. The weld face is the exposed surface of the finished bead.

- The weld is made from the upper side of the joint.

Horizontal (2G) Welding Position

Groove welds made in the ***horizontal (2G) welding position*** must meet these conditions:

- The weld axis must be within 15° of horizontal.
- The weld face must be between 80°–150° or 210°–280°. See **Figure 4-18**. Angles are measured clockwise with 0° at the bottom.

Vertical (3G) Welding Position

A weld on a groove joint in the ***vertical (3G) welding position*** must meet either of these sets of conditions:

Condition A

- The weld axis is 80°–90° from horizontal.
- The weld face is between 0°–360° from horizontal. See **Figure 4-19A**.

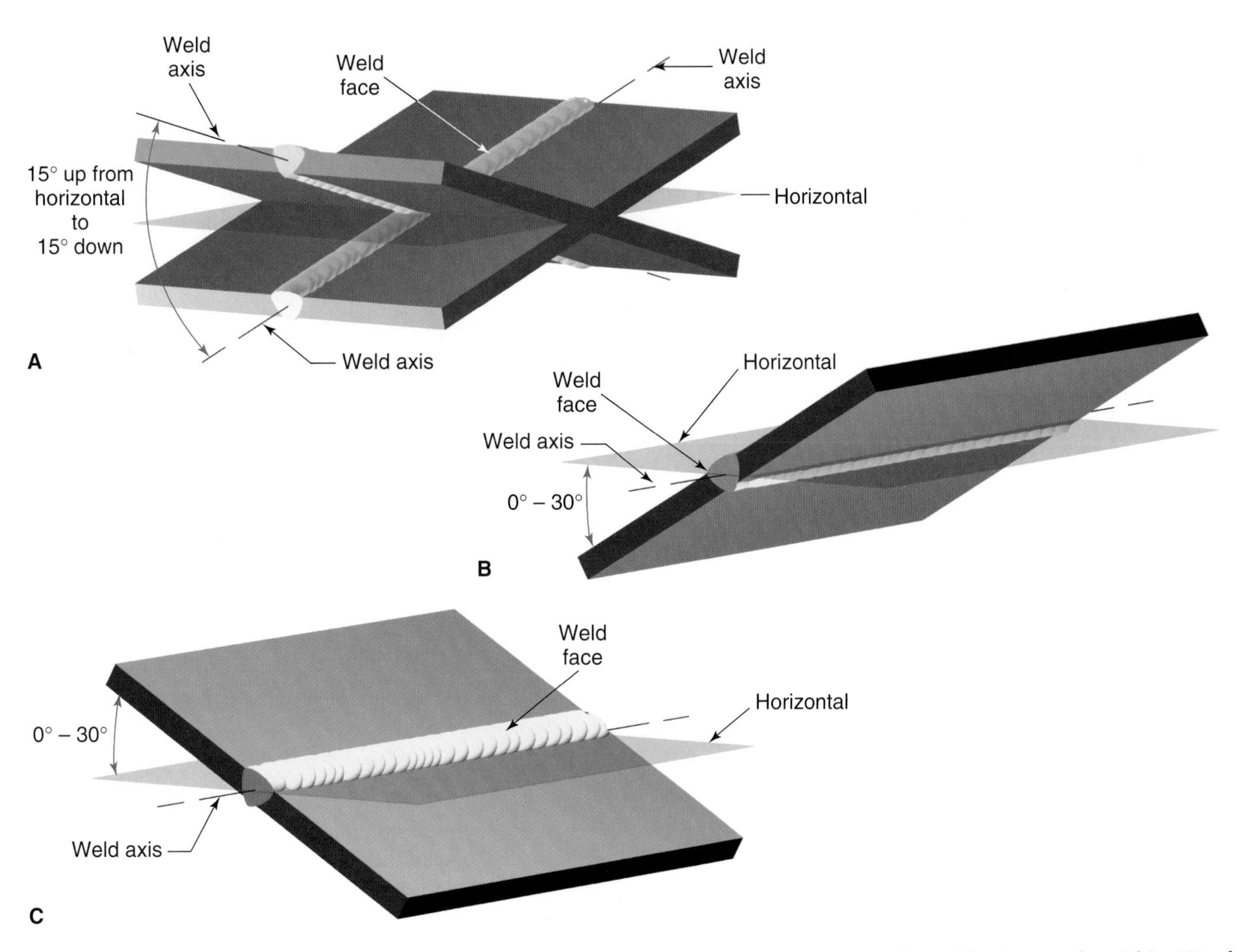

Figure 4-17. These are the specifications for AWS 1G or flat groove welding position. A—The weld axis must be within 15° of horizontal. B and C—The weld face must be within 30° of horizontal.

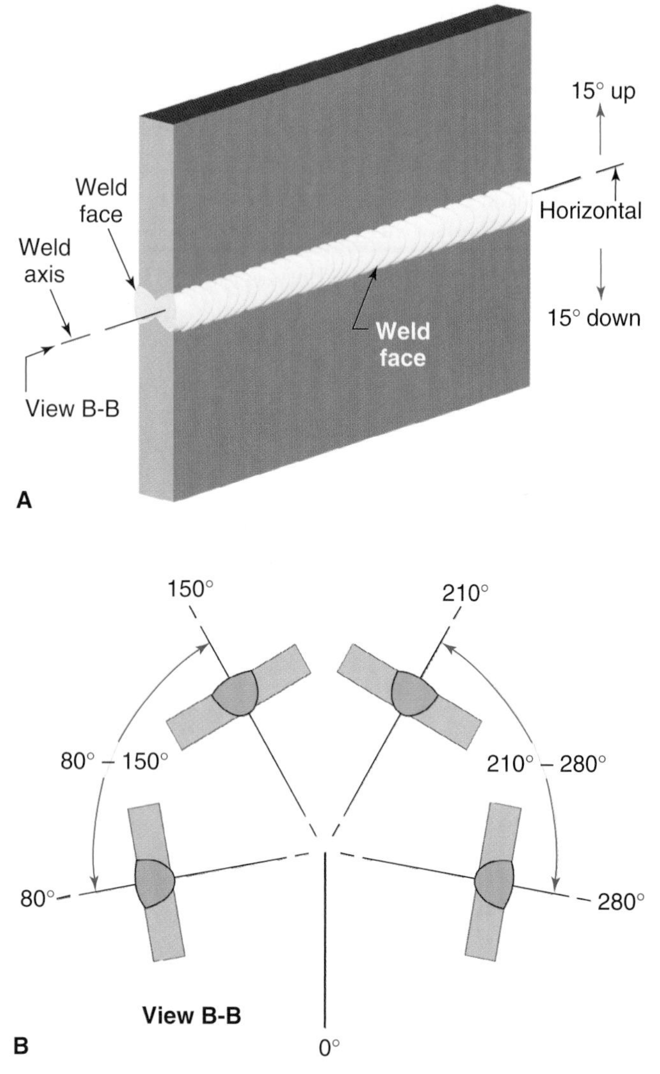

Figure 4-18. These are the specifications for the AWS 2G or horizontal groove welding position. A—The weld axis must be within 15° of horizontal. B—An end view (View B) with the weld shown in blue. The weld face must be within 80°–150° or 210°–280°. All angles are measured clockwise with 0° at the bottom.

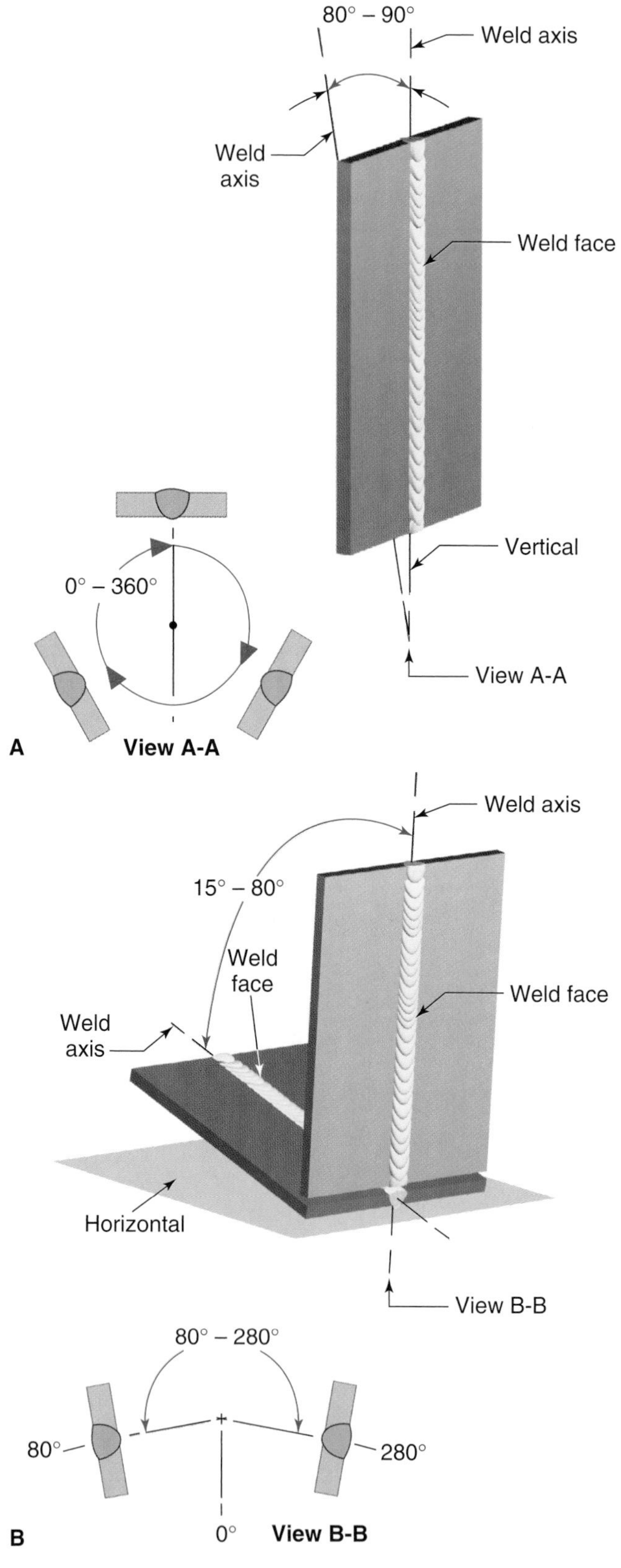

Figure 4-19. These are the specifications for the AWS 3G or vertical groove welding position. A vertical weld must meet either of the following sets of conditions. A—The weld axis must be between 80° and 90° from horizontal. Look at View A. The weld face may be rotated from 0°–360°. B—The weld axis is 15°–80° from horizontal. Look at View B. The weld face must be within 80°–280°.

Condition B

- The weld axis is 15°–80° from horizontal.
- The weld face is within 80°–280° from horizontal.
- The weld is made from the upper side of the joint. See **Figure 4-19B**.

Overhead (4G) Welding Position

Welds made on groove joints in the ***overhead (4G) welding position*** are made under these conditions:

- The weld axis is between 0°–80°.
- The weld face is between 0°–80° or 280°–360°.
- The weld is made from the lower side of the joint. See **Figure 4-20**.

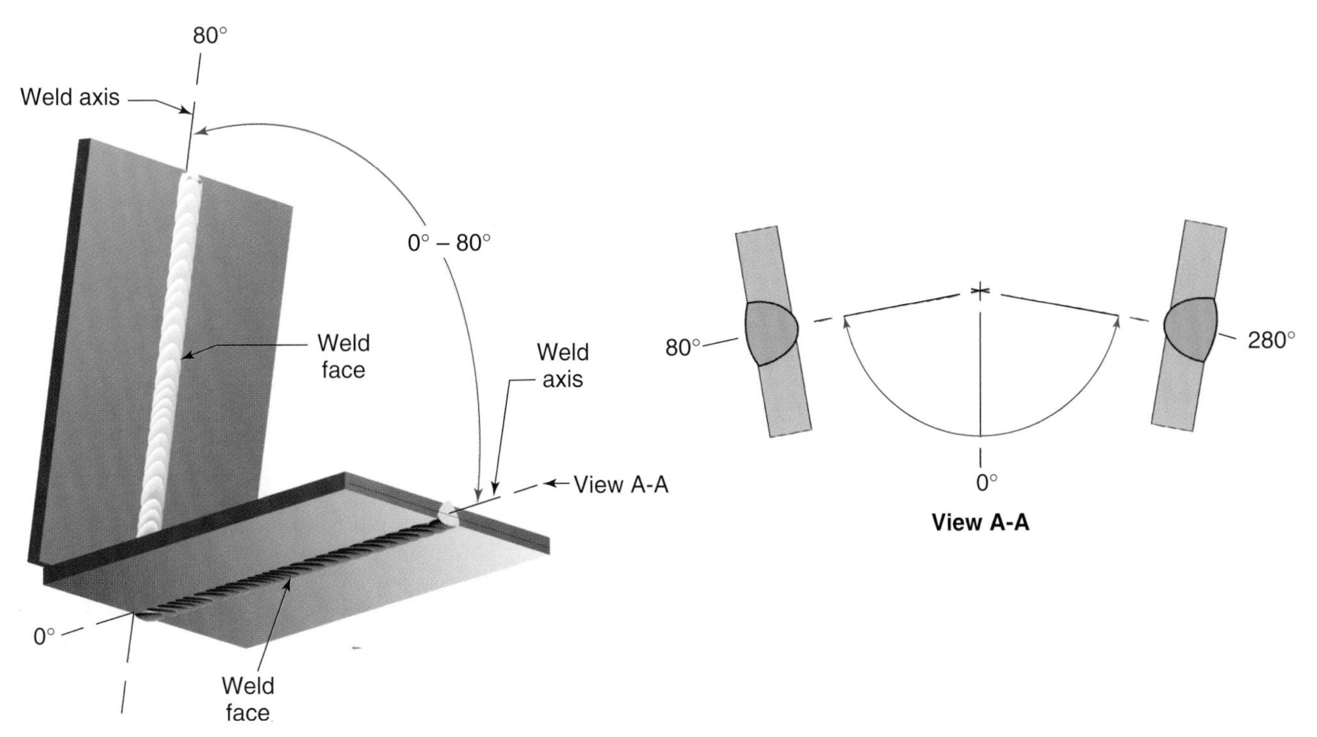

Figure 4-20. *These are the specifications for the AWS 4G or overhead groove welding position. The weld axis must be between 0° and 80°. Look at View A. The weld face must be between 0° and 80° or 280° and 360°. The weld is made from the lower side of the base metal.*

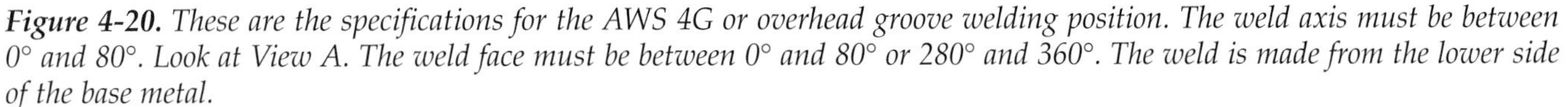

Summary

- A weld joint refers to the way parts are assembled prior to welding. There are five basic types of joints used in welding: butt, lap, corner, T-, and edge.
- A butt joint is used to join parts edge-to-edge.
- A lap joint is formed by two overlapping pieces of base metal. The top surface of one piece is in contact with the bottom surface of the other.
- A corner joint is formed by placing the edge of one piece of base metal against the surface of another piece near or along its outer edge. The pieces may be joined at any angle, but they are commonly welded at a 90° angle.
- A T-joint is formed by two pieces of base metal that are at an angle of approximately 90° to one another. A T-joint is formed at the edge of one piece and away from the edge of the second piece.
- An edge joint is formed when the surfaces of two pieces are in contact and their edges are flush (even).
- A flange joint is formed when the edge of one or more pieces of the joint is bent to form a flange.
- Flare-groove joints are formed when the flanged edges of one or both pieces are placed together to form a single-flare-bevel or double-flare-V-groove.
- A weld bead is one weld pass of filler metal that is added to a weld joint. A weld pass occurs each time a welder lays one weld bead across a weld joint.
- Stringer beads are made by moving the torch or electrode holder along the weld without any side-to-side motion. A weave bead is formed by moving the torch or electrode holder from side-to-side as the weld pass progresses along the weld joint.
- Welds may be made in the flat, horizontal, vertical, or overhead welding positions.

Review Questions

Write your answers on a separate sheet of paper. Please do *not* write in this book.

1. List the five basic weld joints.
2. The metal to be welded is called the _____ metal.
3. The metal added to a weld joint is the _____ metal.
4. Which joint does *not* require any special edge preparation?
5. Sketch the end view of a single-V-groove joint and label the bevel angle and the groove angle.
6. _____, _____, and _____ joints are commonly welded without using filler material.
7. What type of weld bead is formed by moving the torch or electrode from side to side as the weld progresses?
8. Name three methods of preparing the edges of metal for welding.
9. List two reasons why it costs more to make a weld if the groove angle is too large.
10. Name the four welding positions and state the conditions of the AWS 2G position.

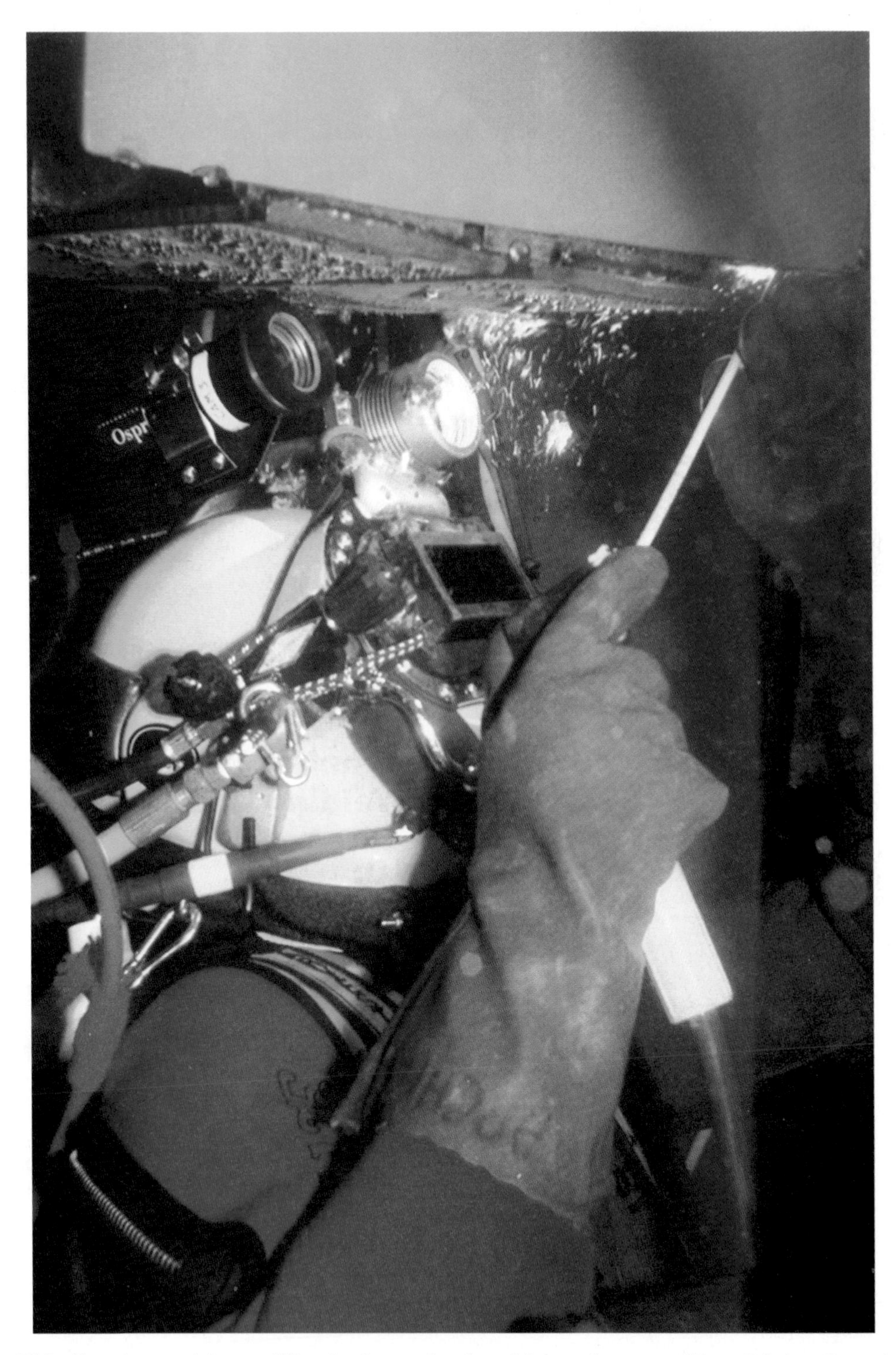

This diver is practicing welding in the overhead position underwater. (Navy Joining Center)

Section 2

Shielded Metal Arc Welding

Jackson Products, Inc.

Chapter 5

SMAW: Equipment and Supplies

Learning Objectives

After studying this chapter, you will be able to:

- Describe the differences between direct current (dc) and alternating current (ac).
- Interpret American Welding Society (AWS) abbreviations regarding welding current polarity.
- Identify the equipment and accessories used in shielded metal arc welding (SMAW).
- Identify factors to consider when selecting an arc welding machine.

Technical Terms

alternating current (ac)
bridge rectifier
chipping hammer
constant current (CC)
constant voltage (CV)
cover plate
cycle
diode
direct current (dc)
direct current electrode negative (DCEN)
direct current electrode positive (DCEP)
direct current reverse polarity (DCRP)
direct current straight polarity (DCSP)
droop curve machines
droopers
duty cycle
electrode holder
electrode lead
filter lens
flash goggles
frequency
gauntlet gloves
helmet
hertz
input power
inverter
lug
open circuit voltage (OCV)
polarity
primary current
rated output current
secondary current
shielding gas
shielded metal arc welding (SMAW)
silicon-controlled rectifier
slag
step-down transformer
terminals
voltage drop
welding outfit
welding station
wire brush
work booth
workpiece lead

Shielded Metal Arc Welding Principles

Shielded metal arc welding (SMAW) is a welding process in which the base metals are heated to fusion or melting temperature by an electric arc. The arc is created between a covered metal electrode and the base metal. A ***shielding gas*** protects the base metals, arc, electrode, and weld from the atmosphere during welding. The shielding gas is created as the flux covering on the electrode melts. When the flux solidifies, it forms a protective slag over the weld bead. The melting electrode wire furnishes filler metal to the weld. See **Figure 5-1**.

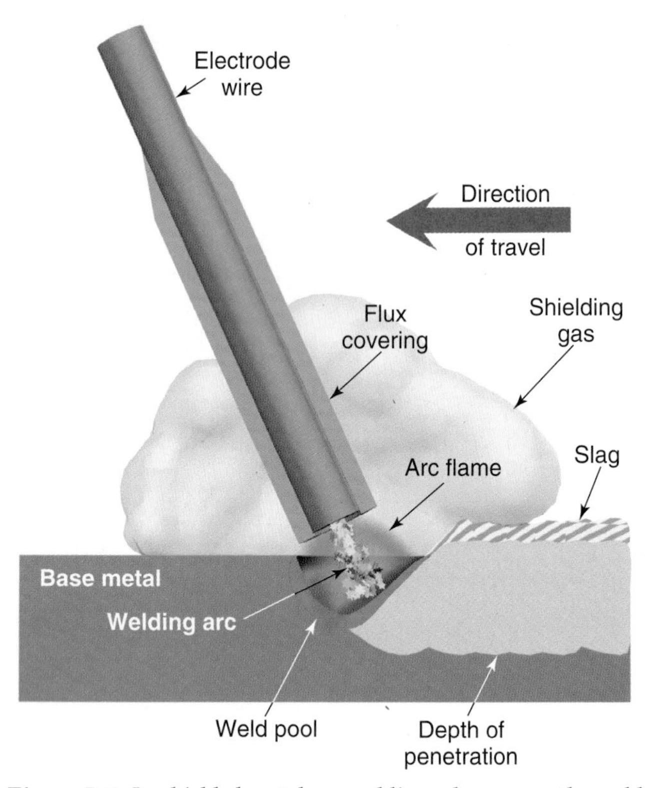

Figure 5-1. In shielded metal arc welding, slag covers the weld bead while it cools. Shielding gas is formed as the flux covering is burned.

The temperature of the arc in SMAW can be over 9000°F (5000°C). With the correct diameter electrode, the heat created by the arc is enough to melt any weldable metal. Mild carbon steel melts at slightly above 2800°F (1540°C).

Shielded metal arc welding is used to construct buildings, ships, truck chassis, pipelines, and other weldments. SMAW equipment is fairly inexpensive. It is widely used on farms and in small repair shops.

SMAW Current and Polarity

In SMAW, electrical current flows across an air gap between the covered electrode and base metal. An electric arc is formed as the current flows across the air gap. The arc creates the heat required for welding.

The electrical current for welding is supplied by an arc welding machine. Welding machines, also called power sources, produce two types of current: direct current (dc) and alternating current (ac).

An electrical current is actually the flow of electrons within a circuit. The direction that the electrons flow is referred to as the ***polarity***. Electrons flow from a negatively charged (polarized) body to a positively charged body. Direct current may flow from the electrode to the base metal. When the current flows in this direction, the electrode has *negative* polarity and the base metal has *positive* polarity. It is called ***direct current electrode negative (DCEN).*** Direct current electrode negative is also known as ***direct current straight polarity (DCSP).*** See **Figure 5-2**. The current direction may be reversed to flow from the base metal to the electrode. When the current flows in this direction, the base metal has *negative* polarity, and the electrode has *positive* polarity. It is called ***direct current electrode positive (DCEP).*** Direct current electrode positive is also known as ***direct current reverse polarity (DCRP).*** See **Figure 5-3**.

The direction selected is determined by the metal thickness, joint position, and type of electrode used. The selection of polarity is covered later in this chapter.

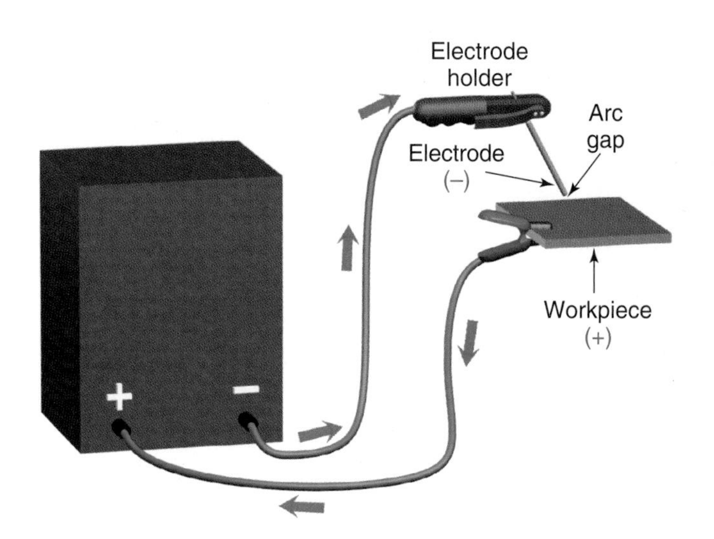

Figure 5-2. A diagram for a direct current electrode negative (DCEN) arc welding circuit. Notice the current is traveling from the negative electrode to the positive base metal.

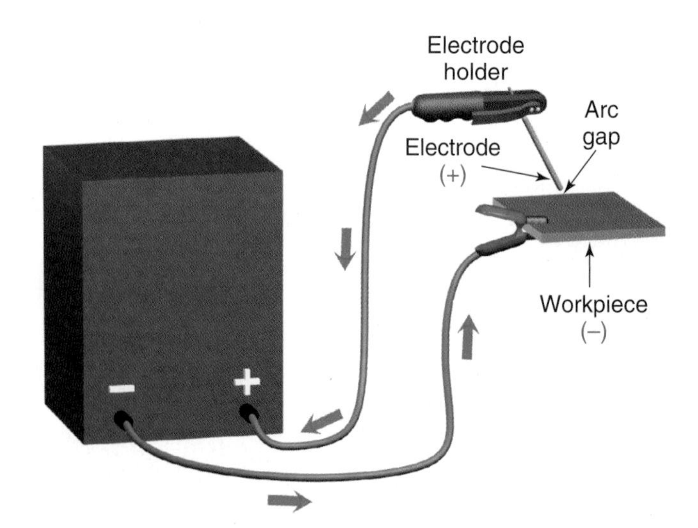

Figure 5-3. A diagram for a direct current electrode positive (DCEP) arc welding circuit. In DCEP, the current travels from the negative base metal to the positive electrode.

Direct Current

Direct current (dc) flows in only one direction. Current flows from one terminal of the welding machine to the base metal. It then flows back to the second terminal of the welding machine. On some welding machines, the electrical leads are manually reversed on the machine terminals. This changes the direction of current flow. On other machines, a polarity switch is flipped to electrically reverse the terminal leads within the machine. See **Figure 5-4**. **Welding must be stopped and the electrode insulated from the circuit whenever the polarity is changed.**

The circuit polarity affects the selection of the covered electrode and how the covered electrode will perform. Direct current electrode positive (DCEP) used with the proper electrode produces deeper penetration than direct current electrode negative (DCEN). Most electrodes use DCEP.

Direct current electrode negative (DCEN) is used with electrodes that do not produce deep penetration. A small electrode using DCEN and a low-current setting will produce a soft arc with only minimal penetration. Thus, DCEN, with proper electrodes, is used for welding thin metals. DCEN is also used with some electrodes to deposit weld metal quickly. DCEN can melt an electrode more rapidly and deposit filler metal at a faster rate than DCEP.

When limited to a certain polarity, the selection of the electrode will depend on the type of polarity. If only certain electrodes are available, the electrode will determine the type of polarity. The polarity and the electrode must always agree.

Alternating Current

Alternating current (ac) generally operates at 60 cycles per second in the United States. A ***cycle*** is a set of repeating events, such as the flow and reversal of flow in an electrical circuit. Alternating current flows in one direction for 1/120 of a second, then reverses its direction and flows for 1/120 of a second to form one complete cycle. See **Figure 5-5**. The total time for a cycle is 1/60 of a second:

1/120 sec + 1/120 sec = 1/60 sec

Frequency is a term used to describe the number of cycles per second there are in ac power. Frequency is measured in ***hertz (Hz)***. One cycle per second is the same as one hertz. Thus, ac power at 60 cycles per second is also called 60-hertz power.

Welding with ac and with ac electrodes produces a medium depth of penetration. Many electrodes (but not all) will work with alternating current. See Chapter 7 and manufacturer's recommendations for more information on electrodes.

SMAW Outfit

SMAW requires the use of proper protective clothing and various tools, supplies, and equipment. The ***welding outfit*** includes equipment required to actually create a weld. The ***welding station*** includes tools, supplies, and other items required to make welding safe and comfortable. A complete SMAW outfit includes the following:

- Welding machine/power source (dc, ac, or ac/dc).
- Electrode and workpiece leads.
- Electrode holder.

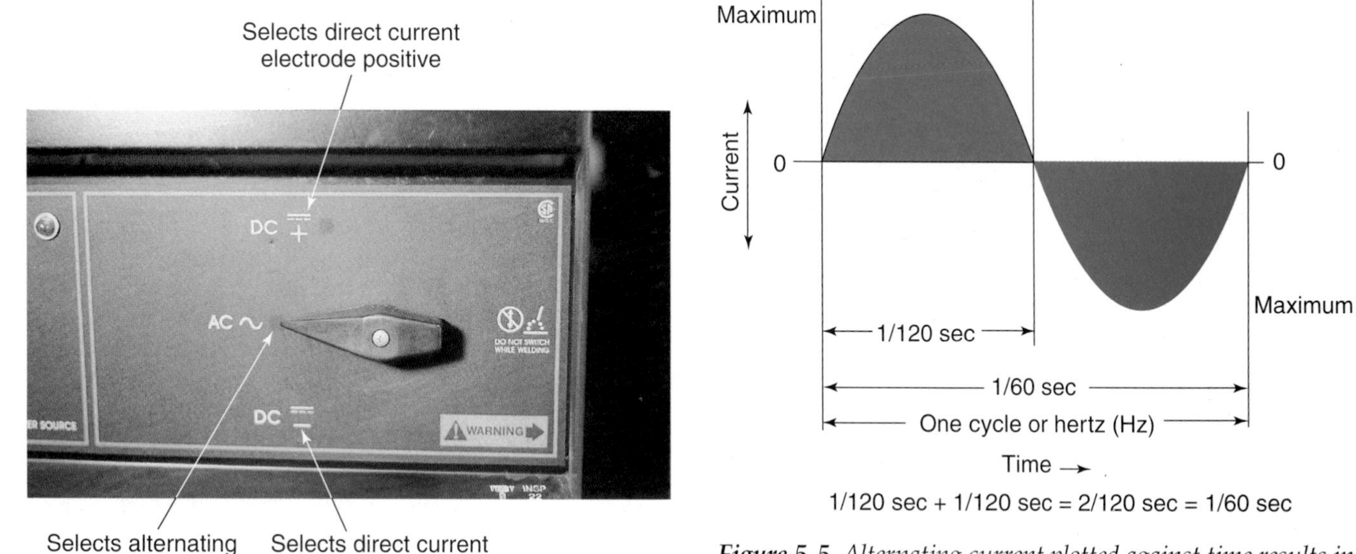

Figure 5-4. *This polarity switch has positions for DCEP, AC, and DCEN.*

Figure 5-5. *Alternating current plotted against time results in an ac sine wave. The current increases from zero and returns to zero. It then increases in the opposite direction and returns to zero. The complete process is one cycle and takes 1/60 second.*

A complete SMAW station, **Figure 5-6**, includes the SMAW outfit plus the following:

- Ventilation.
- Welding table.
- Welding booth with an opaque (solid) or filtered transparent plastic screen.
- Covered electrodes.
- Chipping hammer and wire brush.

Arc Welding Machine

Arc welding machines produce either a ***constant current (CC)*** or a ***constant voltage (CV)***. Manual welding processes, such as SMAW, require a constant current (CC) welding machine. If a CC welding machine is not used, large changes in current occur whenever a welder changes the arc length slightly. Since it is impossible to manually maintain a constant arc length without any variation, a CC welding machine is used.

CC welding machines are also called ***droopers*** or ***droop curve machines***. This name is derived from the voltage versus amperage curve produced by the machines. **Figure 5-7** shows the curve shape for a CC welding machine.

Examine Figure 5-7 closely. An increase from 20 volts to 25 volts is a 25% increase in voltage. This change in voltage results in an increase in current from 125 amps to 130 amps, only a 4% increase. If the welder varies the arc length, it causes a change in voltage. However, the amperage changes very little with a CC welding machine. Although the current varies slightly, these welding machines are considered constant current. Both ac and dc welding machines are available as constant current machines.

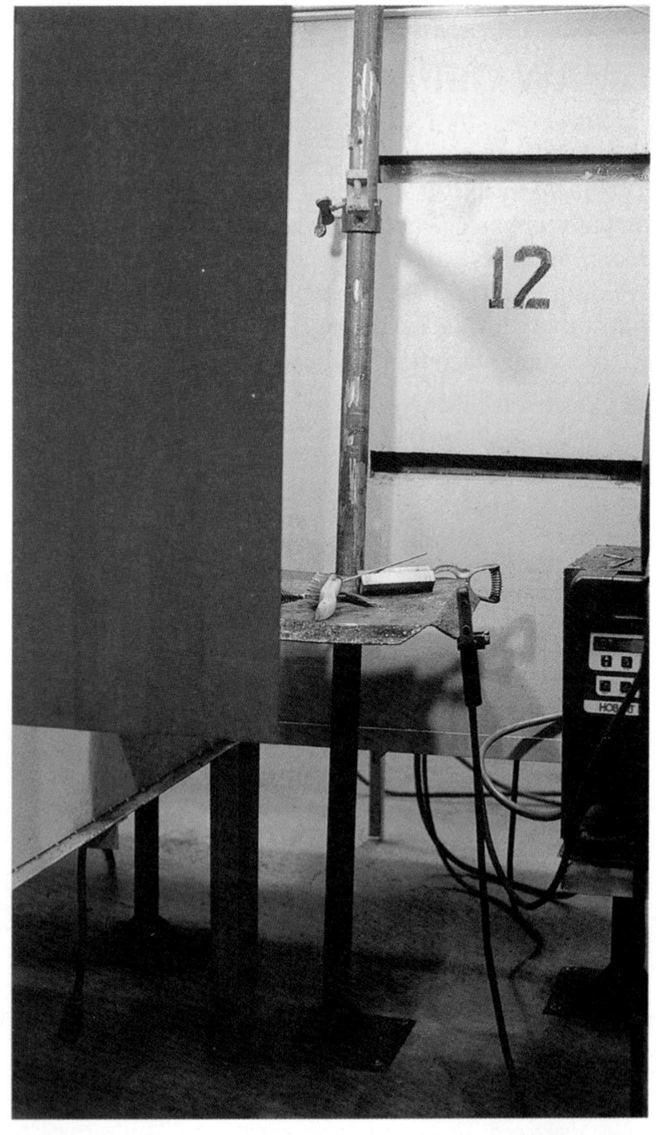

Figure 5-6. *This is a well-equipped shielded metal arc welding station.*

Selecting an Arc Welding Machine

Ac welding machines meet most welding requirements. They are easy to use and generally cost less than comparable dc welding machines. Therefore, they are widely used on the farm and in the home shop. See **Figure 5-8**.

Because the polarity can be changed, dc welding machines are more versatile than ac machines. The ability to change the polarity allows the welder to make out-of-position welds and to weld thin metal more easily. It also allows the welder to vary the heat applied to the metal. DCEP provides deep penetration for welds on thick metal; DCEN deposits filler metal faster. DCEN is used more easily on thin metal.

Some arc welding machines are combination ac/dc machines. These machines can be used as ac welding machines and then can easily be switched to

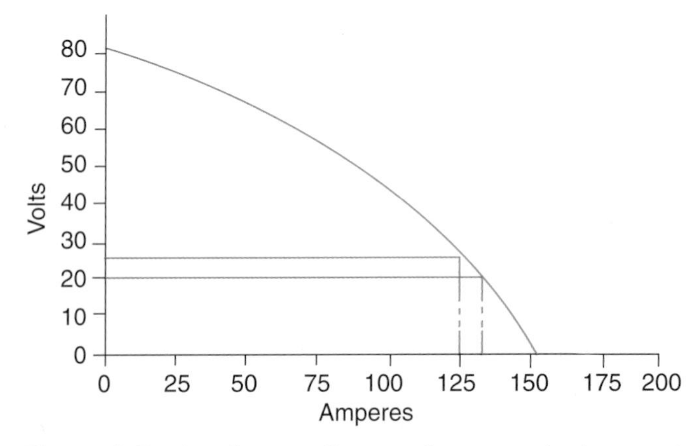

Figure 5-7. *A volt-amp diagram for a constant current machine. The constant current (CC) welding machine is called a drooper because of this curve. A 25% change in voltage results in only a 4% change in amperage. The current change is so slight that the current is considered constant.*

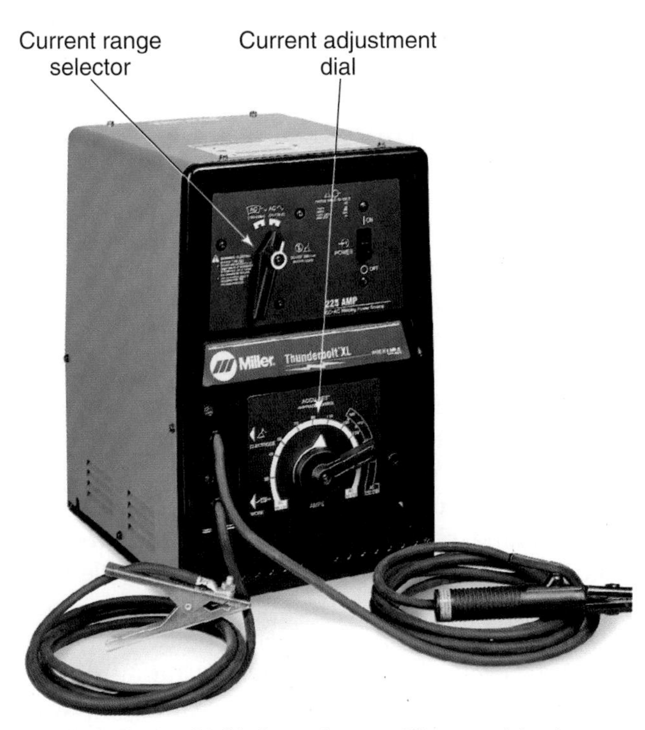

Figure 5-8. An ac shielded metal arc welding machine has two current ranges. The desired current is set by selecting a current range and setting the desired current using the adjustment dial. (Miller Electric Mfg. Co.)

Figure 5-9. A portable engine-driven ac/dc welding power source has a switch used to select ac or either dc polarity. (Lincoln Electric Company)

dc welding machines. Combination machines have a greater flexibility in their use. However, they are more expensive than ac or dc machines.

Machine designs can be different types. The most common are:

- Transformer—used for ac.
- Transformer-rectifier—used for ac and dc.
- Inverter power supply—provides ac and dc options.
- Motor generator—used to create dc power.
- Motor alternator—used to create ac power; with addition of a rectifier, can produce dc power. See **Figure 5-9**.

The function of a power supply is to take the high-voltage, low-current supplied power and change it into the low-voltage, high-current power needed for welding. Welding transformers are called ***step-down transformers*** because the voltage is reduced.

Alternating current is supplied to one side of a transformer. This is called the ***primary current***. The current coming out of the transformer, which is used for welding, is called the ***secondary current***. Alternating current is used as the input to the transformer. The output of the transformer is also alternating current, but at a higher amperage and lower voltage.

To obtain direct current, the alternating current is passed through a set of diodes or silicon-controlled rectifiers (SCRs). ***Diodes*** and ***silicon-controlled rectifiers*** are electrical devices that allow current to flow in only one direction. (Remember, alternating current flows in *two* directions.) A ***bridge rectifier*** is a group of diodes or SCRs used to transform ac into dc. **Figure 5-10** shows a bridge rectifier circuit.

Another type of power supply is an inverter. Inverter machines are very small and lightweight.

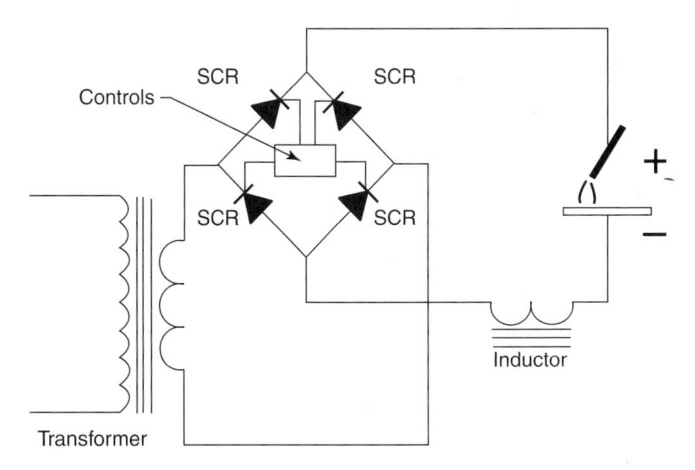

Figure 5-10. This is a bridge rectifier that uses SCRs to change ac current into dc current. SCRs conduct current in only one direction.

An *inverter* power supply uses electrical circuits to change the input frequency to the transformer. See **Figure 5-11**.

First, an inverter machine takes single-phase or three-phase power and passes it through a bridge rectifier. This creates dc current. Next, this dc current is chopped into very high-frequency ac current. The frequency can be between 1,000 and 50,000 cycles per second or 1kHz to 50kHz (the *k* means 1000). This high-frequency current is passed through a small, efficient transformer. Low-voltage, high-current power useful for welding is produced. This power still has a very high frequency. Another bridge rectifier and an inductor are used to create smooth dc output. Inverter power supplies can be used for all arc welding processes and for plasma arc cutting.

Some welding machines taken to job sites where no electricity is available have a gasoline or diesel engine as the power supply. The engine turns a generator or alternator. These create dc and ac power, respectively.

After selecting an ac, dc, or ac/dc machine, the following variables must also be considered:

- Input power requirements.
- Rated output current rating.
- Duty cycle.
- Open circuit voltage.

Input power requirements. Welding machines used in school shops, trade schools, and industry are connected to commercially available electric power. Power requirements for a welding machine must be specified to the electrician when the machine is wired into a building. The *input power* of a welding machine must correspond with the type of power that is available. It is fairly expensive to rewire existing 120V power to 240V or 440V.

The input voltage to a welding machine may be 120V, 240V, 440V or higher. These high voltages are reduced by a transformer within the welding machine to the required welding voltage. Welding voltage ranges from 5V to 30V.

Engine-driven welding machines are used in the field for pipelines, construction, and other welding operations where electric power is not available. Engines are connected to the welding machine to turn an alternator for ac welding current or a generator for dc welding current.

Rated output current. A nameplate on each welding machine shows the *rated output current*. See **Figure 5-12**. A welding machine must be able to supply the current required for the welds being made. The rated output current depends on the duty cycle.

Duty cycle. The *duty cycle* is a rating that indicates how long a welding machine can be used at its maximum output current without damaging it. Duty cycle is based on a ten-minute time period. A welding machine with a 60% duty cycle can be used at its maximum rated output current for six out of every ten minutes. The welding machine may overheat if the duty cycle is exceeded. At lower current settings, the duty cycle may be increased and the power source used for a longer period of time.

A duty cycle chart should be provided with a new welding machine. In **Figure 5-13**, the duty cycle is 20% at the maximum rated output current of 200 amperes (A). The duty cycle is 100% at a rated output current of 100A.

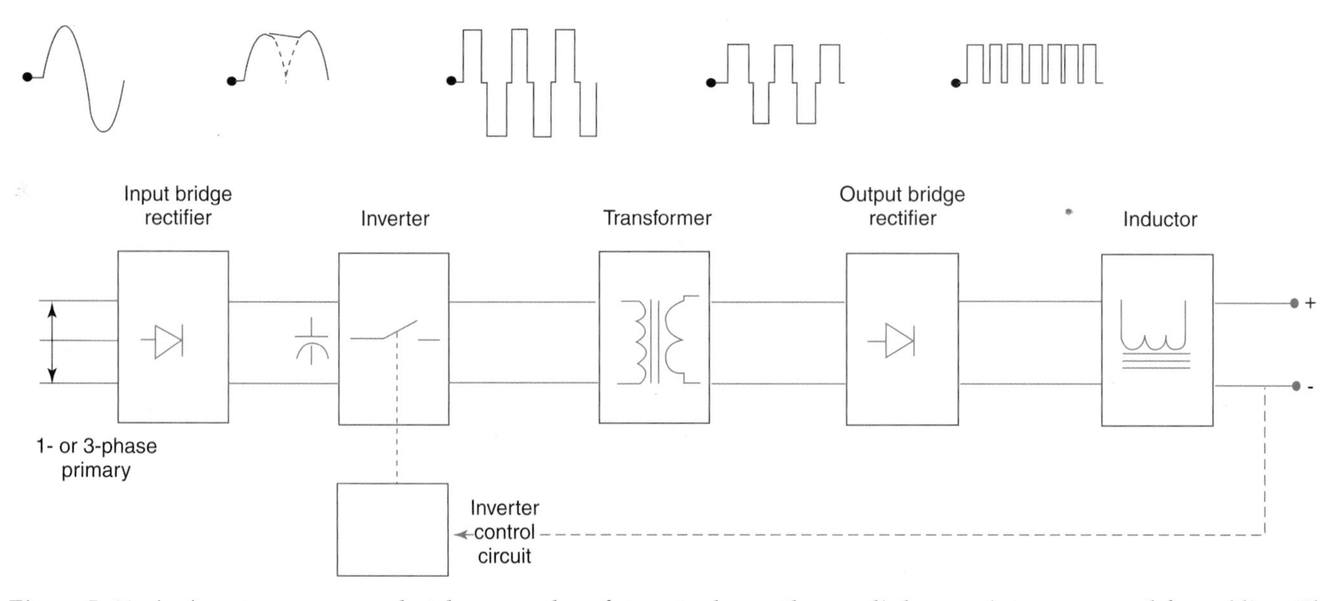

Figure 5-11. *An inverter power supply takes a number of steps to change the supplied power into power used for welding. The different steps and the resulting waveform are illustrated.*

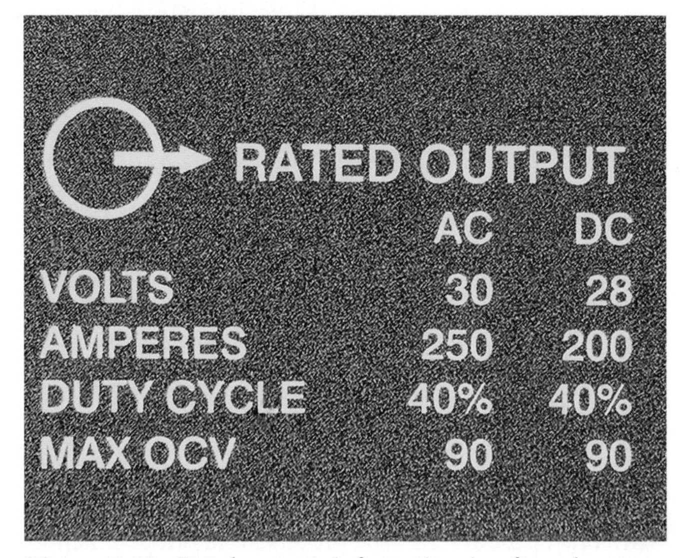

Figure 5-12. Rated output information is often shown on welding machines. During ac operation, the output is 250 amperes at 30 volts. In dc, the output is 200 amperes at 28 volts.

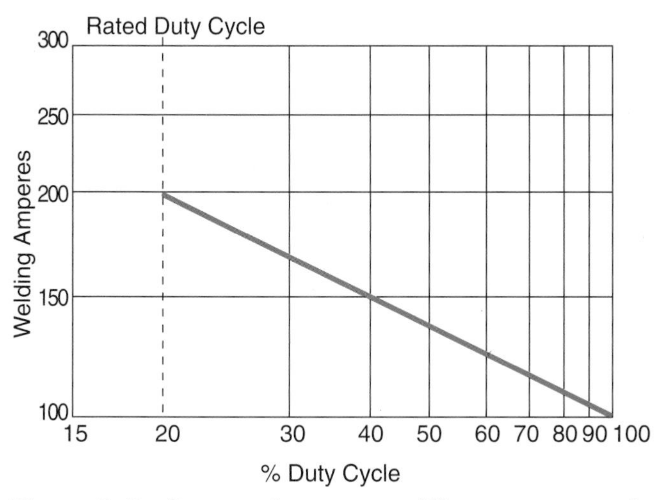

Figure 5-13. Duty cycle versus welding current graphs should be provided with each new arc welding machine. The duty cycle for this machine is 20% at 200A. At 100 amperes, the machine may be used 100% of the time.

Open circuit voltage. Welding machines have a maximum voltage, or open circuit voltage. ***Open circuit voltage (OCV)*** is the voltage of the welding machine when it is on, but is not being used. **The maximum OCV for most manual ac or dc machines is 80V. This relatively low OCV protects workers from electrical shock.** The 80V OCV is still high compared to the 5V-30V closed circuit welding voltage. An OCV of 80V is required to start the arc easily. It is also necessary to maintain the arc during ac welding.

Welding Leads

The electrical cable that connects the electrode holder to a welding machine is the ***electrode lead***. The ***workpiece lead*** (ground) is the electrical cable that connects the base metal to the welding machine.

On large welding machines, leads may be required to carry 600A or more. Leads must have a large diameter to carry such high current. They must also be flexible so a welder can easily move them. As many as 2500 fine conductors are used in a welding lead to produce the flexibility required. Welding leads must also be well insulated to prevent electrical shock. They are usually insulated with a heavy rubber or neoprene covering. See **Figure 5-14**.

Voltage and current are affected when the welding leads are the wrong diameter. Electrical resistance in the lead increases as its diameter decreases or its length increases. ***Voltage drop*** (voltage loss) occurs when electricity travels a long distance from the welding machine. A larger diameter lead is used to counteract the voltage drop when long leads are required. A greater amount of current flows in a larger diameter lead when voltage drop is held to a minimum.

Figure 5-14. This arc welding lead shows the large number of fine copper wires used. When twisted into larger bundles, these fine wires give the lead its flexibility. The lead is then covered with insulating material, cord reinforcement, and an outer layer of rubber or plastic.

Welding leads are available in a variety of diameters. The diameter is referred to by number size. These number sizes and their actual diameters are shown in **Figure 5-15**. Figure 5-15 also shows what size lead should be used to carry a given amperage to the weldment and back. The amperages shown are the maximum that can be carried in the various leads over the stated distances.

Lead Connections

Welding leads are connected to the welding machine and base metal with ***lugs***, clamps, or special ***terminals***. **Figure 5-16** shows several lugs that may be used on the machine end of the workpiece and electrode leads. Lug connections can also be used to connect the workpiece lead to the welding table. Special push-and-turn connectors are shown in **Figure 5-17**. These quick-connect terminals may be used on both welding leads.

Various other types of clamps or special connectors may be used to connect the workpiece lead to the base metal. The workpiece lead may be clamped to the base metal using a spring-loaded clamp, as shown in **Figure 5-18**. Occasionally, a part must be welded while it is being rotated. A special rotating workpiece clamp is used for this application, **Figure 5-19**. Lugs, spring clamps, or other special connectors are mechanically connected or soldered to the bare end of the leads.

Electrode Holder

The ***electrode holder*** is held by the welder during the welding operation. The well-insulated handle of the electrode holder protects the welder from electrical shock. An electrode is clamped in the copper alloy jaws of the electrode holder. The jaws provide a good electrical contact for the electrode. See **Figure 5-20**. The electrode lead is clamped into the electrode holder. The cable clamp is under the insulated handle.

Protective Clothing

Pockets on shirts, pants, or coveralls should be covered to prevent sparks from being caught in them. The top collar button should be fastened, especially when welding out of position. Pant legs should not have cuffs.

A cap should be worn to protect your hair. Leather gloves should be worn. *Gauntlet gloves,* leather gloves with long cuffs, are the type preferred for out-of-position welding. High-top shoes offer added safety from sparks. Steel-toed shoes are recommended when working around heavy metal parts.

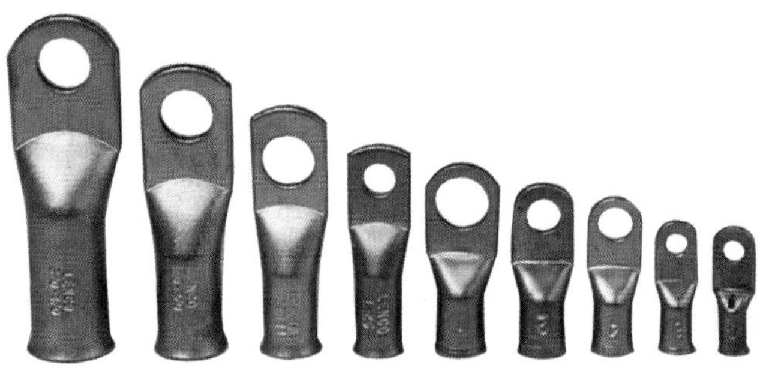

Figure 5-16. *These lugs are for arc welding leads. They are usually connected to the lead by means of a mechanical crimp. (Lenco)*

Welding Lead Current Capacities

Lead Number	Lead Diameter		Length 0 – 50 ft. 0 – 15.2 m	Length 50 – 100 ft. 15.4 – 30.5 m	Length 100 – 250 ft. 30.5 – 76.2 m
	in.	mm	Amperage	Amperage	Amperage
4/0	.959	24.4	600	600	400
3/0	.827	21.0	500	400	300
2/0	.754	19.2	400	350	300
1/0	.720	18.3	300	300	200
1	.644	16.4	250	200	175
2	.604	15.3	200	195	150
3	.568	14.4	150	150	100
4	.531	13.5	125	100	75
Note: Lengths given are for the total combined length of the electrode and work leads.					

Figure 5-15. *Welding lead size recommendations. Lead sizes range from 4/0 to 4. If the work is 100′ (30.5 m) from the machine, 200′ (61 m) of leads are needed: 100′ (30.5 m) for the workpiece lead and 100′ (30.5 m) for the electrode lead. Size 1/0 leads are required to carry 200A through 200′ (61 m) of leads.*

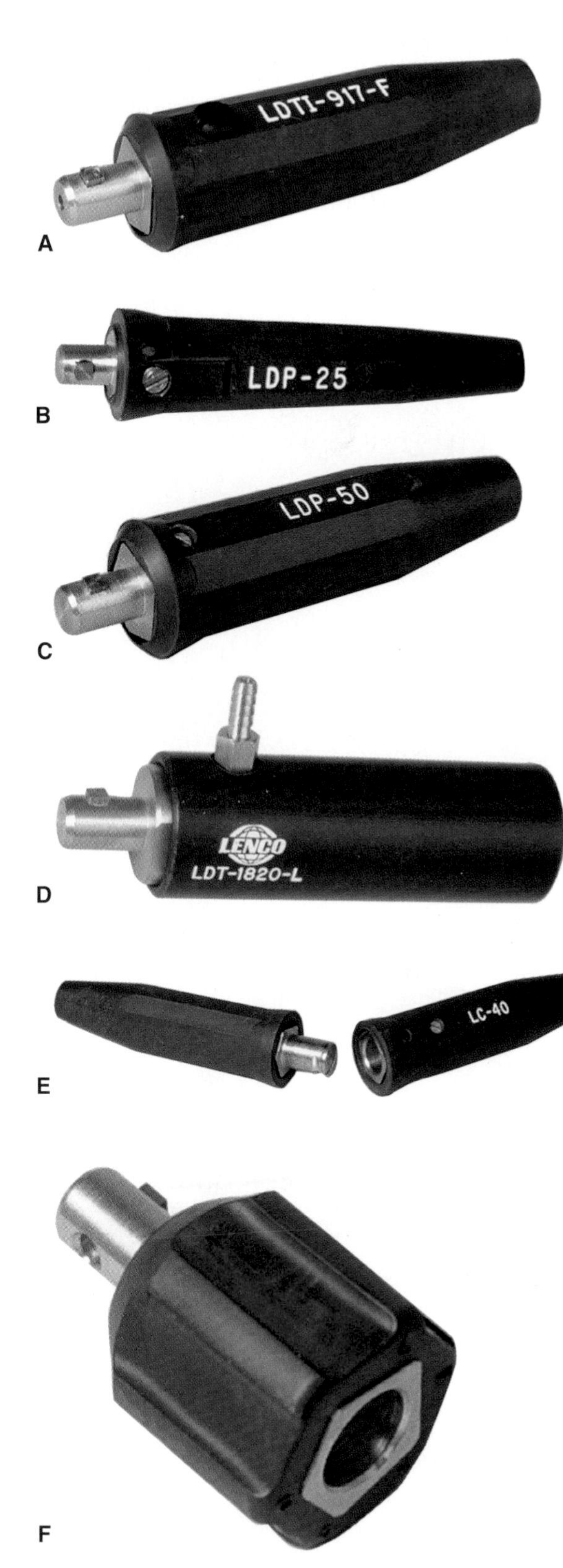

Figure 5-17. A–C—These are all examples of European connectors called DIN connectors. Each connector has a small raised square on the round shaft. DIN connectors are most often used for making lead connections. D—This DIN connector also has a gas connection. E—This is an American-type connector, which has a split round shaft. F—An American to DIN adapter. (Lenco)

Figure 5-18. Two types of spring-loaded workpiece lead clamps. (Lenco)

Figure 5-19. This rotating workpiece lead connector has sockets for four workpiece leads. (Lenco)

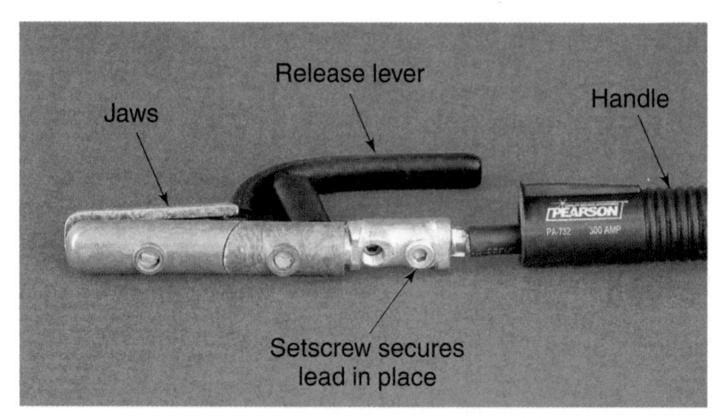

Figure 5-20. This SMAW electrode holder has its insulated handle slid back to show the lead connection. An electrode is held tightly in the jaws by pressure from a heavy coil spring under the release lever.

Arc Welding Helmets and Lenses

An electric arc creates ultraviolet and infrared rays. These light rays are harmful to the eyes and cause burns, like sunburn, on bare skin. A welder should cover all bare skin areas while welding.

An arc-welding ***helmet*** serves several purposes. It shields the face and neck from the harmful rays and protects these areas from molten metal that may spatter from the weld area. The filter lens and cover plate are also held by the helmet. A #10–#14 ***filter lens*** should be used for SMAW. A darker lens is required as the electrode diameter increases. See **Figure 5-21**. Filter lenses must conform to the American National Standards Institute standard Z87.1, which covers the requirements for welding eye protection.

Many filter lenses are auto-darkening. When there is no arc, an auto-darkening lens allows you to view the weld area and position the electrode. The auto-darkening lens senses the arc and immediately (1/12,000th to faster than 1/20,000th of a second or faster) darkens to the preset lens shade number. Many auto-darkening lenses are variable and can be set to different shade numbers.

Some welders use a fixed shade lens. When welding, the helmet is flipped down over the face with a nod of the head. When welding is completed, the helmet is lifted up. **Figure 5-22** shows both types of helmets.

Clear plastic or glass ***cover plates*** protect the filter lens from arc spatter. Always use cover plates. One is used in front of the filter lens and one behind the filter lens. Filter lenses and especially auto-darkening filter lenses are more expensive than cover plates and should be protected. Replace cover plates regularly when they get scratched or speckled with spatter.

Application	Lens Shade Number
SMAW (Shielded metal arc welding) Up to 5/32 in. electrodes 3/16 - 1/4 in. electrodes 5/16 - 3/8 in. electrodes	 10 12 14
GMAW (Gas metal arc welding) (nonferrous) Up to 5/32 in. electrodes	 11
GMAW (Gas metal arc welding) (ferrous) 1/16 - 5/32 in. electrodes	 12
GTAW (Gas tungsten arc welding)	10 to 14

Figure 5-21. *The suggested filter lens shade numbers for various arc welding applications.*

Filter lenses and cover plates are available in two sizes: 2″ × 4 1/4″ (51 mm × 108 mm) and 4 1/2″ × 5 1/4″ (114 mm × 133 mm). An arc-welding helmet can be adjusted for various head sizes. Adjustable screws secure the headband to the helmet. These screws should be tight enough to prevent the helmet from falling down when it is raised. However, the helmet should fall into position over the face when the welder's head is nodded.

A

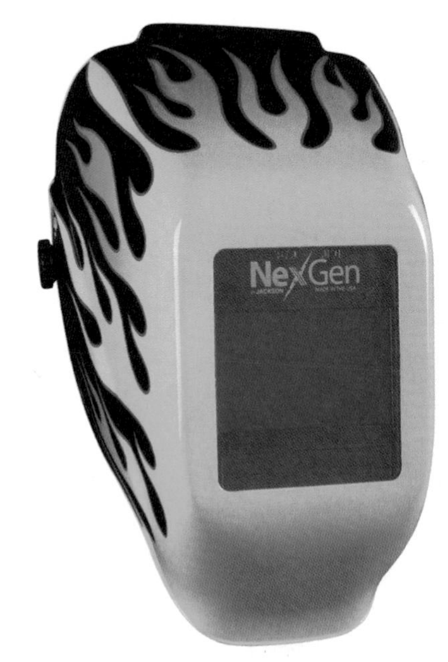

B

Figure 5-22. *A—A fiberglass helmet with a large passive or fixed shade. B—The auto-darkening filter lens in this helmet has a variable shade setting. (Jackson Safety, Inc.)*

Arc welders often wear flash goggles under their helmets. *Flash goggles* protect the welder's eyes from flashes from the rear. A #1–#3 lens is often used in these goggles.

Ventilation

Adequate ventilation must be provided when performing any type of welding. The size or capacity of the ventilation system for a given area should be calculated by a safety engineer.

Exhaust fumes from an arc welding area can be toxic to a welder. A flexible exhaust pickup tube is often used to remove the fumes. The welder positions the pickup tube for maximum performance. Fumes should be picked up and exhausted before they cross the welder's face. See Figure 5-23.

Work Booth and Table

The welding table is part of the arc welding station. The workpiece lead may be bolted to the table with a lug or attached to the table with a spring clamp. A weld positioner may be welded or clamped to the table. A shop-fabricated weld positioner is shown in **Figure 5-24**.

Welding may be done in a ***work booth*** or in an open area. In either case, the electric arc must be shielded from the view of the workers who are not wearing eye protection. Solid walls, canvas curtains, or filtered, transparent plastic may be used. Filtered, transparent plastic curtains must filter more than 99% of the ultraviolet and infrared rays from the electric arc. See Figure 5-25.

Figure 5-23. *This welder is using a welding fume exhaust system. The flexible duct allows the welder to position the pickup vent close to the weld area. (Nederman, Inc.)*

Figure 5-24. *A fixture used to hold weldments for out-of-position welding. This fixture is adjustable in three ways. A C-clamp holds the weldment.*

Figure 5-25. *The curtains in this work area are a filtered, transparent plastic. Note also the flexible fume exhaust system. (Nederman, Inc.)*

Chipping Hammer and Wire Brush

Slag must be removed after each arc weld bead or weld pass is completed. ***Slag*** is the hard, brittle material that covers a finished shielded metal arc weld. The next weld bead will have slag trapped in it if the previous weld bead is not cleaned thoroughly. Slag is chipped away using a ***chipping hammer***. See **Figure 5-26**.

After chipping the slag, the weld bead is further cleaned using a ***wire brush***. The bristles of the brush are steel. See **Figure 5-27**. A wire brush is sometimes combined with a chipping hammer. A rotary wire wheel in a portable electric drill motor may also be used to clean the weld. Always wear goggles when chipping or wire brushing the slag from a weld. This will help to prevent eye injuries.

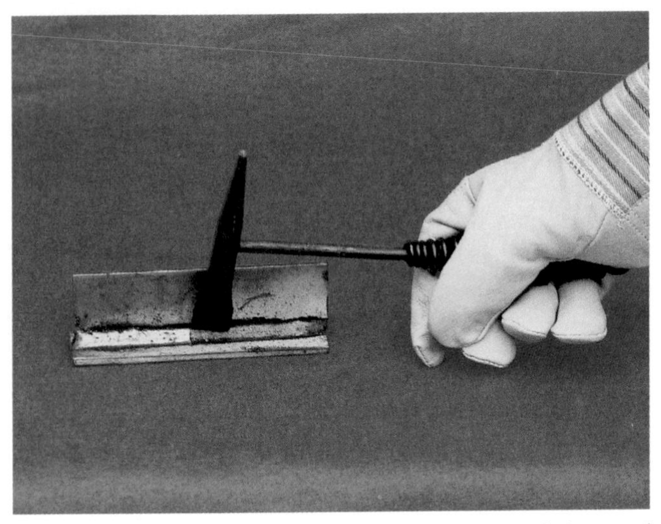

Figure 5-26. *The chisel end of a chipping hammer being used to remove the slag from a finished SMAW weld.*

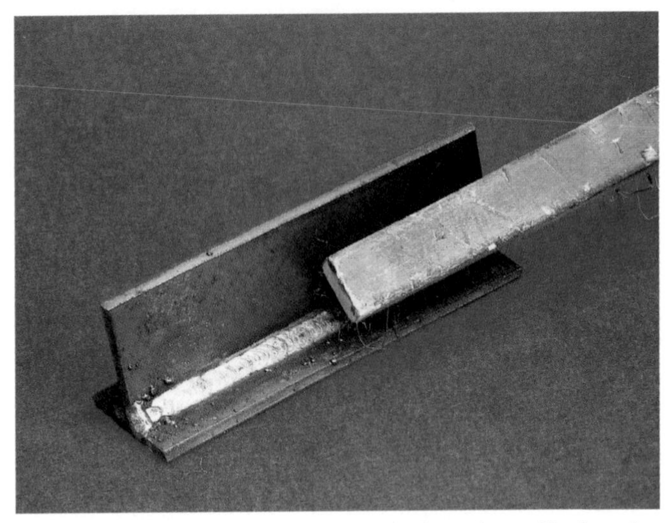

Figure 5-27. *A wire brush is used to clean the weld after slag has been removed.*

Summary

- Shielded metal arc welding (SMAW) is a welding process in which the base metals are heated to fusion or melting temperature by an electric arc. The arc is created between a covered metal electrode and the base metal. The shielding gas is created as the flux covering on the electrode melts. The flux solidifies and forms a slag that protects the weld metal while it cools. The melting electrode wire furnishes filler metal to the weld.
- When direct current flows from the electrode to the base metal, it is referred to as direct current electrode negative or direct current straight polarity.
- When direct current flows from the base metal to the electrode, it is referred to as direct current electrode positive or direct current reverse polarity.
- Direct current flows in only one direction. Alternating current reverses direction at a set frequency, usually 60 cycles per second.
- A welding outfit consists of the equipment required to actually create a weld. A welding station also includes tools, supplies, and other items required to make welding safe and comfortable.
- Arc welding machines used for SMAW produce a constant current (CC).
- The electrical cable that connects the electrode holder to a welding machine is the electrode lead. The workpiece lead (ground) is the electrical cable that connects the base metal to the welding machine.
- A #10–#14 ***filter lens*** should be used for SMAW. A darker lens is required when using larger diameter electrodes.

Review Questions

Write your answers on a separate sheet of paper. Please do *not* write in this book.

1. Identify the parts of this SMAW in progress.

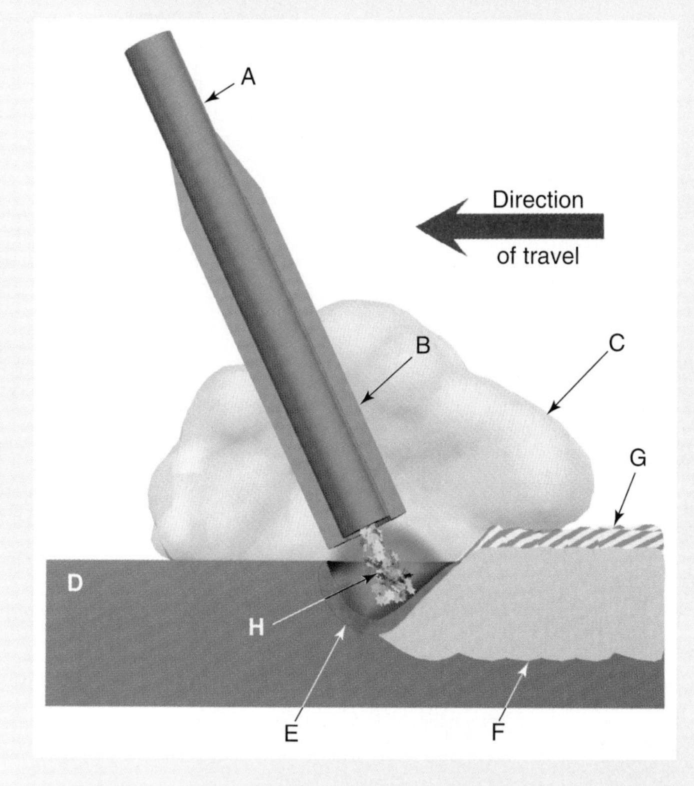

2. Which electrode polarity should be used when welding thin metals?
3. Current that travels from the electrode to the base metal is known as _____ or _____.
4. Ac electrodes provide _____ penetration when used for SMAW.
5. List eight items found in an SMAW station.
6. What do the letters CC and CV stand for? Is a drooper a CC or CV welding machine?
7. Refer to Figure 5-13. The duty cycle for this welding machine at 125 amperes is _____%.
8. Refer to Figure 5-15. A #_____ electrode lead and workpiece lead should be used to carry 150 amperes of electricity 75 feet to the workpiece and 75 feet back to the welding machine.
9. The temperature of an electric arc in SMAW can be over _____°F (_____°C).
10. What is the OCV of most manual ac and dc welding machines?

A welder using the SMAW process on a large weldment. Note how the flexible exhaust tube is positioned to effectively remove fumes. (Nederman, Inc.)

Chapter 6

SMAW: Equipment Assembly and Adjustment

Learning Objectives

After studying this chapter, you will be able to:

- Explain the assembly of a welding machine, leads, and electrode holder.
- List the steps in connecting an electrode holder to the lead.
- Describe the procedure for inspecting a shielded metal arc welding (SMAW) outfit.
- Set the proper amperage and polarity on a welding machine.

cam lock–style cable connector
coarse amperage range
quick-connect terminal
range switch
rectifier

Welding Machine

Ac and dc welding machines can be connected to a variety of input power voltages. **Input power leads should not be repaired by a welder. A qualified electrician should make needed repairs.** However, a welder may need to change the output power leads. The polarity of some dc welding machines is changed by reversing the workpiece and electrode leads.

Input Power

Input power for ac and dc welding machines generally is supplied by commercial sources. Commercial power voltages of 120V, 240V, or higher may be required.

Welding machines using 240V or higher voltages are usually permanently wired (hardwired) into the building's electrical system. However, some electrical codes permit 240V machines to be plugged into special wall receptacles. See **Figure 6-1. Wiring a 120V or 240V wall receptacle or permanently wiring any welding machine must be done by a qualified electrician.**

Engine-driven alternators may be used to produce 120V or 240V input power for arc welding machines. The alternators are driven by gasoline or diesel engines.

Figure 6-1. An approved wall-mounted receptacle used with a special 240V plug on low output welding machines.

Figure 6-2. This engine driven ac/dc arc welding machine can be mounted on wheels to make it more easily moved around the job site. (Lincoln Electric Company)

Engines may also be connected directly to the welding machine, **Figure 6-2**. The engines are used to drive an ac alternator or dc generator. A ***rectifier*** is used to convert alternating current to direct current in the welding machine.

Output Connections

Both ac and dc welding machines have at least two terminals: one for the electrode lead and another for the workpiece lead. These terminals may be threaded brass studs, as in Figure 6-2. Hex or wing nuts are used to secure the leads to these threaded terminals. Tighten the fasteners with the proper size wrench to ensure a good connection. Do *not* put a washer between the lug on the cable and the base of the threaded terminal.

Terminals can also be a socket or internal connection. The machines in **Figures 6-3** and **6-4** have sockets. **Figure 6-5** shows a close-up of an internal connection terminal on a machine and the connector of a lead. The electrode and workpiece leads are quickly attached using a ***cam lock–style cable connector***.

Amperage range terminals may be used on welding machines. Some welding machines can generate welding current in several amperage ranges, known as ***coarse amperage ranges***. Coarse amperage ranges usually extend over 50A or more, depending on the output current of the machine. To select a coarse amperage range, the electrode lead is plugged into the appropriate amperage range terminal.

Figure 6-3. An inverter power supply has a very small, portable design. It has two plugs for its two different operating voltage levels of 120 V and 240 V. The electrode and work leads are plugged into terminals on the front of the machine. (Miller Electric Mfg. Co.)

Figure 6-4. *This is a digitally controlled multiprocess inverter power supply. Welding process, ac/dc, polarity, and current values are set using digital selections. (Thermal Arc, a division of Thermadyne Industries, Inc.)*

Electrode and Workpiece Leads

The electrode and workpiece leads should be the same diameter. Two variables are used to determine the diameter of the electrode and workpiece leads: maximum amperage and total distance to the work and back. Higher amperage requires larger diameter leads. A longer distance to the work and back also requires larger diameter leads. Refer to Figure 5-15 for a table used to determine lead diameter.

External quick-connect terminals or lugs are attached to one end of each lead. A ***quick-connect terminal*** is a heavy-duty electrical terminal that is easy to connect and disconnect from a welding machine. See Figure 6-5. A cam lock device is used to engage the quick-connect terminal. A lug is a heavy-duty electrical terminal that is cylindrical at one end and flat on the other end, **Figure 6-6**. There is a hole in the flat end. This end is connected to a welding machine terminal with a hex or wing nut. A short section of insulation is removed from the end of the lead. The connector or lug is then attached to the bare wires by mechanical means or soldering.

Many workpiece leads contact the workpiece using a spring-loaded clamp. Other workpiece leads use a magnet or C-clamp on the working end, **Figure 6-7**. Another option is to attach a lug to the workpiece lead. In this case, the workpiece lead is bolted to a worktable or to the workpiece.

Figure 6-5. *Cam lock–style cable connectors and quick-connect terminals are often used to connect workpiece and electrode leads to welding machines. Note that this connector is a DIN type.*

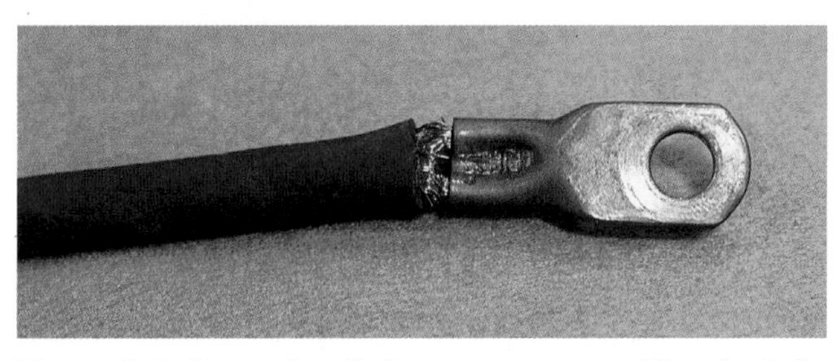

Figure 6-6. *Copper lug fittings connect to welding leads by means of a crimped, swaged, or soldered joint.*

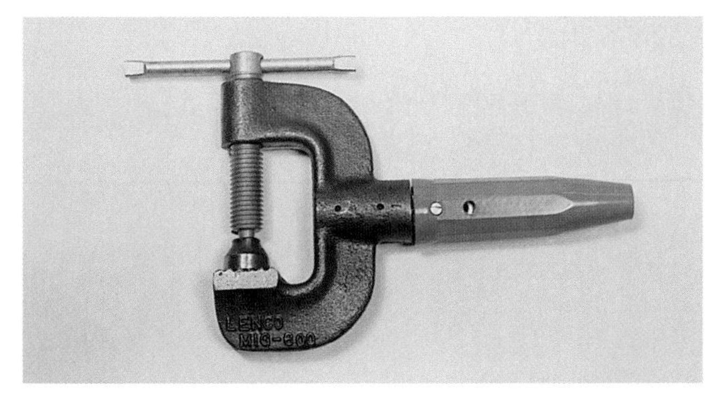

Figure 6-7. *A special C-clamp–type ground clamp for the workpiece lead. (Lenco–NLC, Inc.)*

The electrode lead is connected to the machine and to the electrode holder. The procedure for connecting the electrode holder is discussed later in the chapter.

Leads should not be laid across aisles or other heavy traffic areas. They may be damaged and cause injuries. For permanent installations, run the leads overhead in conduit or below the floor. For temporary installations on a job site, pass the leads under a large section of C-channel for protection.

Connecting the Electrode Holder

One end of the electrode lead is attached to a lug or quick-connect terminal. The opposite end is connected to the electrode holder by mechanical means. See **Figure 6-8**. The procedure for making the mechanical connection is as follows:

1. Remove the insulated handle from the electrode holder.
2. Loosen the hex socket cap screws (Allen screws) in the copper or brass socket.
3. Remove the insulation from the electrode lead. The amount of insulation removed should be equal to the depth of the socket.
4. Wrap copper foil around the bare conductors. This prevents the conductors from unwrapping and also provides a good electrical contact.
5. Insert the wrapped end of the electrode lead with the curved cable pressure plate into the socket.
6. Tighten the hex socket cap screws. This secures the lead tightly between the pressure plate and the socket walls.
7. Reinstall the insulated handle.

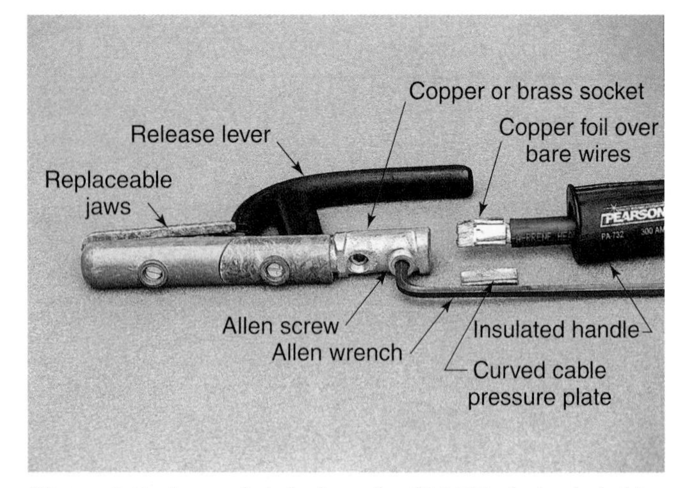

Figure 6-8. An exploded view of a SMAW electrode holder. Notice the insulated handle and the parts required to connect the welding lead to the electrode holder.

The electrode and/or holder should not be in contact with the worktable or workpiece when the welding machine is started, or damage may occur to the welding machine. Also, do not allow them to be in contact with the workpiece when changing current. To prevent the electrode and/or holder from contacting the workpiece, hang the holder from an insulated hanger in the welding booth when not in use.

Inspecting the SMAW Outfit

The SMAW outfit should be inspected before welding to ensure that all connections are tight, all insulation is intact, and that all equipment is safe to use. **The welding machine must be turned off to prevent electrical hazards when inspections are made.**

1. **Check to see that the electrode and workpiece leads are tightly attached to the welding machine terminals. Use the proper size wrench to tighten the connections, if required.**
2. **Visually check the entire length of each lead for damage or wear.**
3. **Ensure that the electrode lead is connected tightly to the electrode holder.**
4. **Verify that the workpiece lead is connected tightly to the worktable or workpiece.**
5. **Make sure that the electrode holder or electrode is not in contact with the worktable or workpiece.**
6. **Verify that booth sides and curtains protect people outside the booth from the arc and flying slag.**
7. **Remove any moisture or standing water from the floor.**
8. **Make sure the ventilation system is working properly and all ventilation ports and ducts are clear of obstructions.**

Adjusting the Machine

The correct polarity and amperage must be determined before adjusting the welding machine. The type of metal, metal thickness, joint design, welding position, and type of electrode should be considered when selecting polarity and amperage.

Polarity

A dc welding machine can be changed from DCEN to DCEP by flipping a switch. See **Figure 6-9**. On some machines, the polarity can also be changed by reversing the workpiece and electrode leads. **The electrode holder must be hung on an insulated hook so no current is flowing when the polarity switch is changed. Another option is to turn the welding machine off,**

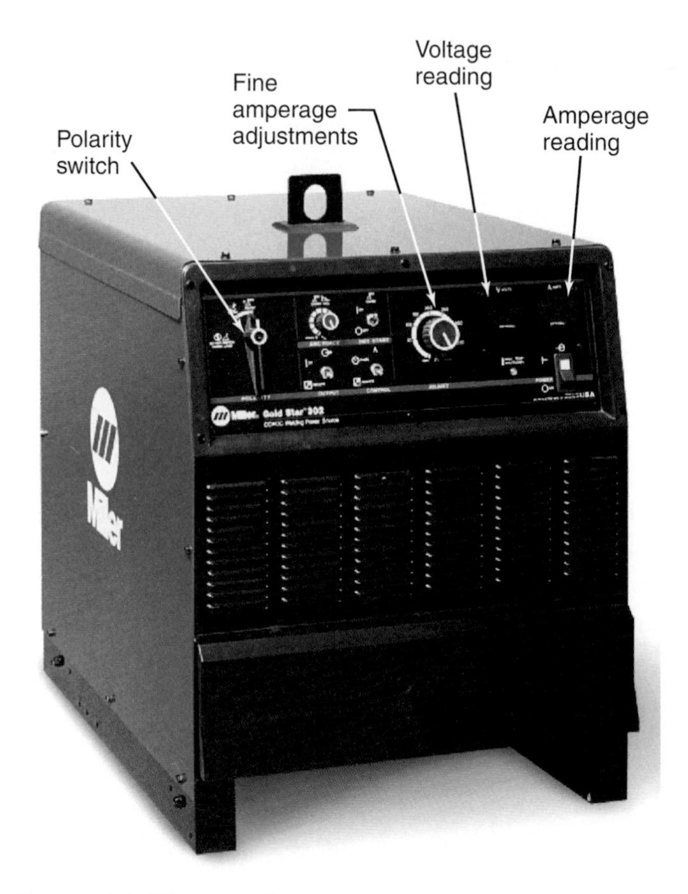

Figure 6-9. The controls on a dc arc welding machine. The lever at the left is used to change the polarity. Fine amperage adjustments are made with the largest knob in the middle. The voltage and amperage may be read on the digital displays at the upper right. (Miller Electric Mfg. Co.)

then change polarity. This prevents internal damage to the welding machine.

Ac and dc welding machines may use coarse and fine adjustment controls to select the required amperage, **Figure 6-10**. The adjustment controls function in the same way, although they may be located in a different place on the welding machine or operate differently. Coarse adjustment provides ranges of amperage settings. Coarse current changes must not be made while welding. Before changing current ranges, stop welding. Hang the electrode holder on an insulated hook or turn the welding machine off. The coarse adjustment lever on a welding machine is called a ***range switch***.

Fine amperage adjustments provide precise control of the current. Fine adjustments are made with a knob, crank, or digital settings. These controls change the amperage within the selected coarse range. The coarse and fine adjustments are generally shown on the face of the welding machine. When equipment is properly adjusted, it will deliver the necessary constant current to make a quality weld.

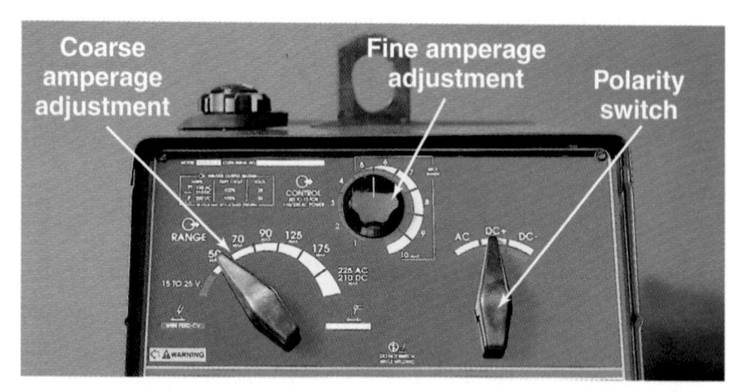

Figure 6-10. Coarse amperage adjustments may be made using the handle at the left. Fine adjustments within each range can be made using the knob at the upper center. Polarity is changed using the handle at the right. (Lincoln Electric Company)

Summary

- Input power for ac and dc welding machines generally is supplied by commercial sources. Commercial power voltages of 120V, 240V, or higher may be required. Engine-driven alternators may be used to produce 120V or 240V input power for arc welding machines.
- Ac and dc welding machines have at least two terminals. One terminal is for the electrode lead and the other is for the workpiece lead. To select a coarse amperage range, use a switch on the machine.
- Two variables are used to determine the diameter of the electrode and workpiece leads: maximum amperage and total distance to the work and back.
- The SMAW outfit should be inspected before welding to ensure that all connections are tight, all insulation is intact, and that all equipment is safe to use.
- The correct polarity and amperage must be determined before adjusting the welding machine. The type of metal, metal thickness, joint design, welding position, and type of electrode should be considered when selecting polarity and amperage.

Review Questions

Write your answers on a separate sheet of paper. Please do *not* write in this book.

1. Although _____V welding machines are usually hardwired into a building's electrical system, some electrical codes permit them to be plugged into special wall receptacles.
2. What is the function of a rectifier in a welding machine?
3. A qualified _____ should permanently wire the welding machine to the commercial power source.
4. Lugs are attached to welding leads using _____ or _____.
5. List two methods used to connect welding leads to a welding machine.
6. What two variables are used to determine the diameter of the welding leads?
7. List three methods of attaching a workpiece lead to a worktable or workpiece.
8. Why should welding leads *not* be laid across an aisle or other heavy-traffic area?
9. Name two methods used to change dc polarity on a welding machine.
10. List three types of controls that might be used to set the fine amperage on ac and dc welding machines.

Chapter 7

SMAW: Electrodes

Learning Objectives

After studying this chapter, you will be able to:

- Identify carbon and low alloy SMAW electrodes.
- List six purposes of an electrode covering.
- Interpret the AWS electrode identification system.
- Determine the trial amperage of a welding machine using the rule-of-thumb method.
- Select an electrode to meet the requirements of a weld.
- Identify two means of storing electrodes.

Technical Terms

AWS electrode specifications
deposition rate
electrode covering
electrode dispenser
electrode drying oven
electrode identification system
flux covering
hydrogen (H_2)
low hydrogen electrodes
welding procedure specification (WPS)

Covered Electrodes

SMAW electrodes are solid, round metal wires that are covered (coated). Electrode wires are made from metal alloys. The wire used for most steel electrodes is a low-carbon steel. The wire melts when an arc is struck. The melting wire provides filler metal for the weld.

Electrode wires and their coverings for SMAW electrodes are produced under strict manufacturing specifications. These manufacturing specifications were developed by the American Welding Society. ***AWS electrode specifications*** are published and revised about every five years. **Figure 7-1** shows the AWS electrode specification numbers for several metals.

Electrode Coverings

When covered electrodes are manufactured, the electrode wire is first cut to the desired length. The

AWS Electrode Specification Number	Metal Referred to in Specification
AWS A5.1	Carbon steels
AWS A5.3	Aluminum and aluminum alloys
AWS A5.4	Corrosion-resistant steels
AWS A5.5	Low alloy steels
AWS A5.6	Copper and copper alloys
AWS A5.11	Nickel and nickel alloys
AWS A5.15	Gray and ductile cast iron

Figure 7-1. *The AWS electrode specification numbers for several commonly welded metals.*

electrode covering materials are combined into a thick clay-like mixture. This mixture is usually extruded onto the electrode wire in a very exact thickness.

An ***electrode covering,*** sometimes called the ***flux covering,*** may serve any or all of the following purposes:

- Add filler metal to the weld.
- Create a protective gas shield around the arc and molten metal.
- Create a flux to clean impurities from the molten metal.
- Create a hard slag covering to protect the molten weld bead as it cools.
- Improve mechanical and chemical properties of the weld by adding alloying elements to the weld metal.
- Determine the current type and polarity specifications of the electrode.

When an arc is struck, the heat from the arc melts the electrode wire and an area of the base metal. Molten electrode wire transfers across the arc and enters the weld pool. When the weld pool solidifies, it is called a weld bead.

The heat from the arc also causes the covering to melt and perform its functions. Some of the materials in the covering are used to create the shielding gas that surrounds the weld area. The shielding gases keep oxygen, nitrogen, dirt, and other airborne contaminants out of the weld area.

Some of the electrode covering materials form a flux as they melt. The flux removes impurities from the weld. The impurities and flux float to the surface of the weld bead.

As the flux and impurities cool, they form slag (hard crust) over the surface of the weld bead. The weld bead cools under the slag, protected from contamination by the thick covering.

The mechanical and chemical properties of the weld metal can also be improved using certain flux coverings. Iron powder may be used in electrode covering compounds. This type of electrode deposits more filler metal into the weld in a given length of time. This reduces welding time and decreases welding costs. Some electrode coverings contain metallic salts. The metallic salts add alloying elements to the molten weld metal as the covering melts. The alloying elements combine with the weld metal to make the weld stronger, more corrosion resistant, or more ductile.

Hydrogen in a completed weld lowers its strength. ***Low hydrogen electrodes*** deposit a minimum amount of hydrogen into the weld. The electrode identification number for low hydrogen electrodes is discussed later in the chapter.

Chemicals in the electrode covering determine the polarity of the electrode. Certain electrode coverings work well with ac. Other electrode coverings are used with DCEN only or DCEP only. Still, other coverings produce good weld beads with ac, DCEN, or DCEP.

Electrode Sizes

Covered electrodes are available in a variety of lengths and diameters. They are available in 9″ (229 mm), 12″ (305 mm), 14″ (356 mm), and 18″ (457 mm) lengths.

The diameter of an electrode is the diameter of the uncoated wire. Flux covering thickness is disregarded when determining electrode diameter. AWS covered electrodes are available in the following diameters: 1/16″ (1.6 mm), 5/64″ (2.0 mm), 3/32″ (2.4 mm), 1/8″ (3.2 mm), 5/32″ (4.0 mm), 3/16″ (4.8 mm), 7/32″ (5.6 mm), 1/4″ (6.4 mm), 5/16″ (7.9 mm), and 3/8″ (9.5 mm).

Not all AWS electrodes are produced by every electrode manufacturer in the diameters or lengths described. Each electrode manufacturer distributes an electrode guide showing the lengths and diameters they produce. These guides also provide a variety of information regarding electrodes and their suggested uses.

Corrugated cartons or metal cans are used to ship 14″ (356 mm) and 18″ (457 mm) electrodes. They are available in 50-pound (22.7 kg) containers. The 9″ (229 mm) and 12″ (305 mm) electrodes are shipped in 25-pound (11.4 kg) containers. Some electrodes are available in one-pound packages.

Electrode Identification

The proper electrode diameter and classification must be determined in order to produce a good weld. The diameter of the electrode can be measured. The electrode is identified (classified) by a number located near its bare end. See **Figure 7-2**.

The American Welding Society has developed an ***electrode identification system.*** Covered electrodes for SMAW low alloy steels and carbon steels are identified

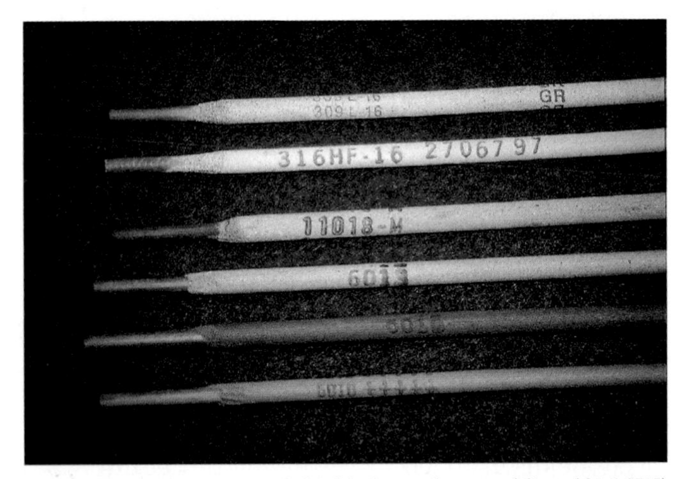

Figure 7-2. *This group of shielded metal arc welding (SMAW) electrodes shows how the AWS identification numbers are imprinted on each electrode.*

with the letter E (for electrode) followed by four or five numbers. Specifications for electrodes used in low-alloy steel SMAW are listed in AWS 5.5. Specifications for electrodes used in carbon steel SMAW are listed in AWS 5.1. See **Figures 7-3** and **7-4**.

The first two (sometimes three) numerals from the left indicate the minimum tensile strength. Minimum tensile strength is measured in thousands of pounds per square inch or in megapascals. In **Figure 7-5A**, the minimum tensile strength is 60,000 psi (414 MPa). The

AWS Classification[a]	Type of Covering	Weld Positions[b]	Type of Current[c]
E7010-X	High cellulose sodium	F, V, OH, H	DCEP
E7011-X	High cellulose potassium	F, V, OH, H	AC or DCEP
E7015-X	Low hydrogen sodium	F, V, OH, H	DCEP
E7016-X	Low hydrogen potassium	F, V, OH, H	AC or DCEP
E7018-X	Iron powder, low hydrogen potassium	F, V, OH, H	AC or DCEP
E7020-X	High iron oxide	H-fillets F	AC or DCEN AC or DC, either polarity
E-7027-X	Iron powder, high iron oxide	H-fillets F	AC or DCEN AC or DC, either polarity
E8010-X	High cellulose sodium	F, V, OH, H	DCEP
E8011-G	High cellulose potassium	F, V, OH, H	AC or DCEP
E8013-G	High titania potassium	F, V, OH, H	AC or DC, either polarity
E8015-X	Low hydrogen sodium	F, V, OH, H	DCEP
E8016-X	Low hydrogen potassium	F, V, OH, H	AC or DCEP
E8018-X	Iron powder, low hydrogen potassium	F, V, OH, H	AC or DCEP
E8045-P2	Low hydrogen sodium	F, OH, H, V-down	DCEP
E9010-G	High cellulose sodium	F, V, OH, H	DCEP
E9010-X	High cellulose sodium	F, V, OH, H	DCEP
E9011-G	High cellulose potassium	F, V, OH, H	AC or DCEP
E9013-G	High titania potassium	F, V, OH, H	AC or DC, either polarity
E9015-X	Low hydrogen sodium	F, V, OH, H	DCEP
E9016-X	Low hydrogen potassium	F, V, OH, H	AC or DCEP
E9018-X	Iron powder, low hydrogen potassium	F, V, OH, H	AC or DCEP
E9018M	Iron powder, low hydrogen	F, V, OH, H	DCEP
E9045-P2	Low hydrogen sodium	F, OH, H, V-down	DCEP
E10010-G	High cellulose sodium	F, V, OH, H	DCEP
E10011-G	High cellulose potassium	F, V, OH, H	AC or DCEP
E10013-G	High titania potassium	F, V, OH, H	AC or DC, either polarity
E10015-X	Low hydrogen sodium	F, V, OH, H	DCEP
E10016-X	Low hydrogen potassium	F, V, OH, H	AC or DCEP
E10018-X	Iron powder, low hydrogen	F, V, OH, H	AC or DCEP
E10018M	Iron powder, low hydrogen	F, V, OH, H	DCEP
E10045-P2	Low hydrogen sodium	F, OH, H, V-down	DCEP
E11010-G	High cellulose sodium	F, V, OH, H	DCEP
E11011-G	High cellulose potassium	F, V, OH, H	AC or DCEP
E11013-G	High titania potassium	F, V, OH, H	AC or DC, either polarity
E11015-G	Low hydrogen sodium	F, V, OH, H	DCEP
E11016-G	Low hydrogen potassium	F, V, OH, H	AC or DCEP
E11018-G	Iron powder, low hydrogen	F, V, OH, H	AC or DCEP
E11018M	Iron powder, low hydrogen	F, V, OH, H	DCEP
E12010-G	High cellulose sodium	F, V, OH, H	DCEP
E12011-G	High cellulose potassium	F, V, OH, H	AC or DCEP
E12013-G	High titania potassium	F, V, OH, H	AC or DC, either polarity
E12015-G	Low hydrogen sodium	F, V, OH, H	DCEP
E12016-G	Low hydrogen potassium	F, V, OH, H	AC or DCEP
E12018-G	Iron powder, low hydrogen potassium	F, V, OH, H	AC or DCEP
E12018M	Iron powder, low hydrogen	F, V, OH, H	DCEP
E12018M1	Iron powder, low hydrogen	F, V, OH, H	DCEP

a. The letter suffix 'X' as used in this table stands for the designator (in Figure 7-7) it replaces that determines the chemical composition of the deposited weld metal.
b. Abbreviations: F = Flat, H = Horizontal, H-fillet = Horizontal fillet; [for electrodes 3/16" (5.0 mm) and under, except 5/32" (4.0 mm) and under for classifications E(X)XX15-X, E(X)XX16-X, and E(X)XX18-X, and E(X)XX18M(1)] V = Vertical, V-down = Vertical with downward progression, OH = Overhead.
c. DCEP means electrode positive (reverse polarity). DCEN means electrode negative (straight polarity).

Figure 7-3. *A listing of AWS low alloy steel, covered arc welding electrodes. An explanation of the Xs is found in Figure 7-7. (AWS A5.5:2006, Table 1, Low Alloy Steel Electrodes for SMAW, reproduced with permission from the American Welding Society, Miami, FL)*

AWS Classification	Type of Covering	Weld Positions[a]	Type of Current[b]
E60 series electrodes			
E6010	High cellulose sodium	F, V, OH, H	DCEP
E6011	High cellulose potassium	F, V, OH, H	AC or DCEP
E6012	High titania sodium	F, V, OH, H	AC or DCEN
E6013	High titania potassium	F, V, OH, H	AC or DC, either polarity
E6018	Low hydrogen potassium, iron powder	F, V, OH, H	AC or DCEP
E6019	Iron oxide titania potassium	F, V, OH, H	AC or DC, either polarity
E6020	High iron oxide	H-fillet	AC or DCEN
E6022[c]	High iron oxide	F, H-fillet	AC or DC, either polarity
E6027	High iron oxide, iron powder	H-fillet F	AC or DCEN AC or DC, either polarity
E70 series electrodes			
E7014	Iron powder, titania	F, V, OH, H	AC or DC, either polarity
E7015	Low hydrogen sodium	F, V, OH, H	DCEP
E7016	Low hydrogen potassium	F, V, OH, H	AC or DCEP
E7018	Low hydrogen potassium, iron powder	F, V, OH, H	AC or DCEP
E7018M	Low hydrogen iron powder	F, V, OH, H	DCEP
E7024	Iron powder, titania	H-fillet, F	AC or DC, either polarity
E7027	High iron oxide, iron powder	H-fillet F	AC or DCEN AC or DC, either polarity
E7028	Low hydrogen potassium, iron powder	H-fillets, F	AC or DCEP
E7048	Low hydrogen potassium, iron powder	F, OH, H, V-down	AC or DCEP

a. Abbreviations: F = Flat, H = Horizontal; H-fillet = Horizontal fillet, V = Vertical, V-down = Vertical with downward progression, OH = Overhead. These are for electrodes 3/16" (5.0 mm) and under, except 5/32" (4.0 mm) and under for classifications E6018, E7014, E7015, E7016, E7018, E7018M, and E7048.
b. DCEP refers to direct current, electrode positive (DC reverse polarity). DCEN refers to direct current, electrode negative (DC straight polarity).
c. Electrodes of the E6022 classification are only for single-pass welds.

Figure 7-4. AWS carbon steel, covered arc welding electrodes. (AWS A5.1:2004, Table 1, Carbon Steel Electrodes for SMAW, reproduced with permission from the American Welding Society, Miami, FL)

minimum tensile strength for the electrode number in **Figure 7-5B** is 100,000 psi (689 MPa).

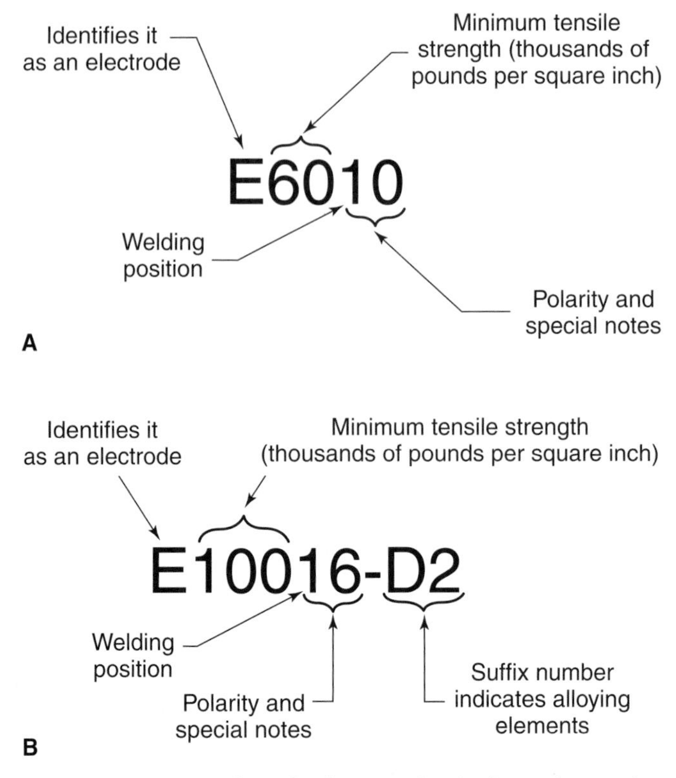

Figure 7-5. Examples of what an AWS electrode number includes. A—E6010 electrode. B—E10016-D2 electrode.

The second numeral from the right indicates the recommended welding position and/or type of weld: 1–all positions, 2–flat welding and horizontal fillet welds, 3–flat position only, 4–flat, horizontal, overhead, or vertical down welding. The electrodes in Figure 7-5 may be used in all positions.

The last two numerals at the right indicate the polarity or any other special notes. **Figure 7-6** shows a list of the last two numerals and their indicated polarities. In Figure 7-5A, the electrode requires DCEP. The electrode in Figure 7-5B requires ac or DCEP. It is also a low hydrogen electrode.

The electrode identification number may also have a suffix to show the alloying elements that are added to the weld metal. **Figure 7-7** shows a list of suffixes and their indicated alloying elements. In Figure 7-5B, D2 indicates that the alloying elements are composed of .25%–.45% molybdenum and 1.25%–2.00% manganese.

Low hydrogen electrodes have AWS identification numbers that end in 5, 6, or 8. See **Figure 7-8**. The numerals 5, 6, or 8 indicate the covering composition and suggested application of the electrode.

Electrodes used to weld other base metals like stainless steel, nickel, aluminum, or copper have their own numbering systems (Figure 7-1).

Electrode Amperage Requirements

The diameter of the electrode wire determines the amount of amperage required to melt it. A larger wire diameter requires higher amperage. Electrode manufacturers provide recommended amperage ranges for the various diameters of AWS electrodes that they produce. **Figure 7-9** suggests amperage ranges for the various diameters of E60XX and E70XX electrodes.

A rule-of-thumb method may be used to find a trial amperage when a manufacturer's electrode guide is not available. This method works well for E6010, E6011, and E6012 electrode diameters between 3/32" (2.4 mm) and 1/4" (6.4 mm). A 1/8" (3.2 mm) diameter electrode requires 90A–120A. Add 30A to the 90A for each 1/32" (.8 mm) of diameter above 1/8" (3.2 mm). Subtract 30A from the 90A for each 1/32" (.8 mm) of diameter below 1/8" (3.2 mm). The amperage that is determined will be near the lower end of the recommended amperage range. For example, the following steps would be taken to determine the amperage that should be set on the welding machine when using 3/16" (4.8 mm) electrodes.

Step 1: 1/8" (3.2 mm) diameter electrode = 90A.

Step 2: Determine the difference between 3/16" and 1/8".

3/16"–1/8" = 1/16".

Step 3: Convert 1/16" to 32nds of an inch.

1/16" = 2/32".

Step 4: Multiply the number of 32nds by 30A.

$2 \times 30 = 60$.

Step 5: Add the 90A from Step 1 to the answer in Step 4.

90A + 60A = 150A.

Answer: The welding machine should be set for 150A when using 3/16" (4.8 mm) electrodes.

The amperage obtained by using the manufacturer's electrode guide, the rule-of-thumb method, or from the tables in Figure 7-9 may need to be modified. The position of the weld and the skill of the welder determine the final amperage setting. Inspect the weld to determine whether the amperage needs to be raised or lowered. The need for amperage changes will be covered in more detail in Chapter 8.

Last Two Numerals	Polarity and Special Notes
EXX10	DCEP
EXX11	AC or DCEP
EXX12	AC or DCEN
EXX13	AC or DCEN
EXX14	DCEN, DCEP, AC (iron powder)
EXX15	DCEP (low hydrogen)
EXX16	AC, DCEP (low hydrogen)
EXX18	AC, DCEP (iron powder and low hydrogen)
EXX20	DCEN, or AC for horizontal fillet welds; DCEN, DCEP, or AC for flat position welding
EXX24	DCEN, DCEP, or AC (iron powder)
EXX27	DCEN, or AC for horizontal fillet welds; DCEN, DCEP, or AC for flat position (iron powder)
EXX28	AC or DCEP (iron powder and low hydrogen)

Figure 7-6. *The recommended polarity to use with SMAW electrodes. The last two numerals indicate the recommended electrical polarity.*

-A1	1/2% Mo
-B1	1/2% Cr, 1/2% Mo
-B2	1 1/4% Cr, 1/2% Mo
-B3	2 1/4% Cr, 1% Mo
-C1	2 1/2% Ni
-C2	3 1/4% Ni
-C3	1% Ni, .35% Mo, .15% Cr
-D1 and D2	.25-.45% Mo, 1.25-2.00% Mn
-G	.50 min Ni, .30 min Cr, .20 min. Mo, .10 min V, 1.00 min Mn, .80 min Si (Only one of the listed elements is required for the G classification).

Figure 7-7. *This chart shows the meanings of each AWS electrode suffix. When a suffix is used, it indicates that alloys have been added to the electrode. Example: E8016-B2.*

Right Hand Digit	Covering Compositions	Application (Use)
5 E-7015	Low hydrogen sodium type.	This is a low hydrogen electrode for welding low carbon, alloy steels. Power shovels and other earthmoving machinery require this rod. The weld files or machines easily. Use DCEP (DCRP) only.
6 E-7016	Same as "5" but with potassium salts used for arc stabilizing.	It has the same general application as (5) above except it can be used on either DCEP (DCRP) or AC.
8 E-7028	Iron powder (low hydrogen), flat position only.	For low carbon alloy steels, use DC or AC.
E-8018	Iron powder plus low-hydrogen sodium covering.	Similar to (5) and (6) DCEP (DCRP) or AC. Heavy covering allows the use of high speed drag welding. AC or DCEP (DCRP) may be used.

Figure 7-8. *Low hydrogen electrode covering compositions and applications. Low hydrogen electrodes withstand higher temperatures and can be used with higher welding amperages.*

Suggested Metal Thickness		Electrode Size		E6010 and E6011	E6012	E6013	E6020	E6022	E6027
in.	mm	in.	mm						
1/16 & less	1.6 & less	1/16	1.6		20-40	20-40			
1/16-5/64	1.6-2.0	5/64	2.0		25-60	25-60			
5/64-1/8	2.0-3.2	3/32	2.4	40-80	35-85	45-90			
1/8-1/4	3.2-6.4	1/8	3.2	75-125	80-140	80-130	100-150	110-160	125-185
1/4-3/8	6.4-9.5	5/32	4.0	110-170	110-190	105-180	130-190	140-190	169-240
3/8-1/2	9.5-12.7	3/16	4.8	140-215	140-240	150-230	175-250	170-400	210-300
1/2-3/4	12.7-19.1	7/32	5.6	170-250	200-320	210-300	225-310	370-520	250-350
3/4-1	19.1-25.4	1/4	6.4	210-320	250-400	250-350	275-375		300-420
1 - up	25.4 - up	5/16	8.0	275-425	300-500	320-430	340-450		375-475

Suggested Metal Thickness		Electrode Size		E7014	E7015 and E7016	E7018	E7024 and E7028	E7027	E7048
in.	mm	in.	mm						
5/64-1/8	2.0-3.2	3/32*	2.4*	80-125	65-110	70-100	100-145		
1/8-1/4	3.2-6.4	1/8	3.2	110-160	100-150	115-165	140-190	125-185	80-140
1/4-3/8	6.4-9.5	5/32	4.0	150-210	140-200	150-220	180-250	160-240	150-220
3/8-1/2	9.5-12.7	3/16	4.8	200-275	180-255	200-275	230-305	210-300	210-270
1/2-3/4	12.7-19.1	7/32	5.6	260-340	240-320	260-340	275-365	250-350	
3/4-1	19.1-25.4	1/4	6.4	330-415	300-390	315-400	335-430	300-420	
1 - up	25.4 - up	5/16*	8.0*	390-500	375-475	375-470	400-525	375-475	

Note: When welding vertically up, currents near the lower limit of the range are generally used.
* These diameters are not manufactured in the E7028 classification.

Figure 7-9. Suggested amperage ranges for use with various E60XX and E70XX electrodes and diameters. Electrode guides furnished by manufacturers also give amperage ranges.

Choosing an Electrode

A ***welding procedure specification (WPS)*** is written to specify welding requirements for commercial jobs, such as bridges, pipelines, or steel structures. A welding procedure specification outlines exactly how each weld is to be made. It also lists the diameter and type of AWS electrode to be used. Welding procedure specifications are generally written by engineers.

Most fabrication, repair, and craft welding does not require a welding procedure. The welder or shop owner selects the electrode to be used for these situations.

Several factors must be considered when selecting an electrode:

- Type of metal.
- Metal thickness.
- Groove design.
- Joint alignment.
- Available welding current.
- Welder skill.
- Welding position.
- Deposition rate (rate of weld deposit).
- Depth of penetration.
- Weld bead finish.
- Additional metal properties.

After considering all the factors, an electrode selection can be made. An electrode manufacturer's guide may also be helpful when selecting electrodes.

Type of Metal

The tensile strength of the deposited weld metal should be as strong as that of the base metal. The tensile strength of the steel being welded must be determined. The steel supplier can provide you with this information. If the steel has a tensile strength of 60,000 psi (414 MPa), an E60XX or E70XX electrode can be used. An E120XX electrode is required for steel with a tensile strength of 120,000 psi (827 MPa).

Metal Thickness

One weld pass may be enough for metal up to 1/4″ (6.4 mm) thick. More than one weld pass is needed for metal over 1/4″ thick. Each weld pass should not be more than 1/4″ (6.4 mm) thick. A 1/4″–3/8″ wide weld bead is required on 1/4″ (6.4 mm) thick steel. Wider weld beads are used on thicker metal.

A weld bead should be 2–3 times the diameter of the electrode. A 1/4″ (6.4 mm) wide weld bead requires an electrode 3/32″–1/8″ (2.4 mm–3.2 mm) in diameter. A 3/8″ (9.6 mm) wide weld bead requires an electrode with a 1/ 8″–3 /16″ (3.2 mm–4.8 mm) diameter.

Certain electrodes work well on sheet metal because they do not penetrate deeply. E6012 and E6013 electrodes work well on thin pieces of metal.

Groove Design

In a wide V-groove joint, a small-diameter electrode may be used for the root pass. Larger diameter electrodes are then used for the filler and cover passes. Smaller diameter electrodes may be required for the root pass and next few passes on narrow bevel-, V-, J-, or U-groove joints.

Joint Alignment

Poorly designed or assembled joints may have large openings that are hard to fill. Certain electrodes are designed to span large root openings. The E6012, E6013, and E7014 electrodes are designed for use on poorly assembled joints with large root openings.

Available Welding Current

A wide variety of electrodes can be used if both direct and alternating welding currents are available. The selection is more limited if only an ac or only a dc welding machine can be used.

Welder Skill

A highly skilled welder may successfully use a large-diameter electrode and high amperage. The same electrode may not provide proper penetration for a less-skilled welder. An inexperienced welder may produce good-quality welds in the flat position using an E7024 electrode.

Welding Position

The welding position must also be considered when selecting an electrode. All electrodes can be used for welds done in the flat welding position. Welds made in the overhead welding position are done with EXX1X or EXX4X electrodes. (The Xs can represent any number.)

Deposition Rate

Deposition rate is the amount of filler metal deposited in the joint in one minute. If a high deposition rate is required, a larger electrode diameter, higher amperage, and DCEP may be required.

Depth of Penetration

E6010 and E6011 electrodes penetrate more deeply than E6012 and E6013 electrodes. E6010 and E6011 electrodes provide good penetration on thick metal.

Weld Bead Finish

An E6013 or E7024 electrode produces a strong weld bead with a good appearance in butt and corner joints. An E6010 electrode produces strong welds in the same joints. However, the appearance of the weld bead will not be as good.

Additional Metal Properties

Electrodes such as the E10016-D2 contain alloying elements in their coverings. The alloying elements are combined with the weld metal during welding. Alloying elements improve the strength, corrosion resistance, or other characteristics of the finished weld.

Low hydrogen electrode coatings are low in moisture and hydrogen. When used properly, low hydrogen electrodes produce welds with lower hydrogen content. This produces a completed weld that is stronger and tougher.

Care of Electrodes

The electrode covering should not be damaged. Electrodes should not be bent. Keep electrodes in their shipping containers until they are used. The shipping container provides safe storage for the electrodes after it is opened.

Electrode coatings that are damp will not work properly. Water (H_2O) contains ***hydrogen*** ***(H_2)***. Hydrogen in a weld and in the heat affected zone weakens the weld. To help prevent such problems, welders should store their electrodes in an ***electrode drying oven***. An adequate supply of electrodes can be kept under ideal conditions in such an oven. See **Figure 7-10**.

Figure 7-10. *Electrode drying ovens keep electrode moisture content low. Different electrode types and diameters can be stored in separate bins inside the oven.*

Low hydrogen electrodes must be kept especially dry to prevent the absorption of hydrogen into the electrode covering.

All electrodes must be used within a specified time after they are removed from the drying oven. The amount of time varies with the type of electrode. The amount of time out of the oven is extremely critical for high-strength, low hydrogen electrodes such as E10018 and E12018. When the specified time has expired, all electrodes should be returned to the oven and rebaked.

An ***electrode dispenser***, **Figure 7-11**, may be used for temporary storage of electrodes. The dispenser protects the electrodes and keeps them relatively dry.

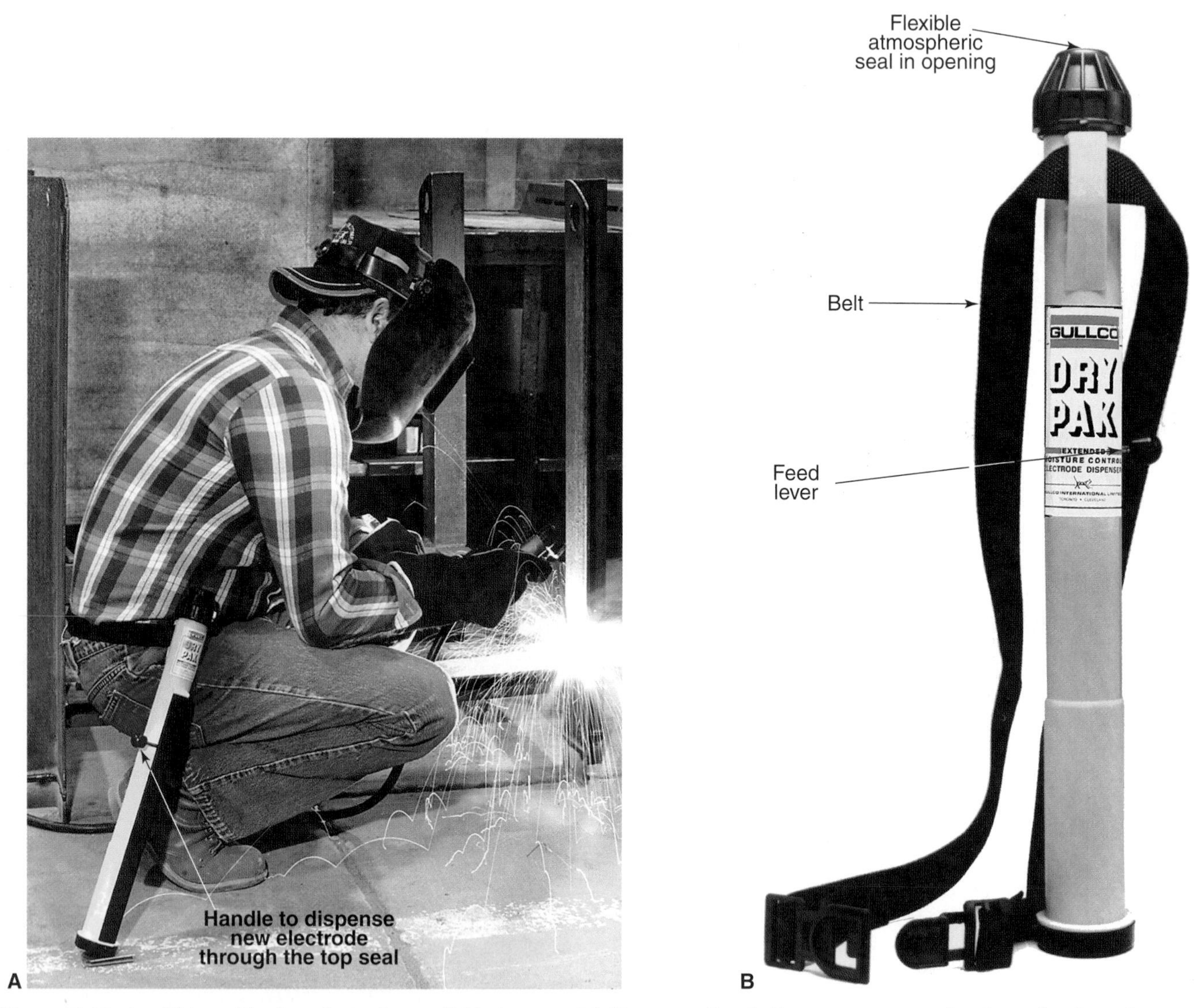

Figure 7-11. *A—This welder has electrodes available in a special dispenser. B—A dispenser protects the electrodes from damage and keeps them relatively dry while they are out of the oven. The electrodes are loaded into the dispenser and the sealing top is installed. Moving the feed lever pushes one electrode through the top seal for easy removal. (Gullco International Limited)*

Summary

- When an arc is struck, the heat from the arc melts the electrode wire and an area of the base metal. Molten electrode wire transfers across the arc and enters the weld pool. The heat from the arc also causes the covering to melt and perform its functions.
- Some of the materials in the covering are used to create the shielding gas that surrounds the weld area. The shielding gases keep oxygen, nitrogen, dirt, and other airborne contaminants out of the weld area. Some of the electrode covering materials form a flux as they melt. The flux removes impurities from the weld. When the flux solidifies, it creates a hard slag covering, which protects the weld bead as it cools.
- Iron powder may be used in electrode covering compounds. This type of electrode deposits more filler metal into the weld in a given length of time. This reduces welding time and decreases welding costs. The electrode covering may add alloying elements to the weld metal to make it stronger, more corrosion resistant, or more ductile.
- Low hydrogen electrodes deposit a minimum amount of hydrogen into the weld.
- Chemicals in the electrode covering determine the polarity of the electrode.
- Covered electrodes for SMAW low alloy steels and carbon steels are identified with the letter E (for electrode) followed by four or five numbers.
- The first two (sometimes three) numerals from the left in the electrode number indicate the minimum tensile strength. The second numeral from the right indicates the recommended welding position and/or type of weld. The last two numerals at the right indicate the polarity or any other special notes. The electrode identification number may also have a suffix to show the alloying elements that are added to the weld metal.
- Low hydrogen electrodes have AWS identification numbers that end in 5, 6, or 8.
- The diameter of the electrode wire determines the amount of amperage required to melt it. A larger wire diameter requires higher amperage.
- A welding procedure specification outlines exactly how each weld is to be made. It also lists the diameter and type of AWS electrode to be used.
- The electrode covering must not be damaged. Electrodes should not be bent. Keep electrodes in their shipping containers until they are used.

Review Questions

Write your answers on a separate sheet of paper. Please do *not* write in this book.

1. Which AWS electrode specification number provides information about electrodes to weld nickel and nickel alloys?
2. List four purposes of the electrode covering.
3. The slag on a SMAW weld bead contains _____ and _____.
4. *True or False?* Metallic salts in the electrode covering may cause the weld metal to become stronger, more corrosion resistant, or more ductile.
5. Describe what is indicated by each of the characters in the E6013 electrode designation.
6. What does B2 represent in an AWS E8016-B2 electrode designation?

(continued)

Review Questions *continued*

7. Use the rule-of-thumb method to determine the trial amperage for a 7/32″ (5.6 mm) electrode. Show your work.
8. *True or False?* The electrode type and diameter is given in a welding procedure specification.
9. *True or False?* An E7013 electrode would be a good electrode to use for steel with a tensile strength of 88,000 psi.
10. A 1/2″ (12.7 mm) wide weld bead requires an electrode with a _____-_____ diameter.

Chapter 8

SMAW: Flat Welding Position

Learning Objectives

After studying this chapter, you will be able to:

- Identify the safety rules required for arc welding.
- Strike an arc.
- Read a weld bead.
- Make a fillet weld on a lap joint in the flat welding position using shielded metal arc welding.
- Make a fillet weld on an inside corner joint in the flat welding position using shielded metal arc welding.
- Make a fillet weld on a T-joint in the flat welding position using shielded metal arc welding.
- Make a square-, J-, and V-grooved weld on a butt joint in the flat welding position using shielded metal arc welding.

arc blow
arc length
backhand welding
backward arc blow
bullet-shaped ripples
chipping
chipping goggles
closed arc
drag angle
drag welding
forehand welding
forward arc blow
keyhole
layer
magnetic field
open arc
phosgene gas
push angle
reading the weld bead
restarting the arc
root pass
running a weld bead
spatter
travel angle
under load
weld pool
work angle

Preparing to Weld

Comfortable fire-resistant clothing should be worn for SMAW. Protect all areas of your skin from burns caused by arc rays or molten metal. Shirts and jackets should be buttoned at the collar. Wear a cap to protect your head and hair. Gloves must be worn to protect your hands and forearms. Hard-toed, high-top shoes or boots should also be worn. Pants without cuffs should be worn to prevent hot metal or sparks from being caught in them. Do not wear clothing with ragged edges or loose threads. The edges and threads may catch fire easily.

Make a visual safety inspection of your welding outfit or station. Refer to Chapter 6 for information detailing the inspection of a welding outfit.

Select the appropriate type and size of electrode for your job. Chapter 7 covers information regarding electrodes, such as suggested amperage and polarity. Guides published by electrode manufacturers also provide polarity and amperage information.

SMAW Safety Precautions

Arc welding presents dangers of electrical shock, fumes and gases, hot metal, arc rays, and fire. The importance of safety precautions cannot be stressed enough. The application of proper safety precautions prevents injury to personnel and damage to equipment.

Safety precautions that may be applied to all forms of arc welding include:

- Always have qualified people perform all installation, maintenance, and repair work on equipment and electrical circuits.
- The electrode and work circuits are HOT (voltage is present) when the welding machine is on. Do not touch the electrode and the workpiece or worktable at the same time with bare skin, wet gloves, or wet clothing.
- Do not weld on damp or metal floors. If welding must be done in these locations, be certain you are insulated from them.
- Keep the welding machine, welding cables, electrode holder, lugs, and clamps in safe working order.
- Connect the worktable or workpiece to a good electrical ground.
- Never make polarity or current range changes on the welding machine while the machine is under load. *Under load* means that current is flowing for welding or that the electrode holder is touching the table or workpiece.
- Avoid breathing hazardous fumes or gases while welding. Fumes should be removed from the weld area before they pass the welder's face.
- When welding on lead, cadmium, or galvanized (zinc-coated) metals, use extra ventilation or wear a supplied air breathing apparatus (positive-pressure respirator). These metals produce toxic or poisonous fumes.
- Do not weld near degreasing or cleaning chemicals that contain chlorinated hydrocarbons. Arc welding rays and the heat of the arc can react with these solvents to produce phosgene gas. *Phosgene gas* is a highly toxic and poisonous gas. Parts cleaned with chlorinated hydrocarbons must be thoroughly rinsed to remove these cleaning agents.
- Always wear clothing that is flame-resistant for protection from fire, molten metal, and harmful arc rays. Clothing must cover all parts of the body. Arc rays can cause skin burns on exposed skin. Wear clean, oil-free clothing, such as leathers. Wear a cap, gloves, high-top shoes, and pants without cuffs.
- Never look at an arc without the proper eye protection. Arc rays can cause permanent damage to the eyes.
- Always use an arc welding helmet with the proper filter lens in place.
- Nonflammable screens should be set up around weld areas. The screens protect others in the area from arc rays and metal spatter.
- Remove all flammable or explosive materials from welding areas. Fire extinguishers should be available.
- Place a person on fire watch. Sparks and hot metal can go through cracks into other areas or fall onto other floors of a building.
- Do not heat, cut, or weld on tanks or other containers until appropriate safety steps have been taken. Containers may have been used to store explosive or chemically toxic materials. Specific steps for cleaning must be followed to ensure that explosions do not occur. The local fire or industrial safety department can advise you on how to properly clean a container prior to welding.
- Wear earplugs to keep sparks out of your ears when welding out of position. Earplugs are also necessary in workplaces where there are very loud noises.
- Wear safety glasses when chipping. Flash goggles should be worn under the welding helmet for protection from reflected rays.
- Engine-driven welders must be used in an open, well-ventilated area. Vent exhaust fumes outside.

Striking an Arc

An electrode holder is generally held with one hand. It can be gripped like a hammer or screwdriver. The electrode lead may be draped over the lower arm to make the electrode holder feel lighter. See **Figure 8-1**.

In order to strike an arc, the electrode must first touch the base metal. The electrode should only remain in contact with the base metal momentarily. This causes electricity to start flowing. The electrode is then pulled away from the base metal a short distance. Current continues to flow across this gap, creating an arc.

Most electrodes have a relatively thin covering of flux. When an arc is struck using a thinly covered

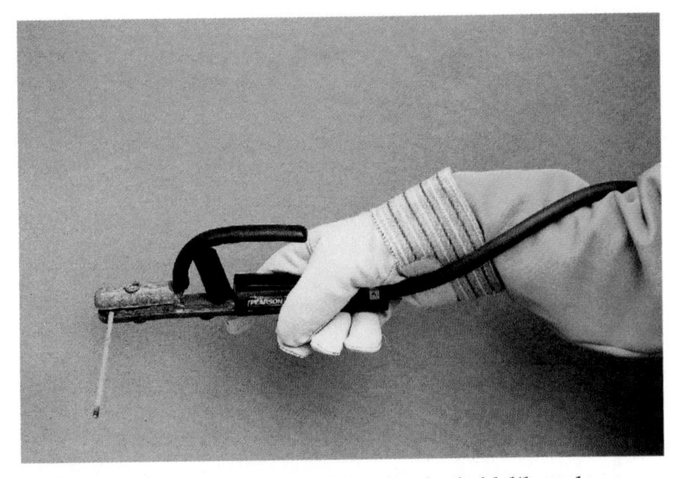

Figure 8-1. The electrode holder can be held like a hammer. Draping the electrode lead over the forearm makes the holder feel lighter.

electrode, the arc can be seen. An ***open arc*** is an arc that can be seen.

Two methods are used to strike an arc. A welder may scratch the electrode on the metal and then withdraw it. A straight up-and-down or pecking motion may also be used. See **Figure 8-2**. After the arc is struck and stabilized, it is brought down to the correct arc length.

The ***arc length*** is the distance between the electrode and the base metal. This distance should be approximately equal to the diameter of the electrode.

The electrode may become welded to the base metal if the arc is not struck correctly or if the arc length is not maintained properly. To release the electrode, proceed as follows:

1. Keep your welding helmet down and release the electrode from the electrode holder.
2. Lift your helmet.
3. Place the electrode holder on an insulated hook.
4. Grasp the electrode near the base metal. Bend the electrode back and forth until it breaks free.

Arc Blow

A ***magnetic field*** is created whenever electricity travels in a wire or electrode. When current travels in an electrode, a magnetic field is created around the electrode. The magnetic field changes direction as the current changes direction. See **Figure 8-3**. The arc is deflected from its normal path by these magnetic forces. This deflection is called ***arc blow***.

When alternating current is used, the magnetic field is continually canceled as the current changes direction. However, when direct current is used, the current flows in one direction and a strong magnetic field may develop. A magnetic field prefers to travel in a conductive material, such as metal. Air causes the magnetic field to weaken. See **Figure 8-4**.

Arc blow is greatest at the ends of the weld joint. The magnetic field tries to stay in the metal. This causes the filler metal to blow toward the center of the joint. See **Figure 8-5**. Arc blow that occurs at the beginning of a joint is ***forward arc blow***. ***Backward arc blow*** occurs at the end of the joint. The least amount of arc blow occurs in the center area of the weld or joint.

Arc blow may affect the quality of a weld. If arc blow does occur, one or more of the following steps may be taken:

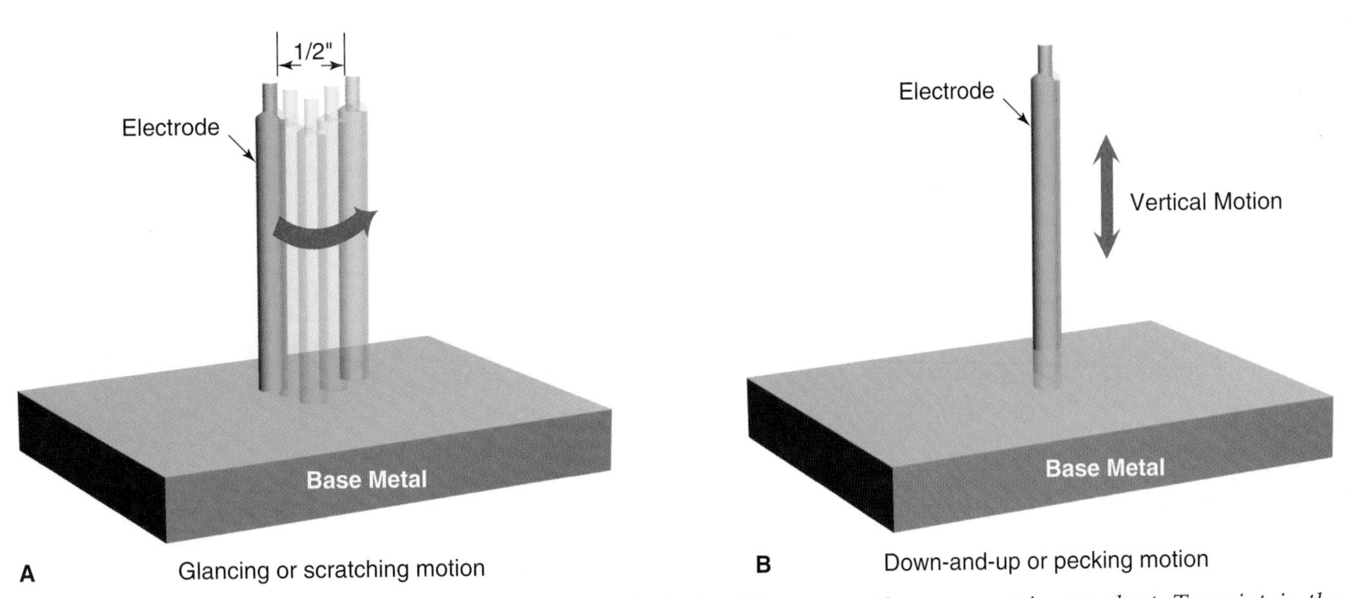

Figure 8-2. A—When using the scratching (glancing) method of striking an arc, the movement is very short. To maintain the arc, the end of the electrode must remain within 1/8″ (3.2 mm) of the base metal. B—When using the straight up-and-down or pecking method of striking an arc, the electrode must remain within 1/8″ (3.2 mm) of the base metal to maintain the arc.

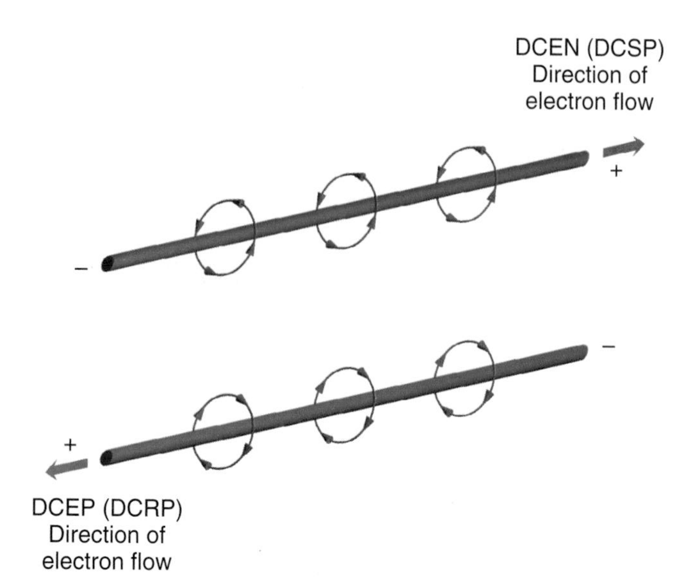

Figure 8-3. The magnetic field around a wire with DCEN (DCSP) rotates in the opposite direction to the field around a wire carrying DCEP (DCRP). The magnetic field is indicated by the arrows around the wire, and the current is indicated by the end arrows.

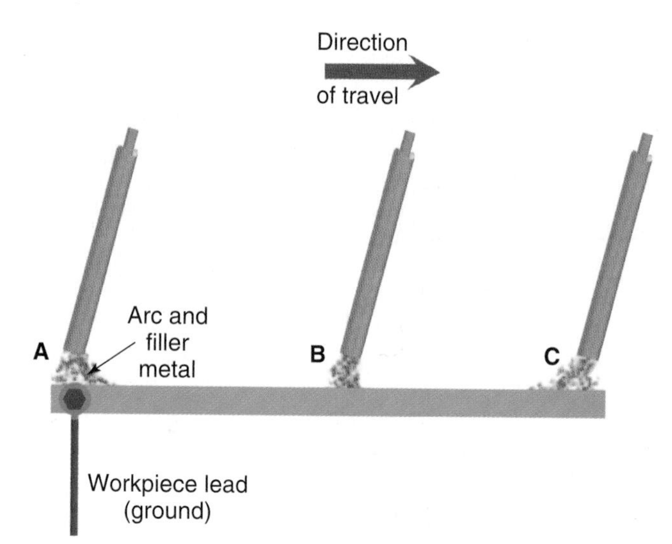

Figure 8-5. The effects of direct current arc blow on the arc and electrode wire. As the arc is started (A), the arc is blown toward the center of the weld joint. In the center (B), the arc travels straight down. As the arc approaches the end of the joint (C), the arc and filler metal are again blown toward the center of the weld joint.

- Place the ground connection as far from the weld joint as possible.
- If forward arc blow is a problem, connect the workpiece lead (ground) near the end of the weld joint.
- If backward arc blow is a problem, place the workpiece lead near the beginning of the weld.
- Reduce the welding current to reduce the strength of the magnetic field.
- Position the electrode so that the arc force counteracts the arc blow force.
- Use the shortest arc possible to produce a good weld bead. A short arc permits the filler metal to enter the weld pool before being effected. A short arc also permits the arc force to overcome the arc blow force.
- Use run-on or run-off tabs on the joint. See **Figure 8-6**.
- Wrap the electrode lead around the base metal in the direction that will counteract the arc blow force.
- Change to an ac machine and electrodes.

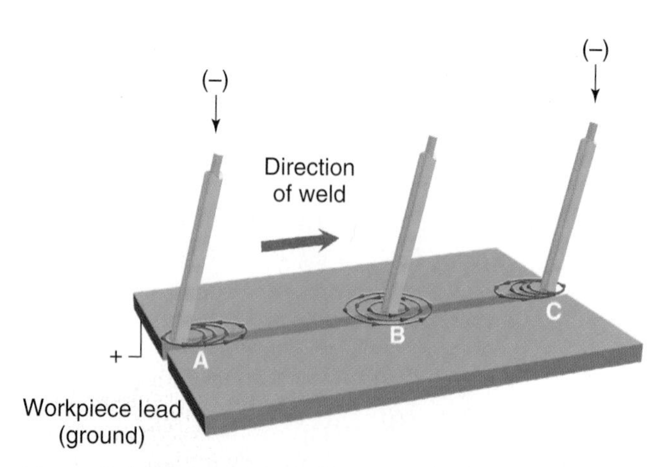

Figure 8-4. The magnetic field around the electrode is deflected at the ends of a weld joint (A and C). The field attempts to flow in the metal and not through the air. The concentration of magnetic flux at the ends of the metal forces the arc toward the center of the base metal. The arc "blows away" from the area directly under the electrode at the ends of the weld. Notice that the magnetic field is not distorted in the center areas of the weld joint at B.

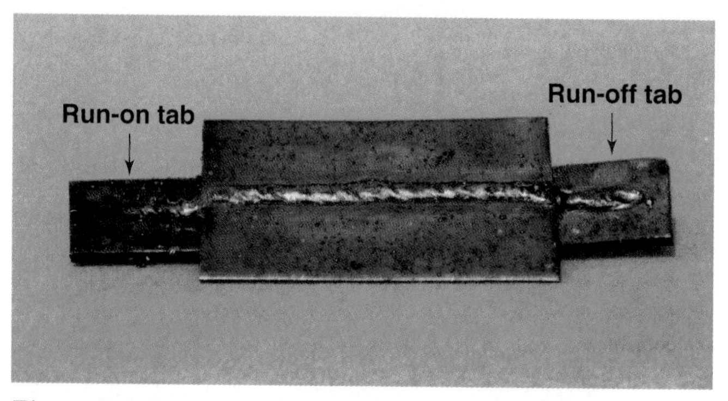

Figure 8-6. Run-on and run-off tabs were used on this square-groove butt joint. The tabs allow the welder to completely stabilize the arc before the actual weldment is reached. The weld bead is continued past the end of the weldment and onto the run-off tab. A strong weld from beginning to end is possible by using such tabs.

Running a Weld Bead

Making a weld bead is known as ***running a weld bead***. A weld pool forms after the arc is started. A ***weld pool*** is the small pool of molten metal that is formed directly below the tip of the electrode. Filler metal melts from the electrode and is deposited in the weld pool. The weld bead now begins to form.

The arc length determines the voltage and amperage across the arc. Therefore, the arc length must remain constant throughout the entire weld. The normal arc length is about equal to the electrode diameter. A beginning welder should practice holding a constant arc length with one hand.

Two types of weld beads are used when welding: the stringer bead and the weave bead. A stringer bead is one made along a line without any side-to-side motion. The width of a stringer bead should be two to three times the electrode diameter. See **Figure 8-7**. Its height should be about one-eighth of the weld bead's width. For example, a stringer bead made with a 1/8″ (3.2 mm) diameter electrode should be about 1/4″–3/8″ (6.4 mm–9.5 mm) wide. The height of the weld bead should be approximately 1/32″–3/64″ (0.8 mm–1.2 mm). As the weld bead reaches the correct width, the electrode is moved forward at a constant speed. A weave bead is one made along a line while the electrode is moved from side to side. See **Figure 8-8**. Weave beads should not be wider than six times the electrode diameter. Weave beads wider than this may cool too rapidly. Impurities may be trapped in the weld bead. Weave beads are often used as the final pass of a multiple-pass weld.

A weld is generally made from left to right by a right-handed welder. A left-handed welder commonly progresses from right to left. The electrode and electrode holder are held perpendicular (90°) to the weld axis. The electrode holder leads the welding end of the electrode. The electrode holder is tilted about 20° from perpendicular in the direction of travel. See **Figures 8-9** and **8-10**.

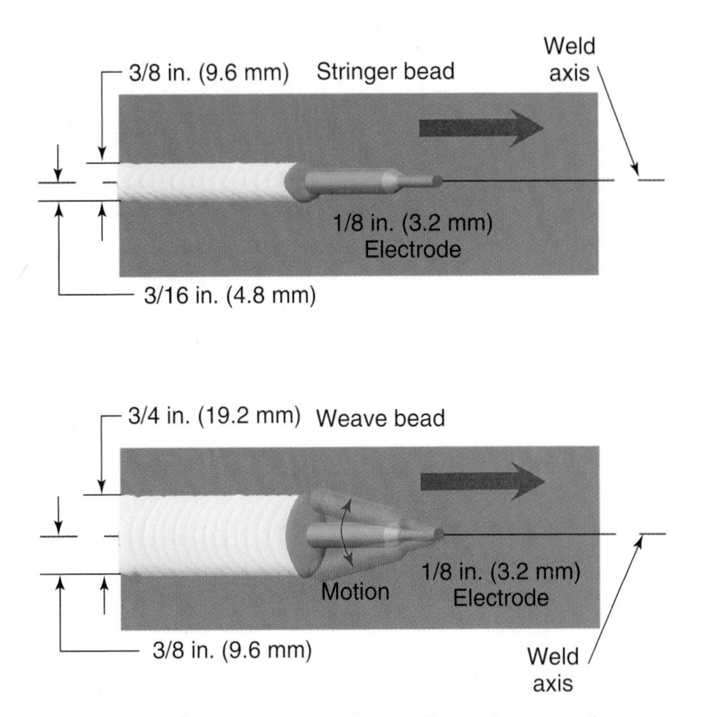

Figure 8-8. Suggested dimensions for stringer and weave beads. The stringer bead is two to three times as wide as the electrode diameter. A weave bead should not be wider than six times the electrode diameter.

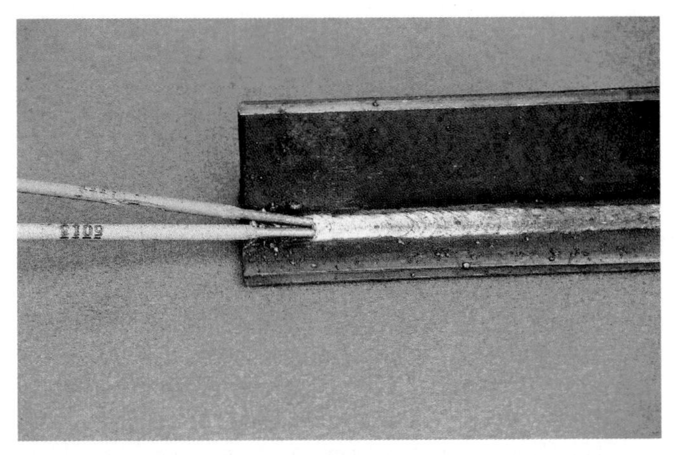

Figure 8-7. The width of a SMAW stringer bead should be two to three times the electrode diameter. After welding for a short distance, check the bead width by comparing it with two of the electrodes being used.

Watch the shape of the ripples that form at the rear of the pool. The ripples should be bullet-shaped and closely, yet evenly, spaced. Pointed ripples indicate that the forward speed is too fast. Flattened ripples indicate that the travel speed is too slow.

The following must be kept constant while welding:

- The arc length.
- The electrode angle.
- The weld bead width.
- The forward welding speed.

The arc will stop if the arc length is allowed to become too large or if the electrode touches the base metal.

Describing the Electrode Angles

Two angles are used to describe how to hold or position an electrode in any arc welding process. These two angles are called the travel angle and the work angle. The travel angle describes the position of the electrode as it is tilted *along* the weld axis. The work angle describes the position of the electrode as it tilts *side to side* about the weld axis.

Travel angle is the angle between a line perpendicular (90°) to the weld axis and the axis of the electrode. This angle is measured in a plane

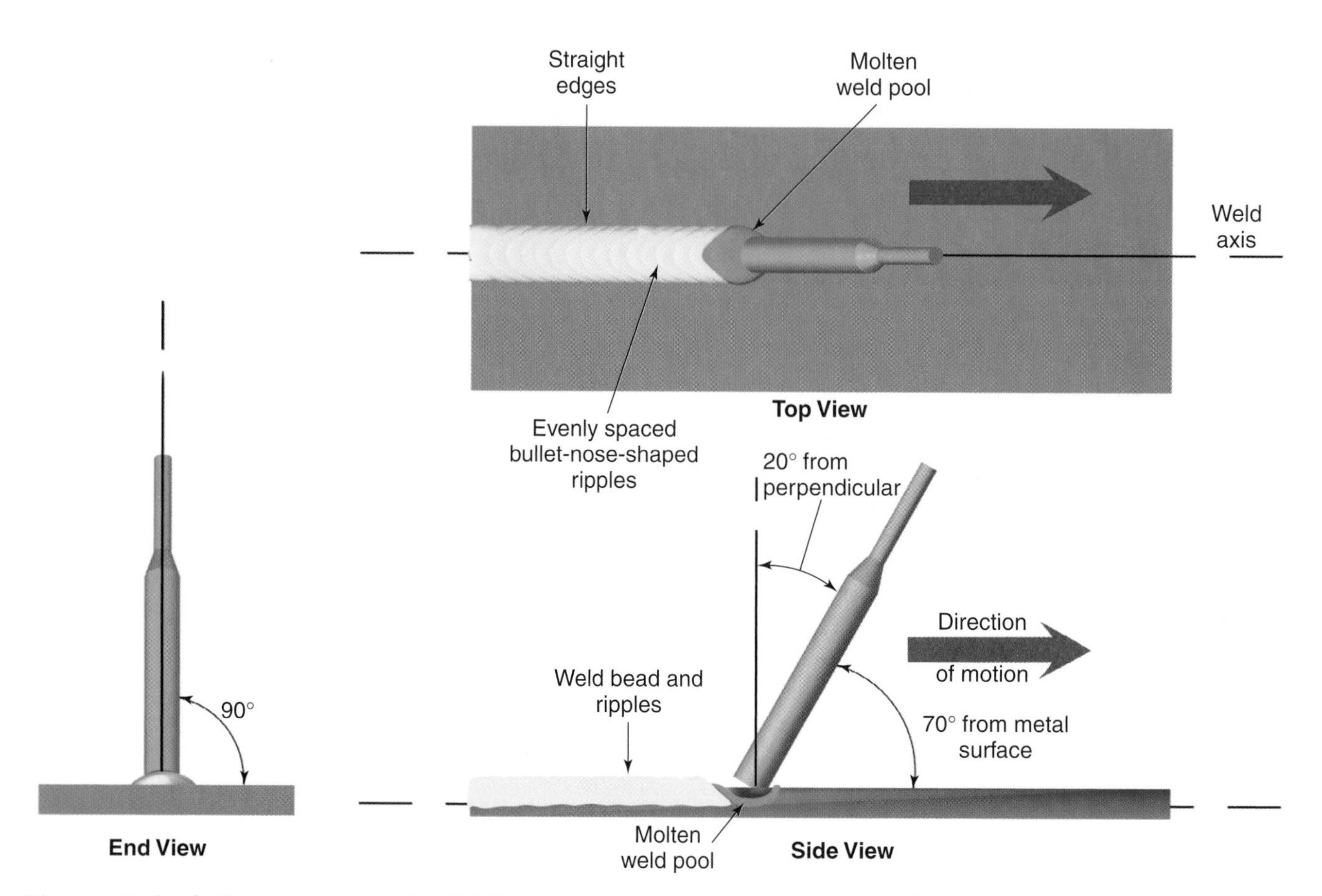

Figure 8-9. Study these three views of a SMAW weld bead in progress. Note that the electrode holder leads the welding end of the electrode at about a 20° angle. This position can also be described as 70° from the surface of the base metal. The end view shows the electrode perpendicular (90°) to the surface of the workpiece. The completed weld bead should have straight edges, evenly spaced ripples, and a uniform height.

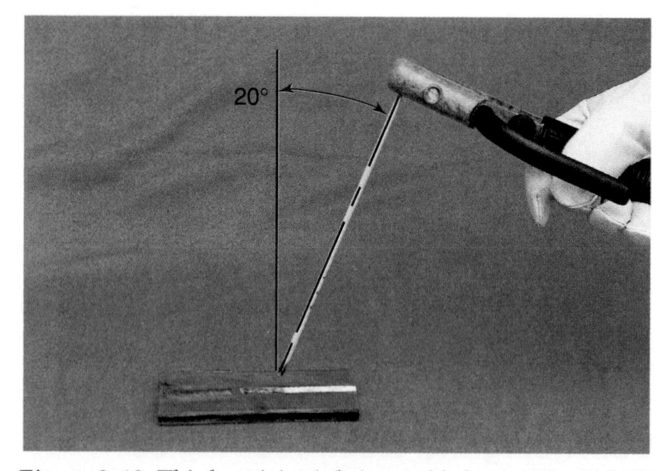

Figure 8-10. *This butt joint is being welded using the SMAW backhand welding method. A 20° drag travel angle is formed. This angle helps to pile up the filler metal behind the weld pool.*

determined by the electrode axis and the weld axis. **Figure 8-11** shows how the travel angle is measured. NOTE: The correct way to describe the electrode position is by using the travel angle. However, sometimes the position of the electrode is described as an angle off the surface of the workpiece. Refer to Figure 8-9.

In SMAW, the backhand welding method is usually used. In ***backhand welding,*** the welding end of the electrode points back toward the weld bead that has been completed. Thus, the electrode tip points opposite the direction of travel. The travel angle used when backhand welding is called a ***drag angle*** or a drag travel angle.

When the ***forehand welding*** method is used, the welding end of the electrode points forward in the direction of travel. The travel angle for forehand welding is called a ***push angle.***

The second angle used to describe the position of the electrode is the work angle. ***Work angle*** is the angle between a line perpendicular (90°) to the major workpiece and a plane determined by the electrode axis and the weld axis. The work angle is shown in **Figure 8-12**. When the electrode is perpendicular to the weld axis, the work angle is zero (0°) degrees.

In a T-joint or a corner joint, the work angle is measured from a line perpendicular to the nonbutting workpiece. The nonbutting workpiece is identified in the illustration.

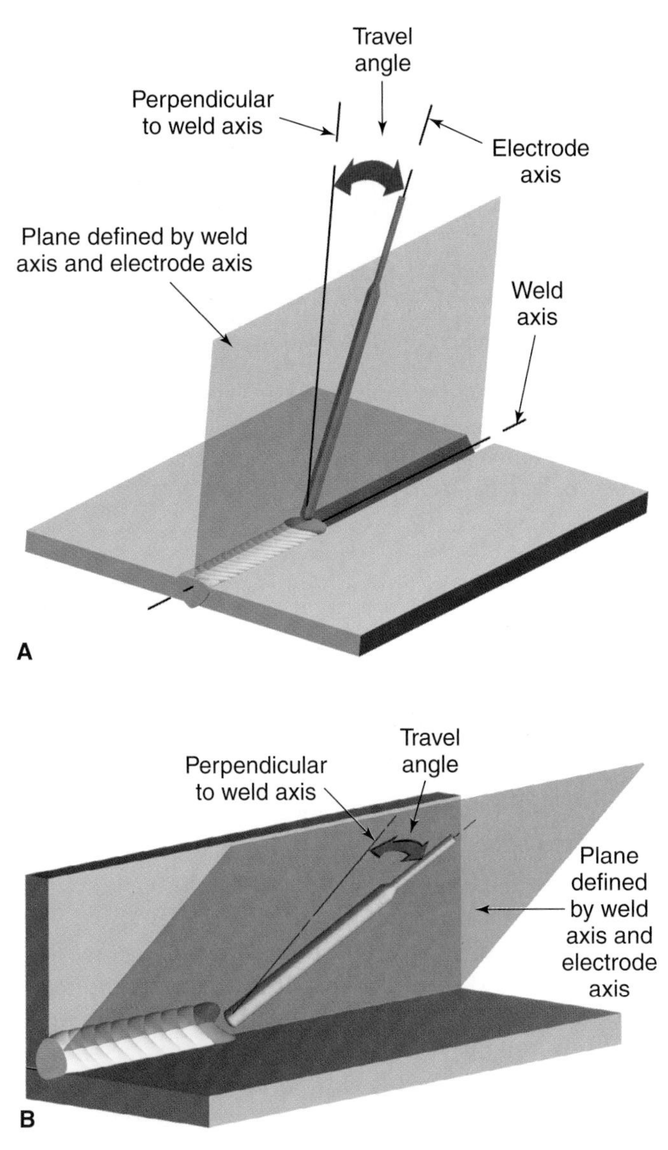

Figure 8-11. Travel angle is measured from the electrode axis to a line perpendicular to the weld axis in the plane defined by the weld axis and the electrode axis. A—Travel angle for a butt joint. B—Travel angle for an inside corner joint.

Restarting the Arc

An arc may stop due to faulty welding techniques. It may also be stopped intentionally to change electrodes or make other adjustments. A weld bead that is stopped for any reason will end with a large arc crater. The arc must be restarted so that the crater is completely filled. The ripples in the weld bead must also appear evenly spaced. When ***restarting the arc***, proceed as follows:

1. Restrike the arc 1/2″–1″ (13 mm–25 mm) ahead of the old crater.
2. Move backward rapidly until the rear of the new weld pool touches the rear of the old crater.
3. Move forward as soon as they touch to complete the weld. See **Figure 8-13**.

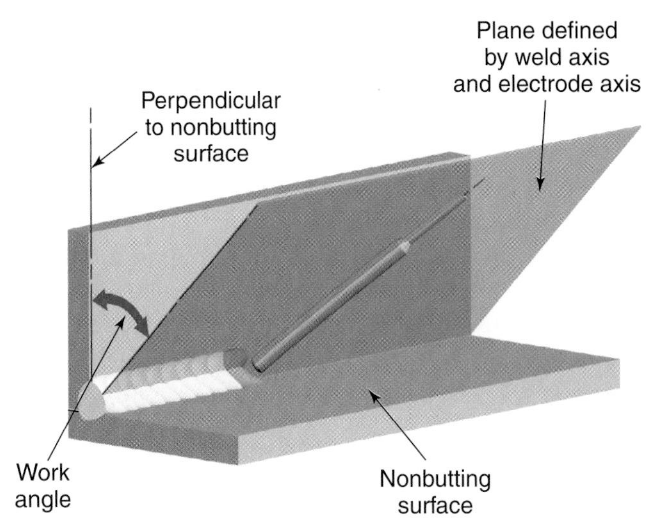

Figure 8-12. Work angle is measured from a line perpendicular to the major or nonbutting surface to the plane containing the weld axis and the centerline of the electrode. The work angle for an inside corner joint is shown here.

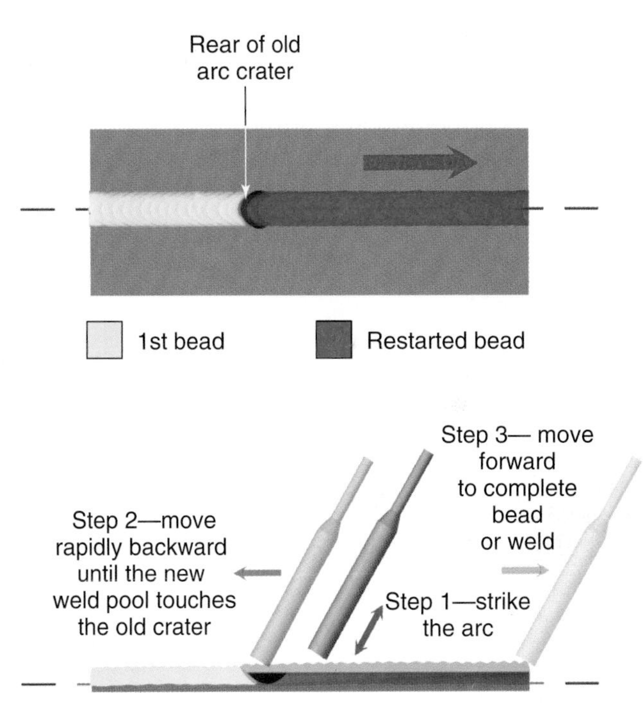

Figure 8-13. Note the three steps in restarting an arc.

Drag Welding

Iron powder electrodes, such as the E7024, have extremely thick coverings. Electrodes with thick coverings can be used for drag welding. ***Drag welding*** is a welding technique in which the electrode covering is in contact with the base metal. Drag welding ensures a constant arc length. The electrode covering does not conduct electricity. As a result, the arc will not stop when the electrode covering touches the base metal.

Striking an arc with a new, thickly covered electrode is the same as with any other shielded electrode.

However, when restarting an arc with a thickly covered electrode, you must hit the base metal hard enough to chip away some of the covering. This exposes the electrode wire and allows the arc to begin. The covering is kept in contact with the base metal after the arc is struck. The arc is enclosed by the thick electrode covering. This type of arc is known as a ***closed arc***. Running a weld bead with the drag method is the same as running a weld bead with an open arc. It is easier, however, because the arc length is kept constant by dragging the electrode on the flux covering. See **Figure 8-14**.

Cleaning the Weld

Each weld bead, or weld pass, must be cleaned by ***chipping*** and brushing. This cleaning prevents slag and dirt from mixing with the following weld passes. Cleaning is also necessary in order to examine a completed weld bead.

Chipping should be done after the weld has cooled for a few minutes. ***Chipping goggles* must be worn while chipping and wire brushing. Chipping should be done where other workers are not endangered by flying particles.** The cooled slag coating is usually hard and brittle. It is removed with a chipping hammer. Small areas of the weld are cleaned with the pick end of the chipping hammer. See **Figure 8-15**. A wire brush is used to clean the weld bead thoroughly after chipping.

Reading the Weld Bead

A welder must be able to visually determine whether a weld is made properly. This can be done by looking at the weld while it is being made and after it is completed. Visual inspection of a weld bead is referred to as ***reading the weld bead***. The process of examining or reading the weld bead can provide information on the following welding variables:

- Amperage setting.
- Arc length.
- Travel speed.

Figure 8-16 shows weld beads that have been made under different welding conditions. The sketches and photographs show the welds as viewed from above and from the end. If the correct amperage is used, a completed stringer bead will have an even width. It will also have evenly spaced, ***bullet-shaped ripples***. The weld bead width and height will also be correct, Figure 8-16A. If the welding amperage is too low, the weld bead will be narrow, built up, and have poor penetration. See Figure 8-16B. Excessive current produces a weld that is wide, low, and has a great deal of ***spatter*** (molten metal droplets) on the metal's surface. See Figure 8-16C.

A built-up weld bead with poor penetration is generally caused by a short arc. Overlap is usually present. See Figure 8-16D. When the arc length is too long, the weld bead will be too low. It will also have poor penetration and have undercutting at the weld toes, Figure 8-16E.

Figure 8-16F shows a weld that was made too slowly. The weld bead is too wide and built up too much. Also notice the flatter shape of the ripples. A weld that is made too fast will be low and narrow as shown in Figure 8-16G. The ripples are usually pointed.

These weld characteristics can be seen while the weld is in progress. If you notice any of these poor characteristics while welding, immediately adjust your arc length and welding speed. The effects of these changes should be seen as a more-perfect weld bead is formed. **If the current needs to be changed, the weld must be stopped. Hang the electrode holder on an**

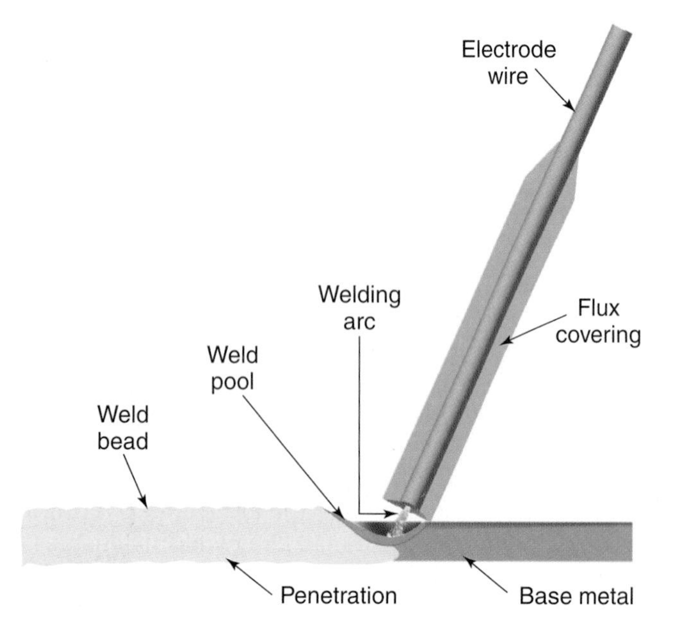

Figure 8-14. When drag welding, the electrode covering is kept in contact with the base metal as the electrode is pulled along the metal surface.

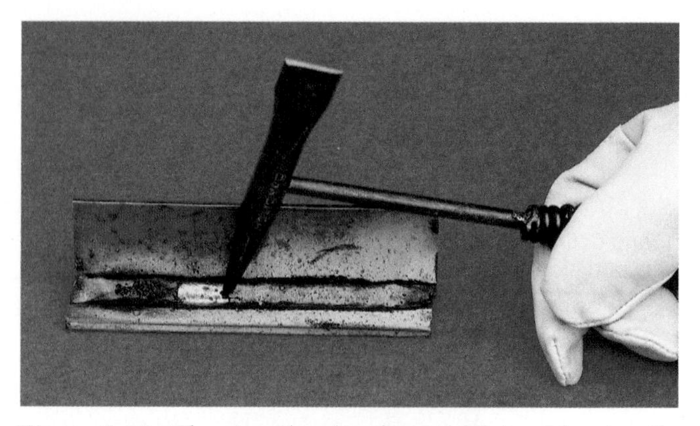

Figure 8-15. Cleaning the slag from a fillet weld using the pick end of the chipping hammer. The chisel end of the hammer can be used on flat weld beads and in larger areas.

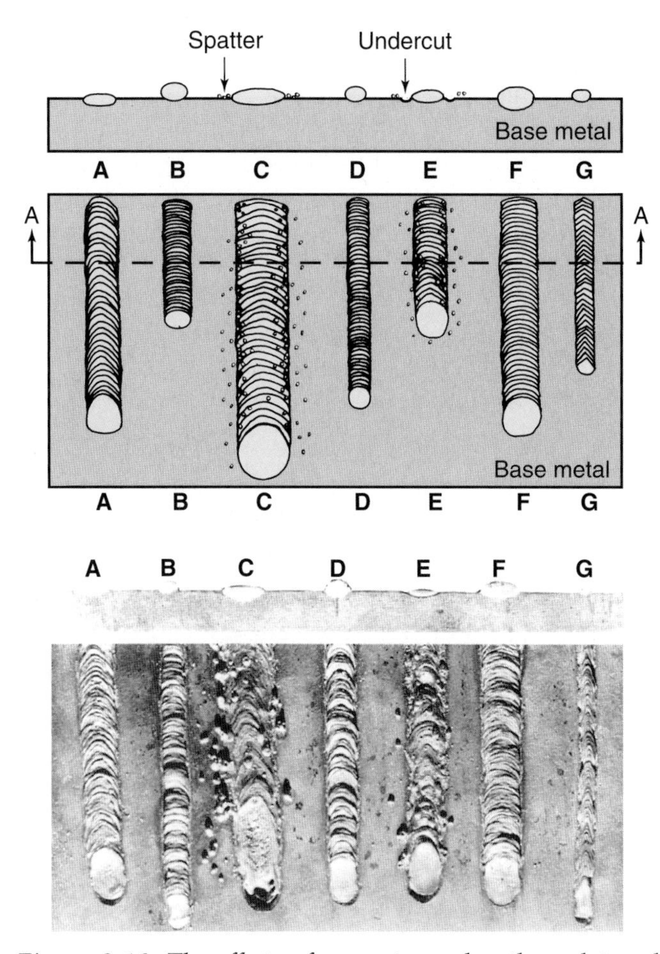

Figure 8-16. The effects of current, arc length, and travel speed on covered SMAW electrode weld beads. A—Correct arc length, travel speed, and current. B—Amperage too low. C—Amperage too high. D—Arc length too short. E—Arc length too long. F—Travel speed too slow. G—Travel speed too fast. (American Welding Society (AWS) Welding Handbook Committee, 2004, Welding Handbook, 9th ed., vol. 2, Welding Processes, Miami: American Welding Society)

insulated hanger before changing the amperage. The weld is then restarted after the amperage is changed.

Fillet Welding

A fillet weld may be made on a lap, inside corner, or T-joint. The legs of the fillet are usually equal. Normally, the size of each leg is at least equal to the thickness of the metal.

Lap Joint

In a lap joint, the weld is made along an edge of one piece and the surface of another piece. The electrode holder end of the electrode is tilted about 20° from vertical in the direction of travel. This is a 20° travel angle. The electrode should point more toward

Exercise 8-1 Running a Weld Bead Using an Open Arc

1. Obtain one low carbon steel plate measuring 1/4″ × 3″ × 6″ (6.4 mm × 75 mm × 150 mm).
2. Also obtain five 1/8″ (3.2 mm) diameter E6013 electrodes.
3. Mark three parallel 6″ (150 mm) long lines on the plate. The lines should be 3/4″ (19 mm) apart.
4. Determine the amperage range and polarity for this electrode. Refer to Chapter 7 or an electrode manufacturer's guide.
5. Make a safety inspection of the arc welding outfit or station.
6. Set the amperage near the low end of the suggested range.
7. Make certain that the workpiece lead is attached to the worktable or practice plate.
8. Strike the arc using the scratching or pecking method.
9. Run a weld bead of correct width and height along the first marked line.
10. Chip and wire-brush the weld bead. **Chipping goggles must be worn.**
11. Read the finished weld bead and compare it to Figure 8-16.
12. Change your amperage, arc length, or travel speed as required. Run two additional weld beads. Read each weld bead while it is being made. Clean the completed welds and make changes in your welding method as required.
13. Make three additional weld beads on the opposite side of the metal about 3/4″ (19 mm) apart.

Inspection:

Each weld bead should be the proper width and height. The ripples should be bullet-shaped and evenly spaced. Each weld bead should improve as errors are read and corrected.

the surface when welding thinner metal. This is not as necessary when welding thick metal. The electrode should be held at a 45° work angle when welding metal over 1/4″ (6.4 mm) thick. See **Figure 8-17**.

When the arc is struck, lay a stringer bead deep into the root of the joint. A *C*-shaped weld pool should be created. A *C* shape at the leading edge of the weld pool indicates that both the edge and surface are melting. Reduce the travel speed if the *C* shape does not form. A welder can determine the correct forward speed by watching the rear of the weld pool. It should be bullet-shaped.

Thick metal may require more than one weld pass. Each weld pass must be cleaned before the next weld pass is made. The correct sequence of the weld passes in a multiple-pass weld is suggested in **Figure 8-18**. A multiple-pass weld generally contains more than one ***layer***. Each layer may contain more than one weld bead or weld pass. The final pass may be a weave bead if the bead width is not too wide.

Inside Corner or T-Joint

A fillet weld for an inside corner or T-joint is made along two adjacent surfaces. The electrode is tilted 20° in the direction of travel. It is held at a 45° work angle. The first weld pass must be made deeply into the weld root.

The leading edge of the weld pool must be *C*-shaped. This ensures that fusion is occurring on both surfaces. Each weld pass or weld bead must penetrate the base metal and/or the previous weld beads that it touches. See **Figure 8-19**. A weave bead may be used for the final weld pass.

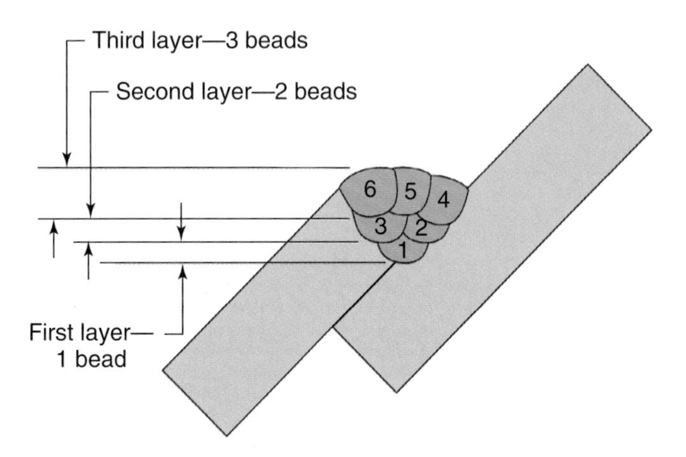

Figure 8-18. This fillet weld shows a good example of the sequence a welder should use when making a multiple-pass weld.

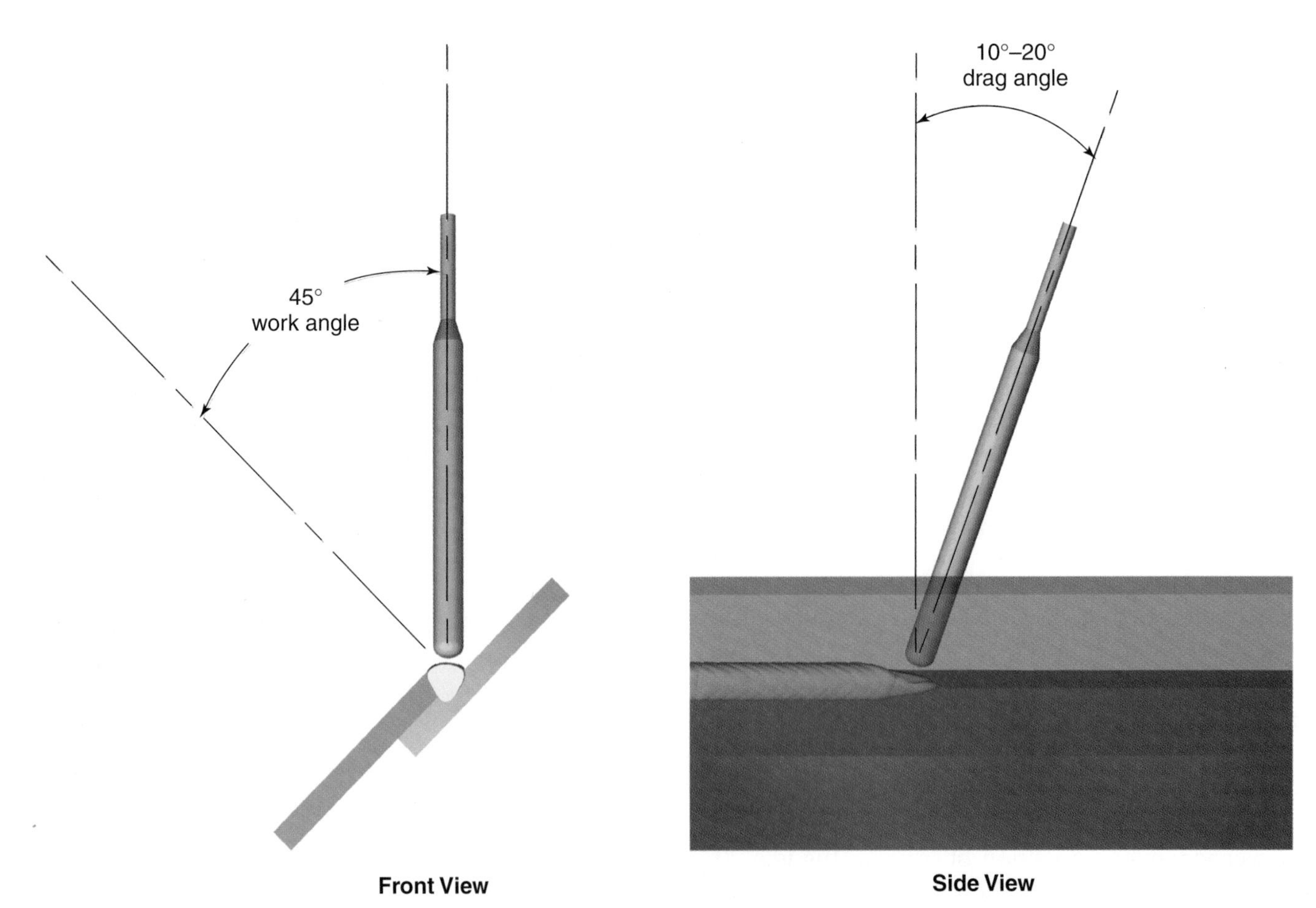

Figure 8-17. For a single weld bead or the first weld bead on a multiple-pass fillet weld on thick metal, the electrode is held at a 45° work angle. A 10°–20° drag travel angle is used.

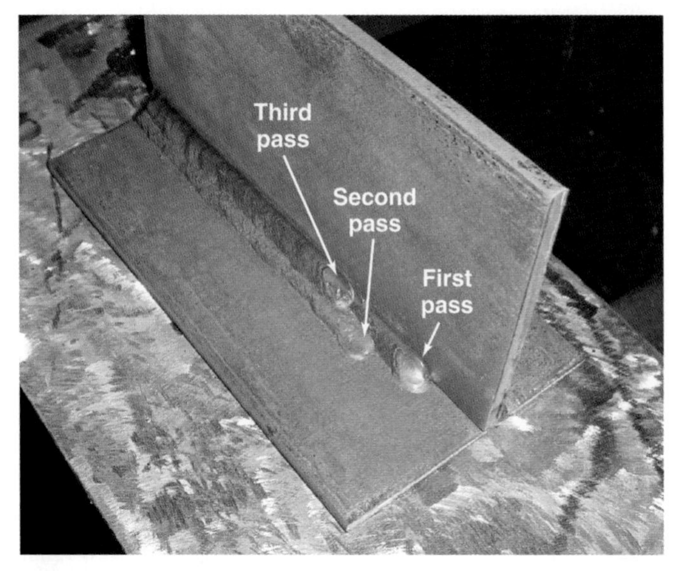

Figure 8-19. *Three weld passes were used on this fillet weld. Each weld bead was stopped at a different point to show their placement. The first weld pass was laid in the corner with equal coverage of both plates. The second weld pass penetrates one-half on the first weld bead and one-half on the horizontal part. The third weld pass was laid from the second weld pass up to the vertical part.*

Exercise 8-2 Fillet Weld on a T-Joint in the Flat Welding Position

1. Obtain three 1/8″ (3.2 mm) diameter E6010 electrodes, six 1/8″ (3.2 mm) E6013 electrodes, and three 5/32″ (4.0 mm) iron powder electrodes, such as E6027.
2. Also obtain four pieces of mild steel measuring 1/4″ × 2″ × 6″ (6.4 mm × 50 mm × 150 mm).
3. Determine and set the correct current type, amperage, and polarity for each electrode.
4. Form two T-joints. Tack weld each T-joint three times on each side using E6010 electrodes.
5. A three-pass weld will be made. Clamp the T-joint so the weld can be made in the flat position. Refer to weld passes 1, 2, and 3 in Figure 8-18.
6. Weld the root pass using an E6010 electrode. Obtain penetration into the root. The weld pool should have a C shape at the leading edge.
7. Clean the slag from the first pass and read the weld. Change the current, arc length, or speed as required.
8. Weld a second pass using an E6010 electrode. Obtain good penetration into both the first pass and the base metal.
9. Clean the slag from the second pass and read the weld.
10. Make any changes in the variables that are needed and weld a third pass to complete the weld joint. This pass must obtain penetration into the first pass, into the second pass, and into the second piece of the base metal. A C-shaped weld pool should be visible.
11. Clean the slag and read the weld. The completed three-pass weld should have a total leg size of about 3/8″ (9.6 mm) and a 1/2″ (13 mm) weld face.
12. Repeat steps 5–11, but use E6013 electrodes. Complete two fillet welds using E6013 electrodes.
13. Repeat steps 5–11 using 5/32″ (4.0 mm) iron powder electrodes, such as an E6027. The fillet weld size will be larger because the electrode diameter is larger. Remember to change the current setting on the welding machine.
14. Compare the ease or difficulty of running a bead with each electrode type. Compare any differences in the appearance of the different welds.

Inspection:

The fillet welds must be the correct size. A convex weld bead must be formed with smooth and evenly spaced ripples. No undercut is permitted.

Butt Joint

Metal less than 1/4″ (6.4 mm) thick is welded using a square-groove joint. A bevel-, V-, J-, or U-groove joint is used on thicker metal. The ***root pass*** is the first and most important weld pass. Complete penetration to the opposite side of the metal can occur only on this weld pass. A ***keyhole***, an enlarged root opening, must be seen throughout the root pass to ensure penetration. Refer to **Figure 8-20**. Additional weld passes must fuse with the base metal and the previous weld beads.

The weld toes must fuse with the surface of the base metal. No undercut should be seen. A butt weld may be made from one or both sides. See **Figure 8-21**. Double-bevel-groove welds are used to ensure 100% fusion and penetration when welding thick sections. Less electrode material and welder time is required to weld a double-bevel-groove than to make a single-bevel-groove weld.

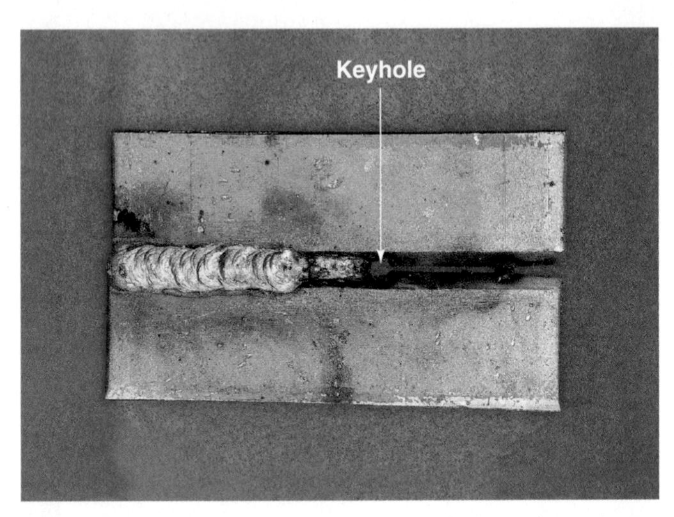

Figure 8-20. This is a keyhole as seen looking down into a bevel-groove butt joint.

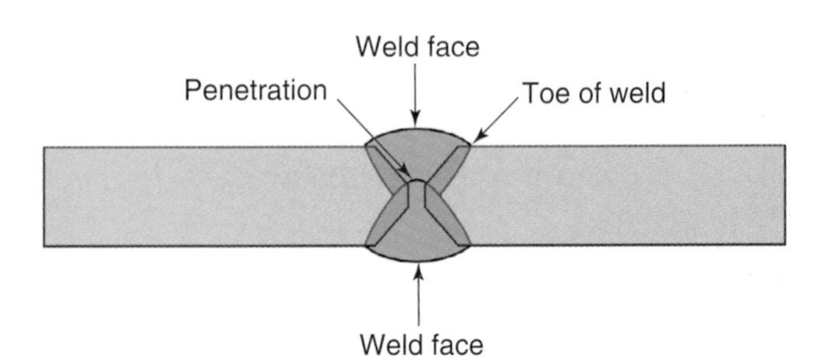

Figure 8-21. Notice the groove preparation on this double-bevel-groove butt joint. The weld on top was made first and then the weld on the bottom was made. Notice how the second weld bead penetrates into the root of the first weld bead.

Exercise 8-3 V-Groove Butt Joint in the Flat Welding Position

1. Obtain four 1/8″ (3.2 mm) diameter E6010 electrodes and four 1/8″ (3.2 mm) diameter E6011 electrodes.
2. Also obtain two pieces of mild steel measuring 3/8″ × 3″ × 6″ (9.5 mm × 75 mm × 150 mm).
3. Flame cut or grind a 45° bevel on the 6″ (152 mm) edge of two pieces.
4. Form a V-groove butt joint with a 1/16″–3/32″ (1.6 mm–2.4 mm) root opening. Tack weld the pieces in three places.
5. Run the root pass using the keyhole method. This ensures 100% penetration.
6. Clean the root pass and run additional weld passes to complete the weld.
7. Repeat steps 3–6 using an E6011 electrode.

Inspection:

The completed weld must have 100% penetration visible on the reverse side. The face of the weld bead must be properly shaped. The bead ripples must be evenly spaced. No undercut should be visible.

Summary

- Arc welding presents dangers of electrical shock, fumes and gases, hot metal, arc rays, and fire. The application of proper safety precautions prevents injury to personnel and damage to equipment.
- An electrode holder is generally held with one hand. It can be gripped like a hammer or screwdriver. The electrode lead may be draped over the lower arm to make the electrode holder feel lighter.
- To start an arc, the welder may scratch the electrode on the metal and then withdraw it. A straight up-and-down or pecking motion may also be used.
- The arc length is the distance between the electrode and the base metal. This distance should be approximately equal to the diameter of the electrode.
- The arc length determines the voltage and amperage across the arc. Therefore, the arc length must remain constant throughout the entire weld.
- Arc blow is a deflection of the arc due to magnetic fields generated by current traveling through the electrode and base metal. Arc blow is greatest at the ends of the weld joint.
- A weld is generally made from left to right by a right-handed welder. A left-handed welder commonly progresses from right to left. The ripples in the weld bead should be bullet-shaped and closely, yet evenly, spaced. Pointed ripples indicate that the forward speed is too fast. Flattened ripples indicate that the travel speed is too slow.
- Two angles are used to describe how to hold or position an electrode in any arc welding process. These two angles are called the travel angle and the work angle. The travel angle describes the position of the electrode as it is tilted *along* the weld axis. The work angle describes the position of the electrode as it tilts *side to side* about the weld axis.
- In SMAW, the backhand welding method is usually used. In backhand welding, the welding end of the electrode points back toward the weld bead that has been completed.
- Drag welding is a welding technique in which the electrode covering is in contact with the base metal. Drag welding ensures a constant arc length.
- Each weld bead, or weld pass, must be cleaned by chipping and brushing.
- Reading the weld bead can provide information about amperage setting, arc length, and travel speed.

Review Questions

Write your answers on a separate sheet of paper. Please do *not* write in this book.

1. Welders must not weld in damp places or with damp gloves. They must not touch the work and electrode at the same time. Why?
2. The heat of the arc will cause chlorinated hydrocarbons used in solvents to form _____, a deadly gas.
3. Why is it advisable to wear earplugs for out-of-position welding?
4. Where is dc arc blow the greatest while welding?
5. Name the two methods used to strike an SMAW arc.
6. What size arc gap is suggested with a 5/32″ (4.0 mm) diameter electrode?
7. How wide should a stringer bead be for a 1/8″ (3.2 mm) diameter electrode? What is the maximum width for a weave bead when using a 1/8″ (3.2 mm) electrode?

(continued)

Review Questions *continued*

8. The arc maintained within the covering during drag welding is called a(n) _____ arc.
9. How far ahead of the previous crater do you strike arc when restarting it?
10. What should be changed if the weld bead is narrow, undercut, and spattered?

Chapter 9

SMAW: Horizontal, Vertical, and Overhead Welding Positions

Learning Objectives

After studying this chapter, you will be able to:

- Identify the proper protective clothing to be worn when welding out of position.
- Weld in the horizontal welding position.
- Identify the characteristics of a good weld.
- Weld in the vertical welding position.
- Describe the procedure for welding uphill and downhill.
- Identify special protective equipment that should be worn for welding in the overhead welding position.
- Weld in the overhead welding position.
- Identify weld defects.

Technical Terms

downhill welding
oscillating
overlap
porosity
slag inclusion
undercut
uphill welding
whip motion

Preparing to Weld

A welder should dress properly for out-of-position welding. Flame-resistant clothing should be worn. All pockets should be covered and/or buttoned. A cap should be worn to protect hair from hot metal and sparks. Earplugs or earmuffs should be worn to prevent hot metal from spattering into your ears. Gauntlet gloves and leathers are recommended. High-top, hard-toe shoes should be worn. Pull your pant legs over the top of your shoes. See **Figure 9-1.** Observe all the arc welding safety precautions discussed in Chapter 8.

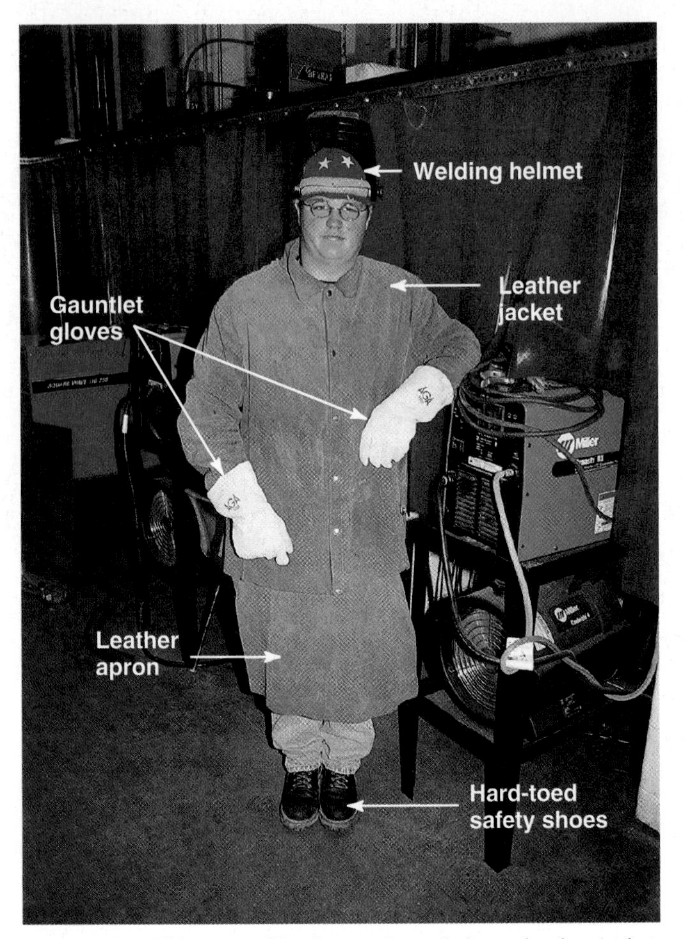

Figure 9-1. *This arc welder is wearing a helmet, leather jacket, leather apron, gauntlet gloves, and hard-toed shoes. Notice that the jacket is buttoned at the collar.*

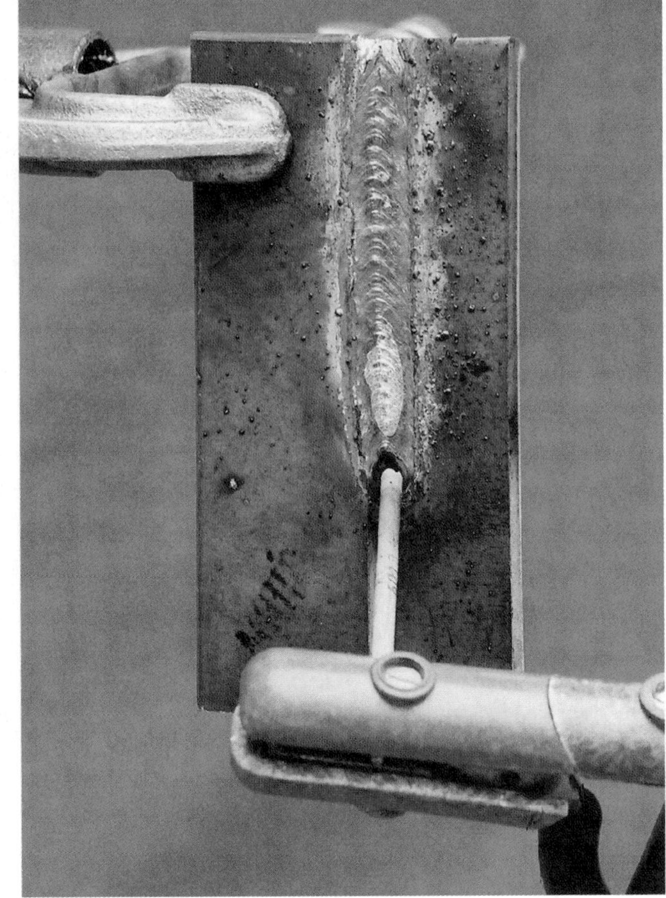

Figure 9-2. *For this fillet weld on a lap joint in the downhill position, the electrode is inserted perpendicular to the electrode holder.*

An electrode used for out-of-position welding should have a smaller diameter than an electrode used for flat position welding. A smaller-diameter electrode provides the following advantages:

- Lower amperage can be used.
- A smaller weld pool is created.
- The molten metal in the pool is easier to control.

The slag deposit from the electrode should be fast-setting (quick-freezing). A fast-setting slag deposit prevents the weld metal from sagging. Some AWS electrodes that have fast-setting slag deposits are E6010, E6011, E6012, and E6013.

The electrode should be placed in the electrode holder at an angle convenient to the welder. Electrodes should not be bent. Their flux covering may chip, resulting in a loss of shielding gas. The electrode may be placed perpendicular to the electrode holder, **Figure 9-2**. It may also be placed at an angle or parallel to the length of the electrode holder, as shown in **Figure 9-3**.

The electrode holder should be held like a hammer or screwdriver. A firm, but not tight, grip should be used. The electrode lead may be draped over your arm to prevent fatigue. Refer to Figure 8-1.

Horizontal Welding Position

When making practice fillet welds in the horizontal position, the weld axis is generally horizontal and the weld face is 45° from horizontal. When making butt welds in the horizontal position, the weld axis is generally horizontal and the weld face is vertical. The molten metal tends to sag downward out of the weld when welding in the horizontal position.

Fillet Welds

Fillet welds are made in the horizontal welding position on the lap, inside corner, and T-joints. A stringer or weave bead can be used to make a fillet weld.

Fillet welds must penetrate deeply into the root of the joint, as shown in **Figure 9-4**. The weld toe should be fused properly into both pieces of the weldment.

Multiple-pass welds may be required on thick metal sections. The lowest weld pass in a layer should be made first to provide a base for the other weld beads. See **Figure 9-5**.

Fillet welds may be made by drag welding. Certain electrodes, such as the E6027, are designed for making fillet welds in the horizontal welding position.

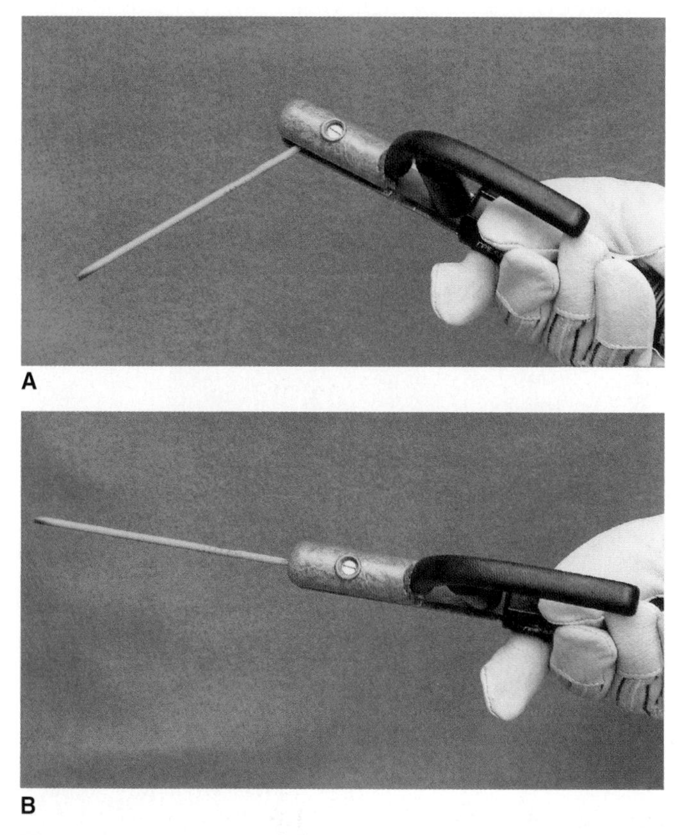

Figure 9-3. *Two ways of clamping the electrode into the electrode holder. Note that the electrode holder is held like a hammer or screwdriver. A—This is recommended for flat, horizontal, and vertical welding positions. B—This is recommended for the overhead welding position.*

Lap Joints

A fillet weld on a lap joint must fuse the edge of one piece and the surface of another piece. Watch the weld toes as they blend into the base metal to ensure proper fusion. Increase the welding amperage or slow the weld speed if the weld is not fused properly. Bullet-shaped ripples at the rear of the weld pool indicate the correct welding speed. There should not be an overlap or undercut on a fillet weld on a lap joint.

Inside Corners and T-Joints

A fillet weld in the horizontal welding position is usually made using the backhand welding method. The electrode points in the opposite direction of travel with about a 20° travel angle. The work angle is about 45°. See **Figure 9-6**. A 20° travel angle ensures that the force of the arc deposits filler metal behind the weld pool. This, in turn, helps to form a uniformly shaped weld bead. The weld bead will build up excessively if more than a 20° angle is used.

A short ***oscillating*** (forward and back) motion also helps in forming a good weld bead. The oscillating motion, combined with a slight crescent-shaped motion as shown in **Figure 9-7**, forces the molten metal upward and backward. This motion prevents the filler metal from sagging. The electrode motion should be stopped briefly each time the electrode reaches the toe of the weld. This helps to prevent undercutting at the weld toes.

Butt Joints

Multiple-pass welds are required for butt welds on metal over 1/4″ (6.4 mm) thick. Small weld pools are used to prevent sagging. The first weld bead laid

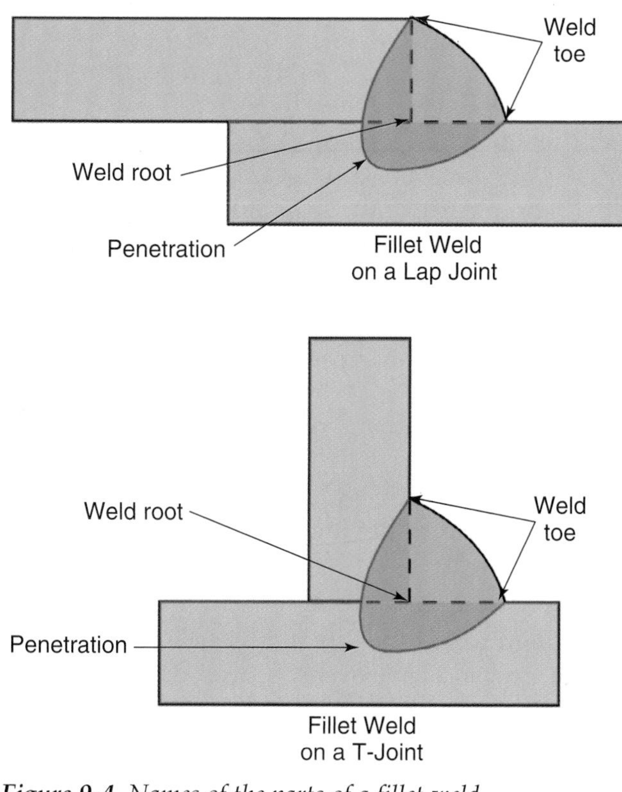

Figure 9-4. *Names of the parts of a fillet weld.*

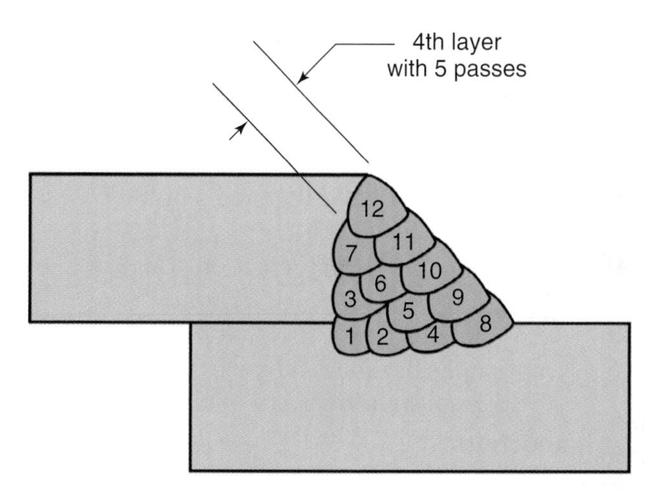

Figure 9-5. *This is the suggested order of welding the various weld beads in a multiple-pass fillet weld. The order of the weld passes is very important. Notice that the second weld pass acts as a step for the third weld pass to rest on. The fourth weld pass likewise is a step for the fifth weld pass and so on.*

down, which fills the root opening, is called the root pass. See **Figure 9-8**. A stringer bead is used for the root pass. The keyhole welding method is used to ensure 100% penetration.

A butt weld in the horizontal welding position is usually made with the tip of the electrode pointing at an upward angle or work angle of 10°–20° from a line perpendicular to the workpiece surface. The upward

Figure 9-6. *Correct electrode angle for welding a fillet weld in the horizontal position. The electrode forms a 20° travel angle and the tip points opposite the direction of travel. A work angle of 45° is used. On a lap joint, the electrode may point more toward the surface and away from the edge.*

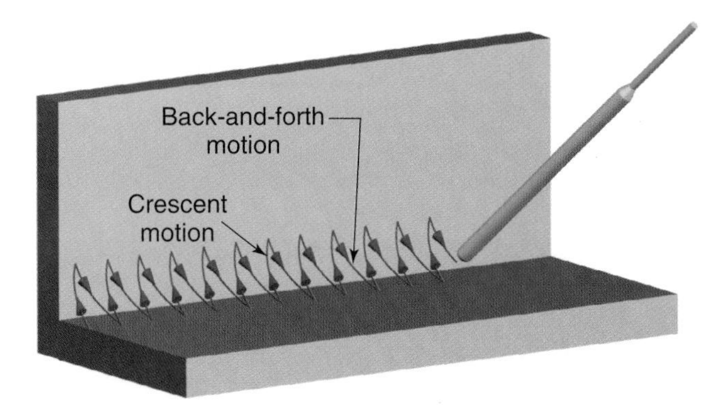

Figure 9-7. *A short back-and-forth arc motion combined with a crescent movement of the electrode may be used to create an even weld bead. This motion will help to push the filler metal up and back in the weld bead.*

Exercise 9-1 Fillet Weld on a Lap Joint in the Horizontal Welding Position

1. Obtain two pieces of mild steel measuring 1/4″ × 1 1/2″ × 6″ (6.4 mm × 40 mm × 150 mm).
2. Obtain four 5/32″ (4.0 mm) diameter E6012 electrodes.
3. Determine the amperage and polarity. Set the welding machine to the correct amperage and polarity.
4. Overlap the two pieces about 3/4″ (19 mm). Tack weld the pieces in three places on each side.
5. Lay a fillet weld on both sides. The legs of the fillets should be 1/4″ × 1/4″ (6.4 mm × 6.4 mm). Use a crescent-shaped motion. Bullet-shaped ripples indicate the correct welding speed.

Inspection:

A convex weld bead is required. The weld beads should be bullet-shaped and have a consistent width. No undercutting should be visible.

Exercise 9-2 Fillet Weld on a T-Joint in the Horizontal Welding Position

1. Obtain two pieces of mild steel measuring 1/4″ × 2″ × 6″ (6.4 mm × 50 mm × 150 mm).
2. Obtain three 5/32″ (4.0 mm) diameter E6013 electrodes and three 5/32″ (4.0 mm) diameter E6027 electrodes.
3. Determine and set the correct amperage and polarity on the welding machine.
4. Form a T-joint with the two pieces and tack weld in three places on each side.
5. Make a normal open arc fillet weld on one side of the T-joint using the E6013 electrodes.
6. Run an additional two weld passes. Clean the slag between each pass.
7. Make a fillet weld on the other side using E6027 electrodes and the drag weld technique.

Inspection:

Both weld beads should be even in width, smooth, and have evenly spaced ripples. The weld beads should be convex with bullet-shaped ripples. Undercutting is not permitted.

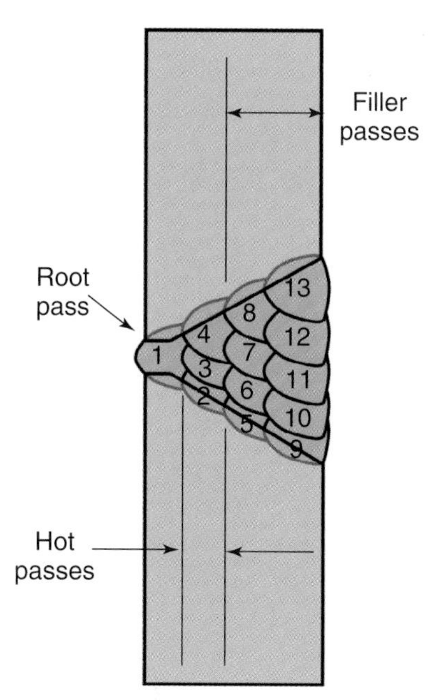

Figure 9-8. The suggested order of welding passes (weld beads) in a multiple-pass V-groove butt weld in the horizontal welding position. The lowest bead in each layer should be welded first to provide a base for the next weld bead to build on.

Exercise 9-3 V-Groove Weld on a Butt Joint in the Horizontal Welding Position

1. Obtain two pieces of mild steel measuring 1/4″ × 2″ × 6″ (6.4 mm × 50 mm × 150 mm).
2. Obtain five 3/32″ or 1/8″ (2.4 mm or 3.2 mm) diameter EXX1X electrodes.
3. Prepare the edges for a 90° V-groove butt joint.
4. Tack weld the pieces in three places in the flat position.
5. Position the joint for horizontal welding.
6. Weld the joint with one or two weld passes using the keyhole welding method. The root pass must penetrate 100%.

Inspection:

The weld must penetrate 100%. A convex weld bead is required. The weld bead should be consistent in width and have evenly spaced, bullet-shaped ripples. The weld face should not sag.

angle prevents the molten metal from sagging. The tip of the electrode points back over the completed weld at an angle of 20°. This travel angle is the same travel angle used when flat welding. See **Figure 9-9**.

Vertical Welding Position

In the vertical position, the weld axis is generally within 10° of vertical. The weld face is also close to being vertical.

Welding in the vertical position is done either downhill or uphill. ***Downhill welding*** begins at the top of the joint and ends at the bottom of the joint. See **Figure 9-10**.

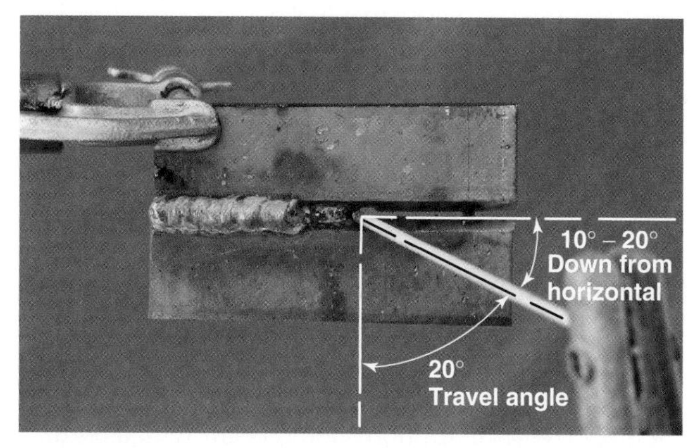

Figure 9-9. Suggested electrode angles for welding a butt joint in the horizontal welding position.

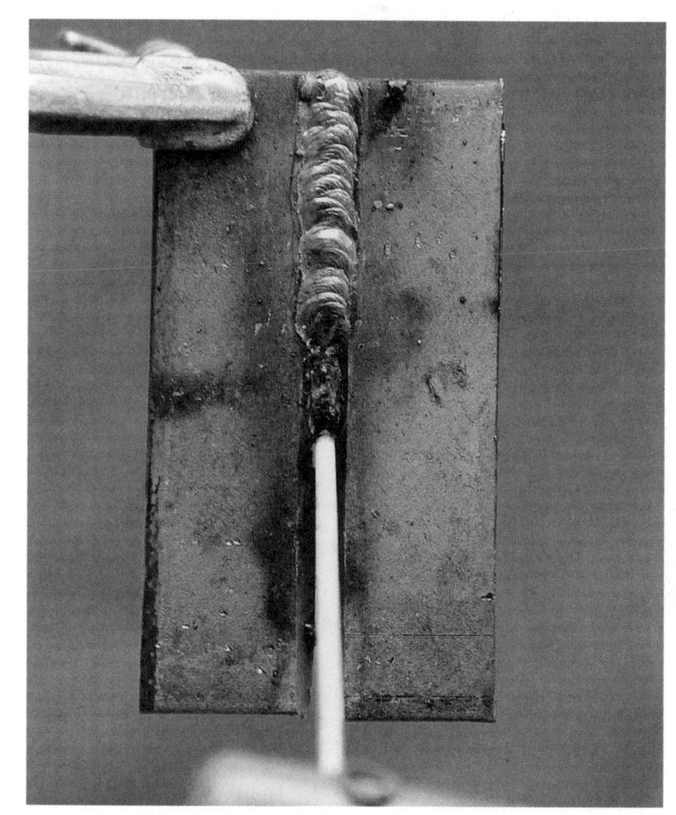

Figure 9-10. This butt joint is being welded using the downhill method. A travel angle of about 10°–20° is used.

The backhand welding method is used when vertical welding. The electrode forms a 10°–20° travel angle. Slag should be prevented from entering the weld pool. The correct welding speed keeps the weld pool ahead of the slag. Slag in the weld causes inclusions (foreign particles) that will weaken the completed weld. The first weld bead laid downhill should be a stringer bead. Weave beads may be used for the following passes.

Uphill welding begins at the bottom of the joint and ends at the top of the joint. A 10°–20° travel angle is used. See **Figure 9-11**. The greatest problem in vertical welding is keeping the molten metal in the weld pool. A ***whip motion*** is used to control the temperature of the pool. A whip motion prevents the weld bead from sagging by allowing the weld pool to cool for a short time. A whip motion can be used for welding any type of joint. Generally, it is only used when welding uphill or overhead.

When using a whip motion, the electrode tip is moved ahead of the weld pool about 1″ (25 mm). The electrode is then pulled away from the base metal while still maintaining a long arc. The electrode is then moved back to the weld pool and a normal arc length. The back-and-forth movement should occur within 1–2 seconds. **Figure 9-12** illustrates the whip motion. The molten weld pool cools during the whip motion. The whip motion should occur with a regular rhythm as the weld bead is laid.

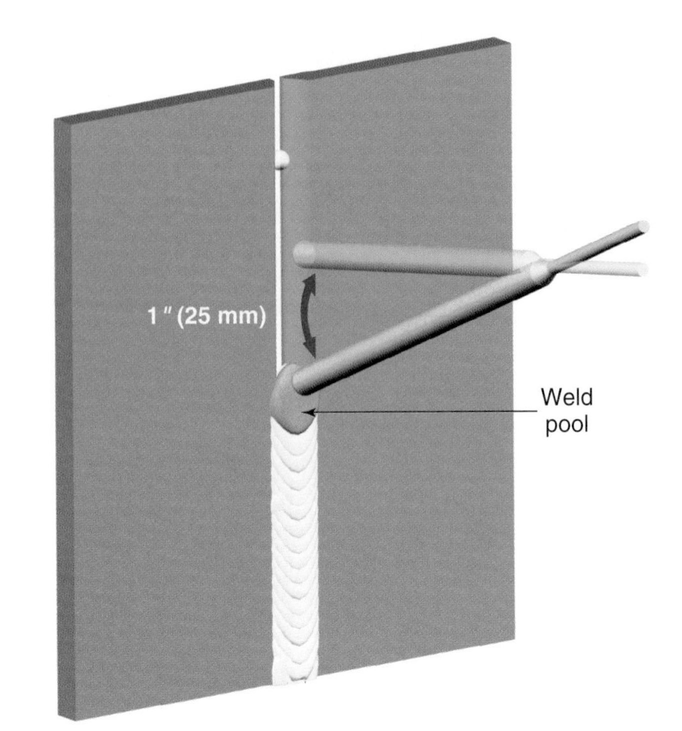

Figure 9-12. *A whip motion is used to control the heat of the weld pool. The electrode is moved ahead of the pool about 1″ (25 mm) and returned to the original weld pool. This entire action should take no more than 1–2 seconds. While the electrode is moving, the weld pool can cool.*

Fillet Welds

Fillet welds may be made using a stringer or a weave bead. A whip motion will help to control weld pool temperature while welding in the vertical welding position.

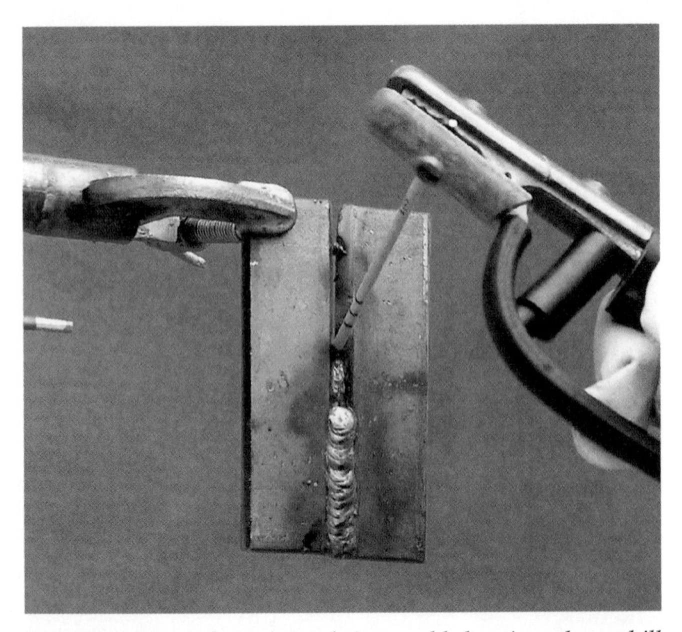

Figure 9-11. *A butt joint being welded using the uphill method. The electrode forms a 10°–20° travel angle and is pointed opposite the direction of travel.*

Lap Joints

On a lap joint, the electrode should be aimed slightly more toward the surface than toward the edge. This will prevent excessive melting of the edge. Greater weld strength is obtained when welding uphill. The welding end of the electrode should be tipped about 10°–20° opposite the direction of travel.

Inside Corner and T-joints

When making a fillet weld in the vertical welding position, the welding end of the electrode should be pointed opposite the direction of travel about 10°–20°. It should be held at a work angle of 45°. A weave bead is commonly used for a fillet weld. A crescent-shaped movement of the electrode is often used. Hesitate at the weld toes when using a weave bead to prevent undercutting. See **Figure 9-13**.

Butt Joints

Butt welds made in the vertical position can be made uphill or downhill. A drag travel angle of 10°–20° is used for both. A work angle of 0° is used.

If the weld pool gets too hot and the molten metal begins to sag, use the whip motion shown in Figure 9-12 to allow the weld pool to cool. The keyhole method is used on the root pass to obtain complete penetration.

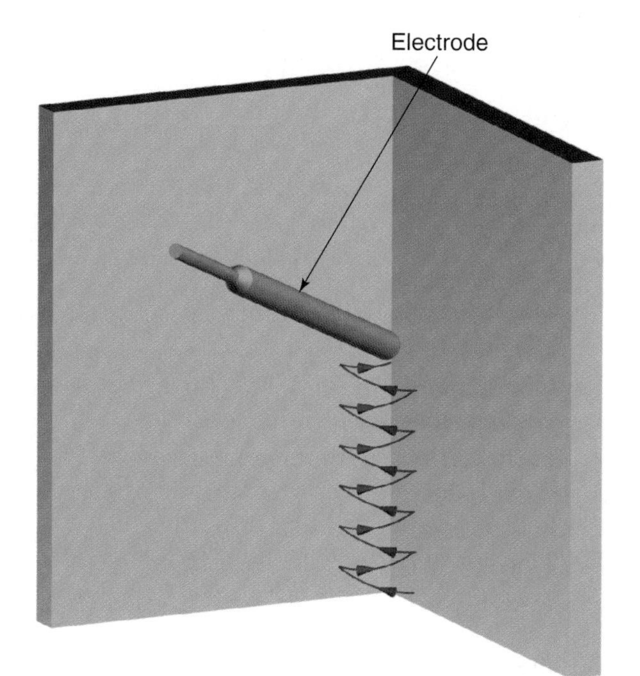

Figure 9-13. A crescent motion is often used with a weave bead on a fillet weld. The angles for the electrode should be the same as flat welding.

Exercise 9-4 Fillet Weld on a Lap Joint in the Vertical Welding Position

1. Obtain two pieces of mild steel measuring 1/8″ × 2″ × 6″ (3.2 mm × 50 mm × 150 mm).
2. Obtain five 1/16″ or 3/32″ (1.6 mm or 2.4 mm) diameter E6011 electrodes.
3. Form a lap joint with a 3/4″ (19 mm) overlap on each side.
4. Tack weld the pieces in three places on each side.
5. Place the weldment in a vertical position.
6. Place a fillet weld on both sides. The fillet leg size should be 1/8″ (3.2 mm). One weld should be made uphill. A whip motion may be used. The other weld should be made downhill.

Inspection:

The completed welds should have convex beads. No overlap or undercut is permitted. The weld bead should have smooth, evenly spaced ripples.

Exercise 9-5 Bevel-Groove Weld on a Butt Joint–Uphill

1. Obtain two pieces of mild steel measuring 1/4″ × 2″ × 6″ (6.4 mm × 50 mm × 150 mm).
2. Obtain five 1/16″ or 3/32″ (1.6 mm or 2.4 mm) diameter E60XX electrodes.
3. Prepare the edges of the metal for a 45° bevel-groove butt joint. Leave a 1/16″ (1.6 mm) root thickness.
4. Form a bevel-groove butt joint with a 1/16″ (1.6 mm) root opening.
5. Tack weld the pieces in three places.
6. Place the weldment in a vertical position.
7. Use the keyhole method on the root pass to ensure 100% penetration.
8. Weld the following passes as stringer or weave beads. Use a whip motion, if necessary, to control the molten weld pool.

Inspection:

The completed weld should be convex with 100% penetration. The bead ripples should be smooth and evenly spaced.

Overhead Welding Position

Special safety precautions must be followed when welding overhead. Hot metal or slag may fall from the weld area, creating unique problems for the welder. Refer to the safety precautions listed in Chapter 8 for specific information.

When welding in the overhead welding position, work is done more comfortably and efficiently across the body. Welding can be done from right to left or left to right. Position yourself so that you can see into the root of the weld.

The electrode should be placed in the holder at a comfortable angle. Some welders clamp the electrode so it is parallel to the electrode holder.

A small weld pool should be maintained to keep the metal from sagging. Welding with a smaller-diameter electrode and lower amperage will help keep a small weld pool. A stringer bead is used for the root pass. The keyhole method ensures 100% penetration. A short oscillating motion is recommended to control the pool and bead shape. See **Figure 9-14**. A whip motion can be used to control the molten metal.

Figure 9-14. A short back-and-forth swing or oscillating motion may be used to control the temperature of the metal in the weld pool.

Fillet Welds

The electrode angle between the base metal and weld axis on all types of joints is the same as when welding in the flat position. The electrode is held at about 45° from the base metal surface. It is tilted opposite the direction of travel at a 10°–20° angle. On a lap joint, the electrode is aimed more toward the surface piece and away from the edge. This prevents excessive melting of the edge as may occur in flat welding.

Butt Joints

When welding a groove-type joint in the overhead position, the electrode is aligned with the weld axis. The backhand welding method is used with the welding end of the electrode pointing 10°–20° in the direction opposite the direction of travel. A multiple-pass weld is cleaned between weld passes. Keeping the weld pool small prevents the molten metal from falling from the weld pool. If overheating occurs, the weld pool can be controlled using an oscillating or whip motion.

Weld Defects

Defects that can be seen include overlap, undercut, voids in the weld face, incorrect weld size and shape, and lack of penetration on a groove-joint weld.

Overlap occurs when the weld toe is not fused into the base metal. **Figure 9-15A** and **B** show examples of overlap.

Undercut occurs when the base metal at the weld toe area is melted but is not filled with filler metal. This causes a weak area in the weldment. See **Figure 9-15C** and **D** for two examples of undercut.

Slag inclusions are caused when slag is trapped during a multiple-pass weld. Slag must be completely removed after each weld pass to prevent slag inclusions.

Porosity can be seen on the surface of some welds. Some porosity can be beneath the surface and not visible on a completed weld. Welding with a long arc or improper electrode motion can cause porosity.

Exercise 9-6 Fillet Weld on a T-Joint in the Overhead Welding Position

1. Obtain two pieces of mild steel measuring 1/4″ × 2″ × 6″ (6.4 mm × 50 mm × 150 mm).
2. Obtain five 1/16″ or 3/32″ (1.6 mm or 2.4 mm) diameter E6012 electrodes.
3. Form a T-joint with the two pieces. Tack weld in three places on each side of the joint in the flat welding position.
4. Place the weldment in an overhead position.
5. **Wear protective clothing recommended for overhead welding.** Make a fillet weld with 3/8″ (9.5 mm) legs, using several weld passes.
6. Make a fillet weld with 3/8″ (9.5 mm) legs on the other side of the T-joint.

Inspection:

The weld should be even in width. Each weld bead should have evenly spaced ripples. The fillet legs must be 3/8″ (9.5 mm) long. The completed weld should have a straight or convex weld face.

Exercise 9-7 V-Groove Weld for a Butt Joint in the Overhead Welding Position

1. Obtain two pieces of mild steel measuring 1/4″ × 2″ × 6″ (6.4 mm × 50 mm × 150 mm).
2. Obtain five 1/16″ or 3/32″ (1.6 mm or 2.4 mm) diameter E6010 electrodes.
3. Prepare the edges of the pieces for a 90° V-groove butt joint. Leave a 1/16″ (1.6 mm) root thickness.
4. Form a butt joint with a 1/16″ (1.6 mm) root opening. Tack weld in three places.
5. Place the joint into an overhead position.
6. Make a root pass using the keyhole method to get 100% penetration. An oscillating or whip motion is suggested to control the weld pool size.

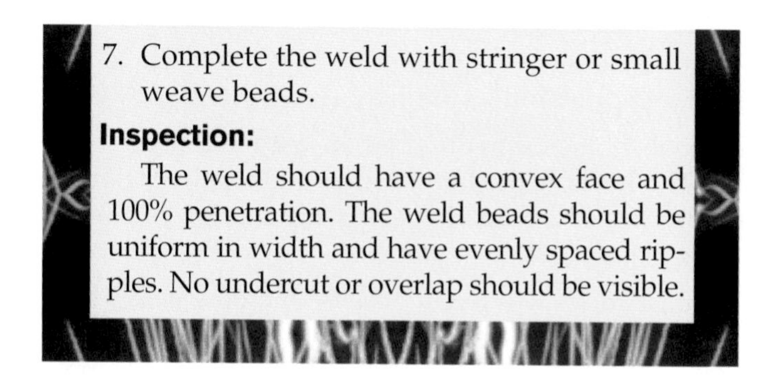

7. Complete the weld with stringer or small weave beads.

Inspection:

The weld should have a convex face and 100% penetration. The weld beads should be uniform in width and have evenly spaced ripples. No undercut or overlap should be visible.

Welds that are too wide, too narrow, have too much buildup, are underfilled, or butt joints that do not have complete penetration can be found by visual inspection. **Figure 9-15E** and **F** shows two defects in bead shape.

Complete (100%) penetration is generally required only on a weld with a groove-type joint. A weld bead with 100% penetration will show a small, uniform bead-like bump on the root side of the weld.

Fillet welds seldom require 100% penetration. They do, however, require enough to penetrate past the intersection of the two pieces being welded. The welding symbol often specifies a leg size and weld size large enough to ensure adequate penetration. Poor penetration on a fillet weld cannot be inspected visually.

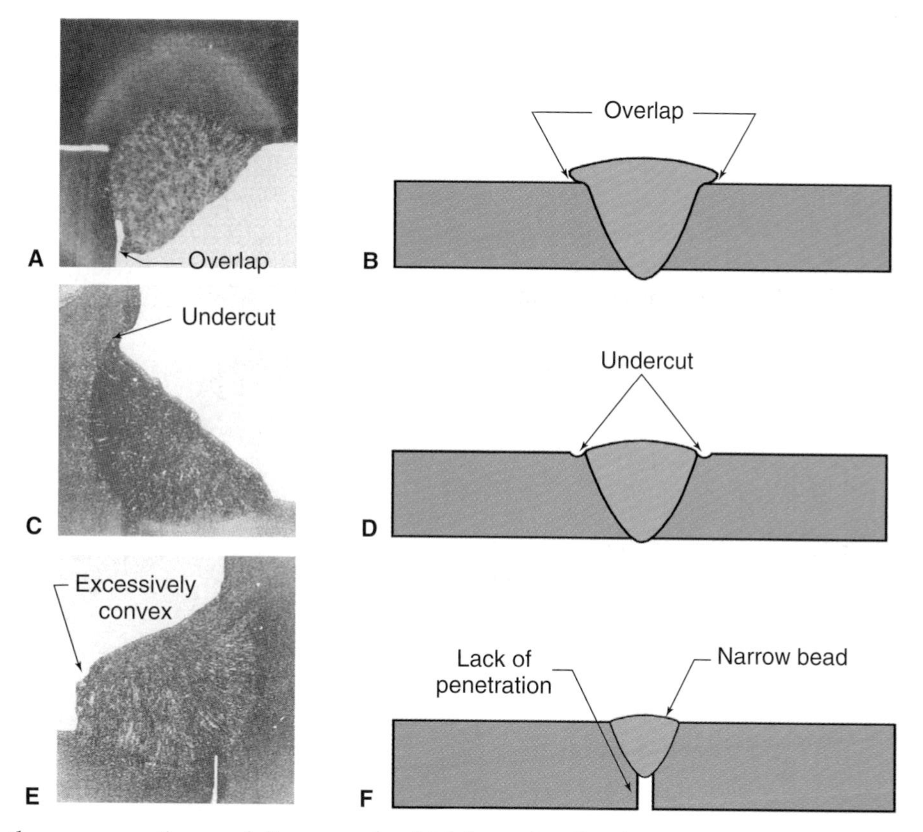

Figure 9-15. *Study these cross sections and diagrams of weld defects. A—Overlap on a fillet weld. B—Overlap on a groove weld on a butt joint. C—Undercut on a fillet weld. D—Undercut on a groove weld on a butt joint. E—A poorly contoured fillet weld. F—A narrow weld bead without complete penetration on a square groove butt joint.*

Summary

- An electrode used for out-of-position welding should have a smaller diameter than an electrode used for flat position welding. The electrode should also be fast-setting (quick-freezing). A fast-setting slag deposit prevents the weld metal from sagging. Some AWS electrodes that have fast-setting slag deposits are E6010, E6011, E6012, and E6013.
- When making practice fillet welds in the horizontal position, the weld axis is generally horizontal and the weld face is 45° from horizontal. When making butt welds in the horizontal position, the weld axis is generally horizontal and the weld face is vertical. If proper technique is not used, the molten metal tends to sag downward out of the weld when welding in the horizontal position.

- In the vertical position, the weld axis is generally within 10° of vertical. The weld face is also close to being vertical. Welding in the vertical position is done either downhill or uphill.
- When welding in the overhead position, a small weld pool should be maintained to keep the metal from sagging. A stringer bead is used for the root pass. The keyhole method ensures 100% penetration. A short oscillating motion is recommended to control the pool and bead shape. A whip motion can be used to control the molten metal.
- Overlap occurs when the weld toe is not fused into the base metal. Undercut occurs when the base metal at the weld toe area is melted but is not filled with filler metal. Slag inclusions are caused when slag is trapped during a multiple-pass weld. Porosity is a defect that can be seen on the surface of some welds.

Review Questions

Write your answers on a separate sheet of paper. Please do *not* write in this book.

1. Name five pieces of protective clothing that are to be worn for out-of-position welding.
2. List three advantages of using smaller diameter electrodes for out-of-position welding.
3. When welding horizontally, why is the lowest weld bead in each layer made first on a multiple-pass V-groove butt weld?
4. Identify the defects shown in the sketch below.

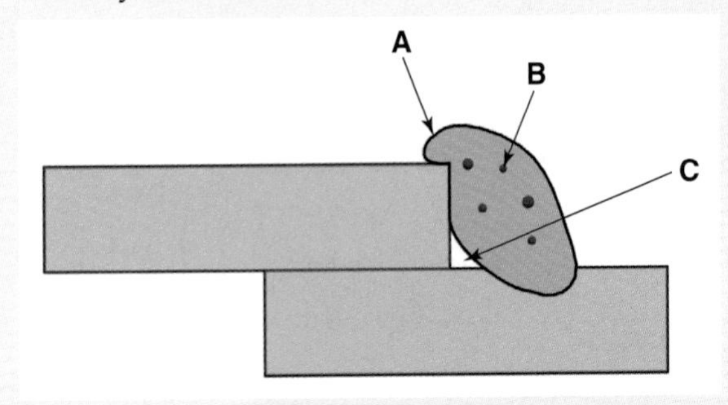

5. Name the part of the weld shown in the sketch.

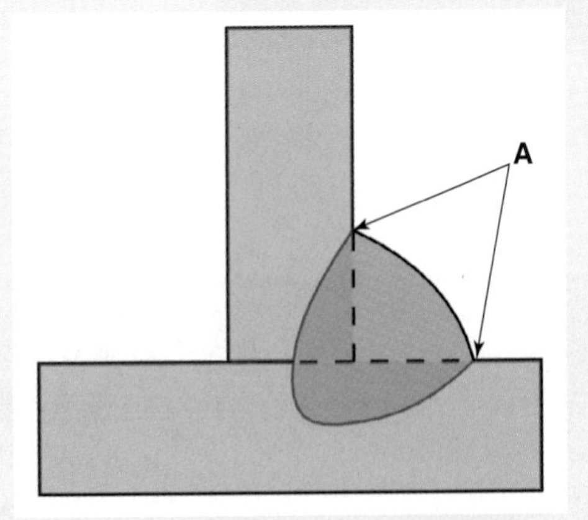

6. An electrode especially designed for making horizontal fillet welds is the _____.
7. Why must the weld pool be kept ahead of the slag when welding downhill?
8. The electrode is moved ahead about _____ (_____ mm) and creates a long arc during the whip motion.
9. The entire whip motion should take only _____ seconds.
10. What type of motion is recommended to control the molten metal during an overhead weld?

Surfacing

Learning Objectives

After studying this chapter, you will be able to:

- List reasons for surfacing a part.
- Identify the various surfacing processes.
- List reasons for wear that occurs in parts.
- Define characteristics of surfacing electrodes.
- List two means of testing material hardness.
- Describe abbreviations used when specifying surfacing electrodes.
- Select the proper surfacing electrode.
- Surface a part.

abrasion
abrasion resistance
Brinell hardness test
buildup process
buttering
chemical corrosion
cladding
corrosion resistance
fatigue
grit blasting
hardfacing
hardness
hot hardness
impact strength
machinability
metal transfer wear
metal-to-metal wear resistance
oxidation resistance
Rockwell C hardness test
surfacing
thermal spraying
vapor degreasers

Surfacing Principles

Surfacing is defined by the American Welding Society as "the application by welding, brazing, or thermal spraying of a layer(s) of material to a surface to obtain desired properties or dimensions, as opposed to making a joint." Surfacing is often less expensive than purchasing a new part. Surfacing also usually reduces the downtime on the equipment. There are several surfacing processes. They include hardfacing, buttering, cladding, and buildup.

Hardfacing is a surfacing process in which hard materials are applied to the surface of a part. This method is used to reduce wear or loss of materials by impact or abrasion. See **Figure 10-1**. Hardfacing a part also results in fewer repairs due to wear.

Hardfacing weld beads may be laid in a basket weave or dot pattern. See **Figure 10-2**. These patterns are used when a sticky material, such as dirt or mud, comes in contact with the part. The sticky material accumulates in the depressed areas of the basket weave pattern. The build-up of dirt or mud helps to protect the metal from further abrasion.

Figure 10-1. The cutting edge of this bean cutter has been hardfaced to increase its useful life. (Stoody Co.)

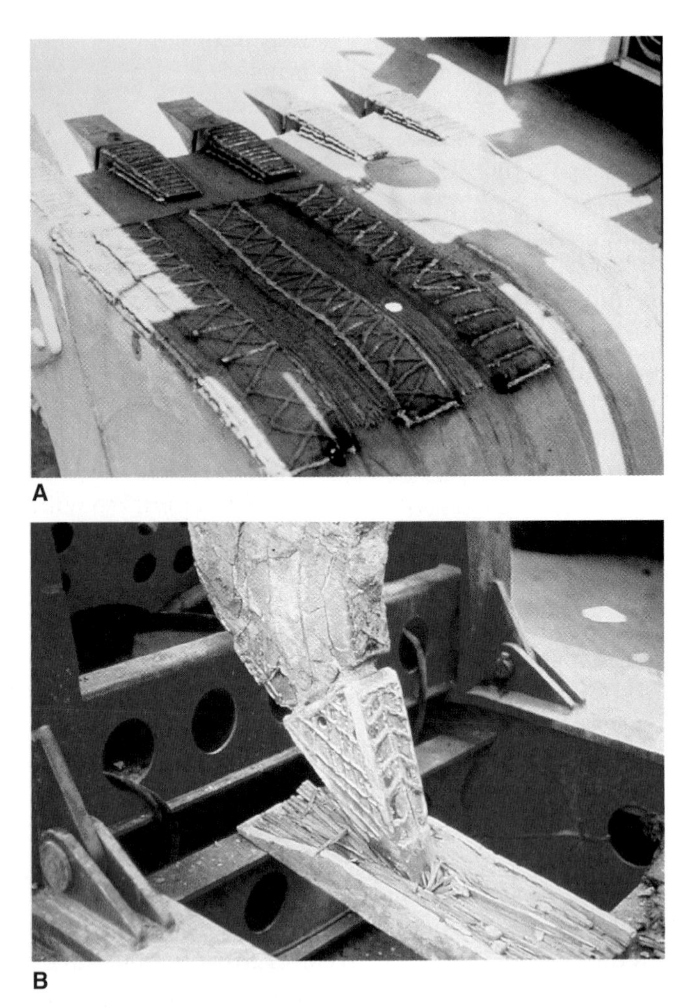

Figure 10-2. Hardfacing patterns. A—Hardfacing material applied to the bottom of a large shovel bucket. An X pattern is used on the bottom and a heavier cross bead is used on the teeth. B—Two different patterns were used to hardface this ripper tooth. (Stoody Co.)

In the ***buttering*** process, one or more layers of easily welded material are applied to the surface of a part that has poor welding characteristics. This process is used to form a transition layer when welding dissimilar metals.

The ***cladding*** process is used to apply surfacing materials that will improve the corrosion or heat resistance of a part. When a part is worn, the surface may be returned to its original dimensions by using the ***buildup process***.

Surfaces of parts wear for several reasons. ***Chemical corrosion*** may eat away the surface. ***Metal transfer wear*** occurs when metal leaves one surface and fuses or sticks to another surface. This commonly occurs on bearing surfaces. Surfaces may be rubbed together constantly and wear down due to friction or ***abrasion***. ***Fatigue*** may occur when surfaces are constantly struck, stretched, compressed, or bent. Metals will eventually weaken and wear down or fracture.

A metal surface is commonly repaired when it wears down. The cause of the failure must be determined to properly repair the part. A surfacing material is then selected to resist the type of wear that occurred on the original part.

Surfacing is applied to a part using solid, powder, or finely ground metallic or nonmetallic materials. These materials may be applied using the oxyfuel gas, arc, or plasma process. Solid surfacing rods may be used when surfacing with oxyfuel gas. Special covered electrodes are used for surfacing with the shielded metal arc process (SMAW). Surfacing materials are enclosed in a hollow wire and applied using the flux cored arc welding (FCAW) process.

Surfacing materials may also be applied by ***thermal spraying***. In this process, powdered or finely ground materials are fed into a flame or arc and carried to the surface of the part. Surfacing materials fed from a hopper into the oxyfuel gas flame may be used. See **Figure 10-3**. Thermal spraying may also be done using the plasma arc. See **Figure 10-4**.

Surfacing Electrodes

Surfacing electrodes are used to apply surfacing materials to metal parts. The surfacing materials reduce wear to a part or build up its dimensions. The following characteristics or qualities should be considered when selecting a surfacing electrode:

- Hardness.
- Hot hardness.
- Impact strength.
- Oxidation resistance.
- Corrosion resistance.
- Abrasion resistance.

- Metal-to-metal wear resistance.
- Machinability.

Hardness is the ability to resist penetration. It is measured by several methods, including the popular Rockwell C and Brinell hardness tests. **Figure 10-5** shows a comparison between the scales used for Rockwell C and Brinell hardness tests.

The ***Rockwell C hardness test*** is used for hard materials. A Rockwell C hardness testing machine forces a pointed diamond into the metal surface. The diamond penetrates deeper into the surface of softer metals than it does the surface of harder metals. A gauge indicates the hardness number. The Rockwell C hardness scale ranges from 0 to approximately 70. A higher number indicates a harder material.

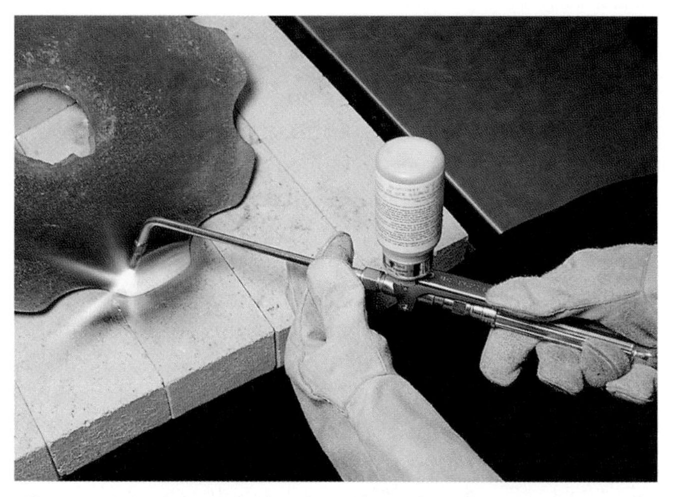

Figure 10-3. *An oxyfuel gas torch equipped with a hopper for surfacing materials. The welder is applying surfacing material to a part used in a farm implement. (Stoody Co.)*

Figure 10-4. *A thermal spraying torch is being used to apply a hardfacing material to a worn shaft. The torch is mounted on a lathe cross-feed for better control of the material thickness. The shaft diameter will be ground to size after the new material is applied.*

Hardness Test Comparisons

Rockwell C	Brinell
69	755
60	631
50	497
40	380
30	288
24	245
20	224
10	179
0	143

Figure 10-5. *A comparison of hardness numbers in the Rockwell C and Brinell scales.*

A ***Brinell hardness test*** is used to evaluate the hardness of soft metals like copper, brass, bronze, and aluminum. In this test, a 10 mm diameter steel ball is forced into the surface of the test piece. The ball penetrates deep into the surface of a soft material. A gauge indicates the hardness number, **Figure 10-6**. Brinell numbers range from approximately 100 to 800. A higher number indicates a harder material.

Hot hardness is the ability of a metal to retain its strength at high temperatures. The hardness of a metal generally decreases as the temperature is increased. A metal that retains its hardness at high temperatures has good hot hardness.

Impact strength is the ability of a material to withstand impact or hammering forces without cracking or breaking. Stone-crushing hammers and forging dies have high impact strength.

Oxidation resistance is the ability of a material to withstand the corrosive effects of atmospheric oxidation. Steel with high oxidation resistance will not rust easily.

Corrosion resistance is the ability of a material to resist corrosion from chemicals, such as salt spray and acids. Corrosion may cause fatigue and failure of parts.

Abrasion resistance is the ability of a material to resist wear from scratching by sharp objects. Rocks, sand grains, or sharp metal are common causes of abrasion.

Metal-to-metal wear resistance is the ability of a material to resist wear from metal-to-metal contact. Bearing surfaces require a high amount of metal-to-metal wear resistance.

Machinability is the ability of a part to be machined or ground to size. The hardness of a material must be considered if the part is to be machined after surfacing. Some surfacing materials are so hard that they may be difficult to machine.

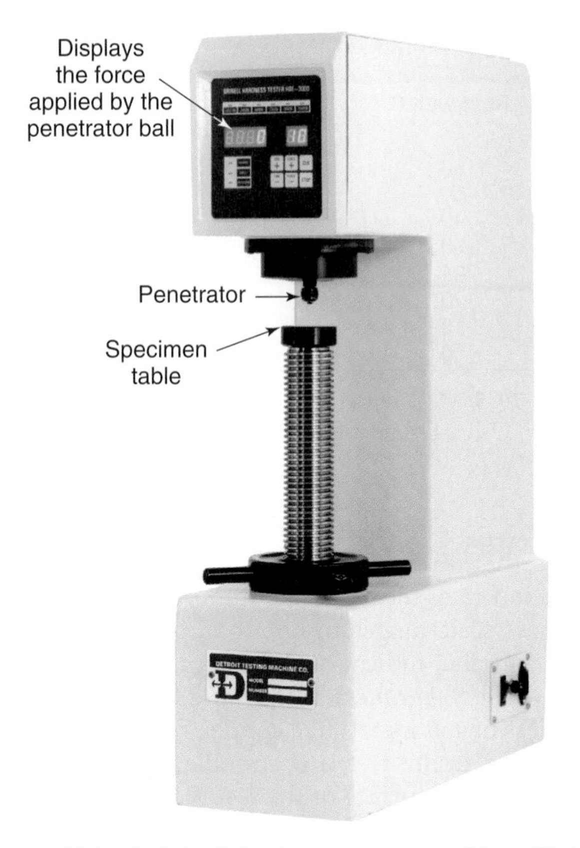

Figure 10-6. A Brinell hardness testing machine. (Detroit Testing Machine Co.)

Electrode Classifications

Surfacing materials can be applied using a variety of welding processes. The oxyfuel gas welding (OFW), flux cored arc welding (FCAW), and gas tungsten arc welding (GTAW) processes may be used. Solid, bare surfacing wires and rods are used with OFW and GTAW. An alloy-filled wire electrode is used with FCAW.

Solid, covered surfacing electrodes are used with the shielded metal arc process. Alloy-filled surfacing electrodes may also be used with this process. The alloying materials may include tungsten carbide crystals, which can create an extremely hard surface when applied to a metal surface. Surfacing electrodes are classified by the primary material contained in the electrode. See **Figure 10-7**.

Surfacing electrodes for the shielded metal arc process are available in the following diameters: 1/8″ (3.2 mm), 5/32″ (4.0 mm), 3/16″ (4.8 mm), 1/4″ (6.4 mm), 5/16″ (8.0 mm), and 3/8″ (9.5 mm). The electrodes are available in 8″–28″ (203 mm–711 mm) lengths. The 14″ (356 mm) and 18″ (457 mm) lengths are the most widely used.

Most surfacing electrodes are designed for use in the flat position. Several are produced for use in all positions.

Classification	Filler Metal
Fe5	High-speed steel
FeMn	Austenitic manganese
FeCr	Chromium steel
CoCr-A	Cobalt and chromium
CoCr-C	Cobalt and chromium
CuZn	Copper and zinc (brass)
CuSi	Copper and silicon
CuAl	Copper and aluminum
NiCr	Nickel and chromium

Figure 10-7. This table displays AWS surfacing filler metal classifications. See Figure 10-8 for the chemical abbreviations.

Specifications for surfacing electrodes are commonly abbreviated. The abbreviations can generally be divided into three parts. See **Figure 10-8**. The first part indicates whether it is to be used as an electrode or a welding rod. The second part indicates the major alloying elements in the filler metal. The third part indicates a subclassification of the filler metal.

Characteristics of various surfacing electrodes and rods are shown in **Figure 10-9**. Suggested applications for the surfacing materials are also shown. For specific corrosion-resistant applications, consult AWS A5.13 or a corrosion resistance expert.

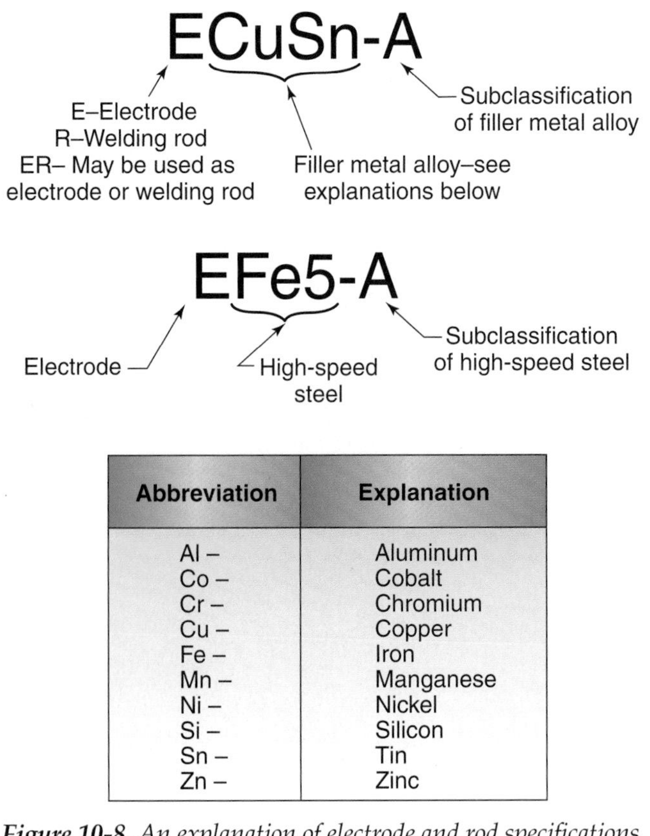

Abbreviation	Explanation
Al –	Aluminum
Co –	Cobalt
Cr –	Chromium
Cu –	Copper
Fe –	Iron
Mn –	Manganese
Ni –	Nickel
Si –	Silicon
Sn –	Tin
Zn –	Zinc

Figure 10-8. An explanation of electrode and rod specifications. The chemical abbreviations used with AWS surfacing filler metal classifications are also shown.

Classification	Hardness	Hot Hardness	Impact Strength	Oxidation Resistance	Corrosion Resistance	Abrasion Resistance	Metal-to-Metal Wear Resistance	Machinability	Applications
Fe5-A,B,C	C55-60[1]	H<1000°F	H[2]	L	L	M	H[3]	H	Cutting tools, dies, guides
FeMn-A & B	450-550BHN[6]	H<500-600°F	H	L	L	M	H	L	Soft-rock-crushing equipment
FeCr-A	C51-62	H<800-900°F	L	H	L	L-H[7]	L-H[7]	L	Farm equipment
CoCr-A	C42	H>1200°F	M	H	H[4]	L	H	L	Engine exhaust valves
-B	C46	H>1200°F	M	H	H[4]	L	H	L	
-C	C55	H>1200°F	L	H	H[4]	M	H	L	
CuAl-A2	115-140BHN[5]		M						Bearings and gear surfaces
CuAl-A3	140-180BHN[5]	H<400°F	L	M	M[4]	L	H	H	
-B	140-180BHN[5]	H<400°F	L	M	M[4]	L	H	H	
-C	180-220BHN[5]	H<400°F	L	M	M[4]	L	H	H	
-D	230-270BHN[5]	H<400°F	L	M	M[4]	L	H	H	
-E	280-320BHN[5]	H<400°F	L	M	M[4]	L	H	H	
CuSi	80-100BHN	H<800°F	M	M	M[4]	L	H	H	Bearing surfaces
CuSn-A	70-85BHN	H<800°F	L	L	M[4]	L	H	H	Bearing surfaces
-C	85-100BHN	H<800°F	L	L	M[4]	L	H	H	
NiCr-A	C28	H<800°F	M	H<1750°F	4	L-H[7]	H	M	Seal rings, cams, screw conveyers
-B	C38	H<800°F	M	H<1750°F	4	L-H[7]	H	M	
-C	C45	H<800°F	L	H<1750°F	4	L-H[7]	H	M	

KEY:
C - Rockwell C hardness number
BHN - Brinell hardness number
L - Low
M - Medium
H - High

1. Anneal to Rockwell C30
2. After tempering
3. After annealing
4. Consult authority for each application
5. Hardness depends on welding process used
6. After work hardening
7. Depends on application

Figure 10-9. Characteristics of various surfacing materials. Refer to AWS A5.13 or a corrosion authority for specific applications.

Surfacing Process

The surfacing process requires a few essential steps. Each part to be surfaced must be thoroughly cleaned. This ensures that the surfacing material can adhere to the part. Preheating some metals is necessary. Preheating can be done using an oven, torch, electrical resistance heater, or any other suitable heat source. The final step is to choose the desired process and equipment and apply the surfacing materials to produce an acceptable new surface.

Cleaning the Surface

Parts must be cleaned to ensure the proper bonding of the base metal and surfacing materials. The type of base metal determines the cleaning method to be used. Mechanical cleaning may be done with abrasive or ***grit blasting***. Aluminum and magnesium are usually grit blasted with aluminum oxide or quartz crystals. Steam, liquid or ***vapor degreasers***, and industrial solvents may be used to chemically clean the parts. Steel may be cleaned with steam or degreasers. Porous metals like cast iron are heated to 400°F–600°F (200°C–315°C). Heating to these temperatures vaporizes grease and oil in the porous areas.

Preheating

Some metals must be preheated prior to surfacing. Preheating to a specific temperature prepares the metal for surfacing. It ensures a complete bond between the base metal and surfacing material. Preheating recommendations for several metals are listed below. If the type of base metal cannot be determined, preheat to 500°F–700°F (260°C–370°C). Be certain, however, that the part is not made of manganese steel.

- Cast iron: 500°F–700°F (260°C–370°C).
- Low and mild carbon steel: preheating is not required except for heavy sections.
- Manganese steel (steel containing 12–14% manganese): 200°F (93°C). Do not heat above 500°F (260°C).

Surfacing with a Shielded Metal Arc

Direct current electrode positive (DCEP) polarity is generally used when surfacing with a shielded metal arc. The equipment used for surfacing is the same as that used for shielded metal arc welding. However, different electrodes are used.

Surfacing materials may be applied as long weld beads over small areas of a part or over the entire surface of a part. They can be applied as stringer or weave beads or sprayed onto the surface of the base metal. Deep penetration of the surface is not required or desired. Figure 10-2 shows examples of patterns used for surfacing materials.

The surface must be thoroughly cleaned to ensure a good bond of the surfacing materials. After cleaning, the metals to be surfaced may require preheating. Preheating ensures a better bond between the base metal and surfacing materials.

The process of depositing surfacing materials with an electrode is the same as SMAW. An arc is struck in the same manner as used in SMAW; however, a long arc is maintained for surfacing. Such an arc is needed to spread the heat over a large area. Penetration is prevented by spreading the heat out. Higher amperage may be required for a surfacing electrode compared to a welding electrode for a given electrode diameter.

A surfacing electrode is held at approximately a 0° work angle (90° to the base metal). A drag travel angle of approximately 45° is used. An oscillating (forward and back) motion is recommended for stringer beads. See **Figure 10-10**. A weaving or circular motion is used to create wider weld beads. See **Figure 10-11**. A weave or circular bead should not be wider than six times the electrode diameter.

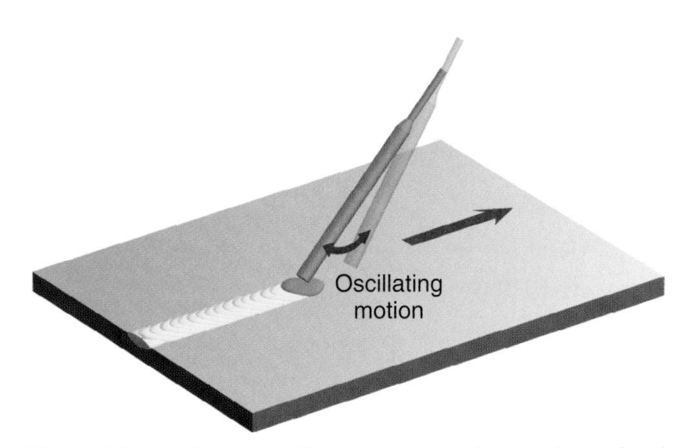

Figure 10-10. *Use an oscillating motion when applying hardfacing material with a SMAW electrode.*

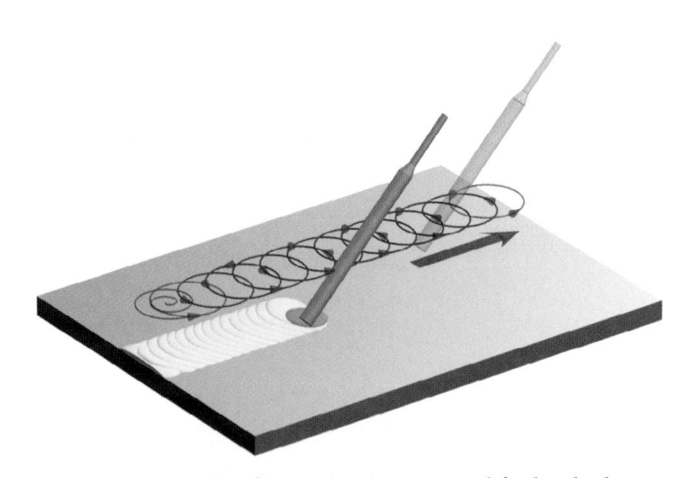

Figure 10-11. *A circular motion is suggested for beads that are from 3/4″–1 1/4″ (19 mm–32 mm) wide.*

Each weld bead of surfacing material should be between 1/16″–1/4″ (1.6 mm–6.4 mm) thick.

Weld beads that are laid side by side should overlap by about 1/4 to 1/3 of the bead width. See **Figure 10-12**. If more than one layer is required, the weld beads should also overlap. Figure 10-12 shows the suggested overlap of weld beads and layers.

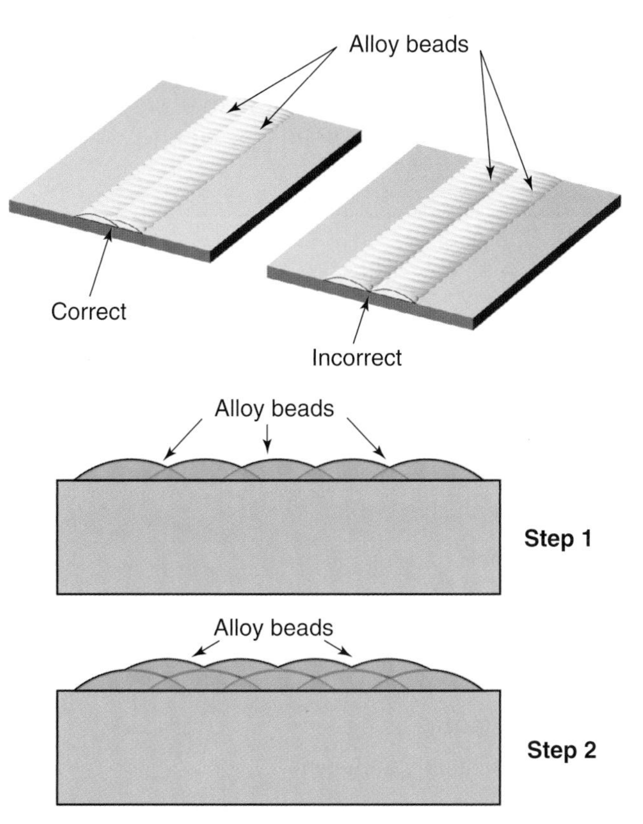

Figure 10-12. *Steps for the suggested method of overlapping surfacing beads across a wide surface.*

Exercise 10-1 Surfacing with a Shielded Metal Arc

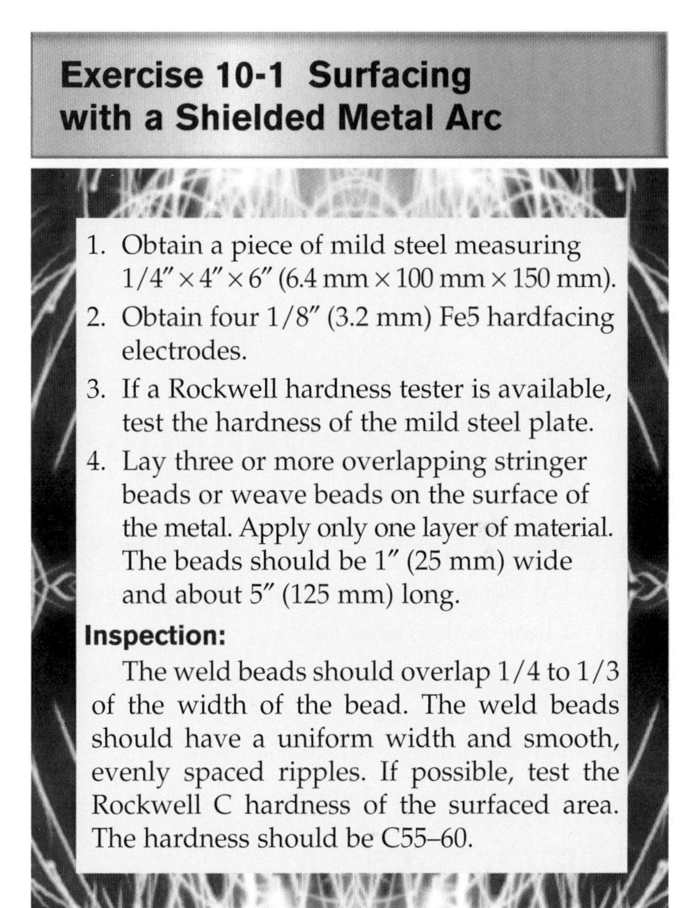

1. Obtain a piece of mild steel measuring 1/4″ × 4″ × 6″ (6.4 mm × 100 mm × 150 mm).
2. Obtain four 1/8″ (3.2 mm) Fe5 hardfacing electrodes.
3. If a Rockwell hardness tester is available, test the hardness of the mild steel plate.
4. Lay three or more overlapping stringer beads or weave beads on the surface of the metal. Apply only one layer of material. The beads should be 1″ (25 mm) wide and about 5″ (125 mm) long.

Inspection:

The weld beads should overlap 1/4 to 1/3 of the width of the bead. The weld beads should have a uniform width and smooth, evenly spaced ripples. If possible, test the Rockwell C hardness of the surfaced area. The hardness should be C55–60.

Summary

- Surfacing is defined by the American Welding Society as "the application by welding, brazing, or thermal spraying of a layer(s) of material to a surface to obtain desired properties or dimensions, as opposed to making a joint."
- Hardfacing is a surfacing process in which hard materials are applied to the surface of a part. This method is used to reduce wear or loss of materials by impact or abrasion.
- Buttering is a process in which one or more layers of easily welded material are applied to the surface of a part that has poor welding characteristics. This process is used to form a transition layer when welding dissimilar metals.
- The cladding process is used to apply surfacing materials that will improve the corrosion or heat resistance of a part.
- When a part is worn, the surface may be returned to its original dimensions by using the buildup process.
- When surfacing, the arc length must remain constant throughout the entire weld.

(continued)

- Surfacing materials may be applied using the oxyfuel gas, arc, or plasma process. Solid surfacing rods may be used when surfacing with oxyfuel gas. Special covered electrodes are used for surfacing with the shielded metal arc process (SMAW). Surfacing materials can be enclosed in a hollow wire and applied using the flux cored arc welding (FCAW) process. Surfacing materials may also be applied by thermal spraying.
- Each part to be surfaced must be thoroughly cleaned. This ensures that the surfacing material can adhere to the part. Preheating of some metals is necessary. The final step is to choose the desired process and equipment and apply the surfacing materials to produce an acceptable new surface.
- Direct current electrode positive (DCEP) polarity is generally used when surfacing with a shielded metal arc. The equipment used for surfacing is the same as that used for shielded metal arc welding. However, different electrodes are used.

Review Questions

Write your answers on a separate sheet of paper. Please do *not* write in this book.

1. List at least two reasons for surfacing a part.
2. List three causes of wear.
3. What type of surfacing process uses powdered surfacing material?
4. A(n) _____ hardness tester forces a pointed diamond into the metal surface.
5. The ability to resist wear from sharp objects is called _____ resistance.
6. Describe the meaning of the abbreviations used in specifying surfacing materials.

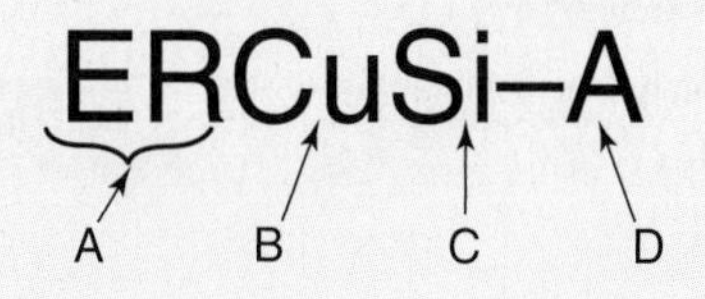

7. Refer to Figure 10-9. What type of surfacing electrode could be used if important considerations were metal-to-metal wear resistance and excellent machinability?
8. Why should cast iron be preheated prior to applying a surface material?
9. What polarity is normally used with SMAW surfacing electrodes?
10. Why is a long arc used when surfacing?

This GMAW outfit is being used to make a fillet weld on a T-joint. (Miller Electric Mfg. Co.)

Section 3

Gas Metal Arc Welding

Stoody Co., a Thermadyne company

Chapter 11

GMAW: Equipment and Supplies

Learning Objectives

After studying this chapter, you will be able to:

- Define GMAW and FCAW.
- Identify the correct polarity to use for GMAW and FCAW.
- List advantages and disadvantages of GMAW as compared to SMAW.
- Define metal transfer.
- Describe the three methods of metal transfer.
- Identify the equipment that makes up a GMAW outfit.
- Summarize the operation of a wire feeder.
- List the parts of a welding gun and cables.
- List four gases used for GMAW and the most common shielding gas used for FCAW.
- Explain the use of a flowmeter for GMAW or FCAW.
- Identify protective clothing and equipment used for GMAW and FCAW.

Technical Terms

active gas
argon (Ar)
background current
carbon dioxide (CO_2)
constant potential
consumable electrode
contact tube
drive rolls
flowmeter
flux cored arc welding (FCAW)
gas metal arc welding (GMAW)
gas nozzle
globular transfer
globule
helium (He)
inch switch
inductance
inert gas
inverter
liner
metal inert gas (MIG)
metal transfer
microprocessor
peak current
pinch force
pulsed spray transfer
pulse
purge switch
push-pull system
self-shielding electrode
short arc
short circuit
short circuiting transfer
slope adjustment
spray transfer
Teflon® liner
transformer-rectifier
transition current
welding gun
wire feeder
wire tension control knob

Gas Metal Arc Welding Principles

Gas metal arc welding (GMAW) is a welding process in which metals are joined by heating them with a welding arc between a continuous consumable electrode and the base metal. A shielding gas or gas mixture is used to prevent the atmosphere from contaminating the weld. Another term that is often used to describe GMAW is MIG welding. ***MIG*** stands for ***metal inert gas***. However, because some of the gases used in GMAW are not inert, MIG is an incorrect term. In the welding trade, both GMAW and MIG are used to describe this welding process.

Gas metal arc welding uses a wire as an electrode. A welding arc is struck between the electrode and the base metal. The electrode melts as it is continuously fed to maintain the welding arc.

Direct current is always used for GMAW and FCAW. Direct current electrode positive (DCEP), or direct current reverse polarity (DCRP), is used for GMAW. Direct current electrode negative (DCEN), or direct current straight polarity (DCSP), is seldom used.

A constant voltage power supply is used for GMAW. On this type of machine, the welder sets the required voltage. A constant voltage welding machine tries to maintain a constant voltage during the welding operation. Although the voltage is constant, the current varies. **Figure 11-1** illustrates the GMAW process. The electrode is melted and becomes part of the weld. For this reason, the electrode is called a ***consumable electrode***. The welding arc, also simply called the "arc," and the molten base metal are protected from air by a shielding gas. Flux is not used in this process.

Flux cored arc welding (FCAW) is similar to GMAW. Flux cored arc welding is a welding process in which metals are joined by heating them with an electric arc between the base metal and a continuous consumable electrode. A main difference between FCAW and GMAW is the electrode. An FCAW electrode is a steel tube containing flux in the center, or core, of the electrode. The flux can be used to add alloying elements to the weld.

The flux creates a shielding gas as it melts. Some flux cored electrodes require the use of additional shielding gas from a cylinder. See **Figure 11-2**. Although some shielding is created by the flux inside these electrodes, they require the use of additional shielding gas. ***Self-shielding electrodes*** produce their own shielding gas and do not require additional shielding gas. See **Figure 11-3.**

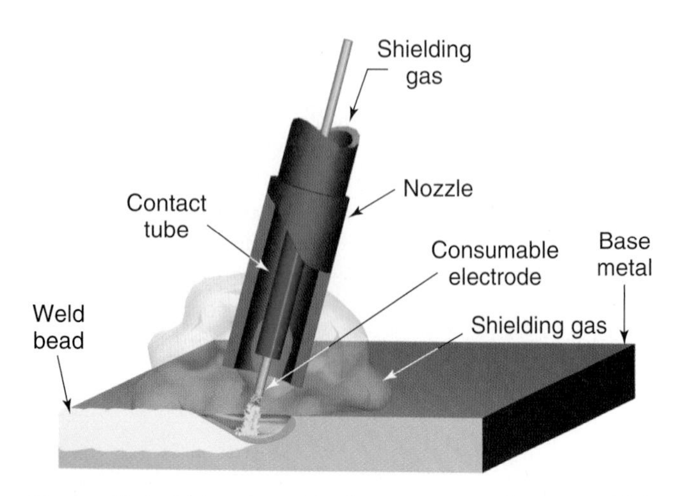

Figure 11-1. *This schematic shows a gas metal arc weld in progress.*

Some of the advantages of GMAW and FCAW include the following:

- A continuous electrode is used so that longer welds can be made without stopping to change electrodes. This eliminates many of the starts and stops that can be a major cause of weld defects.
- Welding speeds are faster than shielded metal arc welding (SMAW).
- No slag is produced in GMAW. Very little spatter is created. More of the electrode becomes part of the weld.
- Deeper penetration is possible with GMAW than with SMAW.
- Less training and skill are required to weld with gas metal arc and flux cored arc welding than with SMAW.

Some disadvantages of GMAW and FCAW include the following:

- GMAW and FCAW equipment costs more than SMAW equipment.
- Some weld joints are hard to reach with the welding gun.
- Rapid air movement, such as a strong wind, can blow the shielding gas away, resulting in a weld defect.

Metal Transfer

Filler metal from the electrode must leave the electrode and enter the weld. This is called ***metal transfer***. Metal transfer can occur in three different ways: short circuiting transfer, globular transfer, and spray transfer. In short circuiting transfer, the electrode contacts the base metal and some molten metal is transferred. In globular transfer and spray transfer, the wire does not touch the base metal. In these transfer methods, drops of molten metal leave the end of the electrode and enter the weld pool through the arc. These drops travel between the end of the wire and the base metal. The size of the drop in globular transfer is much larger than

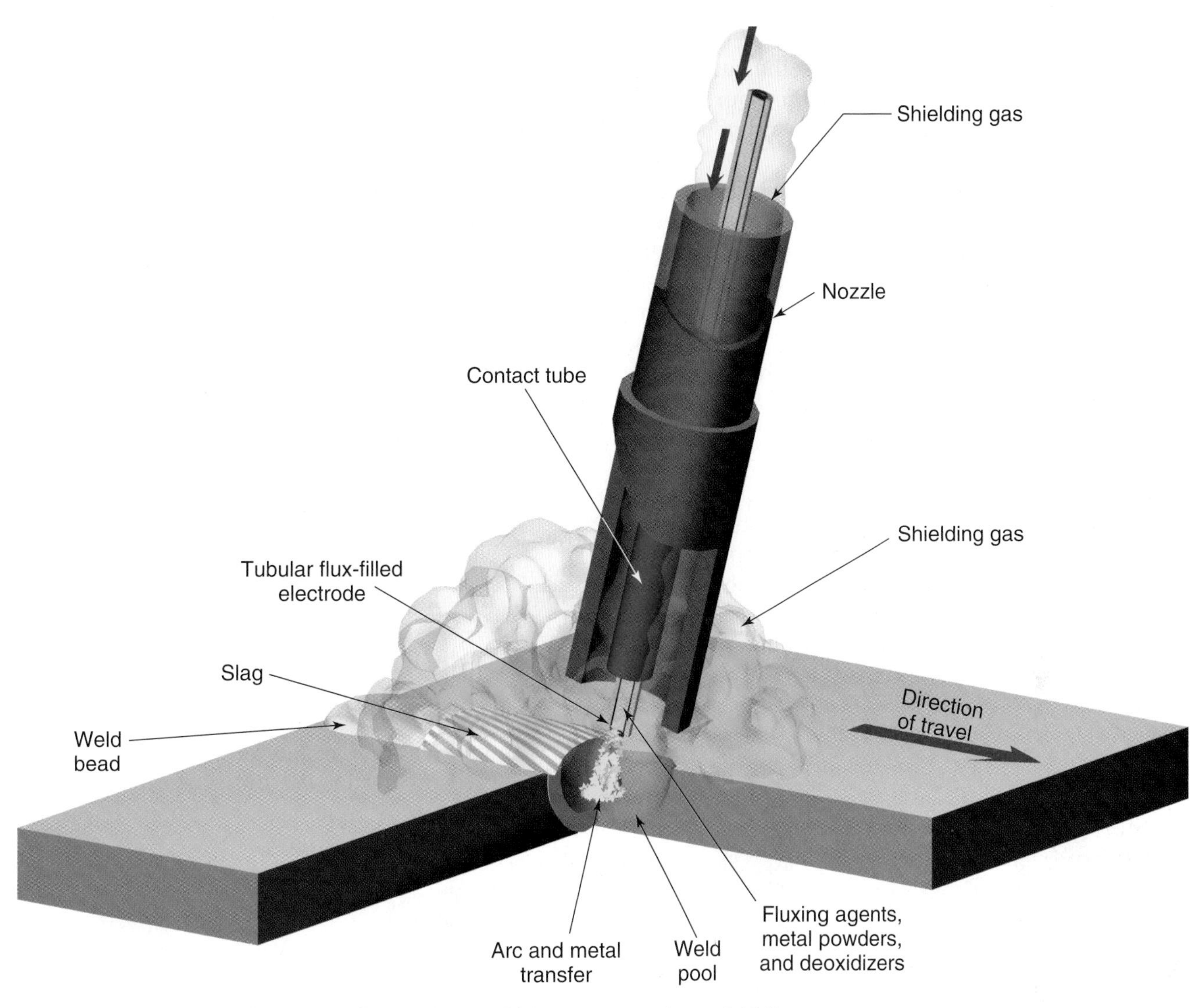

Figure 11-2. This schematic shows a flux core arc weld in progress using a shielding gas.

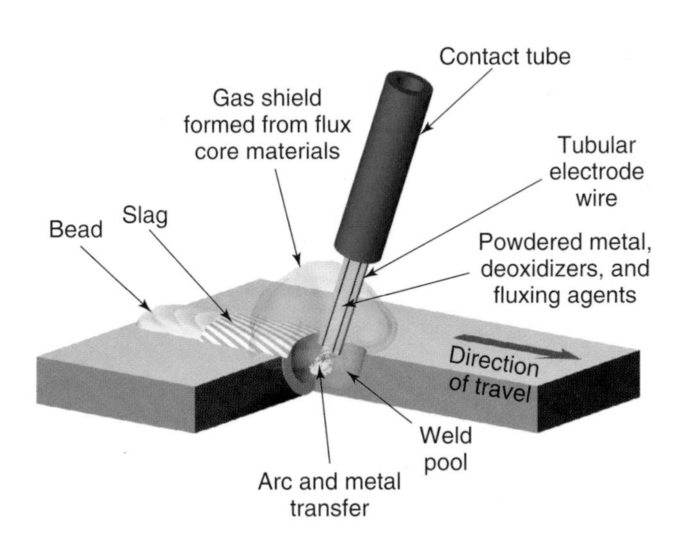

Figure 11-3. This schematic of a flux cored arc weld in progress shows that neither a nozzle nor shielding gas is required when using a self-shielding electrode.

the size of the electrode. The drop size in spray transfer is much smaller than the electrode diameter. Voltage, current, shielding gas, and other factors determine the type of metal transfer you will obtain. Voltage and wire feed speed recommendations are covered in Chapter 12.

Short Circuiting Transfer

Short circuiting transfer is also called ***short arc***. Short circuiting transfer uses a low welding current and voltage. A small-diameter electrode, .030″–.045″ (.76 mm–1.14 mm), is used. The short circuiting transfer method adds the least amount of heat to the base metal of any transfer method. Therefore, short circuiting transfer is used to join thin sheets of metal. It is also used for out-of-position welding and for welding parts that have large gaps.

In short circuiting transfer, the base metal and the end of the electrode are heated by an arc; however, metal is not transferred across the arc. The electrode touches the base metal, but there is no arc when this happens. When the electrode touches the base metal, it is called a ***short circuit***. While the short circuit exists, liquid or molten metal from the end of the electrode enters the weld pool.

A constant voltage welding machine tries to maintain the preset voltage. However, the voltage drops to zero when the wire short circuits to the base metal. When the voltage drops to zero, the current increases rapidly. The increased current creates a ***pinch force*** that causes the end electrode to separate from the base metal. After the end of the electrode separates from the weld pool, the arc reignites and the steps are repeated. **Figure 11-4** shows the short circuiting transfer process. The electrode is continuously fed into the weld where it shorts to the base metal and re-establishes the arc 20–200 times per second. The correct settings to use for short circuiting transfer are covered in Chapter 12.

The end of an electrode explodes if the welding machine increases the welding current too fast. ***Inductance*** is an electrical means of controlling the rate of current change in a circuit.

In a welding machine, inductance controls how fast the welding machine increases the welding current. Some welding machines allow the welder to control the inductance. Inductance is very important when using short circuiting transfer. The inductance should be increased if the end of the electrode appears to explode and creates a lot of spatter. Increasing the inductance slows the rate of current increase. This reduces the pinch force, which helps to prevent the end of the wire from exploding. Increasing the inductance also helps to make the weld pool flatter and more fluid.

Globular Transfer

Globular transfer uses slightly higher voltage and current settings than those used for short circuiting transfer. In ***globular transfer***, the arc melts the end of the electrode and the base metal. The size of the molten metal, ***globule*** (ball) on the end of the electrode increases until it is larger than the diameter of the electrode. It continues to grow in size until the globule falls off the electrode due to its own weight. **Figure 11-5** illustrates globular transfer.

The molten globule does not fall straight into the weld. Forces acting on the arc cause the globule to

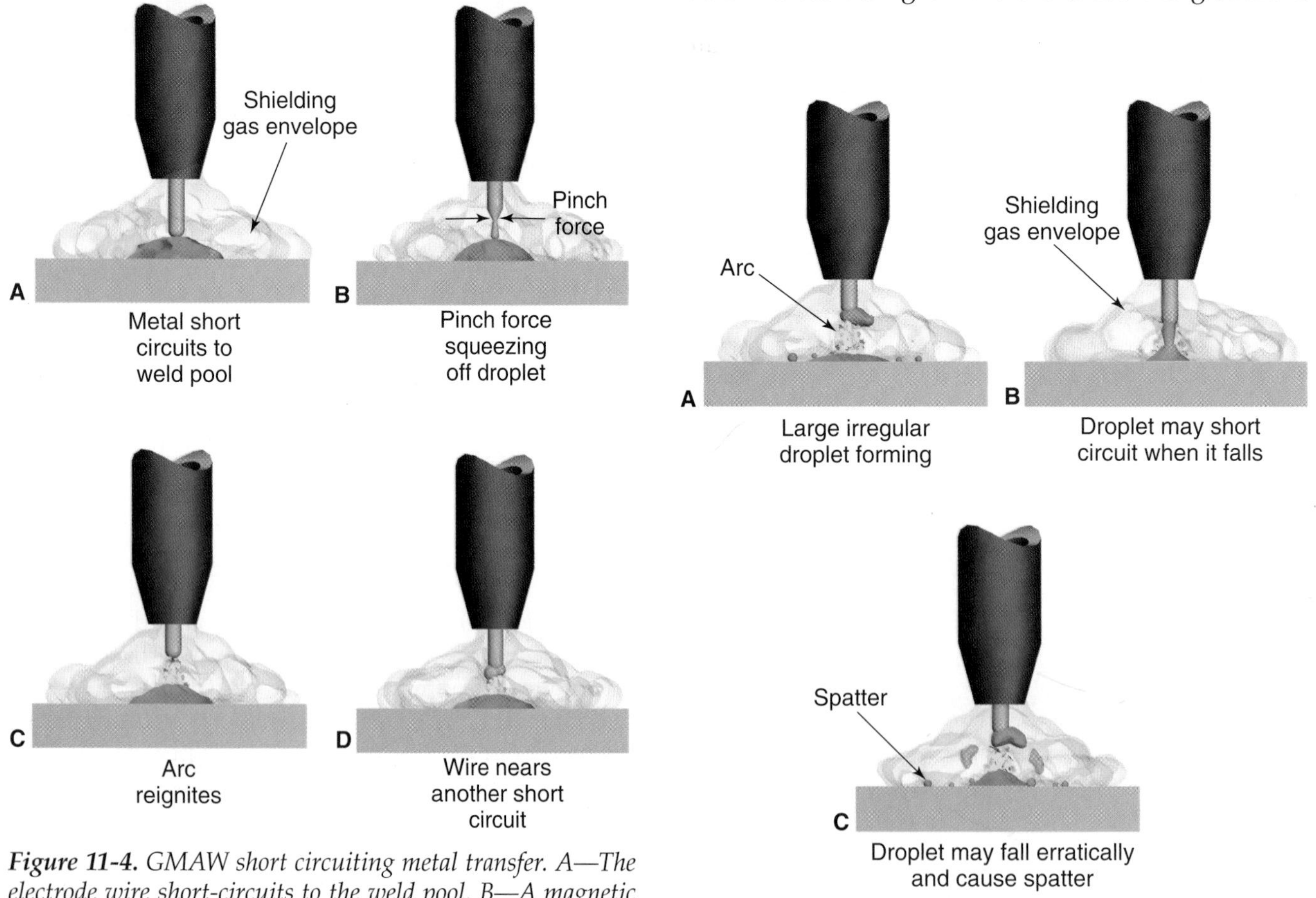

Figure 11-4. *GMAW short circuiting metal transfer. A—The electrode wire short-circuits to the weld pool. B—A magnetic pinch force squeezes off a droplet of molten electrode metal. C—The welding arc reignites. D—The electrode nears another short circuit condition and the process repeats itself.*

Figure 11-5. *GMAW globular metal transfer. A—Large irregular droplets form. B—Droplets may short circuit when they fall. C—Droplets may fall erratically and cause spatter.*

travel across the arc and land on the base metal in random patterns. This usually results in a large amount of spatter. Because gravity is used to transfer the molten globules to the base metal, globular transfer is difficult to perform for out-of-position welding.

Spray Transfer

Spray transfer requires higher voltage and current settings than globular transfer. In spray transfer, hundreds of small droplets are formed every second. They travel at a high rate of speed directly into the weld. See **Figure 11-6**. Good penetration is obtained with spray transfer. Very little spatter is produced, and the weld pool is very fluid. Spray transfer is used primarily for welding in the flat position or on horizontal fillet welds. Metal that is 1/8″ (3.2 mm) and thicker can be welded using the spray transfer process.

A shielding gas mixture containing at least 90% argon must be used for spray transfer. Spray transfer will only occur when the current setting is above the transition current. The ***transition current*** is the amount of current required to convert from globular transfer to spray transfer. **Figure 11-7** lists the transition currents for various electrodes.

Pulsed Spray Transfer

Pulsed spray transfer is very similar to standard spray transfer. However, two different currents are used: background current and peak current. The ***background current*** is an amount of current that is less than the transition current. Background current maintains the arc, but no metal is transferred across the arc during the background current period. ***Peak current*** is a preset current

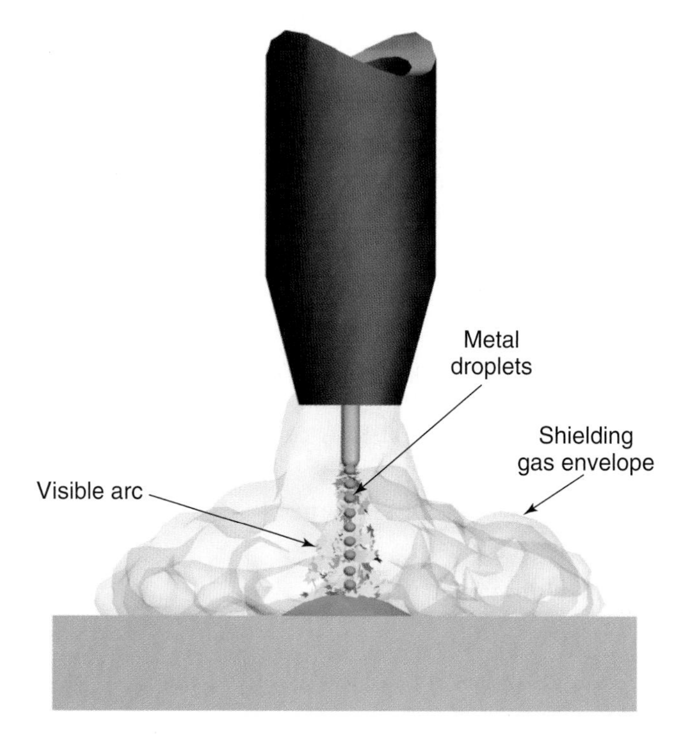

Figure 11-6. In GMAW spray transfer, hundreds of small droplets leave the end of the electrode every second.

Wire electrode type	Wire electrode diameter		Shielding gas	Minimum spray arc current, (A)
	in.	mm		
Mild steel	0.030	0.8	98% argon-2% oxygen	150
Mild steel	0.035	0.9	98% argon-2% oxygen	165
Mild steel	0.045	1.1	98% argon-2% oxygen	220
Mild steel	0.062	1.6	98% argon-2% oxygen	275
Stainless steel	0.035	0.9	98% argon-2% oxygen	170
Stainless steel	0.045	1.1	98% argon-2% oxygen	225
Stainless steel	0.062	1.6	98% argon-2% oxygen	285
Aluminum	0.030	0.8	Argon	95
Aluminum	0.045	1.1	Argon	135
Aluminum	0.062	1.6	Argon	180
Deoxidized copper	0.035	0.9	Argon	180
Deoxidized copper	0.045	1.1	Argon	210
Deoxidized copper	0.062	1.6	Argon	310
Silcon bronze	0.035	0.9	Argon	165
Silcon bronze	0.045	1.1	Argon	205
Silcon bronze	0.062	1.6	Argon	270
Note: Spray transfer will only occur when high percentages of argon or helium are used.				

Figure 11-7. *This table shows the approximate transition current levels for various electrodes. (American Welding Society (AWS) Welding Handbook Committee, 2004, Welding Handbook, 9th ed., vol. 2, Welding Processes, Miami: American Welding Society)*

level that is above the transition current. Spray transfer occurs during periods of peak current. See **Figure 11-8**. The peak current and background current alternate, with current flowing for a short time at each level. Less heat is applied to the base metal in pulsed spray transfer because the weld pool cools during the period of background current.

Peak current ***pulses*** (turns on and off) approximately sixty to several hundred times per second. Pulsed spray transfer allows for good penetration. It can be performed in all positions, since the weld pool has time to cool during the background current period.

GMAW Equipment

GMAW requires more equipment than shielded metal arc or gas tungsten arc welding. In addition to the power supply, a wire feeder and a ***welding gun*** are required. See **Figure 11-9**. A complete GMAW setup includes the following:

- A welding machine.
- A wire feeder.
- A welding gun.
- Electrode wire.
- Shielding gas supply and controls.

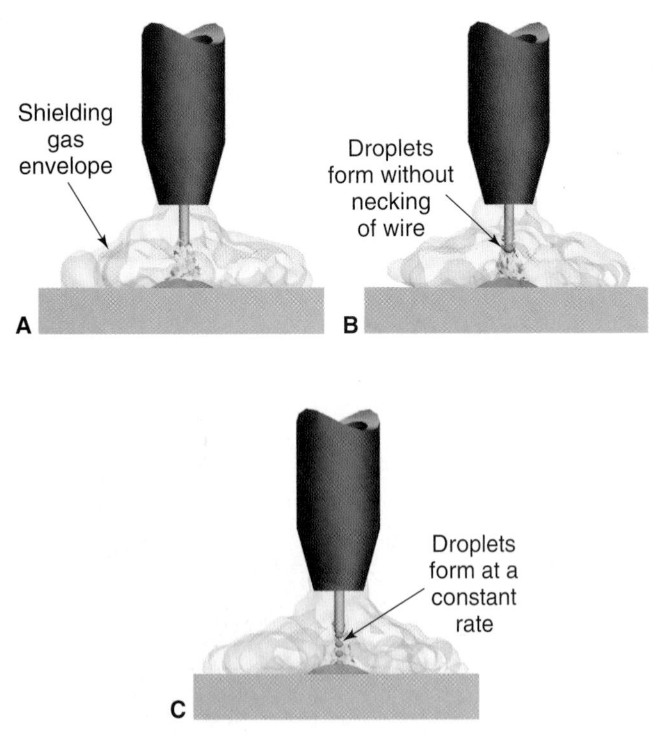

Figure 11-8. Pulsed spray transfer method. A—The shielding gas envelops the weld area. B—Droplets form at the tip of the electrode without any necking down of the wire. C—Droplets form at a constant rate and are transferred to the weld pool.

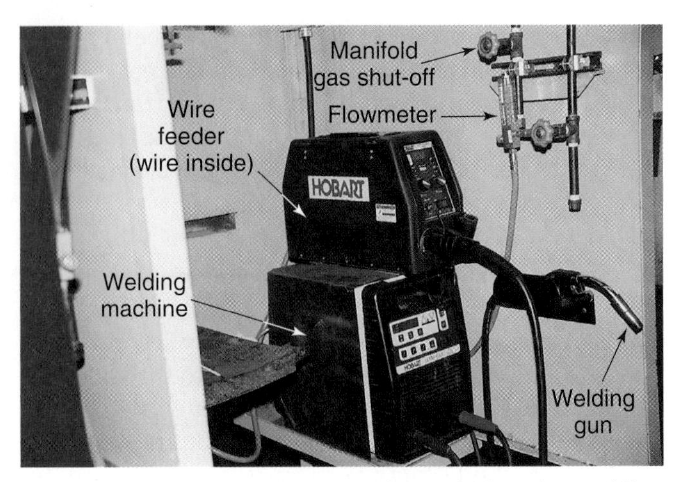

Figure 11-9. The GMAW outfit includes the welding machine, wire feeder, welding gun, electrode wire, and shielding gas supply with flowmeter. The electrode wire is in the wire feeder and the shielding gas is supplied through pipes to the welding booth.

GMAW Power Supply

A constant voltage, or ***constant potential,*** power supply is used for gas metal arc welding. This type of power supply is different from one used for shielded metal arc or gas tungsten arc welding (GTAW). A constant current welding machine is used for SMAW or GTAW. The constant voltage power supply can be a ***transformer-rectifier*** or an ***inverter.*** An inverter welding machine is much smaller and lighter weight than a transformer-rectifier machine. Some welding machines have a built-in microprocessor. A ***microprocessor*** can be considered a small computer. These microprocessor machines can store welding conditions.

A constant voltage power supply has a flat volt-amp curve or flat slope, **Figure 11-10**. The voltage stays about the same, even though the current varies. The relationship between voltage and current is the slope of the power supply. Remember that a constant current power supply has a steep or drooping volt-amp curve.

A constant voltage power supply tries to maintain the preset voltage while welding. The power supply tries to maintain the same arc voltage and arc length, even though the welder may try to change it. The welding machine makes the necessary adjustments very quickly to compensate for any changes that the welder makes.

A welding machine for GMAW is considered to be self-adjusting, since it automatically makes adjustments. If the welding gun is *moved closer* to the base metal while welding, the arc length shortens and the voltage decreases. The power supply will automatically increase the current so the electrode melts faster. This, in turn, increases the arc length. When the electrode burns back to the original arc length and original arc voltage, then the current returns to its original setting or value.

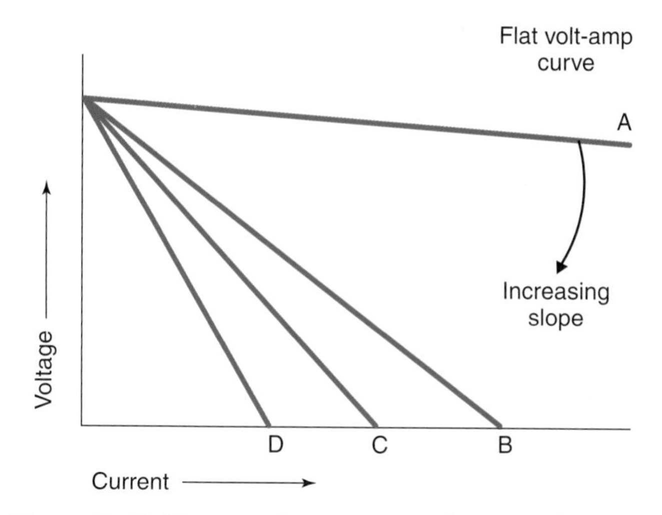

Figure 11-10. These are four power supply curves for a constant voltage welding machine. Increased slope reduces the maximum current during short circuiting transfer. With a flat volt-amp curve, the short circuiting amperage is almost limitless. As the slope of curves of B, C, and D increase, their maximum short circuiting amperage decreases.

If the welding gun is *pulled* away from the base metal, both the arc length and voltage increase. The power supply automatically reduces the current. Less metal will melt off the electrode. This reduces the arc length. Since the electrode now melts more slowly, the arc becomes shorter and the arc voltage decreases. When the former arc length and voltage are obtained, the power supply will return the current to its original value.

A crank or knob is used to adjust the voltage on a GMAW machine. See **Figure 11-11**. There is no adjustment for current. Current is controlled by adjusting the wire feed speed. A faster wire feed speed requires the electrode to melt faster. This requires higher current. Thus, increasing the wire feed speed increases the welding current.

On some machines, the welder can adjust the amount of inductance. Inductance controls how fast the welding machine increases the welding current. Increasing inductance helps to reduce spatter during short circuiting transfer. A certain amount of inductance provides for a better arc start during spray transfer. On some machines, the slope of the volt-amp curve can be changed. ***Slope adjustment*** is used to change the maximum short circuit current. The maximum current decreases as the slope is increased. See Figure 11-10. An increased slope reduces spatter during short circuiting transfer.

A welding machine also includes connections for the wire feeder, shielding gas, water (when used), and ground cable. Some machines have a built-in wire feeder, so they do not have wire feeder connections. See **Figure 11-12**. They do, however, have a place to attach a welding gun. Some machines have a voltmeter and an ammeter. Machines used for pulsed spray transfer have additional adjustments to control the peak current and background current.

GMAW or FCAW machines have a 60% or 100% duty cycle. Most machines used in schools and industry

Figure 11-11. When making GMAW machine adjustments, a knob or crank is used to adjust the required voltage. The wire speed adjustment automatically adjusts the amperage. (Hobart Brothers Co.)

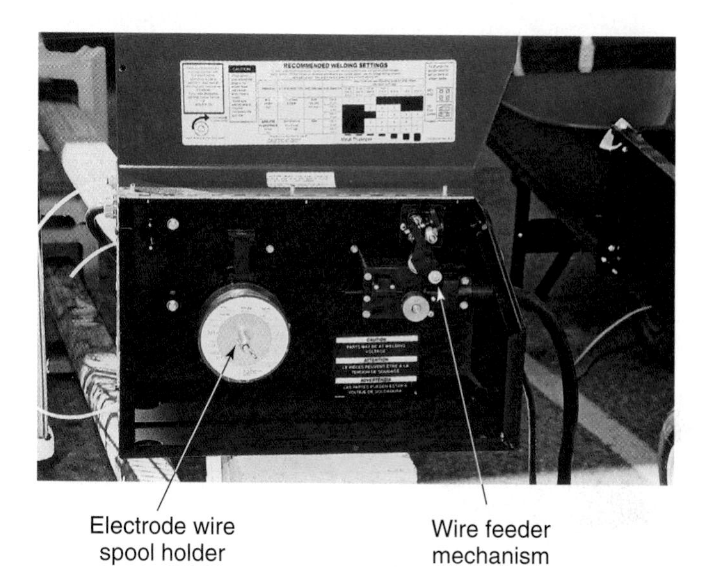

Figure 11-12. The wire feeder and electrode wire spool holder are built into this GMAW machine.

are 240V single-phase or 240V three-phase machines. Some small, light-duty welding machines used in auto body repair are 120V machines.

Wire Feeders

A ***wire feeder*** is used to feed electrode wire to the welding gun. GMAW and FCAW electrode wire is available in coils or spools. The coils or spools generally contain several hundred feet of wire. The wire feeder smoothly pulls the wire from the spool and pushes it to the attached welding gun.

Wire feed speed is set on the wire feeder by the welder. A dc motor attached to the knob controls the speed of the drive rolls. A wire feeder has two or four drive rolls that contact the wire. The ***drive rolls***, or drive wheels, push the wire to the gun at a preset rate. The amount of pressure that each pair of rolls or wheels applies to the wire is adjustable. A ***wire tension control knob*** or screw is used to adjust the pressure. **Figure 11-13** shows a wire feeder with two drive rolls.

On a wire feeder that has two drive rolls, only one drive roll is driven. On a wire feeder with four rolls, two rolls are driven. A center drive gear engages two drive rolls, which in turn drive the remaining two rolls. The roll or drive gear is powered by a dc motor.

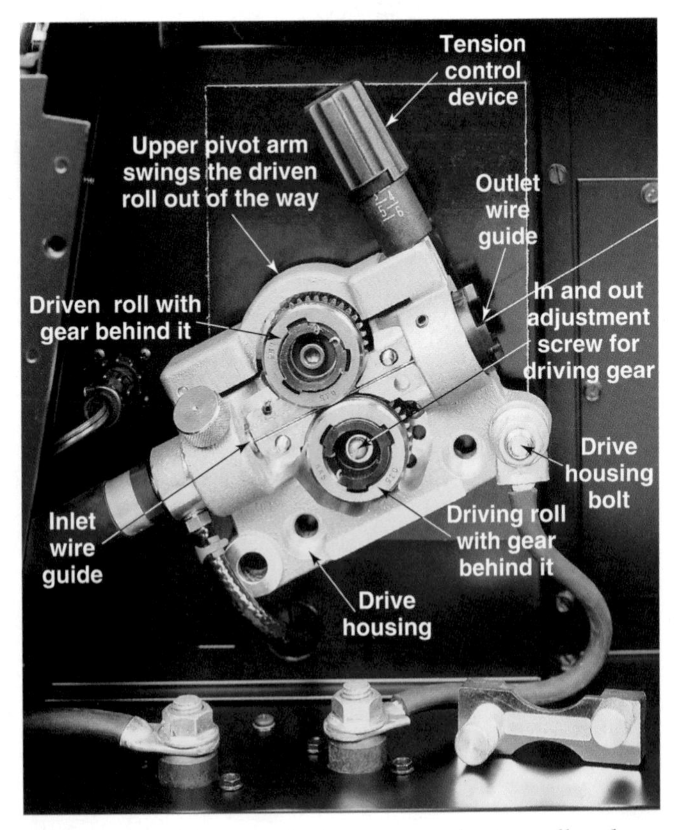

Figure 11-13. This wire feeder uses one driving roll and one driven roll. The upper drive roll can be pivoted out of the way when the pressure adjusting device is completely loosened. (Lincoln Electric Company)

Two other controls commonly found on a wire feeder are an inch switch and a purge switch. Pressing the ***inch switch*** causes the electrode to feed out. The electrode will feed slowly as long as the inch switch is pressed. It is often used when loading a new coil or spool of wire. No welding current flows when the inch switch is pressed. The ***purge switch*** is used to manually control the flow of shielding gas. When the purge switch is pressed, shielding gas will flow. It is used to set the flow rate of the shielding gas and to purge any air in the hoses. See **Figure 11-14**.

Welding Guns and Cables

A GMAW or FCAW gun must perform the following functions:

- Make electrical contact with the electrode.
- Direct the shielding gas to the workpiece (not used in all FCAW).
- Remove heat from the gun.
- Have a switch to start and stop the welding operation.

Electrical contact is made between the welding gun and an electrode by a ***contact tube***. A contact tube made of copper alloy is screwed into the welding gun. See **Figure 11-15**. The electrode wire passes through the contact tube and makes electrical contact.

The hole in the center of the contact tube becomes worn as the electrode passes through it. When the hole becomes larger, it does not make good electrical contact with the electrode. Contact tubes must be replaced when they become worn.

A ***gas nozzle*** is used to direct the shielding gas to the weld area. They are usually made of metal and come in various shapes and diameters. A gas nozzle is

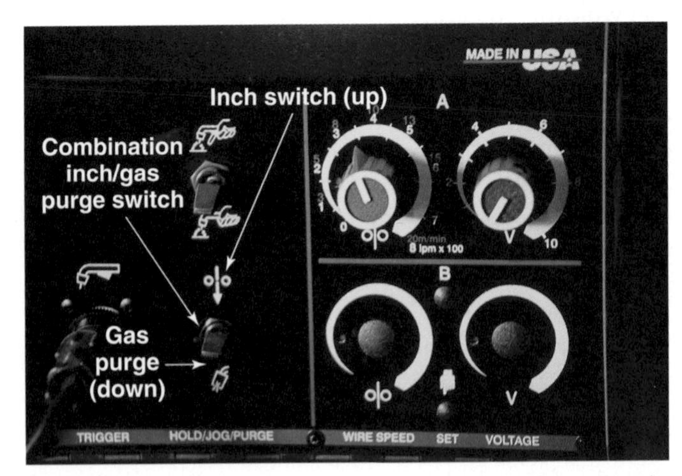

Figure 11-14. The inch or wire jog switch is used to move the electrode wire slowly through the cable to the welding gun. The gas purge switch is used to allow shielding gas to flow through the hose to remove any trapped air.

Figure 11-15. Here are several different sizes of copper contact tubes.

used for GMAW and for FCAW when shielding gas is used. When a shielding gas is not used with FCAW, the gas nozzle is not required. A welding gun with the gas nozzle removed is shown in **Figure 11-16**. Often, the term "nozzle" is used in place of the term "gas nozzle."

GMAW guns can be cooled by using shielding gas or water. Shielding gas-cooled guns are usually limited to 200A (amps). These guns can be up to 600A when CO_2 is used as a shielding gas. A gas-cooled gun may have a combination cable assembly or a cable assembly and a hose. A combination cable assembly on a gun carries the electricity, electrode, and shielding gas. See **Figure 11-17**. For guns that use a cable assembly and a hose, the hose carries the shielding gas. The cable assembly carries the electricity and the electrode.

A water-cooled gun has two hoses and a cable assembly. One hose carries the water to the gun, and the other hose carries the shielding gas. The cable assembly carries the electricity and electrode to the gun and carries cooling water away from the gun. See **Figure 11-18**.

The cable assembly that the electrode wire travels through has a liner in it. See **Figure 11-19**. The ***liner*** keeps the electrode traveling smoothly through the cable. A liner is generally a flexible tube made of coiled wire. When welding with soft electrodes, such as aluminum, a ***Teflon® liner*** is used.

All GMAW and FCAW guns have switches to start and stop the welding operation. When the

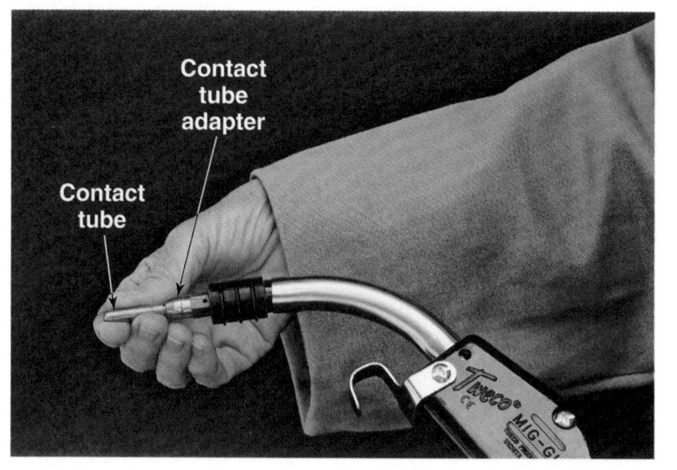

Figure 11-16. This GMAW gun has its gas nozzle removed. The contact tube screws into the contact tube adapter.

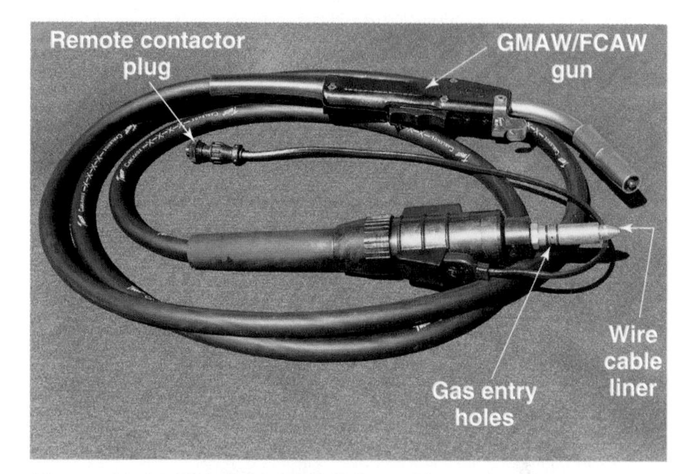

Figure 11-17. This is a combination cable used with a gas-cooled GMAW gun. A combination cable carries the electrode wire, shielding gas, and the remote contactor control cable.

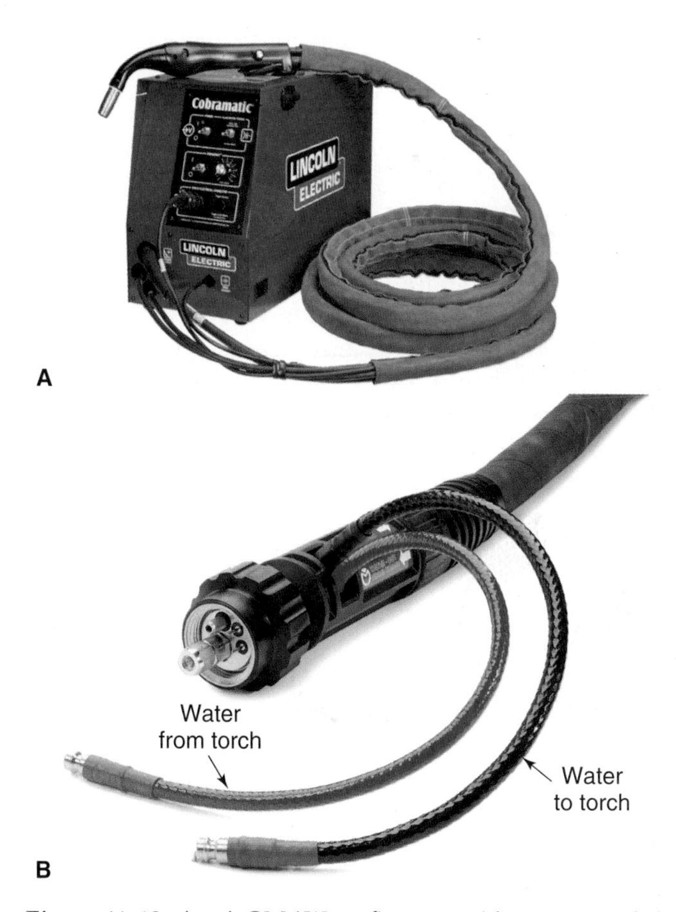

Figure 11-18. A—A GMAW outfit setup with a water-cooled gun. B—A close up of the end of a combination cable for one water cooled gun. (Lincoln Electric Company)

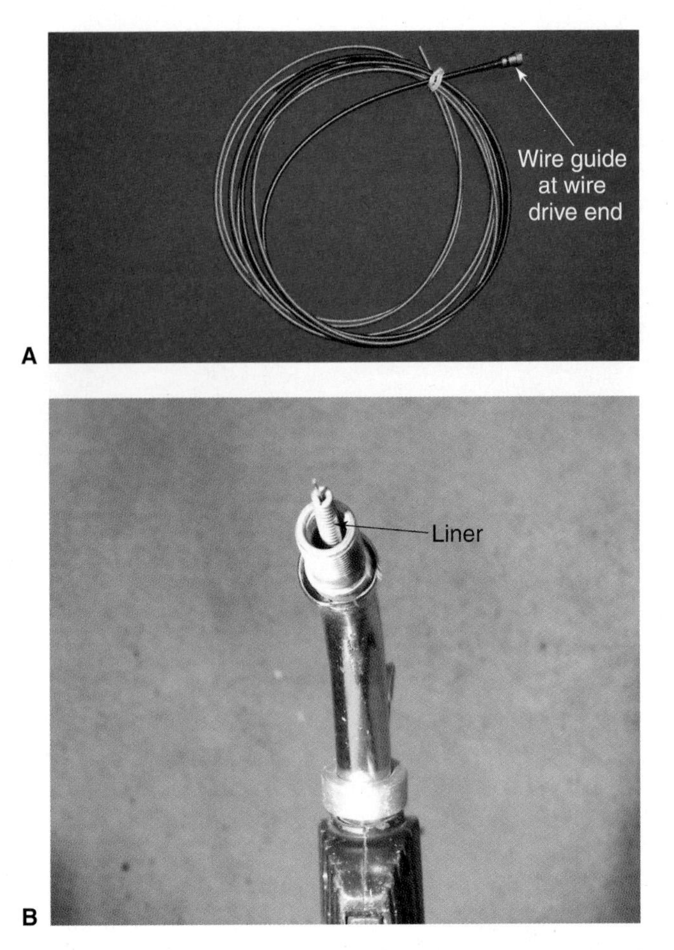

Figure 11-19. *A cable liner is used to carry the electrode through the combination cable to the welding gun. A—This is the cable liner with a wire guide on one end. The wire guide fits into the wire feeder. B—This is the cable liner within the welding gun.*

switch is pressed, the electrode wire begins to feed, and shielding gas and electrical current are also turned on. When the switch is released, the wire and current stop immediately. Shielding gas may continue to flow for a short time to cool the gun and continue shielding the weld and weld area. Water flows continuously in a water-cooled gun.

Welding guns are available in a variety of shapes, as shown in Figures 11-16, 11-17, 11-18, and **Figure 11-20**. Welding guns that are commonly used for steel electrodes and large diameter electrodes have the electrode pushed through them by the wire feeder. However, the electrode may not travel smoothly through the electrode cable when welding a long distance from the wire feeder or when using a small soft electrode, such as aluminum. The electrode wire may get bent and jam the machine. A pull-type gun is used to prevent this. A pull-type gun has a motor that pulls the wire through the cable and prevents the electrode from being bent. A wire feeder is used to push the

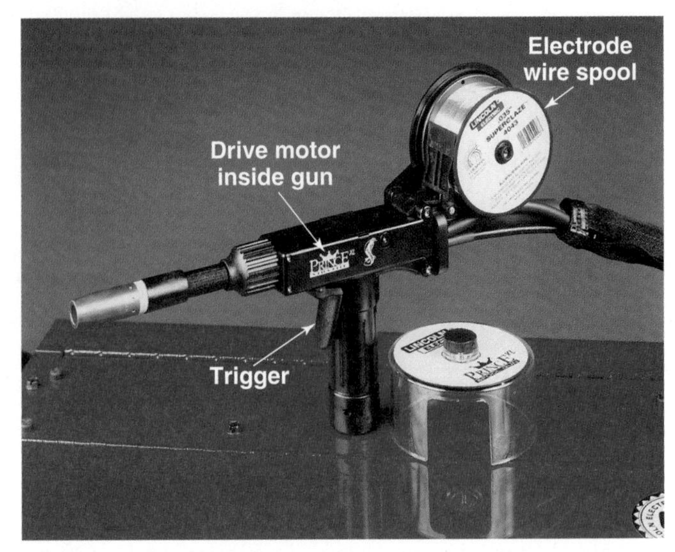

Figure 11-20. *The wire drive motor and a small electrode spool are built onto this GMAW gun. (Lincoln Electric Company)*

electrode while the welding gun pulls the electrode. This is a ***push-pull system***.

Some welding guns allow for a small spool of wire to be placed in the gun. That type of gun has a motor in it to feed the electrode. See Figure 11-20. It is almost impossible to kink or jam the wire with this type of gun. The gun can be easier to handle, but it does weigh more than a standard gun. This is due to the weight of the motor and electrode wire.

Shielding Gases

Shielding gases used for GMAW are primarily argon and carbon dioxide. Mixtures of these gases, along with the addition of helium and oxygen, may also be used. ***Argon (Ar)*** and ***helium (He)*** are inert gases. ***Inert gases*** do not combine with the weld metal. ***Carbon dioxide (CO_2)*** and oxygen (O_2) are active gases. ***Active gases*** combine with the weld metal if given the chance. If carbon dioxide or oxygen enters the weld, the weld will become oxidized. Precautions are taken to prevent oxidation. Carbon dioxide and oxygen are only used when welding on ferrous metals (various steels, including stainless). The type of shielding gas selected for a job affects the following:

- Type of metal transfer.
- Penetration and shape of weld bead.
- Speed of welding.
- Mechanical properties of the weld.

Argon is the most common shielding gas for GMAW. It is used to weld almost all types of metals. Helium, oxygen, or carbon dioxide may be added. When argon is used in spray transfer to weld steel, the weld has very deep penetration in the center. The edges, however, are undercut. Oxygen or carbon dioxide is added to eliminate undercutting. The penetration is reduced, and there is very little spatter.

Typical shielding gas mixtures consist of argon with 1%–5% oxygen or argon with 3%–50% carbon dioxide. Remember, spray transfer requires at least 90% argon. **Figure 11-21** shows the amount of penetration obtained using various shielding gases. See Chapter 12 (Figures 12-11 and 12-12) for a list of the shielding gases that are recommended for welding on different base metals.

Carbon dioxide can be used as a shielding gas alone without any argon or helium; however, it is only used to weld mild steel and low-alloy steel. Carbon dioxide provides more heat to the base metal and gives very good penetration. This allows a higher travel speed while welding. Another advantage of using 100% carbon dioxide is that this gas is cheaper than argon. When using 100% carbon dioxide, only short circuiting or globular transfer can be used. Globular transfer can cause a lot of spatter on the base metal. It is important to select a proper electrode when using carbon dioxide as the shielding gas.

Argon and carbon dioxide (CO_2) displace oxygen. Excellent ventilation is required to remove the shielding gases and supply fresh air when welding in a closed area. Suffocation can result if adequate ventilation is not provided.

Carbon dioxide is the most common shielding gas used for FCAW. It provides globular transfer of the electrode. Carbon dioxide is inexpensive and creates a weld with deep penetration. A common gas mixture for FCAW is 75% argon and 25% carbon dioxide, which allows metal transfer that approaches spray transfer. This shielding gas mixture produces a weld with better strength than one made with 100% carbon dioxide. The shielding gas that is selected is often determined by the type of electrode being used. Remember that not all FCAW electrodes require a shielding gas.

Regulators and Flowmeters

Shielding gases are available in cylinders similar to those used for fuel gases. The gases in the cylinders are at very high pressures. **Care must be taken when handling or moving any gas cylinder. The cylinder valve must be protected by a properly installed safety cap when moving a cylinder or whenever the cylinder is not in use.**

Secure a cylinder to a hand truck when moving it. If a hand truck is not available, tilt the cylinder and roll it along on the bottom edge. Use one hand to support the cylinder. Use the other hand to rotate the cylinder. Cylinders should be stored in the upright position. Each cylinder must be secured to a wall, column, or hand truck using chains or straps.

Shielding gases in cylinders have pressures that may exceed 2000 psi (13.79 MPa). A regulator is used to reduce the cylinder pressure to a much lower working pressure. A regulator used for oxyfuel gas welding often has a high-pressure and a low-pressure gauge. These gauges indicate the cylinder pressure of the gas and the working pressure.

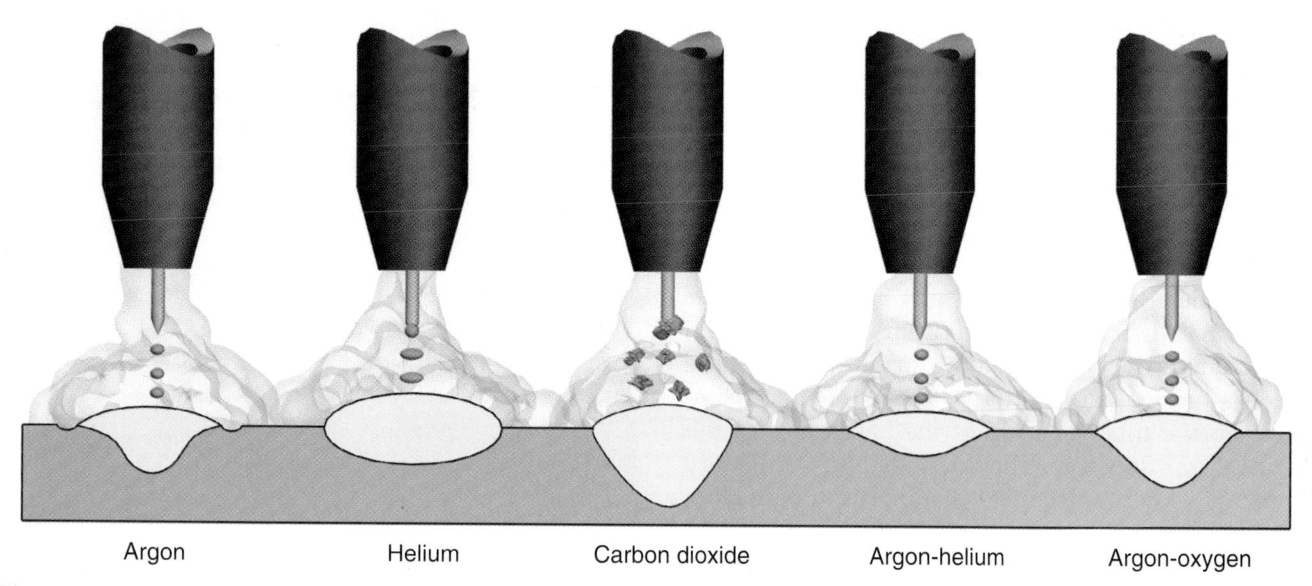

Figure 11-21. *Notice the different GMAW bead shapes and penetration patterns with various shielding gases and gas combinations. These shapes are typical for welds made with DCEP (DCRP). Notice the deep penetration and undercut when using pure argon.*

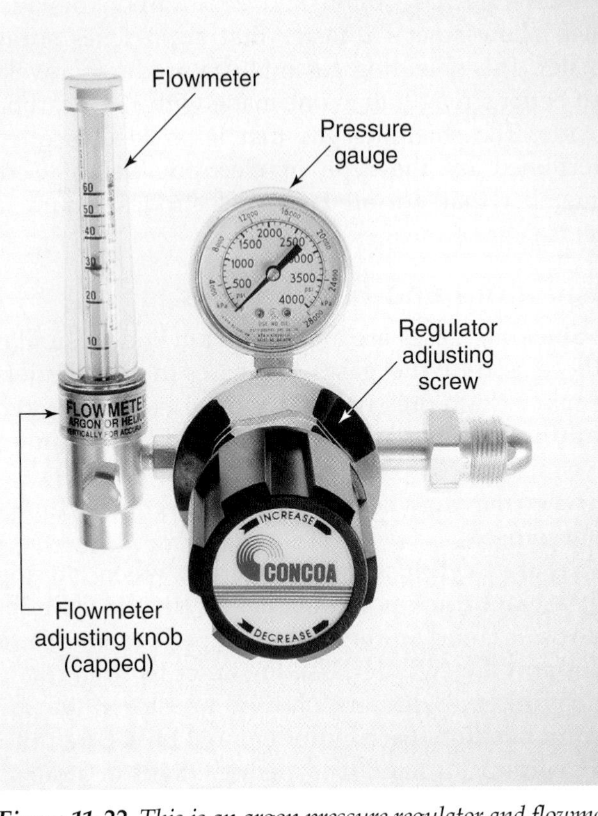

Figure 11-22. This is an argon pressure regulator and flowmeter. Notice that the flowmeter adjusting screw is preset and capped so that the setting cannot be changed. (CONCOA)

When using a shielding gas, it is important to control the flow of gas rather than the pressure of the gas. A ***flowmeter*** is used to control the amount of gas that goes to the welding gun. A combination regulator and flowmeter is used to regulate the pressure *and* control the gas flow. See **Figure 11-22**. The regulator reduces the pressure of the shielding gas to 50 psi (345 kPa). A high-pressure gauge may be used, but it is not required. A knob or adjusting screw on the flowmeter controls the flow of shielding gas.

One type of flowmeter has a ball enclosed in a clear plastic or glass tube. The tube is calibrated to indicate the flow of shielding gas in cubic feet per hour (cfh) or liters per minute (Lpm). Gas flows into the tube from the bottom. As the gas flows, it lifts the ball. A larger gas flow is indicated by the ball rising higher in the tube. The top of the ball indicates the shielding gas flow rate. The flow rate is controlled by adjusting a knob on the flowmeter. The correct flow rate for each welding job varies and is covered in Chapter 13.

A different type of flowmeter must be used for each type of shielding gas. An argon flowmeter is calibrated differently from one for helium. In some school and industrial settings, the shielding gas is piped to the welding area using a manifold system. In such a situation, the pressure of the shielding gas has already been reduced to working pressure so that the welding station does not need a regulator, but it does require its own flowmeter.

Protective Clothing and Equipment

The same type of protective clothing should be worn for GMAW as is required for SMAW. Fire-resistant or leather clothing must be worn. Your pant legs should not have cuffs. Pockets on your shirt should be buttoned to prevent spatter from getting caught in them. A #12 lens is recommended for your welding helmet. A hat and leathers must be worn when welding out of position. Leather gloves capable

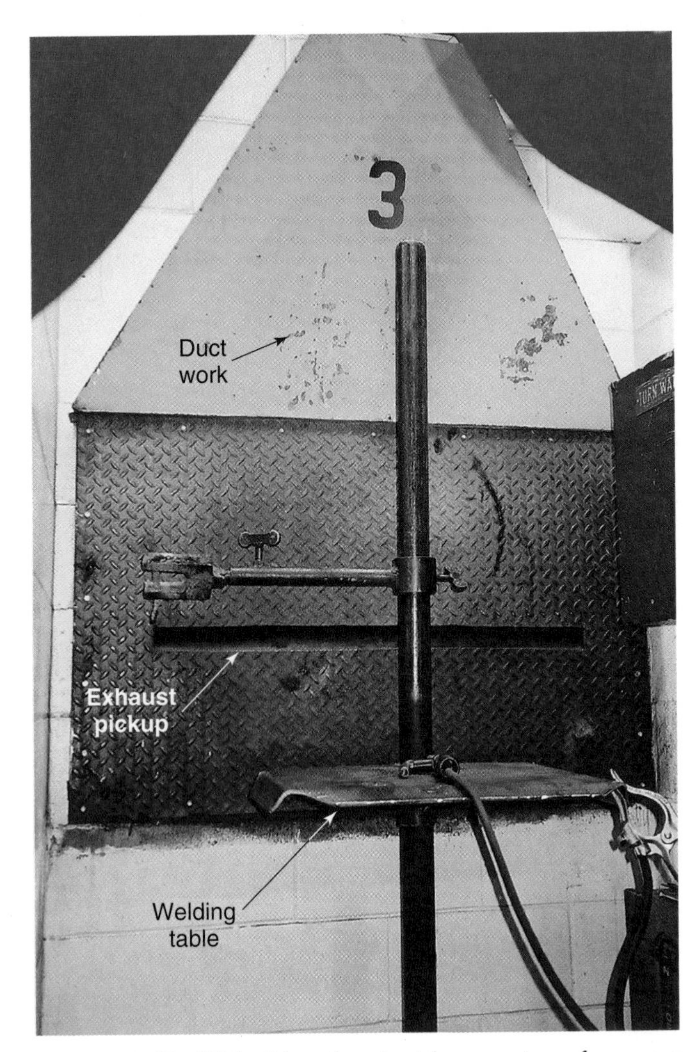

Figure 11-23. With this exhaust pick-up system, fumes are withdrawn through the long rectangular opening above the workbench.

of withstanding high temperatures should be worn for GMAW or FCAW.

A small amount of fumes is generated by the GMAW operation. A large amount of fumes is created when performing FCAW. **When welding in an enclosed area, some type of fume exhaust should be used. See Figure 11-23. A fume exhaust should be positioned to capture and remove as much of the fumes as possible. It should be placed over and slightly to the rear of the welding area. The exhaust should pull the fumes away from your face.** **Figure 11-24** shows a type of fume remover that is attached to the welding gun.

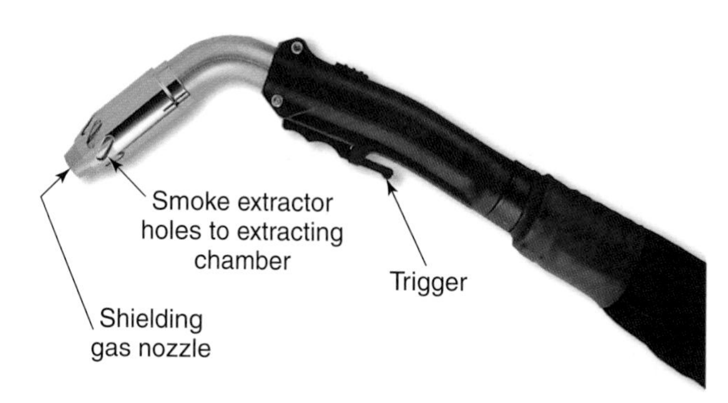

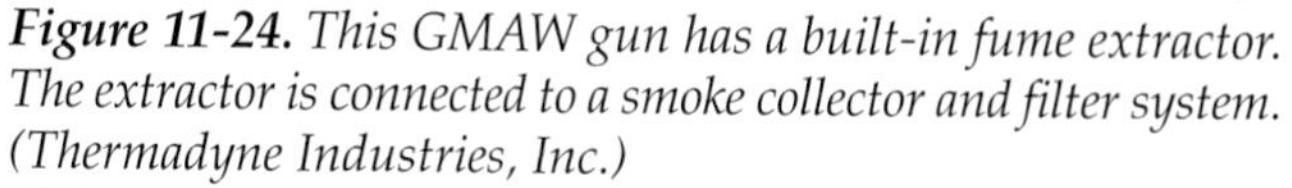
Figure 11-24. *This GMAW gun has a built-in fume extractor. The extractor is connected to a smoke collector and filter system. (Thermadyne Industries, Inc.)*

Summary

- Gas metal arc welding (GMAW) is a welding process in which metals are joined by heating them with a welding arc between a continuous consumable electrode and the base metal. A shielding gas or gas mixture is used to prevent the atmosphere from contaminating the weld.
- Direct current electrode positive (DCEP), or direct current reverse polarity (DCRP), is used for GMAW. Direct current electrode negative (DCEN), or direct current straight polarity (DCSP), is seldom used. Alternating current is not used for GMAW or FCAW.
- Flux cored arc welding is a welding process in which metals are joined by heating them with an electric arc between the base metal and a continuous consumable electrode. A main difference between FCAW and GMAW is the FCAW electrode contains flux in its center, or core.
- Metal transfer in GMAW can occur in three different ways: short circuiting transfer, globular transfer, and spray transfer. In short circuiting transfer, the electrode contacts the base metal and some molten metal is transferred. In globular transfer and spray transfer, the wire does not touch the base metal.
- A complete GMAW outfit includes a welding machine, a wire feeder, a welding gun, electrode wire, and a shielding gas supply and controls.
- A constant voltage, or constant potential, power supply is used for gas metal arc welding. A constant voltage power supply tries to maintain the preset voltage while welding. The power supply tries to maintain the same arc voltage and arc length. A welding machine for GMAW is considered to be self-adjusting, since it automatically makes adjustments.
- A wire feeder is used to feed electrode wire to the welding gun. GMAW and FCAW electrode wire is available in coils or spools. The coils or spools generally contain several hundred feet of wire. The wire feeder smoothly pulls the wire from the spool and pushes it to the attached welding gun.
- A GMAW or FCAW gun must make electrical contact with the electrode wire, direct shielding gas to the workpiece, have a means for removing heat from the gun, and have a switch to start and stop the welding operation.
- Shielding gases used for GMAW are primarily argon and carbon dioxide. Mixtures of these gases, along with the addition of helium and oxygen, may also be used. A flowmeter is used to control the amount of gas that goes to the welding gun. A combination regulator and flowmeter is used to regulate the pressure *and* control the gas flow.
- **The same type of protective clothing should be worn for GMAW as is required for SMAW.**

Review Questions

Write your answers on a separate sheet of paper. Please do *not* write in this book.

1. What does GMAW stand for? Can the terms GMAW and MIG be interchanged? Explain.
2. In gas metal arc welding, the arc is struck between the _____ and the _____.
3. *True or False?* Flux and shielding gas are used to protect the weld from the atmosphere during GMAW.
4. What type of welding machine is used for GMAW, a constant voltage machine or a constant current machine?
5. List four advantages and two disadvantages of GMAW as compared to SMAW.
6. List the equipment that makes up a complete GMAW outfit.
7. In short circuiting transfer, when does the metal from the electrode enter the weld?
8. In globular transfer, what forces the molten metal to leave the electrode?
9. Spray transfer requires the use of shielding gas that contains at least _____% argon.
10. During pulsed spray transfer, molten metal is transferred to the base metal during the _____ current period. This current is greater than the _____ current.
11. Which GMAW metal transfer methods are recommended for out-of-position welding?
12. *True or False?* The volt-amp curve for a constant potential machine is a flat curve.
13. Will the current of the welding machine increase or decrease if the arc voltage increases while welding?
14. *True or False?* If metal explodes from the end of the electrode, the inductance should be increased.
15. What part of a wire feeder is used to pull the electrode wire off the spool and push it to the gun?
16. Which of the following is not a function of a GMAW gun?
    A. Make electrical contact.
    B. Direct the shielding gas.
    C. Maintain the correct voltage.
    D. Control the welding with a switch.
17. In a water-cooled welding gun, what do the two hoses carry? What does the cable assembly carry?
18. List three shielding gases commonly used in GMAW.
19. List two gases that are added to argon to eliminate the undercutting that occurs when pure argon is used to weld steel.
20. Why is a flowmeter used for GMAW and not a low-pressure gauge?

Chapter 12

GMAW: Equipment Assembly and Adjustment

Learning Objectives

After studying this chapter, you will be able to:

- Assemble a GMAW welding outfit.
- Adjust the drive mechanism for the proper pressure and alignment.
- List the proper sequence for removing a bird's nest.
- Adjust the shielding gas flowmeter for the proper pressure and flow rate.
- Identify the electrode wire designation for GMAW electrodes.
- Identify the two adjustments that are made to the welding machine for GMAW.
- Identify safety precautions for GMAW and FCAW.

bird's nest
dash number
open circuit voltage
post flow adjustment
pressure roll
remote contactor control

Assembly and Setup

GMAW or FCAW outfits must be assembled properly to operate properly. Shielding gas must be connected to the welding machine, and the gas flow must be set correctly. The wire feeder must be connected to the machine. It should be adjusted so the electrode wire feeds properly to the welding gun. Also, the welding gun and its parts must be set up properly.

Cylinder, Regulator, and Flowmeter

Shielding gas is stored in cylinders at very high pressures. Care must be taken when handling cylinders so they are not dropped or damaged. A cylinder must always be secured to a wall, column, or hand truck. A safety cap should be secured over the cylinder valve when the cylinder is not in use or when transporting a cylinder. See Chapter 20 for additional information regarding the handling of cylinders.

The outlet on the cylinder must be cleaned before attaching a regulator to it. This is done by quickly opening and closing the cylinder valve. **The cylinder valve must not be pointed at any workers in the area.** The pressure of the gas escaping from the cylinder cleans the outlet. The regulator is then attached to the cylinder. Use an open-end wrench to secure the regulator nut so it is tight and leakproof. However, do *not* overtighten the nut.

A flowmeter is used for GMAW. A ball-float flowmeter must be installed in a vertical position to operate properly. See **Figure 12-1**.

Shielding gas can also be supplied through a manifold system. Shielding gas cylinders are stored in a different area from the welding area. Pressure from the cylinders is reduced by a regulator. Delivery pipes carry the low-pressure shielding gas to the welding area. Each welding station requires a flowmeter so that each welder can set the desired gas flow rate.

Welding Machine

A constant voltage welding machine is used for GMAW. The machine is either plugged into an electrical outlet or hardwired (permanently wired) into the electrical source.

Shielding gas is connected to either the welding machine or the wire feeder. Connect the hose coming from the flowmeter to the proper place on the welding machine or wire feeder.

When a separate wire feeder is used, it is usually placed on top of the welding machine. The wire feeder is then connected to the welding machine. Connect the wire feeder to the ***remote contactor control*** on the welding machine, using the correct cable. See **Figure 12-2**. The wire feeder requires 120V or 240VAC or 24VDC for operation. Connect the power cord to the front of the welding machine or to an electrical outlet with the required voltage.

The welding gun is then connected to the wire feeder. The gun may be connected to the welding machine if the wire feeder is inside of the welding machine. If the welding gun has a combination hose, connect it to the welding machine. If there is an additional shielding gas hose, connect it to its proper fitting on the welding machine or wire feeder. Other welding guns have multiple connections. Be careful to connect all hoses and cables properly. See **Figure 12-3**.

Wire Feeder

A wire feeder should pull the electrode wire smoothly from the spool and push it to the welding gun. Several adjustments must be made on the wire feeder to ensure proper operation.

First, select the type of electrode wire to be used. Mount the spool of electrode wire on the wire spool hub. Secure the spool on the hub so it does not fall off during operation. A lock pin is generally used to keep the spool in place. Next, check that the grooves in the drive rolls are the correct size for the diameter of the electrode wire. Some drive rolls have two grooves. See **Figure 12-4**. Install the proper size drive rolls so that the correct size groove aligns with the electrode wire.

Swing the ***pressure roll*** (top drive roll) up out of the way. Push the end of the electrode wire through the inlet wire guide directly behind the drive rolls. Feed the wire into the grooves of the drive roll and through the outlet wire guide directly in front of the drive roll. Swing the pressure roll back into position to complete the assembly.

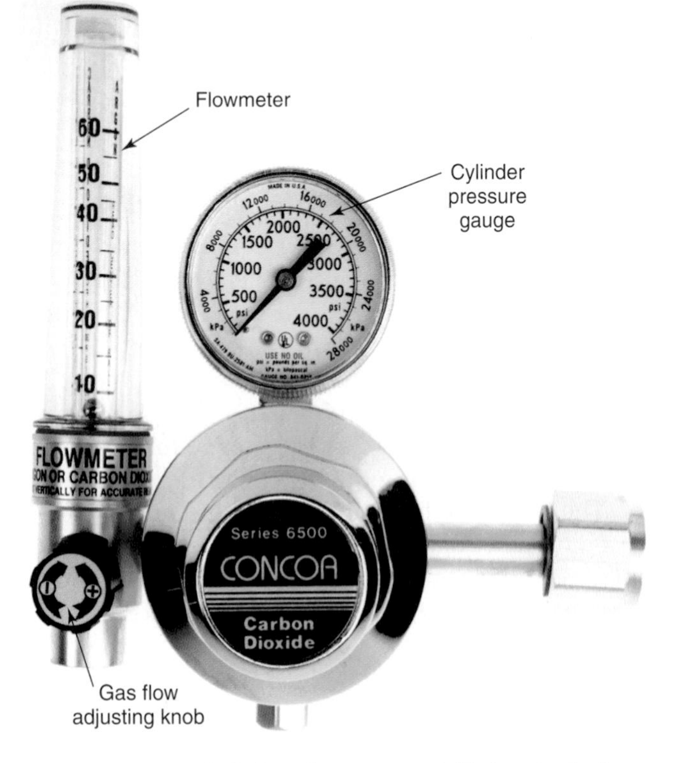

Figure 12-1. *In a ball-float flowmeter, a ball rises in the inner tube of the flowmeter when gas is flowing to indicate the flow in cubic feet per hour. (CONCOA)*

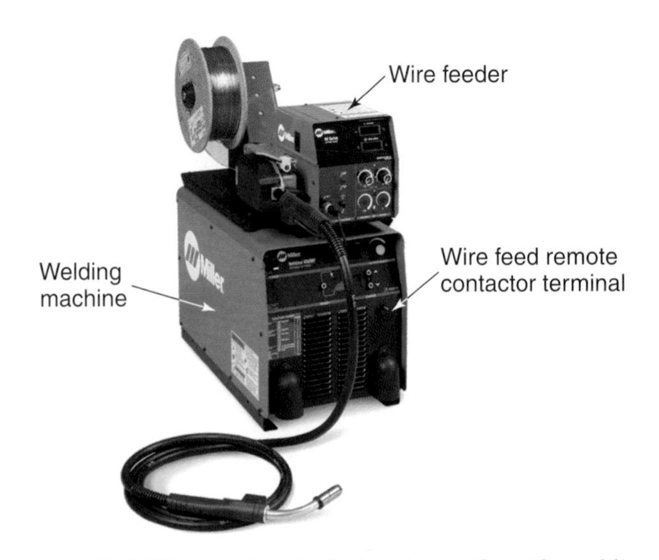

Figure 12-2. *The remote wire feed contactor from the welding gun is plugged into the remote contactor terminal on the machine. (Miller Electric Mfg. Co.)*

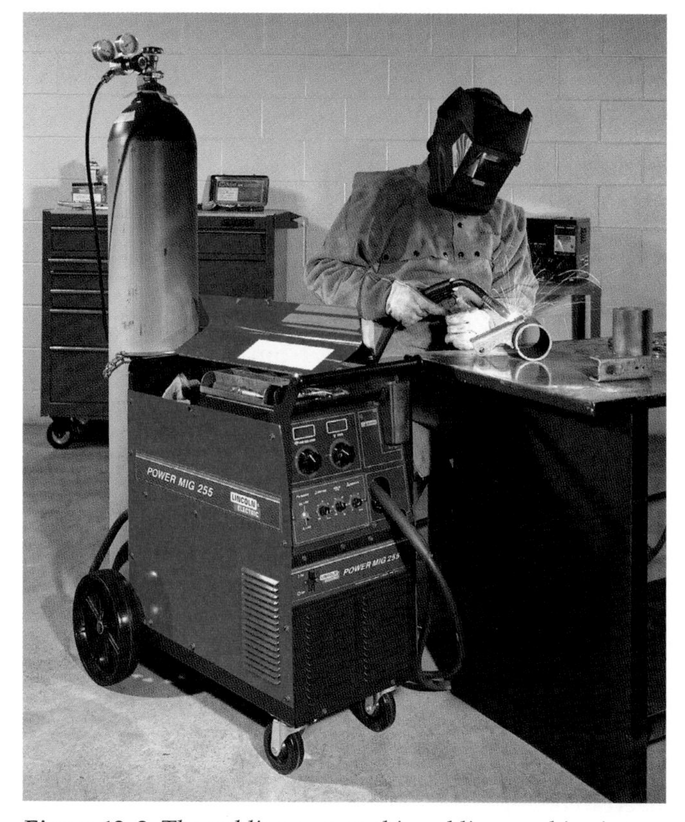

Figure 12-3. The welding gun on this welding machine is connected to the wire feeder through the front of the machine. The shielding gas and electrical leads are connected at the back of the machine. A drawing on the inside of one of the machine covers shows the location of the gas and electrical connections. (Lincoln Electric Company)

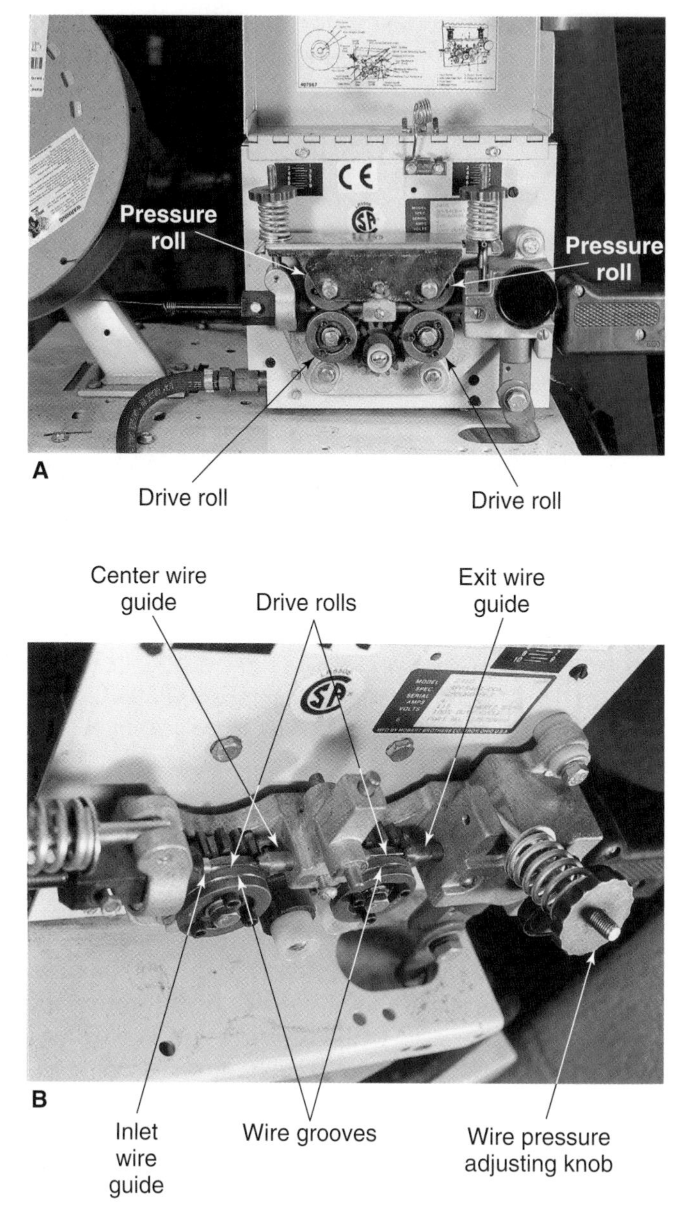

Figure 12-4. A—On this four roll wire feed machine, the electrode wire on the spool is fed through the wire guides as wire enters and exits the wire feeder. The two upper rolls on this wire feeder may be removed to more easily feed new wire through the wire feeder. B—Two wire grooves are cut into the pressure and drive rolls. Each groove may be the same or may be a different size. When the grooves are worn, the rolls may be reversed or replaced to obtain a new set of grooves. Notice that the two upper rolls have been removed to allow easy access to the wire and guides.

A wire feeder that has four rolls is assembled and connected in the same way as a two-roll wire feeder. One difference is that there is a third guide tube between the two sets of rolls. The electrode wire must pass through this guide tube. Another difference is the drive rolls are turned by a main drive gear located between them. See **Figure 12-5**.

After the electrode wire has been inserted, look closely at where it passes between the drive and pressure rolls. The wire should travel in a straight line between the inlet wire guide to the outlet wire guide, **Figure 12-6**. If the drive rolls and guide tubes are not properly adjusted, the wire will not feed smoothly. Adjust the drive housing if the wire bends up or down as it passes between the drive rolls. The grooves in the drive and pressure rolls must also be properly aligned. The drive roll on most wire feeders is adjustable. See **Figure 12-7**.

Once the rolls have been properly adjusted and aligned, the correct amount of pressure must be applied to the electrode wire by the drive and pressure rolls. If not enough pressure is applied, the drive rolls will slip and the electrode wire will not be fed smoothly to the gun. If excessive pressure is applied, the rolls will flatten the electrode wire. The knob(s) on top of the drive housing is used to adjust the pressure exerted on the electrode wire. Refer to Figures 12-4 and 12-5.

When all adjustments have been made to the wire feeder, pull the trigger on the welding gun or press the inch switch on the welding machine or wire feeder. Stop feeding electrode wire when it is visible at the end of the gun. Electrode wire will feed from the wire feeder to the gun whenever the trigger is pressed.

Figure 12-5. Two drive rolls and two pressure rolls are used in this wire drive mechanism. The center drive gear moves the drive rolls. (Lincoln Electric Company)

Removing a Bird's Nest

Occasionally, the electrode wire in the wire feeder will become tangled and produce a ***bird's nest***. See **Figure 12-8**. A bird's nest commonly results from improper feeding as the wire feeder tries to push the wire through the cable to the welding gun. A blockage in the cable, misaligned guide tubes and rolls, or stubbing the electrode wire on the base metal causes the electrode wire to bunch up in the wire feeder. Stubbing the electrode on the base metal is usually a result of one of the following:

- Voltage setting on the machine is too low.
- Wire feed speed is too fast.
- Welding gun is held too close to the work when starting an arc.

A bird's nest is removed by raising the pressure roll and cutting the wire on both sides of it. See **Figure 12-9**. **Put your hand over the wire as you cut to prevent it from flying and causing injury.** Pull all the electrode wire out of the cable connected to the gun. Check the wire coming from the spool to make sure it is not bent or twisted. Remove any bent or twisted wire.

Feed new electrode wire through the wire guides and into the cable connected to the welding gun. Secure the pressure roll into position. Pull the gun trigger or push the inch switch to feed the electrode wire to the gun. When the wire comes out of the contact tube, you will be ready to begin welding again.

Welding Gun

The welding gun and cable assembly have many parts, as discussed in Chapter 11. Some parts may need to be replaced or cleaned due to wear and buildup of debris. The contact tube will need to be changed most often. Select the correct size contact tube for the diameter electrode you will use. The contact tube is threaded into the end of the welding gun. The hole in the center of the contact tube will get larger due to wear. Eight hours of constant welding with a steel electrode will wear a contact tube so that it will need to be replaced. **Figure 12-10** shows a worn contact tube and several usable contact tubes.

After installing the proper contact tube, place a nozzle on the end of the gun. The position of the nozzle may be adjustable.

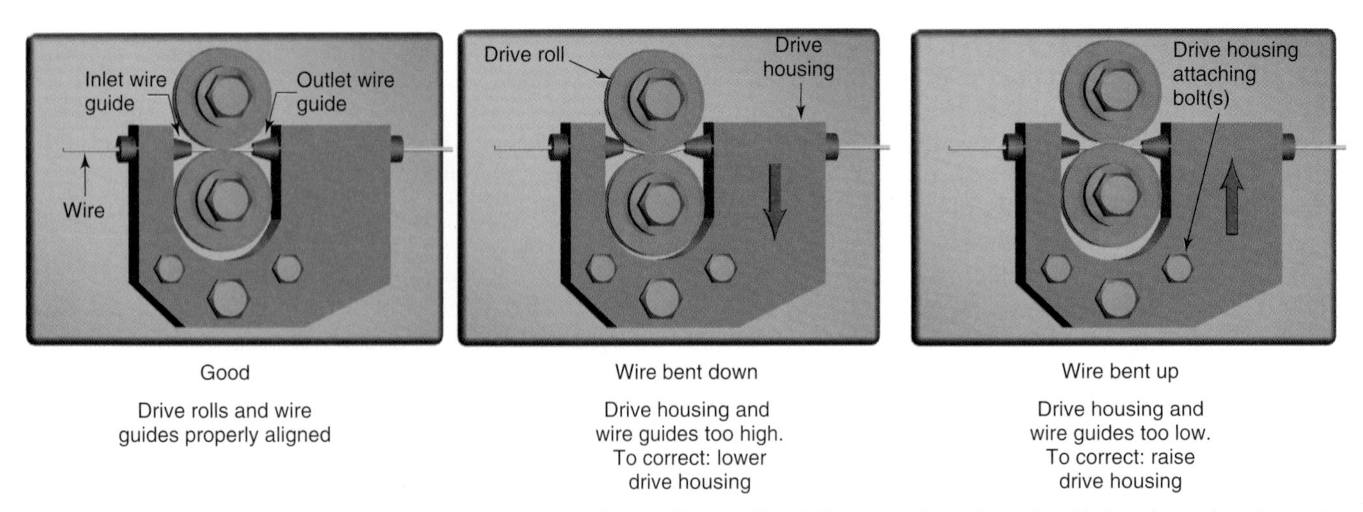

Figure 12-6. Look at these properly and improperly aligned drive rolls. Adjust the drive housing if the electrode wire going through the pressure and drive rolls is bent. To do so, loosen the drive housing bolts, align the rolls, and retighten the bolts.

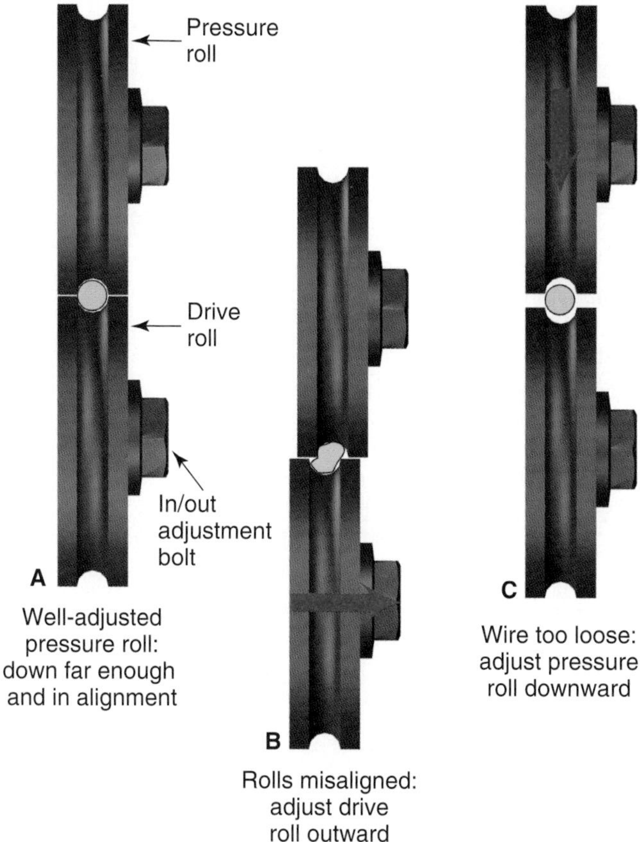

Figure 12-7. A—A properly adjusted pressure roll. B—Rolls misaligned. To correct this problem, the lower drive roll is adjusted in and out by means of an adjustment bolt in the center of the drive roll. C—Wire too loose. To correct this problem, the pressure roll is adjusted up and down by means of the pressure-adjusting wing nut or knob.

Figure 12-8. This is a bird's nest in the wire feed mechanism.

Figure 12-9. Remove the bird's nest by cutting the tangled wire between the inlet wire guide and before the center guide or between the center and exit guides.

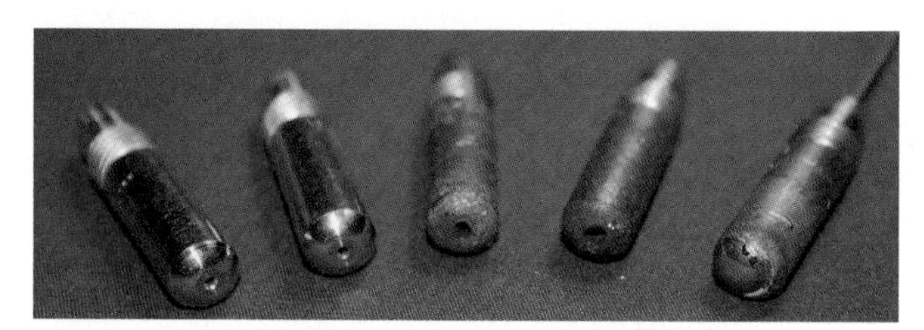

Figure 12-10. The two electrode contact tubes on the left are new. The three tubes on the right are worn and need to be replaced. Note that the tube on the far right had the wire melt onto the end of the tube.

Metal may spatter and get into the nozzle and onto the contact tube while welding. This spatter must be removed. Special tools are available to remove the spatter, but a small wire brush will usually do the job. Antistick compounds that help prevent spatter from sticking to the nozzle and contact tube may also be applied. These compounds make it easier to remove spatter that does stick to the tube and nozzle.

Selecting the Shielding Gas

Shielding gases are chosen after considering a number of factors. Two important factors to consider are the type of base metal being welded and the metal transfer method to be used. Refer to Chapter 11 for information regarding these two factors. **Figures 12-11** and **12-12** list the recommended shielding gases to use for various base metals.

Adjusting the Shielding Gas Flowmeter

Shielding gas is stored in cylinders. Often, a cylinder is used to provide shielding gas for GMAW operations. Cylinders or bulk tanks of shielding gas

Metal	Shielding Gas	Advantages
Aluminum, copper, magnesium, nickel, and their alloys	Argon and argon-helium	Argon satisfactory on sheet metal; argon-helium preferred on thicker sheet metal.
Steel, carbon	Argon-15-25% CO_2	Less than 1/8 in. (3.2 mm) thick; high welding speeds without melt-through; minimum distortion and spatter; good penetration.
	Argon-50% CO_2	Greater than 1/8 in. (3.22 mm) thick; minimum spatter; clean weld appearances; good weld pool control in vertical and overhead positions.
	CO_2 a	Deeper penetration; faster welding speeds; minimum cost.
Steel, low alloy	60-70% Helium-25-35% argon-4-5% CO_2	Minimum reactivity; good toughness; excellent arc stability, wetting characteristics, and bead contour; little spatter.
	Argon-15-25% CO_2	Fair toughness; excellent arc stability; wetting characteristics, and bead countour; little spatter
Steel, stainless	90% Helium-7.5% argon-2.5% CO_2	No effect on corrosion resistance; small heat-affected zone; no undercutting; minimum distortion; good arc stability.
a - CO_2 is used with globular transfer also.		

Figure 12-11. *These are the suggested gases and gas mixtures for use in GMAW short circuiting transfer.*

Metal	Shielding Gas	Advantages
Aluminum	Argon	0.1 in. (0.25 mm) thick; best metal transfer and arc stability; least spatter.
	75% Helium-25% argon	1-3 in. (25-76 mm) thick; higher heat input than argon.
	90% Helium-10% argon	3 in. (76 mm) thick; highest heat input; minimizes porosity.
Copper, nickel, & their alloys	Argon	Provides good wetting; good control of weld pool for thickness up to 1/8 in. (3.2 mm).
	Helium - argon	Higher heat inputs of 50 and 75% helium mixtures offset high heat conductivity of heavier gages.
Magnesium	Argon	Excellent cleaning action.
Reactive metals (Titanium, Zirconium, Tantalum)	Argon	Good arc stability; minimum weld contamination. Inert gas backing is required to prevent air contamination on back of weld area.
Steel, carbon	Argon-2-5% oxygen	Good arc stability; produces a more fluid and controllable weld pool; good coalescence and bead contour, minimizes undercutting; permits higher speeds, compared with argon.
Steel, low alloy	Argon-2% oxygen	Minimizes undercutting; provides good toughness.
Steel, stainless	Argon–1% oxygen	Good arc stability; produces a more fluid and controllable weld pool, good coalescence and bead contour, minimizes undercutting on heavier stainless steels.
	Argon-2% oxygen	Provides better arc stability, coalescence and welding speed than 1% oxygen mixture for thinner stainless steel materials.

Figure 12-12. *These are the suggested gases and gas mixtures for use in GMAW spray transfer.*

can be connected to a manifold and piped to a welding area. A regulator is used to reduce the high pressure in a cylinder to a pressure usable for welding. See Chapters 20 and 21 for information on cylinders and regulators and how to connect them.

The shielding gas flow rate is set after the regulator and flowmeter have been attached to the cylinder. The wire feeder and welding gun must also be connected. An adjusting screw or adjustment knob is used to set the shielding gas flow rate.

Open the cylinder valve before setting or adjusting the shielding gas flow rate. Turn the adjusting screw or knob on the flowmeter so no gas flows. The screw or knob is usually turned counterclockwise. Then, slowly open the cylinder valve. If the flowmeter is open when the cylinder valve is opened, high pressure from the gas in the cylinder may damage the flowmeter.

Set the flow rate after opening the cylinder valve. Shielding gas must be flowing when adjusting the flowmeter. Press the purge switch if the welding machine or wire feeder has one. This causes the gas to flow. Another way to start gas flowing is to set the ***post flow adjustment*** for 15–20 seconds and pull the gun trigger for one second or less. Do not strike an arc. After you pull the trigger, shielding gas will flow for a while so the flow rate can be adjusted.

While gas is flowing, turn the adjusting screw or knob until the flowmeter indicates the correct flow rate. The flow rate is indicated by the top of the ball on a ball-float flowmeter or by a needle on a gauge flowmeter. See **Figure 12-13**. When you are finished setting the flow rate, reset the post flow time to the correct value.

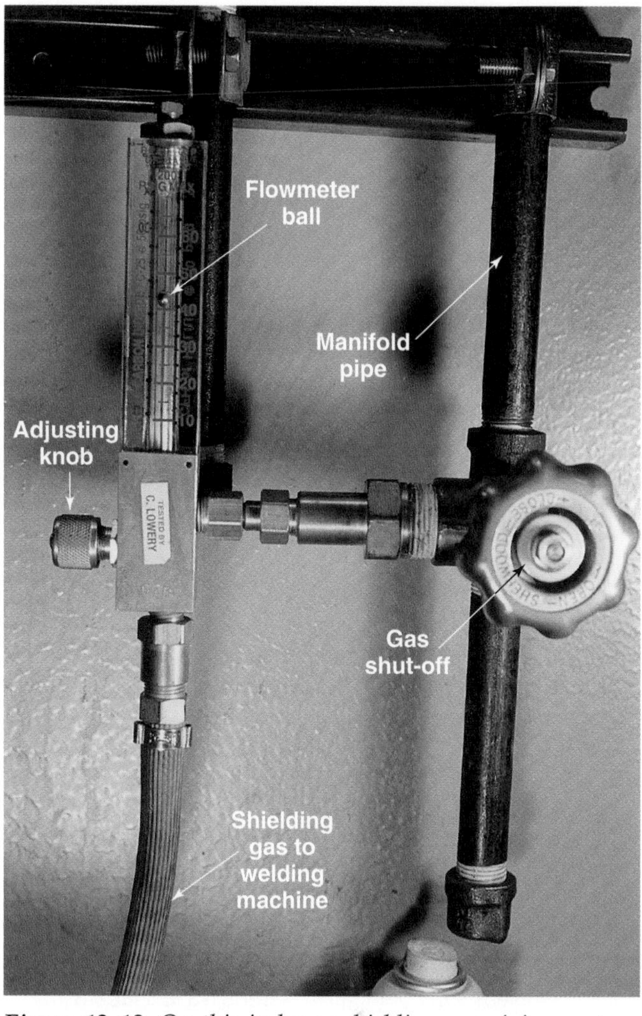

Figure 12-13. *On this in-house shielding gas piping system, shielding gas flowing through the flowmeter raises the ball float in the vertical tube. The correct flow rate is obtained when the top of the ball lines up with the desired flow rate on the gauge.*

Selecting the Electrode Wire

Choice of the electrode wire used for GMAW is based on a number of factors. They include the following:

- Base metal composition.
- Base metal properties.
- Cleanliness of the base metal.
- Shielding gas.
- Metal transfer method.
- Welding position.

Figure 12-14A lists some GMAW electrodes commonly used for welding different base metals. **Figure 12-14B** lists common FCAW electrodes. FCAW electrodes are limited to welding steel and stainless steel base metals. **Figure 12-14C** identifies the letters and numbers used in the electrode designations.

Most flux cored electrodes are designed to be used in the flat position or to make horizontal fillet welds. These electrodes are designated as E70T-X, E80T-X, E90T-X, etc. Electrodes designed to be used in all welding positions are designated as E71T-X, E81T-X, E91T-X, etc. The "1" indicates that the electrode is used for all positions.

Electrode Wire Composition

Many different alloy electrodes are available. An electrode wire that has a chemical composition similar to the base metal is commonly used, but there are important exceptions. The following AWS publications should be referenced for information on the electrodes listed in Figure 12-14.

- AWS A5.9—Stainless steel electrodes for GMAW.
- AWS A5.10—Aluminum alloy electrodes for GMAW.
- AWS A5.14—Nickel alloy electrodes for GMAW.
- AWS A5.18—Carbon steel electrodes for GMAW.
- AWS A5.19—Magnesium alloy electrodes for GMAW.

- AWS A5.20—Carbon steel electrodes for FCAW.
- AWS A5.22—Stainless steel electrodes for FCAW.
- AWS A5.28—Low alloy steel electrodes for GMAW.
- AWS A5.29—Low alloy steel electrodes for FCAW.

FCAW can be done with or without the addition of shielding gas. Electrodes used for welding with a shielding gas have a different composition than those that do not require shielding gas. Flux core electrodes can produce some gases to protect the welding area. The flux also contains elements to prevent the weld metal from oxidizing.

GMAW	
Base Metal	**Recommended Electrodes**
aluminum	ER1100, ER4043, ER5356
copper and copper alloys	ERCu, ERCuSi-A, ERCuAl-A1
carbon steel	ER70S-2, ER70S-6
low alloy steel	ER80S-B2, ER80S-D2
stainless steel	ER308, ER308L, ER316, ER347
nickel and nickel alloys	ERNi-1, ERNiCr-3, ERNiCrMo-3
magnesium	ERAZ61A, ERAZ92A
titanium	ERTi-1

A

FCAW	
Base Metal	**Recommended Electrodes**
carbon steel	E70T-1, E71T-1, E70T-2
low alloy steel	E80T1-B2, E80T1-Ni2
stainless steel	E308T-3, E308LT-3
	E316LT-3, E347T-3

B

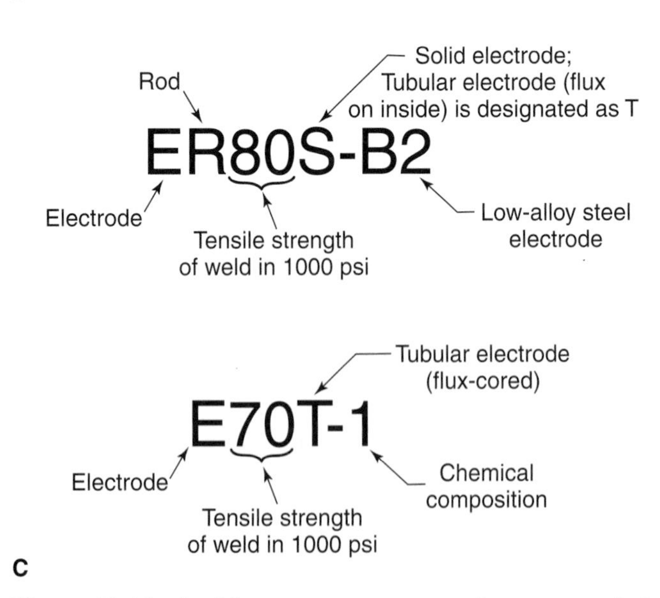

Figure 12-14. A—These are some commonly recommended GMAW electrodes for use with various base metals. There are many more electrodes to choose from. The electrode must be selected for the particular welding job. B—These are some commonly recommended FCAW electrodes for use with various base metals. C—This is an explanation of electrode designations.

Figure 12-15 lists FCAW electrodes used for welding carbon steel. Each FCAW electrode has a ***dash number*** (numerals or letters at the end of the electrode designation, following a dash). The dash number indicates the following information about the electrode:

- The current to use.
- The polarity to use.
- What shielding gas to use (if one is necessary).
- Single-pass or multiple-pass use.

Electrodes that are used to make multiple-pass welds can also be used to make single-pass welds. However, electrodes designed for single-pass welds cannot be used to make multiple-pass welds.

Electrode Wire Diameter

The diameter of electrode is selected after the type of electrode is chosen. Common electrode diameters range between .030″–.0625″ (.76 mm–1.59 mm). Large-diameter electrodes are not used for out-of-position welding. Small-diameter electrodes deposit large amounts of metal using spray transfer. Soft electrodes like aluminum and magnesium with small diameters cannot be pushed through a welding cable as far as a large-diameter electrode can. A push-pull wire feeder and welding gun must be used with soft, small-diameter electrodes. A Teflon® liner should also be used with soft electrodes.

Welding Machine Settings

Two settings must be made on the welding machine, voltage and wire feed speed. Adjusting these two variables determines the arc length and current.

AWS Electrode Classification	External Shielding Medium	Current and Polarity
EXXT-1 (Multiple-pass)	CO_2	DC, electrode positive
EXXT-2 (Single-pass)	CO_2	DC, electrode positive
EXXT-3 (Single-pass)	None	DC, electrode positive
EXXT-4 (Multiple-pass)	None	DC, electrode positive
EXXT-5 (Multiple-pass)	CO_2	DC, electrode positive
EXXT-6 (Multiple-pass)	None	DC, electrode positive
EXXT-7 (Multiple-pass)	None	DC, electrode negative
EXXT-8 (Multiple-pass)	None	DC, electrode negative
EXXT-10 (Single-pass)	None	DC, electrode negative
EXXT-11 (Multiple-pass)	None	DC, electrode negative
EXXT-G (Multiple-pass)	a	a
EXXT-GS (Single-pass)	a	a

a. As agreed upon between supplier and user.

Figure 12-15. These are some of the recommended welding variables for FCAW electrodes used to weld carbon steel.

The metal transfer method is set by these two variables and the type of shielding gas selected. The electrode diameter, electrode composition, and desired metal transfer method determine the proper settings. **Figures 12-16, 12-17,** and **12-18** list the voltage and wire feed speeds required for welding various metals.

Welding Machine Settings for Mild and Low Alloy Steel

Electrode Diameter		Wire Feed Speed		Arc Voltage
in.	mm	in./min.	M/min.	
Short Circuiting Transfer[1]				
.030	.76	150-340	3.81-8.64	15-21
.035	.89	160-380	4.06-9.65	16-22
.045	1.14	100-220	2.54-5.59	17-22
Spray Transfer[2]				
.030	.76	390-670	9.91-17.02	24-28
.035	.89	360-520	9.14-13.21	24-28
.045	1.14	210-390	5.33-9.91	24-30
1/16	1.59	150-360	3.81-9.14	24-32
3/32	2.38	75-125	1.91-3.17	24-32

[1] Values are based on using CO_2 for mild steel and argon–CO_2 for low alloy steel.
[2] Values are based on using argon with 2% to 5% oxygen shielding gas.

Figure 12-16. *Use these suggested arc welding machine settings for short circuiting transfer and spray transfer with GMAW on mild and low alloy steel.*

Welding Machine Settings for Series 300 Stainless Steel

Electrode Diameter		Wire Feed Speed		Arc Voltage
in.	mm	in./min.	M/min.	
Short Circuiting Transfer[1]				
.030	.76	150-430	3.81-10.92	17-22
.035	.89	120-400	3.05-10.16	17-22
.045	1.14	100-240	2.54-6.10	17-22
Spray Transfer[2]				
.030	.76	440-650	11.18-16.51	24-28
.035	.89	430-500	10.92-12.70	24-29
.045	1.14	220-400	5.59-15.16	24-30
1/16	1.59	110-210	2.79-5.33	24-32
3/32	2.38	50-80	1.27-2.03	24-32

[1] Values are based on shielding gas mixture of 90% helium, 7 1/2% argon, and 2 1/2% CO_2 with a flow rate of 20 cfh (566 Lpm).
[2] Values are based on a shielding gas mixture of argon and 1%–5% oxygen.

Figure 12-17. *Use these suggested arc welding machine settings for short circuiting transfer and spray transfer GMAW on series 300 stainless steel.*

Welding Machine Settings for Aluminum and Aluminum Alloys

Electrode Diameter		Wire Feed Speed		Arc Voltage
in.	mm	in./min.	M/min.	
Short Circuiting Transfer[1]				
.030	.76	300-580	7.61-14.73	15-18
.035	.89	250-450	6.35-11.43	17-19
3/64	1.19	200-350	5.08-8.89	16-20
Spray Transfer[1]				
.030	.76	470-680	11.94-17.27	22-28
.035	.89	350-475	8.89-12.06	22-28
3/64	1.19	235-375	5.97-9.52	22-28
1/16	1.59	180-300	4.57-7.63	24-30
3/32	2.38	100-210	2.54-5.33	24-32

[1] Values are based on the use of argon shielding gas.

Figure 12-18. *Use these suggested arc welding machine settings for short circuiting transfer and spray transfer GMAW on aluminum and aluminum alloys.*

The voltage set on the welding machine is the ***open circuit voltage***. The open circuit voltage is higher than the actual welding voltage. The voltage on the welding machine will have to be set higher than the voltage required for welding. The actual difference in voltages varies with the slope of each machine. Some machines have controls for slope and inductance. The functions of these controls are covered in Chapter 11.

Microprocessor welding machines are set differently than non-microprocessor machines. As discussed above, voltage and wire feed speed are normally set on a welding machine.

Often on a microprocessor machine, the base metal type (steel), electrode diameter, and shielding gas or gas mixture are chosen, and the information is entered into the welding machine. The microprocessor then selects correct values for voltage and wire feed speed. These values can be adjusted by the welder and saved into the welding machine. Each time this set of information is entered into the welding machine, the machine sets the same welding conditions.

Preparing the Base Metal

Metal to be welded by GMAW and FCAW should be chemically or mechanically cleaned. A stainless steel brush is commonly used to mechanically clean the metal. Cleaning removes oxides, oil, and other debris.

The edges of some joints may require edge preparation, such as beveling. Refer to Chapter 4 for information on joints and edge preparation. The groove angle required for GMAW is less than the angle required for SMAW. See **Figure 12-19**. A smaller groove angle requires less weld metal and less welder time to fill the joint.

Safety Precautions

GMAW and FCAW can result in spatter flying through the air. They also produce intense light. **It is important to wear proper clothing and eye protection.** Refer to Chapter 11 for a discussion of proper protective clothing. **It is also important to check that the welding equipment is assembled and operating properly.** Each day prior to welding, check the following:

- **The welding gun, wire feeder, and welding machine should be properly connected.**
- **The cables are in good condition (no bare or exposed wire) and are properly connected to the machine and welding gun.**
- **The regulator and flowmeter are correctly installed on the cylinder.**
- **The contact tube is in good condition.**
- **An excessive amount of spatter is not built-up on the inside of the nozzle.**
- **A #12 or darker filter lens is installed in your helmet.**
- **Appropriate protective clothing and gloves are being worn.**
- **Adequate ventilation for removing fumes and shielding gases is present.**

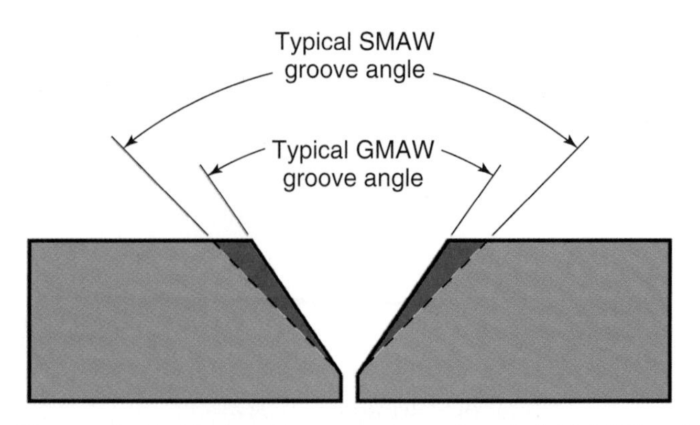

Figure 12-19. *Notice this comparison of typical SMAW and GMAW groove angles. The red area represents the amount of weld metal that is saved. The smaller angle used in GMAW also saves a great deal of welding time.*

Shutting Down a GMAW or FCAW Station

The proper sequence must be followed to properly shut down a GMAW or FCAW welding station. Damage to equipment and supplies can be avoided by following the proper sequence.

1. Adjust the flowmeter so that shielding gas does not flow.
2. Close the cylinder valve by turning it clockwise or close the valve on the manifold.
3. Readjust the flowmeter to allow shielding gas to flow.
4. Press the purge switch on the wire feeder to remove all shielding gas from the hose.
5. Turn the wire feeder off.
6. Turn the welding machine off.
7. Coil the cable connected to the welding gun, then the ground cable. Place cables where they will not get damaged.

Summary

- The outlet on the shielding gas cylinder must be cleaned before attaching a regulator to it. This is done by quickly opening and closing the cylinder valve.
- Shielding gas is connected to either the welding machine or the wire feeder. Connect the hose coming from the flowmeter to the proper place on the welding machine or wire feeder.
- When a separate wire feeder is used, it is usually placed on top of the welding machine. The wire feeder is connected to the remote contactor control on the welding machine
- The welding gun is connected to the wire feeder. The gun may be connected to the welding machine if the wire feeder is inside of the welding machine. If the welding gun has a combination hose, it is connected to the welding machine. If there is an additional shielding gas hose, it is connected to its proper fitting on the welding machine or wire feeder.
- The wire feeder must be set up with proper size drive rolls so that the correct size groove aligns with the electrode wire. Next, the pressure roll is swung up out of the way. The end of the electrode wire is pushed through the inlet wire guide directly behind the drive rolls. The wire is fed into the grooves of the drive roll and through the outlet wire guide directly in front of the drive roll. The pressure roll is swung back into position to complete the assembly. Once the rolls have been properly adjusted and aligned, the correct amount of pressure must be applied to the electrode wire by the drive and pressure rolls.
- A bird's nest is removed by raising the pressure roll, cutting the wire, pulling the wire out of the cable, feeding new electrode wire through the wire guides and into the cable, and then securing the pressure roll into position. The gun trigger or the inch switch is depressed. When the wire comes out of the contact tube, the process is complete.
- The shielding gas flow rate is set after the regulator and flowmeter have been attached to the cylinder. The wire feeder and welding gun must also be connected. An adjusting screw or adjustment knob is used to set the shielding gas flow rate.
- When selecting electrode wire for GMAW, the proper composition is selected first. Next, the electrode diameter is determined.
- Two settings must be made on the welding machine, voltage and wire feed speed. Adjusting these two variables determines the arc length and current. The metal transfer method is set by these two variables and the type of shielding gas selected. The electrode diameter, electrode composition, and desired metal transfer method determine the proper settings.
- Metal to be welded by GMAW and FCAW should be chemically or mechanically cleaned.
- **It is important to wear proper clothing and eye protection. It is also important to check that the welding equipment is assembled and operating properly.**

Review Questions

Write your answers on a separate sheet of paper. Please do *not* write in this book.

1. When shielding gas is supplied to a welding station through a manifold system, what piece of equipment is used at the welding station to control the shielding gas?
2. When a separate wire feeder and welding machine are used, the cable from the wire feeder is plugged into the _____.
   A. remote voltage receptacle
   B. remote contactor receptacle
   C. remote current receptacle
   D. flow control
3. How can you adjust the pressure that the drive rolls exert on the electrode wire?
4. What part of the welding gun needs to be replaced most often?
   A. Nozzle.
   B. Liner.
   C. Switch.
   D. Contact tube.
5. During welding, spatter gets into the _____ and onto the _____.
6. How is the correct shielding gas flow rate set on the flowmeter?
7. Select the factors below that should be considered when selecting an electrode.
   A. Base metal composition.
   B. Manufacturer of the power supply.
   C. Base metal properties.
   D. Shielding gas used.
   E. Welding position.
   F. Flow rate of shielding gas.
   G. Cleanliness of the base metal.
   H. Metal transfer method.
8. What do the letters and numbers in the electrode designation ER70S–3 represent?
9. The type of shielding gas recommended for welding low alloy steel using spray transfer is _____.
10. The type of shielding gas recommended for welding stainless steel using short circuiting transfer is _____.

Chapter 13

GMAW: Flat Welding Position

Learning Objectives

After studying this chapter, you will be able to:

- Describe the GMAW process.
- Determine the appropriate electrode to use with GMAW or FCAW in the flat welding position.
- Identify the correct electrode extension to use with GMAW or FCAW using different metal transfer methods.
- Lay a weld bead on a plate using GMAW or FCAW.
- Make a fillet weld on a lap joint in the flat welding position.
- Make a fillet weld on a T-joint in the flat welding position.
- Weld a butt joint in the flat welding position.
- Describe how to weld aluminum using GMAW.
- Identify various weld defects.

Technical Terms

backhand welding
backing
drag angle
electrode extension
electrode stick out
forehand welding
keyhole
pull gun
push angle
push-pull system
root opening
root pass

Gas Metal Arc Welding Principles

In GMAW and FCAW, a welding arc is struck between the electrode and the base metal. The welder presses the trigger on the welding gun to start the process. Shielding gas begins to flow. Electrode wire feeds out of the gun and a welding arc is struck. The arc melts the base metal and forms a weld pool. Metal from the electrode melts and enters the weld pool. The gun is moved forward to keep the weld pool the correct size.

Welding continues until the welder releases the trigger. Electrode wire stops feeding, and the arc also stops. Shielding gas flows for a short time after the welding stops to protect the weld and base metal as they cool.

Preparing to Weld

For GMAW or FCAW, begin by setting up the equipment. Connect the welding machine, wire feeder, welding gun, and shielding gas as described in Chapter 12. **Check that all electrical, shielding gas, and water connections are tight and leakproof. Be certain that the pressure on the wire is accurately calibrated in the wire feeder. See Figure 13-1. Also, make sure that you are wearing the proper protective clothing for welding.**

After checking the equipment, decide on the type of welding you will be doing. Answer the following questions:

- What type of base metal is to be welded?
- How thick is the base metal?
- In what position will the welding be done?
- What type of metal transfer will be used?

Select an electrode recommended for the type of base metal you will be welding. Refer to Chapter 12 for information regarding the type and diameter of GMAW or FCAW electrodes to use with that base metal. Install a spool of electrode wire on the hub of the wire feeder. Select a contact tube for the diameter of electrode wire you have chosen. Examine the contact tube. The hole in the center should not be worn or contain spatter. Install the contact tube into the gun. Attach the gas nozzle to be used to the end of the gun.

Next, select the shielding gas and the flow rate recommended for the base metal you will be welding and the type of metal transfer to be used. Chapter 12 describes the type of shielding gas that should be used for various base metals and types of metal transfer. **Figure 13-2** lists recommended flow rates for shielding gases. Set the recommended flow rate on the flowmeter as discussed in Chapter 12.

Finally, set the voltage and wire feed speed recommended for the type of metal transfer you have selected. Chapter 12 lists recommended voltages and wire feed speeds.

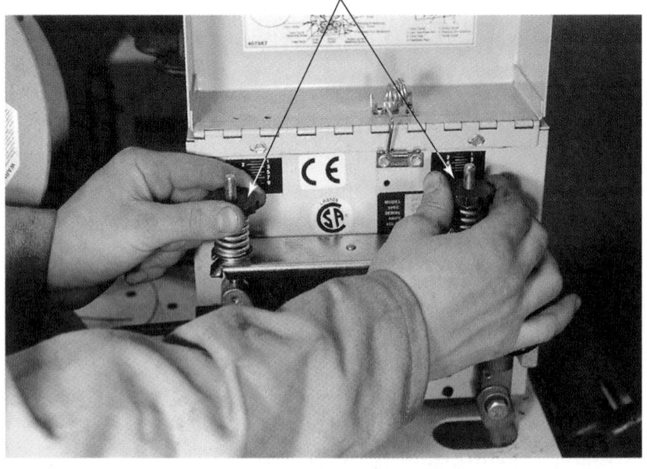

Figure 13-1. Rotating the wire tension adjusting knobs will increase the pressure on the electrode wire as it passes through the rolls.

Metal	Thickness		Argon flow	
	in.	mm	ft^3/hr	L/min
Mild steel	1/16	1.6	25	12
	1/8-3/16	3.2-4.8	30	14
	1/4-1/2	6.4-12.7	35	16
Stainless steel (98%Ar-2%O_2)	1/16	1.6	30	14
	1/8-3/16	3.2-4.8	35	16
	1/4-1/2	6.4-12.7	40	19
Aluminum and aluminum alloys	1/16	1.6	30	14
	1/8-3/16	3.2-4.8	35	16
	1/4	6.4	40	19
	3/8	9.5	50	24
Nickel and nickel alloys	Up to 3/8	Up to 9.5	30	14
Magnesium and magnesium alloys	.025-3/16	0.6-4.8	50	24
	1/4-1/2	6.4-12.7	60	28

Figure 13-2. Suggested shielding gas flow rates for use with various metals and thicknesses are listed. Shielding gas flow rates are shown for the flat welding position. When welding in the horizontal or vertical position, increase the flow about 10%. When welding in the overhead position, increase the flow rate about 30%.

Electrode Extension

In GMAW and FCAW, the voltage set on the welding machine determines the arc length. The wire feed speed determines the current. The welder must control the electrode extension. ***Electrode extension*** or ***electrode stick out*** is the distance from the end of the contact tube to the end of the electrode. See **Figure 13-3**. The electrode must extend beyond the contact tube to preheat the electrode wire. Some preheating is desirable. Preheating too much results in shallow penetration. Proper electrode extension is different for GMAW than for FCAW, **Figure 13-4**.

The contact tube usually cannot be seen when welding because it is inside the nozzle. Therefore, to control the electrode extension, you must control the nozzle-to-work distance. Hold the nozzle off the base metal to obtain the correct electrode extension.

Holding the Welding Gun

Pick up and hold the welding gun. Use a grip that is comfortable, so you can weld for long periods of time without fatigue. Hold the gun so your finger or thumb (depending on the type of gun) is on the switch. Find a

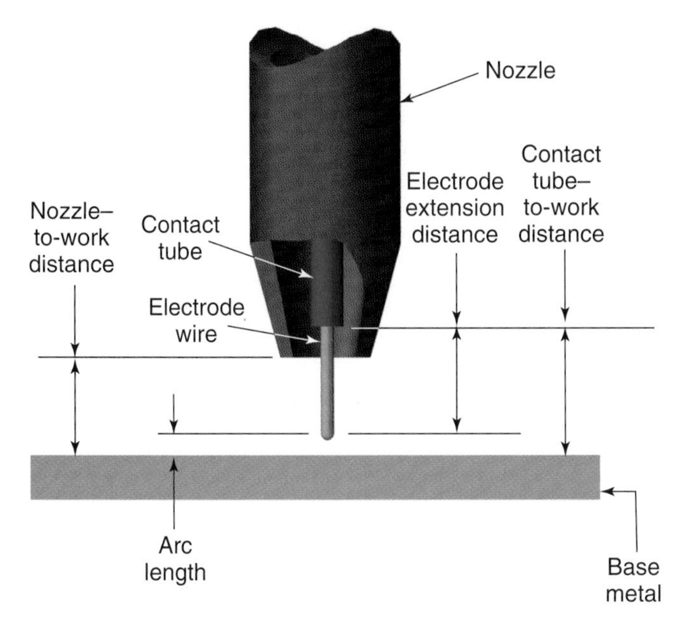

Figure 13-3. Note the various distances associated with GMAW or FCAW. Notice the electrode extension distance.

Process	Method	Electrode Extension
GMAW	Short circuiting	1/4" – 1/2" (6 mm – 13 mm)
GMAW	Spray transfer	1/2" – 1" (13 mm – 25 mm)
FCAW	Gas-shielding	3/4" – 1 1/2" (19 mm – 38 mm)
FCAW	Self-shielding	3/4" – 3 3/4" (19 mm – 95 mm)

Figure 13-4. These are the proper electrode extensions of the different metal transfer methods for GMAW and FCAW.

position that is comfortable. You may want to place the cable across your arm. See **Figure 13-5**.

Press the inch switch on the wire feeder or press the trigger on the welding gun until the electrode sticks out about 2". Using a pair of wire cutters, cut the electrode wire to obtain the proper electrode extension. **Caution: Place one hand over the electrode wire as it is cut to prevent the excess piece from flying through the air. The piece of electrode wire could hit someone in the eye, resulting in injury.**

Laying a Weld Bead

Welding is performed using a backhand or forehand welding method or with the gun held vertically. The ***backhand welding*** method pulls or drags the weld pool along the weld axis. The tip of the electrode points opposite the direction of travel. Backhand welding uses a travel angle of about 25°. A backhand travel angle is called a ***drag angle***. See **Figure 13-6A**.

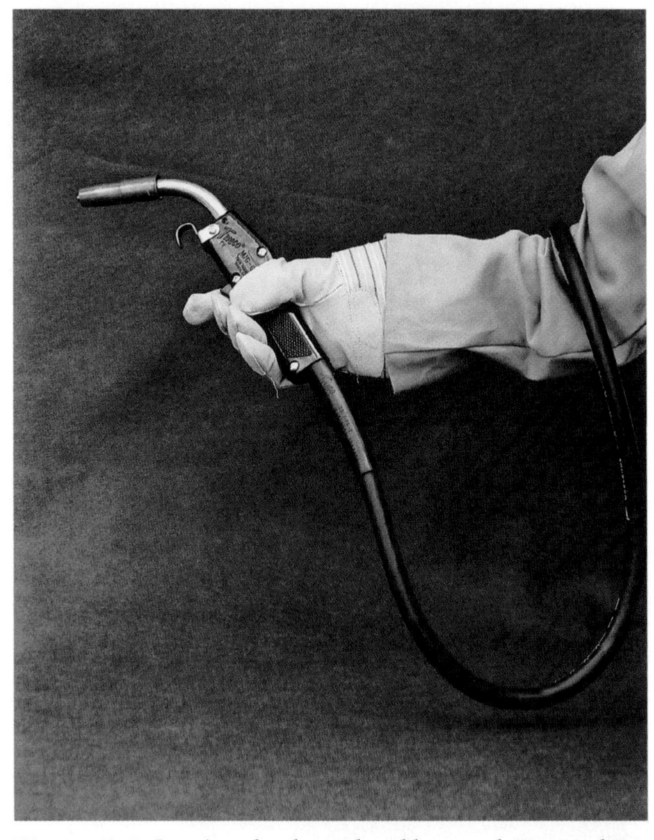

Figure 13-5. Looping the electrode cable over the arm reduces the weight of the gun and makes it more comfortable to hold the gun.

The electrode can be held perpendicular to the workpiece. In such instances, the travel angle is 0°. See **Figure 13-6B**.

The forehand welding method pushes the weld pool, as seen in **Figure 13-6C**. The tip of the electrode points in the direction of travel. The ***forehand welding*** travel angle is called ***push angle***. The electrode and gun form a push travel angle of about 25°. In all these situations, a work angle of 0° is used. The backhand welding method is used to obtain better penetration than forehand or vertical. It also has a more stable arc and produces less spatter than the other two methods.

Position the gun over the area where you will begin welding. Tilt the gun about 20°–25° in the backhand position. Touch the electrode wire to the base metal, but do *not* press the switch. Raise the gun about 1/16" (1.6 mm) or less. Lower your helmet and press the switch. See **Figure 13-7**. A welding arc will then form. The arc rapidly melts the base metal and forms a weld pool (depression). The depth of this weld pool is the amount of penetration of the weld. **Figure 13-8** shows these parts of the weld.

Move the welding gun in the direction of travel. Metal from the electrode will constantly be added to the pool. The weld pool shape is round to oval. A wider weld bead is obtained by moving the welding

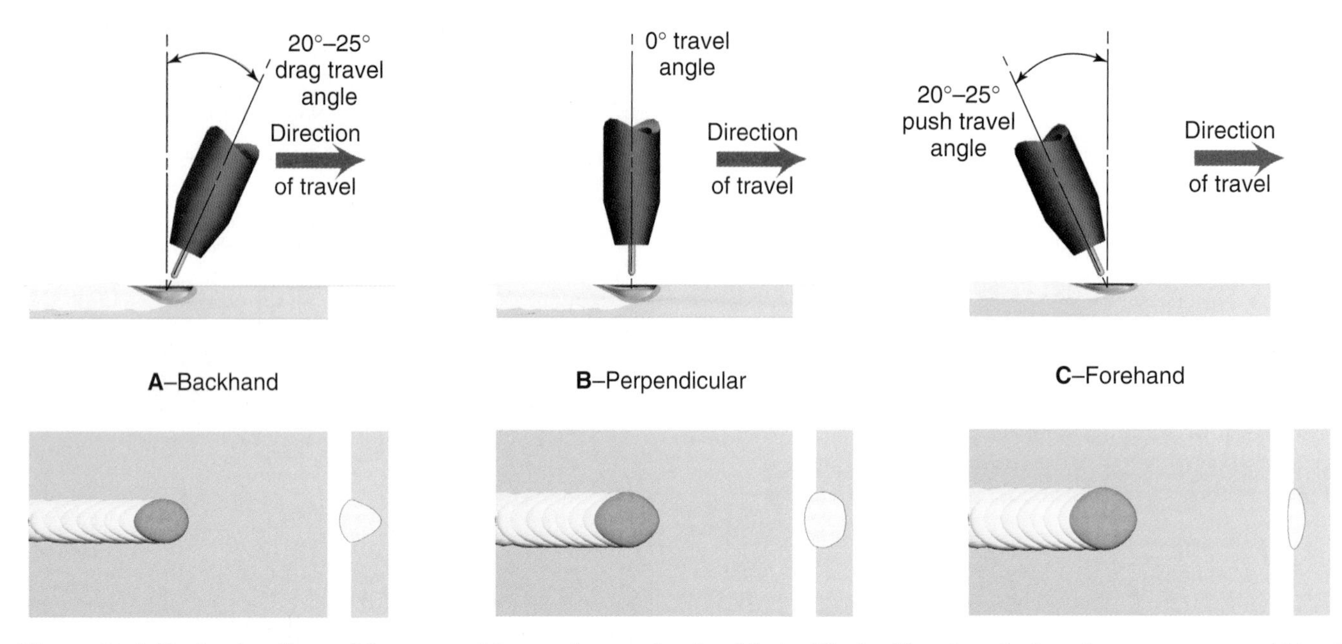

Figure 13-6. Notice the effects of the gun position on the completed weld. A—The backhand method produces a narrow weld bead with deep penetration. B—The perpendicular position produces a medium-width weld bead and medium penetration. C—The forehand welding position produces a wide weld bead with shallow penetration. Note the travel angles shown in each position.

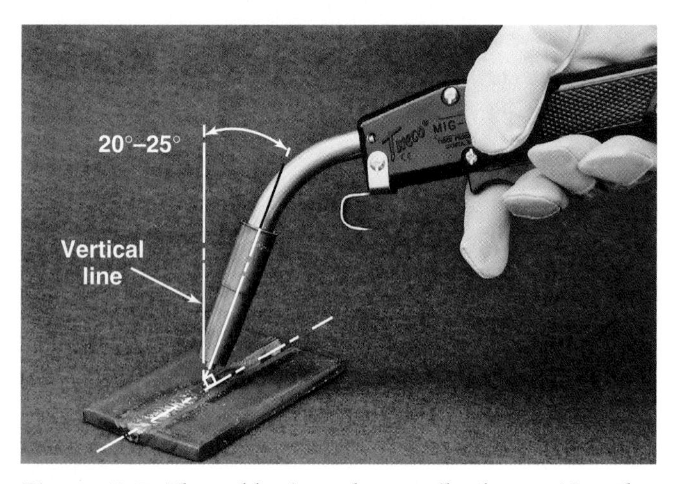

Figure 13-7. The welder is ready to strike the arc. Note that the torch is held at 20°–25° from vertical. This weld will be made in the backhand position.

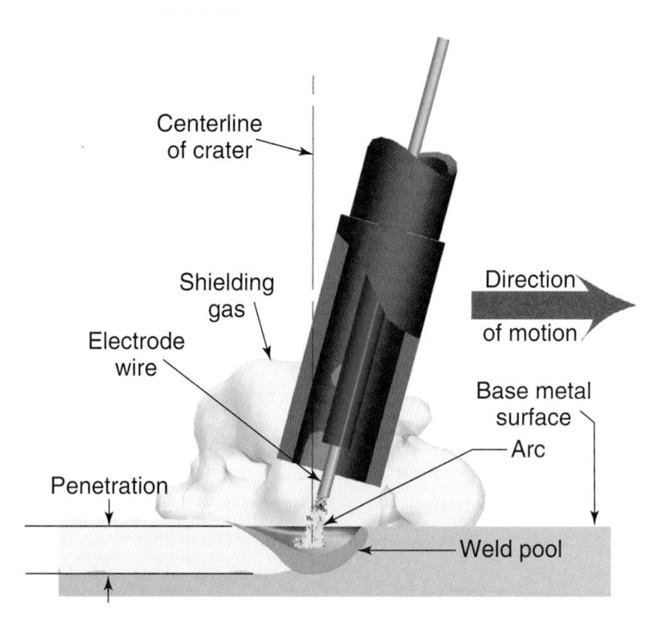

Figure 13-8. This diagram shows the depth of penetration with regard to the surface of the base metal and the weld bead. Note that the electrode wire is kept ahead of the centerline of the weld crater.

gun from side to side. The electrode should be in front of the centerline of the pool. See Figure 13-8. Continue moving the arc and pool until you reach the end of the weld. Keep the nozzle-to-work distance and the angle of the welding gun as constant as possible.

The weld pool must be filled at the end of the joint or when you are ready to stop. Move the gun backwards over the weld pool 1/4″–1/2″ (6 mm–13 mm), and then release the switch. The electrode wire will stop feeding and the welding arc will also stop. Hold the gun in the same position until the shielding gas stops flowing.

For the same diameter electrode, more metal is deposited using spray transfer than globular or short circuiting transfer. The welder must move the welding gun much faster when using spray transfer than with the other two transfer methods. Regardless of the transfer method, watch the weld pool size. It determines the travel speed. If the pool gets too large, move the gun faster. If you are moving the arc in front of the pool, slow the travel speed down. Keep the electrode wire directly in front of the center of the pool.

A good weld bead should have evenly spaced ripples. The edges of the weld should be the same width. There should be about 1/16″–1/8″ (1.6 mm–3.2 mm) of reinforcement. **Figure 13-9** shows an example of a good fillet weld bead on a T-joint. If the ripples are not evenly spaced, practice moving the gun at a constant speed. If the edges are not the same width along the entire weld, practice holding the gun steady. Try to avoid moving the gun from side to side.

The amount of reinforcement is controlled by the travel speed. A wide, shallow weld bead is produced by too slow a travel speed. A narrow weld bead with undercutting along the edges is made by too fast a travel speed. At the correct travel speed, a weld bead with good penetration and no undercutting is produced.

Spatter is reduced by changing the angle of the welding gun, adjusting the inductance and slope on the machine, or lengthening the electrode extension. A longer electrode extension reduces spatter. However, an electrode that extends too far results in poor penetration. Too long of an extension may cause porosity because the shielding gas is not able to cover the welding area. A low flow rate of shielding gas may also cause weld porosity.

Figure 13-9. This is a well-formed fillet weld on a T-joint. Note that the weld bead is slightly convex.

Exercise 13-1 Laying a Weld Bead on a Plate

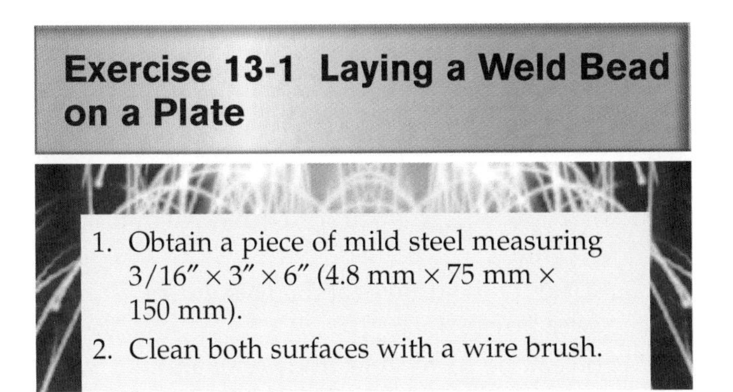

1. Obtain a piece of mild steel measuring 3/16″ × 3″ × 6″ (4.8 mm × 75 mm × 150 mm).
2. Clean both surfaces with a wire brush.
3. Using a ruler and a piece of soapstone, mark out five straight lines. You will lay your weld beads along those lines.
4. Set up the welding machine for short circuiting transfer. Refer to Chapter 12 for details. Select the appropriate shielding gas and set the flowmeter. Figures 12-11 and 13-2 list the correct gases and flow rates.
5. Use the backhand method. Figure 13-7 shows the correct angle between the welding gun and the plate. This angle is the same for most GMAW and FCAW in any position.
6. Watch the metal as the welding arc starts. A molten weld pool develops quickly. Move the arc toward the forward edge of the pool. Note the width of the pool and the amount of penetration. Keep the same width and penetration as the weld progresses.
7. Fill the weld pool at the end of the weld bead.
8. Examine the weld bead. Change the voltage, wire feed speed, travel speed, or shielding gas flow rate as required.
9. Weld beads on the other four lines on the plate.

Inspection:

Each weld bead should be straight with evenly spaced ripples and continuous reinforcement. The crater at the end of the weld should be filled. No evidence of porosity should be seen.

Turn the plate over and repeat this exercise. Set the welding machine for spray transfer. Change the shielding gas to argon or argon with oxygen. See Figures 12-12 and 12-16.

Making a Fillet Weld

A fillet weld made with GMAW or FCAW looks similar to one made with other welding processes. The two legs should be equal in size. The weld face should be convex or flat with evenly spaced ripples. Undercutting, porosity, or cracks should not be visible. **Figure 13-10** shows a good fillet weld made with GMAW. **Figure 13-11** shows cross sections of three good fillet welds.

Figure 13-10. This is a well-formed fillet weld on a lap joint. The weld bead is slightly convex.

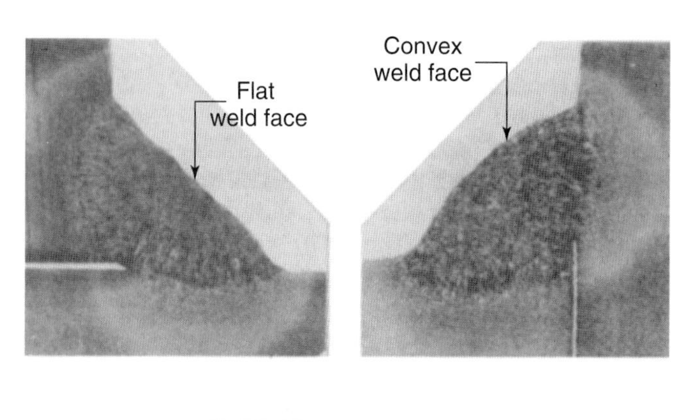

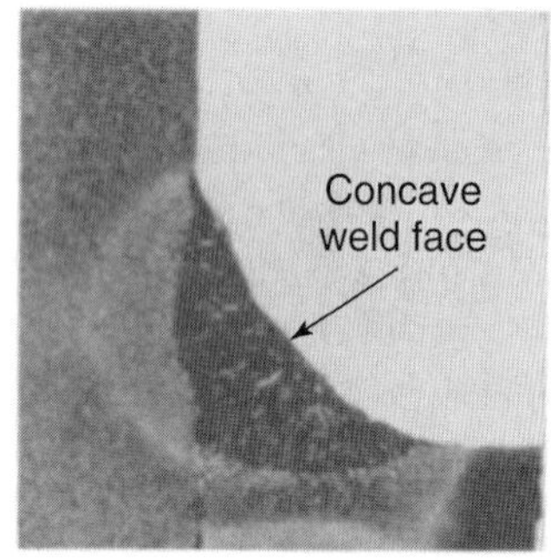

Figure 13-11. Cross sections of three acceptable fillet welds made on an inside corner joint. These macrographs were etched and magnified four times. (US Steel)

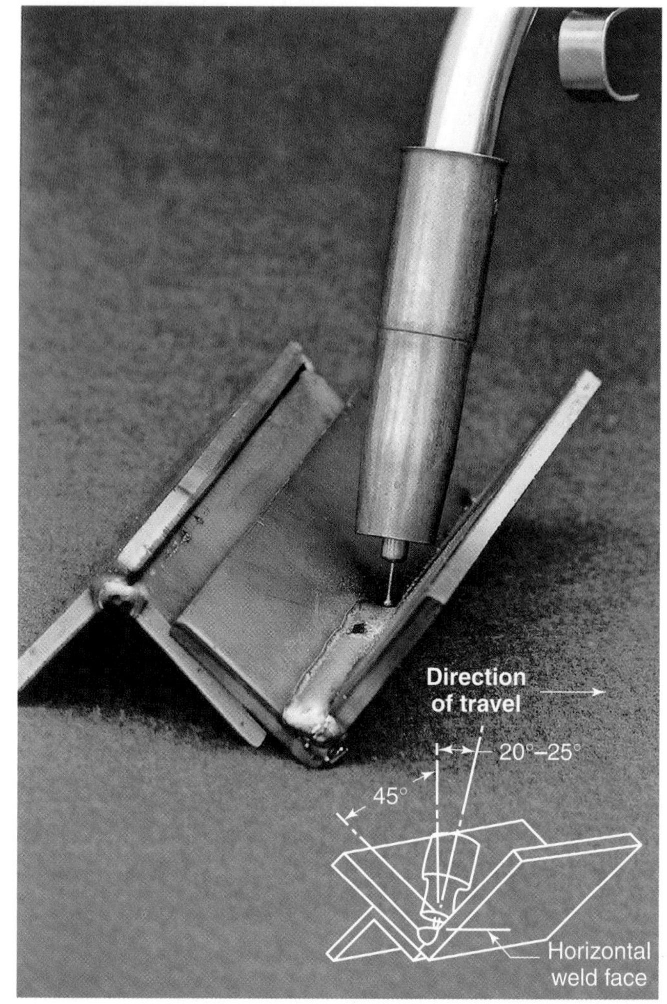

Figure 13-12. These are the correct angles for the GMAW gun for making a backhand fillet weld on a joint in the flat welding position. The gun is held at a 45° work angle to the metal surfaces and a 20°–25° drag travel angle.

A fillet weld is started in the same manner as starting to lay a weld bead on a plate. The electrode and the arc are centered over the weld root. The welding gun should be held at about a 45° work angle to the base metal. The gun can be held in a forehand, perpendicular, or backhand position. See **Figure 13-12**. Be sure to use the proper electrode extension.

Once the welding arc is started, a pool will develop under it. Some side-to-side movement of the gun may be required to form a C-shaped weld pool. The arc should be centered over the weld root and kept toward the front of the pool. Watch the pool as you move along the weld joint. The pool should flow into both pieces of the base metal. Keep the pool the same width as the weld progresses. Weld defects and ways to prevent them are discussed later in this chapter.

Exercise 13-2 Making a Fillet Weld on a Lap Joint

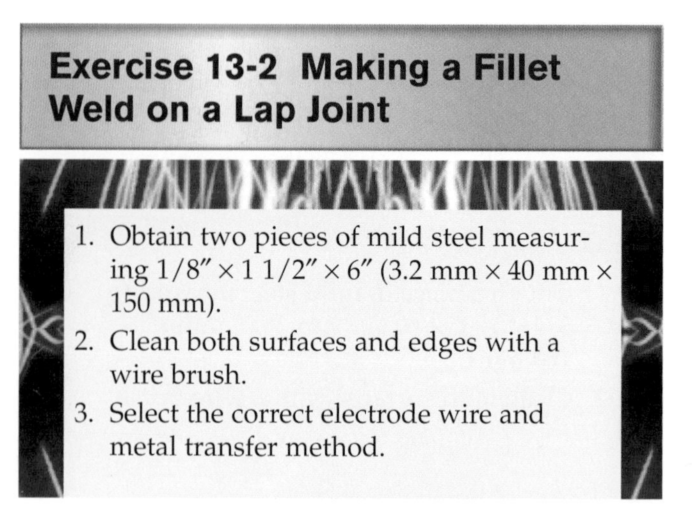

1. Obtain two pieces of mild steel measuring 1/8″ × 1 1/2″ × 6″ (3.2 mm × 40 mm × 150 mm).
2. Clean both surfaces and edges with a wire brush.
3. Select the correct electrode wire and metal transfer method.

4. Set the shielding gas flow rate and the machine adjustments.
5. Place the two pieces together to form a lap joint.
6. Tack weld the pieces in three places on each side.
7. Place the weldment so the weld will be made in the flat position.
8. Position the gun over one end of the joint using the correct angles. It may be necessary to point the gun more toward the surface than the edge when welding an edge to a surface.
9. Press the switch and begin welding.
10. Make a fillet weld along the seam. Fill the crater at the end of the weld bead.
11. Make a second fillet weld on the other side.

Inspection:

The completed welds should be convex with evenly spaced ripples. The edges of the weld bead should be straight and should not overlap.

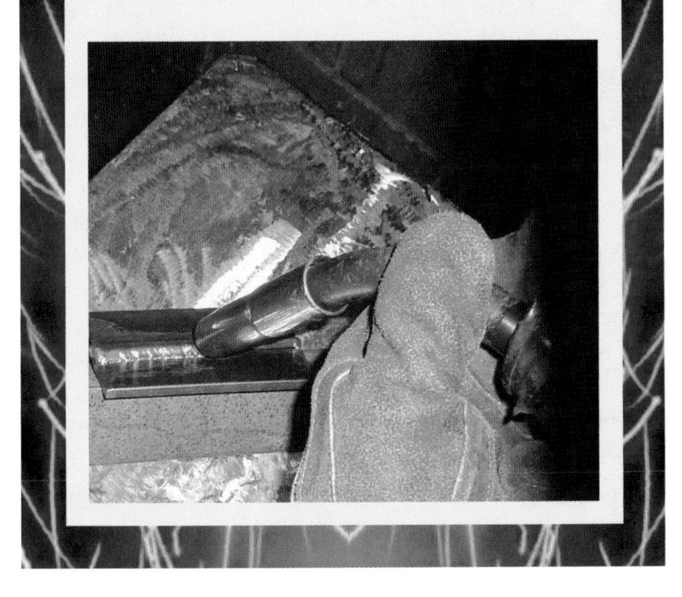

Exercise 13-3 Placing a Fillet Weld on a T-Joint

1. Obtain two pieces of mild steel measuring 1/8″ × 1 1/2″ × 6″ (3.2 mm × 40 mm × 150 mm).
2. Clean one edge of one piece and one surface of the other piece.
3. Select the correct welding electrode and shielding gas. Set up the machine for the metal transfer method you will be using.
4. Tack weld the clean edge to the clean surface in three places on each side.
5. Place the weldment so the weld is in the flat position, Figure 13-12.
6. Make a fillet weld on each side of the T-joint. Watch the weld pool as it forms. Keep a C-shaped pool.

Inspection:

The weld should be straight with evenly spaced ripples. Undercutting or overlap should not be visible.

Welding a Butt Joint

A butt joint can be formed with a square groove or the edges can be prepared with a groove, such as a V-groove. Steel that is 3/16″ (4.8 mm) or thicker usually requires a prepared groove. Grooves used in GMAW and FCAW have a smaller angle than grooves used in SMAW.

Full penetration of butt joints is usually required. A ***root opening*** (gap) is left between the two pieces of base metal when tack welding them. The arc must melt the edges of both pieces when welding. The root of the weld pool looks like a ***keyhole*** when welding. The round portion of the keyhole is smaller than the keyhole in oxyfuel gas welding or SMAW. The keyhole is only seen on the ***root pass*** (first weld pass) of a multiple-pass weld. A keyhole at the root opening indicates that both pieces of base metal are being melted. It also indicates that the molten metal is flowing through the joint, resulting in full penetration. See **Figure 13-13**. To produce consistent penetration, the size of the pool and the keyhole must be kept constant as the weld progresses.

A backing may be used to help produce consistent penetration. ***Backing*** is attached to the root side of the joint to control penetration. Backings can be used in all welding positions. Some backings become part of the completed joint while most are removed after welding. Removable metal backings can be flat or have a radius machined in them. See **Figure 13-14**. Removable metal backings are not melted while welding. A copper backing is used when welding mild or stainless steel. A stainless steel backing is used when welding aluminum.

When welding a butt joint, the electrode should be directly over the centerline of the weld. The welding arc should melt both pieces being welded equally. If one piece is melting more than the other, adjust the

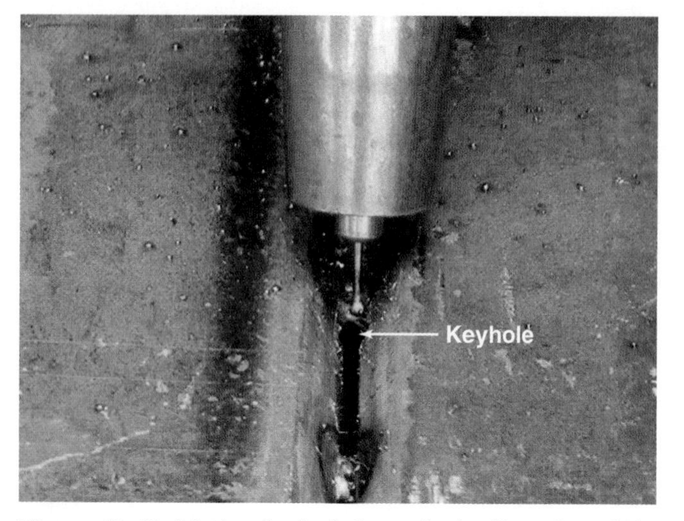

Figure 13-13. Notice the keyhole at the leading edge of the weld crater.

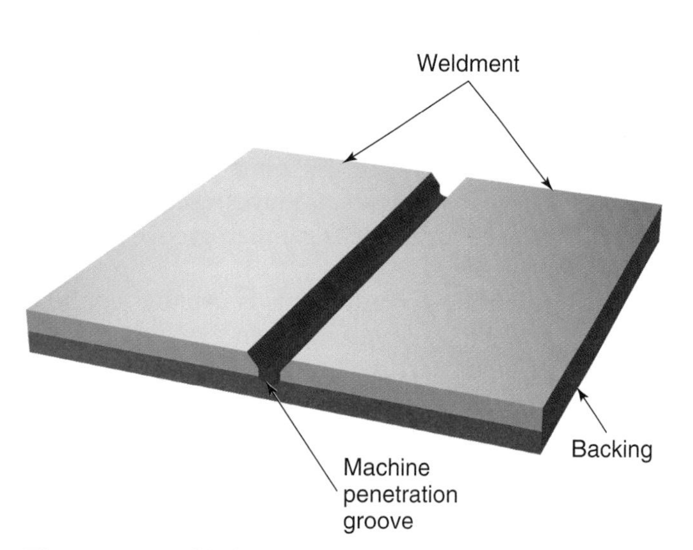

Figure 13-14. This backing has a round groove milled into it. The groove ensures uniform penetration. A copperplate is used for welding steel. The metal to be welded is shown in gray.

location or angle of the gun, so both pieces melt equally. A work angle of 0° is used, as shown in **Figure 13-15**. Use the backhand welding method so the tip of the electrode points opposite the direction of travel. A travel angle of 20°–25° should be used. This gun position can also be described as 65°–70° from the surface of the workpiece. See Figure 13-15.

Multiple passes or weld beads are required to fill the grooves for thick base metals. The keyhole method is used only on the root pass. After the root pass, each weld pass is similar to laying a weld bead on a plate. Good penetration into the base metal and the previous weld bead is required. **Figure 13-16** shows a multiple pass groove weld. Each weld bead must be thoroughly cleaned before another weld pass is made. When FCAW is used, all slag must be removed after each weld pass is made.

Exercise 13-4 Welding a Square Groove Butt Joint

1. Obtain two pieces of mild steel measuring 1/8″ × 1 1/2″ × 6″ (3.2 mm × 40 mm × 150 mm).
2. Clean one edge on each piece.
3. Select the correct welding electrode and shielding gas. Set up the machine for the metal transfer method you will be using
4. Position the two pieces so there is a 3/32″ (2.38 mm) gap between them. Tack weld the pieces together in three places.
5. Support the pieces so a full penetration weld can be made. A backing is optional.
6. Align the gun and electrode with the center of the joint before welding. Press the switch on the gun. Move the gun forward when the pool forms. Keep the keyhole and the pool as consistent as possible.
7. Continue welding the butt joint to the end of the plate. Fill the weld pool at the end of the weld.

Inspection:

A butt weld should have complete penetration. The weld root should indicate complete penetration. This penetration and the face of the weld should have evenly spaced ripples. The weld bead should be consistent in width. Signs of defects should not be visible. The weld bead should be built up slightly higher than the base metal. Excessive spatter on the surface is not acceptable.

Welding Aluminum

A few problems are encountered when welding aluminum that are not encountered when welding mild steel. These include frequent bird's nests and the forming of aluminum oxide. The electrode wire in the wire feeder may become tangled, producing a bird's nest. The number of bird's nests may be reduced by:

- Selecting the correct electrode wire.
- Adjusting the equipment properly.

Aluminum electrode wire is very soft. Bird nests can be avoided by selecting the largest possible diameter electrode wire when welding aluminum. A large-diameter wire is stiffer than a smaller-diameter

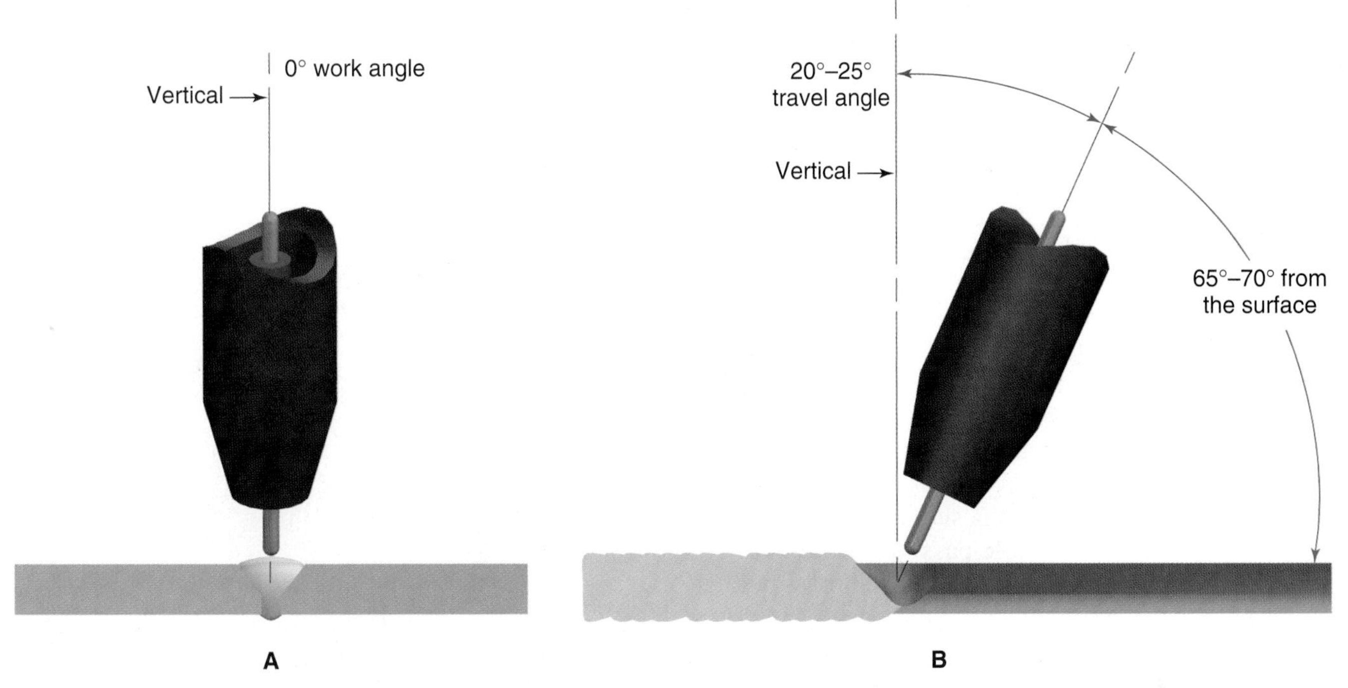

Figure 13-15. These are the suggested gun angles for welding a butt joint. A—The gun is aligned vertically with the weld axis. This is a 0° work angle, as shown in the end view. B—Welding is done using the backhand welding method. A 20°–25° drag travel angle is used.

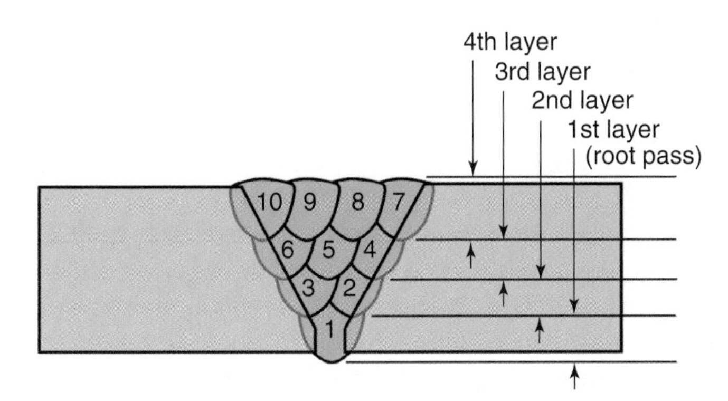

Figure 13-16. Notice the equal size and penetration of each weld bead on this multiple-pass weld. This weld has four layers and ten weld passes in it.

wire. Use an aluminum alloy electrode that is stiffer. This allows the electrode to be pushed through the cable to the welding gun without kinking. Two common aluminum electrode wires are ER5356 and ER4043. ER5356 is stiffer than ER4043 and will result in fewer bird's nests. Each of these is a different alloy and must be selected correctly, depending on the base metal being welded and the desired weld properties.

Proper adjustment of the equipment is an important consideration in reducing the number of bird's nests. One necessary adjustment is to position the guide tubes in the wire feeder as close to the drive rolls as possible. The guide tubes should also be aligned with each other.

Aluminum electrode wire is soft and cannot be easily pushed through a long cable to the welding gun. Use the shortest cable assembly possible from the wire feeder to the welding gun. Also use a Teflon® liner in the cable. A ***pull gun***, which pulls the electrode wire through the cable, may be used. A ***push-pull system*** is formed when a pull gun pulls the electrode wire through the cable as the wire feeder pushes the wire. A push-pull system helps to prevent bird's nests from forming.

Aluminum oxidizes very easily, forming aluminum oxide. Aluminum oxide has a higher melting temperature than aluminum. The base metal should be chemically or mechanically cleaned before welding. A stainless steel wire brush is recommended to mechanically clean the joint. The tip of the electrode wire may also oxidize. Use a pair of wire cutters to cut the tip off the electrode wire. This should be done each time before starting a new weld. **Place your hand over the area to be cut to prevent the piece of wire from flying through the air and injuring someone.**

The forehand welding method helps to remove oxides on unwelded areas. However, penetration may be less with the forehand welding method than with the backhand welding method. Aluminum conducts heat very well. The metal you are welding is already preheated as you move along the weld joint. Therefore, it may be necessary to move the welding gun faster to prevent melt-through.

Weld Defects

Defects produced in GMAW and FCAW are similar to those found in work done using other welding processes. These defects include incomplete penetration, lack of fusion, slag inclusions, undercut, overlap, and porosity. Defects cause a weld to be weaker than its design requirements. **Figure 13-17** shows a test sample of a fillet weld on a T-joint. The sample was tested until it failed. It had no apparent inclusions.

Two common defects in a fillet weld are undercut and overlap. It is important to use the correct gun angles and gun motion to avoid these defects. Make sure both pieces of metal are being melted. You may need to move the gun toward the toe of the fillet weld to melt the base metal.

Another common defect is slag in the weld. This often happens on a multiple-pass weld. Be sure to completely remove all slag from one weld pass before starting another pass.

Most defects can be avoided by taking the following steps:

- Selecting the correct shielding gas and electrode.
- Setting the correct shielding gas flow rate, voltage, and wire feed speed.
- Maintaining the proper electrode extension (contact tube–to-workpiece distance).
- Using the proper gun angles.
- Moving the gun properly.
- Removing slag between weld passes.

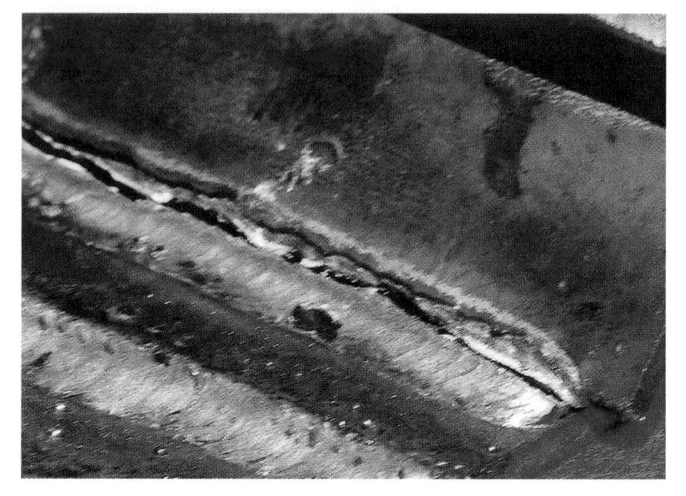

Figure 13-17. *This is a test sample of a fillet weld on a T-joint. The weld was bent until it failed. The break occurred within the weld and not at the weld toes. This indicates a good fillet weld.*

Summary

- In GMAW and FCAW, the welder presses the trigger on the welding gun to start the process. Shielding gas begins to flow. Electrode wire feeds out of the gun and a welding arc is struck. The arc melts the base metal and forms a weld pool. Metal from the electrode melts and enters the weld pool. The gun is moved forward to keep the weld pool the correct size. Welding continues until the welder releases the trigger.
- Before welding, install a spool of the electrode wire recommended for the type of base metal you will be welding. Install a contact tube that matches the diameter of the selected electrode wire. Attach the gas nozzle to be used to the end of the gun. Select the proper shielding gas and set the proper flow rate. Finally, set the voltage and wire feed speed recommended for the type of metal transfer you have selected.
- In GMAW and FCAW, the voltage set on the welding machine determines the arc length. The wire feed speed determines the current.
- GMAW is performed using a backhand or forehand welding method or with the gun held vertically.
- A good weld bead should have evenly spaced ripples. The edges of the weld should be the same width. There should be about 1/16″–1/8″ (1.6 mm–3.2 mm) of reinforcement.
- A fillet weld is started in the same manner as starting to lay a weld bead on a plate. The electrode and the arc are centered over the weld root. The welding gun should be held at about a 45° work angle to the base metal. The gun can be held in a forehand, perpendicular, or backhand position.
- For a butt joint, the electrode should be directly over the centerline of the weld. The welding arc should melt both pieces being welded equally. A work angle of 0° and a drag angle of 20°–25° should be used.

(continued)

- Multiple passes or weld beads are required to fill the grooves for thick base metals. The keyhole method is used only on the root pass. After the root pass, each weld pass is similar to laying a weld bead on a plate.
- Aluminum electrode wire is very soft. Bird's nests can be avoided by selecting the largest possible diameter electrode wire when welding aluminum.
- When aluminum is being welded, the forehand welding method helps to remove oxides on unwelded areas.
- Defects produced in GMAW and FCAW are similar to those found in work done using other welding processes. These defects include incomplete penetration, lack of fusion, slag inclusions, undercut, overlap, and porosity.

Review Questions

Write your answers on a separate sheet of paper. Please do *not* write in this book.

1. The molten metal under the welding arc is called the _____.
2. List four factors that a welder must consider before selecting the electrode and shielding gas.
3. Refer to Figures 12-11, 12-14, and 13-2. What electrode, shielding gas, and flow rate are recommended for gas metal arc welding of 1/8″ 308 stainless steel using short circuit transfer?
   Electrode: __________
   Shielding gas: __________
   Flow rate: __________
4. Refer to Figures 12-18 and 13-2. What voltage, wire feed, and flow rate should be used to weld 1/8″ aluminum using 3/64″ (1.2 mm) diameter electrode wire and spray transfer?
   Voltage: __________
   Wire feed: __________
   Flow rate: __________
5. Which of the following does the welder actually observe and control to obtain a good-quality weld bead?
   A. Nozzle-to-work distance.
   B. Electrode extension.
   C. Contact tube–to-work distance.
   D. Arc length.
6. The electrode should extend _____ beyond the end of the contact tube for spray transfer with GMAW.
   A. 1/4″–1/2″ (6 mm–13 mm)
   B. 1/2″–1″ (13 mm–25 mm)
   C. 3/4″–1 1/2″ (19 mm–38 mm)
   D. 3/4″–3 3/4″ (19 mm–95 mm)
7. List two advantages of backhand welding over forehand welding.
8. _____ transfer deposits more metal for a given size electrode wire.
   A. Short circuiting
   B. Globular
   C. Spray
   D. Pulsed-spray
9. List two ways to reduce spatter.
10. The weld pool shape for a fillet weld is a(n) _____ shape. The opening at the root pass of a butt weld is a(n) _____ shape.

This welder is using the GMAW process to join two thick plates. (Thermadyne Industries, Inc.)

Chapter 14

GMAW: Horizontal, Vertical, and Overhead Welding Positions

Learning Objectives

After studying this chapter, you will be able to:

- Explain why flat position welding is preferred over out-of-position welding.
- Identify the correct welding gun angle for out-of-position welding.
- Weld in the horizontal welding position using GMAW or FCAW.
- Weld in the vertical welding position using GMAW or FCAW.
- Weld in the overhead welding position using GMAW or FCAW.

downhill welding
overlap
underfill
uphill welding

Out-of-Position GMAW or FCAW

Both GMAW and FCAW can be done out of position. When performing GMAW out of position, only short circuiting transfer and pulsed spray transfer methods can be used. Pulsed spray transfer requires a welding machine with pulse controls. Since not all machines have pulse spray capabilities, short circuiting transfer is more often used for out-of-position welding.

The flat welding position should be used whenever possible. Welding in the flat position is preferred over out-of-position welding for the following reasons:

- The pool will not sag or fall from the weld joint.
- All metal transfer methods can be used.
- The flat welding position is more comfortable for a welder.
- Better welds may be produced because welders are usually better at welding in the flat position.

However, some industrial weldments require out-of-position welding to be performed. Weldments such as pipelines and structural welding are too large to move. See **Figure 14-1**. Most welding qualification tests are performed on out-of-position weld joints. A welder who can make out-of-position welds successfully can also produce good welds in the flat welding position. Welds made out of position should look as good and should be as strong as welds made in the flat welding position.

Figure 14-1. Gas metal arc welding is being used to repair a leak in a piping system. Welds may be required in any welding position. (Miller Electric Mfg. Co.)

Welding in the Horizontal Welding Position

The weld axis and weld face are horizontal or close to horizontal when making a weld in the horizontal position. The weld axis can vary from 0°–15°. The weld face can vary from 80°–150° or from 210°–280°.

Welding in the horizontal welding position can be more difficult than welding in the flat welding position. Gravity pushes the molten weld pool down. This causes the weld bead to sag and not properly fill the joint. It is important to use proper gun angles and techniques to obtain high-quality welds.

Fillet Welding

A fillet weld in the horizontal welding position is the most common out-of-position weld. A horizontal fillet weld is made on inside corner joints, lap joints, and T-joints. All transfer methods can be used.

The process for making a horizontal fillet weld is similar to making a fillet weld in the flat position. Notice the gun angles in **Figure 14-2**. The gun points toward the root of the weld. A C-shaped weld pool must be formed when welding. The best penetration is produced using the backhand welding method. The same electrode extension used for making flat welds is used for welding in the horizontal position. The recommended electrode extension for short circuiting transfer with GMAW is 1/4″–1/2″ (6 mm–13 mm). Chapter 13 describes the proper electrode extension.

Thick base metals require more than one pass to fill the weld joint. Each weld pass must melt into the base metal and into the previous weld pass. **Figure 14-3** shows a multiple-pass weld.

A common weld defect encountered when making a fillet weld in the horizontal welding position is undercutting on the vertical plate. To prevent undercutting, use a slight weave bead by moving or pointing the arc at the vertical piece for a short period of time (less than a half second). This allows more electrode metal to be deposited to the vertical piece. The gun angle can also be changed to a 50° work angle (40° from the horizontal), instead of 45°. This would help to eliminate undercutting.

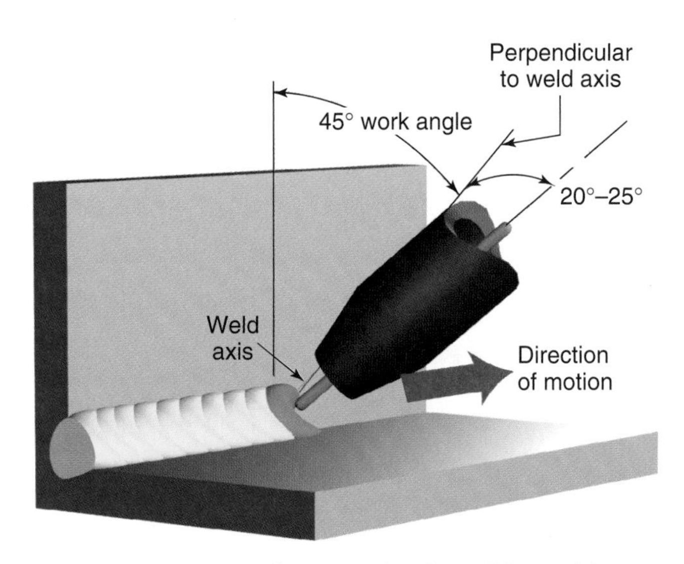

Figure 14-2. Suggested gun angles for a fillet weld on an inside corner joint in the horizontal welding position.

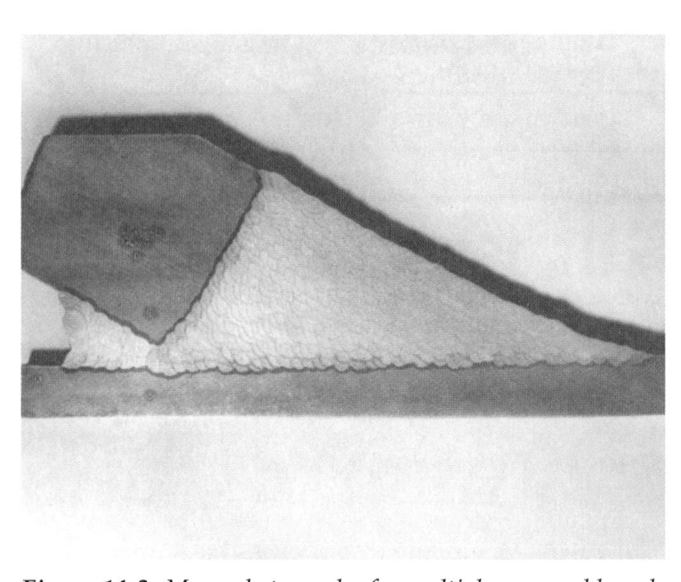

Figure 14-3. Macrophotograph of a multiple-pass weld made on very thick metal using flux cored arc welding (FCAW). Note the number of weld passes and layers required to complete this weld. (Highland Fabricators, Scotland)

Exercise 14-1 Making a Fillet Weld in the Horizontal Welding Position

1. Obtain two pieces of mild steel measuring 3/16″ × 1 1/2″ × 6″ (4.8 mm × 40 mm × 150 mm).
2. Clean one edge on one of the pieces and one surface on the other piece.
3. Set the proper voltage, wire feed speed, and shielding gas flow.
4. Position the pieces to form a T-joint. Tack weld the pieces together in three places on each side of the joint.
5. Make a horizontal fillet weld on each side of the T-joint. Be sure to control the electrode extension.

Inspection:

The completed weld should have a flat-to-convex weld bead shape. The ripples should be even. There should be no evidence of undercutting or other defects.

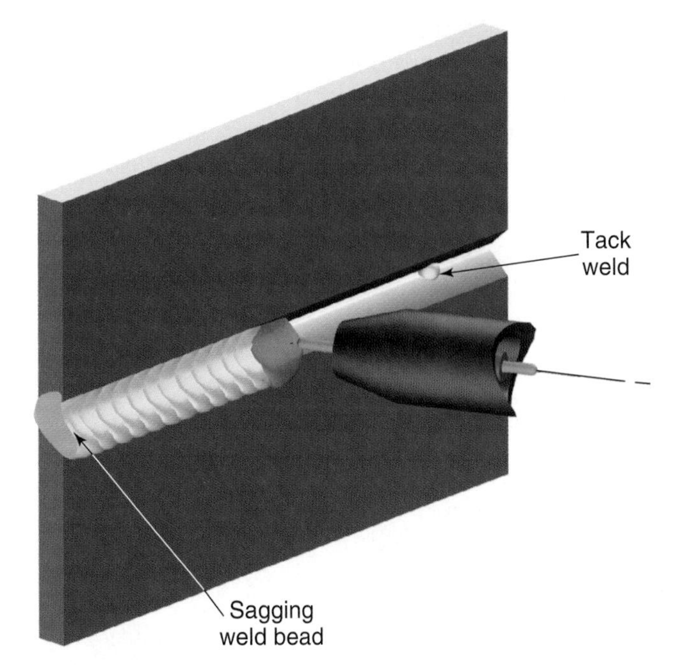

Figure 14-4. The weld bead on this horizontal butt joint was too hot and sagged downward. This resulted in an underfilling of the weld bead on the upper plate. The weld becomes weaker at this point due to the thinning.

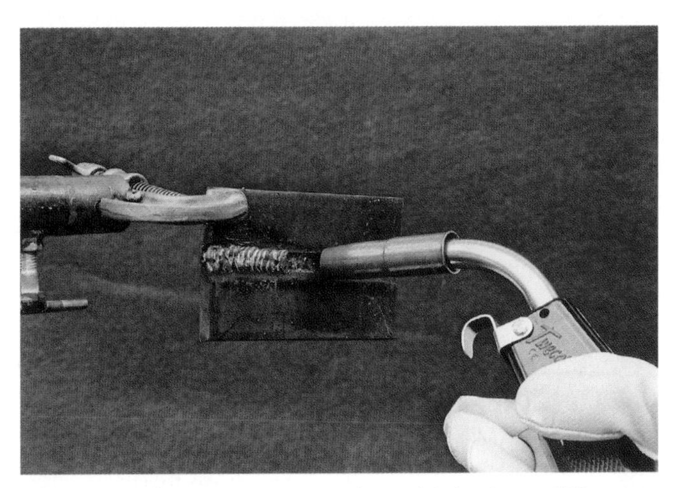

Figure 14-5. Notice the gun angle on this horizontal V-groove butt joint.

Welding a Butt Joint

Short circuiting or pulsed spray transfer is used when welding a butt joint. The weld pool remains fairly small and not too fluid in the horizontal position. When using these methods, a pool will sag downward if it gets too large. A slight sag is acceptable. However, overlap and underfill are not acceptable. ***Overlap*** is a condition in which the metal sags downward and overlaps the lower piece without melting the lower piece. ***Underfill*** is a condition in which the molten metal sags downward and does not fill the top part of the weld joint. See **Figure 14-4**.

The opening at the bottom of the weld bead should look like a keyhole to obtain a full penetration weld. This indicates that both pieces are melting and flowing together. It also shows that the metal is flowing through the joint to the backside for proper penetration. The round part of the keyhole should not be allowed to get too large. Backing may be used to control the penetration. Refer to Chapter 13 for more information on backing.

The angle of the welding gun and welding arc should be considered when making a horizontal butt weld. See **Figure 14-5**. The arc must melt both pieces. Excessive melting and undercutting may result if the electrode is pointed too much toward the upper piece. If the electrode is not directed to the upper piece, then it will not melt enough and will result in a lack of fusion. Watch the pool carefully. Adjust the gun angle as required to maintain a pool that is equally melted on both pieces. **Figure 14-6** shows the angles to use when making a horizontal butt weld. The weld pool should remain molten; however, it should not sag and cause overlap or underfill. Point the welding gun upward at a 5°–10° work angle to help control any sagging of the weld bead. Using a backing helps to control the penetration.

Higher shielding gas flow rates are used on horizontal and vertical position welds than on flat position welds. Since argon, CO_2, and other shielding gases tend to fall away from the weld area, more gas needs to be used to protect these out-of-position welds.

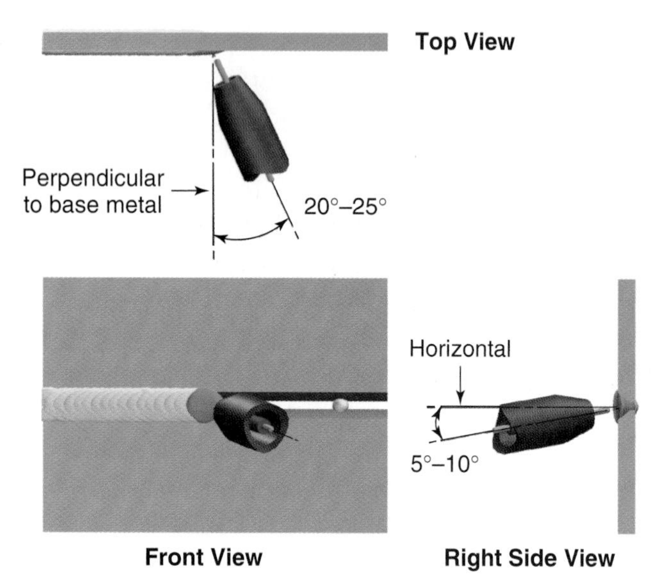

Figure 14-6. The suggested gun angles for welding a butt joint in the horizontal welding position.

Exercise 14-2 Welding a Butt Joint in the Horizontal Position

1. Obtain two pieces of mild steel measuring 1/8″–3/16″ × 1 1/2″ × 6″ (3.2 mm–4.8 mm × 40 mm × 150 mm).
2. Clean one long edge on each piece.
3. Set the proper voltage and wire feed speed for short circuiting transfer. Set the shielding gas flow.
4. Position the pieces to form a butt joint. Tack weld the pieces in three places along the joint. There should be a 1/8″ (3.2 mm) root opening.
5. Position the pieces for welding in the horizontal welding position.
6. Make a horizontal butt weld. Watch and control the weld pool. Fill the weld pool at the end of the weld. It may be necessary to weld more than one pass to fill the joint.

Inspection:

The completed weld should have a flat-to-convex weld bead shape. The ripples should be even. There should be no evidence of undercutting or other defects. Even penetration should be achieved on the back of the plates.

Welding in the Vertical Welding Position

Vertical position welding can be done traveling up (called ***uphill welding***) or down (called ***downhill welding***). The forehand welding method is used when welding uphill. The backhand method is used when welding downhill. When traveling up or down, the gun is usually pointed up. When welding in the vertical position, the weld pool should not become too large. Proper gun angle and control of the weld pool size will help prevent the pool from sagging.

The electrode is directed toward the root of the weld when welding in the vertical position. The electrode should be centered over the root of the weld. This heats both pieces of metal equally. A side-to-side weaving motion is used to obtain a wider weld bead. Hesitate at the edges of the weld to prevent undercutting when using a weaving motion. The gun angles used to make a vertical weld are the same as those used for flat welding. Gun angles for fillet and butt welds are discussed in Chapter 13.

A C-shaped weld pool is used when making a fillet weld. This indicates that both surfaces are melting. If one plate is thicker than the other, the electrode may have to be directed more toward the thicker plate. When welding a lap joint, the electrode may need to be pointed more toward the surface than the edge.

A fillet weld made in the vertical welding position should appear similar to and have the same strength as a weld made in the flat welding position; however, more skill is required to make the vertical weld. **Figure 14-7** shows two fillet welds being made in the vertical position.

More than one weld pass or weld bead may be required to complete a weld. The weld is cleaned after making each weld pass. Remove all slag from the weld and wire-brush the weld. The second pass uses a slightly different angle of the gun. The drag angle remains the same (20°–25°), but the gun is pointed between the first pass and the base metal. A third pass requires the same angle but in the opposite direction. Each pass must melt the base metal and the previous pass. All slag must be removed from the weld before making an additional weld pass. See **Figure 14-8**.

The same gun angles are used for making butt welds in the vertical and flat welding positions. A keyhole shaped opening should be seen on the first pass on a butt joint to obtain complete penetration. Backing can be used to control the penetration.

When making a multiple-pass weld, each weld must be cleaned prior to making the next pass. Each pass must melt into the base metal and the previous pass. Either stringer beads or weave beads can be used to fill and build up the weld.

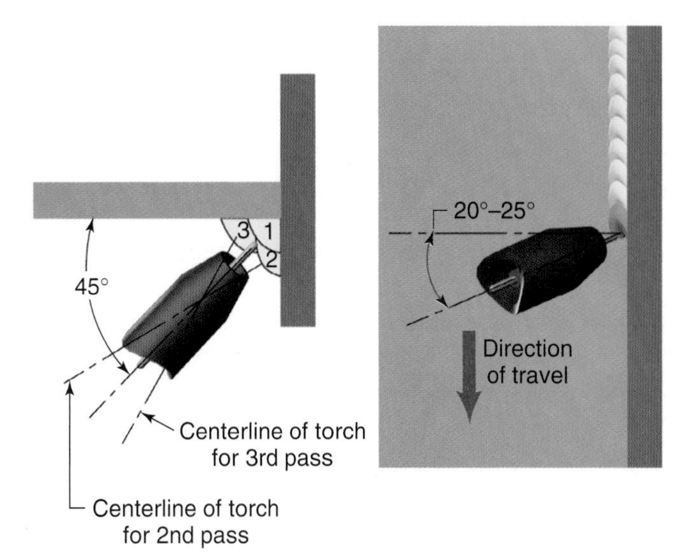

Figure 14-8. The suggested gun angles for welding a multiple-pass fillet weld on a lap joint in the vertical position. Notice that the work angle must change slightly from 45° for the second and third passes.

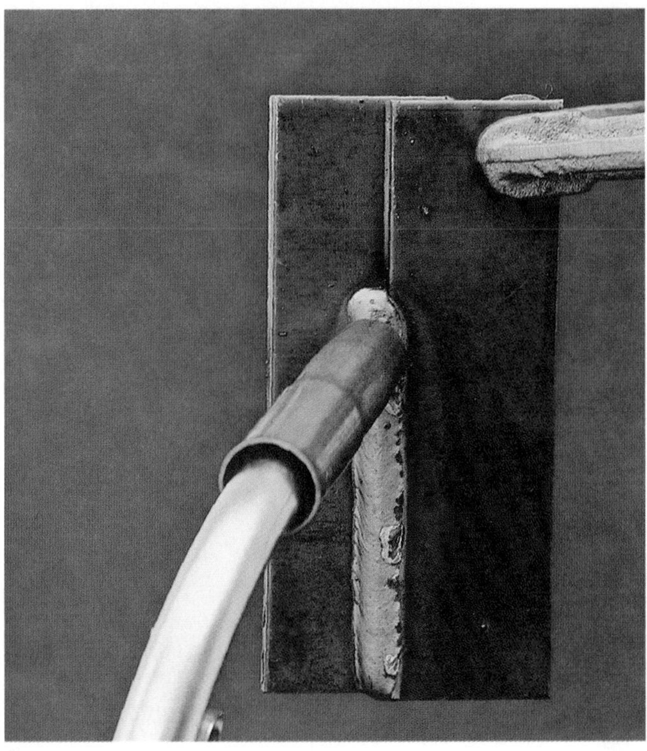

A

B

Figure 14-7. A—A fillet weld being made on a vertical lap joint. This weld is being made uphill using the forehand welding method. B—A fillet weld being made on a vertical T-joint. The weld is being made uphill using the forehand welding method.

Exercise 14-3 Making a Fillet Weld on a Lap Joint in the Vertical Welding Position

1. Obtain two pieces of mild steel measuring 1/8″–3/16″ × 1 1/2″ × 6″ (3.2 mm–4.8 mm × 40 mm × 150 mm).
2. Clean an edge and one surface on each piece.
3. Set the proper voltage, wire feed speed, and flow rate for short circuiting transfer.
4. Clamp the pieces together to form a lap joint.
5. Tack weld the pieces in three places along the joint. Tack welding can be done in the flat position.
6. Place the pieces in a welding positioner for welding in the vertical position.
7. Make a vertical fillet weld traveling downhill. Maintain the proper electrode extension. Make sure the weld penetrates into the surface of the lower piece.
8. Weld the joint on the other side of the metal. This weld can be made uphill or downhill. See Figure 14-7A.

Inspection:

The weld should have evenly spaced ripples on a convex or flat weld face. The width of the weld bead should be consistent. There should be no signs of defects.

Exercise 14-4 Making a Fillet Weld on a T-Joint in the Vertical Welding Position

1. Obtain two pieces of mild steel or stainless steel measuring 3/16″–1/4″ × 1 1/2″ × 6″ (4.8 mm–6.4 mm × 40 mm × 150 mm).
2. Clean a long edge on one piece and a surface on the other piece.
3. Set the voltage, wire feed speed, and shielding gas flow rate for short circuiting transfer.
4. Place the clean edge on the clean surface to form a T-joint. Tack weld the pieces in three places on each side of the joint. Tack welding can be done in the flat position.
5. Place the pieces in a welding positioner for welding in the vertical position.
6. Make a multiple-pass vertical fillet weld with three passes on each side of the T-joint. Weld one side uphill and the other downhill. Do not allow the weld pool to get too large when welding downhill. Reduce the wire feed speed or increase the travel speed if it gets too large.

Inspection:

The completed welds should not penetrate through the base metal. The weld beads should be convex and even in width with evenly spaced ripples. Spatter on the plates should be kept to a minimum. Refer to Figure 14-7B.

Exercise 14-5 Welding a Butt Joint in the Vertical Welding Position

1. Obtain two pieces of mild steel measuring 3/16″ × 1 1/2″ × 6″ (4.8 mm × 40 mm × 150 mm).

2. Set the proper voltage, wire feed speed, and shielding gas flow rate for short circuiting transfer.
3. Tack weld the pieces in three places along the joint to form a square-groove butt joint. There should be a root opening of 1/8″ (3.2 mm), as shown in the following drawing.
4. Mount the tack-welded pieces in a vertical position for welding.
5. Make a vertical butt weld using the keyhole method. Watch and control the weld pool. Do not let it sag.

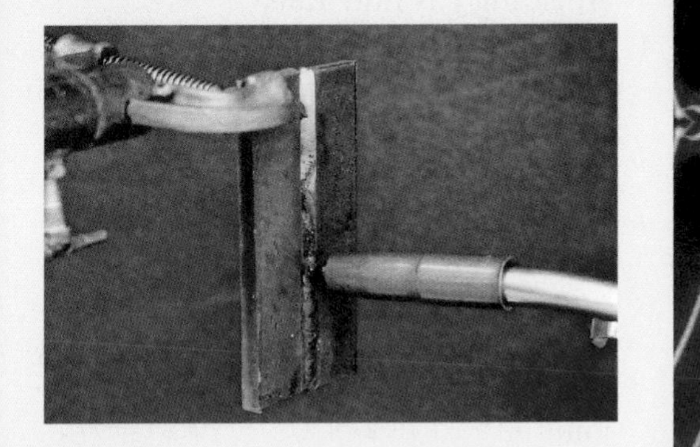

Inspection:

The completed weld should have a flat-to-convex weld bead shape. The ripples should be even. There should not be evidence of undercutting or other defects. There should be even penetration on the backside of the plates.

Welding in the Overhead Welding Position

The overhead position is considered to be the most difficult of all welding positions. The voltages and wire feed speeds used for overhead welding are lower than those used in the flat position. Refer to Chapter 12 for the proper voltage and wire feed speed. Use the values that are on the lower end of the charts. Short circuiting transfer and pulsed spray transfer are the only types of metal transfer that can be used. The weld pool must be kept to a small and controllable size.

Shielding gas flow rates are higher for welding in the overhead position than for other welding positions. Argon and carbon dioxide fall away from the weld area, thus more shielding gas must be used.

Welding gun angles used in the overhead welding position are the same as those used in the flat position. The gun should be pointed at the root of a fillet weld. See **Figure 14-9**. The gun is held at a work angle of 0° when welding a butt joint. A 5°–15° drag angle or backhand angle should be used. See **Figure 14-10**.

Figure 14-9. *A fillet weld being made on an overhead T-joint. The gun angles are the same as those used for the flat or vertical welding positions.*

Figure 14-10. *A V-groove butt joint being welded in the overhead welding position. The gun angles are the same as when flat welding or vertical welding.*

Hold the gun in a comfortable position. Sit or stand so that you can see the joint while welding.

Be sure to wear the proper protective clothing when welding in the overhead welding position. Cover all pockets and button your shirt collar. A cap and leathers are strongly recommended. Both GMAW and FCAW produce a lot of spatter. Very little spatter sticks to the base metal; instead, it falls downward.

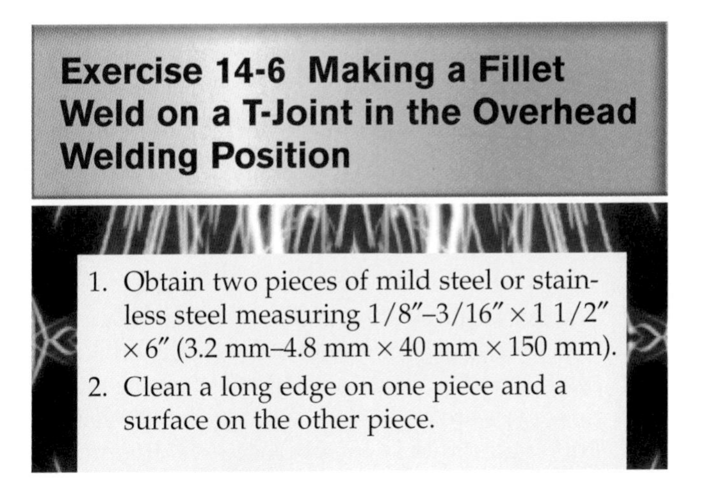

Exercise 14-6 Making a Fillet Weld on a T-Joint in the Overhead Welding Position

1. Obtain two pieces of mild steel or stainless steel measuring 1/8″–3/16″ × 1 1/2″ × 6″ (3.2 mm–4.8 mm × 40 mm × 150 mm).
2. Clean a long edge on one piece and a surface on the other piece.

3. Set the voltage, wire feed speed, and shielding gas flow rate.
4. Place the clean edge on the clean surface to form a T-joint. Tack weld the pieces in three places on each side of the joint.
5. Place the pieces in a welding positioner for overhead welding.
6. Make a fillet weld in the overhead welding position. Keep the electrode pointed at the root of the weld. Some weaving motion can be used. Watch the weld pool so that it does not get too large. Increase the travel speed or reduce the wire feed speed if the weld sags. Maintain the correct electrode extension.

Inspection:

The completed welds should not penetrate through the base metal. They should not show evidence of a large weld pool that sags. The weld should be convex, have a consistent width, and have evenly spaced ripples.

Exercise 14-7 Welding a Butt Joint in the Overhead Welding Position

1. Obtain two pieces of mild steel or stainless steel, measuring 1/8″–3/16″ × 1 1/2″ × 6″ (3.2 mm–4.8 mm × 40 mm × 150 mm).
2. Clean one long edge on each piece.
3. Set the voltage, wire feed speed, and shielding gas flow rate.
4. Align the pieces to form a butt joint. Tack weld the pieces in three places along the joint. There should be a 3/32″–1/8″ (2.4 mm–3.2 mm) root opening.
5. Position the pieces for welding in the overhead position.
6. Make an overhead butt weld. Weld the root pass and watch for the keyhole. A second pass may be required to fill the joint.

Inspection:

The completed weld should have a flat to convex weld bead shape. There should be no evidence of any defects.

Summary

- When performing GMAW out of position, only short circuiting transfer and pulsed spray transfer methods can be used.
- The flat welding position should be used whenever possible. Welding in the flat position is preferred over out-of-position welding because the pool will not sag or fall from the weld joint, all metal transfer methods can be used, it is more comfortable for a welder, and better welds may be produced.
- To make a fillet weld in the horizontal position with GMAW, a 20°–25° travel angle and a 45° work angle are used.
- To make a butt weld in the horizontal position with GMAW, a 20°–25° travel angle and a 5°–10° work angle are used.
- For welding in the vertical position, the forehand welding method is used when welding uphill. The backhand method is used when welding downhill. The gun angles used to make a vertical weld are the same as those used for flat welding.
- The overhead position is considered to be the most difficult of all welding positions. The voltages and wire feed speeds used for overhead welding are lower than those used in the flat position. Short circuiting transfer and pulsed spray transfer are the only types of metal transfer that can be used. Shielding gas flow rates are higher for welding in the overhead position than for other welding positions. Welding gun angles used in the overhead welding position are the same as those used in the flat position.

Review Questions

Write your answers on a separate sheet of paper. Please do *not* write in this book.

1. Which out-of-position weld can be made with all metal transfer methods?
   A. Horizontal fillet weld.
   B. Horizontal butt weld.
   C. Downhill butt weld.
   D. Downhill fillet weld.
2. Where should you point the gun and electrode when making a fillet weld?
3. What can be done to prevent undercutting when making a horizontal fillet weld?
4. Select three of the following defects that are more common in a horizontal butt weld than in a flat butt weld.
   A. Slag inclusions.
   B. Weld pool sag.
   C. Overlap.
   D. Rough surface on weld bead.
   E. Underfill.
5. What weld pool shape is used when welding a fillet weld in the vertical and overhead welding positions?
6. When using the backhand method for making a vertical weld, in which direction will you be traveling, uphill or downhill?
7. What two types of weld beads can be used to fill or build up a weld?
8. Are out-of-position welds stronger, weaker, or the same strength as those welded in the flat welding position?
9. Why must more shielding gas be used when welding out of position than when welding in the flat position?
10. What type of protective clothing is strongly recommended when welding in the overhead welding position?

Section 4

Gas Tungsten Arc Welding

Miller Electric Mfg. Co.

Chapter 15

GTAW: Equipment and Supplies

Learning Objectives

After studying this chapter, you will be able to:

- Define gas tungsten arc welding.
- Describe the principles of gas tungsten arc welding.
- Identify the equipment and supplies involved with GTAW.
- List the three different types of GTAW machines.
- Identify the parts of a GTAW torch.
- Describe the functions of the cables and hoses.
- Identify safety considerations when gas tungsten arc welding.

background current
collet
collet body
end cap
foot pedal
gas-cooled torch
gas lens
gas nozzle
gas tungsten arc welding (GTAW)
heliarc welding
nonconsumable electrode
partially rectified current
peak current
post flow
pulsed current
rectified
thumb switch
torch body
tungsten inert gas (TIG)
water-cooled torch

Technical Terms

The American Welding Society defines ***gas tungsten arc welding (GTAW)*** as "a process that joins metals together by heating them with an arc between a nonconsumable tungsten electrode and the work. The electrode and weld area are shielded with an inert gas or gas mixture. Filler metal may or may not be used." In industry, gas tungsten arc welding is commonly referred to as ***TIG***, which stands for ***tungsten inert gas***. Some people also refer to this process as ***heliarc welding***.

Gas Tungsten Arc Welding Principles

Gas tungsten arc welding is a welding process in which an arc is struck between a tungsten electrode and the base metal. See **Figure 15-1**. The tungsten electrode does not melt or become part of the weld. Therefore, it is referred to as a ***nonconsumable electrode***.

The tungsten electrode and the molten weld pool must be protected from the atmosphere while welding. Oxygen in the atmosphere causes the electrode and base metal to oxidize. An inert gas is used to shield the arc area from the atmosphere. Inert gases do not react with the electrode, arc, or molten weld

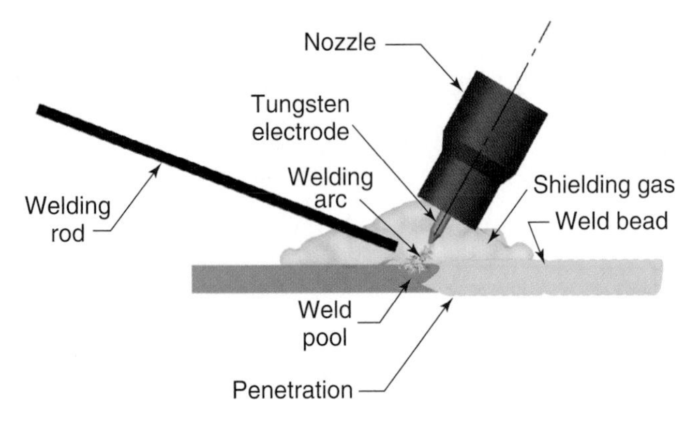

Figure 15-1. *A schematic drawing showing a gas tungsten arc weld in progress. The tungsten electrode is not consumed. A welding rod is often used to supply the filler metal.*

metal. GTAW is considered to be a very clean process, since oxygen is kept away from the electrode and the molten metal.

GTAW Equipment

The gas tungsten arc welding process requires the use of equipment designed for this type of welding. The equipment must be able to start and maintain a welding arc and to withstand the heat generated by welding. The equipment must also provide shielding gas to protect the electrode and the base metal. The basic parts of a GTAW setup include the following:

- A welding machine.
- A torch.
- An electrode.
- An inert gas supply and controls.

Other equipment that is regularly used includes:

- Filler metal.
- Remote control (usually a foot pedal).
- Water cooling equipment for the torch.

A person welding using GTAW must wear the following proper protective clothing: a welding helmet with a number 10 filter (minimum) and gloves. **Figure 15-2** shows a properly dressed welder working at a GTAW station.

GTAW Machine

A constant current welding machine is used for GTAW. This is the same type of machine used in shielded metal arc welding (SMAW). Constant current machines allow the welder to slightly vary the welding current by changing the arc length. Chapter 5 provides detailed information about constant current machines. Most GTAW welding machines provide for the following:

Figure 15-2. *GTAW being used to weld the intersection of two small, stainless steel pipes. Notice the welding hood and gloves. (Miller Electric Mfg. Co.)*

- Selection of current type (ac, DCEP, or DCEN).
- Selection of current range and setting.
- Selection and adjustment of the high-frequency voltage.
- Connection of welding torch.
- Connection of ground lead.
- Connection of shielding gas.
- Connection of water (only on water-cooled machines).
- Controls to allow shielding gas to flow after the current has stopped.
- Current control on the machine panel or from a remote control.

Most of the controls that a welder uses are on the front panel of a welding machine. See **Figure 15-3**. Electrical, water, and shielding gas connections are usually on the bottom front or the back of a machine. A GTAW machine can be an alternating current (ac) machine, a direct current (dc) machine, or a machine that can provide either ac or dc current.

Direct Current (dc) Welding

Two types of direct current are used for GTAW, direct current electrode negative (DCEN) and direct current electrode positive (DCEP). DCEN is also known as direct current straight polarity (DCSP). DCEP is also referred to as direct current reverse polarity (DCRP).

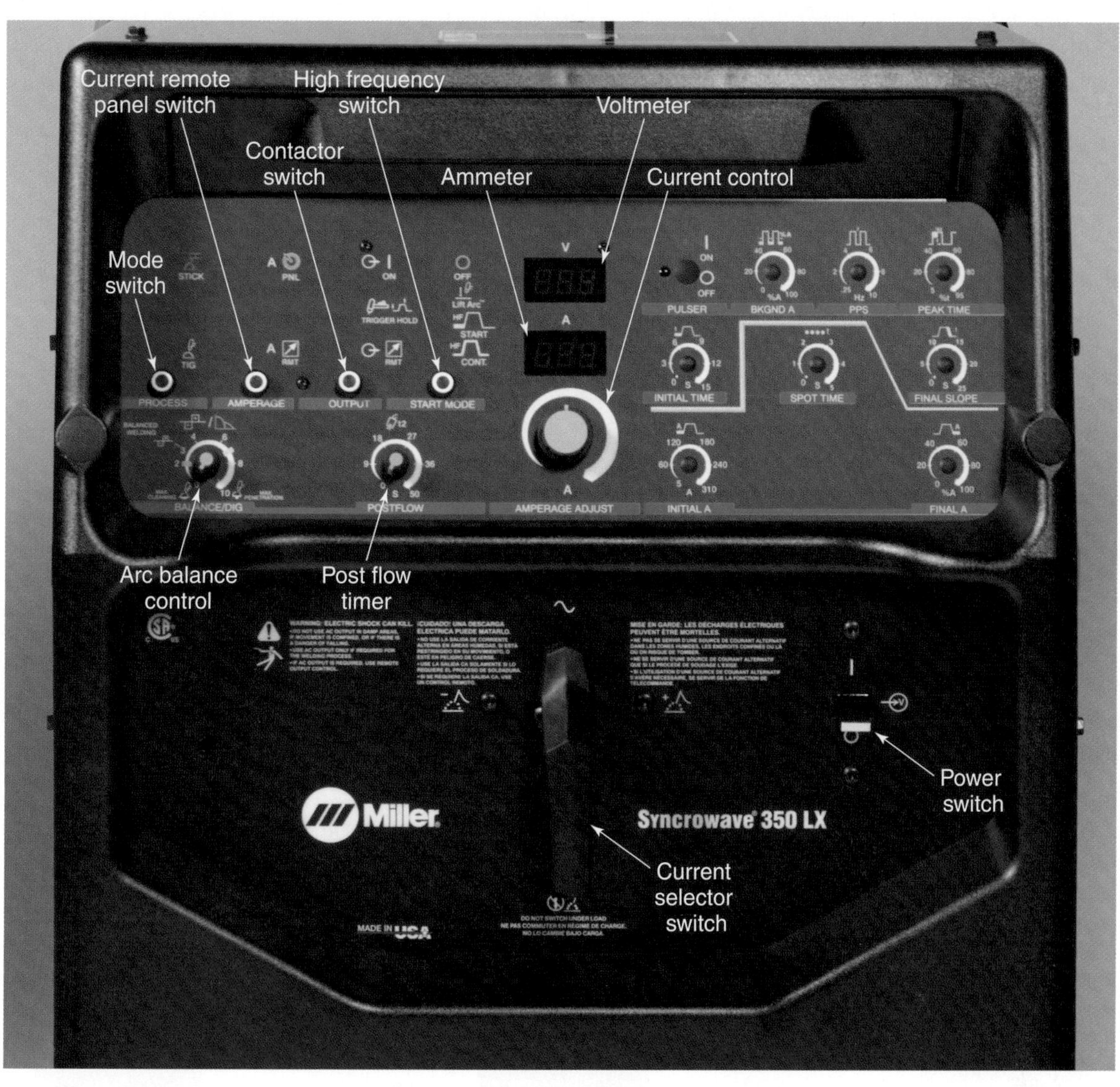

Figure 15-3. The control panel on one type of GTAW machine. (Miller Electric Mfg. Co.)

DCEN provides the deepest penetration. In DCEN, the electrons leave the negative electrode and flow to the base metal. Most of the heat, approximately 70%, is developed at the base metal. Only 30% of the heat is developed in the electrode. See **Figure 15-4**.

When DCEP is used, the electrode is positive, and the base metal is negative. Electrons leave the base metal and flow toward the positive electrode. In this case, approximately 70% of the heat is developed in the electrode, and only 30% is developed at the base metal. The weld bead is wide and shallow. See Figure 15-4. The electrode used during DCEP welding is larger than one used during DCEN for the same current. This allows the electrode to handle the higher heat of DCEP. A small-diameter electrode will melt using DCEP if there is too much current.

One major advantage of DCEP is the cleaning action it has on certain metals, such as aluminum and magnesium. Oxides on the base metal surface are removed. This results in better quality welds, **Figure 15-5**.

Alternating Current (ac) Welding

Alternating current welding combines the characteristics of DCEN and DCEP. The electrode alternates between negative and positive. This change occurs 120 times a second, which means the arc stops and restarts 120 times a second. There is an equal amount of heat on the base metal and the electrode. Penetration in ac welding is less than it is in DCEN but greater than DCEP. A cleaning action occurs, as in DCEP. See Figure 15-5.

Alternating current should flow equally in both directions. However, current in the electrode *positive* half cycle is less than in the electrode *negative* half cycle. In some cases, current does not flow at all in the electrode positive half cycle, **Figure 15-6**. When current does not flow in the electrode positive half cycle, it is said to be ***rectified***. When the current in the electrode positive half cycle is less than desired, it is referred to as ***partially rectified current***.

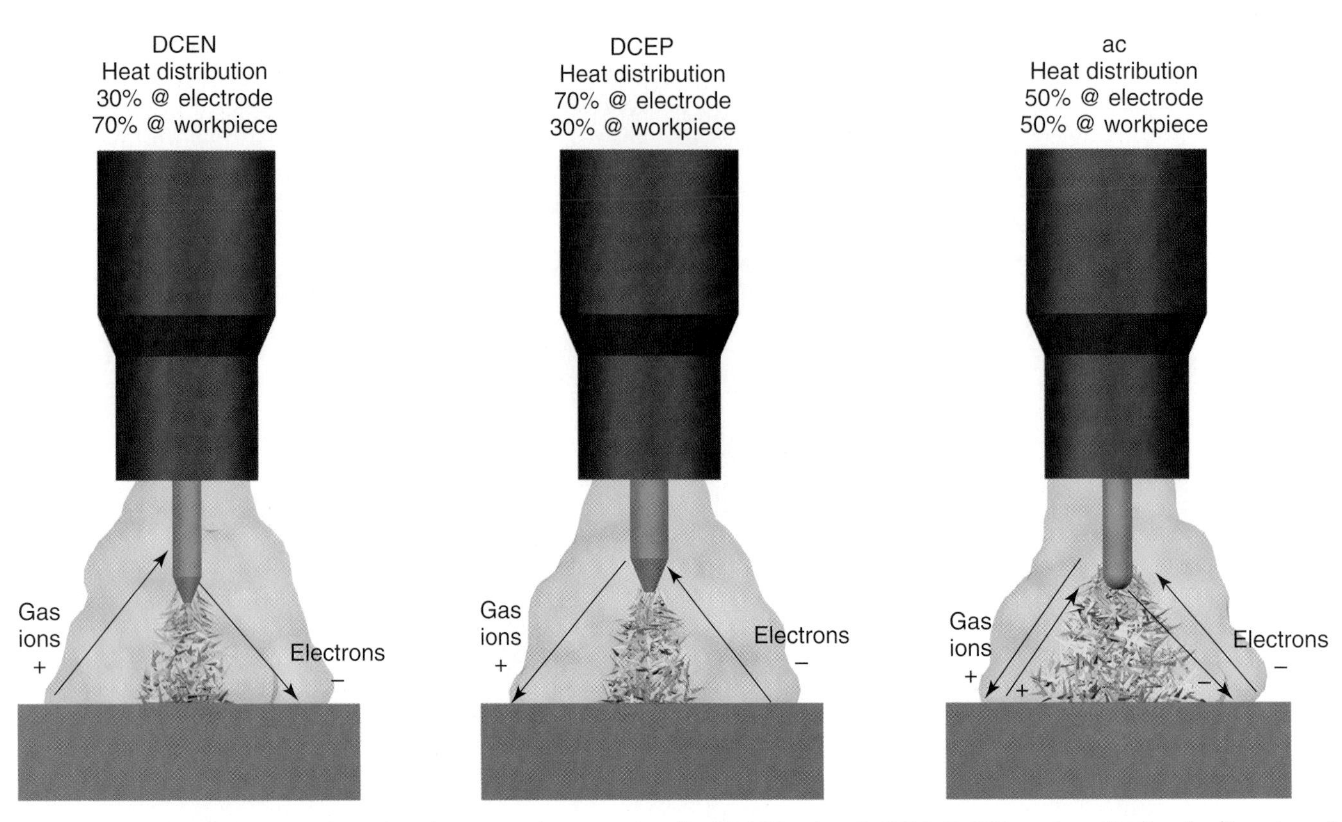

Figure 15-4. *The approximate heat distribution and penetration for GTAW using DCEN, DCEP, and ac. Notice the direction of travel for the gas ions and the electrons in each case. In ac, the gas ions and electrons are constantly reversing direction. Notice the larger diameter electrode needed for DCEP.*

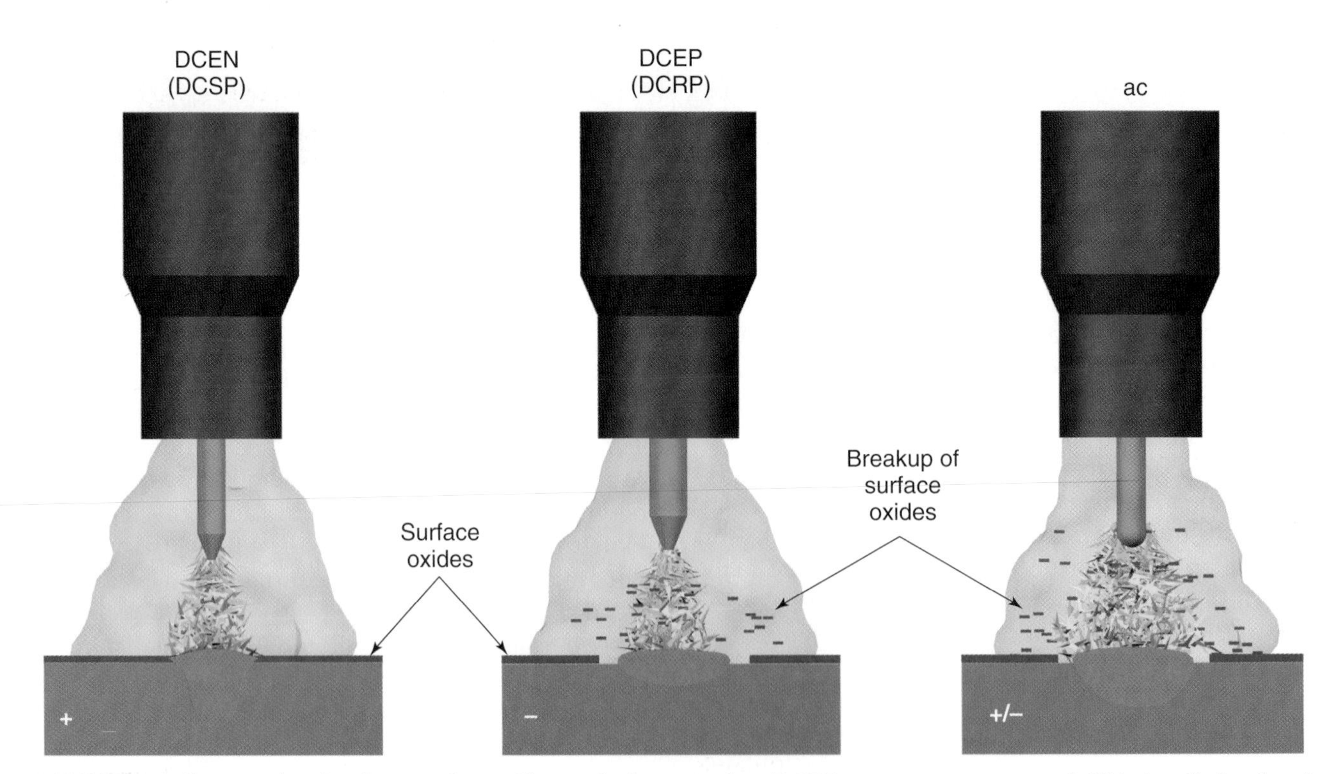

Figure 15-5. *A diagram showing how surface oxides are broken up when DCEP or ac current are used. This is called a cleaning action. The heavy gas ions strike the surface and cause the oxides to break up near the weld bead. No cleaning action takes place when DCEN is used.*

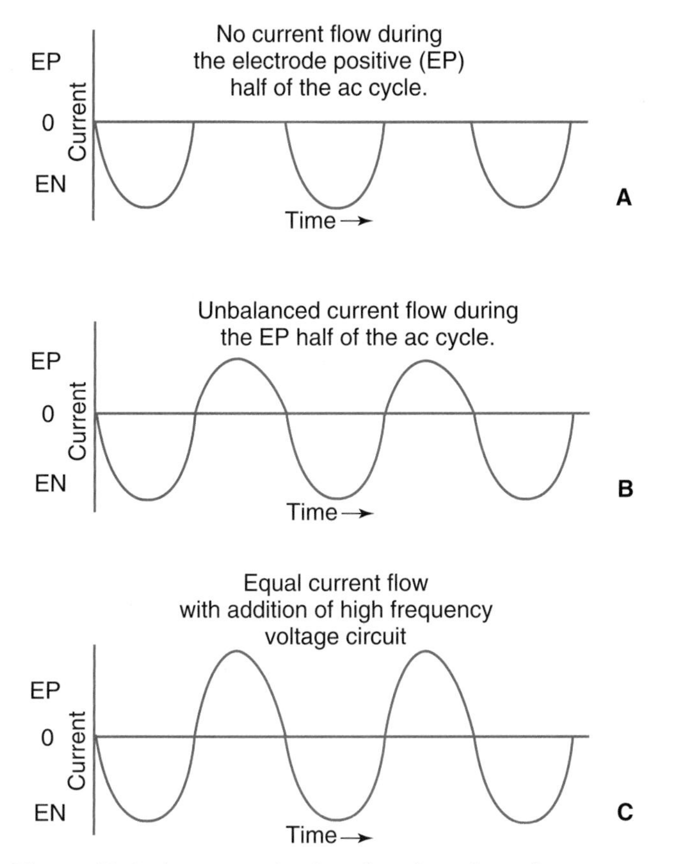

Figure 15-6. *As current is plotted against time, the ac waveform can be seen. A—In this curve, the electrode positive half cycle is rectified (stopped). B—An unbalanced curve where partial rectification is occurring. C—A balanced curve that results from the use of capacitors in the welding circuit.*

Positive half cycle current may be kept from being totally rectified by using a high-frequency, high-voltage, very-low-current generator in series with the welding transformer. This several-thousand-volt source of power is superimposed on the welding current and jumps the gap each time the arc stops. The arc stops 120 times per second. This high-frequency arc ionizes the gas. Ionized gas creates a path for the arc and allows the arc to restart more easily, especially during the positive half cycle.

New square wave power supplies have the ability to switch the current direction very rapidly. The quick current reversal allows the arc to reestablish easily. There is a higher voltage and current delivered from the power supply at the time the current reverses direction. The ionized gas is still present. Square wave power supplies can deliver the same amount of current in both the DCEN and DCEP part of the cycle. These power supplies can be adjusted with a knob to change the amount of current and/or vary the duration of the DCEN and DCEP half cycles.

A balanced ac cycle occurs when the current during the positive and negative half cycles are equal. A balanced wave is beneficial for some welding applications. Electronic circuits are used to balance the wave. Square wave technology allows for a balanced output. Older power supplies used capacitors, which discharged during the DCEP portion of the wave. This additional current was used to create a balanced wave.

Pulsed GTAW

Certain GTAW machines produce a pulsed current. ***Pulsed current*** is like direct current except that it has two current settings. There is a high current period followed by a low current period. The high current is the ***peak current***, and the low current is the ***background current***. See **Figure 15-7**.

Pulsed current is excellent for out-of-position welding. The metal becomes molten during the peak current and forms a spot. The weld then has a chance to cool as the welder moves the torch a short distance during the background current period. Another spot forms during the following peak current period. This pulsed current produces a weld that is actually a series of overlapping welded spots. See **Figure 15-8**.

Torches

GTAW torches perform many functions. The four major functions are:

- Hold the electrode.
- Make good electrical contact with the electrode.
- Direct shielding gas to the weld area.
- Provide a means to cool the electrode.

Figure 15-9 shows an exploded view of a GTAW torch. A ***collet*** is used to hold the electrode and make good electrical contact. Collets are made in different sizes. Each size is for one specific size of electrode. A

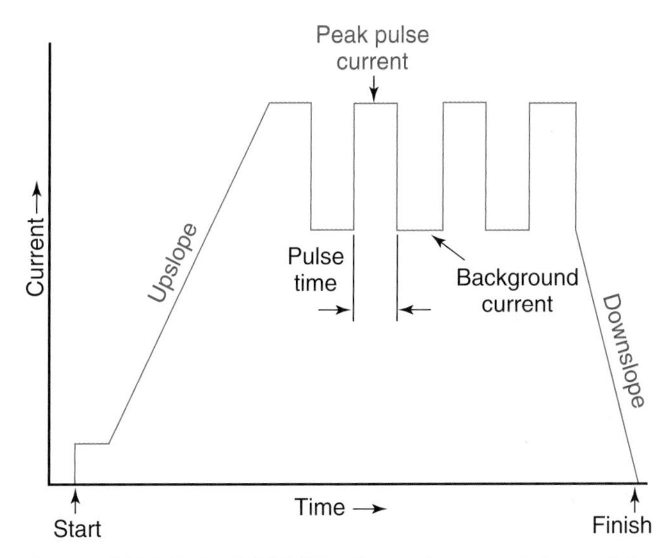

Figure 15-7. *Pulsed GTAW. The peak current is used for welding. The lower background current maintains the arc and allows the weld to cool.*

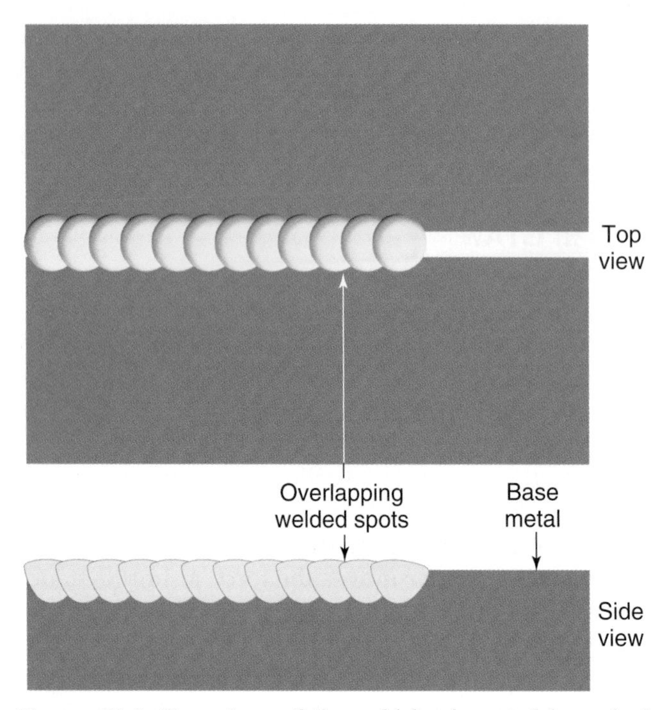

Figure 15-8. Two views of the weld bead created by pulsed current GTAW. A series of overlapping welded spots makes up the weld bead.

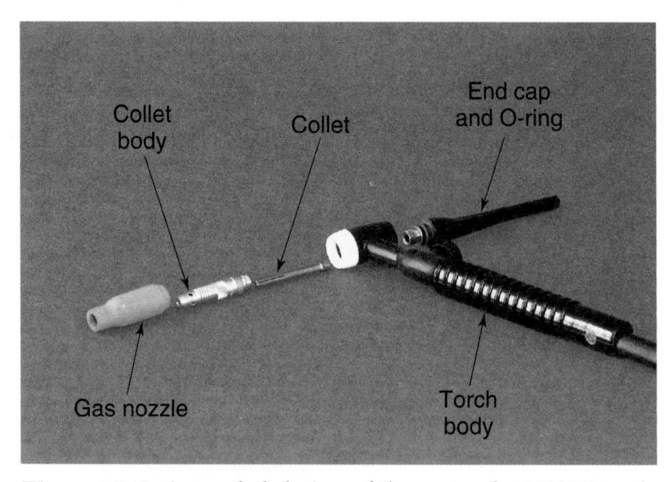

Figure 15-9. An exploded view of the parts of a GTAW torch.

collet body is used to support the collet. Collet bodies and collets are usually available as a set. The set is used for only one specific electrode size. Whenever the electrode size is changed, the collet and collet body must also be changed. Collets have a slot in them. When the ***end cap*** is tightened, the collet tightens around the electrode. A tight fit holds the electrode in its proper place for welding. It also provides a good electrical contact.

The ***torch body*** contains connections for shielding gas, water, and electricity. An end cap and O-ring are used to prevent shielding gas from escaping. They also prevent the atmosphere from getting into the torch. End caps are available in different lengths.

The ***gas nozzle*** is used to direct the shielding gas to the weld area. A number of gas nozzle sizes are available. See **Figure 15-10**. If the selected gas nozzle is too small, there will not be enough shielding gas to cover the weld area. If a too-large gas nozzle is used, shielding gas will be wasted. The gas nozzle must be able to withstand the high temperatures produced by the arc. Most gas nozzles are made of ceramic or metal. Often, the gas nozzle is simply called the "nozzle," or it may be called a "cup."

A gas lens may be used in a GTAW torch. A ***gas lens*** is a series of very fine stainless steel wire screens. It makes the shielding gas exit the nozzle in a column. When using a gas lens, the electrode can stick out of the cup farther, and the welder can see the weld better. Hard-to-reach areas, like inside corner joints, can be welded more easily because the electrode sticks out farther.

The heat that develops in the electrode, cables, and nozzle must be removed, or the torch will overheat. GTAW torches are either gas- or water-cooled. See **Figure 15-11**. ***Gas-cooled torches*** are used for light duty work and are not recommended for use with over 200A. ***Water-cooled torches*** are used to carry currents over 200A. They have a continuous flow of water passing through the torch body to remove heat.

Cables and Hoses

Cables convey the electrical current to the torch. A hose carries the shielding gas to the torch. The cables also carry the water to and from the torch when a water-cooled torch is used. Checking the cables and hoses for wear or leaks once a day is a good idea. When cables are installed properly, little or no adjustment is needed.

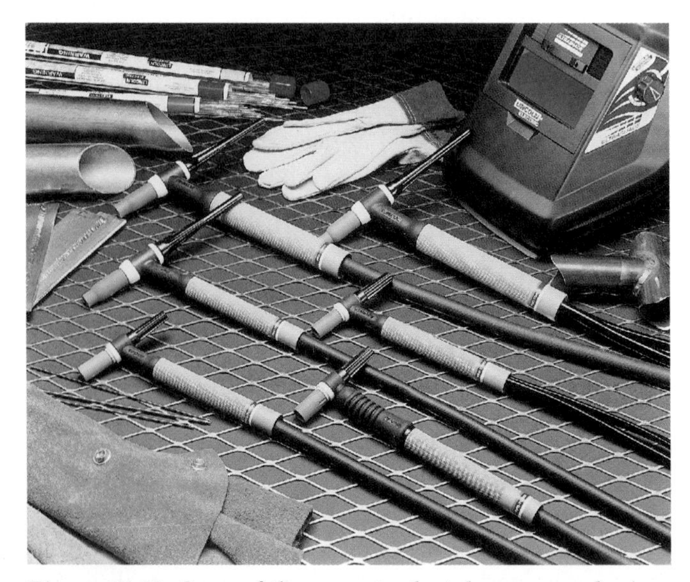

Figure 15-10. Some of the many torch styles, gas nozzle sizes, and end cap lengths used for GTAW. (Lincoln Electric Company)

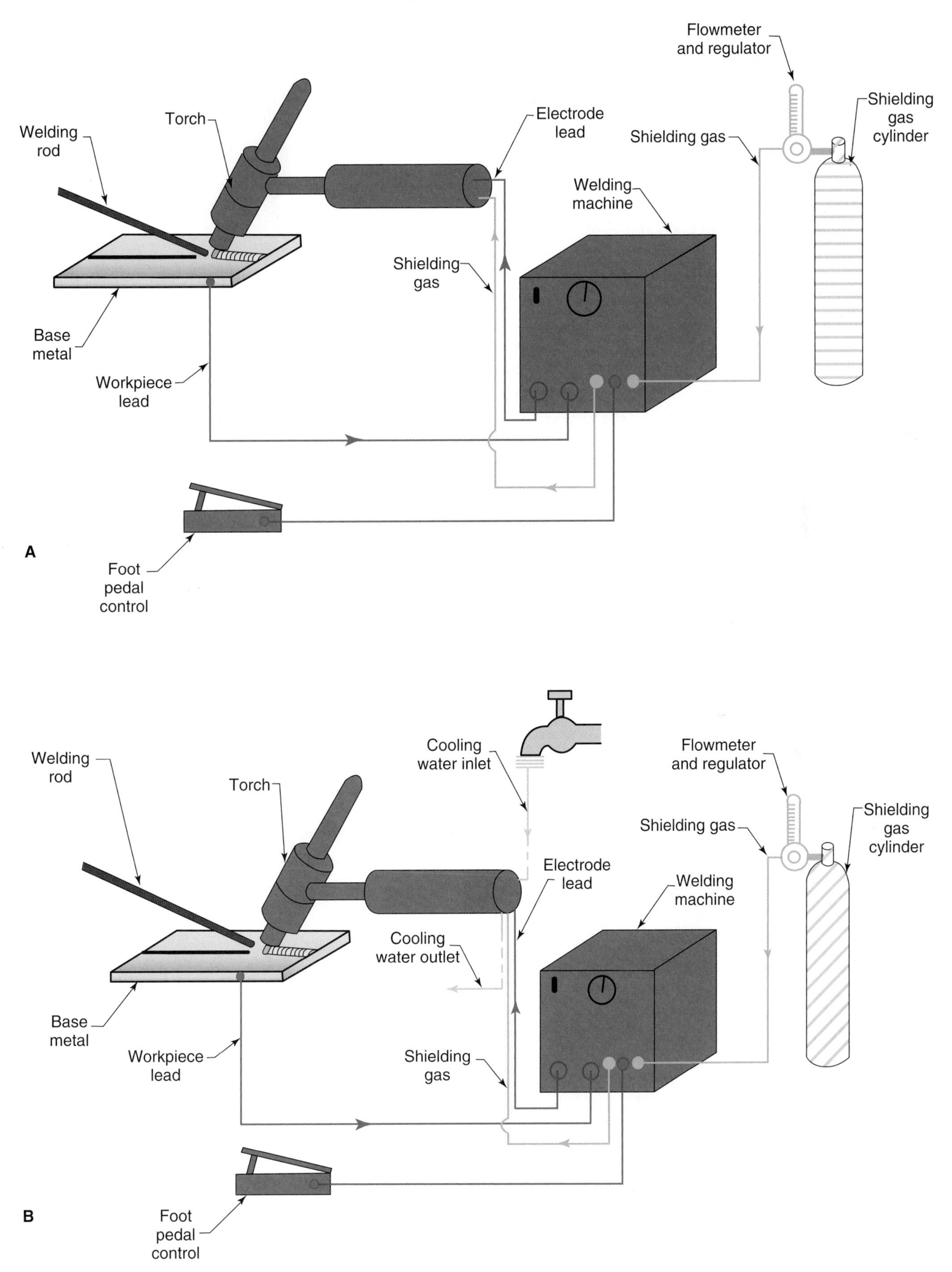

Figure 15-11. *A—A drawing of a gas-cooled GTAW outfit. Notice that the shielding gas hose and electrode lead enter the torch. B—A drawing of a water-cooled GTAW outfit. Note that the electrode lead and hoses for cooling water and shielding gas enter the torch.*

In **Figure 15-12**, the shielding gas and water supply to the torch have their own hoses. The power cable, however, has a dual purpose. It carries the electrical current to the torch and is also used to take water away from the torch. The water cools the power cable. Each fitting shown is attached to the welding machine.

Shielding Gases

Inert shielding gases are used in GTAW. An inert gas does not react with the weld. Two types of shielding gas are used: argon and helium. Each gas can be used by itself, or a combination of the two can be used. Each gas has advantages and disadvantages. It is easier to start and maintain an arc with argon. Helium provides a hotter arc and provides for faster travel speeds. It is commonly used in automatic GTAW. Helium, however, is more expensive than argon. Less shielding gas is required when using argon because it is heavier than air. Argon covers the weld area better when welding in the flat position. See **Figure 15-13**.

Caution: Argon will displace oxygen. When welding in a closed area, there must be good ventilation to remove the shielding gas and supply fresh air. Suffocation can occur if proper ventilation is not provided.

Regulators and Flowmeters

Argon and helium are available in cylinders. Care must be taken when handling, moving, and storing gas cylinders. Chapter 20 provides detailed information regarding cylinders.

A regulator is used to reduce the cylinder pressure to a usable pressure. A flowmeter controls the amount

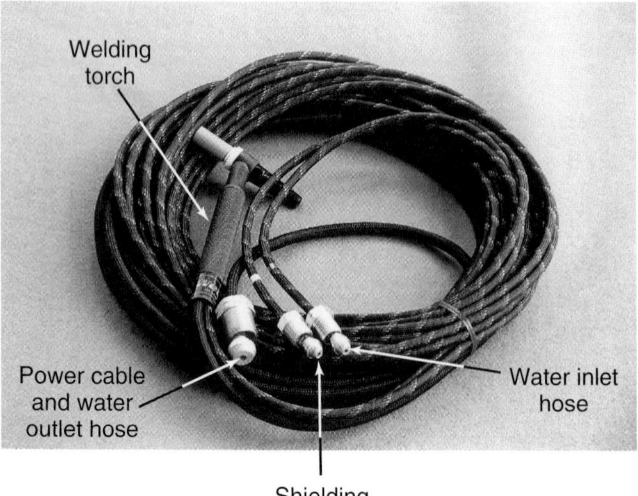

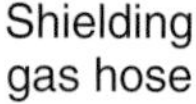

Figure 15-12. A water-cooled GTAW torch with its hoses and fittings. The water outlet hose also carries the current to the torch and electrode.

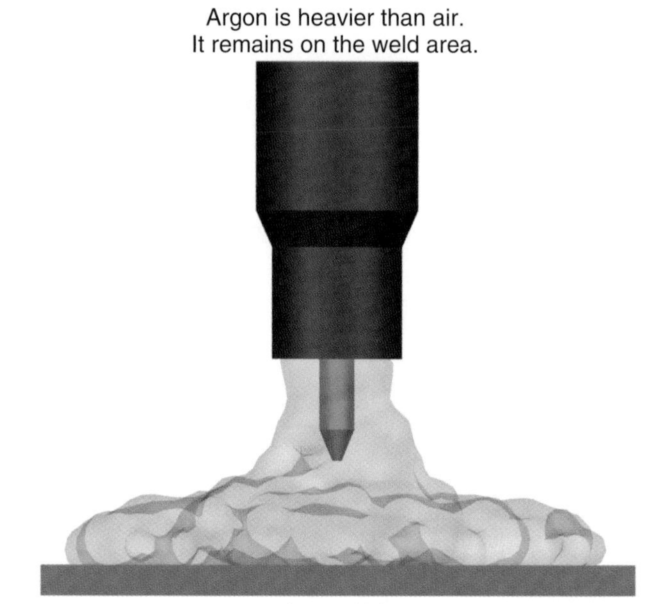

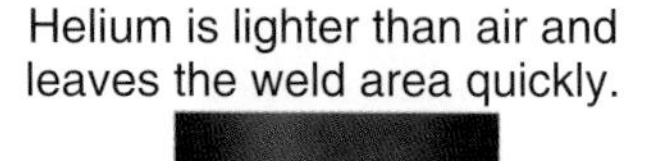

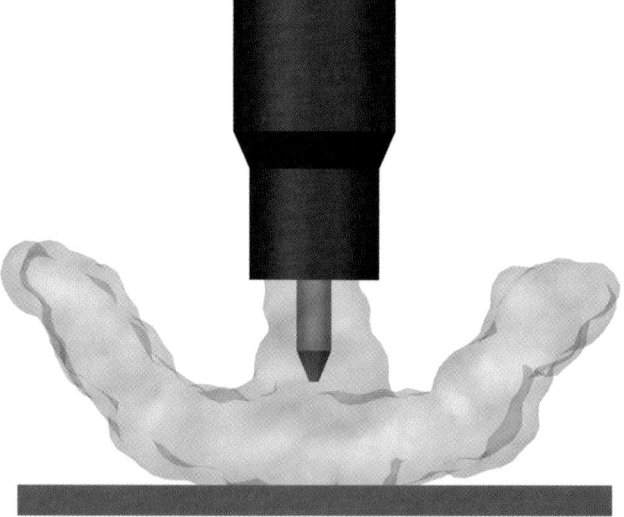

Figure 15-13. The efficiency of argon and helium as shielding gases. Argon is heavier and stays on or near the weld area for a relatively long period of time. Helium is a lighter gas. It rises rapidly and leaves the weld area.

of shielding gas that goes to the welding torch. Flowmeters measure gas flow in cubic feet per hour (cfh) or liters per minute (Lpm). Argon is heavier than helium, so an argon flowmeter is calibrated differently from a helium flowmeter. Sometimes there are different scales on opposite sides of a single flowmeter. Each scale is calibrated for a different shielding gas. **Figure 15-14** shows a typical flowmeter.

The gas flow is adjusted using a screw or knob on the flowmeter. See Chapters 11 and 20 for additional information on regulators and flowmeters.

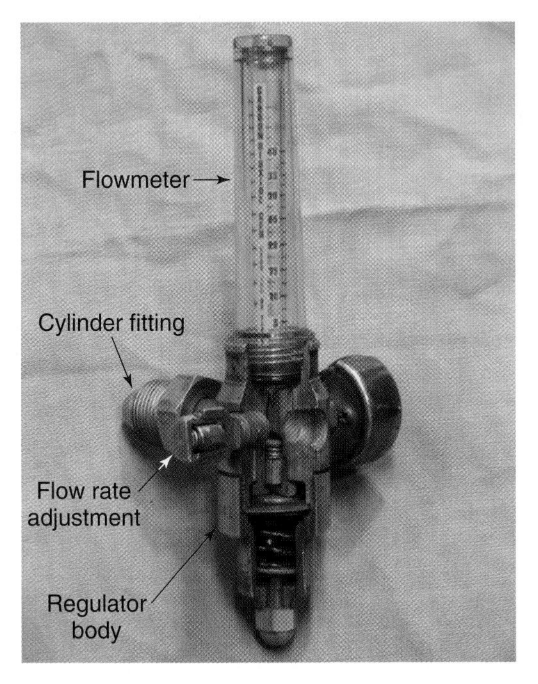

Figure 15-14. *A cutaway view of a ball-type flowmeter with a built-in pressure regulator.*

As discussed earlier, a shielding gas covers and protects the electrode and the base metal during GTAW. If you stopped welding and the shielding gas also stopped, oxygen and other gases could contaminate the weld. The tungsten electrode would then become oxidized. To prevent this, shielding gas flow continues for a short time after the weld current stops. This gas flow is called ***post flow***. The length of time that the gas flows is set on the welding machine.

Remote Current Control

A foot pedal or thumb switch is commonly used in GTAW. The ***foot pedal*** and ***thumb switch*** are remote control devices. They control the current and turn shielding gas on and off. These devices can be used to turn the current on or off and also to control the current within the range on the welding machine. See **Figure 15-15**.

A button or switch on the welding machine controls the foot pedal or thumb switch. The button or switch has two positions; one labeled "control panel" and the other labeled "remote." With the switch or button in the control panel position, the foot pedal or thumb control is used as an on/off switch. The current that is preset on the welding machine is delivered to the torch.

In the remote position, the foot pedal or thumb switch is used to vary the current within the range preset on the welding machine. The foot pedal allows more current to flow as it is pressed down. This gives the welder additional control of the welding operation.

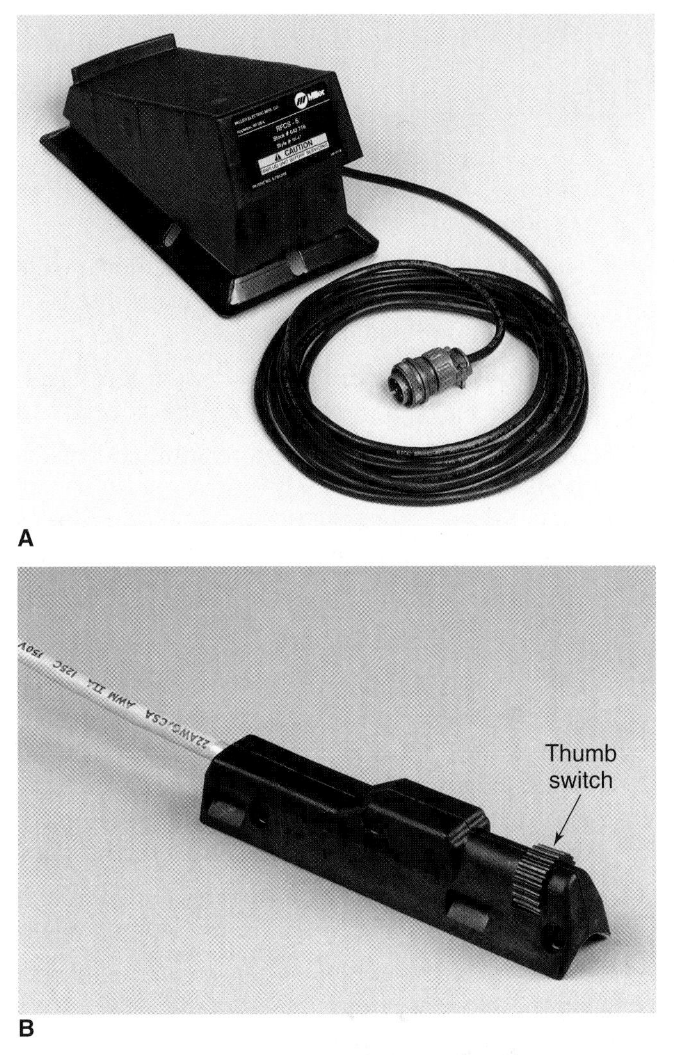

Figure 15-15. *A—A foot pedal for remote current control. Pressing down the foot switch also turns on the water and shielding gas. (Miller Electric Mfg. Co.) B—A thumb-operated current control switch for use on a GTAW torch. The amperage is varied by turning the rolling thumb switch. (Miller Electric Mfg. Co.)*

Protective Equipment

Proper protective clothing must be worn for GTAW. Proper clothing includes a shirt with long sleeves, pants, and gloves. Clothing should be fire-resistant. Flammable objects like matches and lighters should not be carried in your pockets. Gloves worn for GTAW are usually thinner than those used for SMAW or GMAW. GTAW requires good skill in handling the torch and filler rod. Thin gloves are worn so the welder can get a good feel for the torch and filler metal.

A GTAW arc is very intense. It is not shielded by flux or fumes, so a number 10, 12, or 14 lens is recommended in your welding helmet. A clear cover lens should be placed in front of the shade lens. Flash goggles with a number 2 lens should be worn under your helmet.

Summary

- Gas tungsten arc welding is a welding process in which an arc is struck between a tungsten electrode and the base metal. The tungsten electrode does not melt or become part of the weld. An inert gas shields the arc area from the atmosphere.
- The basic parts of a GTAW setup include a welding machine, a torch, an electrode, and an inert gas supply and controls. Other equipment that is regularly used includes filler metal, a remote control (usually a foot pedal), and water cooling equipment for the torch.
- A GTAW machine can be an alternating current (ac) machine, a direct current (dc) machine, or a machine that can provide either ac or dc current.
- DCEN provides the deepest penetration because most of the heat, approximately 70%, is developed at the base metal.
- When DCEP is used, approximately 70% of the heat is developed in the electrode, and only 30% is developed at the base metal. One major advantage of DCEP is the cleaning action it has on certain metals, such as aluminum and magnesium.
- When ac is used, there is an equal amount of heat on the base metal and the electrode. Penetration in ac welding is less than it is in DCEN but greater than DCEP. A cleaning action occurs, as in DCEP.
- In a GTAW torch, a collet is used to hold the electrode and make good electrical contact. A collet body is used to support the collet. An end cap is used to tighten the collet around the electrode and also prevents shielding gas from escaping. The torch body contains connections for shielding gas, water, and electricity. The gas nozzle directs the shielding gas to the weld area. A gas lens may be used in a GTAW torch to make the shielding gas exit the nozzle in a column.
- Cables carry the electrical current to the torch. A hose carries the shielding gas to the torch. The cables also carry the water to and from the torch when a water-cooled torch is used.
- Two types of shielding gas, argon and helium, are used in GTAW. Each gas can be used by itself, or a combination of the two can be used.
- The foot pedal and thumb switch are remote control devices that control the current and turn shielding gas on and off.
- **Proper protective clothing must be worn for GTAW. Proper clothing includes a shirt with long sleeves, pants, and gloves. Clothing should be fire-resistant. Flammable objects like matches and lighters should not be carried in your pockets. A GTAW arc is very intense, so a number 10, 12, or 14 lens is recommended. A clear cover lens should be placed in front of the shade lens. Flash goggles with a number 2 lens should be worn under your helmet.**

Review Questions

Write your answers on a separate sheet of paper. Please do *not* write in this book.

1. List two names other than GTAW that are used to refer to the gas tungsten arc welding process.
2. A tungsten electrode is called a(n) _____ electrode since it does not become part of the weld.
3. A(n) _____ welding machine is used in GTAW. This type of machine allows the welder to vary the _____ slightly by changing the arc length.
4. Which type of welding current applies most of the heat to the base metal and gives the best penetration?
5. What is used to prevent the current from stopping during the electrode positive half cycle when ac welding?

(continued)

Review Questions *continued*

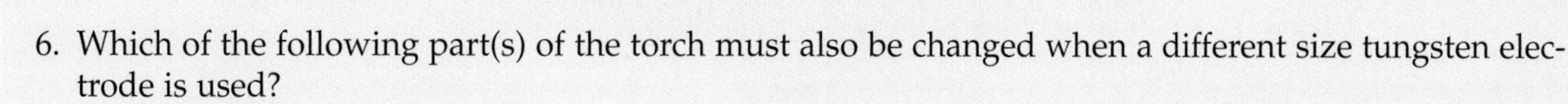

6. Which of the following part(s) of the torch must also be changed when a different size tungsten electrode is used?
   A. Torch body.
   B. Collet.
   C. End cap.
   D. Collet body.
7. Which of the cables on a water-cooled torch serves two purposes? What are those two purposes?
8. A(n) _____ is used to reduce the cylinder pressure of shielding gas to a usable level. A(n) _____ is used to control the amount of shielding gas used.
9. Why is a post flow of shielding gas used with GTAW?
10. *True or False?* Gloves worn for GTAW are thick leather to protect the welder from the intense heat and rays of the arc.

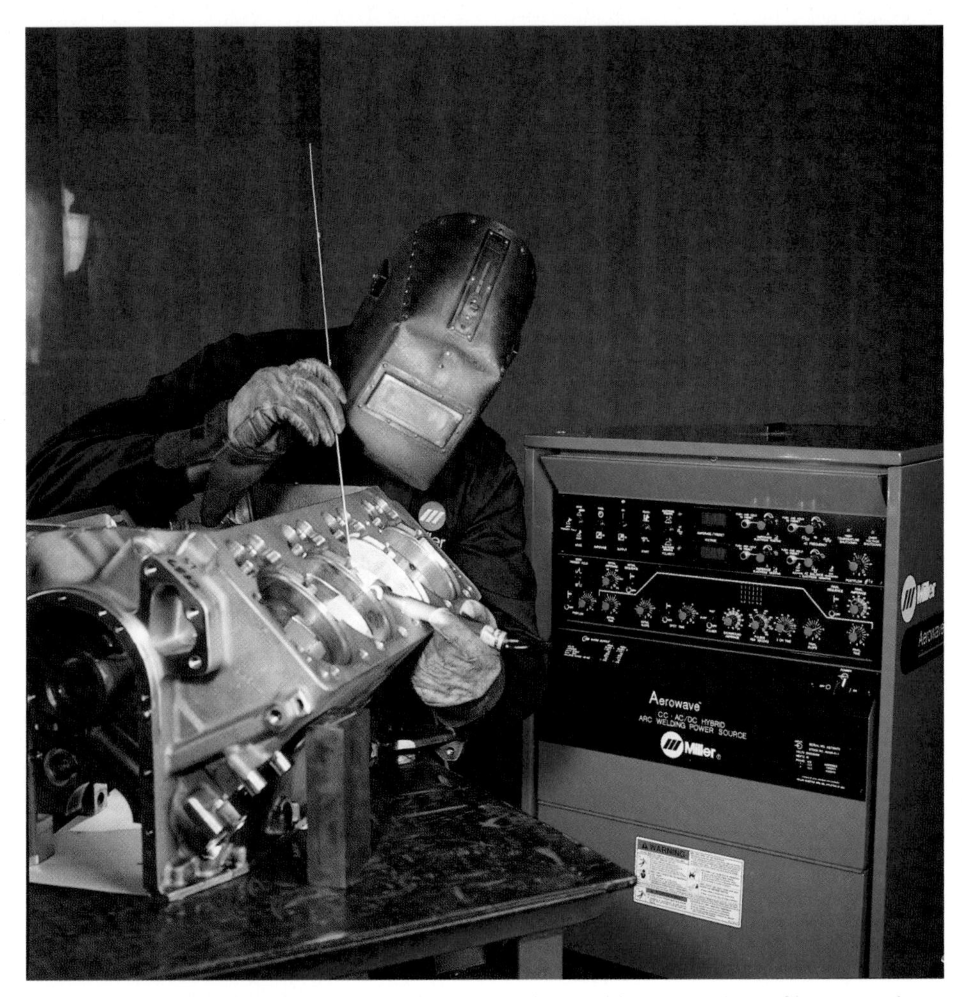

Because it is capable of creating precise, high-quality welds in a variety of base metals, the GTAW process is well-suited for such tasks as repairing aluminum engine blocks. (Miller Electric Mfg. Co.)

Chapter 16

GTAW: Equipment Assembly and Adjustment

Learning Objectives

After studying this chapter, you will be able to:

- Assemble a GTAW welding outfit.
- Prepare an electrode for GTAW.
- Adjust the shielding gas flowmeter for the proper flow rate.
- Identify electrode type designations for GTAW electrodes.
- Select the proper current amount and type for the metal to be welded.
- Know and be able to use the metal cleaning processes used in GTAW.

ceria	electrode tip
circuit breaker	lanthana
contactor switch	thoria
current switch	zirconia

Equipment Assembly

Equipment used in a GTAW outfit must be assembled properly. The main piece of equipment in a welding station is the welding machine. All cables and hoses, the shielding gas supply, and the welding torch are connected to the welding machine. Equipment used to control the flow of shielding gas must be assembled and adjusted. The torch must also be assembled correctly.

Electrical Connections

GTAW machines are connected to 120V or 240V electrical circuits. Machines are often wired into a ***circuit breaker*** panel or a fuse box. **Only a qualified electrician should connect, inspect, or repair the wiring inside a circuit breaker panel or fuse box. A welder should *not* try to repair wiring inside a welding machine.**

All welders should know where the circuit breaker panel for the welding shop is located. **In case of an emergency, all people working in the space should know how to turn off power at the circuit breaker.** In addition to turning off the welding machine at the end of each shift, some welding shops turn off the power at the circuit breaker at the end of the day. A circuit breaker is shown in **Figure 16-1**.

Figure 16-1. A disconnect switch or circuit breaker for an arc welding machine. This switch is mounted on the wall in the welding booth. It may be used to turn the welding machine off in an emergency.

Figure 16-2. Connections for the shielding gas, water, and electrode and workpiece leads are made under the hinged panel of this welding machine.

Cable and Hose Connections

A welding machine has places to connect all needed cables and hoses. Electrical connections for the electrode and workpiece lead are mechanical. They must be tight to make good contact. Water and shielding gas connections are threaded to help prevent leaks. Use the proper size wrench to tighten these connections. **Figure 16-2** shows the cable and hose connections on a welding machine. See Figure 15-12 for connections on a torch.

Most machines have a remote control connection. This is where the foot pedal, thumb switch, or other control feature is connected.

Cylinder, Regulator, and Flowmeter

Cylinders containing compressed gases must be handled with great care. They must be secured to a wall, column, or welding hand truck with chains or steel bands. This keeps the cylinder from falling over and damaging the cylinder valve. See Chapter 20 for more information.

Once the cylinder is secure, a regulator and flowmeter can be attached to it. **First, clean the outlet of the cylinder valve. Make sure the outlet is not pointing toward yourself or anyone else. Quickly open and close the cylinder valve. This cleans the outlet.**

After the outlet is clean, a regulator and flowmeter can be attached. Thread the regulator nut onto the cylinder valve. If a ball float-type flowmeter is used, the flowmeter must be in a vertical position. See **Figure 16-3**. Tighten the regulator nut with an open-end wrench of the correct size. Do not overtighten or you may damage the threads.

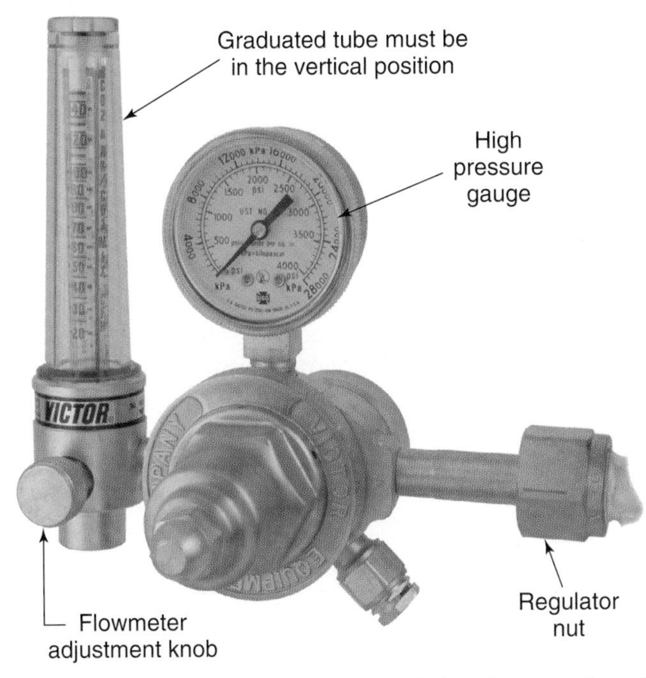

Figure 16-3. *Flowmeters must be installed with the graduated clear tube in the vertical position. This flowmeter is calibrated for three different shielding gases. (Victor Equipment Co.)*

Assembling the Torch

A separate collet and collet body are needed for each electrode diameter. Choose the correct-size collet and collet body for the electrode's diameter. Assemble the torch as follows:

1. Thread the collet body into the torch body.
2. Slide the collet into the collet body from the top of the torch, as shown in **Figure 16-4**.
3. Thread an end cap with an O-ring loosely into the torch body. Do not tighten the end cap yet.
4. Select the gas nozzle you will be using. Thread it into the torch body. A nozzle is also called a cup.

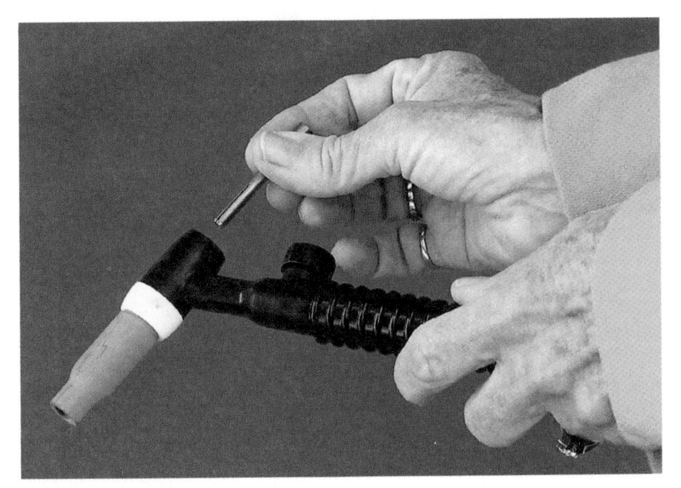

Figure 16-4. *A collet being inserted into the collet body of GTAW torch.*

5. Insert the electrode into the collet. When correctly inserted, the electrode will extend past the end of the gas nozzle only a short distance. The distance depends on the type of joint being welded, as shown in **Figure 16-5**.
6. When the electrode extends the right amount, tighten the end cap. See **Figure 16-6**. This will make the collet clamp the electrode and secure it. To remove or adjust the electrode, loosen the end cap.

Adjusting the Shielding Gas Flowmeter

After the regulator and flowmeter are attached to the cylinder, the correct rate of gas flow can be set.

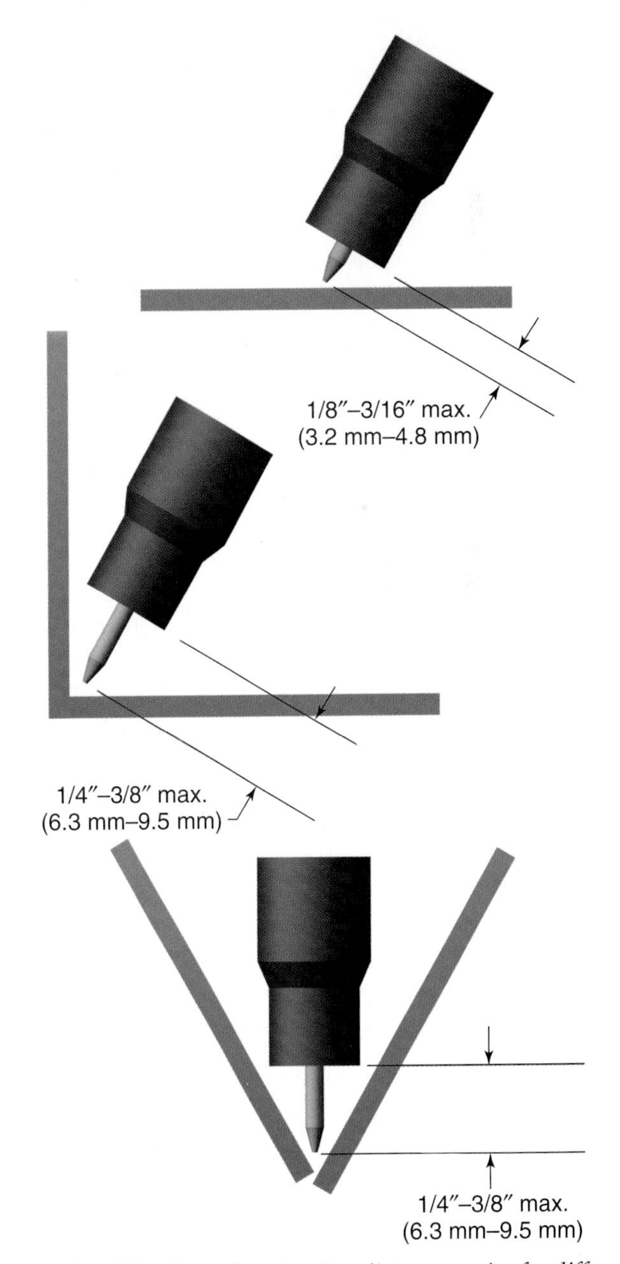

Figure 16-5. *The electrode extension distance varies for different types of joints. It should never be greater than the nozzle diameter.*

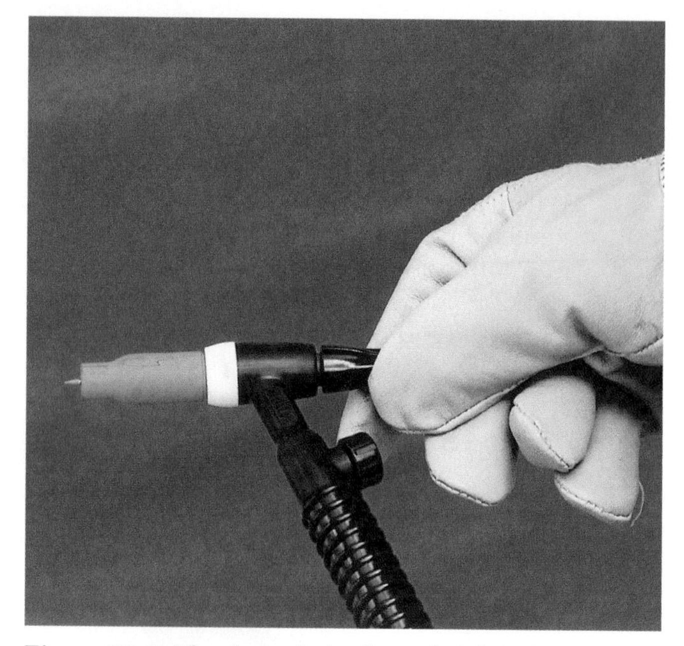

Figure 16-6. The electrode is clamped tightly by the collet when the end cap is tightened.

Flowmeters use a knob or adjusting screw to control gas flow. Some flowmeters use a gauge to indicate the gas flow. However, the most common type of flowmeter uses a ball float in a clear tube. The gas flow is read at the top of the ball. **Figure 16-7** shows how to read this type of flowmeter properly.

To set the gas flow rate, shielding gas must be flowing. Open the cylinder valve, then use the foot pedal or thumb switch to make the shielding gas flow. Do not strike an arc. Another way to make the shielding gas flow is to set the post flow timer for about 10 seconds. Then, press the foot pedal or thumb switch for a second or less. While the shielding gas is flowing, adjust the flowmeter knob to obtain the proper rate for the welding job you will be doing. **Figures 16-8** through **16-12** list gas flow rates for use on different base metals.

Do not change the regulator pressure setting. Shielding gas flowmeters are calibrated for a given pressure. If this pressure is changed, flow rate readings will not be accurate.

Welding Machine Settings

Each GTAW machine has controls on its front panel. You must be able to properly adjust these controls to obtain a good weld. The first step is to select the type of current and the amount of current you will use.

Selecting the Weld Current

Before selecting the weld current, these questions must be answered:

- What type of metal will be welded?
- How thick is the metal?
- What type joint will be used?
- In what position will the welding be done?

After these questions are answered, you can choose the type of welding current. Chapter 15 discussed the three different types of current used in GTAW. The types are: direct current electrode positive (DCEP), direct current electrode negative (DCEN), and alternating current (ac). **Figure 16-13** lists which current is preferred for welding various metals.

Next, you can select the *amount* of current you will use for welding. Figures 16-8 through 16-12 list the recommended currents for welding different base metals.

To choose the desired current type, set the selector switch to ac, DCEP, or DCEN. Set the current range selector to the range (low, medium, high) you need for the amount of current you will be using. Then set the desired current, using the current control knob.

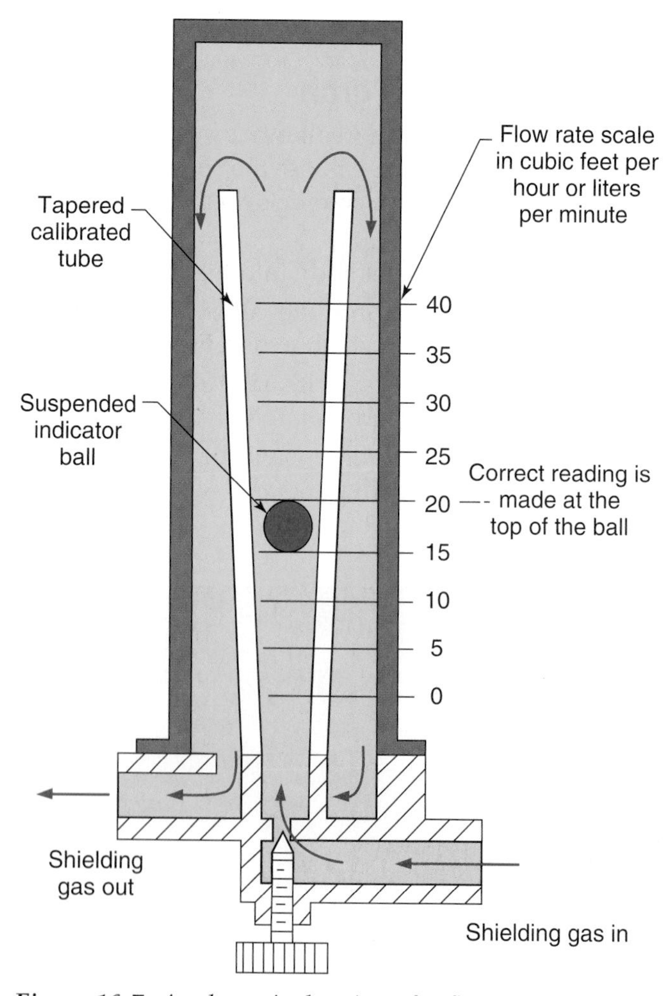

Figure 16-7. A schematic drawing of a flowmeter. The calibration tube is tapered so that a greater flow of gas is required to hold the ball at a higher level. The flow reading is made at the top of the ball.

Mild Steel – DCEN (DCSP)							
Metal thickness	Joint type	Tungsten electrode diameter	Filler rod diameter (if req'd.)	Amperage	Gas		
					Type	Flow (cfh)	L/min.*
1/16″ (1.6mm)	Butt	1/16″ (1.6mm)	1/16″ (1.6mm)	60-70	Argon	15	7.0
	Lap	1/16″	1/16″	70-90	Argon	15	
	Corner	1/16″	1/16″	60-70	Argon	15	
	Fillet	1/16″	1/16″	70-90	Argon	15	
1/8″ (3.2mm)	Butt	1/16″-3/32″ (1.6mm-2.4mm)	3/32″ (2.4mm)	80-100	Argon	15	7.0
	Lap	1/16″-3/32″	3/32″	90-115	Argon	15	
	Corner	1/16″-3/32″	3/32″	80-100	Argon	15	
	Fillet	1/16″-3/32″	3/32″	90-115	Argon	15	
3/16″ (4.8mm)	Butt	3/32″ (2.4mm)	1/8″ (3.2mm)	115-135	Argon	20	9.0
	Lap	3/32″	1/8″	140-165	Argon	20	
	Corner	3/32″	1/8″	115-135	Argon	20	
	Fillet	3/32″	1/8″	140-170	Argon	20	
1/4″ (6.4mm)	Butt	1/8″ (3.2mm)	5/32″ (4.0mm)	160-175	Argon	20	9.0
	Lap	1/8″	5/32″	170-200	Argon	20	
	Corner	1/8″	5/32″	160-175	Argon	20	
	Fillet	1/8″	5/32″	175-210	Argon	20	
*Liters per minute							

Figure 16-8. Variables for manually gas tungsten arc welding mild steel using DCEN (DCSP).

Aluminum and Aluminum Alloys — High-Frequency AC							
Metal thickness	Joint type	Tungsten electrode diameter	Filler rod diameter (if req'd.)	Amperage	Gas		
					Type	Flow (cfh)	L/min.
1/16″ (1.6mm)	Butt	1/16″ (1.6mm)	1/16″ (1.6mm)	60-85	Argon	15	7.0
	Lap	1/16″	1/16″	70-90	Argon	15	
	Corner	1/16″	1/16″	60-85	Argon	15	
	Fillet	1/16″	1/16″	75-100	Argon	15	
1/8″ (3.2mm)	Butt	3/32″-1/8″ (2.4mm-3.2mm)	3/32″ (2.4mm)	125-150	Argon	20	9.0
	Lap	3/32″-1/8″	3/32″	130-160	Argon	20	
	Corner	3/32″-1/8″	3/32″	120-140	Argon	20	
	Fillet	3/32″-1/8″	3/32″	130-160	Argon	20	
3/16″ (4.8mm)	Butt	1/8″-5/32″ (3.2mm-4.0mm)	1/8″ (3.2mm)	180-225	Argon	20	9.0
	Lap	1/8″-5/32″	1/8″	190-240	Argon	20	
	Corner	1/8″-5/32″	1/8″	180-225	Argon	20	
	Fillet	1/8″-5/32″	1/8″	190-240	Argon	20	
1/4″ (6.4mm)	Butt	5/32″-3/16″ (4.0mm-4.8mm)	3/16″ (4.8mm)	240-280	Argon	25	12.0
	Lap	5/32″-3/16″	3/16″	250-320	Argon	25	
	Corner	5/32″-3/16″	3/16″	240-280	Argon	25	
	Fillet	5/32″-3/16″	3/16″	250-320	Argon	25	

Figure 16-9. Variables for manually gas tungsten arc welding aluminum and aluminum alloys using high-frequency ac.

When ac welding, you must use continuous high-frequency voltage. When dc welding, you can use high-frequency voltage to start the arc, but it does not have to be used at all. Set the high-frequency control to the desired position.

The ***contactor switch*** has two positions, standard and remote. In the standard position, open circuit voltage is applied to the welding torch at all times. To strike an arc, the electrode is touched to the work. This is similar to shielded metal arc welding. In GTAW, however, the contactor switch is usually placed in the remote position. A foot pedal or thumb switch is used for on/off control of the current and on/off control of the shielding gas.

The current can be set to a specific amount or can be varied by the welder. To set a specific amount, the ***current switch*** is placed in the *panel* position. In this position, the current setting on the machine will be

Stainless Steel — DCEN (DCSP)							
Metal thickness	Joint type	Tungsten electrode diameter	Filler rod diameter (if req'd.)	Amperage	Gas		
					Type	Flow (cfh)	L/min.
1/16″ (1.6mm)	Butt	1/16″ (1.6mm)	1/16″ (1.6mm)	40-60	Argon	15	7.0
	Lap	1/16″	1/16″	50-70	Argon	15	
	Corner	1/16″	1/16″	40-60	Argon	15	
	Fillet	1/16″	1/16″	50-70	Argon	15	
1/8″ (3.2mm)	Butt	3/32″ (2.4mm)	3/32″ (2.4mm)	65-85	Argon	15	7.0
	Lap	3/32″	3/32″	90-110	Argon	15	
	Corner	3/32″	3/32″	65-85	Argon	15	
	Fillet	3/32″	3/32″	90-110	Argon	15	
3/16″ (4.8mm)	Butt	3/32″ (2.4mm)	1/8″ (3.2mm)	100-125	Argon	20	9.0
	Lap	3/32″	1/8″	125-150	Argon	20	
	Corner	3/32″	1/8″	100-125	Argon	20	
	Fillet	3/32″	1/8″	125-150	Argon	20	
1/4″ (6.4mm)	Butt	1/8″ (3.2mm)	5/32″ (4.0mm)	135-160	Argon	20	9.0
	Lap	1/8″	5/32″	160-180	Argon	20	
	Corner	1/8″	5/32″	135-160	Argon	20	
	Fillet	1/8″	5/32″	160-180	Argon	20	

Figure 16-10. *Variables for manually gas tungsten arc welding stainless steel using DCEN (DCSP).*

Magnesium — High-Frequency AC								
Metal thickness	Joint type	Tungsten electrode diameter	Filler rod diameter (if req'd.)	Amperage[1] with backup	W/O backup	Gas		
						Type	Flow (cfh)	L/min.
1/16″ (1.6mm)	All	1/16″ (1.6mm)	3/32″ (2.4mm)	60	35	Argon	13	6.0
3/32″ (2.4mm)	All	1/16″ (1.6mm)	1/8″(3.2mm)	90	60	Argon	15	7.0
1/8″(3.2mm)	All	1/16″ (1.6mm)	1/8″(3.2mm)	115	85	Argon	20	9.0
3/16″(4.8mm)	All	1/16″ (1.6mm)	5/32″(4.0mm)	120	75	Argon	20	9.0
1/4″(6.4mm)	All	3/32″ (2.4mm)	5/32″(4.0mm)	130	85	Argon	20	9.0
3/8″(9.5mm)	All	3/32″ (2.4mm)	3/16″ (4.8mm)	180	100	Argon	25	12.0
1/2″(13.0mm)	All	5/32″(4.0mm)	3/16″ (4.8mm)	—	250	Argon	25	12.0
3/4″(19.0mm)	All	3/16″ (4.8mm)	1/4″(6.4mm)	—	370	Argon	35	17.0
1 - Use alternating current with a constant high frequency (AC-HF)								

Figure 16-11. *Variables for manually gas tungsten arc welding magnesium using high-frequency ac.*

Deoxidized Copper — DCEN (DCSP)							
Metal thickness	Joint type	Tungsten electrode diameter	Filler rod diameter (if req'd.)	Amperage[1]	Gas		
					Type	Flow (cfh)	L/min.
1/16″ (1.6mm)	All	1/16″ (1.6mm)	1/16″ (1.6mm)	110-150	Argon	15	7.0
1/8″(3.2mm)	All	3/32″ (2.4mm)	3/32″ (2.4mm)	175-250	Argon	15	7.0
3/16″(4.8mm)	All	1/8″(3.2mm)	1/8″(3.2mm)	250-325	Argon	18	9.0
1/4″(6.4mm)	All	1/8″(3.2mm)	1/8″(3.2mm)	300-375	Argon	22	10.0
3/8″(9.5mm)	All	3/16″ (4.8mm)	3/16″ (4.8mm)	375-450	Argon	25	11.0
1/2″(13.0mm)	All	3/16″ (4.8mm)	1/4″ (6.4mm)	525-700	Argon	30	14.0
1 - Use DCEN (DCSP)							

Figure 16-12. *Variables for manually gas tungsten arc welding deoxidized copper using DCEN (DCSP).*

Base Material	Direct Current DCEN	DCEP	Alternating Current
	DCSP	DCRP	ac
Aluminum up to 3/32″	P	G	E
Aluminum over to 3/32″	P	P	E
Aluminum bronze	P	G	E
Aluminum castings	P	P	E
Beryllium copper	P	G	E
Brass alloys	E	P	G
Copper base alloys	E	P	G
Cast iron	E	P	G
Deoxidized copper	E	P	P
Dissimilar metals	E	P	G
Hard facing	G	P	E
High alloy steels	E	P	G
High carbon steels	E	P	G
Low alloy steels	E	P	G
Low carbon steels	E	P	G
Magnesium up to 1/8″	P	G	E
Magnesium over 1/8″	P	P	E
Magnesium castings	P	G	E
Nickel & Ni-alloys	E	P	G
Stainless steel	E	P	G
Silicon bronze	E	P	P
Titanium	E	P	G
E-Excellent G-Good P-Poor			

Figure 16-13. Suggested current and polarity choices for use when gas tungsten arc welding on various base metals. (Welding and Fabrication Data Book)

the current used to weld with; it cannot be changed while welding. If the current switch is in the *remote* position, the current can be changed while welding. This is done by pressing a foot pedal or a thumb switch. Place the contactor and current switches in the positions you have selected.

After the arc stops, shielding gas will continue to flow. This is called post flow. The post flow timer on the machine controls how long shielding gas will flow. A general rule is one second of gas flow for each 10 amps of current. Even when welding with less than 100 amps, a minimum of 10 seconds of post flow should be used.

Some machines have a voltmeter and an ammeter. These show the voltage and current used when welding. All welding machines have a power switch. This is used to turn the machine on and off.

When you set up a GTAW machine, use this checklist:

- Set the type of current.
- Set the current range.
- Set the desired current amount.
- Set the high-frequency switch to the desired position.
- Set the contactor switch to the desired position.
- Set the current switch to the desired position.
- Set the post flow timer for the desired post flow.

Selecting and Preparing the Electrode

The process of selecting and preparing the electrode for GTAW includes the following steps:

- Selecting the type of electrode.
- Selecting the diameter of electrode.
- Selecting the shape for the tip.

The electrode used in GTAW may be one of the following types:

- Pure tungsten.
- Tungsten with 1% or 2% ***thoria*** (thorium dioxide).
- Tungsten with .15%–.40% ***zirconia*** (zirconium oxide).
- Tungsten with 2% ***ceria*** (cerium oxide).
- Tungsten with 1% ***lanthana*** (lanthanum oxide).

Selecting the correct type of tungsten electrode to use is very important. Each electrode type has certain advantages.

The following are characteristics of a pure tungsten electrode:

- Forms a ball on the end when heated.
- Recommended for ac welding.
- Can be used for dc welding.

The characteristics of a thoria tungsten electrode include the following:

- Makes starting the arc easier.
- Allows electrons to flow easier.
- Can carry more current than a pure tungsten electrode of the same size.
- Recommended for dc welding.

Zirconia tungsten electrodes have the following characteristics:

- Similar to pure tungsten, but with easier arc-starting abilities.
- Carries the same current as thoria tungsten.
- Used for ac welding.

Note that zirconia tungsten electrodes are recommended only for ac welding. Thoria tungsten is best for dc welding, but it can be used for either ac or dc welding. When welding steel or stainless steel, a tungsten electrode with thoria is normally used. When ac welding aluminum, use either a pure tungsten electrode or an electrode with zirconium.

Welding Electrode Classifications

Each tungsten electrode type is given a classification by the American Welding Society. **Figure 16-14** lists the different classifications. Each letter and number in the classification has a meaning. The letters can be interpreted as follows:

- E—electrode.
- W—tungsten.
- P—pure.
- Th—electrode contains thoria.
- Zr—electrode contains zirconia.
- Ce—electrode contains ceria.
- La—electrode contains lanthana.

The diameter of an electrode is determined by the amount of current used to weld. As the amount of current increases, so must the diameter of the electrode. Tungsten electrodes are available in diameters from .010″–.250″ (.25 mm–6.4 mm). The most common sizes and the current ranges of each electrode type are listed in **Figure 16-15**. Select an electrode with the

AWS Classification	Defined	Color
EWP	Pure tungsten	Green
EWCe-2	2% Ceria	Orange
EWLa-1	1% Lanthana	Black
EWLa-1.5	1.5% Lanthana	Gold
EWLa-2.0	2.0% Lanthana	Blue
EWTh-1	1% Thoria	Yellow
EWTh-2	2% Thoria	Red
EWZr-1	0.15-0.40% Zirconia	Brown
EWG	Other	Gray

Figure 16-14. The AWS tungsten electrode classifications and the color codes applied in the form of bands or dots on the surface of the electrode. (ANSI/AWS A5.12:2007, Table 1, Chemical Composition of Tungsten Electrodes, reproduced with permission from the American Welding Society, Miami, FL)

	Current Range – Amperes			
Pure tungsten diameter (mm)	**DCEN or DCSP argon**	**DCEP or DCRP argon**	**Ac-unbalanced wave-argon**	**Ac-balanced wave-argon**
.010″ (.25)	Up to 15	*	Up to 15	Up to 10
.020″ (.50)	5-20	*	5-20	10-20
.040″ (1.0)	15-80	*	10-60	20-30
1/16″ (1.6)	70-150	10-20	50-100	30-80
3/32″ (2.4)	125-225	15-30	100-160	60-130
1/8″ (3.2)	225-360	25-40	150-210	100-180
5/32″ (4.0)	360-450	40-55	200-275	160-240
3/16″ (4.8)	450-720	55-80	250-350	190-300
1/4″ (6.4)	720-950	80-125	325-450	250-400
Thorium (1% and 2%), cerium, and lanthanum alloyed tungsten				
.010″ (.25)	Up to 25	*	Up to 20	Up to 15
.020″ (.50)	15-40	*	15-35	5-20
.040″ (1.0)	25-85	*	20-80	20-60
1/16″ (1.6)	50-160	10-20	50-150	60-120
3/32″ (2.4)	135-235	15-30	130-250	100-180
1/8″ (3.2)	250-400	25-40	225-360	160-250
5/32″ (4.0)	400-500	40-55	300-450	200-320
3/16″ (4.8)	500-750	55-80	400-550	290-390
1/4″ (6.4)	750-1000	80-125	600-800	340-525
Zirconium alloyed tungsten				
.010″ (.25)	*	*	Up to 20	Up to 15
.020″ (.50)	*	*	15-35	5-20
.040″ (1.0)	*	*	20-80	20-60
1/16″ (1.6)	*	*	50-150	60-120
3/32″ (2.4)	*	*	130-250	100-180
1/8″ (3.2)	*	*	225-360	160-250
5/32″ (4.0)	*	*	300-450	200-320
3/16″ (4.8)	*	*	400-550	290-390
1/4″ (6.4)	*	*	600-800	340-525
Special alloyed tungsten	Refer to manufacturer for specific information about alloy content and recommended current ranges.			
* Not recommended				
The figures listed are intended as a guide, and are a composite of recommendations from American Welding Society and electrode manufacturers.				

Figure 16-15. Suggested current ranges for various types and sizes of tungsten electrodes.

ability to carry the current needed for the job you will be welding.

Electrodes come in 3″–24″ (76.2 mm–609.6 mm) lengths. The most common lengths for manual gas tungsten arc welding electrodes are 3″, 6″, and 7″ (76.2 mm, 152.4 mm, and 177.6 mm). Electrodes are purchased with either a ground or a chemically cleaned surface.

Preparing the Tip of the Electrode

Finally, the ***electrode tip*** must be prepared for welding. The tip of an electrode can be a blunted point or a ball end. For dc welding, grind the electrode to a point. After forming the point, blunt it slightly. The grinding marks on the electrode should run lengthwise. See **Figure 16-16**.

Most welding shops will have one or more pedestal grinders or grinding wheels that are used strictly for grinding tungsten electrodes. Aluminum and other metals should never be ground on these wheels. Metal particles left on these wheels may become embedded in a tungsten electrode. A contaminated electrode will give poor results when welding. **Figure 16-17** shows a grinder made for grinding tungsten electrodes. **Caution: Always wear safety glasses when you use a grinding wheel.**

For ac welding, the electrode tip is formed into a ball. This is done by striking an arc on a piece of steel or copper. When an arc is struck, the electrode tip will melt and form a ball shape. Use ac or DCEP to form the ball. The ball should be no larger than the diameter of the electrode. For a ball slightly smaller in diameter than the electrode, taper the electrode first on a grinding wheel, then form the ball. **Figure 16-18** shows an electrode with a correctly formed taper and one with a correctly formed ball.

Whenever you use a new electrode, you must first grind or form it into the correct shape. Also, whenever you dip an electrode into the molten weld pool, it must be reground before being used again. This removes any contamination from the electrode.

Never allow a tungsten electrode to become so hot that part of it falls into the weld pool. If this happens, change to a larger size electrode or use less welding current.

Figure 16-17. *A special purpose grinder for tapering the GTAW electrode. The electrode is held in a collet and rotated by means of a small crank handle as it contacts the grinding wheel. (Intercon Enterprises, Inc.)*

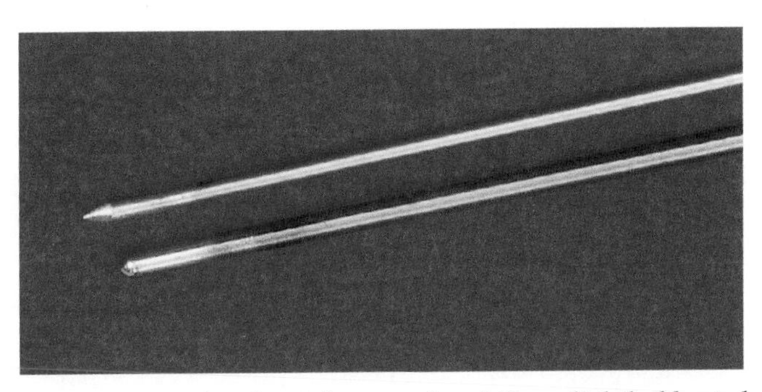

Figure 16-18. *An electrode tapered and then slightly blunted for GTAW with dc and a balled electrode for use with ac.*

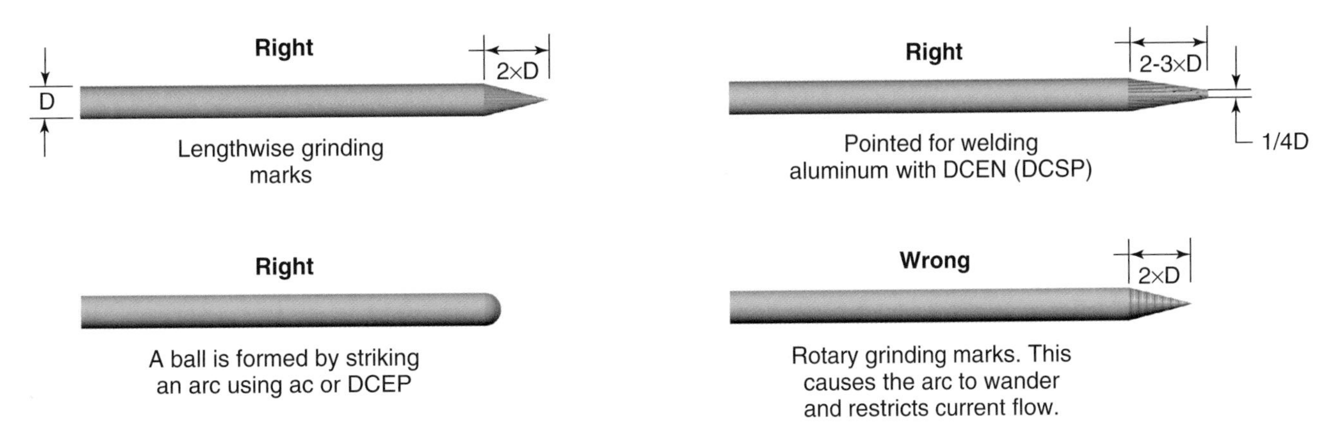

Figure 16-16. *Electrode tip shapes for welding with dc and ac. Grinding marks should run lengthwise on the electrode for better arc control. The point of a tapered tip should be centered on the electrode.*

Filler Metal

Filler metal is needed to make most welding joints. Filler metal used in GTAW is solid wire. It comes in precut lengths or in large spools or coils. Precut lengths, or "rods," are usually 36" (914 mm) long. Spools or coils are used for automatic GTAW, but they can also be used for manual welding. The welder cuts the wire to the length needed. The wire diameter can be as small as .015" (.38 mm). The most common *precut* wire diameters are from 1/16"–3/16" (1.6 mm–4.8 mm).

For a given welding job, filler wire with a composition similar to the base metal is usually used. There are many special alloy filler metals available for specific types of welds. The most common filler metals used in GTAW are shown in **Figure 16-19**. Note that each type begins with the letters ER. The E stands for electrode and the R stands for rod.

NOTE: The filler metals used when GTAW carbon steel are not copper-coated, as are the filler metals used for oxyfuel gas welding of carbon steel.

Preparing Metal for Welding

Gas tungsten arc welding is a very clean process. Prior to welding, the base metal should be cleaned. In industry, chemicals may be used to remove grease and other dirt from the surface of metals. Chemicals are also used to remove the oxides from aluminum and magnesium. If chemical cleaning is not used, the base metals must have a mechanical cleaning.

Mechanical cleaning is done by scrubbing a metal surface with an abrasive cloth or a wire brush. A grinding wheel can also be used. All rust, grease, oil, and oxides must be removed before welding. If not removed, they could cause a weak or defective weld.

Cleaning is done after all edge preparations (chamfers, grooves, beveled edges) are completed.

Base Metal	Recommended Filler Metal
Aluminum	ER1100, ER4043,ER5356
Copper and copper alloys	ERCu, ERCuSi-A, ERCuAl-A1
Magnesium	ERAZ61A, ERAZ92A
Nickel and nickel alloys	ERNi-1, ERNiCr-3, ERNiCrMo-3
Carbon steel	ER70S-2, ER70S-6
Low alloy steel	ER80S-B2, ER80S-D2
Stainless steel	ER308, ER308L, ER316, ER347
Titanium	ERTi-1
Zirconium	ERZr-2

Figure 16-19. *The most-often-recommended filler metals for use when gas tungsten arc welding different base metals.*

Summary

- Electrical connections for the electrode and workpiece lead are mechanical. They must be tight to make good contact. Water and shielding gas connections are threaded to help prevent leaks. Use the proper size wrench to tighten these connections. **Cylinders containing compressed gases must be secured to a wall, column, or welding hand truck with chains or steel bands.** Once the cylinder is secure, a regulator and flowmeter can be attached to it. Thread the regulator nut onto the cylinder valve. If a ball float-type flowmeter is used, the flowmeter must be in a vertical position.
- To assemble a GTAW torch, begin by selecting the correct-size collet and collet body for the electrode's diameter. Thread the collet body into the torch body. Slide the collet into the collet body from the top of the torch. Thread an end cap with an O-ring loosely into the torch body. Select the gas nozzle you will be using. Thread it into the torch body. Insert the electrode into the collet. Tighten the end cap.
- To set the gas flow rate, shielding gas must be flowing. Open the cylinder valve, then use the foot pedal or thumb switch to make the shielding gas flow. While the shielding gas is flowing, adjust the flowmeter knob to obtain the proper rate for the welding job you will be doing.
- When setting up a GTAW machine, perform the following tasks in order: set the type of current, set the current range, set the desired current amount, set the high-frequency switch to the desired position, set the contactor switch to the desired position, set the current switch to the desired position, and set the post flow timer for the desired post flow.

(continued)

- The electrode used in GTAW may be pure tungsten, tungsten with 1% or 2% thoria, tungsten with .15%–.40% zirconia, tungsten with 2% ceria, or tungsten with 1% lanthana. Zirconia tungsten electrodes are recommended only for ac welding. Thoria tungsten is best for dc welding, but can be used for either ac or dc welding. When welding steel or stainless steel, a tungsten electrode with thoria is normally used. When ac welding aluminum, a pure tungsten electrode or an electrode with zirconium is recommended.
- Each tungsten electrode type is given a classification by the American Welding Society. The letters can be interpreted as follows: E—electrode, W—tungsten, P—pure, Th—electrode contains thoria, Zr—electrode contains zirconia, Ce—electrode contains ceria, and La—electrode contains lanthana.
- The electrode tip must be prepared for welding. The tip of an electrode can be a blunted point for dc or a ball end for ac welding. The grinding marks on the electrode should run lengthwise.
- Prior to welding, the base metal should be cleaned. In industry, chemicals may be used to remove grease and other dirt from the surface of metals. Chemicals are also used to remove the oxides from aluminum and magnesium. If chemical cleaning is not used, the base metals must have a mechanical cleaning.

Review Questions

Write your answers on a separate sheet of paper. Please do *not* write in this book.

1. What is a welding machine often wired into?
2. What type of connection is used for shielding gas and water hoses?
3. In what position must a ball float–type flowmeter be mounted?
4. In a ball float–type flowmeter, the flow is read from what part of the ball?
5. List three of the major types of electrodes used in GTAW.
6. When grinding an electrode, in what direction should the grind marks go?
7. What shape should an electrode tip used for DCEN welding have?
   A. Tapered with a sharp point.
   B. Tapered with a blunted point.
   C. Balled on the end.
   D. Square cut.
8. How is the electrode held in place?
9. What type of current requires the high-frequency voltage to be used continuously?
10. If you want to set the current to a certain amount and not vary it during welding, what position should the current switch be in?

The GTAW process is being used to assemble a junior dragster frame from 4130 chrome-moly tubing. (Lincoln Electric Company)

Chapter 17

GTAW: Flat Welding Position

Learning Objectives

After studying this chapter, you will be able to:

- Describe the GTAW process.
- Determine the appropriate filler rod to use when gas tungsten arc welding.
- Lay a bead on a plate using GTAW.
- Make a fillet weld on a lap joint in the flat welding position.
- Make a fillet weld on a T-joint in the flat welding position.
- Weld a butt joint in the flat welding position.
- Describe the use of a backing when welding aluminum using GTAW.
- Identify various weld defects.

backing
filler metal
hot shortness
oxidized
porosity
tungsten inclusion
unstable arc

Gas Tungsten Arc Welding Principles

Gas tungsten arc welding is done by striking an arc between the tungsten electrode and the base metal. The welding arc melts a small spot on the base metal. This small melted area of base metal is called the weld pool. ***Filler metal*** is often used, but not always required. Filler metal is added to the molten weld pool to fill the weld joint. **Figure 17-1** shows a welder using the GTAW process.

In GTAW, the torch is moved along the weld joint and filler metal added. This continues to the end of

Figure 17-1. GTAW being used to join small parts. Note the welding rod.

the joint. The speed of GTAW is much slower than SMAW or GMAW.

The electrode, arc, and base metal are protected during welding by a shielding gas. Even after the welding arc stops, shielding gas continues to flow. This post flow protects the electrode and base metal from contamination as they cool.

Preparing to Weld

Before beginning to weld, you must assemble the equipment. Connect the torch cables to the welding machine. Connect a source of shielding gas. Check that all electrical, shielding gas, and water connections are tight. Refer to Chapter 16 for further information on assembling equipment.

Select and set the current you will use on the machine, as described in Chapter 16. **Figure 17-2** shows the controls you may need to adjust on a welding machine. Next, choose the correct type and diameter of tungsten electrode for the welding you will be doing. Prepare the tip of the electrode as described in Chapter 16 and install it in the torch. Finally, adjust the shielding gas flow rate. See Chapter 16.

Make sure that the filter lens in your welding helmet is a #10, #12, or #14. Check that you are wearing the proper clothes for welding. Make sure all parts of your body are protected from the intense rays of the welding arc. Remove all combustible materials from the area where you will be welding.

Selecting a Filler Rod

Welding (filler) rods are available in many different alloys. Usually a filler rod with a composition similar to the base metal is used. Figure 16-18 lists filler rods recommended for use with different base metals.

The diameter of the rod is determined by the size of the weld. Filler metal should have to be added to the weld only about once every two seconds to obtain the correct weld size.

If the rod diameter is too small, metal will have to be added very often, and the rod will be used up quickly. If the rod is too large, metal will be added in too-large amounts. The large amount of filler metal will cool the weld pool, causing large bumps in the weld. These bumps will remain unless you take time to smooth them out by reheating the area.

When using filler metal, keep the end of the rod close to the arc. Holding the filler metal close to the arc keeps the end of the rod heated and also protects the end of the rod with shielding gas.

Using the GTAW Torch

Hold the torch as shown in **Figure 17-3**. This should be a comfortable grip. It is similar to holding a pencil. You can also hold the torch like a hammer, **Figure 17-4**. This is used on larger torches. When welding out of position, this grip allows for greater flexibility and control of the torch.

If you use a foot pedal, place it on the floor where your foot can reach it comfortably. Sit or stand at the welding station. Place the metal to be welded in front of you on the table or in a welding fixture. You must be able to move the torch and filler metal along the joint without hitting anything. See **Figure 17-5**.

Make sure the ground cable is securely attached to the table or to the workpiece. Now you are ready to begin welding.

Starting the Arc

There are two different ways to start, or strike, an arc. One is to use a high-frequency generator. The other is to touch the electrode to the base metal.

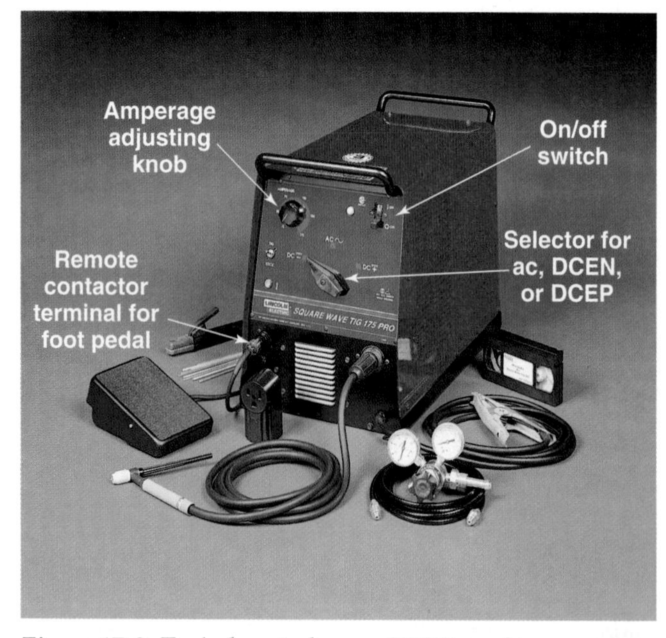

Figure 17-2. *Typical controls on a GTAW machine. (Lincoln Electric Company)*

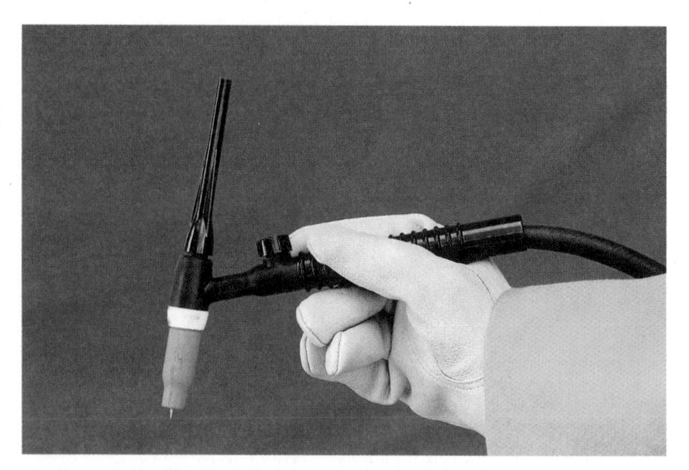

Figure 17-3. *Light duty torches may be held like a pencil.*

The most common way to start the arc is to use a high-frequency voltage generator. When you weld with ac, you use high-frequency voltage continuously. When this feature is used to start the arc for dc welding, it will shut off automatically a few seconds after the arc stabilizes.

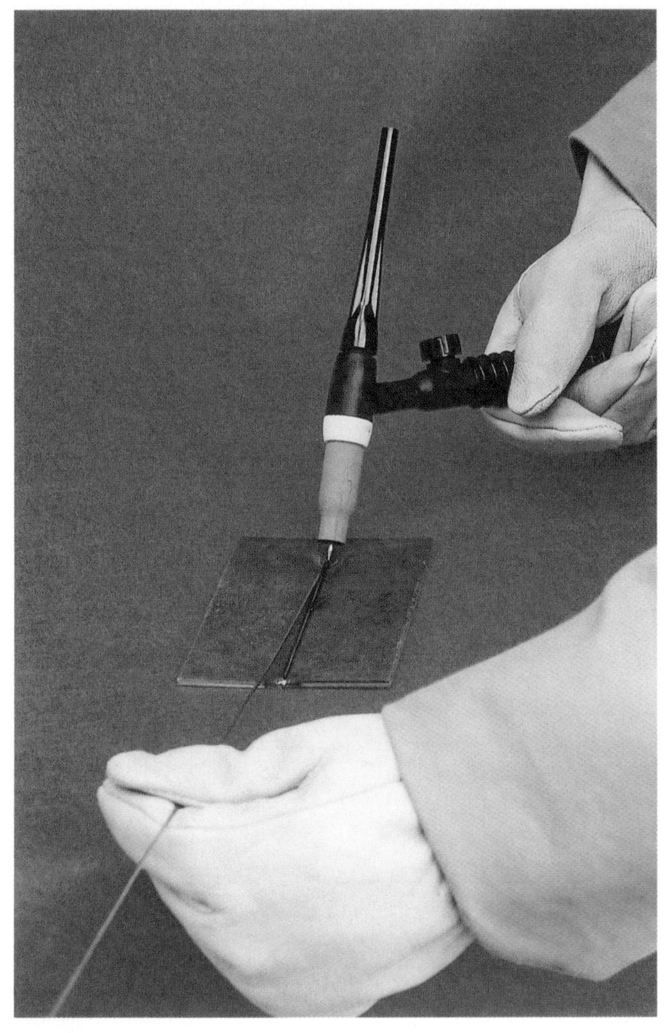

Figure 17-4. This GTAW torch is being held like a hammer. Note the torch and welding rod angles.

One way to start the arc with high-frequency voltage is to hold the electrode about 1/8″ (3.2 mm) above the base metal. Press the foot pedal or thumb switch. The high-frequency arc will jump the gap. Wait about one or two seconds and the arc will stabilize. A high-frequency arc can actually jump a gap of more than 1/2″ (13 mm), but the main current will not flow with a gap this large.

You can also start the arc with high-frequency voltage by holding the electrode about 2″ (51 mm) above the base metal. Press the foot pedal or thumb switch, then quickly move the torch down toward the work. This can be a swinging motion, as shown in **Figure 17-6**. When the electrode gets close enough to the work, the high-frequency voltage will cause an arc to jump the gap.

Another way to start an arc is to touch the electrode to the work. Touch starting is used only when dc welding. When using this method, very quickly touch the electrode to the work and then pull it back to about 3/32″–1/8″ (2.4 mm–3.2 mm). Wait for the arc to become stable. Touch-starting the arc can contaminate the electrode. A contaminated electrode will produce an unstable arc. An unstable arc makes welding difficult and can cause a poor quality weld. Refer to the *Weld Defects* heading in this chapter for additional information.

After you start the arc, hold the torch so the electrode is about 1/16″–1/8″ (1.6 mm–3.2 mm) above the base metal. This is the correct arc length for GTAW. Wait as a weld pool forms, then begin welding.

Stopping the Arc

When the weld is complete, you must stop the arc. Stop the current by removing your foot from the pedal or turning off the thumb switch. Keep the torch over the area where you stopped welding. Shielding

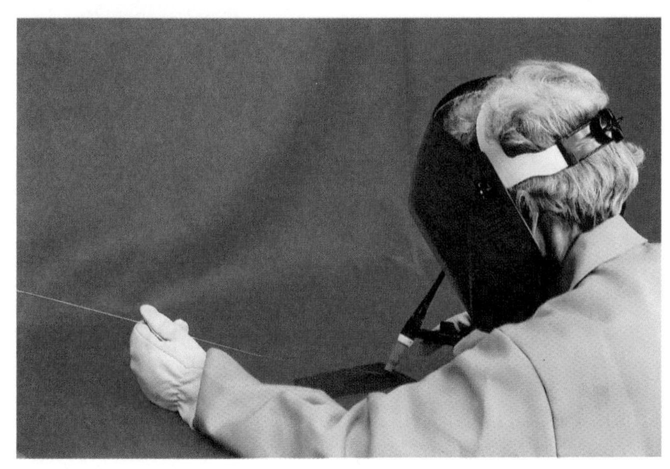

Figure 17-5. Welding with nothing to block or disturb torch and rod movement.

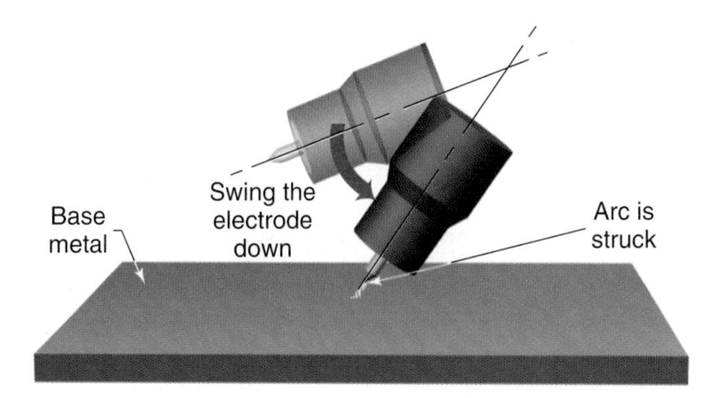

Figure 17-6. Starting the GTAW arc with a swinging motion. The foot control is depressed and the electrode is swung down toward the base metal until the arc is struck.

gas will continue to flow (post flow). Post flow will protect the weld metal as it cools.

If the current switch on the welding machine is in the remote position, you can use the foot pedal or the thumb switch to control the amount of current. You can slowly reduce the current near the end of the weld. This gives you time to fill the weld pool with filler metal, if necessary.

Welding a Bead on Plate

In GTAW, the torch and electrode tip almost always point in the direction of travel. This is the forehand welding method. When using the forehand welding method, the travel angle is called a push angle. A push angle of 15°–30° is used. See **Figure 17-7**.

If the torch were pointed opposite the direction of travel, this would be the backhand welding method. The backhand method uses a drag angle, which means the electrode points opposite the direction of travel.

The electrode is held at a work angle of 0°, which is perpendicular (90°) to the work surface. See the heading titled *Describing the Electrode Angles* in Chapter 8 for additional information on travel and work angles.

Start the welding arc as described in the preceding section. After the arc stabilizes, hold the electrode about 1/16″–1/8″ (1.6 mm–3.2 mm) above the work. A molten weld pool will form from the heat of the arc.

Hold the torch in one place until a weld pool forms. When welding on thin metals, the molten pool will sag slightly. This will occur when the amount of penetration is equal to the thickness of the metal. On thicker metals, no sag will occur. Wait until the pool is the proper size. Usually, the weld pool diameter should be two to three times the diameter of the electrode. Then, begin to move forward. Keep the pool the same size as you move along the length of the weld joint.

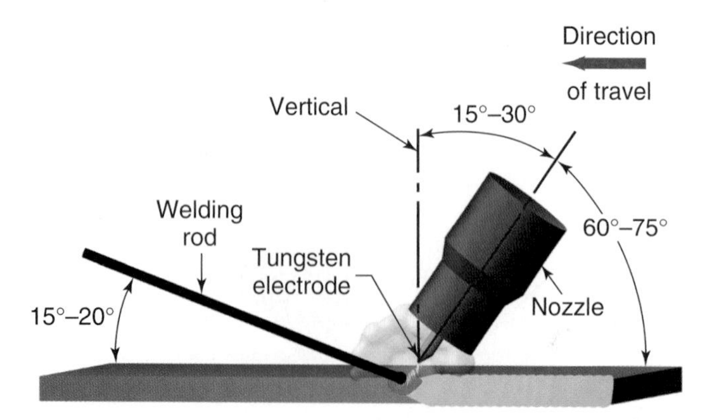

Figure 17-7. The suggested electrode and welding rod angles for welding a bead on plate. The same angles are used when making a butt weld. The torch is held at a 15°–30° travel angle using the forehand method (60°–75° from the base surface). A work angle of 0° is used (90° to the base surface).

You can change the current slightly by moving the electrode toward or away from the base metal. If you need more current (for more penetration or more heat), move the electrode closer to the base metal. See **Figure 17-8**. If the weld pool is too large and you want to reduce the current, pull the electrode away from the base metal. Do not pull the electrode so far away that the arc becomes unstable. A long arc length and especially an unstable arc will cause ***porosity*** (trapped gas bubbles or holes) in the weld.

Another way to control current is to set the current switch on the machine to the remote position. This allows the current to be changed by using a foot pedal or thumb switch. These changes can be large, compared to moving the electrode, which can cause only small changes of up to 10 amps.

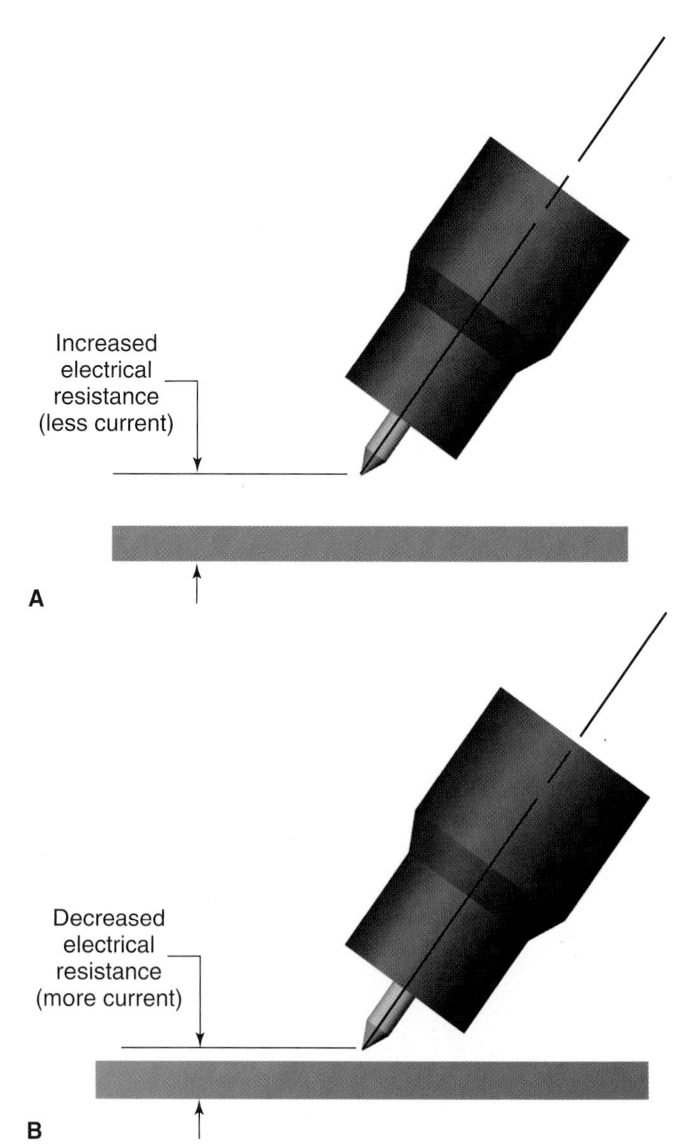

Figure 17-8. The current can be changed slightly by moving the electrode toward or away from the base metal. A—A larger gap will decrease the current, heat, and penetration. B—A smaller gap will increase these factors.

Once you have formed a good weld pool, continue moving the electrode and the pool until you reach the end of the plate. GTAW is a fairly slow process, but it produces high-quality welds.

Exercise 17-1 Laying a Bead on Plate without Filler

1. Obtain a piece of mild steel measuring 1/16″ × 3″ × 6″ (1.6 mm × 75 mm × 150 mm).
2. Clean both surfaces with a wire brush.
3. Mark five straight lines lengthwise on each side of the metal. Place the first line 1/2″ (12.7 mm) in from one edge, and the last line 1/2″ (12.7 mm) in from the opposite edge. Place the metal in the flat position.
4. Set up the welding machine for use with mild steel. Use DCEN (DCSP) and 70 amp current. Place the current switch in the panel position, so the foot pedal or thumb switch can act as an on/off switch. Set the high-frequency switch to the start position.
5. Set the shielding gas flow rate. Refer to Figure 16-9 for shielding gas flow rates. The shielding gas should be argon.
6. Obtain a 2% thoria tungsten electrode 1/16″ (1.6 mm) in diameter. Form the tip into a blunted point. Make sure the grinding marks are lengthwise, as shown in Figure 16-15.
7. Select the correct size collet and collet body. Install the collet, collet body, and electrode in the torch. The electrode should extend 1/8″–3/16″ (32 mm–4.8 mm).
8. Hold the torch at the correct angle about 1/8″ (3.2 mm) above the base metal.
9. Press the foot pedal or thumb switch. The welding arc will start.
10. Move the electrode closer to the base metal, so it is about 3/32″ (2.4 mm) above the surface. Watch the weld pool as it forms. When it reaches about 3/16″ (4.8 mm) in width, it will begin to sag. Move the torch forward along the line marked on the plate.
11. Maintain the same size weld pool as you move along the plate. Keep the electrode a constant distance above the surface.
12. Stop welding at the end of the line.
13. Repeat the process until you have welded five beads on each side of the plate.

Inspection:

Each weld bead should be straight and have even ripples. There should be signs of slight penetration on the back side of the plate. There should not be any porosity in the weld.

Welding a Bead on Plate with Filler Metal

When welding a butt joint or a bead on plate, hold the torch so it has a push angle, or travel angle, of 15°–30°. Also, hold the torch in line with the weld axis, creating a work angle of 0°.

When welding a fillet weld, the travel angle should be 15°–30°. The work angle should be 45°. The position of the welding torch is the same, whether or not filler metal is used. As shown in Figure 17-4, the filler rod is held in the opposite hand from the torch and grasped like a pencil. The filler metal is held at an angle of 15°–20° to the base metal and nearly in line with the weld axis. See Figure 17-7.

Start the arc, as described earlier in this chapter. When the weld pool reaches the correct size (two to three times the diameter of the electrode), add the filler metal at the front of the weld pool. At the same time that you add the filler metal, move the torch toward the back of the pool. This will prevent the filler metal from touching and contaminating the electrode. Before the filler metal is added, the weld pool has a slight depression. After it is added, the pool is filled and slightly crowned (convex).

After adding the filler metal, remove the rod from the weld pool. Keep the end of the filler rod close to the arc and the pool. The rod end should be about 1/8″–3/16″ (3.2 mm–4.8 mm) away from the pool. This will keep the filler metal in the area protected by the shielding gas.

Move the torch forward. As the weld pool moves, it will again form a slight depression. Watch the weld pool. When filler metal is needed, add the rod to the pool's leading edge. **Figure 17-9** illustrates the steps used to add filler metal. Keep the amount of reinforcement or build-up equal for the full length of the bead.

To form a wider bead, move the torch from side to side. Filler metal is still added to the weld pool's leading or front edge. Instead of being added to the center

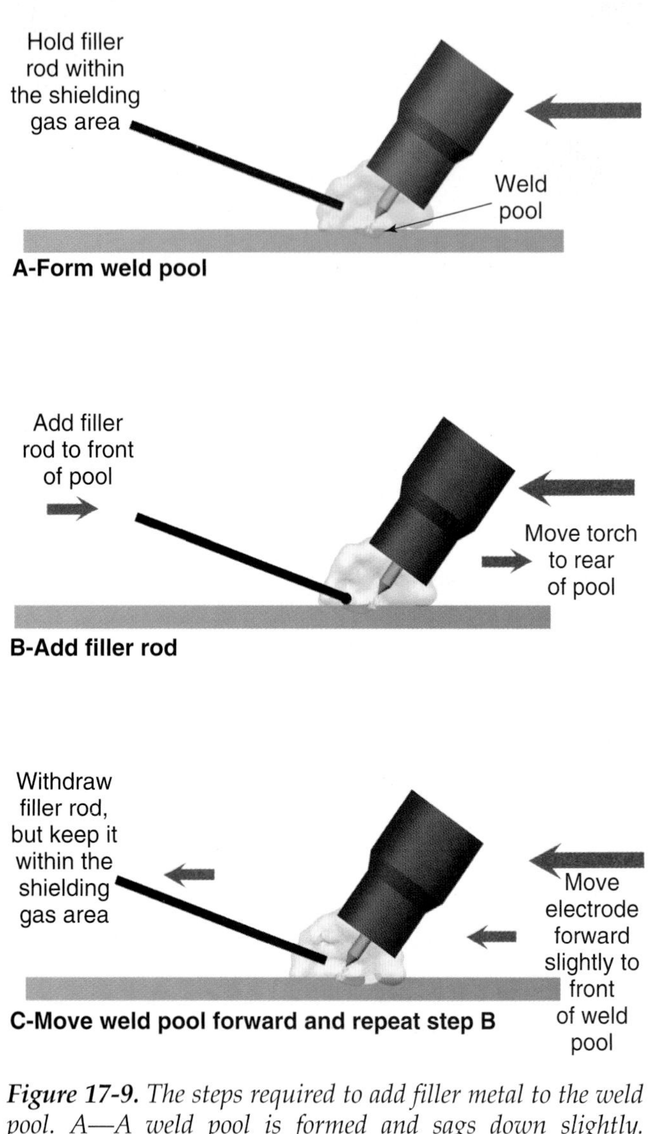

Figure 17-9. *The steps required to add filler metal to the weld pool. A—A weld pool is formed and sags down slightly. B—The torch is moved to the rear of the pool as the rod is added to the front of the pool. C—The rod is withdrawn and the electrode is moved to the front of the weld pool again. This process is repeated to the end of the weld.*

of the pool, it is added to one side or the other. When the torch moves to the left, filler metal is added on the right front side of the weld pool. The next time filler metal is needed, it is added to the left front side of the pool, while the torch is on the right side.

As you finish a weld, add filler metal to the weld pool to fill the crater. Move the electrode to the rear of the pool, and add filler metal to the front edge. Remove the filler metal, and move the torch over the center of the pool again. This will smooth out the weld pool. Then release the foot pedal to stop the arc. Hold the torch in place for a few seconds to let the shielding gas protect the electrode and weld, while they are cooling.

Exercise 17-2 Laying a Bead on Plate with Filler Rod

1. Obtain a piece of mild steel measuring 1/16″ × 3″ × 6″ (1.6 mm × 75 mm × 150 mm).
2. Clean both surfaces with a wire brush.
3. Mark five straight lines lengthwise on each side of the metal. Place the first line 1/2″ (12.7 mm) in from one edge, and the last line 1/2″ (12.7 mm) in from the opposite edge. Place the metal in the flat position.
4. Set up the welding machine for use with mild steel. Use DCEN (DCSP) and 70 amp current. Place the current switch in the panel position, so the foot pedal or thumb switch can act as an on/off switch. Set the high-frequency switch to the start position.
5. Set the shielding gas flow rate. Refer to Figure 16-9 for shielding gas flow rates. The shielding gas should be argon.
6. Obtain a 2% thoria tungsten electrode 1/16″ (1.6 mm) in diameter. Form the tip into a blunted point. Make sure the grinding marks are lengthwise.
7. Select the correct size collet and collet body. Install the collet, collet body, and electrode in the torch. The electrode should extend 1/8″–3/16″ (32 mm–4.8 mm).
8. Obtain a mild steel filler rod 1/16″ (1.6 mm) in diameter.
9. Hold the torch at the correct angle about 1/8″ (3.2 mm) above the base metal.
10. Press the foot pedal or thumb switch. The arc will start.
11. Move the electrode closer to the base metal, so it is about 3/32″ (2.4 mm) above the surface. Watch the weld pool as it forms.
12. When the weld pool reaches about 3/16″ (4.8 mm) in width, it will begin to sag. Add the filler rod to the front edge of the pool. Begin moving the torch forward along the line marked on the plate.
13. Keep the same size weld pool as you move along the plate. When the pool begins to sag, add filler rod. Keep the electrode a constant distance above the surface. Keep the filler rod close to the weld pool.

14. Stop welding at the end of the line. Fill the weld pool at the end of the weld.
15. Repeat the process until you have welded five beads on each side of the plate.

Inspection:

Each weld bead should be straight and have even ripples. The reinforcement should be about 1/16″ (1.6 mm) or less. There should be signs of slight penetration on the back side of the plate. There should not be any porosity along the weld.

Welding without Filler Metals

Some joints do not require filler metal. These include the edge joint, the lap joint, and the outside corner joint. When welding these joints, the base metal is melted until the two metals flow together. **Figures 17-10** through **17-12** show GTAW on these joints.

The processes of starting the arc and controlling the weld pool are like those used to weld a bead on plate. When welding an edge joint, the torch is held in line with the weld axis, which is a work angle of 0°. When welding a lap joint or outside corner joint, a work angle of 45° is used. The torch is held at a travel angle of 15°–30°. The travel angle is a push angle.

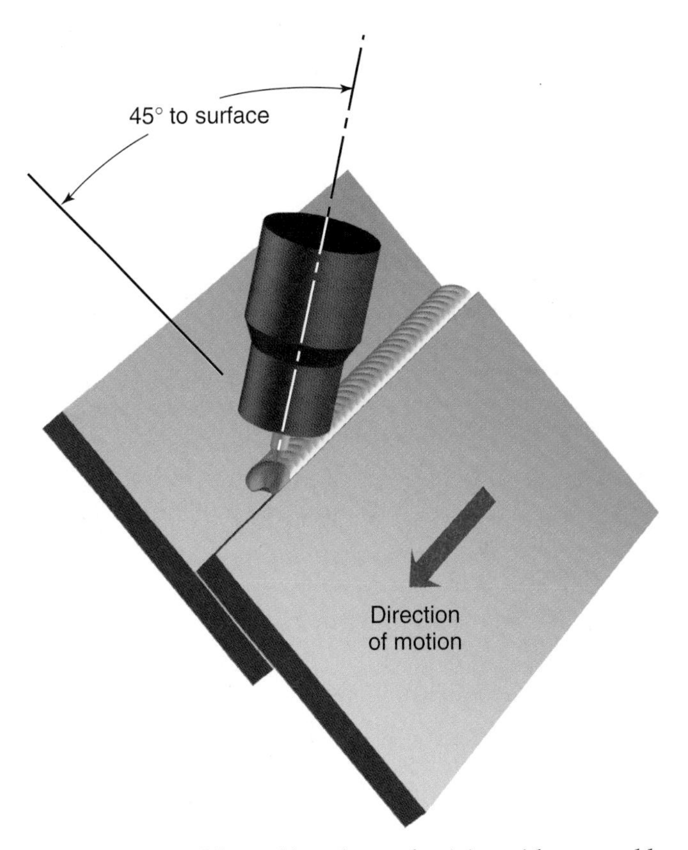

Figure 17-11. *A fillet weld made on a lap joint without a welding rod. The edge is melted into the surface piece to form a weld bead. The weld face should be horizontal.*

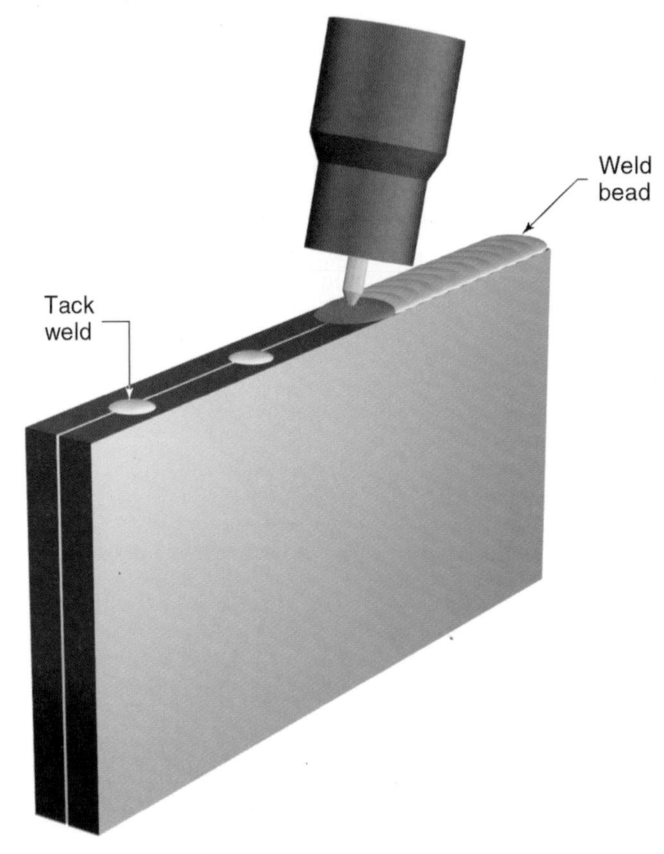

Figure 17-10. *An edge weld made without a welding rod. The base metal is melted by the arc to form a weld bead.*

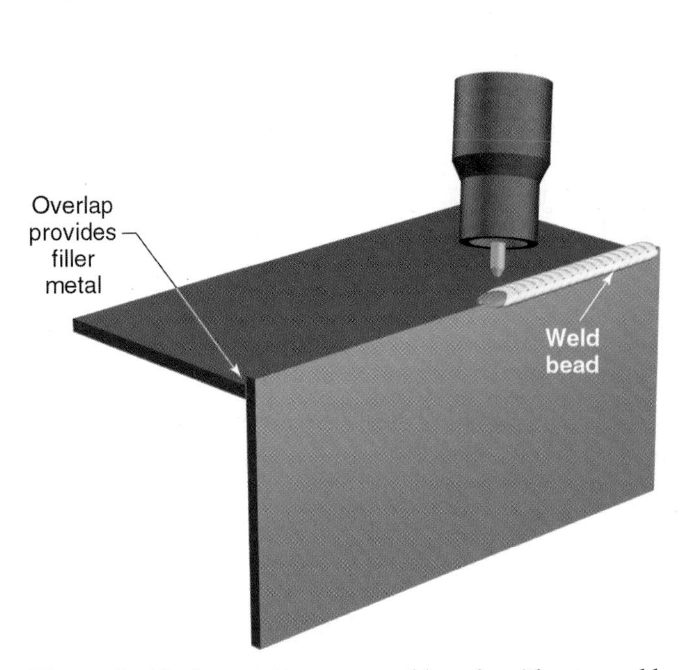

Figure 17-12. *An outside corner weld made without a welding rod. One edge is overlapped one metal thickness. This edge is then melted down to form the weld bead.*

A side-to-side motion may be needed to obtain a bead that will flow and join the two metals. When you come to a tack weld, continue to move forward at a constant speed. Do not skip over it. You must have a continuous weld bead from the start to the end of weld joint.

The welding current, electrode diameter, and shielding gas flow rates are selected on the basis of the thickness of one piece of metal. Refer to Figures 16-8 through 16-12 for correct settings.

Exercise 17-3 Making an Edge Weld without Filler Rod

1. Obtain two pieces of mild steel measuring 3/32″ × 3″ × 6″ (2.4 mm × 75 mm × 150 mm).
2. Clean the edges with a wire brush.
3. Set up the welding machine for welding mild steel.
4. Set the shielding gas flow rate.
5. Select an electrode of the proper type and correct diameter. Prepare the electrode for welding.
6. Select the correct size collet and collet body. Install the collet, collet body, and electrode in the torch. Make sure the electrode extends the proper amount.
7. Tack weld the two pieces in three places to form an edge weld.
8. Strike an arc and weld the edge joint. Make the weld bead as wide as the two pieces of metal are thick. No filler metal is required. A weaving motion may be necessary, depending on the diameter of electrode you are using.
9. Keep the electrode a constant distance above the work.
10. Stop welding at the end of the joint.
11. Repeat the process and weld the other side of the joint.

Inspection:

Each weld bead should be as wide as the two pieces are thick.

Fillet Welding a Lap Joint

A lap joint can be made with or without filler metal. When welding a lap joint, hold the torch at a travel angle of 15°–30° and a work angle of 45°. The weld pool has a C shape, as shown in **Figure 17-13.** This C-shaped weld pool shows that both pieces of metal are being melted by heat from the arc.

Often, the edge will melt more than the surface. Only half of a C-shaped weld pool will form. To correct this, point the torch more toward the surface to reduce the melting of the edge. Change the angle of the torch as needed so both pieces melt evenly and the C-shaped weld pool forms.

A filler rod may not be required when welding thin metals. When filler metal is not used, the metal from the upper piece flows down and forms a fillet. When using filler metal, add it to the upper edge of the weld pool, and hold the filler rod at an angle of 15°–20° to the base metal. A completed weld will have evenly spaced ripples. When filler metal is added, the bead should have a slight convex surface. See **Figure 17-14**. If filler metal is not used, the bead usually will have a flat surface. Remember to weld with a constant forward motion, even when welding over a tack weld. Proper technique is needed to obtain a weld that is free of defects. Weld defects are discussed near the end of this chapter.

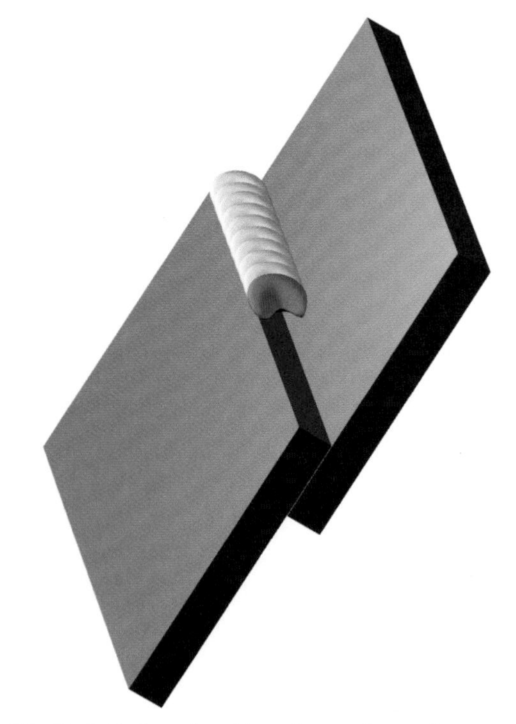

Figure 17-13. *A C-shaped weld pool must form whenever a fillet weld is made on a lap, T, or inside corner joint. The C shape indicates that both pieces of metal have melted and the filler metal has been added.*

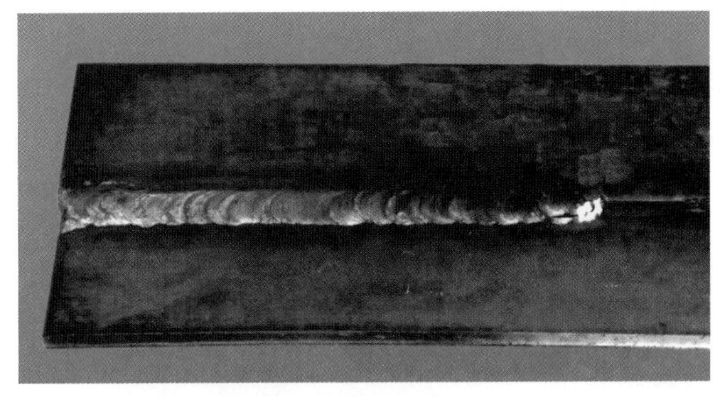

Figure 17-14. *A well-formed fillet weld on a lap joint. The weld bead has an even width, evenly spaced ripples, and is convex (curved outward). There is no undercutting or underfilling.*

Exercise 17-4 Making a Lap Joint without Filler Rod

1. Obtain two pieces of mild steel measuring 3/32″ × 3″ × 6″ (2.4 mm × 75 mm × 150 mm).
2. Clean one long edge and one surface on each piece with a wire brush.
3. Set up the welding machine for welding mild steel.
4. Set the shielding gas flow rate.
5. Select and prepare the proper type and diameter electrode. Select the correct size collet and collet body.
6. Install the electrode in the torch with the proper amount extending.
7. Tack weld the two pieces together to form a lap joint. Make sure there is no gap between them. Place the pieces at an angle so the weld is made in the flat welding position.
8. Strike an arc and weld the joint. Use a C-shaped weld pool. As the overlapping piece melts, it will form a fillet. No filler metal is needed.
9. Hold the torch steady and complete the weld.
10. Repeat the process and weld the other side of the joint.

Inspection:

Each weld should be even in width. It should have a flat face with evenly spaced ripples. The weld should be free of defects. Refer to the heading *Weld Defects* at the end of this chapter.

Exercise 17-5 Making a Lap Joint with Filler Rod

1. Obtain two pieces of mild steel measuring 3/32″ × 3″ × 6″ (2.4 mm × 75 mm × 150 mm).
2. Clean one edge and one surface on each piece with a wire brush.
3. Set up the welding machine for welding mild steel. Place the current control switch in the remote position.
4. Set the shielding gas flow rate.
5. Select and prepare the proper type and diameter electrode. Select the correct size collet and collet body.
6. Install the electrode in the torch with the proper amount extending.
7. Tack weld the two pieces together to form a lap joint. Make sure there is no gap between them. Place the pieces at an angle so the weld is made in the flat welding position.
8. Select the correct filler rod type and diameter.
9. Strike an arc and weld the joint. Use a C-shaped weld pool. Add the rod as needed to form a convex bead. When adding filler metal, the overlapping plate should not be melted as much as it is when no filler metal is used.
10. Hold the torch steady and complete the weld. When you reach the end of the weld, slowly reduce the welding current. Add filler metal to fill the weld pool.
11. Repeat the process and weld the other side of the joint.

Inspection:

Each weld should be even in width. Each weld should have a convex face with evenly spaced ripples. Compare your weld beads to Figure 17-14.

Welding Inside Corner and T-Joints

Fillet welds are made on inside corner joints and T-joints. The arc must heat both surfaces to be welded. A C-shaped weld pool indicates that the surfaces are being melted equally. Remember that the electrode must extend about 1/4″ (6.4 mm) from the nozzle. Refer to Figure 16-5.

The torch is held at a travel angle of 15°–30° and a work angle of 45°. The filler rod is held at an angle of 15°–20° to the metal. See **Figure 17-15**.

To hold the metal in position, tack weld about every 3″ (76 mm) along the joint. When welding, start the arc about 1/8″–1/4″ (3.2 mm–6.4 mm) from the edge of the joint. Once the arc is started, move it back to the beginning of the joint and begin welding.

To make the fillet weld, filler metal is added. When you are welding in the flat position, the rod is added to the center of the weld pool. When you are welding in the horizontal position, the rod is added to the top edge of the weld. This can help reduce undercutting. The proper torch angle is most important in preventing undercutting. **Figure 17-16** shows the correct angles for the torch and filler metal.

Keep the fillet the same size as you move along the joint to the end of the weld. Fill the weld pool before stopping the arc. After the arc stops, hold the torch over the end of the weld to protect it with shielding gas while it cools.

To obtain the proper size weld, add filler metal as needed. As a rule of thumb, a fillet weld should be as thick as the metal being welded. The size of a fillet weld is measured by the lengths of its legs. See **Figure 17-17**. Thick metals will require more than one weld pass to complete a properly sized fillet weld.

Welding Butt Joints

Full-penetration butt welds can be made on a square groove joint up to 3/16″ (4.8 mm). Thicker metals are beveled to allow for complete penetration.

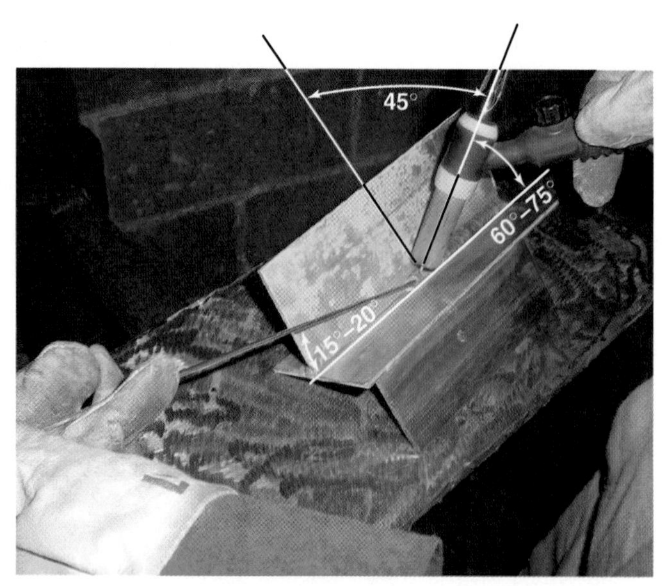

Figure 17-15. A fillet weld on a T-joint. Notice the electrode extension and the electrode and welding rod angles.

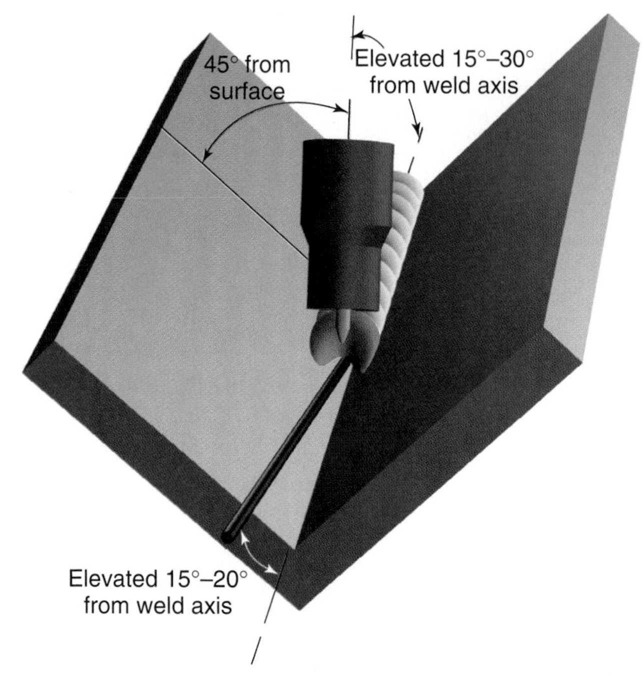

Figure 17-16. The suggested electrode and welding rod angles for making a fillet weld using GTAW in the flat position.

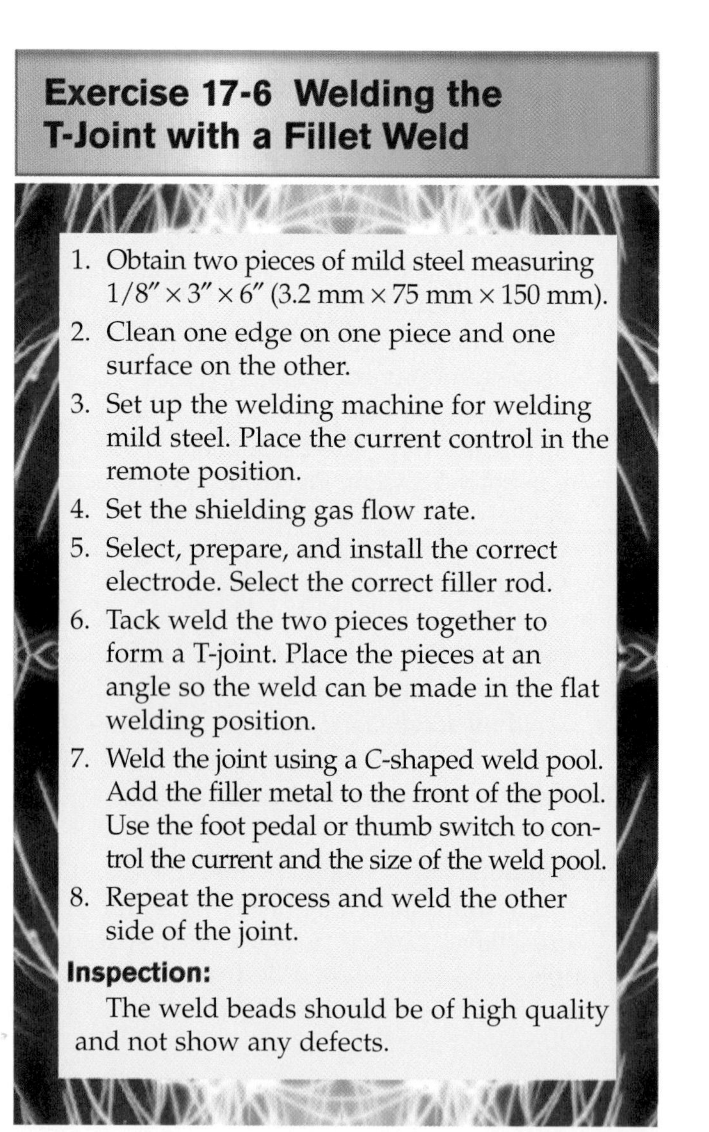

Exercise 17-6 Welding the T-Joint with a Fillet Weld

1. Obtain two pieces of mild steel measuring 1/8″ × 3″ × 6″ (3.2 mm × 75 mm × 150 mm).
2. Clean one edge on one piece and one surface on the other.
3. Set up the welding machine for welding mild steel. Place the current control in the remote position.
4. Set the shielding gas flow rate.
5. Select, prepare, and install the correct electrode. Select the correct filler rod.
6. Tack weld the two pieces together to form a T-joint. Place the pieces at an angle so the weld can be made in the flat welding position.
7. Weld the joint using a C-shaped weld pool. Add the filler metal to the front of the pool. Use the foot pedal or thumb switch to control the current and the size of the weld pool.
8. Repeat the process and weld the other side of the joint.

Inspection:

The weld beads should be of high quality and not show any defects.

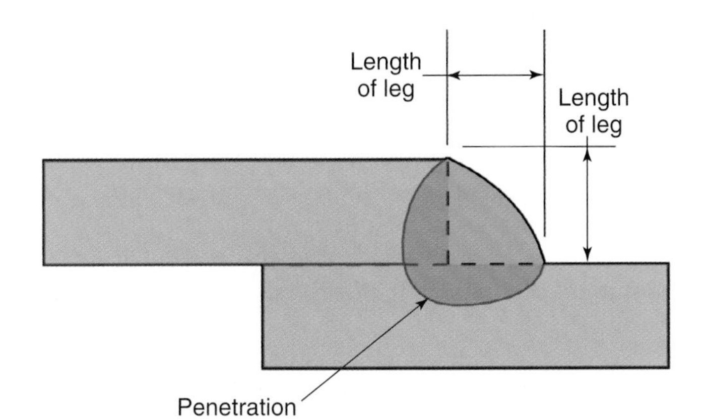

Figure 17-17. This drawing shows how the legs of a fillet weld are measured. The size of a fillet weld is usually about the same size as the metal thickness.

Even 3/16″ (4.8 mm) metal is sometimes beveled for this reason. Full penetration is obtained by using the keyhole method. It is important to tack weld the metals to hold them in position for welding.

The torch and electrode are centered over the root of the joint. The electrode has a 15°–30° push angle and a 0° work angle. The filler rod is held at 15°–20° above the work surface, as shown in **Figure 17-18**.

Start the welding arc as described earlier. When the weld pool is properly formed, add filler metal to the front of the pool. While the filler metal is being added, the torch is moved to the rear of the weld pool. After the rod is removed, the torch is moved forward again. Keep the rod close to the weld pool, adding it as needed. **Figure 17-19** shows a welder making a butt weld.

On thick metals, more than one weld pass is needed to fill the groove. After the root pass, either stringer beads or weave beads can be used to fill the groove. These weld beads are made the same way as welding a bead on plate. You must be sure to get good penetration into the previous weld bead and into the sides of the joint.

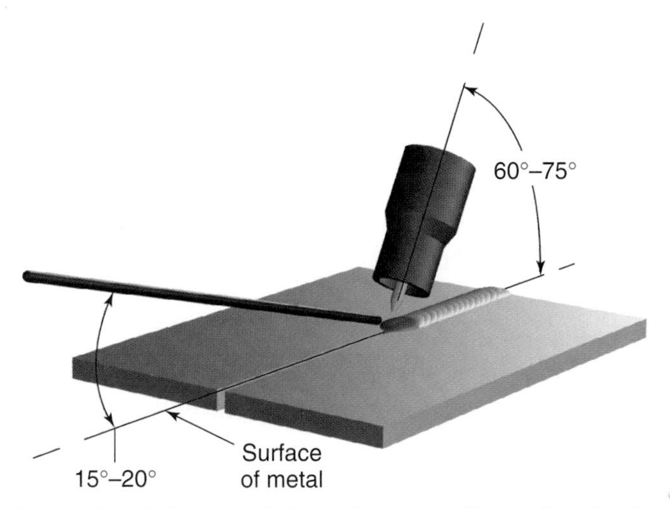

Figure 17-18. Suggested electrode and welding rod angles for welding a butt joint in the flat welding position.

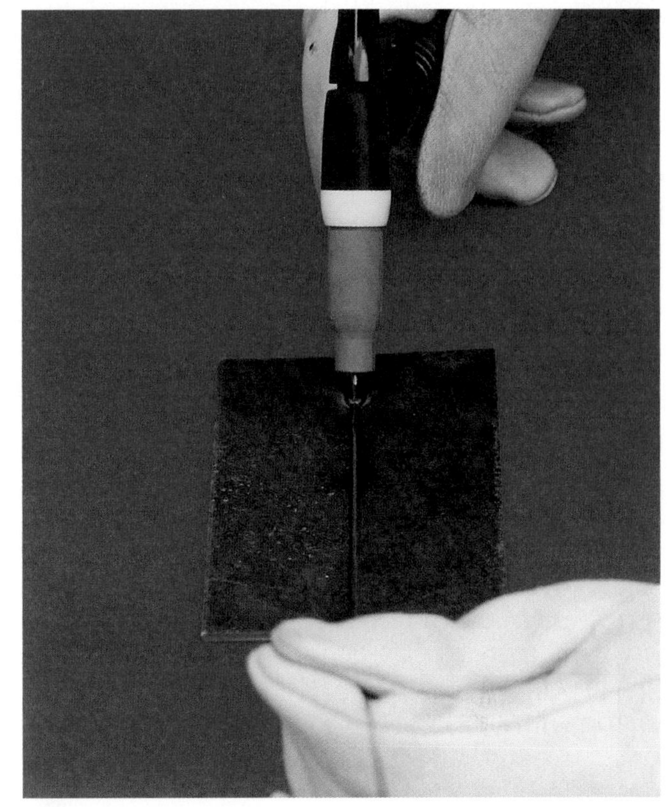

Figure 17-19. A butt weld in the flat position. Note that the welding rod is held in line with the weld axis and at a low angle.

A completed butt weld should show even ripples on the face of the metal. There should be even ripples of penetration on the back side, as well. No defects should be seen on either side.

Welding Other Metals

Base metals other than mild steel are often welded with the gas tungsten arc welding process. The most commonly welded metals include stainless steel and aluminum. Magnesium, titanium, and other types of metals require the high-quality welds that GTAW produces. The inert gas shielding prevents oxidation and contamination of the weld.

Stainless steel is welded much like mild steel. Aluminum and magnesium, however, are somewhat different. In these metals, the molten weld pool tends to fall through if it gets too large. If the weld pool falls through the metal, a large hole results. These metals do not have much strength when they are very hot. This is called ***hot shortness***. If the pool begins to sag more than it should when welding these metals, either add welding rod to cool it or reduce the current.

One way to prevent the weld pool from falling through the plate in hot-short metals is to use a backing. A ***backing***, often referred to as a "back-up," is a piece of metal, positioned behind the weld, that will

Exercise 17-7 Welding a Butt Joint

1. Obtain two pieces of mild steel measuring 3/32″ × 3″ × 6″ (2.4 mm × 75 mm × 150 mm).
2. Clean one long edge on each piece.
3. Set up the welding machine for welding mild steel. Place the current control in the remote position.
4. Set the shielding gas flow rate.
5. Select and prepare the correct electrode.
6. Select and install the correct size collet and collet body. Install the electrode and nozzle.
7. Select the correct diameter and type of filler rod.
8. Align the two pieces to form a butt weld with a 1/32″–1/16″ (0.8 mm–1.6 mm) root opening. Tack weld the two pieces together.
9. Start the arc. Move the torch until both pieces of metal begin to melt and a keyhole forms. Push on the foot pedal or adjust the thumb switch to adjust the current as required. When both pieces are molten, add the welding rod. Move the torch ahead, making sure that both pieces melt equally. Also watch that the size and shape of the keyhole remains constant. Complete the butt joint.

Inspection:

The weld beads should be of high quality and not show any weld defects. If you do not get enough penetration, the keyhole must be made larger.

not melt and will not join to the metal being welded. A stainless steel backing is used when welding aluminum, magnesium, or copper. On metals that do not have hot shortness, a backing is used to control penetration. For example, copper is used when welding steel or stainless steel.

When welding other base metals, refer to Figure 16-19 for the recommended current type. Figures 16-9 through 16-12 give the recommended current setting, electrode diameter, filler rod diameter, and shielding gas flow for aluminum, stainless steel, magnesium, and copper.

Weld Defects

Common weld defects in GTAW include porosity, undercutting, lack of penetration, overlap, and tungsten inclusions. Two problems that may occur while welding are metal oxidation and an unstable arc. These two problems are not weld defects. Oxidation shows there is a problem in the equipment or in the technique. An unstable arc is not a defect but can cause defects.

Porosity is caused when shielding gas does not protect the weld area. If you find porosity in the weld area, check the flow rate of the shielding gas. Also, check all shielding gas connections for leaks. Keep a correct arc length. Too long an arc will allow oxygen from the air to get into the weld area.

Undercutting is caused by using incorrect torch angles. Use the proper torch angles and add filler metal to fill any undercutting.

Lack of penetration is caused when the base metal is not heated properly for fusion to occur well below

Exercise 17-8 Welding Other Base Metals

Each of the welding exercises in Chapter 17 can be repeated using different base metals. You should learn to weld stainless steel and aluminum with GTAW. The techniques of starting the arc, forming the weld pool, and ending a weld are the same as those used for mild steel.

1. Clean the metal surfaces to be joined.
2. Set up the welding machine for the type of metal to be welded.
3. Set the shielding gas flow.
4. Select an electrode of the proper type and correct diameter. Prepare the electrode for welding.
5. Select the correct size collet and collet body. Install the collet, collet body, and electrode in the torch. Make sure the electrode extends the proper amount.
6. Make the weld.

Inspection:

Completed welds in stainless steel, aluminum, and other metals should all exhibit high quality weld beads and not have any defects.

the surface of the base metal. Penetration is improved by increasing the amperage, slowing down the travel speed, and using the correct torch angles.

Overlap is caused when a large weld pool flows onto the base metal and becomes solid. The molten weld pool does not melt the base metal; it just lies on top of it. To prevent overlap, keep the weld pool the proper size, use the correct torch angles, and do not add too much filler metal.

One defect that occurs only in GTAW is getting tungsten in the weld. When tungsten gets into the weld, it is called a ***tungsten inclusion***. Tungsten inclusions are caused by the following:

- Dipping the electrode into the weld pool.
- Using too much welding current.
- Using an electrode that is too small.

Tungsten inclusions can occur when too much current is used for the electrode diameter. To correct this, change to a larger diameter electrode or reduce the amount of current you are using.

Neither the weld bead nor the base metal should be oxidized. ***Oxidized*** means that oxygen has combined with the base metal. This is caused by a lack of shielding gas coverage. Check the shielding gas flow rate. Also make sure you use the correct electrode extension and the correct arc length. Remember that you must hold the torch over the end of the weld for the post flow of shielding gas.

When welding, the arc should be straight and concentrated between the end of the electrode and the base metal. When the arc jumps around the base metal or moves up and down on the electrode in an uncontrolled manner, it is an ***unstable arc***. An unstable arc is caused by one of the following:

- Electrode is contaminated.
- Electrode is too large in diameter.
- Weld joint is too narrow.
- Base metal is dirty.

Summary

- Gas tungsten arc welding is done by striking an arc between the tungsten electrode and the base metal. The welding arc creates a small weld pool on the base metal. Filler metal is usually added to the molten weld pool to fill the weld joint.
- Welding (filler) rods are available in many different alloys. Usually a filler rod with a composition similar to the base metal is used. Filler metal should have to be added to the weld only about once every two seconds to obtain the correct weld size.
- There are two different ways to start, or strike, an arc. One is to use a high-frequency generator. The other is to touch the electrode to the base metal.
- When the weld is complete, you must stop the arc. Stop the current by removing your foot from the pedal or turning off the thumb switch. Keep the torch over the area where you stopped welding. The post flow of shielding gas protects the weld as it cools.
- To weld a butt joint or a bead on a plate, a push angle of 15°–30° and a work angle of 0° are used. The filler rod is held at 15°–20° above the work surface. Start the welding arc. After the arc stabilizes, hold the electrode about 1/16″–1/8″ (1.6 mm–3.2 mm) above the work. A molten weld pool will form from the heat of the arc. Hold the torch in the same location until the weld pool is the proper size, usually two or three times the diameter of the electrode. When the weld pool reaches the correct size, add the filler metal at the front of the weld pool. At the same time that you add the filler metal, move the torch toward the back of the pool. After adding the filler metal, remove the rod from the weld pool. Keep the end of the filler rod close to the arc and the pool. Move the torch forward. Keep the pool the same size as you move along the length of the weld joint. As you finish a weld, add filler metal to the weld pool to fill the crater.
- Some joints do not require filler metal. These include the edge joint, the lap joint, and the outside corner joint. When welding these joints, the base metal is melted until the two metals flow together.
- A lap joint can be made with or without filler metal. When welding a lap joint, hold the torch at a travel angle of 15°–30° and a work angle of 45°. The weld pool should have a C shape. Hold the filler rod at an angle of 15°–20° to the base metal and add it to the upper edge of the weld pool. A completed weld will have evenly spaced ripples. When filler metal is added, the bead should have a slight convex surface. If filler metal is not used, the bead usually will have a flat surface.

(continued)

- Fillet welds are made on inside corner joints and T-joints. The arc must heat both surfaces to be welded. A C-shaped weld pool indicates that the surfaces are being melted equally. The torch is held at a travel angle of 15°–30° and a work angle of 45°. The filler rod is held at an angle of 15°–20° to the metal. When you are welding in the flat position, filler rod is added to the center of the weld pool. When you are welding in the horizontal position, filler rod is added to the top edge of the weld.
- Base metals other than mild steel are often welded with the gas tungsten arc welding process. The most commonly welded metals include stainless steel and aluminum. Welding aluminum and magnesium is somewhat different than welding mild steel. In these metals, the molten weld pool tends to fall through if it gets too large. One way to prevent the weld pool from falling through the plate is to use a backing. A backing is a piece of metal, positioned behind the weld, that will not melt and will not join to the metal being welded.
- Common weld defects in GTAW include porosity, undercutting, lack of penetration, overlap, and tungsten inclusions.

Review Questions

Write your answers on a separate sheet of paper. Please do *not* write in this book.

1. List two ways to strike, or start, an arc.
2. What is the correct arc length for GTAW?
3. What are the two ways to increase the current while welding?
4. Why should the welding rod be held close to the arc while welding?
5. Where should the torch be while filler metal is being added to the weld pool?
6. List two ways to prevent porosity from developing in GTAW.

Use the information (charts, figures, and text) provided in this chapter and in Chapter 16 to answer questions 7 through 10.

7. Answer the following for welding a fillet on 1/8″ (3.2 mm) mild steel:
   Type of current: _____
   Amount of current: _____
   Electrode extension (beyond the nozzle): _____
   Type of electrode: _____
   Time of post flow: _____
8. Answer the following for welding a butt joint in 1/16″ (1.6 mm) 6061 aluminum:
   Type of current: _____
   Amount of current: _____
   When high-frequency voltage is used: _____
   Type of filler metal: _____
9. Answer the following for welding on 1/4″ (6.4 mm) stainless steel:
   Type of welding current: _____
   Size of electrode: _____
   Type of shielding gas: _____
   Shielding gas flow: _____
   Shape of electrode tip: _____
10. Answer the following for welding on 1/4″ (6.4 mm) nickel alloys:
   Type of current: _____
   Type of filler metal: _____

Chapter 18

GTAW: Horizontal, Vertical, and Overhead Welding Positions

Learning Objectives

After studying this chapter, you will be able to:

- Explain why out-of-position welding is often an important part of welder qualification tests.
- Identify the correct torch and filler rod angles for out-of-position welding.
- Weld in the horizontal welding position with GTAW.
- Weld in the vertical welding position with GTAW.
- Weld in the overhead welding position with GTAW.

downhill
out-of-position welding
uphill

Out-of-Position GTAW

Welding in a position other than flat is called ***out-of-position welding***. Small assemblies usually can be placed so the welding is done in the flat position. Larger assemblies, which often cannot be moved, require the welder to weld out of position. See **Figure 18-1**.

Figure 18-1. *GTAW being used to weld a V-groove butt joint on pipe. (Miller Electric Mfg. Co.)*

Many welder qualification tests are performed on out-of-position joints. If a welder can make good welds out of position, he or she can also weld properly in the flat welding position. Chapter 4 describes and illustrates the different welding positions.

Welding in the Horizontal Welding Position

The weld axis and weld face are horizontal or close to horizontal when making a weld in the horizontal position. The weld axis can vary from 0°–15°. The weld face can vary from 80°–150° or from 210°–280°.

Welding in the horizontal position can be more difficult than welding in the flat position. Gravity pushes the molten weld pool down. This causes the weld bead to sag and not properly fill the joint. It is important to use proper torch angles and techniques to obtain high-quality welds.

Preparing to Weld

Before beginning any welding, check that the equipment is assembled and set up correctly. Refer to Chapter 16. Select and set the type and amount of current, the electrode type and diameter, and the type and amount of shielding gas you will use. Select the correct filler rod. Refer to Chapters 16 and 17 for more information.

Pulsed GTAW may be used to make high quality welds in out-of-position joints. The pulsed arc allows the weld pool to cool between pulses. Good penetration is still obtained. Pulsed GTAW is discussed in Chapter 15.

Protective clothing worn for out-of-position welding is the same as for flat welding. A long-sleeved shirt with the collar buttoned should be worn. Wearing a welding cap is a good practice. A leather coat or cape is recommended when welding overhead. Bare skin must be well-protected from falling molten metal.

Holding the torch and starting an arc is the same as in flat welding. Weld pools in out-of-position welding are kept the same size or slightly smaller than those used in flat welding. The current setting on the machine can be reduced. This keeps the weld pool smaller. However, penetration is shallower when the current is reduced.

Fillet Weld

Horizontal fillet welds are very common. They are made on lap joints, inside corner joints, and T-joints. **Figure 18-2** shows a welder making a fillet weld in the horizontal welding position.

When making horizontal fillet welds, the travel angle is held at a 15°–30° angle. The work angle is 45°. The torch is angled so the electrode points in the direction of travel. This is a push angle. Filler metal is held 15°–20° from the weld axis and in line with the axis of the weld. The torch and filler metal angles used to make a horizontal fillet weld are shown in **Figure 18-3**. The torch is centered over the root of the weld but pointed more toward the vertical piece. This helps prevent undercutting. Filler metal is added to the upper leading edge of the weld pool. When making a fillet weld in any position, a C-shaped weld pool is used. This shows that both pieces of metal are melting. This is similar to welding in the flat position.

The face of a completed fillet weld should be flat or slightly convex. See **Figure 18-4**. The weld should not sag down onto the lower piece. Sagging of weld metal can cause undercutting, underfill, or overlap. If any of these happen, change your torch angles and add the filler metal to the top edge of the weld pool. There should be no porosity. If there is, check that the shielding gas is set for the proper flow rate and that there are no leaks in the hoses, fittings, and connections.

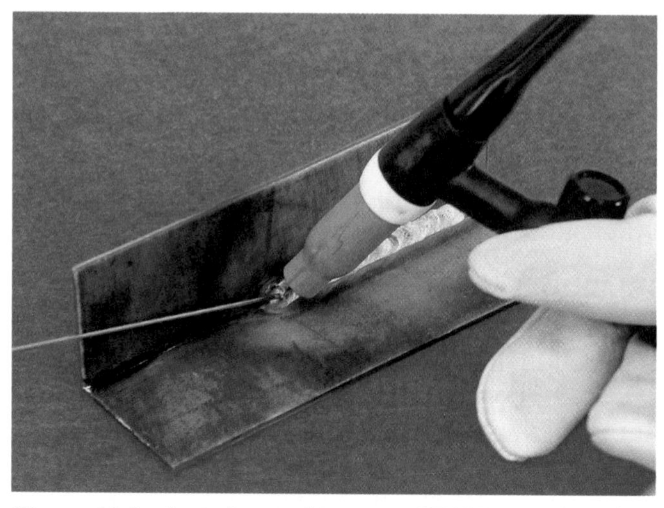

Figure 18-2. A student welder uses GTAW to make a fillet weld on a horizontal T-joint.

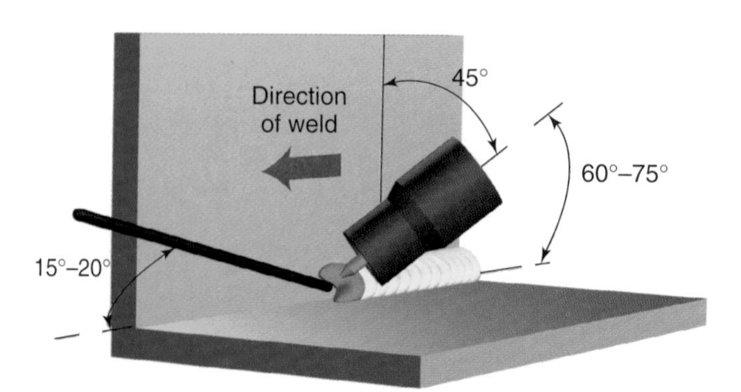

Figure 18-3. The suggested electrode and filler rod angles for gas tungsten arc welding a T-joint in the horizontal position. The angles are the same for forehand or backhand welding.

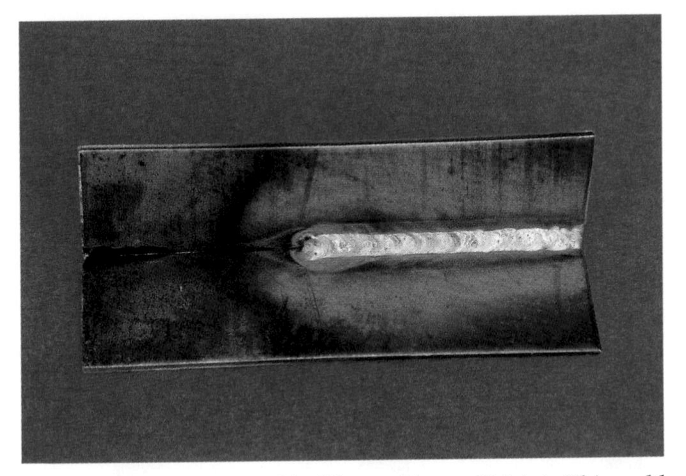

Figure 18-4. An acceptable fillet weld on a T-joint. This weld was made on mild steel. It has a flat bead, even ripples, and an even bead width. There is no apparent undercutting, sag, or overlap.

Exercise 18-1 Making a Horizontal Fillet Weld on a Lap Joint

1. Obtain two pieces of mild steel measuring 1/8″ × 1 1/2″ × 6″ (3.2 mm × 40 mm × 150 mm).
2. Clean one long edge and one face on each piece.
3. Set up the welding machine for welding mild steel. Use DCEN (DCSP) and 100A. Place the current switch in the remote position. Set the high-frequency start switch on.
4. Set the shielding gas flow rate. Refer to Figure 16-8 for shielding gas flow rates.
5. Obtain a 2% thoria tungsten electrode 1/16″ (1.6 mm) in diameter. Grind the tip into a blunted point. Make sure the grind marks are lengthwise, as shown in Figure 16-16.
6. Select the correct size collet and collet body. Install these and the electrode in the torch. The electrode should extend 1/8″–3/16″ (3.2 mm–4.8 mm) beyond the end of the nozzle.
7. Obtain a 3/32″ diameter (2.4 mm) mild steel welding rod.
8. Clamp the pieces to form a lap weld. Tack weld the pieces together at three places on each side of the joint. Place the workpiece in a horizontal position.
9. Hold the torch at the correct angle about 1/8″ (3.2 mm) above the work.
10. Press the foot pedal or thumb switch. The arc will start.
11. Move the electrode closer to the work, so it is about 3/32″ (2.4 mm) above the base metal. Watch the weld pool as it forms into a C shape. When it reaches the correct size and shape, add the filler rod. Begin moving the torch forward.
12. Maintain the same size weld pool, adding welding rod as needed, as you steadily move along the plate. Keep the electrode a constant distance above the work.
13. Stop welding at the end of the joint.
14. Weld the other side of the joint.

Inspection:

The weld bead should be straight with even ripples. There should not be any overlap or porosity.

Exercise 18-2 Making a Horizontal Fillet Weld on a T-Joint

1. Obtain two pieces of mild steel measuring 1/8″ × 1 1/2″ × 6″ (3.2 mm × 40 mm × 150 mm).
2. Clean one long edge on one piece and one face on the other piece.
3. Set up the welding machine for welding mild steel. Set the shielding gas flow rate.
4. Select, prepare, and install the electrode. Select the correct filler rod.
5. Tack weld the two pieces together to form a T-joint.
6. Weld both sides of the T-joint. Watch the weld pool to make sure both surfaces melt. Add welding rod to form a 1/8″ (3.2 mm) fillet weld. The bead should be slightly convex.

Inspection:

The completed weld should have even ripples. The bead should be even in width along the entire joint. There should not be any defects in the weld. Compare your completed weld to the one shown in Figure 18-4.

Butt Weld

Molten metal in a horizontal butt weld will sag downward due to the force of gravity. If the weld sags, it will have a poor appearance. It also can be weakened by undercutting or underfilling of the weld joint. Overlap may also occur.

The angles of the torch and filler rod are shown in **Figure 18-5**. Notice that the welding rod is held 15°–20° above the axis of the weld. The welding rod is added to the front upper part of the weld pool. Adding the filler metal to the top of the pool reduces undercut and underfill. The tip of the torch is angled upward about 15°. Pointing the electrode up helps keep the weld pool from sagging.

Welding with a small weld pool and a short arc length will help prevent these defects. Some common defects are shown in **Figure 18-6**.

Butt joints require full penetration welds. Point the arc at the root of the weld. Heat the weld joint with the arc, so both pieces begin to melt. They should melt all the way to the root of the joint. You can then add the filler rod. Move the torch forward at a uniform speed, keeping both pieces molten.

The root of the weld will look like a keyhole, as shown in **Figure 18-7**. The back and the sides of the weld are molten. Keep the keyhole size the same as the weld progresses.

A completed root pass will have even ripples on both the root side and the face side of the weld. To fill thick joints, use stringer bead or weave bead passes. See **Figure 18-8**.

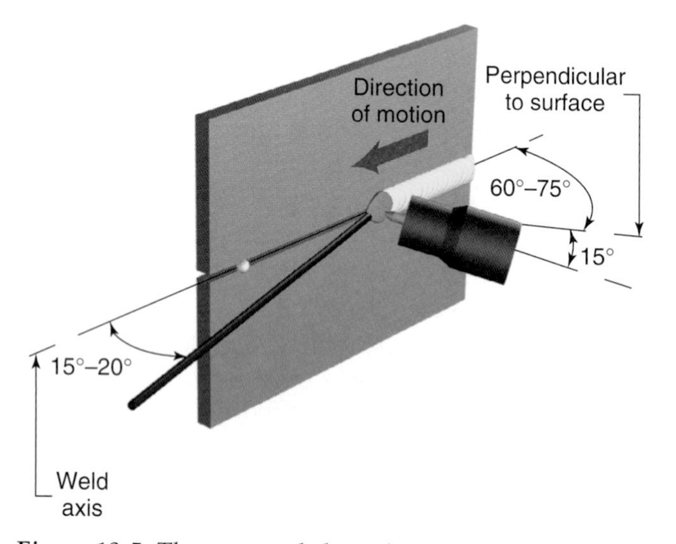

Figure 18-5. The suggested electrode and filler rod angles for making a butt weld in the horizontal position. The electrode should point upward 15°–20° to keep the bead from sagging.

Welding in the Vertical Welding Position

When vertical welding with the gas tungsten arc process, center the electrode over the root of the weld joint. Point the torch upward so that it forms a push angle of 15°–30°. **Figure 18-9** shows a welder welding a joint in the vertical position. Notice the angles of the torch and welding rod.

Thicker metals are welded with the weld pool moving from the bottom of the joint toward the top.

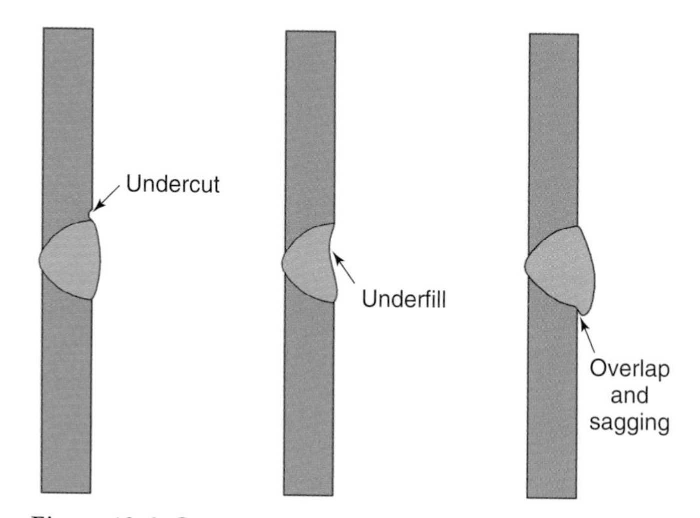

Figure 18-6. Common weld defects that may occur on a horizontal butt weld. Undercut is caused by an incorrect electrode angle. When not enough filler metal is added, underfill occurs. Overlap or sagging occurs when too much filler rod is used or if the bead is overheated.

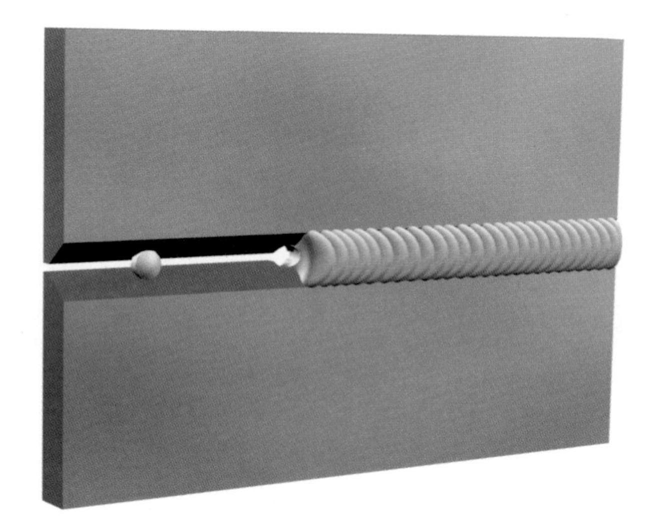

Figure 18-7. The keyhole may be seen at the bottom of the groove weld when there is proper penetration.

Exercise 18-3 Making a Horizontal Butt Weld

1. Obtain two pieces of mild steel measuring 1/8″ × 1 1/2″ × 6″ (3.2 mm × 40 mm × 150 mm).
2. Clean one long edge on each piece.
3. Set up the welding machine. Set the shielding gas flow rate.
4. Select and prepare the electrode. Select the correct filler rod.
5. Tack weld the two pieces together to form a butt joint that has a root opening of 1/16″ (1.6 mm). Place the workpiece in a horizontal position.
6. Weld the butt joint. Add filler metal to obtain a slightly crowned weld face. Watch the weld pool to make sure both pieces are melted to the root of the joint.

Inspection:

The completed weld should have even ripples on both the face side and root side of the joint. The bead should be even in width along the entire joint. There should not be any defects.

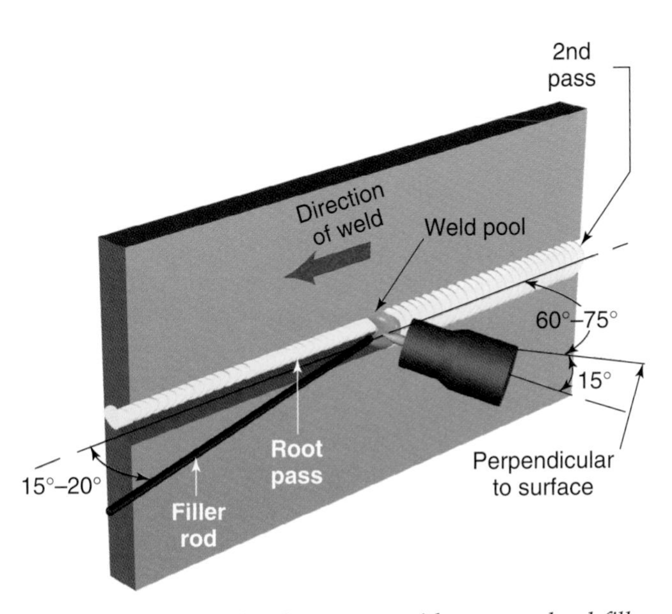

Figure 18-8. A completed root pass with a weave bead filler pass being added to complete a V-groove weld.

This is called ***uphill*** (vertically up) welding. Thin metals are welded ***downhill*** (vertically down). When welding downhill, the filler metal is added to the trailing edge or the side of the weld pool. **Figure 18-10** shows the angles used to weld downhill.

Fillet welds use a weld pool shaped like a C. This is the same as in flat welding. Butt welds use the keyhole method to obtain full penetration. See **Figure 18-11**.

Welding in the Overhead Welding Position

Overhead welding is the most difficult. It can be uncomfortable for the welder. The weld pool must be kept small when welding overhead. Use only stringer beads or very small weave beads.

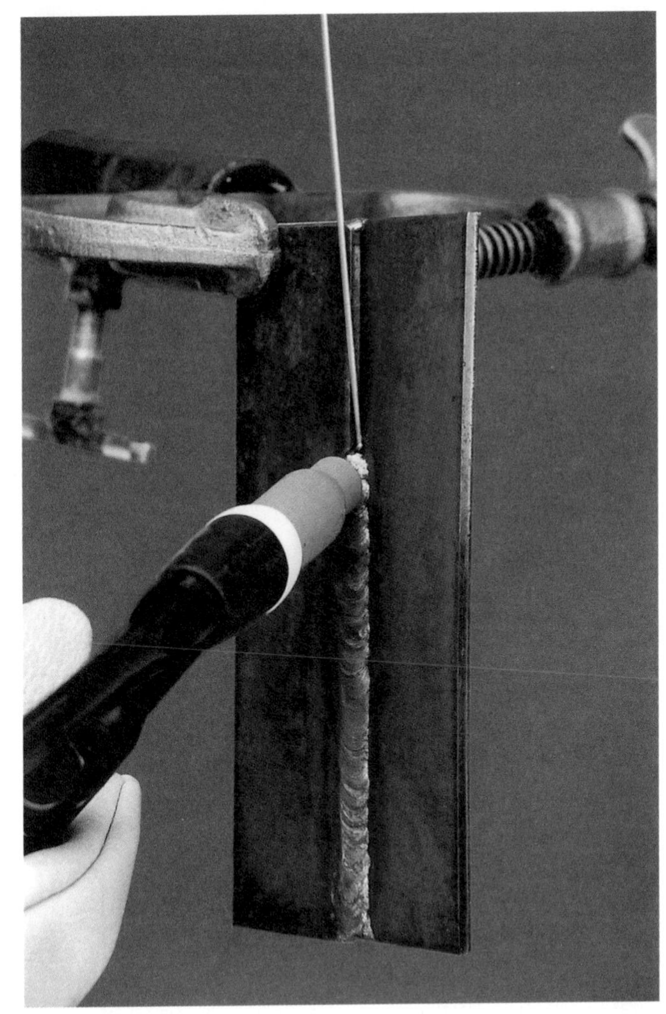

Figure 18-9. A fillet weld on a lap joint being made uphill. The electrode is positioned with a push travel angle of about 15°–30°. It is pointing more at the surface than at the edge. Note that the filler rod is held at about 15°–20° to the weld axis.

Exercise 18-4 Making a Vertical Fillet

1. Obtain two pieces of mild steel measuring 3/16″ × 3″ × 6″ (6.4 mm × 75 mm × 150 mm).
2. Clean one long edge on one piece and one face on the other piece.
3. Set up the welding station for welding mild steel.
4. Tack weld the two pieces together to form a T-joint. Use a welding positioner to hold them in the vertical position.
5. Weld both sides of the T-joint uphill. Each completed bead should be a 3/16″ fillet weld. This will require three stringer bead passes or one stringer bead pass and one weave bead pass. Do not let the weld pool get too large, or it will sag.

Inspection:

The completed weld should be of high quality. The ripples should be evenly spaced, and the completed weld bead should be even in width. There should not be any defects.

Exercise 18-5 Making a Vertical Butt Weld

1. Obtain two pieces of mild steel measuring 1/8″ × 1 1/2″ × 6″ (3.2 mm × 40 mm × 150 mm).
2. Clean one long edge on each piece.
3. Set up the welding equipment for welding mild steel.
4. Tack weld the two pieces together to form a butt joint with a root opening of 1/16″ (1.6 mm).
5. Weld the butt joint uphill with the filler metal. Use the recommended torch and welding rod angles and the keyhole method. The weld must have complete penetration.

Inspection:

Examine the weld for complete penetration. There should be no porosity, underfill, or any other defects present in the weld. Ripples on the face and root of the weld should be even and straight. Compare your completed weld to the one shown in Figure 18-11.

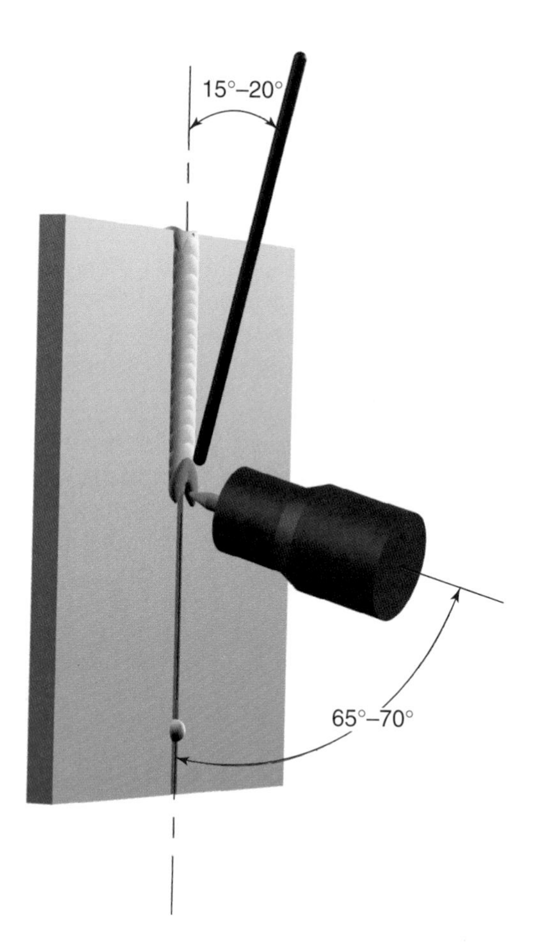

Figure 18-10. *When welding downhill, the electrode is held in line with the weld axis with a 15°–20° travel angle. The filler rod is held at 15°–20° from the base metal. It may be added from the top or side.*

The amount of current set on the welding machine to weld thin metals overhead is usually 10A less than when flat welding. Thicker metals use 20A less current for overhead welding than for flat welding.

Pulsed welding gives very good results. It provides good penetration and gives the weld pool a chance to cool. Reducing current or using a pulsed arc keeps the weld pool smaller than the pool used for flat welding.

Shielding gas flow rates must be higher when overhead welding. Remember that argon gas is heavier than air. It will remain on a plate being welded in the flat position. Refer to Figure 15-13. However, when welding overhead, the gas will fall away from the plate and not give good protection. Flow rates are increased by 3–5 cfh (1.4–2.4 Lpm) for thin metals. On metals 3/16″ (4.8 mm) or thicker, an additional 5 cfh (2.4 Lpm) or more is used. If you get porosity when welding overhead, check to make sure you are keeping the correct arc length. If your arc length is correct and you are still getting porosity, increase the shielding gas flow.

The angles used to weld fillet or butt welds overhead are the same as those used in other positions. Filler metal is added to the leading edge on a fillet weld. **Figure 18-12** shows a welder making a fillet weld in the overhead welding position. **Figure 18-13** shows a butt weld being made in the overhead welding position.

The torch is often held like a hammer when overhead welding. This provides flexibility and control.

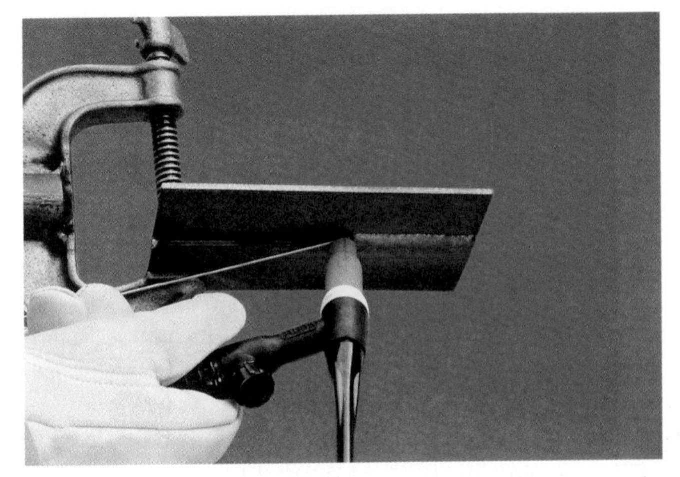

Figure 18-12. *A fillet weld being made on a lap joint in the overhead position. The angles of the electrode and welding rod are the same as those used in the flat position.*

Figure 18-13. *A bead being laid in a butt joint in the overhead welding position.*

Figure 18-11. *An acceptable butt weld made on mild steel that was made welding uphill. The bead is convex and straight, with evenly spaced ripples. There is no apparent undercut, underfill, or sag.*

Welding Other Metals

Stainless steel and aluminum are metals very commonly welded with gas tungsten arc welding. You should develop proper skills for welding both. Chapter 17 discusses how welding these metals differs from welding mild steel.

After learning to weld mild steel out of position, you should practice welding these metals in the horizontal, vertical, and overhead positions. Figures 16-9 through 16-12 list current and shielding gas settings for metals other than mild steel. Each of the exercises in this chapter can be done with aluminum, stainless steel, or any other base metal weldable with GTAW.

Backings are often used to control penetration when welding out of position. See Chapter 17 for more information about backings.

Exercise 18-6 Making an Overhead Fillet Weld on a T-Joint

1. Obtain two pieces of mild steel measuring 3/16″ × 3″ × 6″ (4.8 mm × 75 mm × 150 mm).
2. Clean one long edge on one piece and one face on the other piece.
3. Set up the welding station for welding mild steel. Use 85A–90A current. Use 18 cfh–20 cfh (8.5 Lpm–9.4 Lpm) argon shielding.
4. Tack weld the two pieces together to form a T-joint. Clamp the tack welded pieces in the overhead position.
5. Weld both sides of the T-joint. Keep the molten weld pool small so the metal does not fall or sag.

Inspection:

The completed weld should be of the same quality as a weld made in the flat position. Examine the completed weld for evidence of porosity. Also look for any metal sagging that might cause underfill or overlap.

Exercise 18-7 Making an Overhead Butt Weld

1. Obtain two pieces of mild steel measuring 1/8″ × 1 1/2″ × 6″ (3.2 mm × 40 mm × 150 mm).
2. Clean one long edge on each piece.
3. Set up the welding equipment for welding mild steel. Adjust the current and shielding gas flow rate for overhead welding.
4. Tack weld the two pieces together to form a butt joint with a root opening of 1/16″ (1.6 mm). Clamp the tack welded pieces in the overhead position.
5. Weld the butt joint. Use the correct torch and welding rod angles. Watch the weld pool. It must melt both pieces and form the keyhole shape. The weld must have complete penetration.

Inspection:

The completed weld should have even ripples. The edges of the weld should be the same width along the entire length. The root side of the weld should show consistent penetration. There should not be any porosity along the weld.

Summary

- Many welder qualification tests are performed on out-of-position joints. If a welder can make good welds out of position, he or she can also weld properly in the flat welding position.
- Horizontal fillet welds are very common. They are made on lap joints, inside corner joints, and T-joints. When making horizontal fillet welds, a 15°–30° push angle and 45° work angle is used. Filler metal is held 15°–20° from the weld axis and in line with the axis of the weld. The torch is centered over the root of the weld but pointed more toward the vertical piece.
- When making a horizontal butt weld, a 15° work angle and a 15°–30° push angle are used. The welding rod is held 15°–20° above the axis of the weld. The welding rod is added to the front upper part of the weld pool. Adding the filler metal to the top of the pool reduces undercut and underfill.
- When vertical welding with the gas tungsten arc process, center the electrode over the root of the weld joint. Point the torch upward so that it forms a push angle of 15°–30°. The welding rod is held 15°–20° above the axis of the weld. Thick metal is welded uphill, and thin metal is welded downhill. When welding downhill, the filler metal is added to the trailing edge or the side of the weld pool.

(continued)

- Overhead welding is the most difficult. It can be uncomfortable for the welder. The weld pool must be kept small when welding overhead. Only stringer beads or very small weave beads should be used. The amount of current set on the welding machine to weld thin metals overhead is usually 10A less than when flat welding. Thicker metals use 20A less current for overhead welding than for flat welding.
- Shielding gas flow rates must be higher when overhead welding. Flow rates are increased by 3–5 cfh (1.4–2.4 Lpm) for thin metals. On metals 3/16″ (4.8 mm) or thicker, an additional 5 cfh (2.4 Lpm) or more is used.
- The angles used to weld fillet or butt welds overhead are the same as those used in other positions. The torch is often held like a hammer when overhead welding. This provides flexibility and control.

Review Questions

Write your answers on a separate sheet of paper. Please do *not* write in this book.

1. What safety precautions should a welder take to prevent being burned by falling molten metal?
2. How does the size of the weld pool used for out-of-position welding compare to the size of the weld pool for flat welding?
3. Where is filler metal added to a horizontal fillet weld?
4. What are three common defects that occur when welding in the horizontal position?
5. What method is used to obtain complete penetration on a butt weld?
6. Are thin metals welded uphill or downhill?
7. How much is current increased or decreased when welding thick metals in the overhead welding position?
8. Why is the flow of argon shielding gas increased when welding overhead?
9. When welding overhead, is the torch most often held like a pencil or like a hammer?
10. In out-of-position welding, which of the following can be done to keep the weld pool small?
    A. Change to ac from dc.
    B. Reduce the amount of current.
    C. Change to dc from ac.
    D. Use a pulsed arc.
    E. Just A and B.
    F. Either B or D.

Section 5

Plasma Arc Cutting

Chapter 19:
Plasma Arc Cutting

Thermal Dynamics Corp., a division of Thermadyne

Chapter 19

Plasma Arc Cutting

Learning Objectives

After studying this chapter, you will be able to:

- Define plasma arc cutting.
- Describe the plasma arc cutting process.
- Identify and assemble the equipment and supplies used for PAC.
- Identify the parts of a PAC torch.
- Identify safety considerations for PAC.
- Set up PAC equipment for cutting.
- Perform cuts using PAC equipment.

constricting nozzle
dross
gouging
heat shield
ionized
kerf
nontransferred
pilot arc
plasma
plasma arc cutting (PAC)
swirl ring
transferred arc

Plasma Arc Cutting Principles

Plasma arc cutting (PAC) is a cutting process that uses an arc and a high-velocity, ionized gas coming through a constricting nozzle to cut all metals. The arc is created between the electrode and the base metal. The arc melts the base metal. The high-velocity gas blows the molten metal through the base metal. As the torch is moved along the path to be cut, metal continues to be removed. A slot is created in the base metal. This slot is called the ***kerf***.

Plasma is a state of matter. The most common states of matter are solid, liquid, and gas. Plasma is a superheated gas that is ionized. ***Ionized*** means some of the electrons in the gas have broken away from their atoms. These free electrons in the plasma conduct electricity.

Actually, a plasma gas is present in all arc welding processes. However, the plasma arc cutting process uses a constricting nozzle to create, concentrate, and direct the high velocity plasma. Using a small hole, the constricting nozzle concentrates the arc and plasma flow to create a cutting system. Only processes that use a constricting nozzle to create the plasma arc condition are called plasma arc.

Temperatures used during the plasma cutting process range between 18,000°F–25,000°F (10,000°C–14,000°C). These temperatures will melt all metals. The high velocity gas removes the molten metal from the part being cut. **Figure 19-1** shows the PAC process being used.

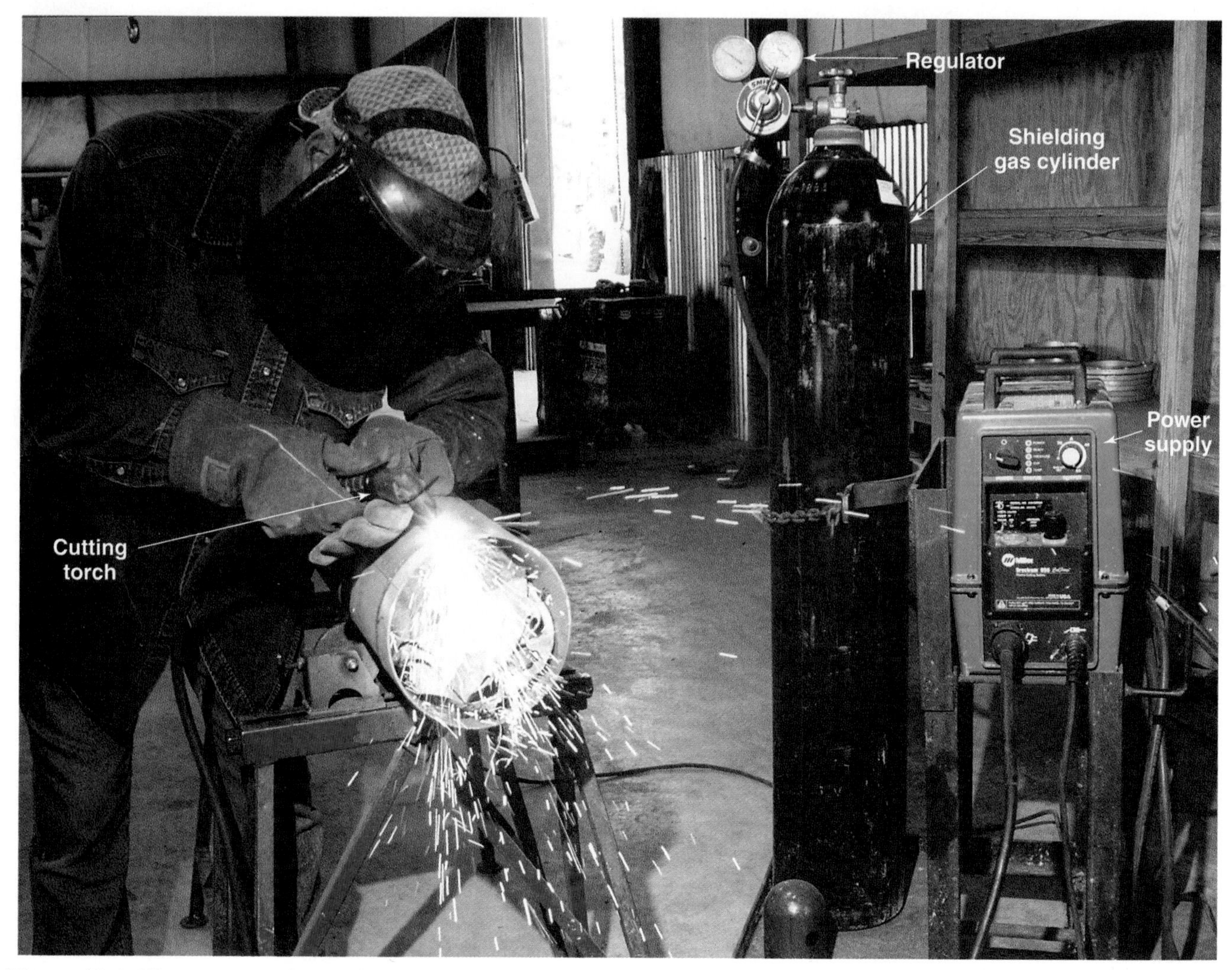

Figure 19-1. Plasma arc cutting equipment. An inverter power supply along with a gas cylinder are used to supply power and shielding gas for the cutting process. Notice the safety equipment being worn by the welder. (Miller Electric Mfg. Co.)

Some of the advantages of PAC include:

- The ability to cut all metals.
- Cutting speeds are faster than oxyfuel cutting, especially on metals less than 1″ (25.4 mm) thick.
- No preheating is required. Cutting begins immediately.
- Distortion and heat-affected zones are minimized.
- No hazardous or explosive gases are used.

Some disadvantages of plasma arc cutting include:

- The cost of PAC equipment can be higher than other processes.
- A source of electricity is needed to use the PAC process.
- The presence of an arc can cause safety hazards.
- The metal fumes created can be a health hazard.

Plasma Arc Cutting Equipment and Supplies

The plasma arc cutting process requires equipment designed for this process. The equipment must be able to start and control the arc. The torch design must create the plasma and properly direct it to perform the cutting operation. The end of the torch must be able to withstand very high temperatures. The main parts of a plasma arc cutting system are:

- A power supply.
- A torch.
- A supply of gas or gases and regulator(s).

Also required is proper safety equipment for the person using the PAC outfit.

Power Supplies

A power supply used for plasma arc cutting is a constant current machine. This is the same type used for shielded metal arc or gas tungsten arc welding. Plasma arc cutting uses direct current electrode negative (DCEN).

Most medium and heavy duty PAC power supplies require 230 or 460 volts and use three-phase input. Some units use single-phase power. Smaller units, including most inverter power supplies, operate using 115 or 230 volts. These units use single-phase power.

Many plasma arc cutting machines are inverter power supplies. An inverter power supply is lightweight. This makes it very portable.

Inverter power supplies electronically change the frequency of the supplied electrical power going to the transformer. The supplied power is alternating current at 60 hertz or 60 cycles per second. Hertz is a unit of measure for frequency. One hertz is one cycle per second.

An inverter power supply first changes the alternating current into direct current. This direct current then goes through an inverter. An inverter chops the current and creates square wave alternating current that will be between 1000 and 50,000 cycles per second. This high-frequency power is passed through a very small and efficient transformer. This small transformer is why inverter power supplies are very lightweight. Figure 19-1 shows an inverter power supply. The inverter process is covered in more detail in Chapter 5.

Plasma arc cutting power supplies have a high-frequency generator. The high-frequency voltage is used to initiate the arc.

Torches and Cables

The torch is an important part of the cutting system. The torch holds the electrode and provides a good electrical connection to the electrode. The torch also provides a path for the plasma (cutting) gas. This cutting gas and the arc pass through the constricting nozzle. A second nozzle is used to direct the shielding (secondary) gas. The torch body has a button or trigger used to start and stop the cutting process. The main parts are shown in **Figure 19-2**. Some torches are liquid (water) cooled. Most torches are cooled by the shielding gas used.

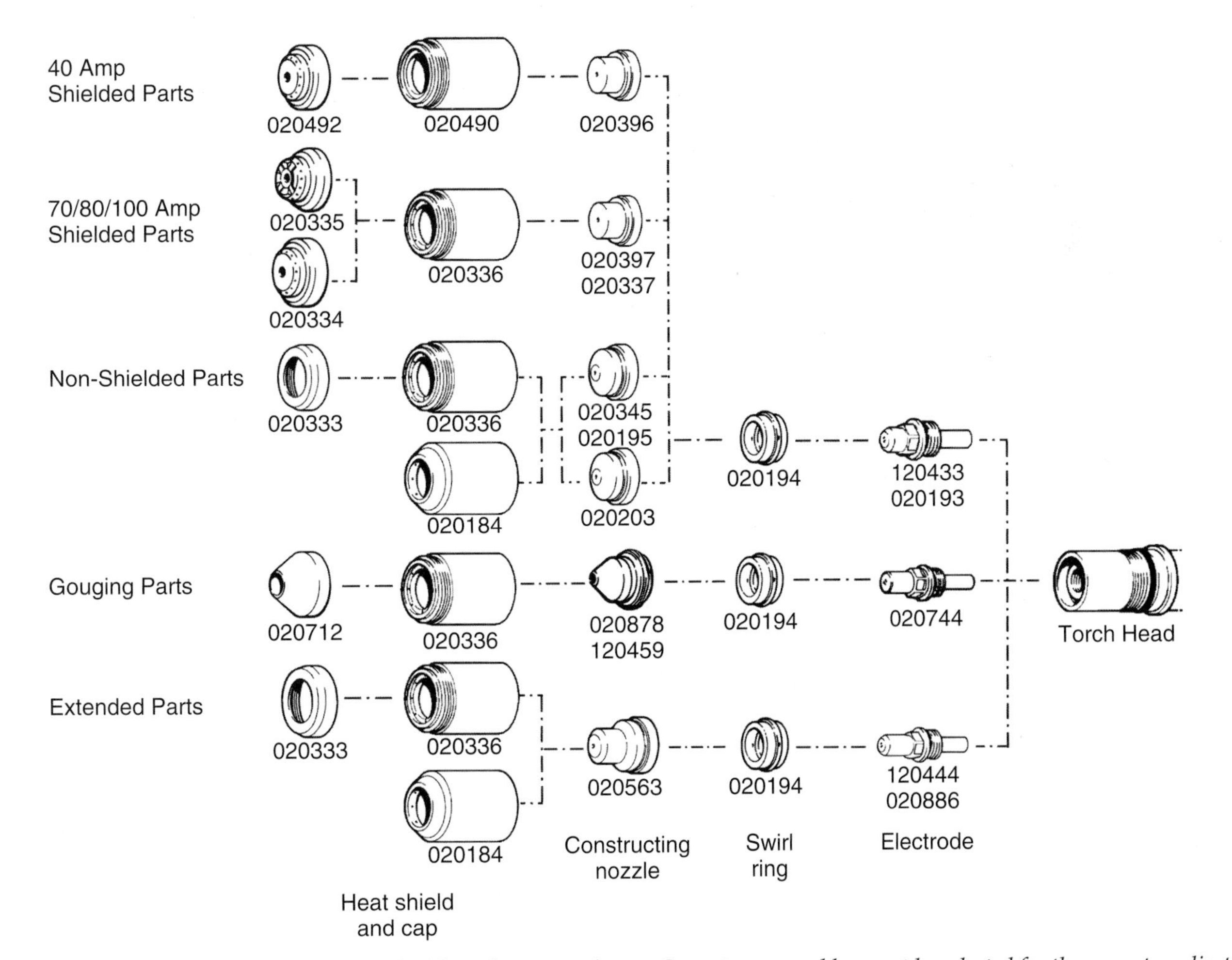

Figure 19-2. *Different electrodes, nozzles, shields and caps are shown. Correct consumables must be selected for the correct application. (Hypertherm)*

Electrodes are made of hafnium, zirconium, or tungsten. Hafnium electrodes are most common. Hafnium electrodes are used when air is used as the plasma gas. Tungsten electrodes are used when argon-hydrogen plasma gas is used.

The electrode installed into the torch body has a copper alloy surrounding the hafnium, zirconium, or tungsten electrode material. The electrode assembly is threaded into the torch body. **Figure 19-3** shows some PAC electrodes.

A nozzle is attached to the torch body. This nozzle is called a ***constricting nozzle***, **Figures 19-4** and 19-2.

Figure 19-3. *Electrodes come in many shapes. The hafnium, zirconium, or tungsten is surrounded by copper alloy. The light grey area in the center of the cutaway electrode is the electode material. (American Torch Tip)*

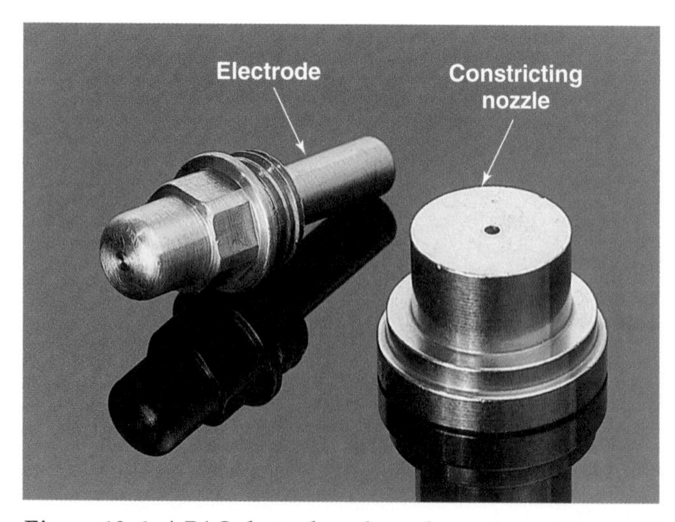

Figure 19-4. *A PAC electrode and nozzle are shown. The small hole in the nozzle aligns with the electrode when assembled. The arc and plasma gas flow through the hole in the nozzle. (American Torch Tip)*

The nozzle has a small hole in it called the constricting hole. The arc and the plasma cutting gas flow through this hole. The small hole in the constricting nozzle force the plasma gas to pass directly in the path of the arc. The arc superheats the gas and ionizes it, turning it into plasma. The gas exits the constricting nozzle at a very high speed and temperature.

The electrode and the nozzle are both considered consumable parts. They will get worn. As they wear, the quality of the cut is reduced. Both need to be replaced when they become worn.

A ***swirl ring*** is another part of most plasma arc cutting torches. The swirl ring works to create a rotating plasma gas. This keeps the flow of the plasma gas more concentrated. Figure 19-2 shows several swirl rings.

A ***heat shield*** is installed on the cutting torch, Figure 19-2. The purpose of a heat shield is to protect the torch from the high heat of the cutting process. Heat shields function as a second nozzle to contain and direct the shielding gas. Shielding gas works to cool the torch and in some applications protects the cutting area from contamination.

The parts of a PAC torch are assembled into the torch body. **Figure 19-5** shows the parts to be installed into a torch body.

Usually a single combination cable assembly from the torch body is used to connect the torch to the power supply. The connections to be made include:

- Electrical cutting power.
- Trigger control wire.
- Pilot arc control wire.
- Gas connection.

Torches used for automated or mechanized cutting, cutting thick material, and high duty cycles can be fluid (water) cooled. Torches for automated or mechanized cutting are straight. The nozzle on most manual torches is nearly perpendicular to the torch body.

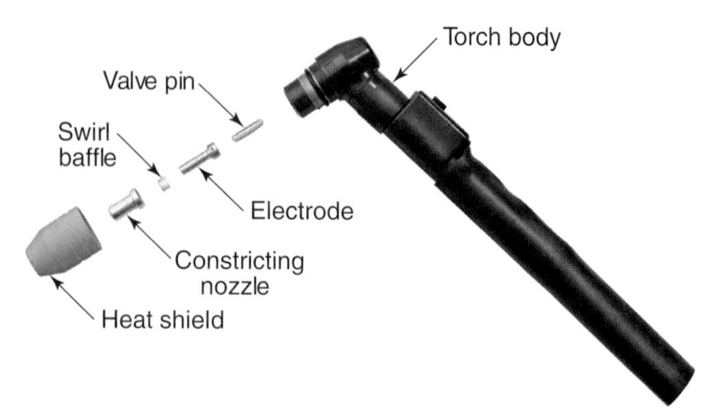

Figure 19-5. *Parts of a plasma arc cutting torch. (ESAB Welding and Cutting Products)*

A workpiece lead is also required. One end is connected to the power supply. The other end is connected to the work. The workpiece lead usually has a clamp on the end to secure it to the part being cut.

Plasma and Shielding Gases

As discussed earlier, gases are used in two areas. One use is the plasma gas. This gas flows through the constricting nozzle. It is heated by the arc and becomes a plasma.

The second use of a gas is as the shielding, or secondary, gas. The shielding gas may be the same as, a mixture of, or different than the plasma gas.

The most common gas used for manual plasma arc cutting is air. Air is used as both the plasma and the shielding gas. Air is used to cut carbon steels, aluminum, and stainless steel. Compressed air is supplied from a gas cylinder or, more often, from a shop manifold system. Some portable plasma arc cutting power supplies have a built-in air compressor.

Air is very low cost. To obtain the best results, air must be filtered and dried to remove dirt, dust, oil, and water vapor before it is used in a plasma cutting operation.

A second gas that can be used to cut carbon steel is oxygen. Oxygen is used as the plasma gas. Air is used as the shielding gas. Cutting speeds are increased when oxygen is used. Oxygen is not used to cut stainless steel or aluminum.

Nitrogen is a good plasma gas choice when cutting stainless steel or aluminum. Air is used as the shielding gas. Carbon dioxide (CO_2) can also be used as the shielding gas. CO_2 slightly improves the cutting speeds and metal surface finish but is more costly than air.

When cutting stainless steel or aluminum that is 1/2″ (12.7 mm) or thicker, the plasma gas used is often argon with 35% hydrogen. This gas combination is called H-35. Nitrogen is used as the shielding gas.

When using air or oxygen as the plasma and the shielding gas, the base metal is not shielded from the atmosphere. The base metal will be slightly oxidized from the cutting process. Only when argon or nitrogen is used as the plasma gas and nitrogen, water, or CO_2 provide the shielding, will the base metal not be oxidized.

Gases can be provided in cylinders. Cylinders are necessary if the cutting operation will be done in the field. Some gases can be supplied through a shop using a manifold system. **Figure 19-6** lists various shielding gases used for different base metals. Information on gas cylinders and regulators are covered in Chapters 20 and 11.

Plasma Gas	Shielding Gas	Mild Steel	Stainless Steel	Aluminum
Air	Air	Good cut quality & speed. Lowest cost.	Good cut quality & speed. Lowest cost.	Good cut quality & speed. Lowest cost.
Oxygen (O_2)	Air	Excellent cut quality & speed. Very little dross.	Not recommended.	Not recommended.
Nitrogen (N_2)	CO_2	Fair cut quality, some dross. Excellent part life.	Good cut quality. Excellent part life.	Excellent cut quality. Excellent part life.
Nitrogen (N_2)	Air	Fair cut quality, some dross. Excellent part life.	Good cut quality. Excellent part life.	Good cut quality. Excellent part life.
Nitrogen (N_2)	H_2O	Fair cut quality, some dross. Excellent part life.	Excellent cut quality. Excellent part life. Best surface preparation for welding.	Excellent cut quality. Excellent part life. Best surface preparation for welding.
Argon/Hydrogen 35%	Nitrogen (N_2)	Not recommended.	Excellent on material over 1/2″ (12.7 mm) thick.	Excellent on material over 1/2″ (12.7 mm) thick.

Figure 19-6. *Chart used to select plasma and shielding gases. Many handheld torches use air plasma and air shielding gas. This makes good quality cuts and is inexpensive to use.*

Equipment Assembly

There are only a few pieces of equipment that need to be assembled before cutting can begin.

The cutting power supply is plugged into an electrical outlet or wired into an electrical source. A qualified electrician should connect the power supply to the electrical source. Equipment to be used away from electrical power may have a generator or alternator run by an engine to create power.

If using a cylinder supply of gas for cutting, attach a regulator to the cylinder. Connect a hose from the regulator to the power supply. If using gas from a shop manifold, connect a hose from the manifold to the power supply. If a power supply has its own air compressor, no connection for shielding gas is required.

Connect the cutting torch to the power supply. On most PAC equipment, there is a combination cable attached to the cutting torch. This combination cable plugs into the power supply. A threaded ring or collar is tightened to secure the cable to the power supply.

Assemble the parts of the cutting torch following the manufacturer's directions and as discussed in the *Torches and Cables* section. Secure each piece, but do not overtighten them, as this will damage the parts or the torch body.

Safety Equipment

Plasma arc cutting is a combination of an arc process and a cutting process. The dangers of arc welding are present. These include the intense light, heat, and radiation of the arc. The dangers of cutting, which include flying particles and fire hazards, are also present.

Select an area for cutting that has good ventilation. Prior to starting the cutting process, inspect the area for any hazards. There should be no flammable items in the area, no paper, cardboard, rags, excessive grease, oil, or fuel. Any welding cables or hoses must be protected from flying metal.

Plasma arc cutting adds another safety concern. The high-velocity gas passing through the constricting nozzle can create very loud noise. High noise is more of a problem with high-current, high-velocity automated cutting than with manual cutting.

To protect the welder, the following safety precautions should be taken. Wear appropriate clothing. Wear long pants and long sleeve cotton shirts. The pants cannot have cuffs that can catch sparks. Do not carry flammable items, such as matches or lighters. Shirts should be buttoned at the collar. A welding jacket may be worn to further protect the welder. Gloves are worn to protect the hands.

Many welders using PAC wear a welding helmet. The filter plate or lens should be a #9 or greater. Some welders wear only welding goggles. These should only be used for light cutting and if only cutting for a short period of time. Arc rays are harmful.

Earplugs, earmuffs, or both are worn to prevent damage to the ears from loud noise. **Figure 19-7** shows earplugs and earmuffs being worn.

Some PAC is done on a water table to greatly reduce the noise. Parts to be cut are placed on a table. The parts are slightly below the surface of the water. Plasma arc cutting is still capable of cutting the parts. The plasma torch has a water jacket or ring of water flowing around the shielding gas. Water is recirculated from the water table through the torch. The purpose of the water is to significantly reduce the noise created. This is most often done on high-amperage, automated cutting processes. See **Figure 19-8**.

Plasma arc cutting and gouging produce fumes. Proper ventilation is required to remove contaminated air from the work environment.

Preparing to Cut

Very few settings are required when plasma arc cutting. Turn on the power supply. Open the shielding gas cylinder valve or the valve to the shop manifold system. Press the gas test button on the power supply, which will cause the shielding gas to flow. Adjust the regulator to the desired pressure. Refer to the manufacturer's recommended pressure setting. Common pressures for air plasma cutting are 65–70 psig.

Set the desired current on the power supply. There is often a dial on the power supply to adjust the current. The manufacturer usually supplies a chart showing the amount of current required to cut various thicknesses of base metal.

If the proper current setting is not known, it is better to set the current higher than needed prior to cutting. More current will ensure a complete cut. However, the travel speed may need to be faster and the cut quality may be poorer. If the current is found to be too high, adjust the current down.

Cutting Procedure

The process of starting the cutting arc occurs very quickly. The cutting process begins by pressing and holding a switch or button on the torch. A preflow of shielding gas will flow for a very short period of time. When air is used as the cutting gas, there is no need for preflow of shielding gas.

After the preflow time, a pilot arc starts. The ***pilot arc*** is an arc between the electrode and the constricting nozzle. The purpose of the pilot arc is to ionize the plasma cutting gas. High-frequency current is used to start and maintain the pilot arc. The pilot arc is ***nontransferred,*** which means the arc does not occur between the electrode and the work.

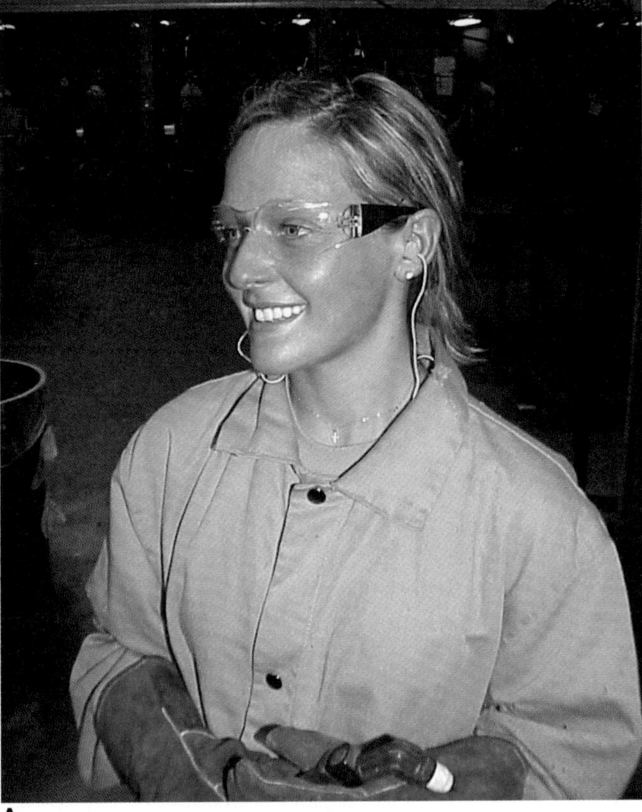

A

B

Figure 19-7. *A—This plasma cutting operator is showing the proper use of earplugs. B—Earmuffs block more noise than do earplugs. In very high noise areas, both should be worn.*

Figure 19-8. *An automated plasma arc cutting application. A water table is used to reduce noise. (Hypertherm)*

When starting the cutting process, it is best to bring the torch to the desired distance from the work before pressing the switch to start the cut. The nozzle must be close enough to the work to allow the arc to jump to the work. For manual cutting, the distance from the nozzle to the work should be 1/16″–1/8″ (1.6 mm–3.2 mm).

Press the switch to start the cut. The high open-circuit voltage and the ionized gas produced by the pilot arc allows the arc to jump from the electrode to the base metal. When the arc occurs between the electrode and the base metal, it is called a ***transferred arc***. Once the transferred arc is established, the pilot arc is no longer needed, and it is stopped.

The base metal is rapidly melted by the arc. High-velocity plasma coming through the constricting nozzle blows the molten base metal away. Cutting will continue so long as the switch is held and the torch is close enough to the work to maintain the cutting arc.

To stop the cutting process, release the switch. The arc will stop. Shielding gas will continue to flow for a number of seconds after the switch is released. This flow of shielding gas cools the torch and, depending on the shielding gas used, may protect some base metals from oxidizing as they cool.

Another way to stop the arc is to pull the torch away from the base metal. The primary cutting arc will stop; however, the pilot arc will restart. Remember, the pilot arc occurs between the electrode and the constricting nozzle. The pilot arc will erode both the electrode and constricting nozzle. Allowing the pilot arc to continue for long periods of time is not desirable.

When nearing the end of a cut on thick metal, point the nozzle forward in the direction of travel. This allows the bottom edge to be cut properly. If you do not do this, the bottom edge may not separate completely.

When cutting, do not let the nozzle come in contact with the work. When this happens, the nozzle and the work are electrically connected. The arc can jump to the nozzle instead of the workpiece. This can rapidly erode the nozzle. Attachments are available that allow the torch to contact the work. These attachments are insulated from the electrical part of the torch, **Figure 19-9**.

Plasma arc cutting can be started from the edge or from the center of the part to be cut. When starting from an edge, hold the torch perpendicular to the workpiece surface. Hold the torch nozzle between 1/16″–1/8″ (1.6 mm–3.2 mm) from the surface. Press the switch and begin the cut.

When starting from the center of a workpiece, the initial start of the cut is a piercing operation. The molten base metal will fly upward from the surface being cut. After the material is pierced, the molten metal will be blown through the base metal.

During the initial pierce, the torch should be held at a 60° angle pointing away from the welder. The molten metal will be less likely to get into the torch nozzle and will not hit the welder. Hold the torch nozzle 1/8″–3/16″ (3.2 mm–4.8 mm) from the surface.

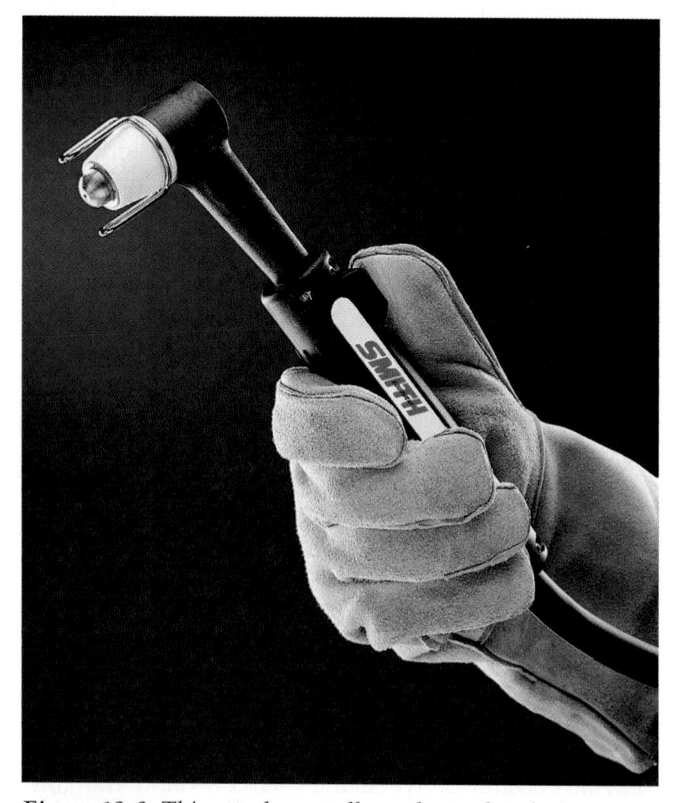

Figure 19-9. This attachment allows the torch to be in contact with the base metal. This maintains a consistent torch-to-work distance for better cut quality. (Arcsmith)

As the pierce begins, rotate the torch toward perpendicular. Once the pierce is complete, hold the torch perpendicular to the workpiece and lower it to 1/16″–1/8″ (1.6 mm–3.2 mm) from the surface. Begin forward motion while maintaining the desired distance for cutting, which is 1/16″–1/8″ (1.6 mm–3.2 mm) from the surface.

Cutting Speeds

Plasma arc cutting speed is fairly quick. **Figure 19-10** lists some cutting speeds. The cutting speed should be fast enough so the cutting takes place in the front of the arc, not on the sides.

If travel speed is too slow or too fast, dross will be found on the bottom side of the workpiece. ***Dross*** is a term used for metal removed during a cut that becomes attached to the metal surface. Dross is most often found on the bottom surface but can also be found on the top surface.

There is a difference between dross from traveling too fast and traveling too slow. When metal is cut too fast, dross is formed because the metal has not been completely cut. This dross is difficult to remove. It requires grinding. When metal is cut too slowly, dross is formed because the metal was melted but not blown completely out of the cut. This dross can usually be removed with a chipping hammer.

If dross is present, adjust the travel speed to eliminate the dross.

Cutting Quality

A properly made plasma arc cut will have a kerf width that is about two times the diameter of the hole

PAC on Mild Steel				
Thickness		Current	Travel Speed	
inch	mm	amps	inch/min	mm/min
1/16	1.6	35	175	4450
1/8	3.2	40	90	2300
1/4	6.4	40	40	1000
1/4	6.4	80	100	2540
3/8	9.5	40	18	460
3/8	9.5	80	55	1400
1/2	12.7	40	20	510
1/2	12.7	80	35	890
3/4	19.0	80	18	460
1.0	25.4	80	10	250

Figure 19-10. Chart showing the relationship between amperage, cutting speed, and material thickness. Manufacturers provide this information for each type of torch they make.

in the constricting nozzle when cutting thin base metals. Kerf is the name for the slot cut in the base metal. There will be little or no dross present. The cut surfaces will be fairly smooth with a slight taper.

As the thickness of the base metal increases, the kerf width also increases. The kerf width on 1″ (25.4 mm) thick base metal can be about 3/16″ (4.8 mm) wide.

The cut surfaces should be smooth, **Figure 19-11**. If the cut surface quality starts to have ripples, the consumable parts of the torch (the electrode and nozzle) may need to be replaced. There should be no dark deposits on the edges of the surface. If there are, there may be oil or moisture in the plasma or shielding gas.

If dross is present and difficult to remove, reduce the travel speed. Dross produced from slower travel speeds is much easier to remove than dross from faster travel speeds. When the travel speed is correct, molten metal is blown through the kerf. This molten metal exits the bottom surface and creates a trail at about a 15° angle opposite the direction of travel.

Dross on the top surface often results from holding the torch too high above the workpiece. Reduce the distance to eliminate dross on the top surface.

The plasma gas in a plasma arc cutting system is made to rotate as it exits the constricting nozzle. This is to help keep the gas in a tight column.

When this rotating column of plasma is used to cut, one side of the kerf will be squarer or more perpendicular than the other side. Learn from experience which side of the kerf will be more square.

Often one side of the part to be cut will remain and one side is scrap. Some examples are beveling a pipe or plate. Another example is cutting a large circle out of a piece of base metal. If the result is to be the base metal with a hole cut in it, the direction of travel when cutting the hole should be counterclockwise. This will leave the squarer edge on the base metal. However, if the desired part is the circular part removed from the base metal, then the torch should travel in a clockwise direction around the hole. This will put the squarer edge on the circular piece removed. This information is true for plasma gas that swirls in a clockwise direction. The plasma gas in some torches swirls in the opposite direction. In those cases, the cutting directions will be opposite. **Figure 19-12** shows an automated cut in progress.

A good cut will have a slight taper on the cut edges. If the taper is too great, adjust the torch-to-work distance. If the cut edge has a larger gap at the top than the bottom, the distance is too small. If the gap is greater at the bottom surface than the top surface, the torch-to-work distance is too great.

Plasma arc cutting equipment has the ability to make hundreds of arc starts and great distances of cuts. However, at some point the cut quality begins to get worse. When the cut quality gets worse, check the electrode and nozzle. Examine the electrode to see if it is getting pitted. If the electrode is pitted or is eroding at the edges, replace it.

Examine the constricting nozzle. The orifice should not be blocked by slag. If there is slag in the

Figure 19-11. *Edge quality from plasma arc cutting. (Hypertherm)*

Figure 19-12. *A mechanized plasma arc cutting torch operating on a cutting table. A circular part is being cut from the base metal. The circular part is going to be used, thus the torch is traveling clockwise to leave a squarer cut edge on the circular part. (Thermal Dynamics, a Thermadyne Company)*

orifice, remove it without enlarging the hole diameter. If the orifice is enlarged, replace the nozzle. Dipping the nozzle in anti-spatter compound helps prevent spatter from sticking to the nozzle.

The electrode and nozzle need to be replaced when they are worn. The most common reason for wear is the pilot arc. A pilot arc occurs between the electrode and nozzle. Keep the amount of time a pilot arc is active, such as while starting and stopping, to a minimum. Hold the nozzle 1/16″–1/8″ (1.6 mm–3.2 mm) from the surface when starting to cut. Even under the best conditions, after numerous arc starts, the electrode and nozzle will wear and must be replaced.

Plasma Arc Gouging

Plasma arc equipment can be used for gouging. ***Gouging*** is the process of removing only a portion of the thickness of a metal. Gouging can be used to form U-groove or J-groove weld preparations.

The equipment used for gouging is the same as that used for cutting, except a different constricting nozzle is used. The nozzle used for gouging usually has a larger orifice diameter. The larger orifice diameter creates a softer arc. Also, the velocity of the cutting gas is reduced. This allows the base metal to be gouged and prevents the metal from being cut.

A higher open circuit voltage is required for power supplies to be used for gouging. The arc length is longer because the torch is not perpendicular to the work surface. A longer arc length requires a higher voltage. The torch-to-work distance for gouging is about 1/8″–1/4″ (3.2 mm–6.4 mm). If the torch-to-work distance is too small, the base metal may be cut and not gouged.

Factors which determine the quality of a plasma arc gouge operation include the following:

- Shielding gas.
- Torch-to-work distance.
- Travel speed.
- Travel angle.
- Current setting.

Plasma gases commonly used for gouging are air or argon with 35% hydrogen added. Helium can be added to argon instead of hydrogen. Using helium instead of hydrogen will provide a shallower gouge. Shielding gases can be air, argon, or nitrogen.

The torch is held at about a 55°–60° travel angle. This is 30°–35° off the base metal surface. The cutting end of the torch is pointed in the direction of travel.

The settings for current are determined by the travel speed. The amount of metal to be removed is also a factor. Do not try to remove too much metal in one pass. More than one pass may be required to create a deep gouge on thick base metal.

Summary

- Plasma arc cutting (PAC) is a cutting process that uses an arc and a high-velocity, ionized gas coming through a constricting nozzle to cut all metals.
- In plasma arc cutting, the arc is created between the electrode and the base metal. The arc melts the base metal. The high-velocity gas blows the molten metal through the base metal.
- The torch is an important part of the plasma arc cutting system. The torch holds the electrode, provides a good electrical connection to the electrode, provides a path for the cutting gas, and directs the secondary gas. The torch body has a button or trigger used to start and stop the cutting process.
- The most common gas used for manual plasma arc cutting is air. Air is used as both the plasma and the shielding gas. Nitrogen is a good plasma gas choice when cutting stainless steel or aluminum. When cutting stainless steel or aluminum that is 1/2″ (12.7 mm) or thicker, the plasma gas used is often argon with 35% hydrogen.
- To assemble a plasma arc cutting outfit, plug the cutting power supply into an electrical outlet. If using a cylinder supply of gas for cutting, attach a regulator to the cylinder. Connect a hose from the regulator to the power supply. If using gas from a shop manifold, connect a hose from the manifold to the power supply. Connect the cutting torch to the power supply.
- Plasma arc cutting is a combination of an arc process and a cutting process. The dangers of arc welding are present. These include the intense light, heat, and radiation of the arc. The dangers of cutting, which include flying particles and fire hazards, are also present.

(continued)

- **Select a work area that has good ventilation. Before cutting, inspect the work area for any hazards. There should be no flammable items in the area, no paper, cardboard, rags, excessive grease, oil, or fuel. Any welding cables or hoses must be protected from flying metal. Earplugs, earmuffs, or both should be worn to protect against loud noise.**
- To set up the PAC equipment for cutting, turn on the power supply, open the shielding gas cylinder valve or the valve to the shop manifold system, press the gas test button on the power supply, and adjust the regulator to the desired pressure.
- The cutting process begins by pressing and holding a switch or button on the torch. A preflow of shielding gas will flow for a very short period of time. After the preflow time, a pilot arc starts. The high open-circuit voltage and the ionized gas produced by the pilot arc allows the arc to jump from the electrode to the base metal. The base metal is rapidly melted by the arc. High-velocity plasma coming through the constricting nozzle blows the molten base metal away. Cutting continues as long as the switch is held and the torch is close enough to the work to maintain the cutting arc.
- When nearing the end of a cut on thick metal, point the nozzle in the direction of travel. This allows the bottom edge to be cut properly. If you do not do this, the bottom edge may not separate completely.
- When cutting, do not let the nozzle come in contact with the work. When this happens, the nozzle and the work are electrically connected. The arc can jump to the nozzle instead of the workpiece. This can rapidly erode the nozzle.
- When starting a cut from an edge, hold the torch perpendicular to the workpiece surface. Hold the torch nozzle between 1/16″–1/8″ (1.6 mm–3.2 mm) from the surface. Press the switch and begin the cut.
- When starting from the center of a workpiece, the initial start of the cut is a piercing operation.
- When cutting thin base metals, a properly made plasma arc cut will have a kerf width that is about two times the diameter of the hole in the constricting nozzle.
- A good cut will have a slight taper on the cut edges. If the taper is too great, adjust the torch-to-work distance. If the cut edge has a larger gap at the top than the bottom, the torch-to-work distance is too small. It the gap is greater at the bottom surface than the top surface, the torch-to-work distance is too great.
- Plasma arc equipment also can be used for gouging. The equipment used for gouging is the same as that used for cutting, except a different constricting nozzle is used. The nozzle used for gouging usually has a larger orifice diameter.
- Plasma gases commonly used for gouging are air or argon with 35% hydrogen added. Helium can be added to argon instead of hydrogen. Using helium instead of hydrogen will provide a shallower gouge. Shielding gases can be air, argon, or nitrogen.
- For gouging, the PAC torch is held at about a 55°–60° travel angle. The current settings are determined by the travel speed.

Review Questions

Write your answers on a separate sheet of paper. Please do *not* write in this book.

1. Which of the following statements is *not* true?
   A. Plasma is an ionized gas.
   B. Plasma conducts electricity.
   C. Plasma is present only in welding and cutting processes that use a constricting nozzle.
   D. Plasma is a state of matter.
2. List four advantages of PAC.
3. *True or False?* Power supplies for PAC are usually constant voltage machines using direct current electrode negative (DCEN).
4. *True or False?* An inverter power supply is usually heavier than a non-inverter power supply.
5. List the two parts of the PAC torch that need to be replaced most often.
6. What gas or gas mixture is the most commonly used for manual PAC?
7. Before cutting begins, the area must be inspected and be free of flammable items. List four items that should not be present in a cutting area.
8. The pilot arc occurs between the electrode and the _____.
9. What distance should the PAC torch nozzle be above the base metal when cutting?
10. To cut 1/4″ (6.4 mm) thick mild steel at a rate of 40 inches per minute, the current would be set to approximately _____. Use the example presented in Figure 19-10.

A PAC torch is being used to cut a beveled edge in thick steel plate. (Hypertherm)

Section 6

Oxyfuel Gas Processes

Victor Equipment Co.

Chapter 20

Oxyfuel Gas Cutting and Welding: Equipment and Supplies

Learning Objectives

After studying this chapter, you will be able to:

- Identify the parts of an oxyfuel gas cutting or welding outfit.
- Describe the function of each of the parts in an oxyfuel gas cutting and welding outfit.
- Identify the safety features of an oxyacetylene cutting or welding outfit.
- Describe protective clothing used for oxyacetylene cutting or welding.
- List safety precautions that must be taken when performing oxyfuel gas cutting or welding.

acetone
acetylene (C_2H_2)
acetylene generator
backfire
check valve
cover plates
cutting outfit
cutting oxygen
cutting oxygen lever
cutting oxygen orifice
cutting station
cutting torch
cutting torch attachment
cylinder
cylinder pressure gauge
cylinder valve
Dewar flask
distill
filter lenses
fittings
flashback
flashback arrestor
fuel gases
fusible plug
hand truck
hose
infrared rays
injector-type torch
leathers
liquefaction
manifold
mill file
mixing chamber
orifice
oxygen (O_2)
positive pressure torch
preheating orifices
pressure regulator
regulator adjusting screw
safety cap
safety valve
single-stage regulator
tip nut
torch tip
torch valves
two-stage regulator
ultraviolet rays
welding outfit
welding rod
welding station
welding torch
working pressure gauge

Oxyfuel Gas Cutting and Welding

Oxyfuel gas cutting and *oxyfuel gas welding* are general terms for a group of cutting and welding processes (methods) that use heat produced by a gas flame to cut and join various metals. Oxyacetylene and oxyhydrogen cutting and welding are examples of oxyfuel gas welding processes. See **Figure 20-1**.

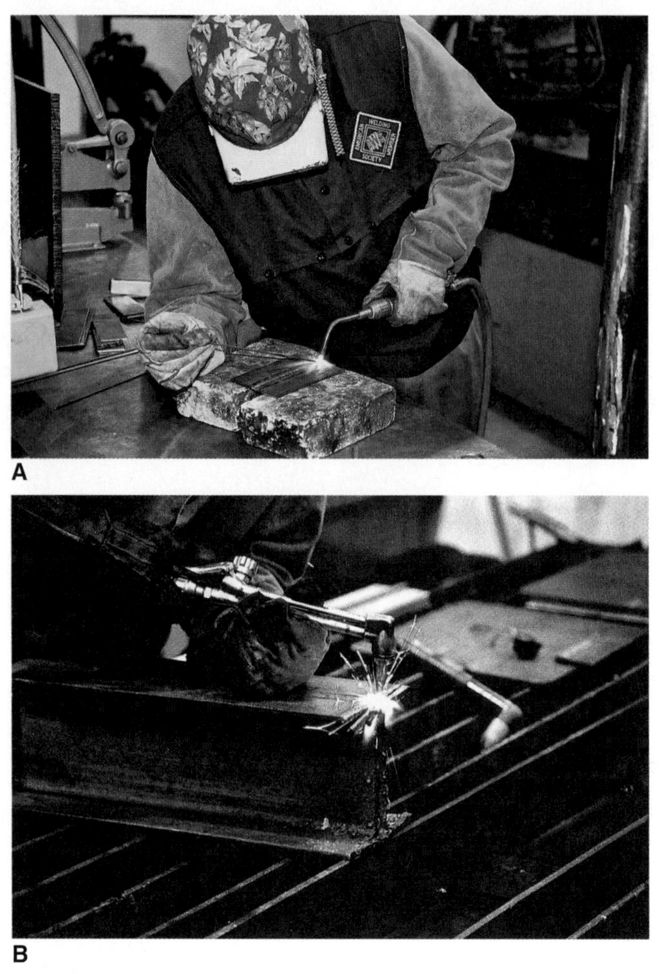

Figure 20-1. Oxyfuel gas welding (OFW) processes include welding and cutting operations. A—Making an oxyacetylene weld in the flat position. B—Using an oxyacetylene cutting torch to cut a steel beam.

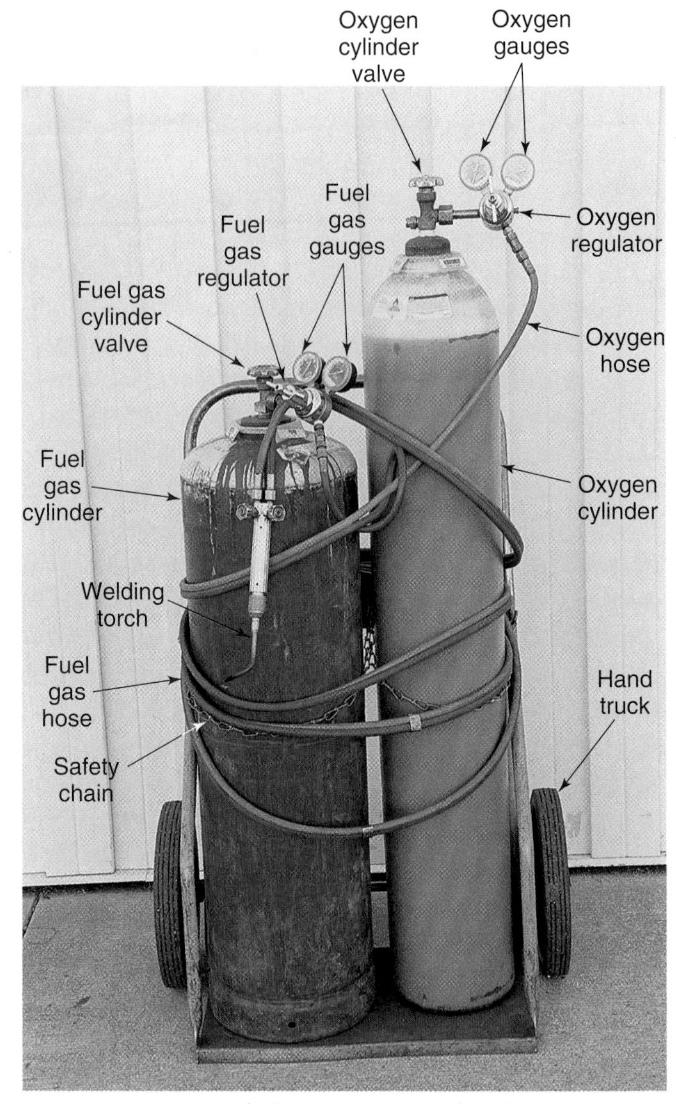

Figure 20-2. An oxyfuel gas welding outfit. The equipment is the same for oxyfuel gas cutting, but a cutting torch is used in place of the welding torch.

Fuel gases are those that will support combustion (burn) when combined with oxygen. These fuel gases include acetylene, propane, butane, hydrogen, natural gas, and MPS (methylacetylene-propadiene) gas, which is commonly sold by the trade name MAPP® gas. Acetylene is commonly used for cutting and welding. When combined with oxygen in a neutral flame, it produces temperatures around 5600°F (3093°C). This is the highest temperature produced by any combination of oxygen and fuel gas. Propane, butane, city gas, and natural gas do not produce enough heat for welding. However, they may be used for soldering and brazing. The ***welding outfit*** includes equipment required to actually create a weld. A ***cutting outfit*** includes the equipment needed to perform cutting operations.

An oxyfuel gas cutting outfit or welding outfit, **Figure 20-2**, consists of the following pieces of equipment:

- Fuel gas cylinder.
- Oxygen cylinder.
- Means of securing the cylinders in an upright position.
- Fuel gas regulator and gauges.
- Oxygen regulator and gauges.
- Fuel gas hose and fittings.
- Oxygen hose and fittings.
- Flashback arrestors.
- Check valves.
- Cutting or welding torch.

The main differences in the cutting and welding outfits are in the oxygen regulator and the torch. When cutting thick metal, a larger volume of higher pressure oxygen is needed than when welding thinner metal. This requires the use of an oxygen regulator that is calibrated for this larger volume and higher pressure. The differences in cutting and welding torches are explained later in this chapter. A complete oxyfuel gas ***cutting station*** or ***welding station*** includes the cutting or welding outfit, a cutting or welding table, means of ventilation, goggles, and a spark lighter.

Acetylene

Acetylene (C_2H_2) is a colorless fuel gas with a distinct garlic-like odor. It is produced by adding calcium carbide to water.

An ***acetylene generator*** is used to produce acetylene in a controlled environment. See **Figure 20-3**. Calcium carbide is placed in a hopper and fed into the water at a controlled rate. The chemical combination produces acetylene.

A low-pressure acetylene generator produces the gas at a gauge pressure of .25 psig (pounds per square inch gauge) (1.72 kPa). A medium-pressure generator produces acetylene at a gauge pressure of 15 psig (103.42 kPa). **Pure acetylene is very unstable at a gauge pressure above 15 psig (103.42 kPa). Do not use acetylene at a gauge pressure above 15 psig (103.42 kPa) because it may explode or burn rapidly.**

Acetylene Cylinder

Acetylene can be stored safely as a gas at pressures above 15 psig (103.42 kPa) if it is kept from collecting in large volume. An acetylene ***cylinder*** is a portable container used to store acetylene. Cylinders are available in sizes to meet the needs of most users, **Figure 20-4**.

An acetylene storage cylinder is constructed of steel and filled with a porous material. The porous material has many small air pockets. **Figure 20-5** shows a cutaway view of an acetylene cylinder. The cylinder is filled with ***acetone***, a colorless and extremely flammable liquid. The acetone fills the air pockets of the porous material. Acetylene is then pumped into the cylinder and absorbed into the acetone. The porous

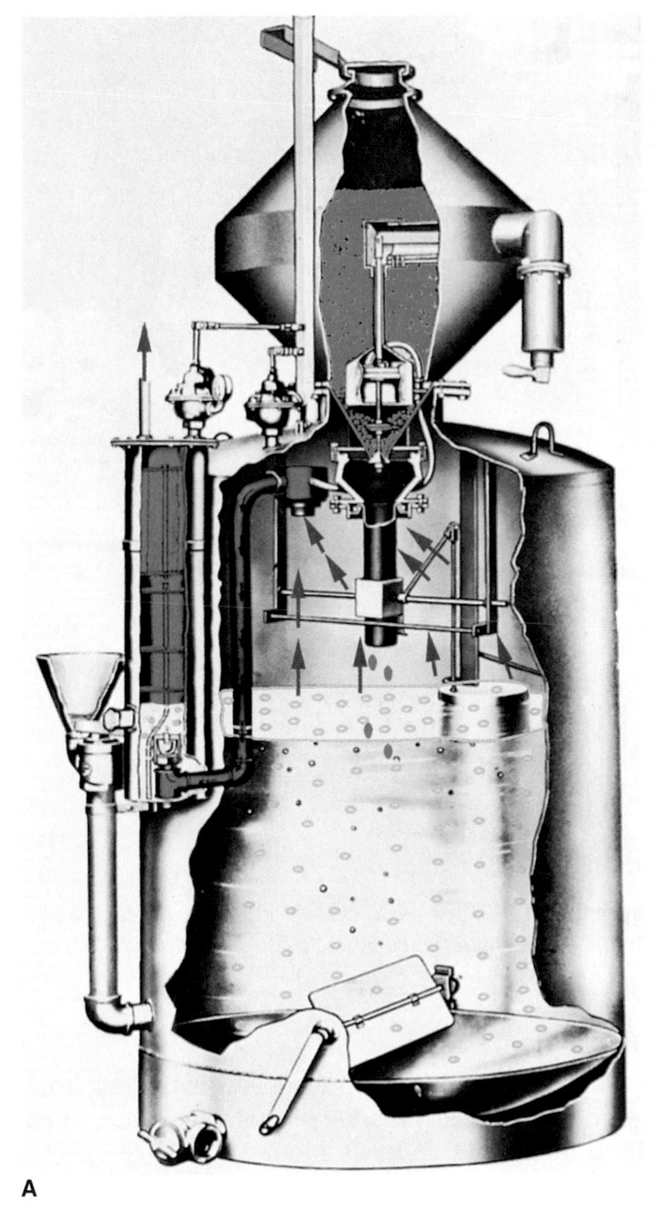

A

B

Figure 20-3. A—A cutaway view of an acetylene generator. Calcium carbide is fed from the hopper at the top into the water to produce acetylene gas. B—A larger, two-hopper acetylene generator. (Rexarc, Inc.)

Figure 20-4. *Acetylene cylinders are available in a variety of sizes. The safety caps were removed to show the type of cylinder valves used on the larger cylinders.*

material and acetone prevent the acetylene from collecting in large volume. Acetylene bubbles out of the acetone when the cylinder valve is opened. Acetylene can be safely stored at pressures up to 250 psig (1724 kPa) using this method.

A ***safety cap*** **must be screwed over the cylinder valve when a cylinder is stored or moved. In case a cylinder falls, the safety cap prevents the cylinder valve from being damaged. Warning: Without the safety cap, the cylinder valve could break, causing the high-pressure gas to be suddenly released. This can result in the cylinder becoming an extremely dangerous and flammable projectile. Cylinders should be carefully moved with a properly designed hand truck or by rolling them along their bottom edge.** A ***hand truck*** is a two-wheel cart with a chain or strap for securing cylinders, **Figure 20-6**. **A cylinder may also be moved by tilting it with one hand on the safety cap. The other hand is used to roll the cylinder in the direction of travel.** **Figure 20-7** shows this method of moving a cylinder. Cylinders filled with gases under pressure must be stored in an upright position. They should be fastened to a wall, column, or hand truck with chains or steel straps.

Figure 20-5. *Cutaway view of an acetylene cylinder. Notice how the porous material completely fills the cylinder.*

Fusible Plugs

Fusible plugs prevent acetylene cylinders from exploding in a fire. A ***fusible plug*** is a steel plug filled with a metal that melts at about 212°F (100°C). The plug is externally threaded and is screwed into an opening at the top or bottom of the acetylene cylinder, **Figure 20-8**. In case of a fire, the center of the fusible plug melts. This allows the acetylene to

Figure 20-6. A hand truck used to hold and transport oxygen and fuel gas cylinders. Note the safety chains used to keep the cylinders secured.

Figure 20-7. A cylinder being moved by tilting it with one hand and rolling it into position with the other hand. Notice that the safety cap is in place.

escape slowly. Although the gas escapes and burns, the tank will not explode.

Acetylene Cylinder Valve

An acetylene ***cylinder valve*** controls the flow of gas from the cylinder. A handwheel or a cylinder valve wrench is used to open and close the valve. See **Figure 20-9**. **Warning: The handwheel or valve wrench should always remain on the cylinder valve when the cylinder is in use.** An acetylene cylinder valve is usually opened only 1/4–1/2 turn for sufficient gas flow. This allows it to be closed quickly in case of an emergency.

Figure 20-8. Fusible plug on an acetylene cylinder. The valve is depressed and has internal threads. This type of cylinder is known as a POL (Prestolite) cylinder.

Acetylene Safety Precautions

Acetylene can be very explosive if not handled safely. Always refer to the manufacturer's instructions when handling or using equipment.

- **Acetylene gas is unstable at pressures above 15 psig (103.42 kPa).**
- **Avoid contact between acetylene and copper, silver, or mercury. Also avoid contact with the salts, compounds, and high concentrations of these metals. Under certain conditions, acetylene and these metals can form explosive compounds.**

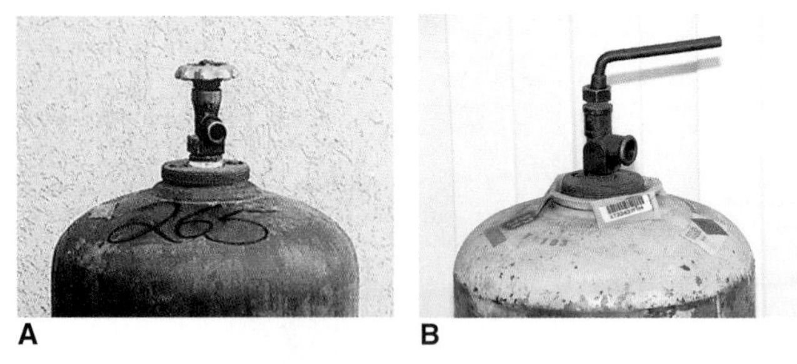

Figure 20-9. *A—Handwheel used to regulate the flow of acetylene. This type of cylinder is known as a commercial cylinder. It has an externally threaded valve and two fusible plugs. B—An acetylene cylinder with the acetylene valve wrench in place.*

- **Fusible plugs melt at approximately 212°F (100°C).**
- **Concentrations of acetylene between 2.5% and 80% by volume in air ignite easily and may cause an explosion.**
- **Acetylene has a garlic-like odor. Acetylene may displace air in a poorly ventilated space. Adequate ventilation is essential. An area must contain at least 18% oxygen to prevent dizziness, unconsciousness, or possibly death.**
- **Smoking, open flames, unapproved electrical equipment, or other ignition sources are not permitted in acetylene storage areas.**
- **Do not place cylinders beneath overhead welding or cutting operations. Hot slag (metal) may fall on them and melt a fusible plug.**
- **Keep the handwheel or valve wrench on the cylinder valve while the cylinder is in use.**
- **Avoid contact between the torch flame and cylinder.**
- **Secure all cylinders in an upright position.**
- **Do not use safety caps for lifting cylinders.**
- **Keep cylinder valves covered with safety caps when cylinders are not in use.**
- **Do not force cylinder valves open.**
- **Do not use any cylinder that does not have a label or that has an illegible label. Never assume a cylinder contains a particular gas. Return the cylinder to the supplier.**
- **Do not use a cylinder with a leaking valve. Return it immediately to the supplier.**

Also, you should be familiar with the following safety standards:

- American National Standards Institute (ANSI) Z49.1, "Safety in Welding, Cutting, and Allied Processes."
- National Fire Protection Association (NFPA) No. 51, "Oxygen-Fuel Gas Systems for Welding, Cutting, and Allied Processes."

Oxygen

Oxygen (O_2) is a colorless, odorless gas contained in the earth's atmosphere. The atmosphere consists of 78% nitrogen, 21% oxygen, and 1% other gases, such as argon and helium. When oxygen is added to a fuel gas flame, an increased flame temperature and rate of combustion result.

Oxygen is produced for the welding industry using the liquefaction process. ***Liquefaction*** is a process that liquefies air and then separates the gases in it at their various boiling points. The process requires large, expensive equipment.

Air becomes a liquid at a very low temperature. When air is liquefied, the gases ***distill*** (boil away) at different temperatures. Nitrogen is distilled at –320°F (–196°C), and oxygen is distilled at –297°F (–183°C). After oxygen is distilled, it is cleaned, and all traces of water are removed. After distilling, oxygen is a gas.

Oxygen is stored as a gas or a liquid. Gaseous oxygen is stored in strong cylinders at approximately 2200 psig (15.168 MPa). Liquid oxygen is stored in a Dewar flask. A ***Dewar flask*** is a pressurized container with an insulated double wall.

Oxygen Cylinder

An oxygen cylinder is a seamless, portable container used to store oxygen. Oxygen is stored in cylinders at a pressure of about 2200 psig (15.168 MPa). The Interstate Commerce Commission (ICC) regulates the construction of oxygen cylinders. Oxygen cylinders are made of forged steel. The minimum thickness of a cylinder is 1/4″ (6.4 mm). **Figure 20-10** shows the typical construction of an oxygen cylinder.

Oxygen cylinders are available in a variety of sizes to meet the needs of the users. **Figure 20-11** shows four common sizes of oxygen cylinders. The available sizes differ slightly by manufacturer. Typical sizes are:

- 20 ft^3 (566 L)
- 55 ft^3 (1557 L)
- 80 ft^3 (2265 L)
- 122 ft^3 (3455 L)
- 220 ft^3 (6230 L)
- 244 ft^3 (6909 L)
- 330 ft^3 (9345 L)

An oxygen cylinder should be transported with a properly designed hand truck. A cylinder may also be moved by tilting it and rolling it along the bottom edge. Do not lift cylinders by their safety caps.

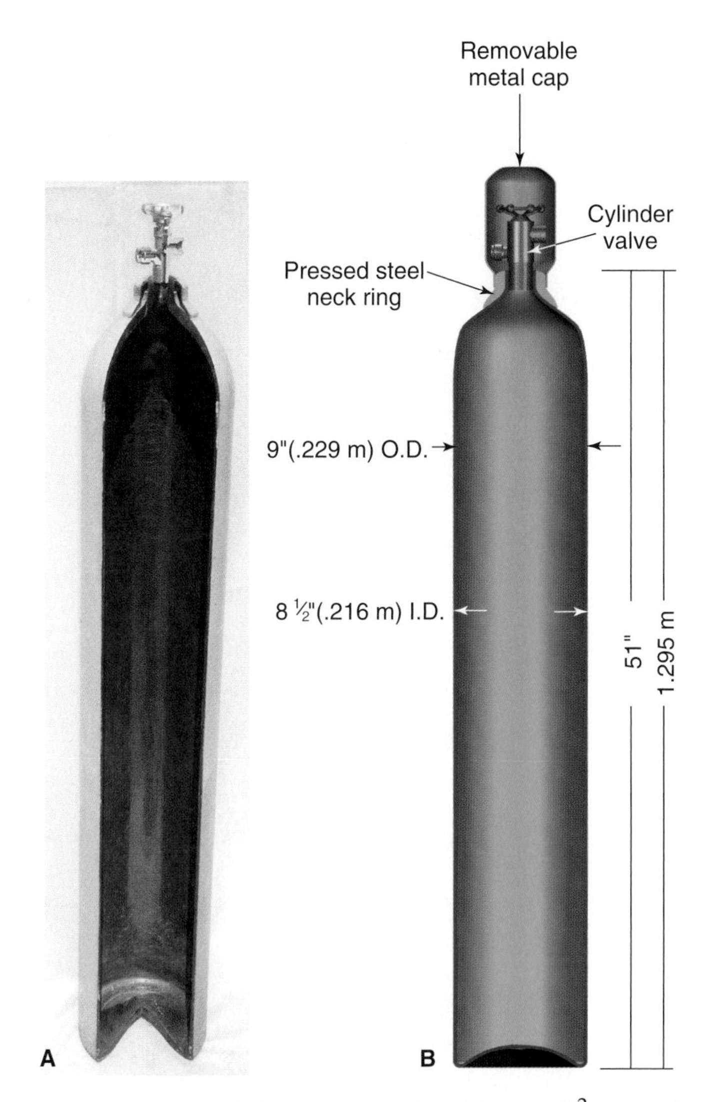

Figure 20-10. *Typical oxygen cylinder with a 244 ft^3 (6909 L) capacity. A—Internal construction of a cylinder. Note the one-piece construction. B—Dimensions of a 244 ft^3 (6909 L) cylinder. (Pressed Steel Tank Co.)*

Figure 20-11. *Four common sizes of oxygen cylinders.*

An oxygen cylinder must be stored in an upright position. It should be secured to a wall, column, or hand truck with chains or straps. Warning: The cylinder valve may break and leak if the cylinder falls. Never attempt to move a cylinder without the safety cap screwed on securely.

Dewar Flask

Weld shops that need a large amount of oxygen commonly use liquid oxygen. Liquid oxygen is stored and shipped in a Dewar flask. The flask is designed to store liquid oxygen at –297°F (–183°C) under pressure. Oxygen is withdrawn from the flask through tubing that passes between the inner and outer walls. This warms the liquid, causing it to vaporize (turn into a gas). See **Figure 20-12**. A standard regulator is used to maintain the correct working pressure of the gas.

Oxygen Cylinder Valve

A forged brass cylinder valve controls the flow of oxygen from the cylinder, **Figure 20-13**. **The valve should be protected with a properly installed safety cap when the cylinder is being moved or is not in use.**

The oxygen cylinder valve should always be completely open when in use. A backseating valve prevents oxygen from escaping around the valve stem. The backseating valve operates only when the cylinder valve is completely open. **Figure 20-14** shows a cutaway view of an oxygen cylinder valve.

Safety Valve

A safety valve is an integral part of the oxygen cylinder valve. A ***safety valve*** prevents an explosion when the cylinder is exposed to high temperatures. As the temperature increases, the cylinder pressure also

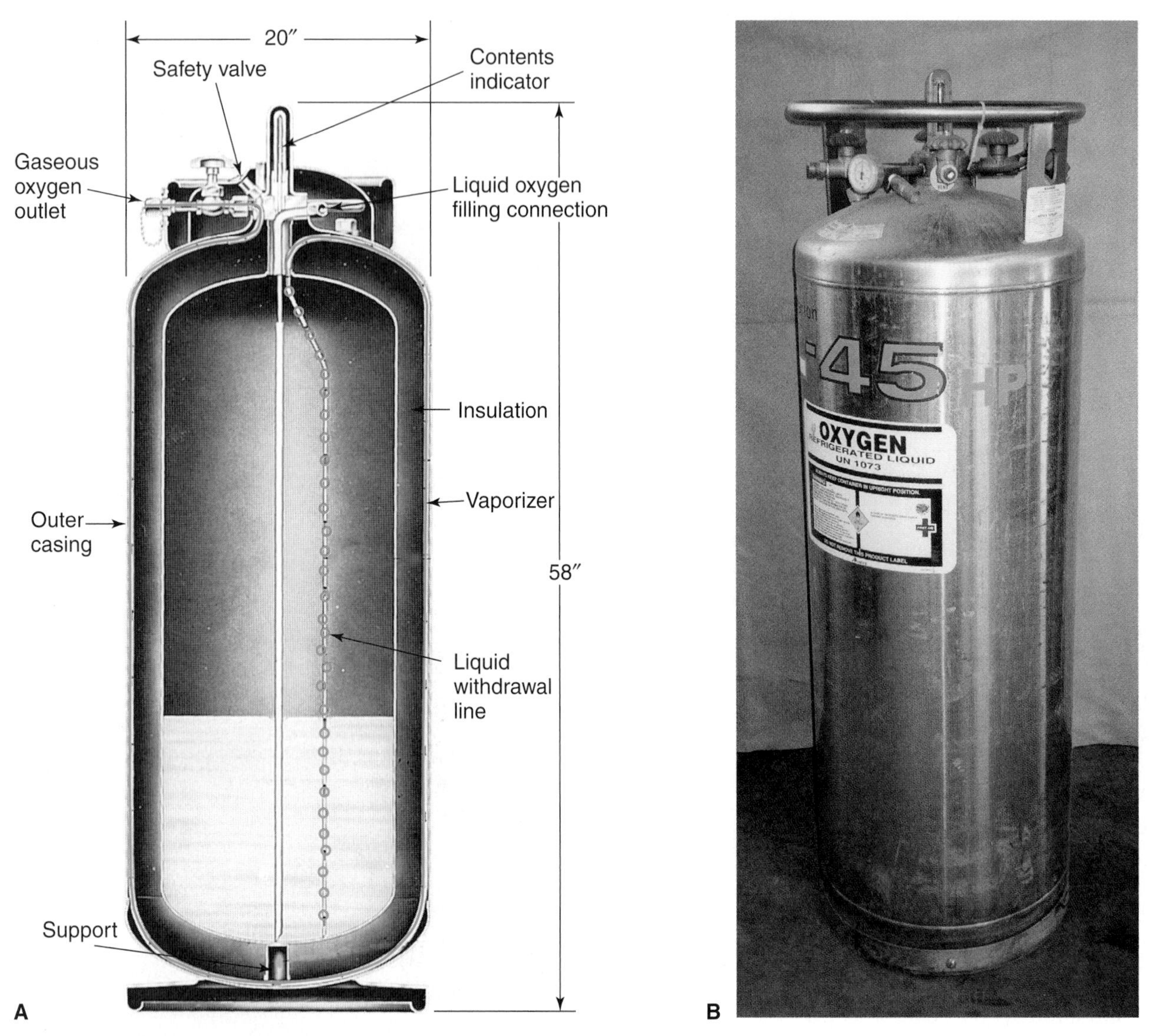

Figure 20-12. A—Cross-sectional view of a Dewar flask. Liquid oxygen vaporizes as it passes between the inner and outer walls. (Linde Div., Union Carbide Corp.) B—This Dewar flask can supply several torches at one time.

Figure 20-13. A forged brass oxygen cylinder valve. The white material is Teflon tape that has been used to seal the threads between the valve and cylinder.

increases. When a predetermined pressure is reached, a safety disc within the valve ruptures and slowly releases the oxygen in the cylinder. See **Figure 20-15.**

Oxygen Safety Precautions

General safety precautions that should be followed when using oxygen include:

- **Do not place liquid oxygen equipment on asphalt or surfaces with oil or grease deposits.**
- **Smoking or open flames are not permitted in areas where oxygen is stored, handled, or used.**
- **Do not use a cylinder that has no label or that has an illegible label. Do not assume a cylinder contains a particular gas. Return the cylinder to the supplier.**
- **Avoid contact between liquid oxygen and your skin or eyes. Liquid oxygen is at –297°F (–183°C) and will cause freeze burns.**

- Remove all clothing that has been splashed or saturated with liquid oxygen. Oxygen-saturated clothing is highly flammable. Allow the clothing to air out at least 30 minutes so that the oxygen dissipates.

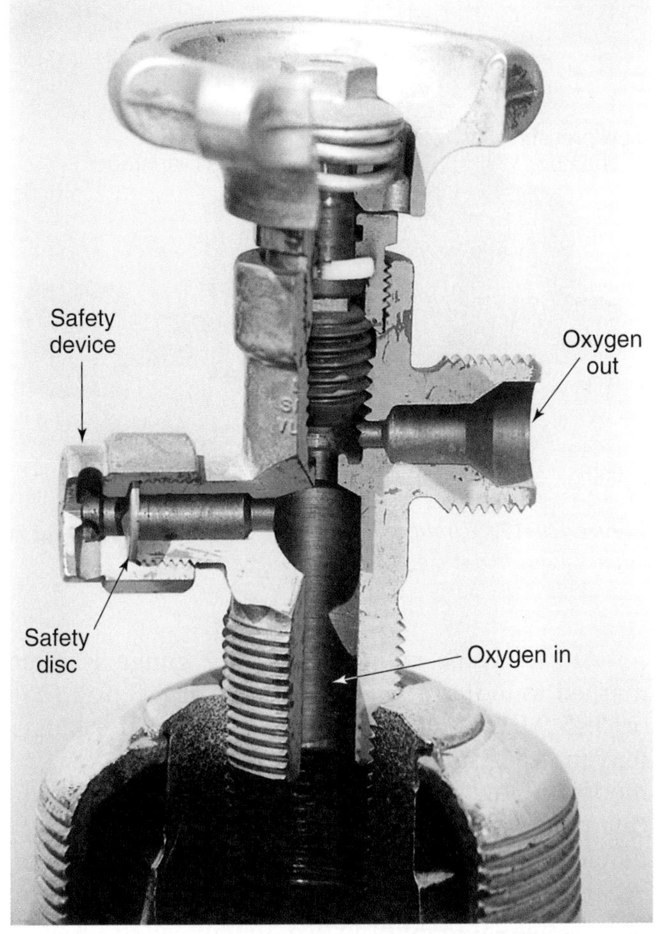

Figure 20-14. Cutaway view of an oxygen cylinder valve. The cylinder valve must be completely opened for the backseating valve to function properly.

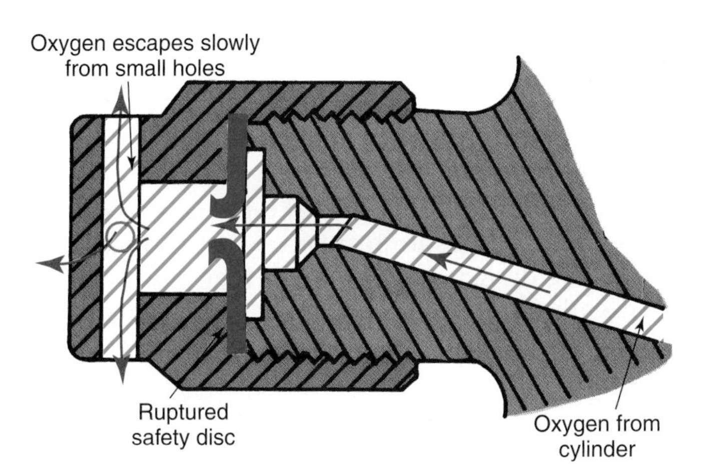

Figure 20-15. A schematic of the safety valve on an oxygen cylinder valve. When the disc ruptures, oxygen escapes through several drilled holes.

- Avoid contact between oxygen or oxygen fittings and organic materials, such as oil, grease, kerosene, cloth, wood, tar, and coal dust. These materials are highly combustible when combined with oxygen.
- Secure all cylinders in an upright position.
- Avoid contact between the torch flame and cylinder.
- Do not use safety caps for lifting cylinders.
- Always transport a cylinder with a hand truck or by rolling it along the bottom edge.
- Keep cylinder valves covered with safety caps when not in use.
- Do not force cylinder valves open. If the handwheel is missing or will not operate properly, return the cylinder to the supplier.

Manifolds

Manufacturers and repair shops that need a large volume of oxygen and fuel gas for welding or cutting may use manifolds. A ***manifold*** is a brazed assembly of pipes that delivers gas from several cylinders into one supply pipe. The pipe distributes gas to several workstations or locations. **Figure 20-16** shows an acetylene manifold.

Oxygen and acetylene manifolds must be kept separate to comply with local fire codes. Their construction also is governed by fire codes. Acetylene may form explosive compounds when combined with copper, mercury, or silver. For this reason, copper tubing or pipe must not be used in an acetylene manifold system. Manifolds must also be enclosed, ventilated, and located outside of the welding or cutting area.

Figure 20-16. An acetylene manifold. Two acetylene cylinders are attached to the manifold piping and supply gas for a number of users. Note the manifold shut-off valves and the pressure regulator.

Cylinders of oxygen and acetylene must never be connected to the wrong manifold. Intermixing gases in the delivery pipe could result in an explosive condition. Connecting oxygen and acetylene cylinders to the wrong manifold is virtually impossible due to the fittings. Acetylene fittings have left-hand threads; oxygen fittings have right-hand threads.

Check the following safety standards for more information on the use of oxygen and fuel gas cylinders and manifolds:

- ANSI Z49.1 "Safety in Welding, Cutting, and Allied Processes."
- NFPA No. 50, "Bulk Oxygen Systems at Consumer Sites."
- NFPA No. 51, "Oxygen-Fuel Gas Systems for Welding, Cutting, and Allied Processes."
- Linde Form 9888, "Precautions and Safe Practices—Liquid Atmospheric Gases."
- Local and national safety codes.

Pressure Regulators

A ***pressure regulator*** is a device used to reduce the pressure at which oxygen or fuel gas is delivered. Gas pressure in acetylene and oxygen cylinders is high. The working pressure for welding, however, may be as low as 1 psig (6.9 kPa). A pressure regulator is used to reduce the pressure of the gas.

A ***single-stage regulator*** reduces cylinder pressure to working pressure in one stage (step). The ***regulator adjusting screw*** controls the working pressure of gas delivered by the regulator. See **Figure 20-17**.

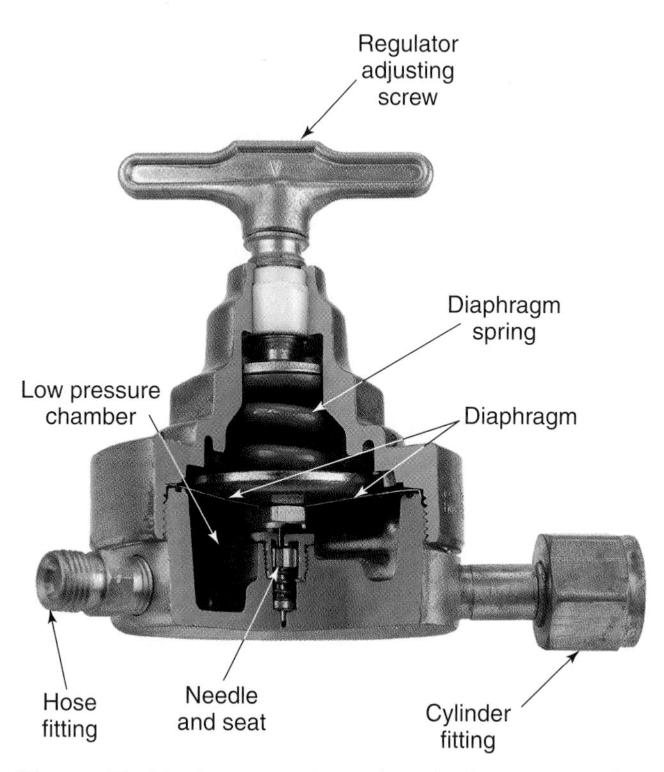

Figure 20-17. *Cutaway view of a single-stage regulator. (Victor Equipment Co.)*

A ***two-stage regulator*** reduces cylinder pressure to working pressure in two stages. Cylinder pressure is first reduced to an intermediate pressure. The intermediate pressure is then reduced to the working pressure in the second stage. See **Figure 20-18**. Two-stage regulators are more expensive than the single-stage regulators but provide more precise control.

Pressure adjustments are made by using the regulator adjusting screw. When the screw is turned clockwise (in), the working pressure increases. Turning the screw counterclockwise (out) decreases the working pressure. The regulator is turned completely off when the screw is turned counterclockwise until it feels loose.

Pressure Gauges

Pressure regulators usually have two gauges. The ***cylinder pressure gauge*** indicates cylinder pressure. The ***working pressure gauge*** shows the working pressure of the gas that flows through the torch. Gauges are marked with pressure readings at least 50% higher than the highest pressure that is expected.

An acetylene cylinder pressure gauge is often marked to indicate pressures of 400 or 500 psig (2.76 or 3.45 MPa). Acetylene working pressure gauges may indicate pressure up to 30 psig (207 kPa). Several models of acetylene working pressure gauges indicate pressure up to 15 psig (103.42 kPa) and identify dangerous pressure levels with a red background. See **Figure 20-19**. **For safe use, acetylene working pressure must be kept below 15 psig (103.42 kPa).**

An oxygen cylinder pressure gauge is commonly marked to indicate pressures up to 3000 or 4000 psig (20.7 or 27.6 MPa). Oxygen working pressure gauges may show pressure up to 200 psig (1380 kPa). See **Figure 20-20**. Special working pressure gauges, with markings as high as 1000 psig (6.89 MPa), may be needed for heavy cutting operations.

Hoses and Fittings

A ***hose*** is a flexible rubber tube used to convey gases from a pressure regulator to a welding torch. Hoses are designed to withstand high pressure. The oxygen hose is green. The fuel gas hose is red. Both single and dual (siamese) hoses are available. See **Figure 20-21**.

Fittings are used to connect the hoses to the regulator and torch. Each end of the hose has a fitting consisting of a brass nut and gland. Oxygen hose nuts have right-hand threads. Fuel gas hose nuts

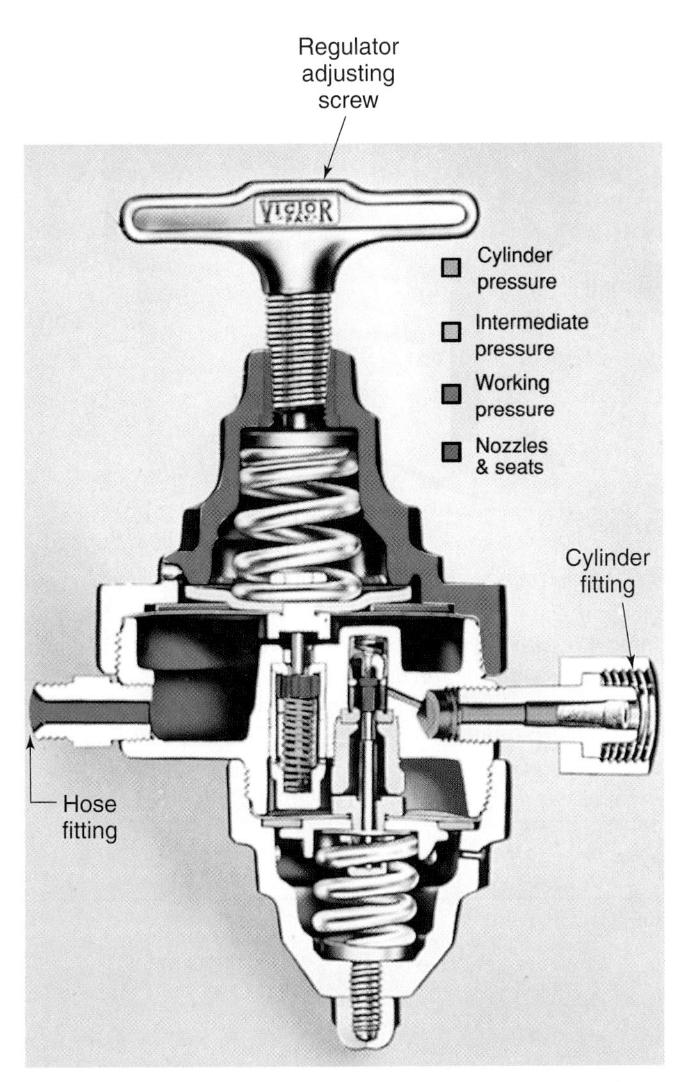

Figure 20-18. *Cross-sectional view of a two-stage regulator. This regulator uses stem-type valves in both stages. (Victor Equipment Co.)*

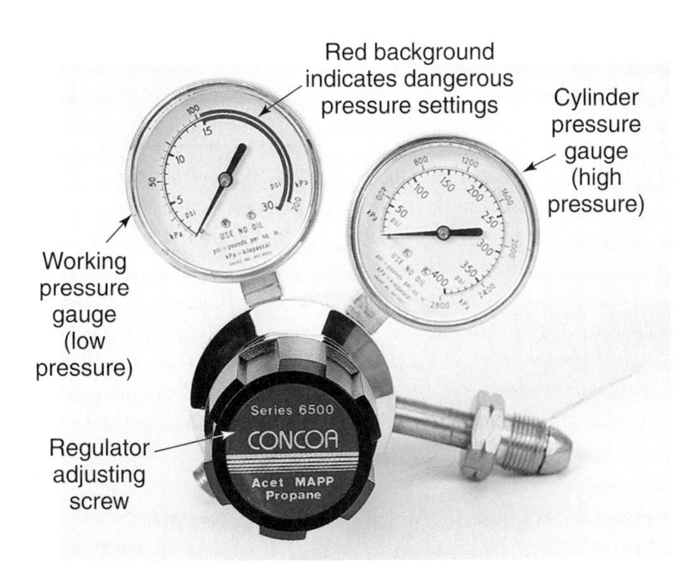

Figure 20-19. *Fuel gas regulator and gauges. The working pressure gauge is shaded in red above 15 psi, because acetylene is dangerous above this point. (CONCOA)*

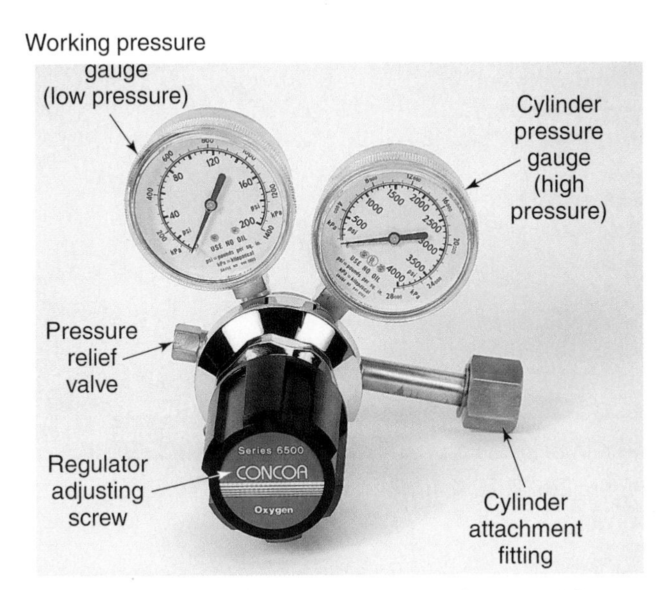

Figure 20-20. *Typical oxygen regulator and gauges. The pressure regulating mechanism is built into the regulator body. (CONCOA)*

Figure 20-21. *A dual (siamese) oxygen and fuel gas welding hose is on the right. Notice that the fuel gas hose is red and the oxygen hose is green. A single oxygen welding hose is on the left.*

have left-hand threads and a groove machined around the nut for identification. See **Figure 20-22**. The left-hand and right-hand threads and different color hoses are safety precautions. They prevent the hoses from being connected incorrectly. Oxygen flowing through a fuel gas hose could cause an explosion.

Hoses should not be laid across an area where they may be damaged or run over by a vehicle. When hoses must be temporarily run across a traffic path, they can be protected by placing them under a section of inverted channel iron or steel angle.

Figure 20-22. Hose fittings.

Backfires and Flashbacks

A ***backfire*** is a small explosion that produces a sharp popping sound. After a backfire, the flame may be extinguished or may continue to burn. A backfire usually remains in the torch head. It occurs when the torch head or tip overheats, when the tip is held too close to the work, or when the tip is dirty.

One of the greatest dangers in oxyfuel gas welding is a flashback. A ***flashback*** occurs when the flame moves into or beyond the ***mixing chamber*** of the torch. A flashback may move into the hoses, regulator, and possibly into the cylinder. **The flame burning back into the torch can result in a damaged torch or hose. It can also cause a violent explosion, which may destroy the regulator or cylinder. The explosion could also cause serious personal injury and fire.**

Flashbacks are controlled by check valves and flashback arrestors. Flashbacks may be caused by a reverse flow of gases or by loose and leaking connections. Leaks reduce the gas flow. The lessened flow of gas may not be able to support the flame at the torch tip. The flame may then burn back into the torch or even farther.

Check Valves and Flashback Arrestors

A ***check valve*** is a valve that allows the flow of gas in only one direction. Such valves are used to prevent the reverse flow of gases through the torch, hoses, and/or regulators. Gas pressure opens the valve to allow gas to flow. The valve closes when the flow stops or when gas tries to flow in a reverse direction. See **Figure 20-23**. Check valves are most often installed at the torch inlet. They sometimes are placed at the regulator outlet or at both the torch and regulator.

A ***flashback arrestor*** is a device that prevents the flow of a burning fuel gas and oxygen mixture from the torch back into the hoses, regulators, and cylinders. It is designed to eliminate the possibility of an explosion in the regulator or cylinder. See **Figure 20-24**. Flashback arrestors are installed between the torch and hose. An arrestor consists of the following safety valves:

- Reverse-flow check valve—prevents the flow of gas in the wrong direction.
- Pressure-sensitive cut-off valve—stops gas flow in case of an explosion.

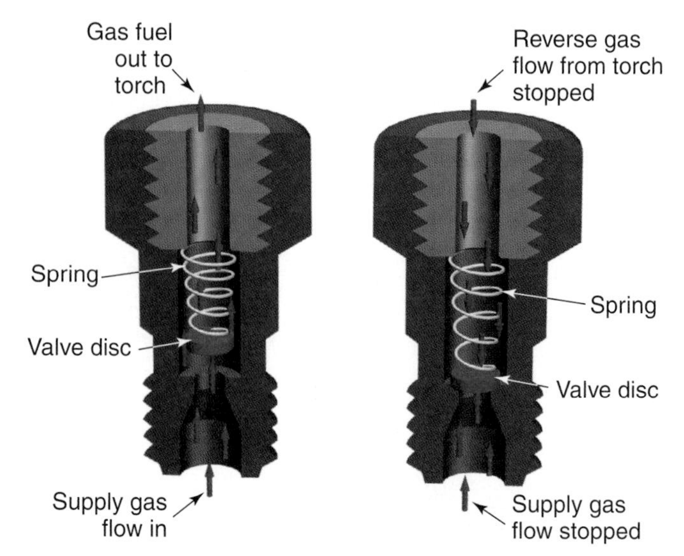

Figure 20-23. Check valve. Left Side—In normal operation, gas from the regulator is at a higher pressure than gases in the torch. Pressure from the regulator lifts the valve disc and allows gas to flow into the torch. Right Side—If the pressure within the torch plus the force of the spring on the valve disc exceeds the pressure from the regulator, the disc seats and does not allow gas to flow from the torch into the hose. A check valve is not reusable after a flashback. (Welding Design & Fabrication)

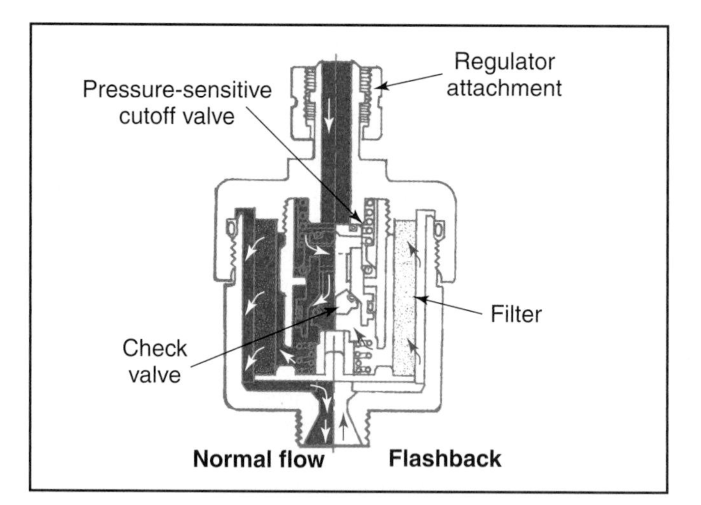

Figure 20-24. Flashback arrestor. Left Side—In normal operation, gas flows through the open cutoff valves, check valve, and flame arrestor filter into the hose. Right Side—In case of a flashback, the stainless steel filter stops the flame, and the pressure fluctuations activate the cutoff valve. The flow of gas is halted, extinguishing the flame. The check valve operates when gas flows toward the cylinder. If the arrestor is exposed to fire, the thermal cutoff valve shuts off the gas supply. A flashback arrestor is reusable after a flashback. (Welding Design & Fabrication)

- Stainless steel filter—prevents the flame from entering the hose.
- Heat-sensitive check valve—stops gas flow if the arrestor reaches 220°F (104°C).

Cutting Torches

An oxyfuel gas ***cutting torch*** is used to control and mix fuel gas with oxygen and to direct the oxyfuel gas flame to the cutting area. A complete cutting torch has many parts, **Figure 20-25**. These include the torch body, torch tube, standard torch valves, cutting oxygen valve, cutting oxygen lever, torch head, tip nut, and a special cutting torch tip.

While the principles of oxyfuel gas cutting will be discussed in detail in Chapter 22, a basic understanding is needed to understand the parts of the cutting torch. The cutting torch is designed to preheat metal until it reaches the temperature at which it can be cut. Then, the torch supplies a stream of oxygen to begin the actual cutting. The stream of oxygen that cuts the metal is referred to as ***cutting oxygen***. Cutting oxygen is released by pressing the ***cutting oxygen lever***. A cutting torch has three passageways: one for oxygen, another for fuel gas, and a third passageway for cutting oxygen.

Two types of cutting torches are widely used. The first is a standard, one-piece cutting torch used only for cutting. The second is called a ***cutting torch attachment*** or two-piece combination torch. The cutting torch attachment can be quickly converted to a welding torch by simply removing the cutting attachment and replacing it with a welding tip and mixer. See **Figure 20-26**. The advantage of the cutting torch attachment is its quick attachment.

Torch Valves

Torch valves control the flow of oxygen and fuel gas into the torch. A mark on the torch body near each valve indicates whether the valve is for fuel gas or for oxygen. On cutting torches, torch valves controlling the oxygen and fuel gas are located at the hose end of the torch body. A one-piece cutting torch has two valves on the torch body. One is the torch acetylene valve; the other is the torch oxygen valve. When the cutting torch attachment is mounted on the torch body, there are four torch valves. They include the two valves mentioned above plus a second torch oxygen valve on the cutting attachment. The fourth valve is the cutting oxygen lever, which regulates the flow of cutting oxygen.

Needle-and-seat or ball-and-seat valve construction is used to control the gas flowing to the mixing chamber. In a needle-and-seat valve, a conical stem and beveled seat are used to obstruct the passageway for gases. See **Figure 20-27**. A hardened steel ball and hemispherical seat are used to control gas flow in a ball-and-seat valve.

Torch valves are opened and closed by hand. Only finger force should be used. Do *not* use a wrench to open and close the valves. One-half to one full turn counterclockwise generally will open the valves completely. The valves are turned clockwise to close them.

Cutting Torch Tips

A cutting ***torch tip*** is the part of the end of the torch where the fuel gas and oxygen are mixed and ignited. **Figure 20-28** shows a cutaway view of a cutting torch attachment. Oxygen and fuel gas flow through separate passageways in the torch, mix together, and then exit through ***preheating orifices*** (holes) in the cutting tip. The pure cutting oxygen is carried through a separate passageway to the ***cutting oxygen orifice*** in the center of the tip. ***Orifices*** (precise holes) are bored into the end of the cutting torch tip when it is made. Cutting torch tips are made with many different configurations of preheating orifices. See **Figure 20-29**.

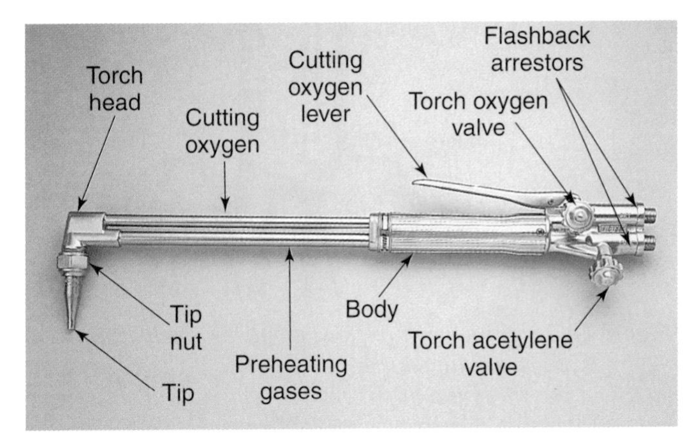

Figure 20-25. A fuel gas cutting torch. The upper tube supplies pure oxygen to the cutting tip. (Victor Equipment Co.)

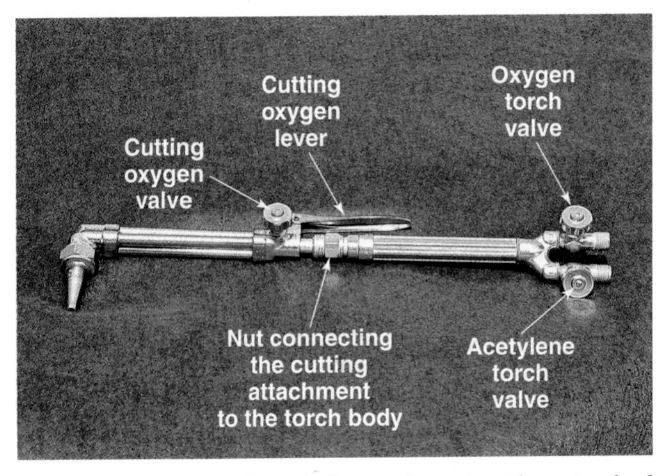

Figure 20-26. A cutting torch attachment with a standard welding torch body.

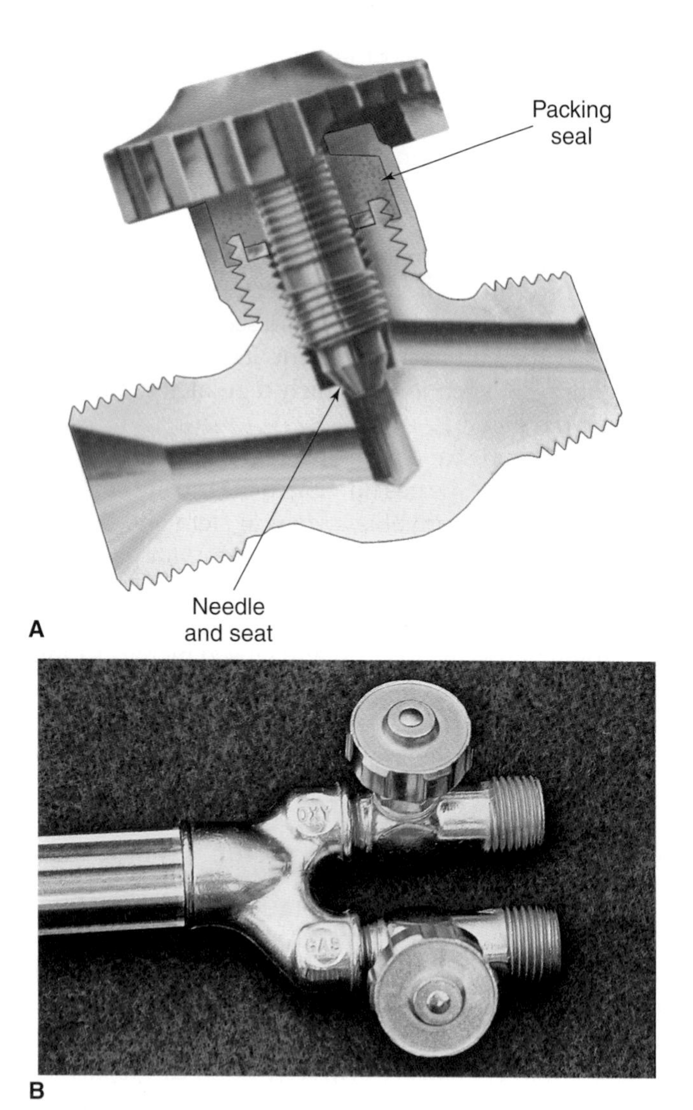

Figure 20-27. Torch valves. A—The valve shown uses needle-and-seat construction. B—Note that the valves are marked (OXY) and (GAS). (Veriflow Corp.)

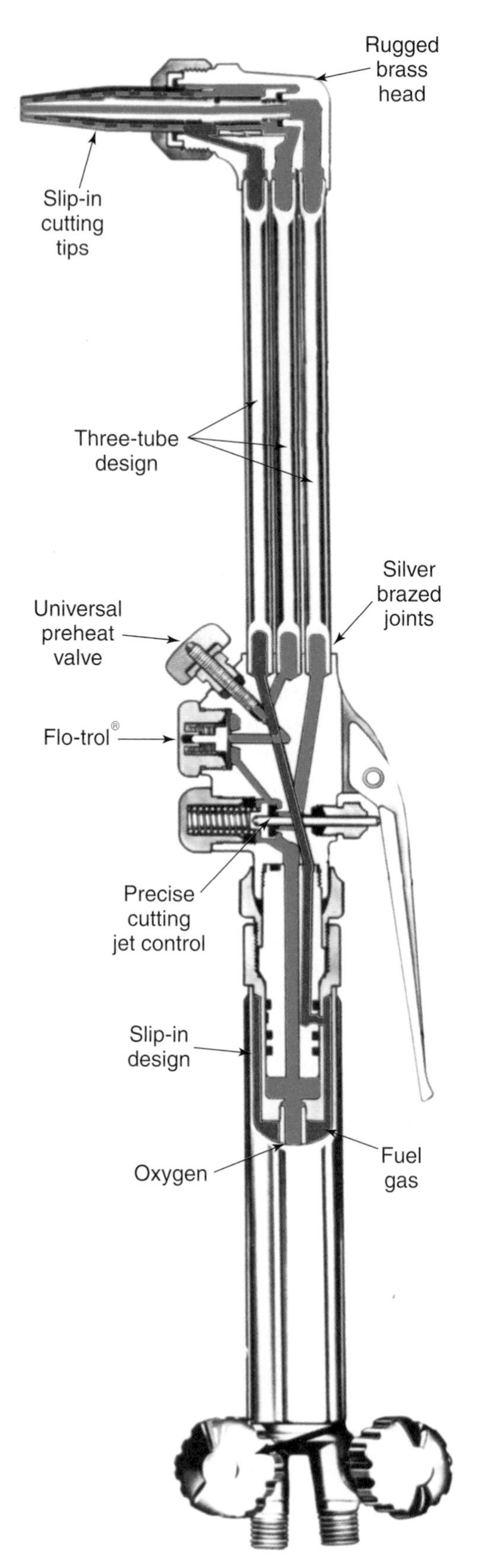

Figure 20-28. A sectioned view of a cutting torch attachment mounted on a welding torch handle. The preheat gases are mixed in the tip of the torch where it connects into the torch head. Depressing the oxygen cutting lever will allow pure oxygen to travel through the center hole in the cutting tip. (Smith Equipment, Division of Tescom Corp.)

Cutting torch tips are generally made of copper. Sometimes they are chrome-plated to reflect heat. Cutting tips have two or more orifices in the end. The cutting oxygen always flows from the center orifice. Gas for the preheating flames flows from the other orifices.

Tips used in cutting outfits are always separate from the torch tube. They are commonly unthreaded and held in the cutting torch head by a ***tip nut*** (large threaded nut). The *mating* (touching) surfaces of the tip and the torch head are machined smooth. The machined surfaces seal and prevent gas leakage when the tip nut is tightened. See **Figure 20-30**.

The flame end of the cutting tip becomes very hot in use. The tips may be easily damaged or pitted. See **Figure 20-31**. Never use the cutting tip to hold or pound on anything.

Number of Preheat Orifices	Degree of Preheat	Application
2	Medium	For straight line or circular cutting of clean plate.
2	Light	For splitting angle iron, trimming plate and sheet metal cutting.
30° 2	Light	For hand cutting rivet heads and machine cutting 30° bevels.
4	Light	For straight line and shape cutting clean plate.
4, 6, 8	Medium	For rusty or painted surfaces.
6	Heavy	For cast iron cutting and preparing welding Vs.
6	Very heavy	For general cutting, also for cutting cast iron and stainless steel.
20° 6	Medium	For grooving, flame machining, gouging, and removing imperfect welds.
6	Medium	For grooving, gouging, or removing imperfect welds.
45° 3	Medium	For machine cutting 45° bevel or hand cutting rivet heads.
6	Heavy	Flared cutting orifices provide a large oxygen stream of low velocity for rivet head removal (washing).

Figure 20-29. A table showing some common cutting torch tips and their uses. (Veriflow Corp.)

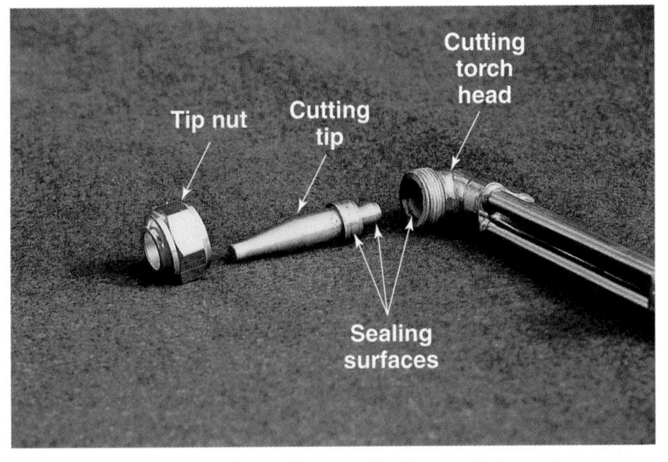

Figure 20-30. Typical cutting torch, tip, and tip nut arrangement. The smoothly machined surfaces in the torch head and on the torch tip form a gas-tight seal when the nut is tightened.

Some damaged cutting tips may be cleaned and re-formed. **Figure 20-32** shows a tool used to re-form the end of a cutting or welding tip. The preheating and oxygen orifices can be cleaned with a set of wire broaches. The correct method for cleaning a tip is described later in this chapter.

Welding Torches

An oxyfuel gas ***welding torch*** controls and mixes the fuel gas and oxygen. It is also used to direct the gas flame to the welding, brazing, or soldering work area. A complete welding torch has several parts. These include the torch valves, torch body, mixer or injector, torch tube, and torch tip. See **Figure 20-33**.

Either an injector-type torch or a positive pressure torch may be used for oxyfuel gas welding. An ***injector-type torch*** is used with a low-pressure acetylene

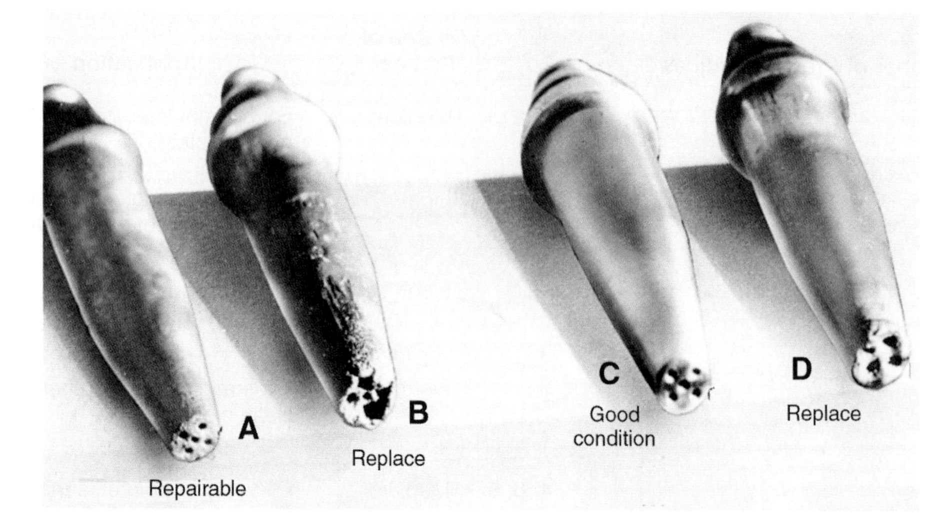

Figure 20-31. *Four cutting tips. The end of tip A can be refaced and the orifices cleaned. Tips B and D are beyond repair. Tip C is in good condition.*

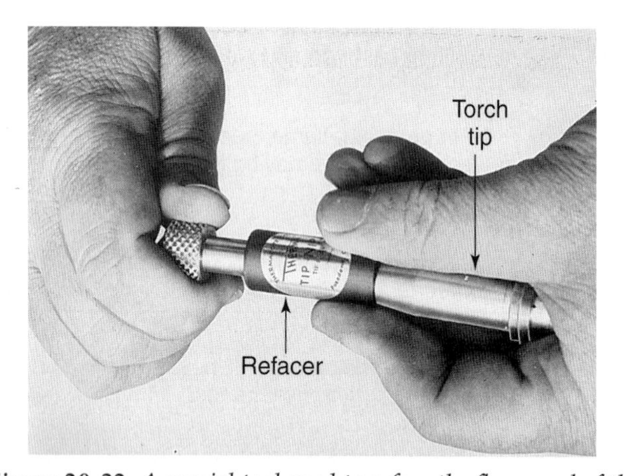

Figure 20-32. *A special tool used to reface the flame end of the cutting tip. This could be used on tip A shown in Figure 20-31. (Thermacote-Welco Co.)*

generator. See **Figure 20-34**. Acetylene pressures as low as .25 psig (1.7 kPa) are used with an injector-type torch. When using the injector-type torch, acetylene is drawn into the injector by oxygen traveling through the injector. The oxygen and acetylene are mixed in the mixing chamber. The mixed gases flow to the tip of the torch where they are burned. The welding torch, like the cutting torch, has two valves, which control the flow of oxygen and fuel gas into the torch.

A ***positive pressure torch*** is used with acetylene pressures above .25 psig (1.7 kPa). The oxygen and acetylene pressures are high enough to force gas through the torch into the mixing chamber. The mixing chamber may be in the torch body or at the torch end of the tip. See **Figure 20-35**. The gases are mixed and then flow to the tip where they are burned.

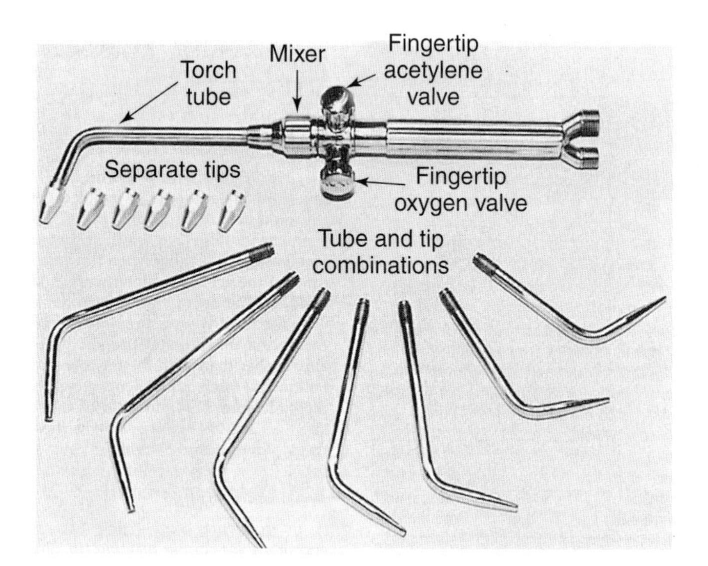

Figure 20-33. *A light-duty positive pressure welding torch. Two different styles of tips can be used with the same mixer. Note the location of the valves. (Smith Equipment, Division of Tescom Corp.)*

Welding Torch Tips

Two types of welding torch tips are available. A welder may choose a tube and tip combination or a separate tube and tip. A single orifice is bored into the end of the welding torch tip when it is made. The orifice size number is stamped on the tip by the manufacturer. Unfortunately, tip identification numbering systems differ among tip manufacturers. Generally, a small orifice number indicates a small hole. A small orifice delivers a small amount of heat to the base metal. A large orifice delivers a greater amount of heat to the base metal.

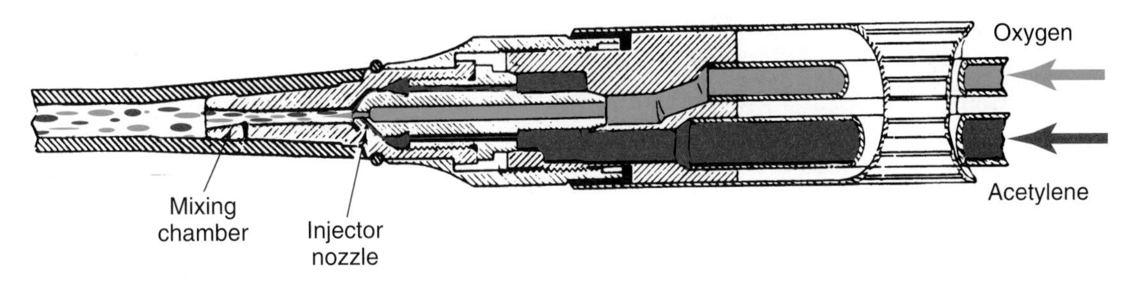

Figure 20-34. Cross-sectional view of an injector-type welding torch. The acetylene is injected (drawn) into the mixing chamber by suction created by the flow of oxygen.

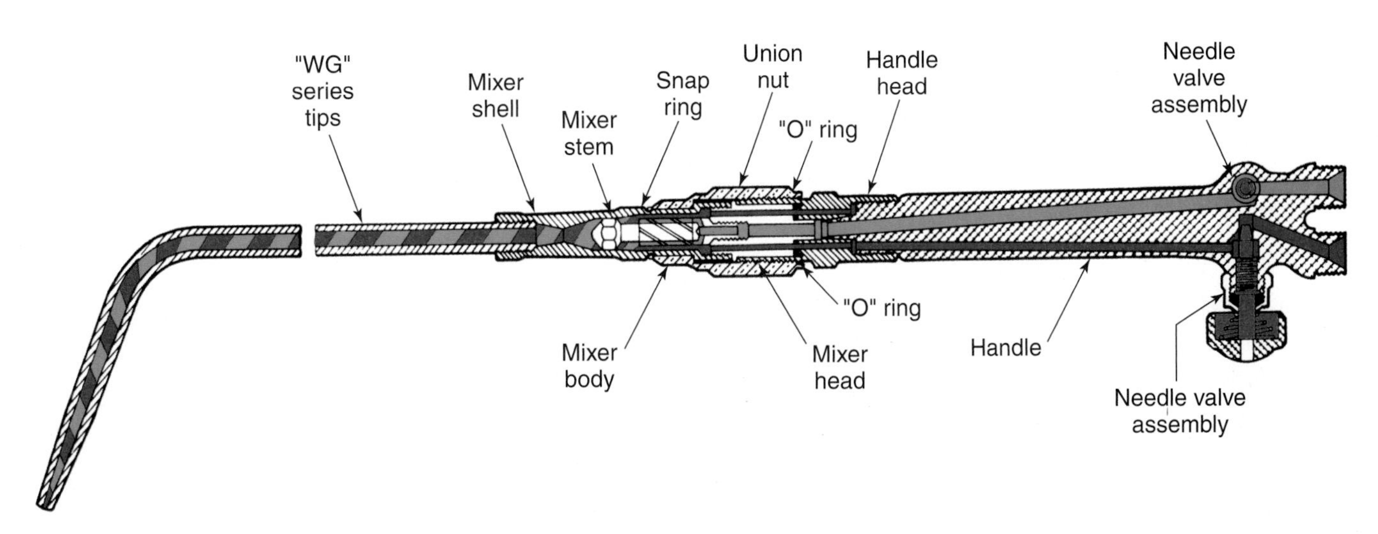

Figure 20-35. A cross-sectional view of a positive pressure torch. The working pressure of the oxygen and acetylene is high enough to force the gases into the mixing chamber. (Modern Engineering Co., Inc.)

The manufacturer's recommendation should be followed when selecting the correct tip size for a welding job. Welding supply companies supply tip size recommendation charts to their customers. **Figure 20-36** is a chart of the approximate tip size to use with metals of various thicknesses. The drill size of an orifice may be found using a set of numbered drill bits. **Care should be taken not to scratch or distort the orifice when using the bits.**

Molten metal or dirt may collect on the end of the tip. This may change the size or shape of the orifice. A drill bit or broaching wire is used to clean the orifice. See **Figures 20-37A** and **20-37B**. The end of the tip may need to be cleaned or filed flat. Dirt, filler metal, weld metal, or other materials may build up on the tip. To prevent these unwanted materials from falling into the torch, the oxygen torch valve can be opened slightly while the torch tip is being cleaned. A ***mill file*** may be used as shown in **Figure 20-37C**. After the tip is filed flat, the orifice may need to be cleaned again. The oxygen torch valve should be closed once the tip is clean.

Welding Rods

The edges of thick metal are commonly ground, flame cut, or machined before welding. Edge preparation removes a portion of the base metal and thins it. Even though metal is removed from the edges, a completed weld must still be at least as strong as the base metal. Filler metal is generally added to the weld bead to increase the thickness of the weld.

A ***welding rod*** is a long, thin rod of a similar type of metal to that being welded. As the rod is melted, it is added to the weld bead as filler metal.

Welding rod is available in 36″ (.91 m) lengths. Diameters range from 1/16″ (1.6 mm) to 3/8″ (9.5 mm). See **Figure 20-38**. The rods are usually packaged in 50-pound (22.7 kg) bundles. Welding rod is made from various metals, such as carbon steel, stainless steel, bronze, aluminum, and magnesium. Carbon steel welding rods are usually copper-coated to protect them from rusting.

Metal Thickness	Size* Welding Tip Orifice	Welding Rod Diameter	Oxygen		Acetylene		Welding Speed Ft./hr.
			Psig Pressure	Cu. ft./hr.	Psig Pressure	Cu. ft./hr.	
1/32″	74	1/16″	1	1.1	1	1	
1/16″	69	1/16″	1	2.2	1	2	
3/32″	64	1/16″ or 3/32″	2	5.5	2	5	20
1/8″	57	3/32″ or 1/8″	3	9.9	3	9	16
3/16″	55	1/8″	4	17.6	4	16	14
1/4″	52	1/8″ or 3/16″	5	27.5	5	25	12
5/16″	49	1/8″ or 3/16″	6	33	6	30	10
3/8″	45	3/16″	7	44	7	40	9
1/2″	42	3/16″	7	66	7	60	8

*Note the tip orifice size as shown is the number drill size. These recommendations are approximate. The torch manufacturer's recommendations should be carefully followed.

Figure 20-36. *Table used for oxyfuel gas welding with a positive pressure torch. Notice the relationships between tip orifice size, gas pressures, welding rod diameter, and metal thickness.*

Figure 20-37. *A—Cleaning the welding tip orifice using broaching wires. B—A broaching tool has many different size broaching wires. C—A small file in the broaching tool can be used to clean the end of the tip.*

Figure 20-38. *Welding rods are available in several diameters. A 1/16″ (1.6 mm), 3/32″ (2.4 mm), 1/8″ (3.2 mm), and 5/32″ (4.0 mm) are shown.*

Torch Lighters

A torch lighter provides safe ignition for oxyfuel gas welding torches. Never use a match or butane lighter to light a torch. Severe burns may occur.

The flint-and-steel spark lighter is the most common torch lighter. A spark lighter produces a spark by squeezing the handles together. As the handles are squeezed together, a replaceable flint rubs across a file segment to create a spark. See **Figure 20-39**.

In large production shops, an economizer and pilot light may be used to light a torch. The pilot light is fueled by a fuel gas. It remains burning during welding or cutting operations. When the torch is not being used, it is hung on the economizer. That device pivots to shut off the supplies of fuel gas and oxygen. When the torch is lifted from the economizer, the gases flow again. See **Figure 20-40**.

Figure 20-39. *A flint-and-steel spark lighter being used to light an oxyfuel gas welding torch.*

Figure 20-41. *This welder is wearing a light flame resistant jacket that is buttoned at the collar. He is also wearing a filter quality face shield. A filter quality face shield is ideal for those who wear glasses.*

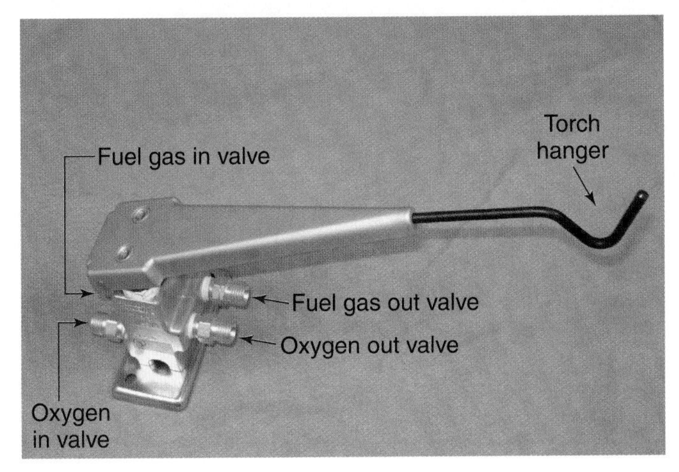

Figure 20-40. *Torch economizer.*

Protective Clothing

A welder must be properly dressed to avoid injury due to flying sparks and metal. **The welder should wear coveralls or a shirt or jacket that buttons at the collar, Figure 20-41.** Dark-colored clothing is preferred because it reflects less light. **Clothing must be flame-resistant and have covered pockets. Pant legs should be cut to length. Cuffs or folds at the bottom of the leg are dangerous. They may catch sparks or hot metal.**

Flammable objects, such as matches and butane lighters, should not be carried in your pockets. They could be ignited by a spark while you are welding.

A cap should be worn to keep your hair clean and free from sparks and hot metal.

Gloves must be worn while welding. Leather gloves or those with leather palms are preferred. Gloves with tight-fitting cuffs can be used for light-duty welding. Gauntlet gloves should be worn for heavy-duty or out-of-position welding.

Leather safety (hard toe) shoes are recommended to protect the welder's feet from falling objects. The shoe tops should be high enough to fit under the pant leg.

Leathers should be worn when overhead welding. The term ***leathers*** refers to the leather coat, hood, cape, leggings, and chaps worn by welders. See **Figure 20-42.**

Goggles

Ultraviolet and ***infrared rays*** are created by most welding processes. These rays are harmful to the eyes. Direct exposure should be avoided.

Welding goggles with protective filter lenses must be worn. ***Filter lenses*** are the lenses in welding goggles or the plates in welding helmets that protect the eyes from infrared, ultraviolet, and visible radiation. Some types can be worn over prescription glasses. Filter-quality face shields are also available.

Filter lenses range in shade from a #1 to #14. A higher number indicates a darker lens. A #4–#6 lens is recommended for oxyfuel gas welding.

Clear glass or plastic cover plates are installed outside the filter lenses. ***Cover plates*** protect the costly filter lenses from damage. Cover plates are inexpensive. They should be replaced when badly spattered or scratched. See **Figure 20-43.**

Figure 20-42. Gauntlet gloves, welding goggles, and a leather, flame-resistant jacket are worn for oxyfuel gas cutting or welding.

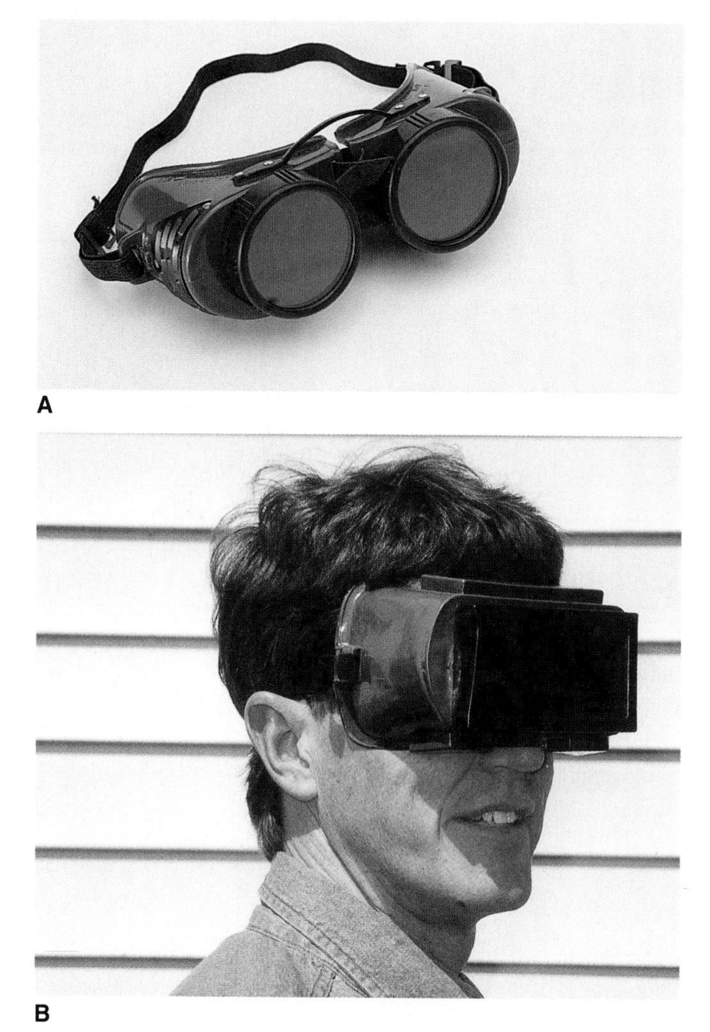

Figure 20-43. Welding goggles. A—These 2″ (50 mm) round welding goggles will fit over prescription glasses. (Jackson Products, Inc.) B—The rectangular filter lens in these goggles measures 2″ × 4 1/4″ (50.8 mm × 108 mm). Rectangular goggles are easily worn over prescription glasses.

Safety Glasses

Shatter-resistant safety glasses are designed to protect your eyes when grinding, chipping, or performing other activities that produce flying particles. For extra protection, safety glasses may be worn under a welding hood or welding goggles. Wear safety glasses whenever you work in an area where eye hazards are present. These include welding, cutting, and grinding areas, as well as areas used for machining, turning, stamping, or similar operations. Safety glasses may also be tinted to help protect your eyes from ultraviolet and infrared rays.

Summary

- Oxyfuel gas cutting and oxyfuel gas welding are general terms for a group of cutting and welding processes (methods) that use heat produced by a gas flame to cut and join various metals.
- An oxyfuel gas cutting outfit or welding outfit consists of a fuel gas cylinder, an oxygen cylinder, a means of securing the cylinders in an upright position, a fuel gas regulator and gauges, an oxygen regulator and gauges, fuel gas hose and fittings, an oxygen hose and fittings, flashback arrestors, check valves, and a cutting or welding torch.

(continued)

- Acetylene (C_2H_2) is a colorless fuel gas with a distinct garlic-like odor. It is produced by adding calcium carbide to water.
- An acetylene storage cylinder is constructed of steel and filled with a porous material. The cylinder is filled with acetone. The acetone fills the air pockets of the porous material. Acetylene is then pumped into the cylinder and absorbed into the acetone. The porous material and acetone prevent the acetylene from collecting in large volume. Acetylene bubbles out of the acetone when the cylinder valve is opened. Acetylene can be safely stored at pressures up to 250 psig (1724 kPa) using this method.
- **A safety cap must be screwed over the cylinder valve when a cylinder is stored or moved. In case a cylinder falls, the safety cap prevents the cylinder valve from being damaged.**
- Oxygen (O_2) is a colorless, odorless gas contained in the earth's atmosphere. When oxygen is added to a fuel gas flame, an increased flame temperature and rate of combustion result.
- An oxygen cylinder is a seamless, portable container used to store oxygen. Oxygen is stored in cylinders at a pressure of about 2200 psig (15.168 MPa).
- Manufacturers and repair shops that need a large volume of oxygen and fuel gas for welding or cutting may use manifolds. A manifold is a brazed assembly of pipes that delivers gas from several cylinders into one supply pipe. The pipe distributes gas to several workstations or locations.
- A pressure regulator is a device used to reduce the pressure at which oxygen or fuel gas is delivered. Gas pressure in acetylene and oxygen cylinders is high. The working pressure for welding, however, may be as low as 1 psig (6.9 kPa). A pressure regulator is used to reduce the pressure of the gas.
- A hose is a flexible rubber tube used to convey gases from a pressure regulator to a welding torch. Hoses are designed to withstand high pressure. The oxygen hose is green. The fuel gas hose is red. Fittings are used to connect hoses to the regulator and torch.
- A check valve is a valve that allows the flow of gas in only one direction. Such valves are used to prevent the reverse flow of gases through the torch, hoses, and/or regulators.
- A flashback arrestor is a device that prevents the flow of a burning fuel gas and oxygen mixture from the torch back into the hoses, regulators, and cylinders. It is designed to eliminate the possibility of an explosion in the regulator or cylinder. Flashback arrestors are installed between the torch and hose.
- An oxyfuel gas cutting torch is used to control and mix fuel gas with oxygen and to direct the oxyfuel gas flame to the cutting area. A complete cutting torch has many parts, including the torch body, torch tube, standard torch valves, cutting oxygen valve, cutting oxygen lever, torch head, tip nut, and a special cutting torch tip.
- Two types of cutting torches are widely used. The first is a standard, one-piece cutting torch used only for cutting. The second is called a cutting torch attachment or two-piece combination torch. The cutting torch attachment can be quickly converted to a welding torch.
- Torch valves control the flow of oxygen and fuel gas into the torch. The torch valves are located at the hose end of the torch body. A cutting torch tip is the part of the end of the torch where the fuel gas and oxygen are mixed and ignited.
- An oxyfuel gas welding torch controls and mixes the fuel gas and oxygen. It is also used to direct the gas flame to the welding, brazing, or soldering work area. A complete welding torch has several parts. These include the torch valves, torch body, mixer or injector, torch tube, and torch tip. An injector-type torch is used with a low-pressure acetylene generator. A positive pressure torch is used with acetylene pressures above .25 psig (1.7 kPa).
- A welding rod is a long, thin rod of a similar type of metal to that being welded. As the rod is melted, it is added to the weld bead as filler metal.
- **The welder should wear coveralls or a shirt or jacket that buttons at the collar. Clothing must be flame-resistant and have covered pockets. Pant legs should be cut to length. Flammable objects, such as matches and butane lighters, should not be carried in your pockets. A cap should be worn to keep your hair clean and free from sparks and hot metal. Gloves must be worn while welding. Leather safety (hard toe) shoes are recommended.** Leathers should be worn when overhead welding. A #4–#6 filter lens is recommended for oxyfuel gas welding. Wear safety glasses whenever you work in an area where eye hazards are present.

Review Questions

Write your answers on a separate sheet of paper. Please do *not* write in this book.

1. List five fuel gases used in oxyfuel gas welding or cutting.
2. A low-pressure acetylene generator produces acetylene gas at _____ psig (_____ kPa).
3. Acetylene becomes unstable and explosive at or above _____ psig (_____ kPa) if it is allowed to collect in large volume as a gas.
4. Which type of gas cylinder uses a fusible plug?
5. Which will boil away first when air is liquefied, oxygen or nitrogen?
6. Why should a safety cap be used on a cylinder when it is moved or stored?
7. What happens in an oxygen cylinder safety valve to prevent an explosion during a fire?
8. Liquid oxygen is stored in a(n) _____ flask.
9. A(n) _____ is used to connect several cylinders to one delivery pipe.
10. Name one metal that should not be used for tubing or pipes that carry acetylene.
11. A(n) _____ is used to reduce and control the pressure of the welding gases.
12. When the regulator adjusting screw feels loose in its threads, is the regulator opened or closed?
13. A high-pressure oxygen gauge is usually marked to indicate pressures up to _____ psig (_____ MPa).
14. Low-pressure acetylene gauges are often shaded in red above _____ psig (_____ kPa).
15. What is a siamese welding hose?
16. Explain the difference between a cutting torch tip and a welding torch tip.
17. *True or False?* Oxygen flowing through an acetylene hose may cause an explosion.
18. Refer to Figure 20-36. List the diameters of four commonly used welding rods.
19. When opening or closing torch valves, how much force should be applied?
20. A correct size _____ or _____ should be used to clean the torch tip orifice.
21. The recommended filter lenses for oxyfuel gas welding are #_____ to #_____.
22. A(n) _____ is a one-way valve that prevents gases from burning back to the regulator.
23. Explain the difference between acetylene and oxygen hose nuts.
24. Pant leg cuffs and pockets are not recommended on welding clothing. Why?
25. Oil, grease, kerosene, cloth, tar, wood, and coal dust must be kept away from all _____ fittings to prevent fires or explosions.

Chapter 21

Oxyfuel Gas Cutting and Welding: Equipment Assembly and Adjustment

After studying this chapter, you will be able to:

- List the steps required to assemble an oxyfuel gas cutting and welding outfit.
- List the steps required to turn on an oxyacetylene cutting and welding outfit.
- Describe the procedure used to check for leaks in an oxyacetylene cutting and welding system.
- List the steps required to light and adjust the flame on an oxyacetylene cutting torch.
- List the steps required to light and adjust the flame on an oxyacetylene welding torch.
- Identify the three types of flames.
- Describe the procedure for shutting off an oxyacetylene cutting or welding outfit.

Technical Terms

carburizing flame
neutral flame
nonpetroleum-based
oxidizing flame
petroleum-based
purging

Assembling the Cutting or Welding Outfit

Proper assembly, care, and security of the oxyfuel gas welding outfit is necessary for its safe and effective use. Care must be taken when assembling the various threaded fittings. When tightened, the fittings must not leak. The fittings are generally made from soft metal like brass. Do not overtighten fittings, or the threads may be damaged.

Petroleum-based **oil, grease, or soap contains flammable material. Petroleum-based products must not be used to lubricate any part of a cutting or welding outfit. These materials could ignite and cause a fire.**

An oxyfuel gas cutting outfit is assembled in the same manner as an oxyfuel gas welding outfit. As noted in Chapter 20, oxyfuel gas cutting and welding equipment is essentially the same. The main differences between the cutting and welding outfits are in the oxygen regulator and the torch. A larger volume of higher pressure oxygen may be needed for cutting heavy metal. A pressure regulator with a larger volume capacity and higher pressure indications is used. Aside from the torch and the oxygen regulator, the assembly of the oxyfuel gas outfit is the same for both cutting and welding.

To properly assemble an oxyfuel gas cutting or welding outfit, follow these procedures:

1. Securely fasten the oxygen and fuel gas cylinders to a wall, column, or hand truck in the vertical position. See **Figure 21-1**. Chains or steel straps are commonly used to fasten the cylinders. Safety caps may be removed after the cylinders are safely secured.
2. Clean the oxygen and fuel gas cylinder outlets before attaching the regulators. This is done by quickly and briefly opening and closing the cylinder valves. **Figure 21-2** shows a cylinder valve being opened. The escaping gas cleans any dust or dirt from the outlet. **Do not point the cylinder outlet toward people in the area. Flames or sparks must not be present when opening the fuel gas cylinder valve. If cylinder valves are damaged or leaking, the cylinders must not be used. Return defective cylinders to the supplier.**
3. Attach the regulators to the cylinder outlets. The regulators should be threaded onto the cylinder nozzles by hand, as shown in **Figure 21-3**. The regulator nuts should then be tightened with a regulator wrench or a proper size open-end wrench so they are snug and leakproof. The brass nuts used on regulators and hoses should not be overtightened, or they may be damaged. The oxygen regulator nut has right-hand threads and can only be threaded onto oxygen cylinders, which also have right-hand threads. Large, commercial fuel gas cylinders have left-hand threads. This is a safety precaution that prevents a welder from being able to screw an oxygen regulator onto a fuel gas cylinder. Smaller, noncommercial fuel gas cylinders with Prestolite valves have right-hand threads.
4. Attach the oxygen hose to the oxygen regulator, **Figure 21-4A**, and attach the fuel gas hose to the fuel gas regulator. A fuel gas hose should be red in color for easy identification. A groove is machined around the nuts used on fuel gas hoses. Fuel gas hose connections

Figure 21-1. *An oxyacetylene welding outfit secured to a cylinder truck by chains.*

Figure 21-2. *A cylinder valve is opened briefly to remove dirt or foreign debris from the outlet. The cylinder outlet should be pointed away from nearby people.*

Figure 21-3. *To attach a regulator to a gas cylinder, start the connector by hand and then tighten it with the proper wrench.*

have left-handed threads. An oxygen hose is green, and the nuts do not have a groove. Also, oxygen hose connections, like those of the oxygen regulator, have right-handed threads. These characteristics make the hoses easy to distinguish.

5. Attach the cutting or welding torch to the hoses. The torch inlet fittings have right-hand and left-handed threads like the regulators and hoses. Attach the oxygen hose to the torch oxygen inlet fitting, **Figure 21-4B**, and the fuel gas hose to the torch fuel gas inlet fitting. After tightening by hand, use a wrench to seal the connections, but do not overtighten.
6. Select the proper size torch tip and place it into the torch or into the torch tube. Tighten it securely.

Turning on an Oxyacetylene Cutting or Welding Outfit

A positive pressure torch is used for many oxyacetylene cutting and welding operations. The cylinder, regulator, and torch valves must be turned on in a given sequence in order to be used safely. Before lighting the oxyacetylene flame, the oxygen and acetylene systems must be purged.

Purging is the process of passing the correct gas through the entire system to remove air or undesirable gases. Purging ensures that the correct gas is flowing in the appropriate regulator, hose, and torch passage. Purging is done by allowing acetylene to flow through the acetylene hose and oxygen to flow through the oxygen hose for a short period of time. When the following steps are carried out, the system will be properly purged and ready to light.

Follow these steps when turning on an oxyacetylene welding or cutting outfit:

1. **Visually check the torch, valves, hoses, fittings, regulators, gauges, and cylinders for damage. Be sure to check that the oxygen and acetylene torch valves are closed by turning them clockwise (to the right).**
2. **Make certain the regulators are closed before opening the cylinder valves. This will prevent damage to the regulators and gauges.** See **Figure 21-5**. Close the regulators by turning the regulator adjusting screws on the oxygen and acetylene regulators *counterclockwise* (to the left). Continue to turn the screws counterclockwise until they feel loose. Be careful not to turn the adjusting screws too far, or they will fall off.

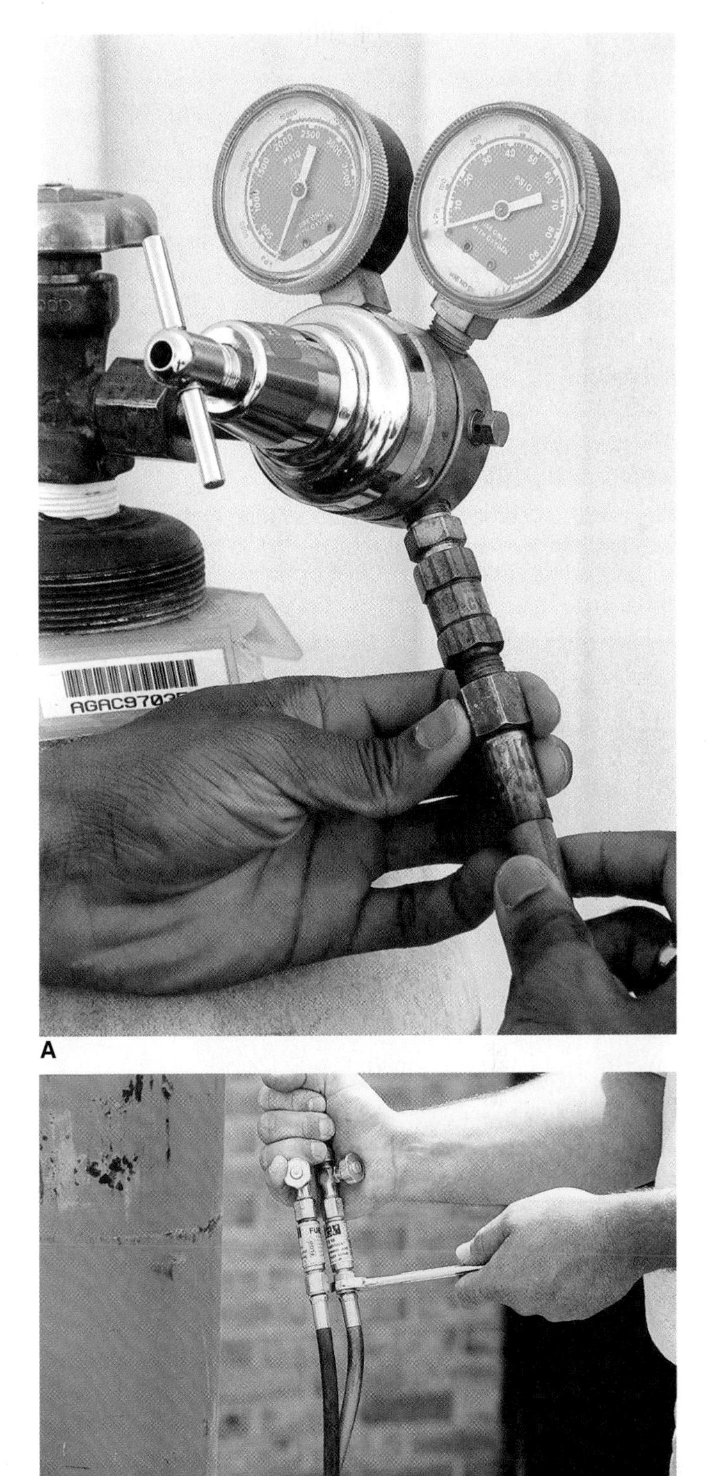

Figure 21-4. *A—When attaching the oxygen hose to the regulator, start the connector by hand and then tighten it with the proper size wrench. B—When attaching the oxygen hose to the torch, start the connector by hand and then tighten it with the proper size wrench.*

Figure 21-5. Loosen the regulator adjusting screw until it feels loose in the regulator threads.

Figure 21-6. Opening the oxygen cylinder valve. Always stand to one side when opening any cylinder valve.

3. **Stand to one side of the regulators while opening the cylinder valves, as shown in Figure 21-6. A regulator or gauge could burst, causing severe injury.**
4. Slowly open the acetylene cylinder valve by turning it counterclockwise until it is open (about 1/4 to 1/2 turn). See **Figure 21-7**. This provides enough acetylene flow for most purposes. Use the proper size wrench for the cylinder valve. **Leave the wrench in place so that the valve can be closed quickly in case of an emergency.**
5. Slowly turn the oxygen cylinder valve counterclockwise until it is fully open (the valve may leak if it is not fully opened). **Remember that cylinder pressure may be 2200 psig (15.17 MPa). A rapid flow of high-pressure gas could rupture the regulator diaphragm or gauges.**
6. Turn the acetylene torch valve one-half to one complete turn counterclockwise.
7. Set the correct working pressure on the acetylene regulator gauge by turning the acetylene regulator adjusting screw *clockwise* (to the right) until the low-pressure gauge shows the correct working pressure. The correct working pressure depends on the tip size used. Refer to Figure 20-36 for a table showing the recommended pressure in relation to the tip orifice size.

Figure 21-7. Opening the acetylene cylinder valve. A special wrench may be required. The wrench should remain on the valve so that it can be turned off quickly in an emergency.

8. Close the acetylene torch valve. Check the acetylene regulator for possible leaks. Checking for leaks is discussed in the next section.
9. Open the oxygen torch valve one-half to one turn counterclockwise.
10. Turn the oxygen regulator adjusting screw clockwise until the desired working pressure is obtained, **Figure 21-8**. Note: When turning on an oxyacetylene cutting outfit, the cutting torch oxygen valve (lever) should also be open while the oxygen cutting pressure is set. After setting the working pressure, close the torch valve and release the oxygen cutting lever.
11. Close the oxygen torch valve. Check the oxygen regulator for leaks. Checking for leaks is discussed in the next section. **Remember that the preceding steps for turning on the oxyacetylene cutting or welding outfit must be performed in the correct order to properly purge the system prior to lighting the torch.**

Each of these steps should be carefully followed. Once the procedure is complete, the outfit is ready to light.

Note: Regarding the need to set the correct working pressure, remember that different manufacturers of welding tips have different charts. If the actual orifice size is unknown or if no manufacturer's table is available, use the alternate method described later in this chapter. The alternate method can also be used if the welding hose is excessively long (over 25 feet). Long hoses can cause the actual gas pressure that reaches the welding tip to be lower than the pressure reading shown back at the gauge.

Figure 21-8. *Adjusting the oxygen working pressure. The torch valve must be open to properly adjust the working pressure. The torch valve is closed after the pressure is set.*

Checking for Leaks

An oxyfuel gas cutting or welding outfit should be checked for leaks each time it is turned on. External leaks may be found using soapsuds. Both internal and external leaks may be located by watching the low-pressure gauge. **A *nonpetroleum-based* soap contains no flammable material and is therefore a safe solution to use in checking for leaks. Petroleum-based products may cause fires when in contact with oxygen.** Completely cover fittings with soapsuds to locate leaks. Look for bubbles, which will occur in the soapsuds when a leak is present.

Regulators

The high-pressure valve under the regulator diaphragm opens and closes to control the working pressure. A leaking regulator valve allows pressure to rise uncontrollably inside the regulator. The regulator or low-pressure gauge may burst if there is excessive pressure below the diaphragm. Check regulators for leaks immediately after setting the working pressure. The regulator may be checked for leaks as follows:

1. Turn on the welding or cutting outfit.
2. Close both torch valves.
3. Carefully watch the acetylene and oxygen low-pressure gauges. The pressure readings should remain constant. **A leaky regulator valve is indicated if the gauge pressure continues to rise. Immediately close the cylinder valves and shut off the outfit.**
4. Send the bad regulator out for repair.

Cylinder-to-Regulator Connections

Cylinder-to-regulator fittings may be checked for leaks by intentionally trapping pressurized gas between the regulator and cylinder. A leak is indicated by a decrease of pressure on the high-pressure gauge. This test must be made on both the oxygen and acetylene systems as follows:

1. Turn on the outfit. Note the pressure on the high-pressure gauge.
2. Close the regulator by turning the adjusting screw counterclockwise. Close the cylinder valve. High-pressure gas is now trapped between the regulator and cylinder valve.
3. The high-pressure gauge reading should remain constant. If it drops, there is a leaking connection between the regulator and the cylinder. Check the fitting for tightness and repeat steps 1 and 2. If a leak is still indicated, the cylinder outlet fitting may be bad. If so, return the cylinder to the vendor.

4. Try the regulator on another cylinder and repeat steps 1, 2, and 3. If there is still a leak, the regulator fitting itself may be bad.

Hoses and Hose Fittings

Gas leaks in the hoses or fittings between the regulators and the torch valves will show as a pressure drop on the low-pressure gauge. Gas must be trapped between the regulator and the torch valves to perform this test. This procedure must be used to test the oxygen and acetylene hoses and fittings as follows:

1. Turn on the outfit.
2. Close the torch valves.
3. Note the reading on the low-pressure gauge.
4. Close the regulator with the adjusting screw.
5. The low-pressure gauge reading should remain constant. If the pressure drops, there is a leak in the hose or fittings. Locate it using nonpetroleum-based soapsuds.

Flames

Three types of flames can be produced with oxygen and acetylene. They are the carburizing, neutral, and oxidizing flames. All three flames are produced by both the oxyacetylene cutting and welding outfits. See **Figures 21-9** and **21-10**, which illustrate the cutting and welding flames.

Carburizing Flame

A ***carburizing flame*** is produced when too little oxygen is present. Three flame areas are visible (Figure 21-9B and Figure 21-10A). The end of the inner flame is ragged or rough. A carburizing flame may add carbon to the weld area. This can cause undesirable hardening of the weld. A carburizing flame is used whenever it is necessary to avoid adding oxygen to weld or braze joints. Slightly carburizing flames are used for brazing, braze welding, soldering, and welding certain metal alloys.

Neutral Flame

A ***neutral flame*** results from the correct balance of oxygen and acetylene. It is used for most welding and cutting operations. There are two flame areas. The end of the inner cone is smooth and shaped like a bullet, Figures 21-9C and 21-10B. A neutral flame will not burn the weld or base metal because it does not have excessive oxygen. It will not add carbon to the weld because all the carbon is burnt in the flame.

Oxidizing Flame

An ***oxidizing flame*** is produced when too much oxygen is present. Two flame areas are seen. The end of the inner flame is smooth and pointed (Figures 21-9E and 21-10C). The flame is loud and makes a hissing noise. An oxidizing flame is usually not desired because it burns the surface of the base metal and weld.

Lighting and Adjusting a Positive Pressure Cutting Torch

Follow the steps for turning on an oxyacetylene cutting outfit earlier in this chapter. Remember to press the cutting oxygen lever since it must be open while the oxygen working pressure is set. After setting the working pressures for oxygen and acetylene, close the torch valves and release the oxygen cutting lever. If the manufacturer's table indicating correct working pressures is not available or if the hose is excessively long, light the torch using the alternate method described later in this chapter.

When the correct working pressures have been set, the cutting torch is ready to light. **A flint-and-steel spark lighter must be used to prevent severe burns. When lighting the torch, hold it so that the tip is facing downward. Strike a spark while holding the spark lighter approximately 1″ (25 mm) from the tip.**

1. Open the acetylene torch valve 1/16 turn.
2. Light the preheating flames with a spark lighter. All flames should light at the same time. If all the flames do not light, shut down the torch and clean any dirty orifices.
3. Continue opening the acetylene torch valve until the smoking stops. The flame must remain in contact with the end of the tip. See Figure 21-9A.
4. Open the oxygen torch valve until a neutral flame is obtained. See Figure 21-9B. The cutting oxygen valve must be closed during this adjustment.
5. Open the cutting oxygen valve (lever). The preheating flames should remain neutral. See Figure 21-9C. If a carburizing flame is obtained, open the torch oxygen valve a little more. Continue to adjust until neutral preheating flames remain while the cutting oxygen valve is open.

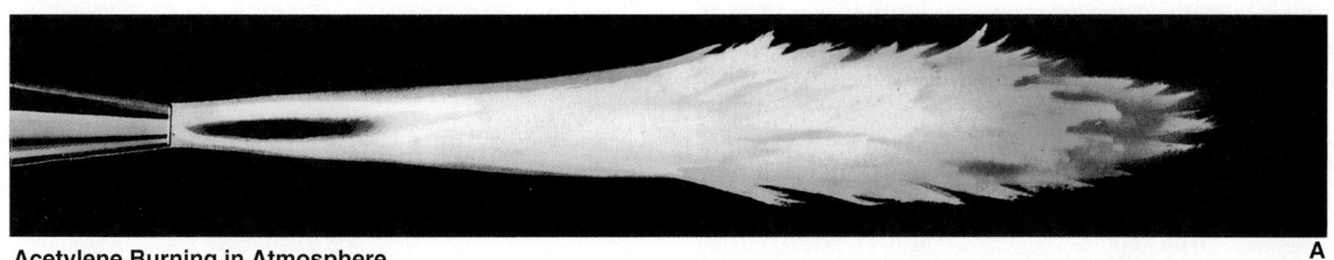

Acetylene Burning in Atmosphere A
Open fuel gas valve until smoke clears from flame.

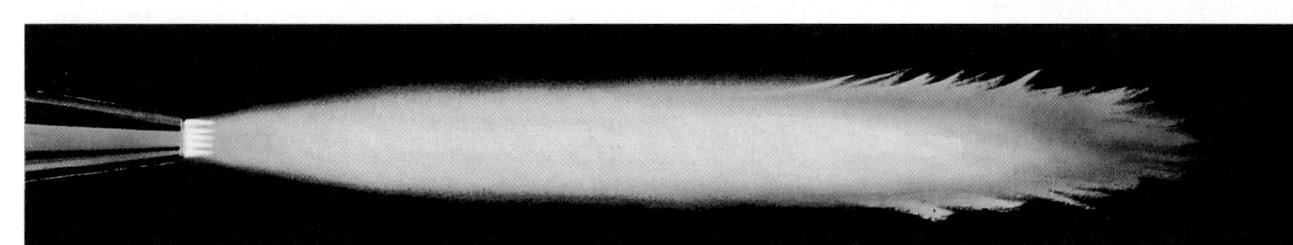

Carburizing Flame B
(Excess acetylene with oxygen.) Preheat flames require more oxygen.

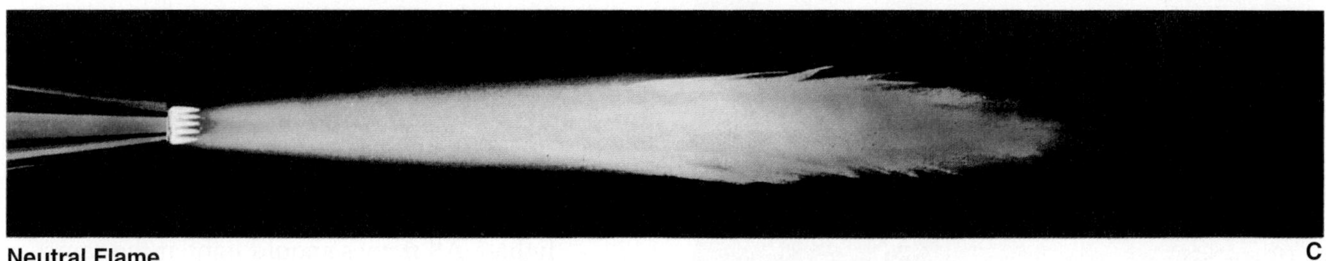

Neutral Flame C
(Acetylene with oxygen.) Temperature 5589°F (3087°C). Proper preheat adjustment for cutting.

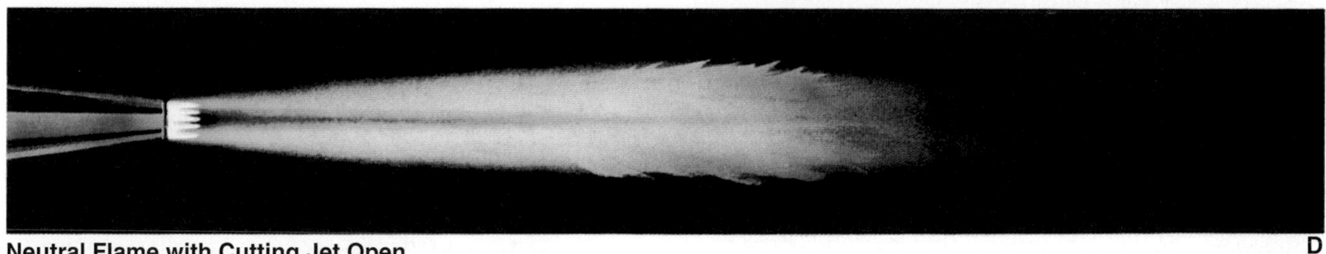

Neutral Flame with Cutting Jet Open D
Cutting jet must be straight and clean.

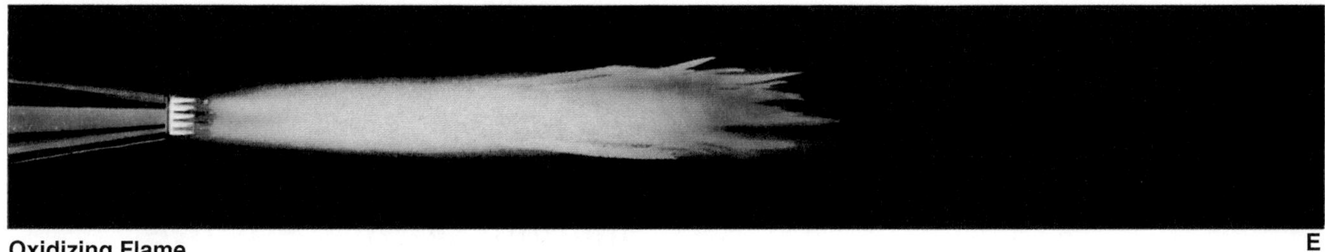

Oxidizing Flame E
(Acetylene with excess oxygen.) Not recommended for average cutting.

Figure 21-9. *Oxyacetylene cutting flame adjustments.*

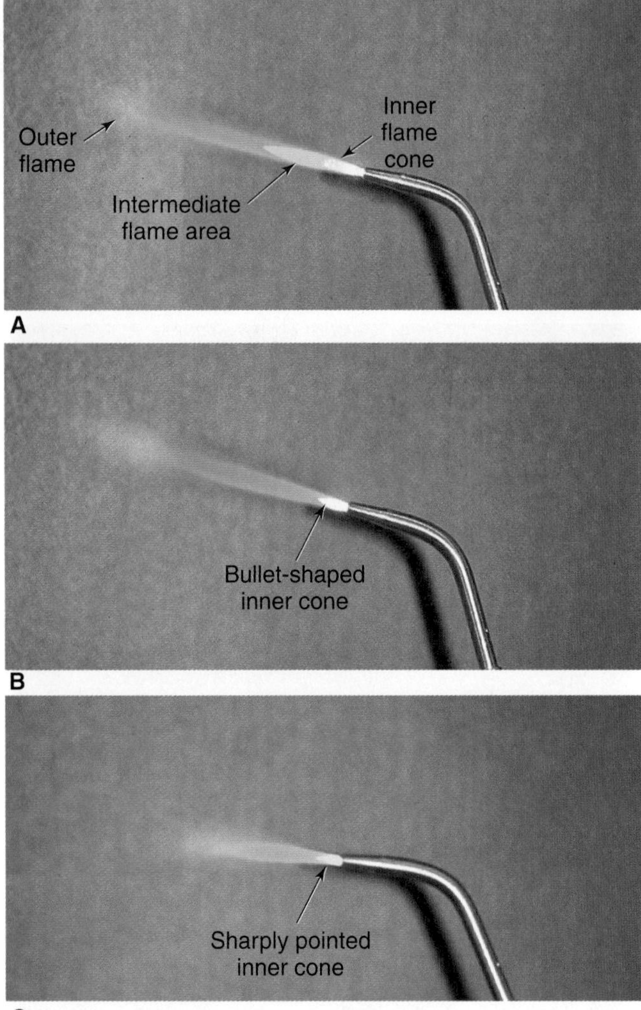

Figure 21-10. *A—Carburizing flame. The flame begins to burn cleanly as oxygen is turned on. Three distinct areas of the flame can be seen. B—Neutral flame. When the oxygen and acetylene are properly adjusted, the inner cone is bullet-shaped or rounded on the end. C—Oxidizing flame. When too much oxygen is present, the inner cone becomes pointed. The flame is noisy and gives off a hissing sound. This flame is undesirable because it causes the weld area to oxidize (rust).*

Shutting Down a Positive Pressure Cutting Torch

The cutting outfit must be shut down when the cutting operation is completed or when the welder is leaving the cutting station. The cutting outfit is shut down using a procedure similar to that used for shutting down a welding outfit. Be certain that the correct procedure is followed so that all gauges read zero when the shutdown is completed. Remember to turn the regulator adjusting screws counterclockwise until they feel loose. Use the following procedure to properly shut down the outfit.

1. Turn off the acetylene torch valve.
2. Turn off the oxygen torch valve.
3. Turn off the acetylene cylinder valve.
4. Turn off the oxygen cylinder valve.
5. Reopen the acetylene and oxygen torch valves.
6. Close both torch valves when all gauges read zero.
7. Close both regulators by turning the regulator adjusting screws counterclockwise until they feel loose.

Lighting and Adjusting a Positive Pressure Cutting Torch Attachment

The flame is ready to light after turning the cutting outfit on and setting the correct working pressures. Before lighting the torch, open the torch oxygen valve on the cutting attachment one full turn. This completely opens the torch oxygen valve on the cutting attachment. All oxygen adjustments will now be made at the torch oxygen valve on the torch body.

A flint-and-steel spark lighter must be used to prevent severe burns when lighting the cutting torch attachment. When lighting, point the tip of the torch downward and hold the spark lighter about 1″ (25 mm) from the tip.

1. Open the acetylene torch valve 1/16 turn.
2. Light the preheating flames with a spark lighter. All flames should light at the same time. If all the flames do not light, shut down the torch and clean any dirty orifices.
3. Continue opening the acetylene torch valve until the smoking stops. The flame must remain in contact with the end of the tip.
4. Open the oxygen torch valve until a neutral flame is obtained. The cutting oxygen valve must be closed during this adjustment.
5. Open the cutting oxygen valve (lever). The preheating flames should remain neutral. If a carburizing flame is obtained, open the torch oxygen valve a little more. Continue to adjust until neutral preheating flames remain when the cutting oxygen valve is open.

The flame must always be turned off if the torch is not in the welder's hands. If the torch will not be used for some time, the outfit must be shut down.

Shutting Down a Positive Pressure Cutting Torch Attachment

The cutting outfit must be shut down after the cutting operation is completed or when the welder is leaving the welding station. The following shutdown procedure must be used. When the procedure is complete, make certain the gauges read zero and the regulator adjusting screws feel loose.

1. Turn off the acetylene torch valve.
2. Turn off the oxygen torch valve.
3. Turn off the acetylene cylinder valve.
4. Turn off the oxygen cylinder valve.
5. Reopen the acetylene and oxygen torch valves.
6. Close both torch valves when all gauges read zero.
7. Close both regulators by turning the regulator adjusting screws counterclockwise until they are loose.
8. Close the oxygen valve on the cutting attachment.

Lighting and Adjusting the Flame of a Welding Torch—Recommended Method

A spark lighter should be used to light an oxyacetylene cutting or welding torch. The flint and steel in the spark lighter create a spark that will safely light the acetylene at the torch tip. A spark lighter should be long enough to keep your hand safely away from the acetylene flame. **Never use a match or butane lighter to light an oxyacetylene torch. Severe burns could result.**

After completing all the steps for turning on the welding outfit and checking for leaks, use the following procedure to light and adjust the flame.

1. Open the acetylene torch valve approximately 1/16 turn.
2. Light the acetylene at the torch tip, using a spark lighter.
3. Open the acetylene torch valve until the flame becomes turbulent (rough) about 3/4″–1″ (19–25 mm) from the end of the tip. Adjust the acetylene so the flame no longer smokes or releases soot. See Figure 21-10A.
4. Turn on the oxygen torch valve slowly after the acetylene is regulated. Adjust the oxygen torch valve until a neutral flame is obtained.

Lighting and Adjusting the Flame of a Welding Torch—Alternate Method

Another method of lighting and adjusting the flame may be used when the actual orifice size is unknown or when the manufacturer's table indicating correct working pressure is not available. This method is also used to compensate for the pressure drop that can result in using an excessively long hose.

1. Visually check the torch, valves, hoses, fittings, regulators, gauges, and cylinders for damage.
2. Make certain the regulators are closed before opening the cylinder valves, Figure 21-5. This will prevent damage to the regulators and gauges. Turn the regulator adjusting screws on the oxygen and acetylene regulators counterclockwise (to the left). Continue to turn the screws counterclockwise until they feel loose.
3. Stand to one side of the regulators while opening the cylinder valves, as shown in Figure 21-6. A regulator or gauge could burst causing severe injury.
4. Slowly turn the acetylene cylinder valve counterclockwise until it is open (about 1/4 to 1/2 turn). See Figure 21-7. This provides enough acetylene flow for most purposes. Use the proper size wrench for the cylinder valve. Leave the wrench in place so that the valve can be closed quickly in case of emergency.
5. Slowly turn the oxygen cylinder valve counterclockwise until it is fully open (the cylinder valve may leak if it is not fully opened). Remember that cylinder pressure may be 2200 psig (15.17 MPa). A rapid flow of high-pressure gas could rupture the regulator diaphragm or gauges.
6. Open the acetylene torch valve one complete turn. Slowly turn the acetylene regulator adjusting screw clockwise until the gas begins to flow.
7. Light the acetylene at the torch tip with a spark lighter.
8. Keep turning the acetylene regulator adjusting screw clockwise until the flame stops smoking or releasing soot. The flame must remain in contact with the end of the tip.
9. Open the oxygen torch valve one turn. (If lighting a cutting torch, press the oxygen lever after opening the oxygen torch valve).
10. Turn the oxygen regulator adjusting screw until the correct flame is obtained. A neutral flame is usually preferred. Note: The cutting flame may need to be readjusted when the oxygen lever is pressed. This is done by readjusting the torch oxygen and acetylene valves until a neutral flame is obtained.

Shutting Down an Oxyacetylene Welding Outfit

The flame must be turned off whenever the oxyacetylene torch is not in your hand. The flame is turned off by closing the acetylene torch valve and then the oxygen torch valve.

The welding outfit should be shut down using the following procedure:

1. To turn off the flame, close the acetylene torch valve. Then, close the oxygen torch valve.
2. Completely close the acetylene and oxygen cylinder valves.
3. Open the oxygen and acetylene torch valves. This allows all the gases in the system to escape.
4. Close the torch valves after the high-pressure and low-pressure gauges read zero. Hang up the torch.
5. Turn the adjusting screws on both regulators counterclockwise until they feel loose.

Note: If the torch manufacturer's procedures differ from those listed here, follow the manufacturer's procedures.

Summary

- To properly assemble an oxyfuel gas cutting or welding outfit, begin by securely fastening the oxygen and fuel gas cylinders to a wall, column, or hand truck. Next, clean the oxygen and fuel gas cylinder outlets. Attach the regulators to the cylinder outlets. Attach the oxygen hose to the oxygen regulator and attach the fuel gas hose to the fuel gas regulator. Attach the cutting or welding torch to the hoses. Finally, select the proper size torch tip and install it into the torch or torch tube.
- The cylinder, regulator, and torch valves must be turned on in a given sequence in order for the oxyfuel cutting or welding outfit to be used safely. **First, inspect the equipment for damage and to make sure the regulators are closed.** Open the acetylene cylinder valve about 1/4 to 1/2 turn. Slowly open the oxygen cylinder valve all the way. Open the acetylene torch valve one half to one complete turn. Set the correct working pressure on the acetylene regulator gauge by turning the acetylene regulator screw. Close the acetylene torch valve. Check the acetylene regulator for leaks. Open the oxygen torch valve one half to one turn. (For cutting torches, the cutting torch oxygen valve (lever) should also be open while the pressure is set.) Set the desired working pressure by turning the oxygen regulator adjusting screw. Close the oxygen torch valve (and release the oxygen cutting lever). Check the oxygen regulator for leaks. The torch is now ready to light.
- An oxyfuel gas cutting or welding outfit should be checked for leaks each time it is turned on. External leaks may be found using soapsuds. Both internal and external leaks may be located by watching the low-pressure gauge.
- A carburizing flame is produced when too little oxygen is present. Three flame areas are visible. The end of the inner flame is ragged or rough.
- A neutral flame results from the correct balance of oxygen and acetylene. It is used for most welding and cutting operations. There are two flame areas. The end of the inner cone is smooth and shaped like a bullet.
- An oxidizing flame is produced when too much oxygen is present. Two flame areas are seen. The end of the inner flame becomes sharply pointed.
- To light a cutting torch, begin by setting the correct working pressures. Point the torch downward and open the acetylene torch valve 1/16 turn. Light the preheating flames with a spark lighter. Continue opening the acetylene torch valve until the smoking stops. With the cutting oxygen valve closed, open the oxygen torch valve until a neutral flame is obtained. Open the cutting oxygen valve (lever). Adjust the torch oxygen valves until neutral preheating flames remain while the cutting oxygen valve is open.
- To shut down a cutting torch, begin by turning off the acetylene torch valve. Next, turn off the oxygen torch valve. Then, turn off the acetylene cylinder valve. Turn off the oxygen cylinder valve next. Reopen the acetylene and oxygen torch valves. When all gauges read zero, close both torch valves. Finally, close both regulators by turning the regulator adjusting screws counterclockwise until they feel loose. (Close the oxygen valve next to the cutting oxygen lever if a cutting attachment is being used.)

(continued)

- To light and adjust a welding torch, open the acetylene valve approximately 1/16 of a turn. Light the acetylene at the torch tip. Open the acetylene torch valve until the flame becomes turbulent about 3/4″–1″ (19–25 mm) from the end of the tip. Adjust the acetylene so the flame no longer smokes or releases soot. Open the oxygen torch valve slowly until a neutral flame is obtained.
- To shut down a welding torch, begin by closing the acetylene torch valve. Next, close the oxygen torch valve. Completely close the oxygen and acetylene cylinder valves. Open the oxygen and acetylene torch valves. Close the torch valves when the gauges read zero. Hang up the torch. Turn both regulator adjusting screws counterclockwise until they feel loose.

Review Questions

Write your answers on a separate sheet of paper. Please do *not* write in this book.

1. How are the cylinder valve outlets cleaned before installing the regulators?
2. The _____ must be closed before the cylinder valves are opened.
3. The regulator is closed when the adjusting screw feels _____.
4. Why must you stand to the side of the regulators while opening the cylinder valves?
5. The torch valve must be _____ before adjusting the working pressure on the regulator.
6. The _____ and _____ are checked for leaks when the welding outfit is turned on and the regulator and torch valves are closed.
7. What type of soap must *never* be used when checking oxygen fittings?
8. How far should you open the acetylene torch valve when lighting the torch?
9. Describe the appearance of a neutral flame.
10. Describe the characteristics of an oxidizing flame.

This image shows oxygen and acetylene regulators properly attached to the cylinders. The left-hand gauge on each regulator displays the working pressure. The right-hand gauge displays the cylinder pressures. You can tell by looking at the gauges that the cylinder valves are open and the regulators are closed. (Thermadyne Industries, Inc.)

Chapter 22

Oxyfuel Gas Cutting

After studying this chapter, you will be able to:

- List fuel gases that are used for oxyfuel gas cutting.
- Turn on and adjust manual oxyfuel gas cutting equipment.
- Perform cuts manually with a cutting torch or cutting torch attachment.
- Describe the procedure for shutting down oxyfuel gas cutting equipment.
- Identify the basic types of cutting machines.
- Turn on and adjust oxyfuel gas cutting machines.
- Perform cuts with an oxyfuel gas cutting machine.
- Describe the procedure for shutting down oxyfuel gas cutting machines.

Technical Terms

bell-mouthed kerf
burning
cutting machine
electric motor-driven carriage
electronic pattern tracer
hard slag
ignition temperature
kerf
motor-driven beam-mounted torch
motor-driven magnetic tracer
oxyfuel gas cutting (OFC)
preheated
slag

Oxyfuel Gas Cutting Principles

Oxyfuel gas cutting (OFC) is a process used to cut metal by rapidly oxidizing it. The heat of a gas flame and pressurized pure oxygen are used in this process. Most materials, including steel, will burn. Oxyfuel gas cutting is also referred to as ***"burning"*** or "flame cutting" by some welders.

If the temperature of paper is raised to its ignition temperature, the paper will burn. The ***ignition temperature*** is the temperature at which a material will burn if enough oxygen is present. Oxygen must be present for burning to occur. When paper is burned, the oxygen comes from the air.

The ignition temperature for steel is 1500°F (816°C). One or more oxyfuel gas flames are used to heat steel to this temperature when oxyfuel gas cutting. Steel is a bright red or orange-red color at this temperature. When the steel attains a temperature of 1500°F (816°C), a jet stream of high-purity (99% +) oxygen is directed at it. This stream of oxygen rapidly oxidizes, or burns, the steel. A continuous cut is made on the steel if the 1500°F (816°C) temperature and oxygen jet are maintained.

Various gases are used as the fuel for oxyfuel gas cutting. These gases include acetylene, hydrogen, natural gas, propane, and MPS (methylacetylene-propadiene). Acetylene and MPS are the most commonly used fuel gases in industry.

Oxyfuel Gas Cutting Equipment

Oxyfuel gas cutting equipment is essentially the same as oxyfuel gas welding equipment. The main differences in the cutting and welding outfits are in the oxygen regulator and the torch. A larger volume of oxygen at a higher pressure is needed for cutting very thick steel. An oxygen regulator with a larger volume capacity and higher-pressure indications is needed for cutting heavy metal.

Preparing to Cut

The step-by-step procedure for assembling an oxyfuel gas cutting outfit is explained in Chapter 21. Select either a cutting torch or a cutting torch attachment and connect it to the hoses. Check the cutting station to ensure it is assembled correctly and all gas connections are tight and leakproof. Be sure to select the proper size cutting tip when preparing to cut. Cutting tip sizes are identified by numbers ranging from 00–8. The numbers are stamped on the tip when they are made. Most cutting tip manufacturers use the same tip numbering system. **Figure 22-1** suggests cutting oxygen and acetylene pressures for use with various cutting tips. Suggested tip sizes for use with various metal thicknesses are also shown.

Proper protective clothing must be worn for oxyfuel gas cutting. Coveralls and a shirt or jacket that buttons at the collar must be worn. **All clothing must be flame resistant. Pant legs should have no cuffs or folds. A cap should be worn to protect hair from sparks and hot metal. Gloves must be worn to protect the hands. Welding goggles with a #5–#6 filter lens should be worn.**

Review Chapter 20 for a thorough overview of oxyacetylene cutting and welding equipment and supplies. Read Chapter 1 to review general safety information. Also, review the procedures listed in Chapter 21 for turning on and shutting down an oxyacetylene cutting outfit. Note the differences in the procedures used for the welding torch and the cutting torch attachment.

Manual Cutting

The edges of metal produced by oxyfuel gas cutting should be smooth and straight. A good cut depends on the welder's skills. Cut quality is also affected by the cutting tip size and the oxygen pressure used. Very accurate circles and straight lines can be cut manually by using a circle cutting attachment, **Figure 22-2**, or a straight line cutting attachment, **Figure 22-3**.

Slag is iron oxide. It may form on the underside of the metal being cut. **Figure 22-4** shows a plate that was turned over to reveal the slag. Slag that is easily removed is acceptable. ***Hard slag*** (slag that cannot be easily removed) is unacceptable. Enough oxygen pressure must be used to cut through the metal without leaving hard slag.

The slot or opening produced in the metal when cutting is the ***kerf***. See **Figure 22-5**. The kerf should be as narrow as possible. A wide kerf requires more oxygen to make a good cut. The width of the kerf is determined by the size and shape of the cutting tip. Too much oxygen pressure may result in a ***bell-mouthed kerf***, **Figure 22-6**.

Material thickness, inches	1/8	1/4	1/2	3/4	1	1 1/2	2	4	5	6	8	10	12
Recommended tip number	00	0	1	1	2	2	3	3	4	6	6	7	8
Oxygen pressure setting, psig	20-25	25-30	30-35	30-35	35-40	40-45	40-45	40-50	45-55	45-55	45-55	45-55	45-55
Acetylene pressure, setting, psig	3-5	3-5	3-5	3-5	3-7	3-7	5-10	5-10	6-12	7-13	8-14	10-15	10-15
Cutting speed range, in./min.	27-30	26-29	20-24	17-21	14-18	13-17	12-15	8-11	7-9	6-8	5-6	4-5	3-4

Figure 22-1. *Suggested tip sizes for cutting various metal thicknesses with a positive pressure torch. The approximate oxygen and acetylene gas pressures are listed. Also shown is the correct cutting speed. (Goss, Inc.)*

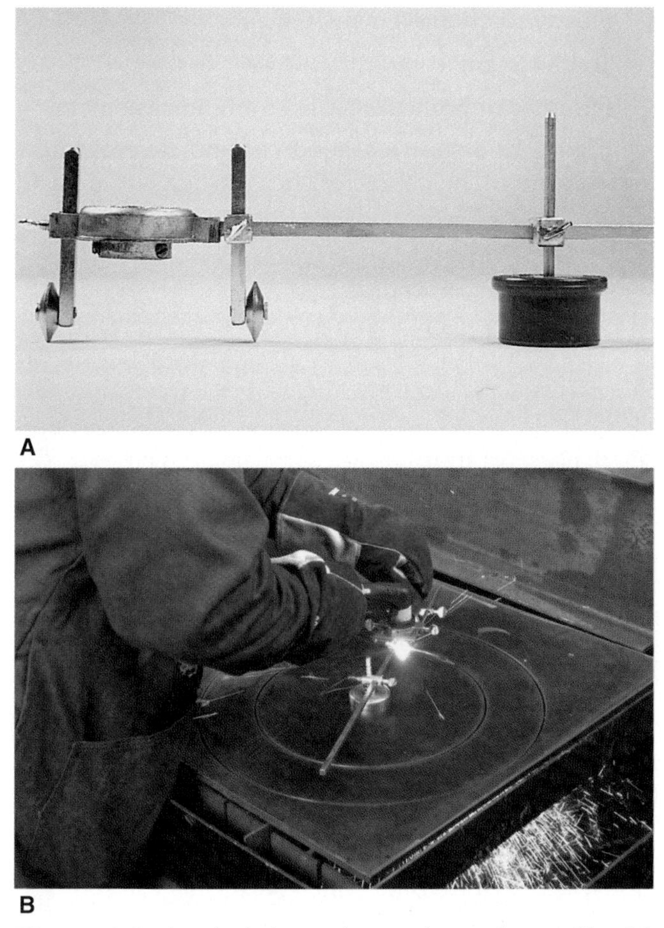

Figure 22-2. A—A circle cutting torch attachment like this one allows welders to turn a torch accurately for precise circular cuts. (Lenco-NLC, Inc.) B—Many circle-cutting guides can be used with different types of cutting torches. (Victor, a division of Thermadyne Industries, Inc.)

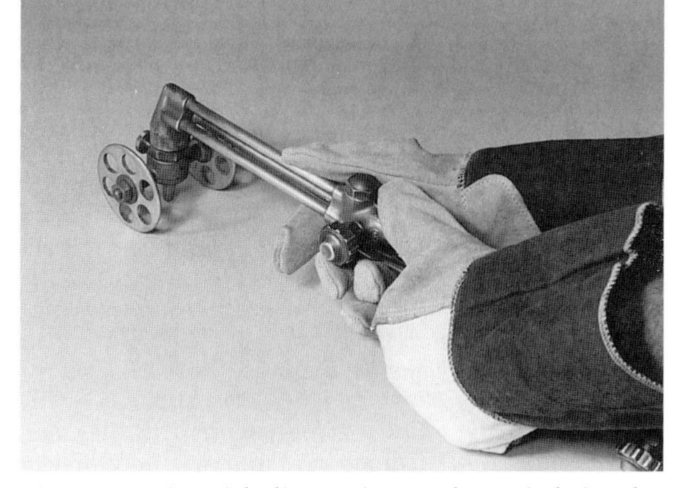

Figure 22-3. A straight-line cutting attachment is designed to hold the torch tip steady for the length of the cut. It can be used freehand or with a straightedge.

Figure 22-4. One edge has a very clean cut with very little slag, while the other side has considerable slag.

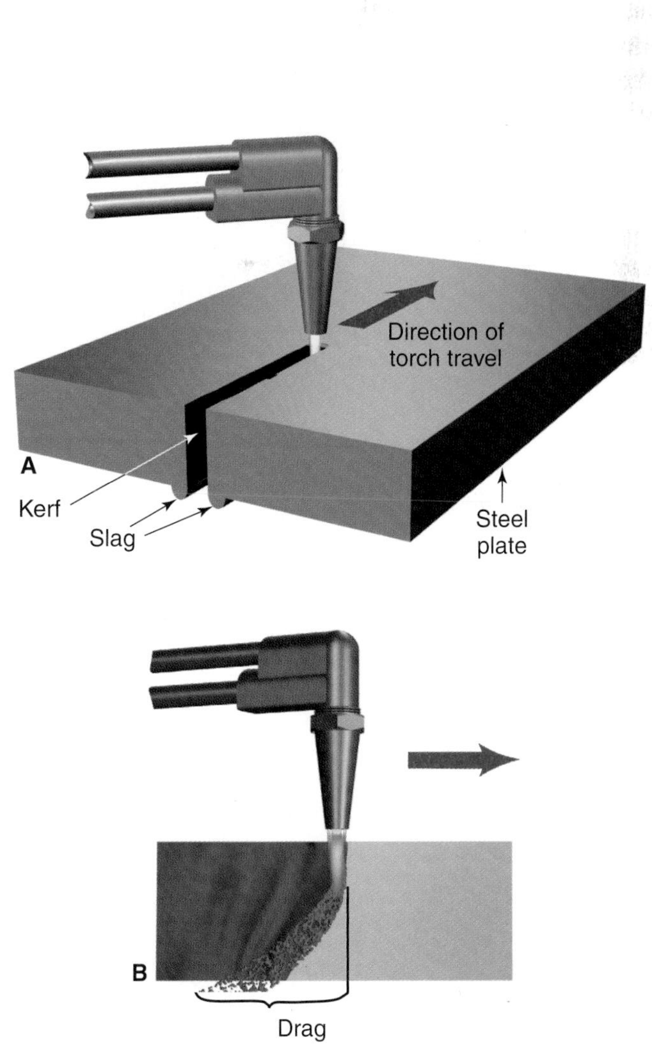

Figure 22-5. An oxyacetylene cut in progress. A—Note the slag at the bottom of the kerf. B—Cutting oxygen slows down as it travels through the metal. Drag results from the oxygen slowing down and the forward motion of the torch.

When oxyfuel gas cutting, the base metal must be ***preheated*** (heated to its ignition temperature) before cutting oxygen is applied. The ignition temperature for steel is about 1500°F (816°C). The small orifices in the cutting tip supply the preheating flames. These flames preheat the metal ahead of the oxygen stream and also heat the sides of the kerf. The cutting tip should be placed into the torch so that it aligns with the cutting line as shown in **Figure 22-7**. A minimum of two preheating orifices should line up with the cutting line. A kerf is formed under the cutting oxygen jet. The quality of the kerf depends on the following:

- Cutting tip size.
- Oxygen pressure.
- Torch forward speed.
- Steady torch movement.
- Torch tip angle.
- Distance of preheat flames from the base metal.

Figure 22-8 shows a good cut and several unacceptable cuts. The characteristics of each cut are also given.

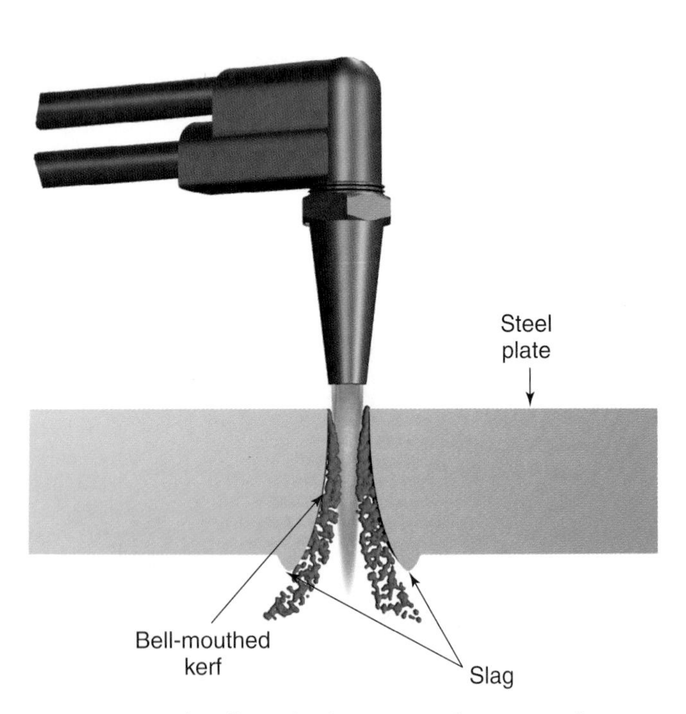

Figure 22-6. The effect of using too much oxygen when cutting steel. Notice how the kerf widens at the bottom of the plate to create a bell-mouthed kerf.

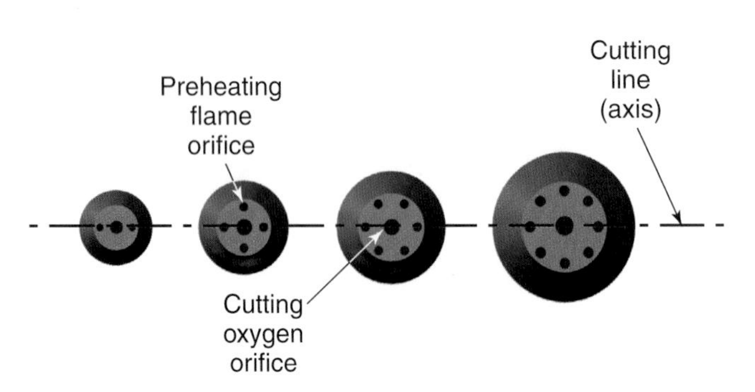

Figure 22-7. Correct alignment of the preheating orifices on the cutting line. At least two preheating orifices should align with the cutting line.

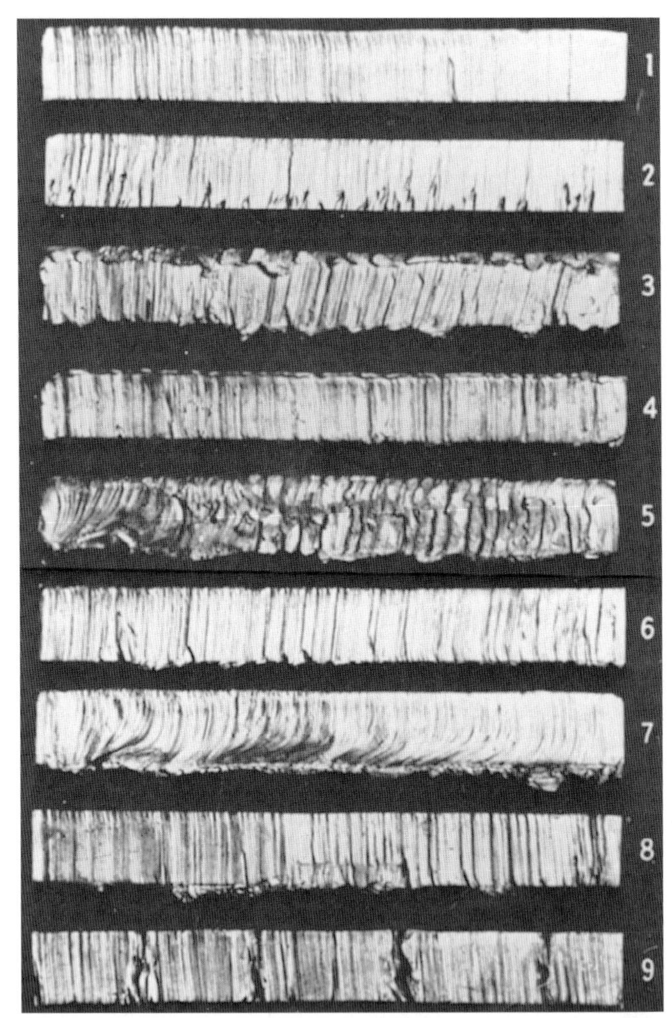

Figure 22-8. Typical edge conditions resulting from oxyfuel gas cutting operation: (1) A good cut in 1" (25 mm) plate. The edge is square, and the drag lines are essentially vertical and not too pronounced. (2) The preheat flames were too small for this cut and the cutting speed was too slow, causing bad gouging at the bottom. (3) The preheating flames were too long, with the result that the top surface melted over. The cut edge is irregular and there is an excessive amount of adhering slag. (4) The oxygen pressure was too low, with the result that the top edge melted over because of the slow cutting speed. (5) The oxygen pressure was too high and the nozzle size too small, with the result that control of the cut was lost. (6) The cutting speed was too slow, with the result that the irregularities of the drag lines are emphasized. (7) The cutting speed was too fast, resulting in a pronounced break in the drag line and an irregular cut edge. (8) The torch travel was unsteady, resulting in a wavy and irregular cut edge. (9) The cut was lost and not carefully restarted, causing bad gouges at the restarting point. (AWS)

Torch Position

The torch angle for cutting will vary with the thickness of the metal. The centerline of the cutting tip should be perpendicular to the surface of the metal to form square edges.

The cutting tip should be tilted backward 5°–20° from vertical when cutting with the correct size tip, **Figure 22-9**. This allows the welder to see into the kerf.

As the thickness of the metal increases, the cutting tip should be held closer to vertical and not tilted backward. When cutting very thick metal, the work angle should approach 0° so that the cutting tip is nearly vertical, **Figure 22-10**.

When cutting very thin metal (under 1/8″ [3.2 mm] thick), even the smallest tip size available may produce too much heat for a quality cut. In this case, lower the torch to 15°–20° from horizontal. This angle increases the amount of material being cut and allows a larger tip to produce a clean cut, **Figure 22-11**.

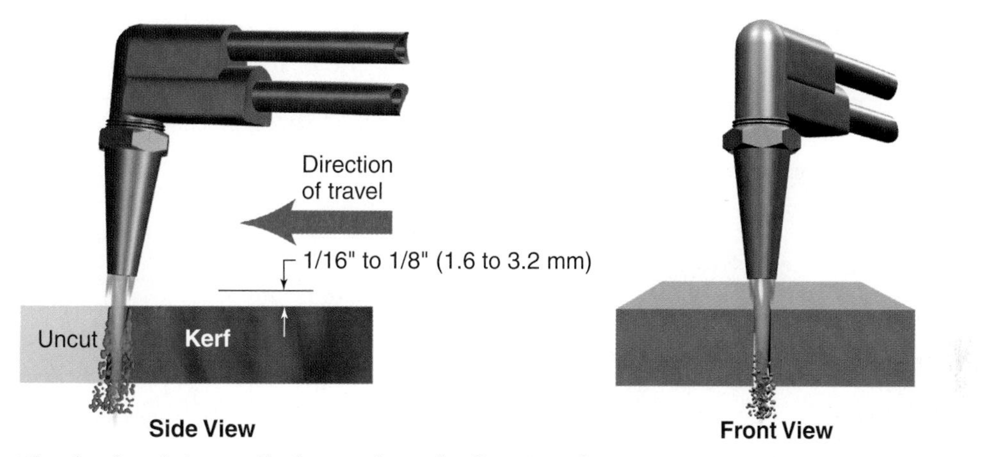

Figure 22-9. *The cutting head and tip are tilted away from the direction of cutting at a 5°–20° angle from vertical.*

Figure 22-10. *A student is practicing oxyfuel gas cutting on thick 1 1/2″ (38 mm) mild steel.*

Hand Position

A cutting torch must be held with two hands. As the cut progresses, the torch is slid through the support hand, as shown in **Figure 22-12**. A 4″–8″ (102 mm–203 mm) cut may be made before moving the support hand.

A piece of steel angle (angle iron) may be used as a guide to make straight or beveled cuts, **Figure 22-13**. The angle may be clamped to the part for use as a straight edge. The torch tip is slid along the angle.

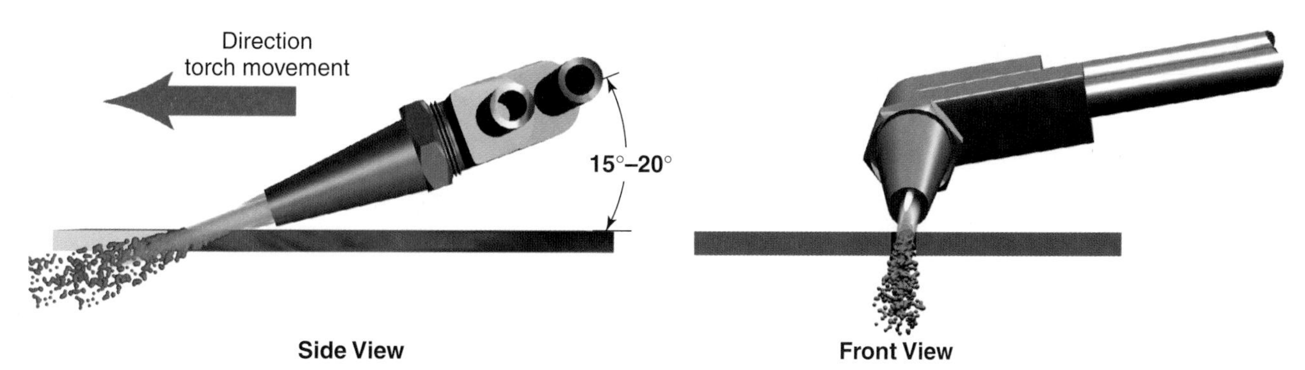

Figure 22-11. *Recommended procedure for cutting thin steel. Notice that the two preheat flames on this tip are in line with the kerf (cut line).*

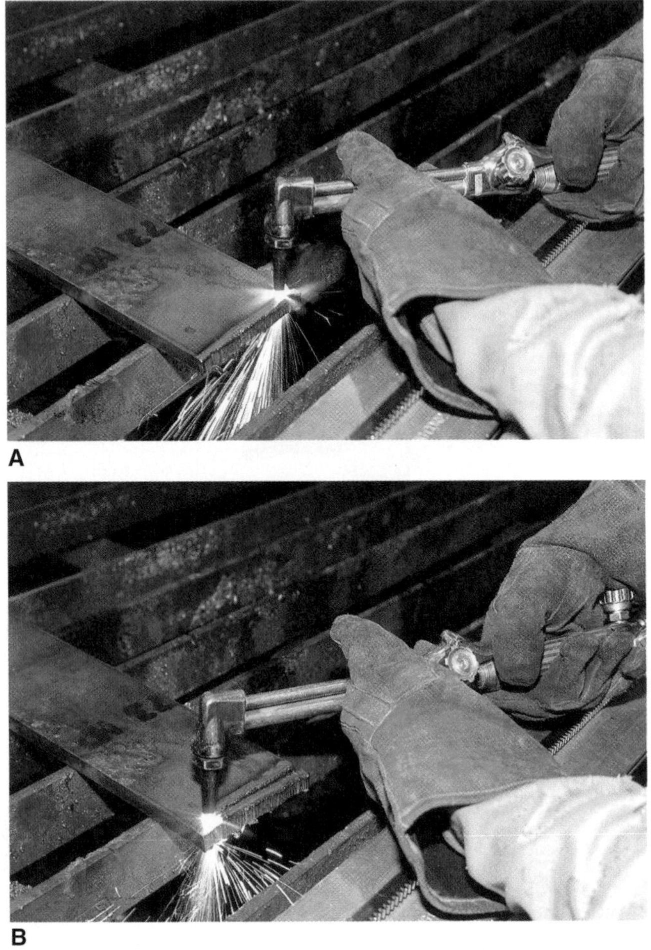

Figure 22-12. Guiding the cutting torch through the left hand. A—Left hand is held near the torch head. B—Torch is slid through left hand as cut progresses.

Figure 22-13. A steel angle (angle iron) used as a guide to make a beveled cut.

Exercise 22-1 Manually Cutting Straight Edges–Freehand

1. Obtain a piece of mild steel that measures at least 1/4″ × 6″ × 6″ (6.4 mm × 150 mm × 150 mm).
2. Using soapstone or chalk, mark out three lines on the surface 1 1/2″ (38 mm) apart. Refer to the following figure.

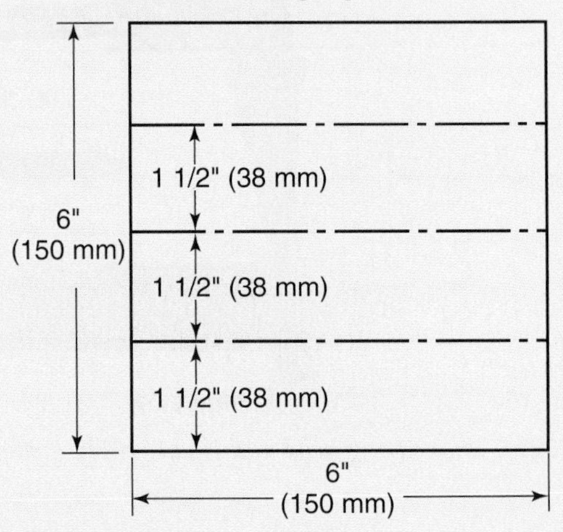

3. Select the correct cutting tip. Use a tip with 4–6 preheating orifices, if possible. Install the tip in the torch.
4. Turn on the station and set the correct pressures. Light the torch and adjust for neutral preheating flames.
5. Place your left hand on the table. Rest the torch in the left hand. As the cut progresses, slide the torch through your left hand. This prevents the torch from shaking.
6. For a right-handed welder, start the cut at the right and progress toward your left.
7. Hold the tip 5°–20° from vertical. The centerline of the tip should be aligned with the line to be cut. The preheating flames should be about 1/16″–1/8″ (1.6 mm–3.2 mm) above the metal.
8. Squeeze the oxygen cutting valve when the metal turns an orange-red color (1500°F/816°C). The cut will begin.
9. Progress at a constant rate to complete the cut.
10. Repeat the process for the next two lines.

Exercise 22-2 Manually Cutting Straight Edges–Using Steel Angle

1. Obtain a piece of mild steel that measures at least 1/4″ × 6″ × 6″ (6.4 mm × 150 mm × 150 mm).
2. Using soapstone or chalk, mark out three lines on the surface 1 1/2″ (38 mm) apart. Refer to the following figure.

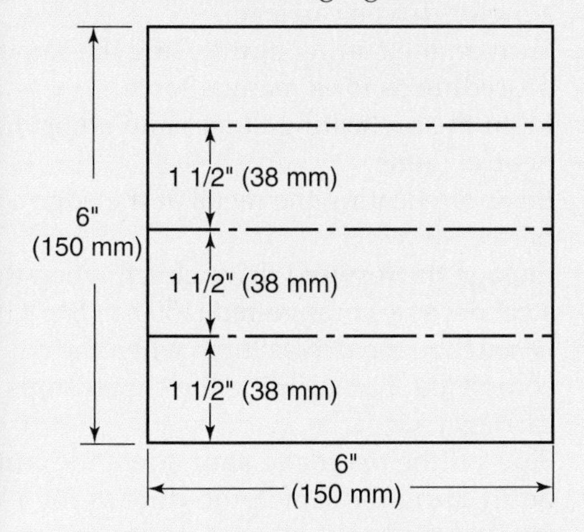

3. Select the correct cutting tip. Use a tip with 4–6 preheating orifices, if possible. See Figure 20-29.
4. Turn on the station and set the correct pressures. Light the torch and adjust for neutral preheating flames.
5. Obtain a piece of angle iron that measures at least 1″ × 1″ × 7″ (25 mm × 25 mm × 178 mm).
6. Lay one leg of the angle flat, with other leg perpendicular to the base metal. Position the angle just off the line to be cut to allow for the tip thickness. Clamp the angle firmly in place.
7. Hold the tip 5°–15° from vertical. The centerline of the tip should be aligned with the line to be cut. The preheating flames should be about 1/16″–1/8″ (1.6 mm–3.2 mm) above the metal.
8. Squeeze the oxygen cutting valve when the metal turns an orange-red color (1500°F/816°C). The cut will begin.
9. Slide the cutting torch tip along the steel angle and progress at a constant rate to complete the cut.
10. Repeat the process for the next two lines.

Exercise 22-3 Manually Cutting Beveled Edges–Using Steel Angle

1. Obtain a piece of mild steel that measures at least 1/4″ × 6″ × 6″ (6.4 mm × 150 mm × 150 mm).
2. Using soapstone or chalk, mark out three lines on the surface 1 1/2″ (38 mm) apart. Refer to the following figure.

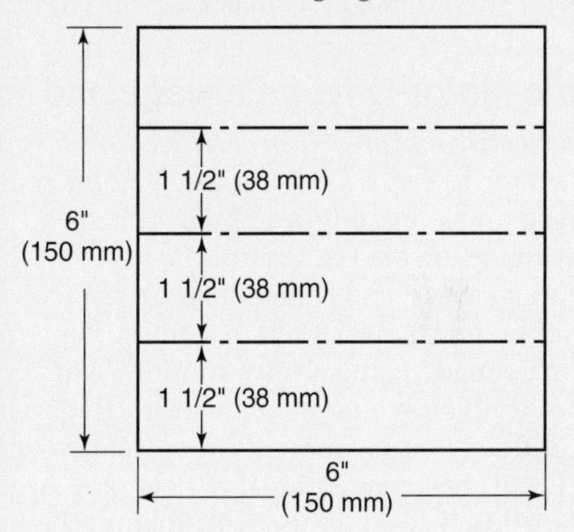

3. Select the correct cutting tip. Install the tip in the torch.
4. Turn on the station and set the correct pressures. Light the torch and adjust for neutral preheating flames.
5. Obtain a piece of angle iron that measures at least 1″ × 1″ × 7″ (25 mm × 25 mm × 178 mm).
6. Position the steel angle on its legs, as in Figure 22-13. Clamp the angle in place.
7. Hold the cutting torch tip on the incline of the clamped steel angle. The centerline of the tip should be aligned with the line to be cut. The preheating flames should be about 1/16″–1/8″ (1.6 mm–3.2 mm) above the metal.
8. Squeeze the oxygen cutting valve when the metal turns an orange-red color (1500°F/816°C). The cut will begin.
9. Progress at a constant rate to complete the cut.
10. Repeat the process for the next two lines.

Cutting Machines and Pattern Tracers

Oxyfuel gas ***cutting machines*** make high-quality cuts at a faster rate than manual flame cutting. Cutting machines are often used to make long cuts. They are also used to cut multiple pieces with close tolerances. The basic types of cutting machines are:

- Electric motor-driven carriage and track.
- Motor-driven beam-mounted torch and electronic tracer.
- Motor-driven magnetic tracer and torch.

Electric Motor-Driven Carriage and Track

An ***electric motor-driven carriage*** uses a variable-speed motor to carry a cutting torch along a straight or curved track. See **Figure 22-14**. The speed of the motor can be adjusted to control the speed of the cut. Torch movement and flame height are consistent because the torch moves along a track. High-quality cuts can be made using a track-mounted carriage.

The cutting torch used on a motor-driven carriage has a different appearance from a manual cutting torch. It still has conventional oxygen and acetylene valves. A small lever is used to operate the cutting oxygen valve. Using the rack-type gear on the cutting machine allows the welder to adjust the flame height by moving the torch up and down. The torch can also be adjusted horizontally to align it with the cutting line.

Only one cutting torch is usually mounted on the carriage. Torch angle, flame height, and cutting speed can be adjusted on the carriage. A clutch switch is used to engage and disengage the carriage drive mechanism.

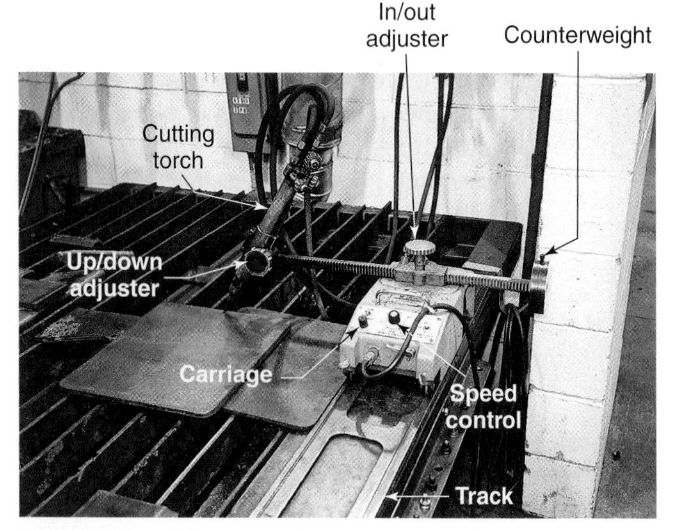

Figure 22-14. The parts of a motor-driven carriage.

An oxyfuel gas cutting torch mounted on a motor-driven carriage is used as follows:

1. Set up the track or circle cutting attachment.
2. Adjust the torch height so the flames are about 1/16″–1/8″ (1.6 mm–3.2 mm) above the base metal. Adjust the torch to the appropriate angle for the metal being cut. Refer to the *Torch Position* subhead under *Manual Cutting*. Adjust the forward speed for the carriage. Figure 22-1 suggests forward speeds to be used with various metal thicknesses. This speed may be increased, when possible, with a motor-driven carriage.
3. Turn on the cutting outfit using the same procedure as for a manual torch.
4. Light the preheating flames and adjust for a neutral flame.
5. Begin preheating the metal at the edge of the part.
6. Engage the forward drive clutch when the steel becomes orange-red (1500°F/816°C).
7. Disengage the drive clutch when the cut is completed. Disengaging the clutch stops the carriage.
8. Shut off the torch and shut down the cutting outfit using the same procedure as for a manual torch.

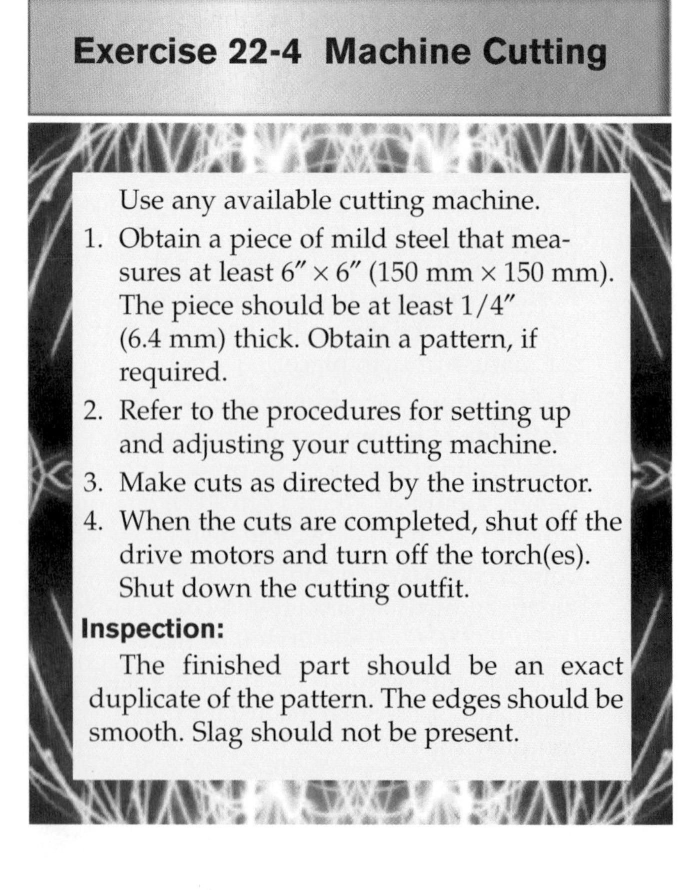

Exercise 22-4 Machine Cutting

Use any available cutting machine.

1. Obtain a piece of mild steel that measures at least 6″ × 6″ (150 mm × 150 mm). The piece should be at least 1/4″ (6.4 mm) thick. Obtain a pattern, if required.
2. Refer to the procedures for setting up and adjusting your cutting machine.
3. Make cuts as directed by the instructor.
4. When the cuts are completed, shut off the drive motors and turn off the torch(es). Shut down the cutting outfit.

Inspection:

The finished part should be an exact duplicate of the pattern. The edges should be smooth. Slag should not be present.

Motor-Driven Beam-Mounted Torches and Electronic Tracer

The oxyfuel gas cutting torch is mounted on a strong beam. Several cutting torches may be mounted on the same beam. This allows the ***motor-driven beam-mounted torch*** machine to cut several parts at the same time. See **Figure 22-15**.

A dense black drawing of the part to be cut is made on a white background. The outline of the drawing is followed by the electronic eye on a motor-driven ***electronic pattern tracer***. As the pattern shape changes, the electric eye follows the edge of the pattern. It will not stray more than .003″ (.076 mm) from the drawing image.

The pattern tracer is electrically connected to two drive motors. As the tracer moves along the outline, it signals the drive motors on the beam. The motors move the beam on which the torches are mounted. All moves that the tracer makes are duplicated by all the torches on the beam.

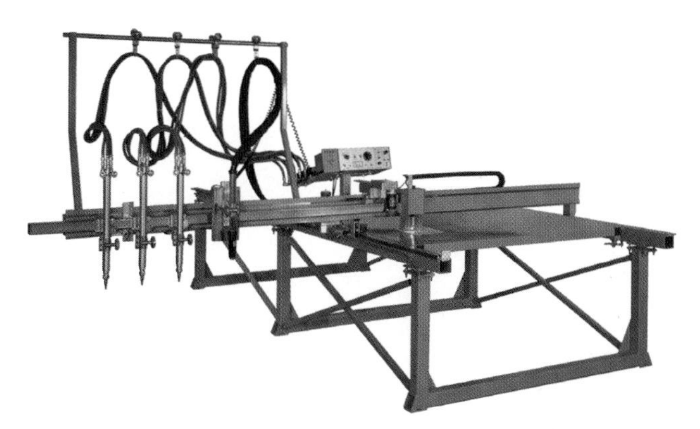

Figure 22-15. A cantilever-style mechanized shape-cutting system. (C&G Systems™, a division of Thermadyne®)

The beam-mounted torch(es) and electronic tracer are used as follows:

1. Place the drawing on the pattern table under the electronic pattern tracer.
2. Turn on the cutting outfit using the same procedure as for manual cutting.
3. Adjust the light beam on the electronic tracer to the line on the drawing.
4. Light and adjust each torch separately. The flame should be 1/16″–1/8″ (1.6 mm–3.2 mm) above the base metal. The torches should also be perpendicular to the surface of the metal. The drive motor speed is adjusted on the electronic tracer control panel.
5. Turn on the cutting oxygen at each torch and pierce a hole in the metal.
6. Start the beam drive motors.
7. Start the tracer. The torches move as the tracer moves.
8. Shut off the tracer and drive motors when the cut is completed. Turn off all the torches.
9. Shut down the cutting outfit in the same manner as for manual cutting.

Motor-Driven Magnetic Tracer and Torch

The ***motor-driven magnetic tracer*** and torch is a relatively inexpensive cutting machine. This type of cutting machine works best with shapes that are not too complex. The tracer will not follow sharp angles very well. One torch is generally mounted on this type of cutting machine.

A motor-driven magnetic tracer and cutting machine is shown in **Figure 22-16**. A steel pattern must be used when duplicating parts. The pattern is firmly mounted directly above the cutting torch. A cylindrical magnetic tracer, or follower, follows the outline of the steel pattern. The tracer is rotated by a small motor and held in contact with the pattern by magnetism. The tracer is mounted directly above and on the exact centerline of the cutting torch. Therefore, the torch follows the movement of the tracer as it moves around the pattern. The part that is cut out will be an exact duplicate of the pattern. Parts cut by this method are not as accurate as those cut with an electronic tracer.

The magnetic motor-driven tracer and torch are used as follows:

1. Mount the steel pattern in position.
2. Place the magnetic tracer in contact with the pattern.
3. Adjust the torch angles.
4. Using the same procedure as for manual cutting, light the torch and adjust the torch height. The tip of the preheat flames should be 1/16″–1/8″ (1.6 mm–3.2 mm) above the base metal.
5. Turn on the machine and start the magnetic tracer motor. The tracer rotates and follows the pattern. The torch follows the tracer movement.
6. After the cut is completed, shut off the tracer motor. Shut down the cutting torch and cutting outfit.

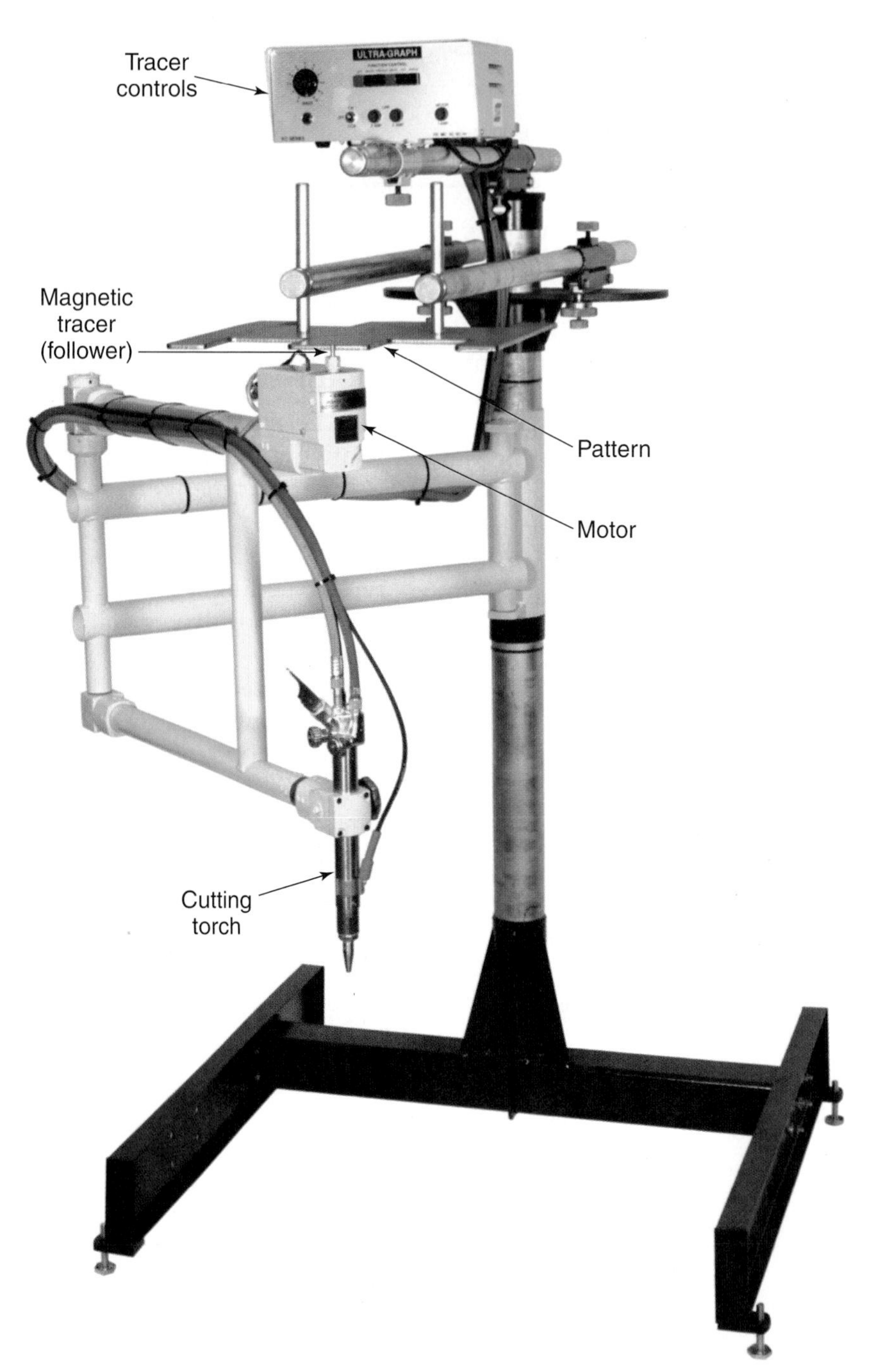

Figure 22-16. *The magnetic pattern tracer is mounted above the cutting torch. As the follower rotates and follows the pattern, the cutting torch travels a duplicate path over the metal. (ESAB Welding and Cutting Products)*

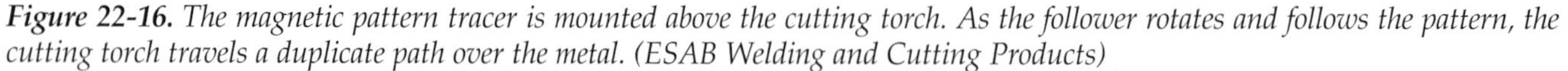

Summary

- Oxyfuel gas cutting (OFC) is a process used to cut metal by rapidly oxidizing or burning it. The heat of a gas flame and pressurized pure oxygen are used in this process.
- The ignition temperature for steel is 1500°F (816°C). When the steel attains this temperature, a jet stream of high-purity (99% +) oxygen is directed at it. This stream of oxygen rapidly burns away all of the steel in the kerf (the slot that is produced during cutting).

(continued)

- Various gases are used as the fuel for oxyfuel gas cutting. These gases include acetylene, hydrogen, natural gas, propane, and MPS (methylacetylene-propadiene). Acetylene and MPS are the most commonly used fuel gases in industry.
- The edges of metal produced by oxyfuel gas cutting should be smooth and straight. A good cut depends on the welder's skills. Cut quality is also affected by the cutting tip selected and the oxygen pressure used.
- When oxyfuel gas cutting, the base metal must be preheated before cutting oxygen is applied. The small orifices in the cutting tip supply the preheating flames. The cutting tip should be installed in the torch so that minimum of two preheating orifices line up with the cutting line.
- The torch angle for cutting will vary with the thickness of the metal. The centerline of the cutting tip should be perpendicular to the surface of the metal to form square edges. The cutting tip should be tilted backward 5°–20° from vertical when cutting with the correct size tip. As the thickness of the metal increases, the cutting tip should be held closer to vertical. When cutting very thin metal, lower the torch angle to 15°–20° from horizontal to increase the thickness of the material being cut.
- A cutting torch must be held with two hands. As the cut progresses, the torch is slid through the support hand.
- Oxyfuel gas cutting machines make high-quality cuts at a faster rate than manual flame cutting. An electric motor-driven carriage uses a variable-speed motor to carry a cutting torch along a straight or curved track.
- A motor-driven beam-mounted torch consists of one or more torches mounted on a strong beam and is capable of cutting multiple parts at one time. A motor-driven electronic pattern tracer allows the beam-mounted torches to cut parts by following a predrawn pattern.
- The motor-driven magnetic tracer and torch is a relatively inexpensive cutting machine that follows a steel pattern to cut parts.

Review Questions

Write your answers on a separate sheet of paper. Please do *not* write in this book.

1. Oxyfuel gas cutting is also known as _____ or _____ in the trade.
2. What color is steel at its ignition temperature? What is its ignition temperature?
3. Name four fuel gases that may be used in oxyfuel gas cutting.
4. What is the minimum number of preheat orifices that should line up with the cutting line?
5. The torch tip is normally held at a _____°–_____° angle from vertical.
6. When cutting metal under 1/8″ (3.2 mm) thick, the torch should be lowered to _____°–_____° from horizontal.
7. Explain how and why manual oxyfuel gas cutting is a two-handed operation.
8. List three devices used to guide a cutting machine along a line or pattern.
9. *True or False?* As the thickness of the metal increases, the cutting tip should be angled further from perpendicular to the surface.
10. Sketch a bell-mouthed kerf.

Many people think of oxyfuel equipment as unavoidably cumbersome. This cutting and welding outfit utilizes small-capacity cylinders, making it highly portable. (Hobart Brothers Company)

Chapter 23

Oxyfuel Gas Welding: Flat Welding Position

Learning Objectives

After studying this chapter, you will be able to:

- Identify the proper protective clothing that must be worn for oxyfuel gas cutting or welding.
- Identify the four positions used in welding.
- Explain how to hold a torch when forehand welding.
- Explain how to hold a torch when backhand welding.
- Carry a weld pool along a weld joint.
- Weld an edge joint without a welding rod.
- Weld a corner joint without a welding rod.
- Weld a flanged butt joint without a welding rod.
- Select a welding rod.
- Lay a weld bead along a weld joint using a welding rod.
- Lay a fillet weld on a lap joint using a welding rod.
- Lay a fillet weld on a T-joint using a welding rod.
- Weld a butt joint using a welding rod.
- Identify weld defects.

Technical Terms

backhand welding
carrying a weld pool
concave weld bead
convex weld bead
creating a continuous weld pool
C-shaped weld pool
downhand welding
drag angle
flat weld bead
forehand welding
freeze
inclusions
keyhole
keyhole welding
laying a weld bead
leading edge
overlap
push angle
travel angle
undercutting
weld pool
work angle

Flat Position Welding

The four basic positions used in welding are the flat, horizontal, vertical, and overhead positions. In flat position welding, the welder's hands and elbows can normally rest on the working surface. This enables the welder to hold the torch and welding rod steady. Welding in other positions often requires the hands and arms to be held above the shoulders, which creates fatigue. As a result, mistakes can more easily occur. Flat position welding, or ***downhand welding*** (a nonstandard term), is the easiest and most efficient position. If the parts to be welded can be moved and aligned, flat position welding is always the position of choice.

Preparing to Weld

The same protective clothing must be worn for oxyfuel gas welding as is worn for oxyfuel gas cutting. See Chapters 20 and 22. Welding goggles with a #4–#6 filter lens should be worn.

Review the safety precautions for oxyfuel gas welding as listed in Chapter 20. Check the oxyfuel gas cutting and welding station to be sure it is assembled correctly (review Chapter 21). All gas connections must be tight and leakproof.

Select a welding tip with the proper orifice size. Refer to Figure 20-36 to select a suitable tip for the thickness of the metal being welded.

Turn on the oxyfuel gas welding outfit. Adjust the oxygen and fuel gas pressures for the orifice size being used.

Holding the Torch

The welding torch may be held like a pencil or a hammer, **Figure 23-1**. With lightweight torches, the pencil grip is comfortable and effective. This method is also effective when welding horizontally, vertically, or overhead. Both methods of holding the torch can be used effectively in all welding positions. Practice and experimentation are needed to determine which grip is best in each position. To reduce fatigue and make the torch feel lighter, the hoses may be hung over the welder's arm.

Two methods may be used to direct the flame to the weld area. In the ***forehand welding*** method, the torch tip is held so that the flame is pointing in the direction of travel, **Figure 23-2A.** For a right-handed welder, forehand welding is begun at the right side of the base metal, and the direction of travel moves to the left. The flame melts the base metal and preheats the welding rod. Excess heat is reflected off the base metal.

The backhand welding method may be used when welding thick metal. In the ***backhand welding*** method, the flame is pointed opposite to the direction of travel, **Figure 23-2B**. For a right-handed welder, backhand welding is begun at the left side. The flame

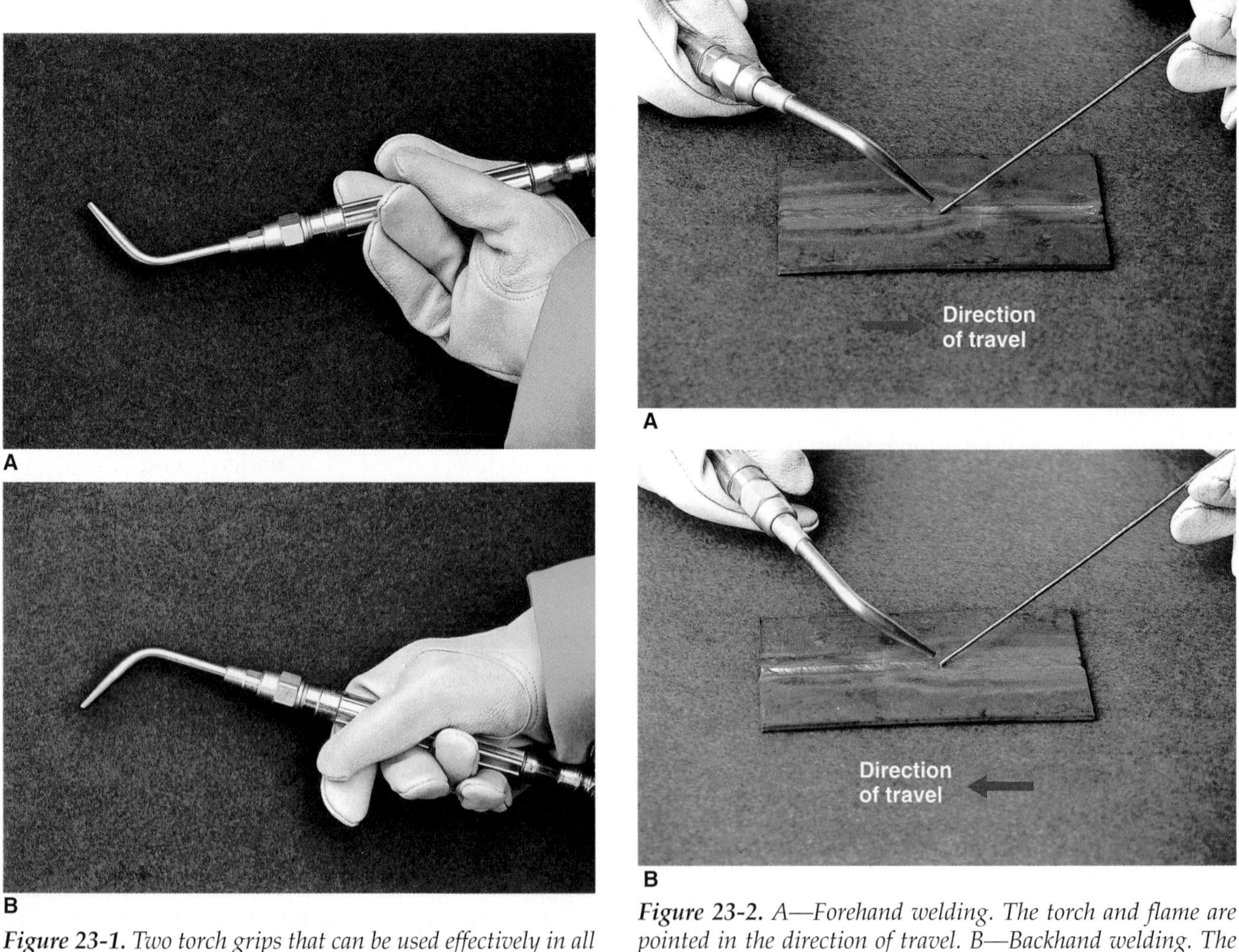

Figure 23-1. *Two torch grips that can be used effectively in all welding positions. A—The torch may be held like a pencil. B—The torch may be held like a hammer.*

Figure 23-2. *A—Forehand welding. The torch and flame are pointed in the direction of travel. B—Backhand welding. The torch and flame are pointed away from the direction of travel. This method is often used on thick metal.*

is pointed toward the beginning of the weld. It is used to create the weld pool, sometimes simply called the "pool." The flame also warms the welding rod. Only a small amount of heat is reflected off the metal and lost. Therefore, more heat is directed at the base metal.

Describing the Torch Angles

Two angles are used to accurately describe the position of the welding torch during a welding procedure. These two angles are called the travel angle and the work angle.

The ***travel angle*** is the angle between a line perpendicular (90°) to the weld axis and the axis of the torch. This angle is measured in a plane determined by the torch or electrode axis and the weld axis. The proper travel angle used in creating a continuous weld pool is 35°–45°, as shown in **Figure 23-3A**. The angle at which the torch is held can also be measured from the surface of the major workpiece to the torch axis. Therefore, the travel angle in Figure 23-3A could also be described as an angle of 45°–55° off the surface.

The ***work angle*** is the angle between a line perpendicular (90°) to the major workpiece and a plane determined by the centerline of the torch tip and the weld axis. When the torch tip is held perpendicular to the major workpiece, the work angle is 0° degrees, **Figure 23-3B**. The angle at which the torch tip is held can also be measured from the surface of the major workpiece to the centerline of the torch tip. In this case, the work angle in Figure 23-3B could be described as an angle of 90° off the surface.

When using the forehand welding method, the torch tip points forward in the direction of travel. The travel angle for forehand welding can also be called the ***push angle***. In backhand welding, the torch tip points back toward the weld bead that has been completed, which is opposite the direction of travel. The travel angle for backhand welding can also be called the ***drag angle***.

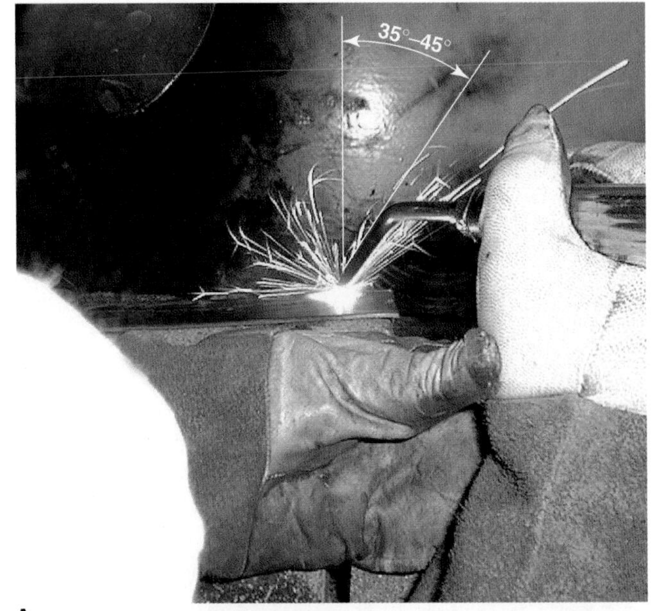

A

B

Figure 23-3. *Creating a continuous weld pool. A—Front view with torch tip held at a 35°–45° travel angle. NOTE: The 35°–45° travel angle may vary, depending on the amount of heat required. B—Left side view with torch tip held at a 0° work angle, which is an angle of 90° off the surface of the workpiece. The 0° work angle keeps the heat equal on both sides of the centerline.*

Creating a Continuous Weld Pool

A ***weld pool*** is the small pool of molten metal that is formed directly below the welding flame. See **Figure 23-4**. ***Creating a continuous weld pool*** is the process of creating a pool and moving it along a line. It provides practice in torch and flame control. Good penetration must be achieved in all welding processes. The amount of penetration cannot always be seen, especially when welding thick metals. Creating a continuous weld pool on thin metal allows the amount of penetration to be seen.

Penetration is controlled by the width of the weld pool. The width of the pool should increase as the thickness of the metal increases. A wider pool results in deeper penetration on the same thickness of metal.

The weld pool must be wide enough to permit the molten metal to sag slightly. The deformation caused by this sag should be visible on the underside of the metal after the weld pass is completed. This deformation shows that total penetration has occurred. The deformation also indicates whether the flame and weld pool have been properly controlled.

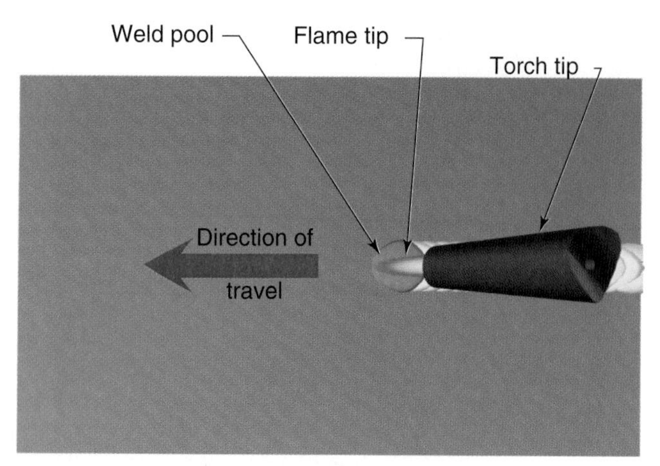

Figure 23-4. The weld pool is shown from above.

Exercise 23-1 Creating a Continuous Weld Pool

1. Obtain a piece of mild steel measuring 1/16″ × 3″ × 6″ (1.6 mm × 75 mm × 150 mm).
2. Clean both surfaces of the metal with steel wool or abrasive paper.
3. Using chalk or a soapstone, mark out five straight lines on the metal.
4. Select the correct tip size and install the tip in the torch. Refer to Figure 20-36.
5. Set the desired welding pressures.
6. Light the torch and adjust to a neutral flame.
7. Begin the pool 1/2″ (12.7 mm) from the edge of the metal. Hold the flame about 1/16″–1/8″ (1.6 mm–3.2 mm) from the metal. Hold the torch at a 0° work angle and a 35°–45° travel angle. See Figure 23-4.
8. Watch the metal melt as the weld pool forms. When the metal sags, quickly note the pool width. Begin to move the flame ahead slowly and carry the weld pool forward. Maintain a steady forward speed and a consistent side-to-side motion. This keeps the pool the same width as it is carried along. Continue the pool to the end of the line. Inspect your first weld pass. Try to prevent any defects in your next weld pass.
9. Carry a weld pool along the length of the other four lines. Inspect each weld pass as it is completed.

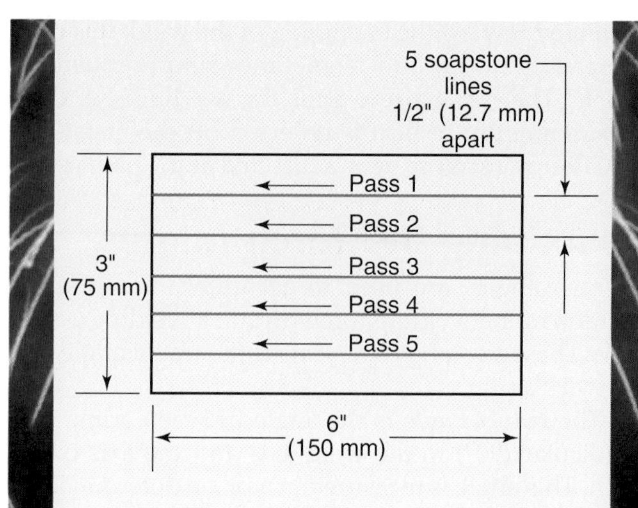

Inspection:

Do a visual inspection of the first weld pass. Turn the metal over using pliers and inspect the penetration. A small continuous deformation (bump) should appear on the underside. The weld pool width should be uniform. This indicates continuous penetration and good torch and flame control. Holes should not be present.

All metal in a weld shop should be considered hot. Use pliers to pick up metal if you are not sure of its temperature. If a piece of hot metal must be left unattended, mark it *HOT* with chalk or soapstone.

Forming and Carrying a Continuous Weld Pool

A welding rod is not needed to carry a weld pool. ***Carrying a weld pool*** refers to creating a pool and moving it along a line. The pool is created with molten base metal. The flame should be held about 1/16″–1/8″ (1.6 mm–3.2 mm) above the surface of the base metal. The torch should be held at a 35°–45° travel angle and a 0° work angle. See Figure 23-3. Use the forehand welding method so the flame is pointed in the direction of travel. The base metal will begin to melt and form a weld pool. The metal starts to sag as the pool increases in diameter. Notice the width of the weld pool when the metal sags. Move the torch flame ahead slowly and the pool will follow. Continue moving the torch ahead slowly, carrying a consistent-width weld pool to the end of the joint.

The end of the metal piece may become overheated as the weld pool approaches the end of the joint. The flame should be raised about 1″ (25 mm) from the surface. This allows the metal to cool. Return the flame to within 1/16″–1/8″ (1.6 mm–3.2 mm) of the metal to complete the weld pass.

A side-to-side torch motion may be required to widen the weld pool. Use a crescent-shaped motion or one similar to those shown in **Figure 23-5** to make a wide weld bead. Some type of flame motion is required to keep the surface of the pool agitated. This prevents the surface from cooling too rapidly.

A weld pool should remain the same width its entire length. This ensures the same amount of deformation on the underside of the metal. See **Figure 23-6**.

If the pool becomes too large, the metal begins to sag excessively. The molten metal may drop through, creating a hole. A hole in the metal indicates that too much time was spent welding in one spot. A hole also indicates that the weld pool was allowed to become too large. If a weld pool becomes too large or too deep, withdraw the flame about 1″ (25 mm). When the metal cools slightly, return the flame to within 1/16″–1/8″ (1.6 mm–3.2 mm) of the metal. The tip size may be too large if the flame must be withdrawn often or if the weld pool must be moved too rapidly.

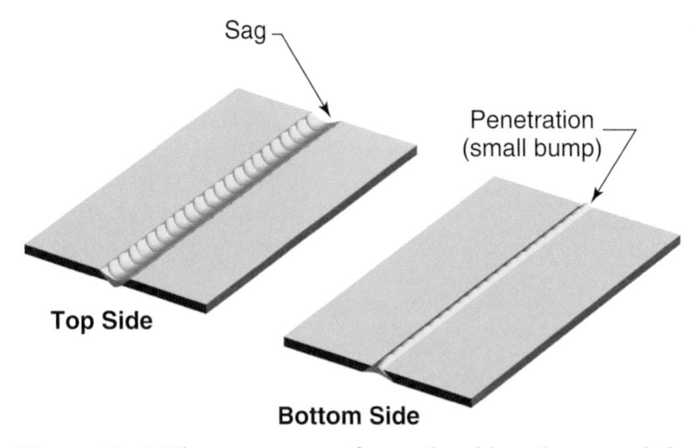

Figure 23-6. *The appearance of a good weld pool pass and the penetration on the underside.*

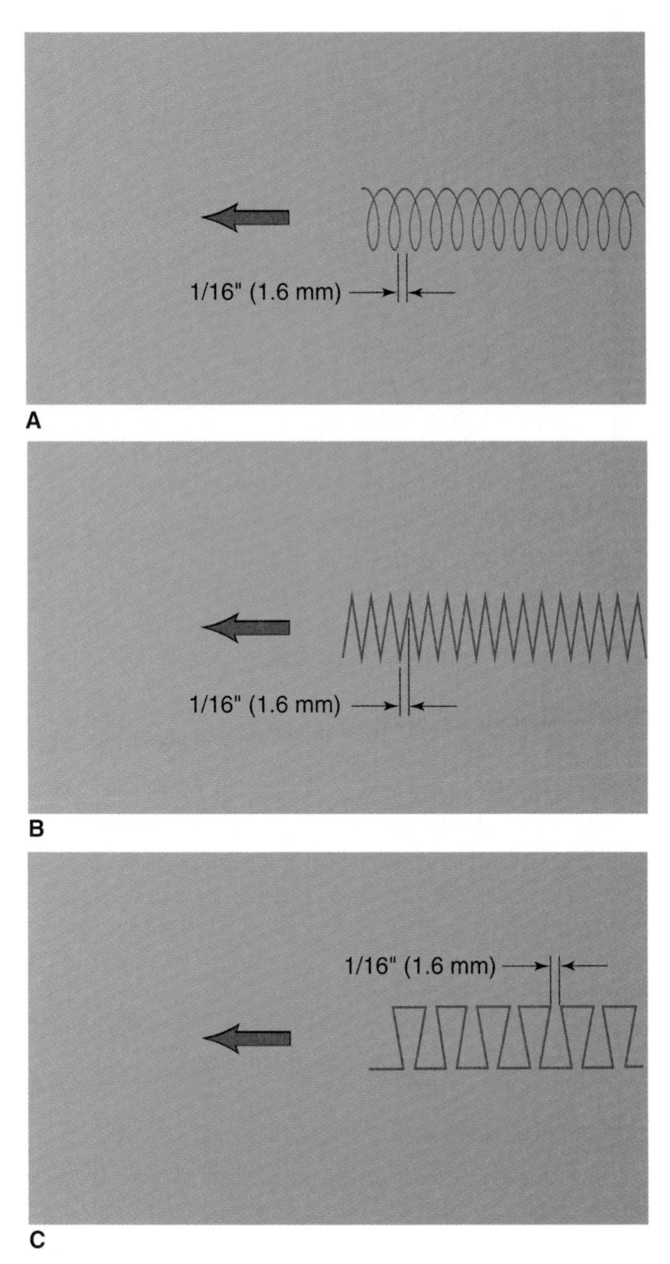

Figure 23-5. *Three suggested flame motions. In each case, the torch should move forward about 1/16″ (1.6 mm).*

A lack of penetration or sag indicates that not enough time was spent welding in an area. A narrow weld pool generally causes little or no penetration. If the weld pool cannot be made wide enough, the tip size may be too small. Select a tip with a larger orifice to produce more heat and create a wider pool.

Welding without a Welding Rod

Several types of welds can be made without using a welding rod. The base metal is used as the filler metal in these welds. When made properly, these welds are as strong as any made using a welding rod.

The edge, outside corner, and edge flanged joint in a butt weld formation can be made without welding rods. See **Figure 23-7**. The flame is used to melt the edges of the metal.

Edge Joint

Welds on edge joints of thin metal assemblies are often made without filler metal. The flame is applied to the edges to melt them and form a weld bead. The resulting weld can have deep penetration and a very uniform weld bead. The joint must be tightly fitted, run from edge to edge, and not overlap. ***Overlap*** is a condition in which the weld bead is not fused into the base metal at the weld toe.

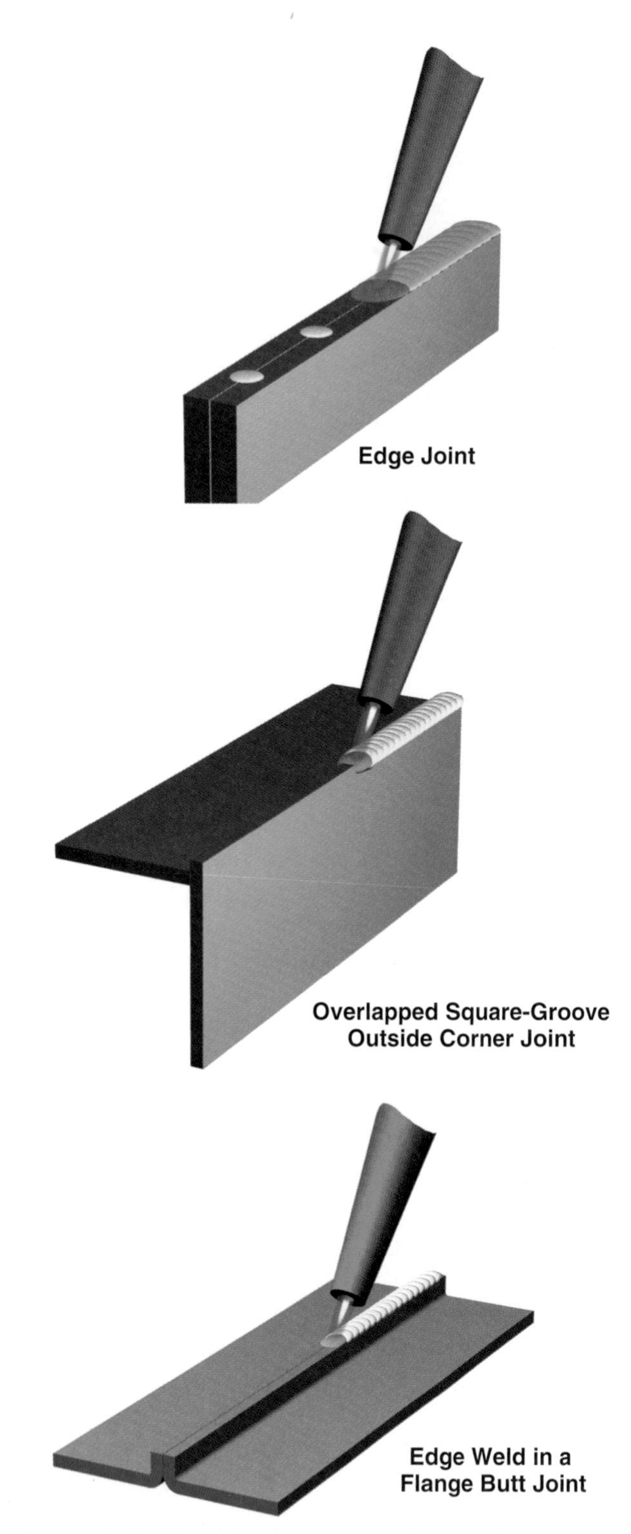

Figure 23-7. Welds in progress on edge, overlapped outside corner, and edge-flange joint welds. A welding rod is not used. The edge of the metal is melted and provides filler metal.

Corner Joint

Inside corner joints are generally not welded without a welding rod. The rod furnishes the additional metal normally needed to reinforce the weld.

Exercise 23-2 Welding an Edge Joint

1. Obtain two pieces of low carbon steel, each measuring 1/16″ × 1 1/2″ × 6″ (1.6 mm × 40 mm × 150 mm).
2. Clean the surface from the edges inward at least 1/2″ (12–15 mm). Position the pieces as shown in the top formation in Figure 23-7.
3. Select the correct torch tip and install it in the torch. See Figure 20-36.
4. Start the welding outfit. Adjust the outfit to obtain the correct pressures.
5. Light the torch and adjust to a neutral flame.
6. Tack weld the pieces in three places.
7. Create a molten pool at the edge of the joint. Be sure to hold the torch at a 0° work angle and a 35°–45° travel angle.
8. Slowly move the molten pool across the joint using a circular or semicircular motion. See Figures 23-5 and 23-7. Form the weld bead by fusing equal amounts of metal from each workpiece into the molten pool.

Inspection:

The weld bead should be smooth. It should fully extend across the joint from one edge to the other.

Several outside corner joints are designed to provide filler metal to reinforce the weld. The pieces used to form an outside corner joint may be overlapped as shown in Figure 23-7. They may also be bent in a brake to form a corner-flange or flare-bevel joint. The overlapped or bent metal edge is melted with the flame to form a uniform weld bead. The bead width should be twice the base metal thickness.

Edge Weld in a Flanged Butt Joint

The flanged butt joint is prepared by bending the metal edge in a brake. These bent edges furnish the filler metal. The edges must fit tightly together, since they are melted to form a weld that runs from outside edge to outside edge. The finished weld has an unusually uniform weld bead. See Figure 23-7.

Exercise 23-3 Welding an Outside Corner Joint

1. Obtain two pieces of low carbon steel, each measuring 1/16″ × 1 1/2″ × 6″ (1.6 mm × 40 mm × 150 mm).
2. Clean the surfaces from the edges inward at least 1/2″ (12–15 mm). Align the pieces as shown in the middle formation of Figure 23-7.
3. Select the correct torch tip and install it in the torch. See Figure 20-36.
4. Start the welding outfit. Adjust the outfit to obtain the correct pressures.
5. Light the torch and adjust to a neutral flame.
6. Tack weld the pieces in three places.
7. Melt the overlapped edge to form a molten pool between the edge and the horizontal piece that is nearly flush with the horizontal surface. See Figure 23-7.
8. Use a circular motion while pointing the torch at the joint. Be sure to melt at least 1/16″–1/8″ (1.6 mm–3.2 mm) of the horizontal surface along with the overlapped edge. Do not point the torch only at the top edge of the overlapped edge as the weld pool is carried along the entire length of the joint. Pointing the torch only at the top edge may cause lack of fusion with the horizontal piece (cold lap defect).

Inspection:

The completed weld should have a smooth bead. The overlapped edge should be flush with the surface of the horizontal piece. The weld bead should be about 1/8″ (3.2 mm) wide. Using pliers, turn the workpiece over. Penetration should show slightly along the inside of the joint.

Exercise 23-4 Welding an Edge Weld in a Flanged Butt Joint

1. Obtain two pieces of low carbon steel, each measuring 1/16″ × 1 1/2″ × 6″ (1.6 mm × 40 mm × 150 mm).
2. Create a 1/4″ (6–7 mm) flange along one edge of each of the pieces. Use a brake press to bend the metal.
3. Clean the surfaces from the edges inward at least 1/2″ (12–15 mm). Align the pieces as shown in the bottom formation of Figure 23-7.
4. Select the correct torch tip and install it in the torch. See Figure 20-36.
5. Start the welding outfit. Adjust the outfit to obtain the correct pressures.
6. Light the torch and adjust to a neutral flame.
7. Tack weld the pieces in three places.
8. Run a weld pool along the edges to form a smooth weld bead. Hold the torch at a 0° work angle and a 35°–45° travel angle. See Figure 23-7.

Inspection:

The completed weld bead should be straight, even in width, and free of holes.

Selecting a Welding Rod

A welding rod is added to a weld pool to accomplish the following:

- Fill a groove weld.
- Form a fillet weld.
- Fill a weld pool that has a depression in it.
- Make a completed weld as strong as the base metal.

Welding rods are available in the following diameters: 1/16″ (1.6 mm), 3/32″ (2.4 mm), 1/8″ (3.2 mm), 5/32″ (4.0 mm), 3/16″ (4.8 mm), 1/4″ (6.4 mm), 5/16″ (7.9 mm), 3/8″ (9.5 mm). They are available in standard 36″ (914 mm) lengths.

The required amount of filler metal should be added to the weld pool each time that the welding rod is dipped into the weld pool.

A welding rod of the correct diameter forms a good ***convex weld bead***. The correct size rod also permits the weld pool to remain fluid as the rod is added. Refer to Figure 20-36 for a table of suggested welding rod sizes.

A rod that is too small does not add enough filler metal to the pool. The rod must be dipped into the weld pool repeatedly to form a good weld bead. A welding rod that is too large in diameter can cause the pool to ***freeze***. This occurs when the welding rod is too large and the attempt to melt it draws too much heat from the pool. The weld pool cools and traps the welding rod. The pool must be reheated to release the rod. A torch tip orifice that is too small in diameter may also cause a welding rod to freeze in the pool.

Your hand position on the welding rod must change as the rod melts and gets shorter, **Figure 23-8**. **To change hand position, place the rod on the worktable with the cool end hanging over the edge and the hot end away from your body. Move your hand to a new position and pick up the rod. Do *not* place the rod against your body to change hand position. Another method is to stand the welding rod upward, with the hot end on the table. Relax your grip on the welding rod and then slide your hand upward to a new position.**

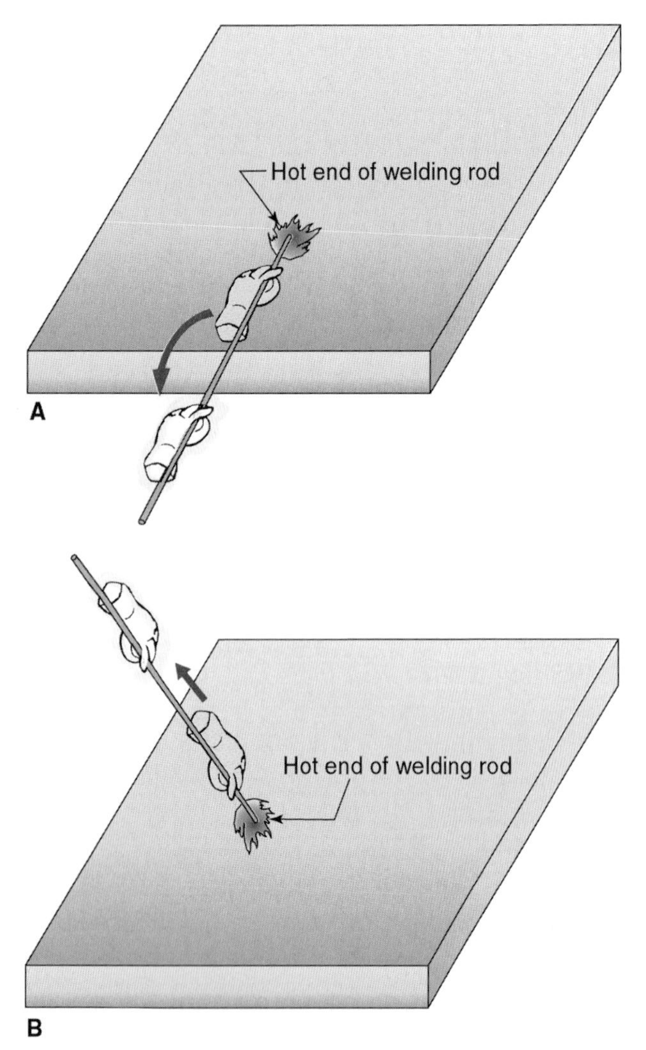

Figure 23-8. *Two methods of changing hand position on the welding rod. A—Place the rod on a table or the weldment with the hot end facing away from you. Then, move your hand to a new position. B—Stand the welding rod on end, with the hot end resting on the table. Relax your grip and slide your hand upward to the new position.*

Laying a Weld Bead

A weld bead results from adding filler metal to a weld pool. In this case, the filler metal is a welding rod. The welding rod is added to the weld pool until the pool is flush with the surface or slightly convex. See **Figure 23-9**. A convex weld bead, one that is raised slightly above the surface, is commonly used. The filler metal increases the strength of the metal in the area of the joint.

A ***flat weld bead*** is used whenever the weld bead is to be ground down after welding. A ***concave weld bead*** is sunk below the surface. It is the weakest type of weld bead because it is thinner. It is used in fillet welds when a blended appearance is more important than strength.

Laying a weld bead requires the use of both hands. A right-handed welder holds the torch in the right hand and the welding rod in the left hand. **Figure 23-10** shows the positioning of the welding rod and torch.

The torch tip is generally inclined with a travel angle of approximately 35°–45°. However, the travel angle can be changed in order to alter the amount of heat. As the tip moves away from perpendicular, less heat is directed onto the base metal. The torch should be held at a 0° work angle (90° to the workpiece surface) as shown in the end view, Figure 23-10. At this angle, the metal is heated equally on each side of the weld axis. If metals of different thickness are being welded, incline the torch toward the thicker piece.

The flame is hottest at a distance of 1/16″–1/8″ (1.6 mm–3.2 mm) from the tip of the flame. Therefore, the tip of the flame should be held at that distance from the base metal. The flame may be drawn away to reduce the heat input to the metal.

The welding rod is normally inclined at 15°–45° to the weld surface. More of the rod is heated by the reflected heat from the torch flame when the rod is held at a low angle.

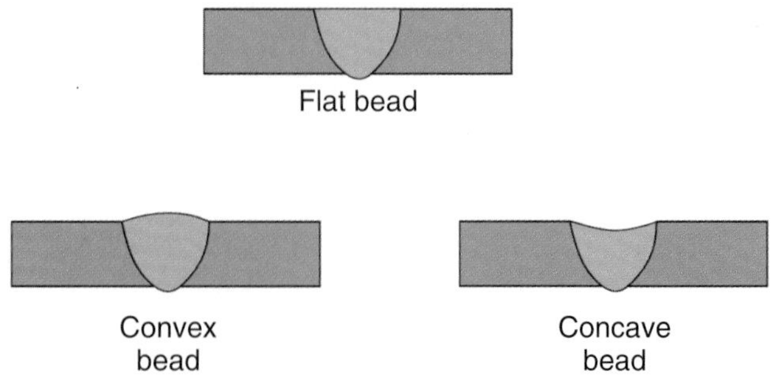

Figure 23-9. *Flat, convex, and concave weld beads. The convex weld bead is generally preferred. A flat weld bead is used if the weld bead is to be ground down after welding. A concave weld bead is weaker because it is thinner.*

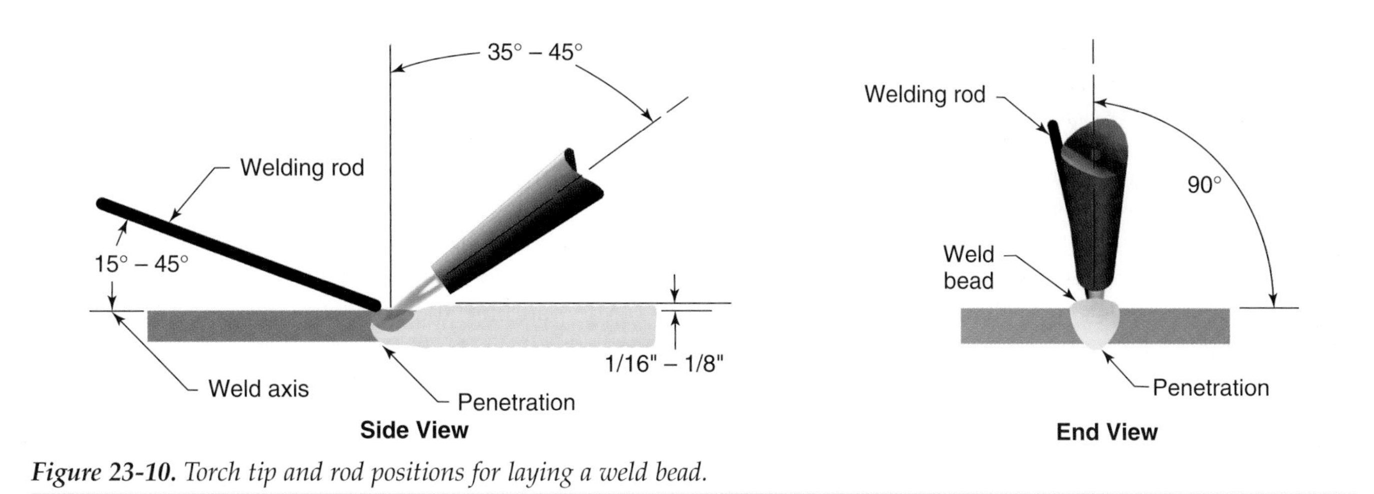

Figure 23-10. *Torch tip and rod positions for laying a weld bead.*

The end of the welding rod must be held near the weld pool to keep it hot. A welding rod will cool if it is held too far from the pool. The pool may freeze if a cool welding rod is placed in it.

A weld pool must be formed first when laying a weld bead. The welding rod is added when the pool sags. The rod is added by dipping it into the ***leading edge*** (the forward edge) of the weld pool. The heat melts the end of the welding rod, filling the pool. A welding rod that is the correct diameter will only need to be dipped into the weld pool every few seconds to form a good weld bead.

The flame movement and the use of the welding rod must be coordinated to produce a good weld bead. Move the flame and weld pool along the weld axis at a constant rate. Control the width and depth of the pool and move the welding rod in and out of the pool consistently. This will produce a uniform ripple in the weld bead. **Figure 23-11** shows a well-formed weld bead with adequate penetration.

Raise the flame slightly toward the end of the weld bead. This allows the weld pool to cool as the welding rod is added to fill the end of the weld bead.

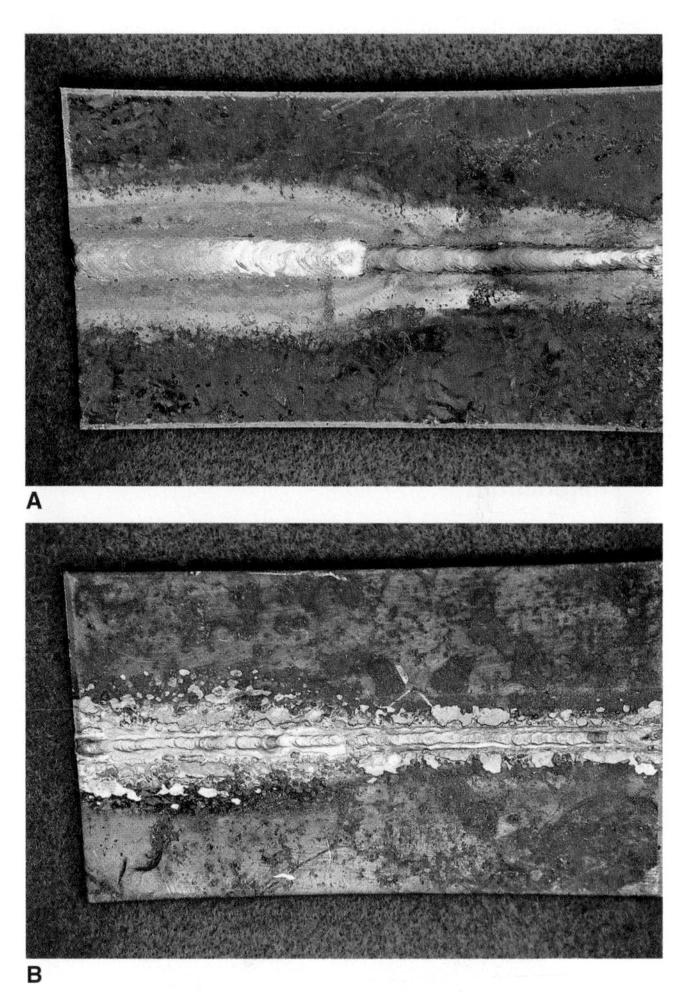

Figure 23-11. *A—A well-formed cover pass bead in a butt weld on 1/4" mild steel. The root pass bead is visible at the right side of the incomplete cover pass bead. B—The underside of the same weld. Note that 100% penetration is indicated by the continuous sag of the metal at the root of the weld.*

Lap Joint

A machined or a flame-cut edge is not required for a lap joint. The two pieces must be in good contact with one another. They should be held tightly together with clamps and then tack welded.

A lap joint is welded along the edge of one piece and the surface of the other piece. The flame should be mainly directed toward the surface, **Figure 23-12**. The surface needs more heat to melt than does the edge.

Exercise 23-5 Running a Weld Bead with Welding Rod

1. Obtain one piece of mild steel measuring 1/16″ × 3″ × 6″ (1.6 mm × 75 mm × 150 mm).
2. Clean the surfaces of the mild steel, using a stainless steel brush.
3. Using chalk or soapstone, mark out four lines on the surface 3/4″ (19.1 mm) apart.
4. Select the proper size welding tip. Obtain several lengths of the proper size welding rod. See Figure 20-36.
5. Start the welding outfit and set the correct pressures. Light the torch and adjust for a neutral flame.
6. Form a weld pool on the first line. Add the welding rod when the pool is the correct width and after it sags. Add enough filler metal to form a convex weld bead.
7. Carry the weld pool, add filler metal, and form a weld bead in a smooth, continuous motion.
8. Continue the weld bead to the end of the metal. Repeat the procedure on the other three lines.

Inspection:

The completed weld bead should be straight, consistent in width, and free of holes. It should have evenly spaced ripples and 100% penetration. On thin metal, penetration is indicated as a continuous sag on the reverse side of the metal.

A ***C-shaped weld pool*** should be formed when welding a lap joint, **Figure 23-13**. The C-shaped weld pool is the indicator that the surface has melted enough to create a good weld. It also indicates good fusion on the surface of the piece. The welding rod is added to the center of the C-shaped weld pool when it forms. Continue to dip the rod into the pool until the correct weld bead shape is formed.

After adding the welding rod, move the flame ahead about 1/16″ (1.6 mm). Wait for the C-shaped weld pool to form again. Add the welding rod and move on. Move ahead at a steady rate, adding the welding rod each time the C-shaped pool is well-formed. Continue this procedure to the end of the weld.

The edge of the lap joint will sometimes melt away rapidly. This is commonly caused when the two pieces of metal are not in good contact. The flame

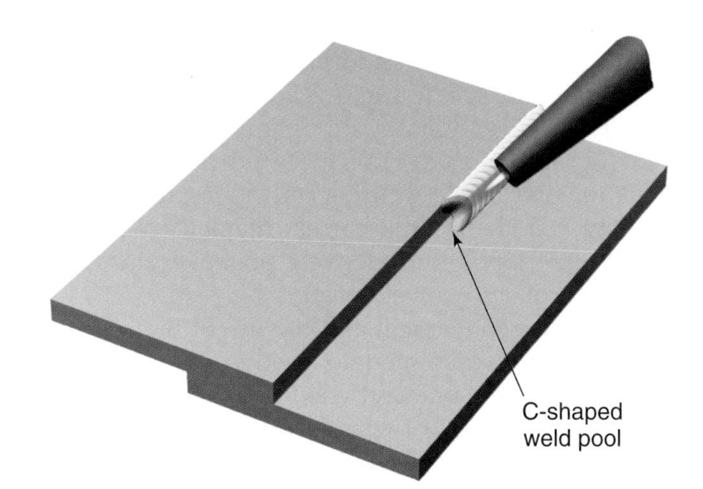

Figure 23-13. A C-shaped weld pool indicates that the surface and edge of the base metal have melted. When making a fillet weld on a lap or corner joint, add the welding rod after the C-shaped weld pool forms.

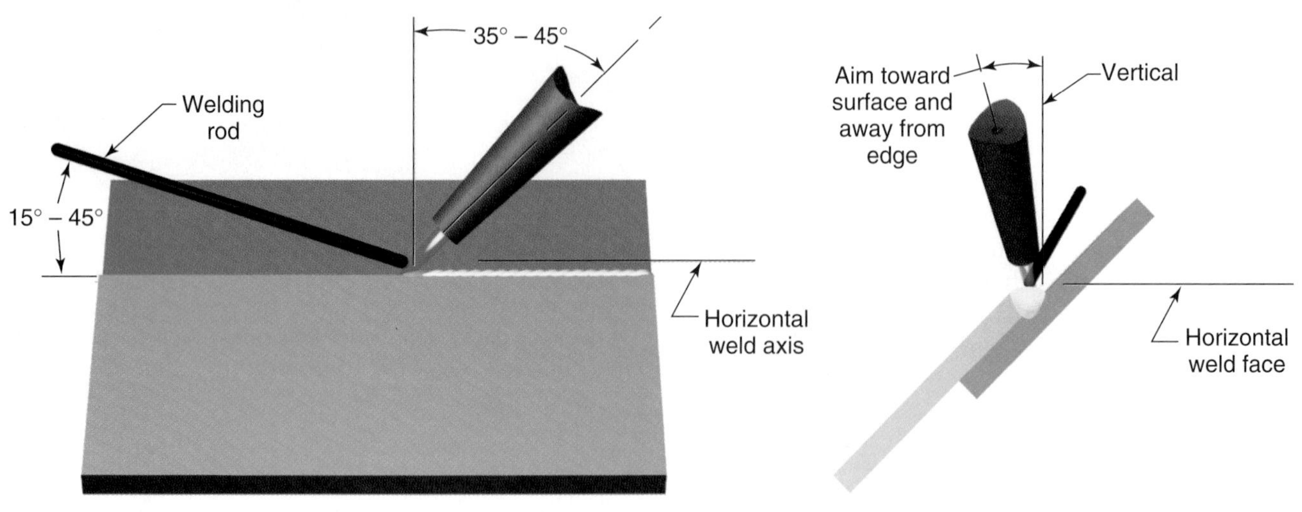

Figure 23-12. A fillet weld on a lap joint. The edge melts before the surface. Aim the flame slightly more toward the surface to apply additional heat there.

travels between the pieces, melting the edge rapidly and uncontrollably. To slow melting of the edge, rest the welding rod on it. The welding rod will draw some heat away from the edge.

Inside Corner and T-Joints

On thin metal, inside corner and T-joints may be welded without edge preparation. A bevel-groove joint may have to be used on metal over 3/16″ (4.8 mm) thick to ensure good penetration. On metal thicker than 3/8″ (9.5 mm), a bevel-groove, double-bevel-groove, single-J-groove, or double-J-groove may be used. The finished weld used in all cases is a fillet weld.

A fillet weld alone is used on square-groove inside corner and T-joints. The inside corner joint is welded from one side only; T-joints may be welded from both sides.

The leg size of a fillet weld is determined by the weld joint designer. However, as a rule of thumb, the leg size should be at least equal to the base metal thickness. For example, 1/8″ (3.2 mm) thick metal requires 1/8″ (3.2 mm) leg sizes. The leg size should be 1/4″ (6.4 mm) for 1/4″ (6.4 mm) thick metal.

The weld for an inside corner or a T-joint is made on two surfaces. The torch flame must heat each piece equally. **Figure 23-14** shows the position of the pieces of metal when welding an inside corner or T-joint in the flat position. The correct angles for the welding tip and welding rod are also shown.

A C-shaped weld pool must be formed when fillet welding an inside corner or T-joint. The weld pool shows that the metal surfaces have melted properly. The welding rod is dipped into the upper edge of the pool after it is formed. Continue dipping the rod into the weld pool until the desired weld bead shape is formed. If the shape is not specified, use a convex weld bead for greatest strength. Move the torch ahead smoothly and dip the welding rod into the weld pool after the C-shape re-forms. Continue the procedure to the end of the joint.

Exercise 23-6 Fillet Weld on a Lap Joint

1. Obtain two pieces of mild steel, each measuring 1/16″ × 1 1/2″ × 6″ (1.6 mm × 40 mm × 150 mm).
2. Clean the surfaces and edges of both pieces.
3. Select the correct torch tip and welding rod size. Refer to Figure 20-36.
4. Start the welding outfit and set the working pressures. Light the torch and adjust to a neutral flame.
5. Position the pieces so that they overlap about 3/4″ (19 mm).
6. Clamp the pieces tightly, using C-clamps.
7. Tack weld the joint three times on the top and bottom sides.
8. Lay the fillet weld with a convex weld bead. See Figure 23-12. Wait for a *C*-shaped weld pool to form before adding welding rod.
9. Make a second fillet weld on the other side.

Inspection:

The completed welds must be straight and have evenly spaced ripples. The weld bead should be convex and have consistent width.

Butt Joint

The edges of metal used to form a butt joint are generally machined or flame cut. Metal thicker than 3/16″ (4.8 mm) must have its edges prepared as a bevel groove, V-groove, J-groove, or U-groove. Thicker metal may be welded from one or both sides.

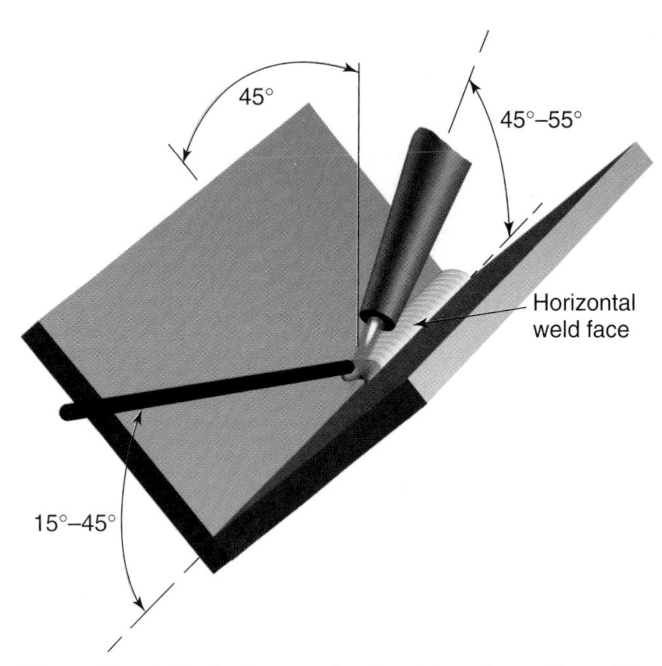

Figure 23-14. Torch, flame, and rod positions for making a fillet weld on an inside corner joint in the flat welding position. The weld face is horizontal. The torch tip is held at a 45° work angle and a 35°–45° travel angle (shown as 45°–55° off the base metal).

Exercise 23-7 Welding a T-Joint with a Fillet Weld

1. Obtain two pieces of mild steel, each measuring 1/16″ × 1 1/2″ × 6″ (1.6 mm × 40 mm × 150 mm).
2. On one piece, clean both sides along one 6″ edge. Clean one surface of the second piece.
3. Select the correct torch tip and welding rod. See Figure 20-36.
4. Start the welding outfit and adjust the pressures. Light the torch and adjust to a neutral flame.
5. Place the clean edge in the middle of the clean surface. Use a brick to prop up the vertical piece.
6. Tack weld the joint three times on each side. Alternate the tack welds from side-to-side and end-to-end. This prevents the vertical piece from pulling in one direction.
7. Weld a fillet weld on one side using the correct torch tip and welding rod positions. Do not dip the welding rod until a good C-shaped pool is formed. See Figure 23-14.
8. Weld a fillet weld on the other side.

Inspection:

The completed welds should be convex and even in width. The weld bead should be straight and have evenly spaced ripples.

Weld beads seldom should be thicker than 1/4″ (6.4 mm). Thicker weld beads will cool before gases and impurities in the weld rise to the surface. Gases and impurities trapped within a weld weaken it.

Welds made on thick metal may require more than one weld bead. Each completed weld bead requires one weld pass. A weld that requires three weld beads needs three weld passes to complete it. Complete penetration is possible on metal 3/16″ (4.8 mm) or less without flame cutting or machining the edges.

Metal in a butt joint will pull together due to the expansion and contraction that takes place during welding. The expansion may be so great that the adjoining edges overlap at the end of a long butt joint. To prevent overlapping or reduction of the weld root, the pieces must be clamped together or tack welded. Tack welds are often placed 3″ (76 mm) apart, **Figure 23-15**. The tack weld melts and becomes part of the main weld as welding proceeds. A tack weld must be well-made so it will not break as the weld is made.

Expansion may be compensated for by yet another method. In this method, the weld root opening tapers from beginning to end, **Figure 23-16**. The amount of taper may be difficult to estimate. Therefore, this method should be used only when identical welds are to be made repeatedly. Several factors must be considered in determining the amount of taper needed to avoid overlap. These include metal thickness, type of metal, tip size, and the length of the weld.

Welds made on butt joints must totally penetrate the base metal. Thick metal may be welded from both sides to ensure 100% penetration. Multiple weld passes may also be required on thick metal.

The edges of a piece of metal that have been prepared by machining or flame cutting are thinner at the root of the joint. This causes the root of the weld to melt more rapidly. The root opening enlarges due to this rapid melting. The enlarged root opening looks like an

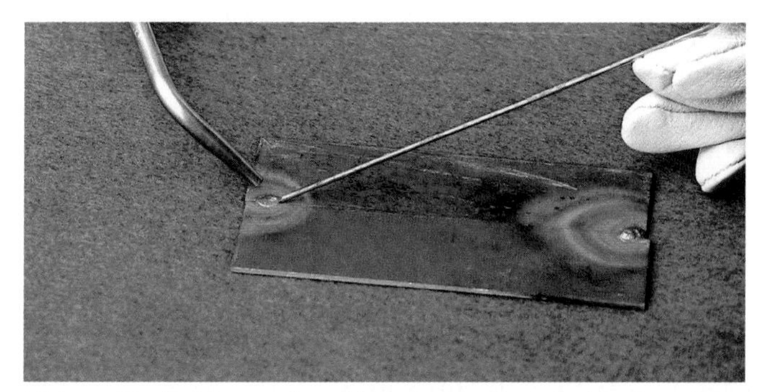

Figure 23-15. *Two tack welds being made to hold a butt joint in alignment for welding.*

Figure 23-16. *The root opening is tapered prior to welding this butt joint. As the weld progresses from the narrow end, the metal pulls together, closing the gap.*

old-fashioned keyhole, **Figure 23-17**. The ***keyhole*** indicates that the metal has been melted through to the reverse side. Total penetration is ensured if the keyhole is allowed to develop before the welding rod is added. This procedure is known as ***keyhole welding*** or keyholing.

Weld Defects

Welders should be able to identify a good weld. They must also be able to recognize defects and know how to prevent them.

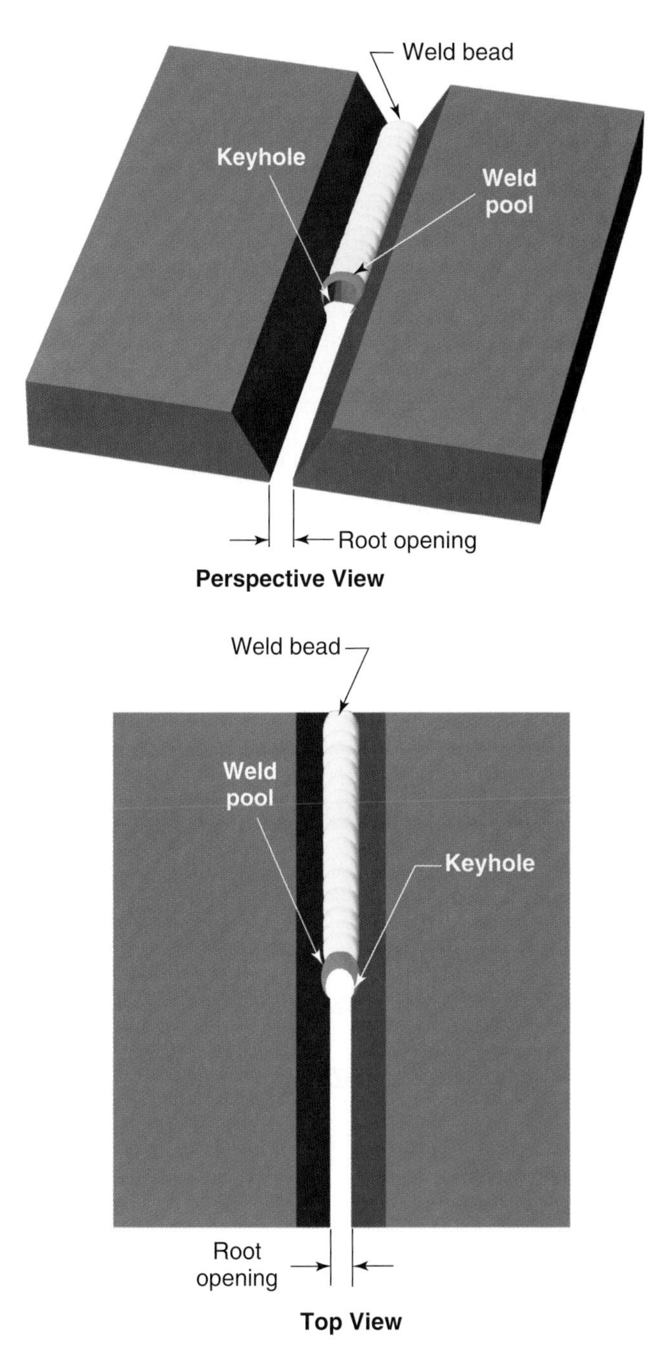

Figure 23-17. Two views of a keyhole. The keyhole indicates that the weld pool just ahead of the weld bead has penetrated through the metal.

Exercise 23-8 Welding a Single Square-Groove Butt Joint

1. Obtain two pieces of mild steel, each measuring 1/16″ × 1 1/2″ × 6″ (1.6 mm × 40 mm × 150 mm).
2. Clean the edges and the top surfaces from the edge inward 1/2″ (12.7 mm).
3. Select the correct torch tip and welding rod. See Figure 20-36.
4. Start the welding outfit and set the pressures. Light the torch and adjust to a neutral flame.
5. Tack weld the pieces near the beginning, middle, and end of the joint. Be sure to maintain a 1/16″–3/32″ (1.6 mm–2.4 mm) root opening.
6. Melt the surfaces of the pieces evenly. Side-to-side torch motion may be needed. The molten metal will fill the root opening. It will also form a weld pool and sag. Notice the width of the pool when this occurs. Add the welding rod to the leading edge of the pool as needed to fill the pool and form a weld bead. Move the torch flame slowly and continuously ahead. Add the welding rod as the correct width weld pool re-forms. Continue the procedure to the end of the joint.

Inspection:

The completed weld should have a uniform width and a consistent ripple pattern. Total penetration should be indicated on the underside.

Most often welds are visually inspected. **Figure 23-18** shows fillet welds with different weld bead contours and some common weld defects. The top row in Figure 23-18 shows well formed fillet welds with different weld bead contours. Each weld has good penetration. In Figure 23-18D, the vertical piece of metal has been thinned and weakened by an undercut weld. A weld bead with overlap is shown in Figure 23-18E. The overlapped weld bead does not add strength to the weld. The actual size of the completed weld as shown by the dashed line, is smaller than desired. The weld in Figure 23-18F will be weak because of poor penetration.

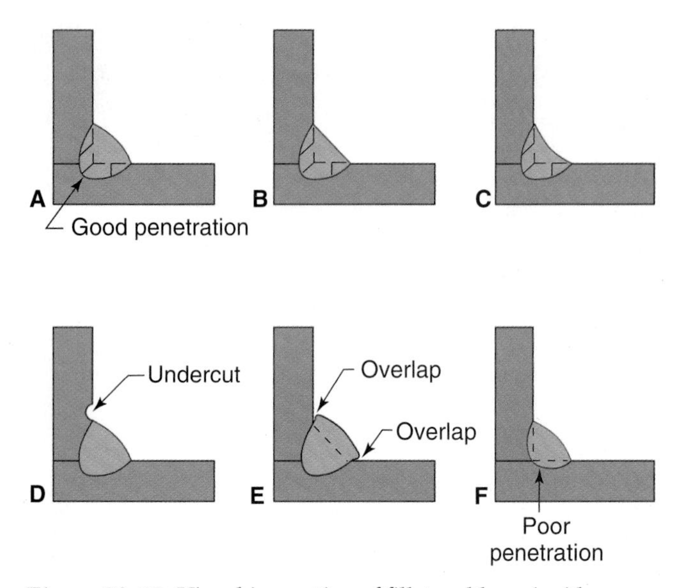

Figure 23-18. Visual inspection of fillet welds on inside corner joints. The penetration of A, B, and C is good. A convex weld bead, as shown at A, is generally preferred. For D, E, and F, the defects are undercut, overlap, and poor penetration.

Figure 23-19 shows one well-formed and two poorly made butt welds. A narrow weld bead, poor penetration, and overlap are caused by insufficient heat. Using too small a welding tip or a welding speed that is too fast results in insufficient heat. If the weld pool is not large enough, the weld will not penetrate, the weld bead will be narrow, and overlap may occur. If the torch tip is too large, a wide weld bead and excessive penetration may result.

Undercutting is the thinning of the base metal along the toe of the weld. Undercutting can occur in fillet welds and in welds made on groove joints. It is eliminated by adding more filler metal to the weld pool. Undercutting occurs when the incorrect torch flame angle is used. On a horizontal fillet weld, the filler metal sags from the vertical piece causing an undercut condition. The problem is corrected by adding filler metal to the upper edge of the weld pool.

Overlap is a condition in which the weld bead is not fused into the base metal. Overlap occurs when the base metal is not molten at the time the welding rod is added.

Incomplete penetration results in a weak weld. When welding thick metal, use a groove-type weld. A larger tip size and carrying a deeper weld pool also will eliminate poor penetration.

Inclusions are foreign materials in the weld pool. Cleaning the metal before welding and keeping a clean welding area will help eliminate inclusions. Air pockets may also occur. The welding flame should be constantly kept in motion to stir the weld pool and eliminate this problem.

Flat or concave weld beads may be required on a weld. However, a convex weld bead is generally preferred since it is stronger. A convex weld bead is obtained by adding more filler metal to the weld pool.

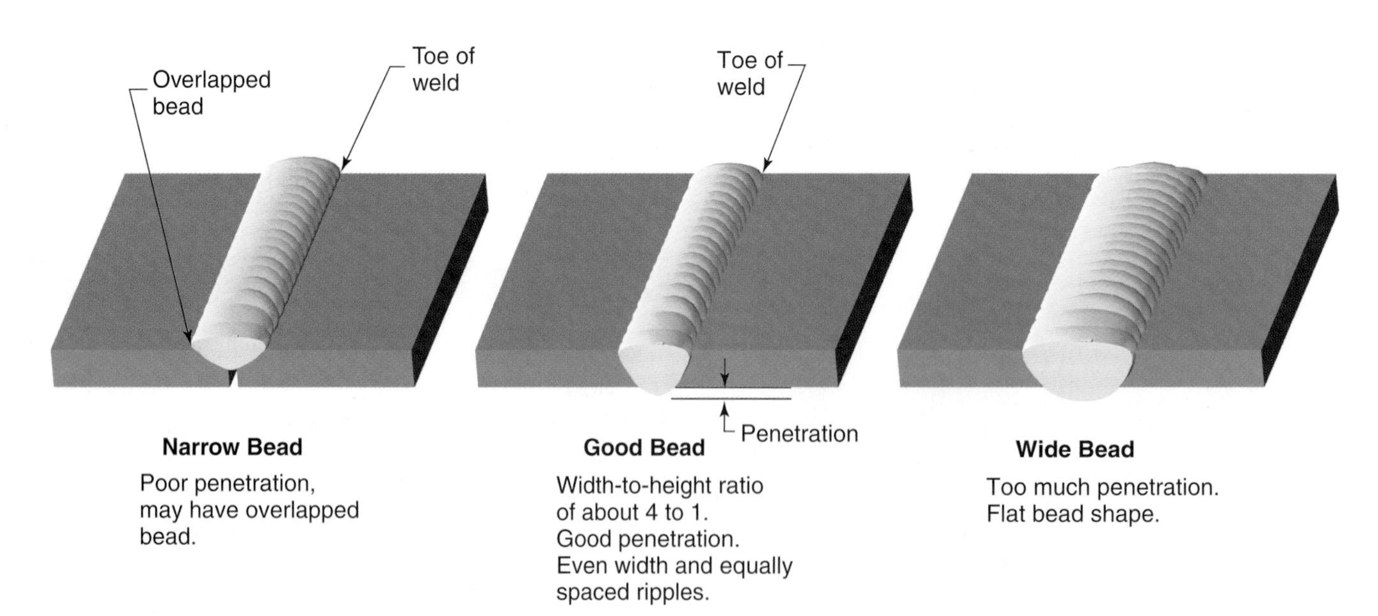

Figure 23-19. Visual inspection of welds on a butt joint. A good weld bead is convex unless a different shape is specified. The weld must penetrate through the entire thickness. A narrow weld bead will not penetrate deeply and may show overlap. Too much heat produces a weld bead that is too wide and has too much penetration.

Summary

- The welding torch may be held like a pencil or a hammer. Both methods of holding the torch can be used effectively in all welding positions.
- In the forehand welding method, the torch tip is held so that the flame is pointing in the direction of travel. In the backhand welding method, the flame is pointed opposite to the direction of travel.
- The travel angle is the angle between a line perpendicular (90°) to the weld axis and the axis of the torch. This angle is measured in a plane determined by the electrode axis and the weld axis. The work angle is the angle between a line perpendicular to the major workpiece and a plane determined by the centerline of the torch tip and the weld axis.
- The phrase "carrying a weld pool" refers to creating a pool and moving it along a line. The pool is created with molten base metal. The flame should be held about 1/16″–1/8″ (1.6 mm–3.2 mm) above the surface of the base metal. The torch should be held at a 35°–45° travel angle and a 0° work angle.
- Welds on edge joints of thin metal, several outside corner joints, and a flanged butt joint can be performed without adding filler metal.
- The correct size welding rod permits the weld pool to remain fluid as the rod is added. A rod that is too small does not add enough filler metal to the pool. A welding rod that is too large in diameter can cause the pool to freeze.
- Laying a weld bead requires the use of both hands. A right-handed welder holds the torch in the right hand and the welding rod in the left hand. The torch tip is generally inclined with a travel angle of approximately 35°–45°. The torch should be held at a 0° work angle. The welding rod is normally inclined at 15°–45° to the weld surface. The end of the welding rod must be held near the weld pool to keep it hot. The welding rod is added when the pool sags. The rod is added by dipping it into the leading edge of the weld pool. The heat melts the end of the welding rod, filling the pool. The flame movement and the use of the welding rod must be coordinated to produce a good weld bead.
- A lap joint is welded along the edge of one piece and the surface of the other piece. The torch flame should be directed more at the surface and less at the edge, using a travel angle of 35°–45°. The welding rod should be raised 15°–45° from the weld axis. A *C*-shaped weld pool indicates that the surface has melted enough to create a good weld. The welding rod is added to the center of the *C*-shaped weld pool when it forms. After adding the welding rod, move the flame ahead about 1/16″ (1.6 mm). Wait for the *C*-shaped weld pool to form again. Add the welding rod and move on.
- The weld for an inside corner or a T-joint is made on two surfaces. The torch flame must heat each piece equally. A travel angle of 35°–45° is used. The welding rod should be raised 15°–45° from the weld axis. As with a lap joint, a *C*-shaped weld pool indicates that the surface has melted enough to create a good weld. The welding rod is dipped into the upper edge of the pool after it is formed.
- Metal in a butt joint will pull together due to the expansion and contraction that takes place during welding. To prevent overlapping or reduction of the weld root, the pieces must be clamped together or tack welded.
- When welding butt joints, a 35°–45° travel angle and a 0° work angle are used. A keyhole in the root opening indicates that the metal has been melted through to the reverse side. Total penetration is ensured if the keyhole is allowed to develop before the welding rod is added.
- Undercutting is the thinning of the base metal along the toe of the weld. It occurs when the incorrect torch flame angles are used. Overlap is a condition in which the weld bead is not fused into the base metal. It occurs when the base metal is not molten at the time the welding rod is added. Inclusions are foreign materials in the weld pool. Cleaning the metal before welding and keeping a clean welding area will help eliminate inclusions.

Review Questions

Write your answers on a separate sheet of paper. Please do *not* write in this book.

1. Refer to Figure 20-36. What welding rod and torch tip size should be used for welding 1/8″ (3.2 mm) thick mild steel?
2. When creating a continuous weld pool, how is penetration indicated?
3. The two angles which must be known in order to accurately describe how to hold the torch while welding or cutting are called the _____ angle and the _____ angle. Explain how these two angles are measured.
4. How does a welder know when the welding rod diameter is too large?
5. List three types of joints that can be welded without using welding rod.
6. Metal that is _____″ (_____ mm) and thicker should have its edges prepared by machining or flame cutting.
7. Why should pieces be tack welded before they are welded?
8. Which of the following will help ensure 100% penetration of a butt joint on thick metal?
   A. Form a bevel or groove.
   B. Weld from both sides.
   C. Use multiple weld passes.
   D. Use bigger tip size.
   E. All of the above.
9. How do you know when to dip the rod in the weld pool while making a fillet weld on a T-joint, lap joint, or inside corner joint?
10. Trace and complete the following sketch to show what a "keyhole" looks like when viewed from above.

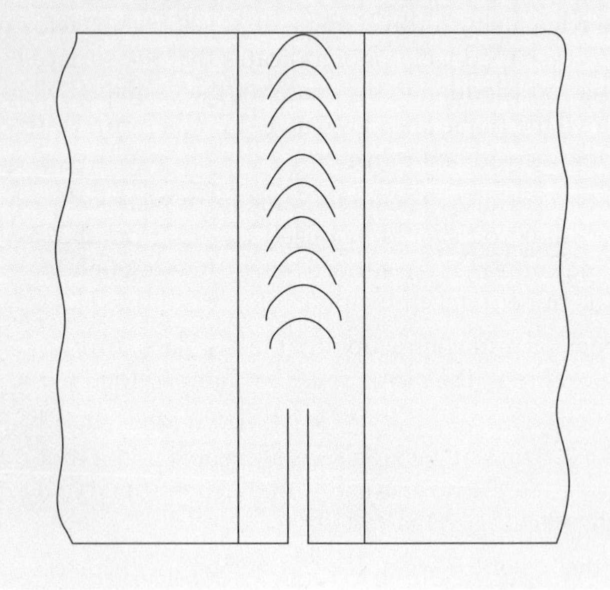

Chapter 24

Oxyfuel Gas Welding: Horizontal, Vertical, and Overhead Welding Positions

After studying this chapter, you will be able to:

- Define out-of-position welding.
- Identify safety measures to be taken when welding out of position.
- Describe methods used to perform welds in the horizontal welding position.
- Weld in the horizontal welding position with OFW.
- Describe methods used to perform welds in the vertical welding position.
- Weld in the vertical welding position with OFW.
- Describe methods used to perform welds in the overhead welding position.
- Weld in the overhead welding position with OFW.

Technical Terms

crescent-shaped motion
out-of-position welding
positioner
sagging

Out-of-Position Welding

Welding is easiest when done in the flat welding position. However, parts that are fixed in position or that are too heavy to move must be welded where they are, which may be in the horizontal, vertical, or overhead positions. See **Figure 24-1**. Welding in a position other than the flat position is commonly referred to as ***out-of-position welding***.

Welds made out of position must be as strong as welds made in the flat welding position. An out-of-position weld should look as good as a weld made in the flat welding position. More training and talent are required to weld out of position. Those who can weld out of position have a chance to earn more money.

Some companies use positioners to rotate large parts. ***Positioners*** are large machines that rotate parts so that welds may be made in the flat welding position. See **Figure 24-2**.

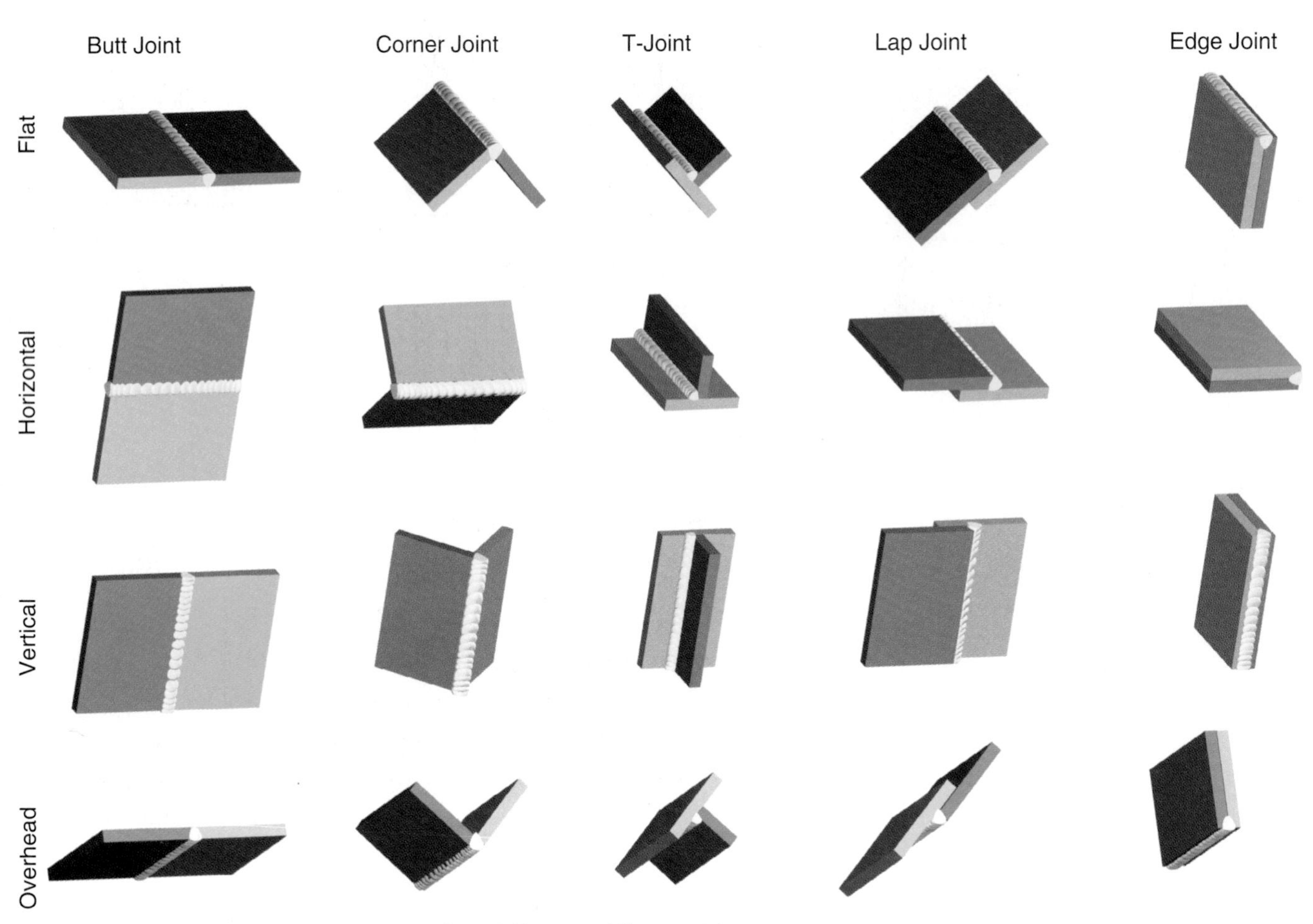

Figure 24-1. *The five basic weld joints in four different welding positions.*

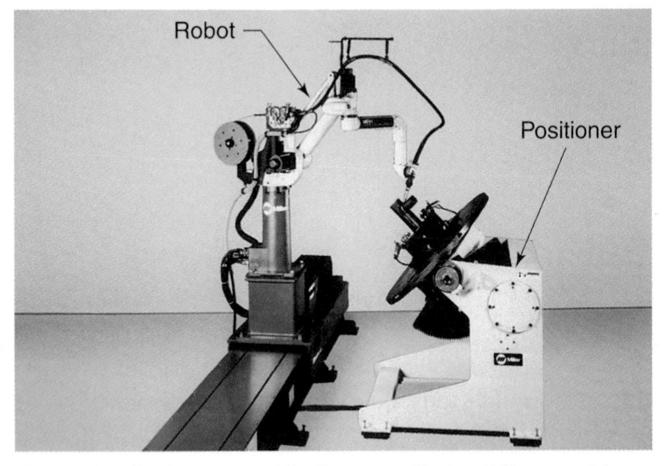

Figure 24-2. *An automatically controlled welding positioner working with a robot. The positioner moves the workpiece to allow good access to the weld joint and to obtain the best welding position. (Miller Electric Mfg. Co.)*

Preparing to Weld

Many out-of-position welds are performed at or above eye level. The protective clothing worn for out-of-position welding is similar to that worn for flat position welding. **Coveralls or jackets should be buttoned at the collar to prevent hot metal from getting inside clothing. Pockets must be buttoned. Flammable items should not be carried in pockets. A cap should be worn when performing overhead welds. Flame-resistant leather clothing may also be worn for additional protection. Welding goggles with #4–#6 lenses are recommended.**

The torch tip and welding rod sizes are selected based on the metal thickness and/or the manufacturer's recommendations. Figure 20-36 shows a table of recommended tip orifice and welding rod sizes. The welding outfit is turned on and the flame is lit and adjusted. A neutral flame should be obtained.

Welding in the Horizontal Welding Position

Welds made in the horizontal (2G) welding position must meet the following conditions:

- The weld axis is within 15° of horizontal.
- The weld face is between 80°–150° or 210°–280°.

Figure 24-3 shows the position of a horizontal weld. For practice welding, the weld face is usually vertical with a horizontal weld axis. Parts for practice welds may be held in a fabricated welding positioner, **Figure 24-4**.

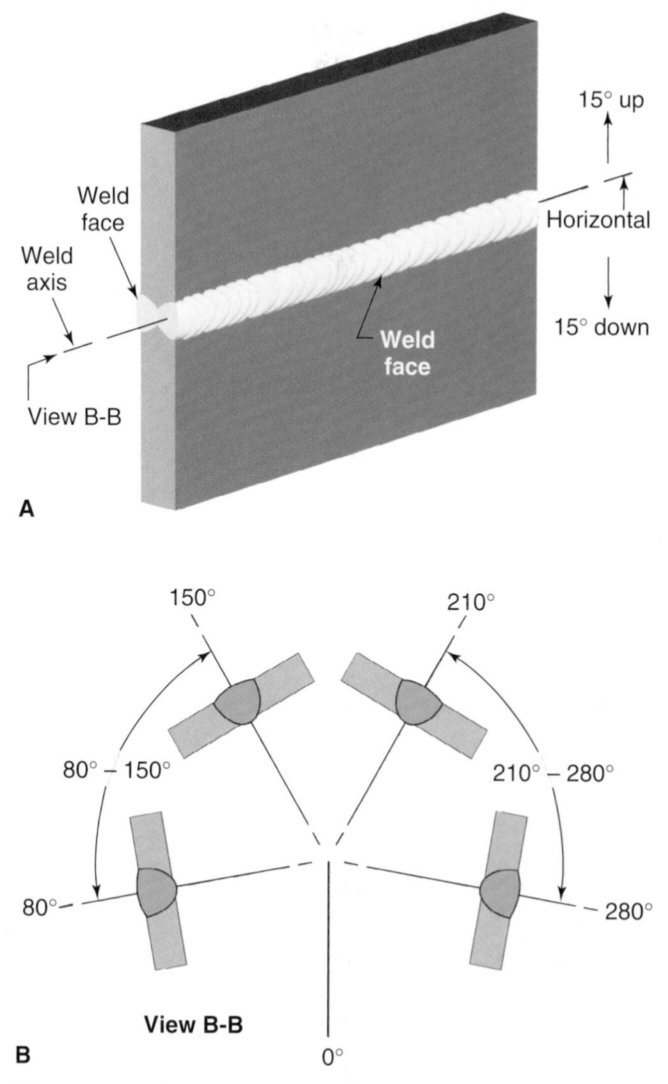

Figure 24-3. The AWS 2G, or horizontal, position. A—The weld axis must be within 15° of horizontal. B—An end view (as viewed along the View B-B axis) of a butt joint in the horizontal position. The weld face must be within 80°–150° or 210°–280°. All angles are measured clockwise with 0° at the bottom.

Lap Joint

The pieces of metal for a lap joint in the horizontal position should be arranged as shown in **Figure 24-5**. The weld face for practice pieces should be at an angle of 135° or 225°.

Torch and Flame Position

When welding a lap joint, the edge of one piece melts more rapidly than the surface of the other. Therefore, the centerline of the tip and flame should point more at the surface than at the edge. Hold the torch at a work angle of 30°–45° (which can also be described as an angle of 45°–60° off the surface). See the *Describing the Torch Angles* section in Chapter 23. This lessens the amount of heat applied to the edge. The torch tip and flame should be held at a travel angle of 45°. See Figure 24-5.

Figure 24-4. A fixture used to hold practice welds while welding out of position. This fixture is adjustable in three ways. A device like a C-clamp is used to hold the work.

Welding Rod Position

The edge of thin metal tends to melt quickly when welded. To prevent this, hold the welding rod on the edge of the metal before inserting it in the weld pool. The rod draws heat away from the pool and decreases the amount of heat applied to the edge. This helps slow the melting of the edge. The rod is held at a 15°–45° angle from the metal surface, **Figure 24-6**. It is usually aligned with the weld axis.

Weld Pool

A C-shaped weld pool should be formed before adding the welding rod. Insert the welding rod at the upper edge of the pool. This helps to overcome gravity and improve the shape of the weld bead. A convex weld bead should normally be formed. A flat or concave weld bead may be specified in the working drawing. A concave weld bead is not as strong as a convex weld bead. However, a concave weld bead may be required when appearance is important, such as on the outside of an appliance.

Corner and T-Joints

A procedure similar to the one used for a fillet weld on a lap joint is used to place a fillet weld in an inside corner or T-joint. The procedure for welding an outside corner joint is similar to welding a butt joint. **Figure 24-7** shows a fillet weld being made on an inside corner joint. The weld face for most horizontal practice welds should be at about 135°.

Torch and Flame Position

The centerline of the torch tip and flame should be centered between the surfaces of the two pieces of metal.

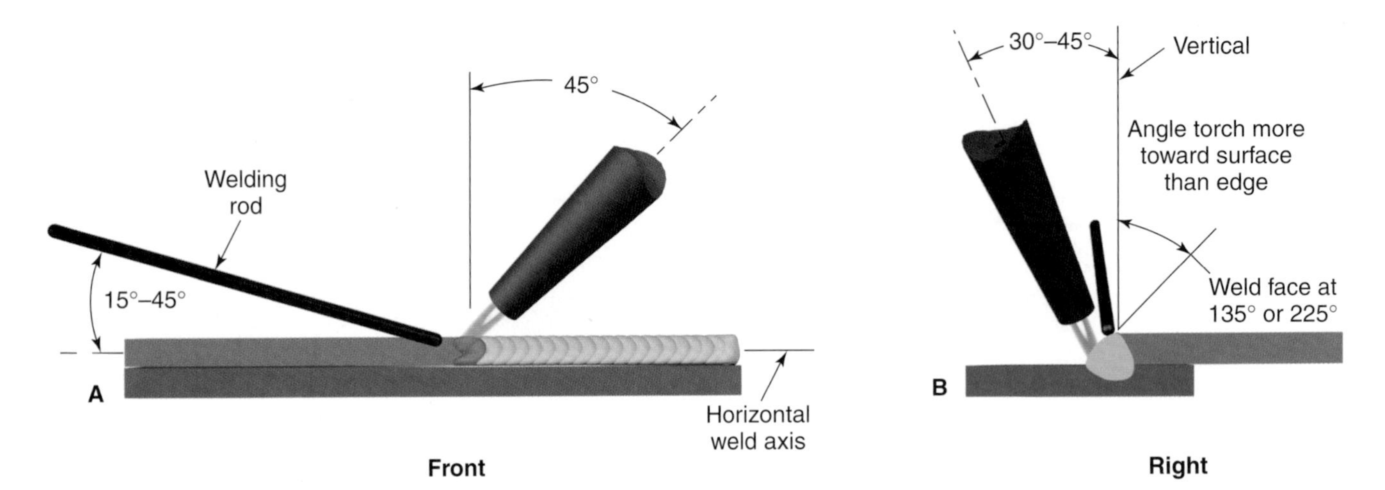

Figure 24-5. Views of torch, rod, and flame positions for lap welding. A—Front view with torch tip held at a 45° travel angle. B—Right side view with torch tip held at a 30°–45° work angle. Note: These angles are typical for lap welding in any position.

Figure 24-6. A lap joint being welded in the horizontal welding position. The torch and flame are directed so that more heat is applied to the surface than to the edge.

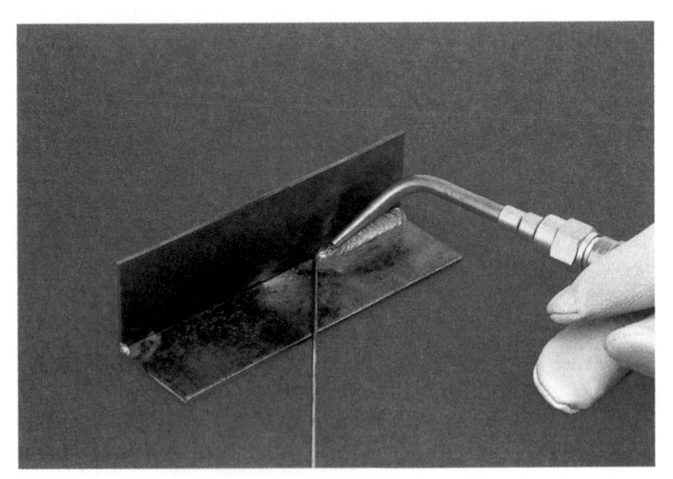

Figure 24-7. A fillet weld being made on an inside corner joint. Note that the tip is held at a 45° work angle so the heat is applied evenly to both surfaces. If undercutting occurs, the flame is aimed more toward the vertical piece.

Exercise 24-1 Fillet Weld on a Lap Joint in the Horizontal Welding Position

1. Obtain two pieces of mild steel that measure 1/16″ × 1 1/2″ × 6″ (1.6 mm × 40 mm × 150 mm).
2. Clean the surfaces of both pieces.
3. Place one piece on the other with a 3/4″ overlap (19.1 mm).
4. Clamp the parts tightly together.
5. Select the correct torch tip and welding rod size. Refer to Figure 20-36.
6. Start the welding outfit and set the working pressure. Light the torch and adjust to a neutral flame.
7. Tack weld the pieces together in three places on both sides.
8. Remove the clamps.
9. Form a C-shaped weld pool before adding the welding rod to the pool. Weld both sides.

Inspection:

The completed welds should have convex weld beads with a consistent width and even ripples.

Hold the torch tip at a 45° work angle and a 35°–45° travel angle. See **Figure 24-8**. Both surfaces must be heated evenly. If you notice one piece heating more than the other, move the tip slightly to compensate.

Welding Rod Position

The welding rod should be placed into the upper edge of the weld pool. This reduces the possibility of undercutting on the vertical surface. The rod should be held at a 15°–45° angle to the weld axis. The welding rod should be used to fill any undercutting. If undercutting does occur, aim the flame a few degrees more toward the vertical piece. The pressure of the welding gases also helps to reduce sagging and undercutting.

Weld Pool

A *C*-shaped weld pool eventually forms as the metal melts. The *C*-shaped pool shows that the surfaces of both pieces are melting. The welding rod is inserted only after the weld pool forms.

Welding a square-groove, bevel, or V-groove outside corner joint is similar to welding a butt joint. The edge of one piece of metal and the surface of another piece are heated and fused together. More heat is required to melt the surface, especially when welding thin metal. The flame is usually pointed more toward the surface of the joint than at the edge, **Figure 24-9**.

Butt Joint

The pieces of metal for a horizontal butt weld are arranged as shown in **Figure 24-10**. The pieces may be tack welded to hold them in correct alignment.

Various techniques are used to create strong welds and well-formed weld beads. Most thicknesses of metal can be welded using the forehand welding

Exercise 24-2 Inside Corner Joint in the Horizontal Welding Position

1. Obtain two pieces of mild steel that measure 1/16″ × 1 1/2″ × 6″ (1.6 mm × 40 mm × 150 mm).
2. Clean the surfaces of both pieces.
3. Place the metal together to form an inside corner joint. Use a welding positioner or metal block to support the vertical piece.
4. Select the correct torch tip and welding rod size. Refer to Figure 20-36.
5. Start the welding outfit and set the working pressure. Light the torch and adjust to a neutral flame.
6. Tack weld the pieces in three places.
7. Hold the torch tip, flame, and welding rod at the suggested angles. Refer to Figure 24-8 for angles.
8. Melt the two pieces evenly and watch for a C-shaped weld pool to form. When the pool forms, add the welding rod to the upper edge of the pool.
9. Continue this procedure to the end of the joint.

Inspection:

The finished weld should not penetrate through either piece. The weld bead should be convex, even in width, and have evenly spaced ripples.

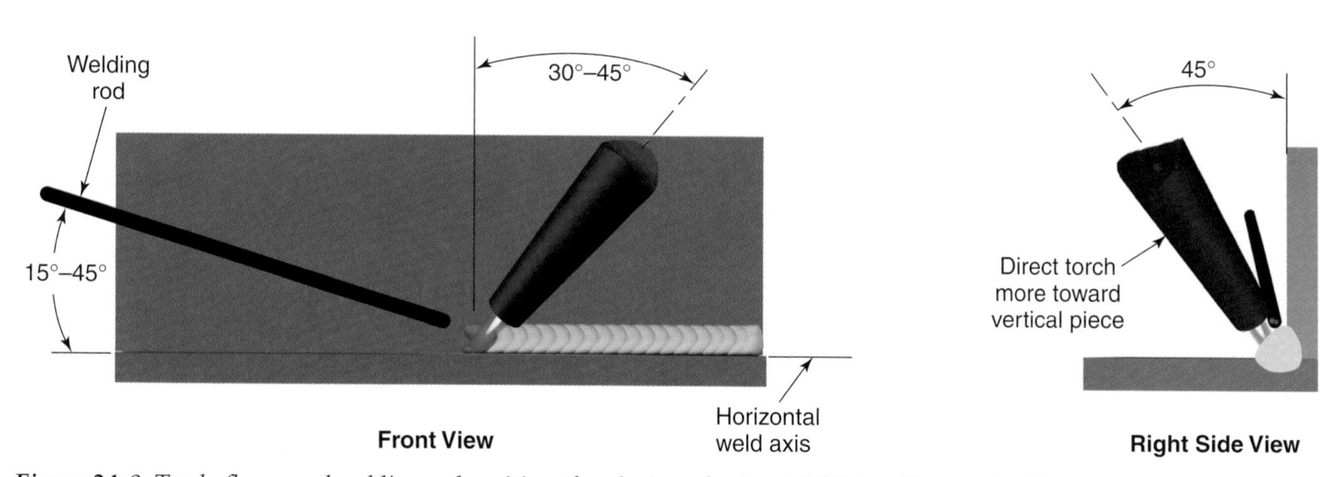

Figure 24-8. Torch, flame, and welding rod positions for placing a horizontal fillet weld on an inside corner or T-joint. The flame should be directed toward the vertical piece to help oppose the force of gravity on the weld pool.

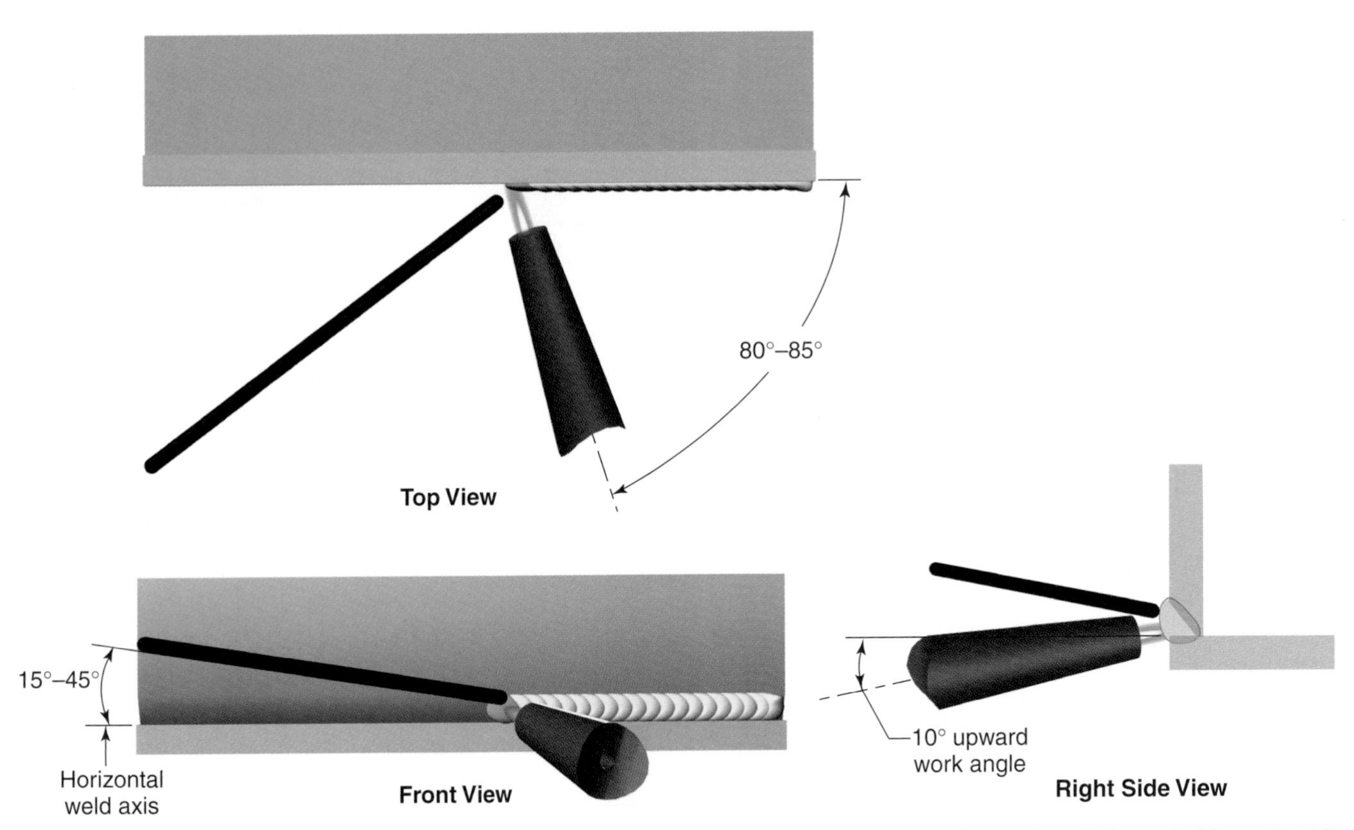

Figure 24-9. Torch, flame, and welding rod positions for welding an outside corner joint. The torch and tip are held at a 5°–10° travel angle from the surface of the metal (80°–85°). The torch should be angled up slightly, at a 5°–10° work angle to counteract gravity. These same angles are also used on a horizontal butt joint.

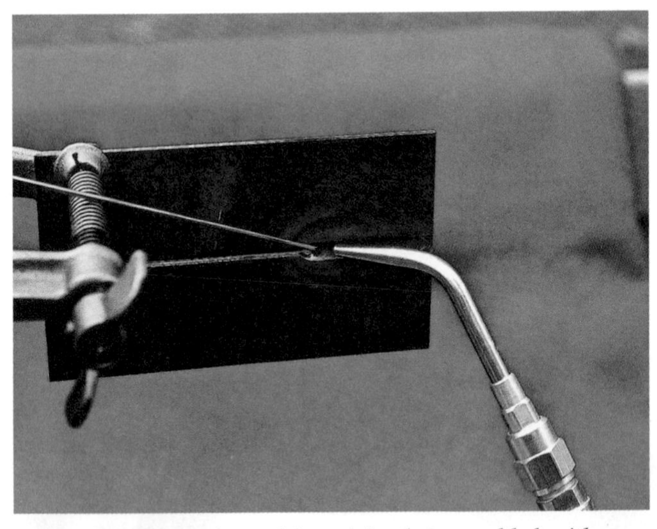

Figure 24-10. Horizontal butt joint being welded with oxyacetylene welding (OAW). Note the slight upward angle of the torch. The pressure of the welding gases is used to oppose the force of gravity on the weld pool.

method. The downward pull of gravity tends to deform the weld bead in all out-of-position welding. The force of the welding gases is used to overcome the downward pull of gravity. Careful placement of the welding rod into the upper edge of the pool also helps to overcome gravity.

Torch and Flame Position

The centerline of the torch tip and flame are pointed upward toward the weld axis. The upward work angle should be 5°–10°, as shown in **Figure 24-11**. The upward angle of the flame and the force of the welding gases will help to prevent the weld bead from ***sagging*** (flowing downward). The angle may be changed to form an acceptable weld bead. The metal near the centerline of the weld on both surfaces must be heated and melted evenly. The torch tip should be held at a 5°–10° travel angle. A 35°–45° travel angle is generally used for welding butt joints in the other positions. Travel and work angles vary depending on setup conditions in order to counteract gravity.

Welding Rod Position

The welding rod is held at a 15°–45° angle from the base metal. This angle is generally used for all welding positions. When horizontal welding, the rod is placed into the upper edge of the molten weld pool. This permits the rod to melt and cool the weld pool before the molten metal sags.

Weld Pool

The weld pool may become too hot while welding. To cool the pool, withdraw the flame tip about 1″ (25 mm) from the base metal and return it to within

1/16″–1/8″ (1.6 mm–3.2 mm). This rapid up-and-down motion allows the weld pool to cool, **Figure 24-12**. If undercutting occurs on the upper piece, insert the welding rod in the pool more often. This fills the undercut (depression). Increasing the upward angle of the flame also helps to force molten metal higher into the weld pool.

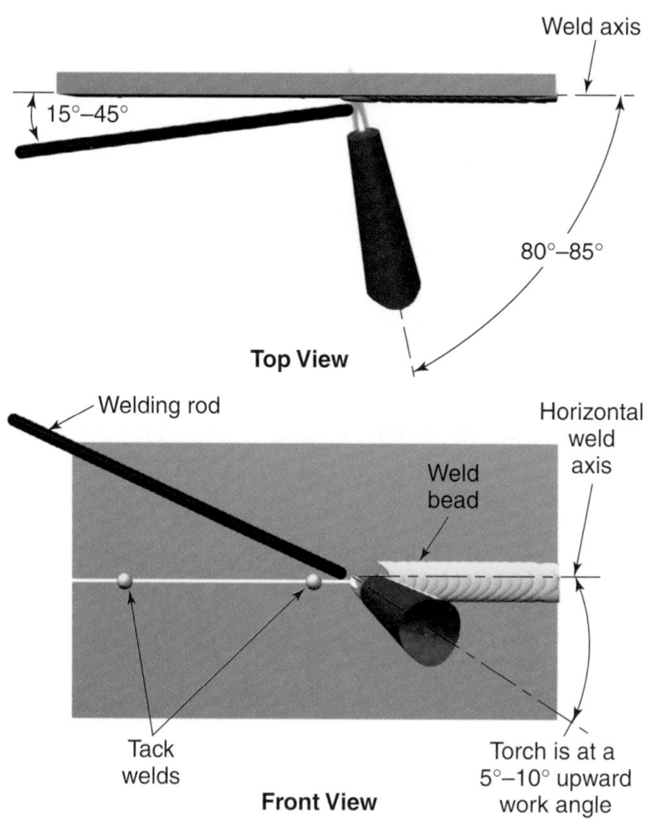

Figure 24-11. Torch, flame, and welding rod positions for welding a horizontal butt joint.

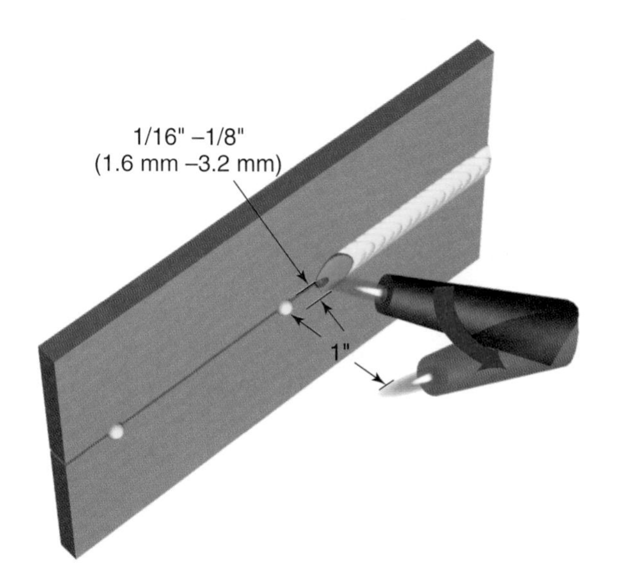

Figure 24-12. The flip motion of the torch and flame. This motion may be used if the metals tend to overheat. It allows time for the weld pool to cool slightly.

Exercise 24-3 Square-Groove Butt Joint in the Horizontal Welding Position

1. Obtain two pieces of mild steel 1/16″ × 1 1/2″ × 6″ (1.6 mm × 40 mm × 150 mm).
2. Clean the edges.
3. Select the correct torch tip and welding rod size. Refer to Figure 20-36.
4. Start the welding outfit and set the working pressure. Light the torch and adjust to a neutral flame.
5. Place the pieces in the flat welding position and tack weld in three places.
6. Position the pieces for horizontal welding.
7. Weld the joint using the suggested torch tip and flame angles.

Inspection:

The completed weld should have a convex weld bead with even ripples, an even bead width, and 100% penetration.

Welding in the Vertical Welding Position

Vertical welding is also called welding in the 3G position. Welds in the vertical welding position must meet either of the following sets of conditions.

Condition A

- Weld axis must be between 80° and 90° from horizontal.
- Weld face must be between 80° and 90° from horizontal.

Condition B

- Weld axis must be between 15° and 80° from horizontal.
- Weld face must be between 80° and 280°.
- Welding is performed from upper side of the joint. For practice welding, place the weld axis as close to vertical as possible.

Vertical welding on any type of joint is generally done using the forehand method. The weld begins at the lowest point on the joint and progresses upward.

The upward pressure of the welding gases keeps the molten metal in the weld pool from sagging. Placing the welding rod into the upper edge of the weld pool also helps to prevent sagging. A pencil-like grip is effective and comfortable for vertical welding. This is an especially good way to hold a light-duty torch.

Lap Joint

A fillet weld is used on a lap joint. For practice, the weld axis is positioned about 80°–90° from horizontal. In a lap joint, the edge of one piece and the surface of another piece are heated.

Torch and Flame Position

The centerline of the torch tip and flame is held at a 60° travel angle (about a 30° angle from the base metal surface or weld axis). The flame is pointed in the direction of motion. See **Figure 24-13**.

A 30° work angle is generally used, with most of the heat directed at the surface. This is to avoid excessive melting away of the edge. The tip of the flame should be about 1/16″–1/8″ (1.6 mm–3.2 mm) from the weld axis.

Welding Rod Position

The welding rod is usually positioned above the torch and flame. It is held about 30°–45° from the base metal or weld axis. The end of the welding rod should be held close to the weld pool. The end is preheated by keeping it in the reflected heat of the torch flame, **Figure 24-14**.

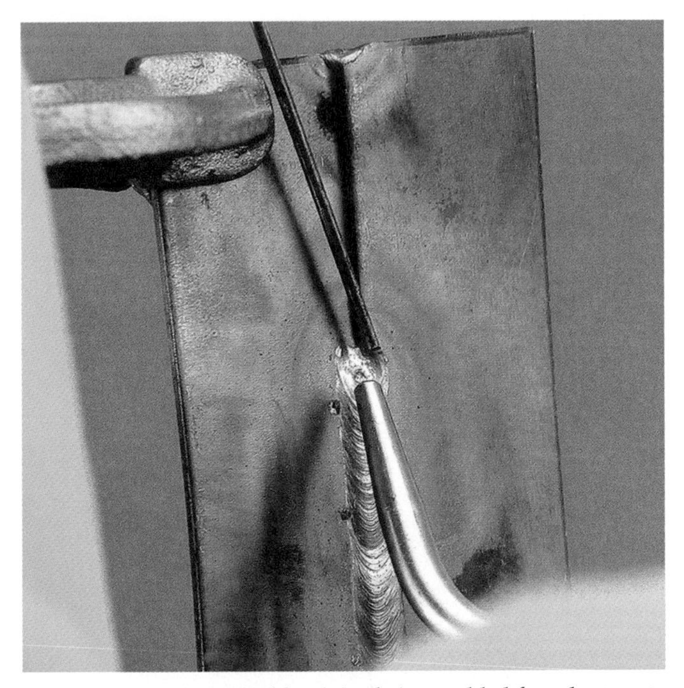

Figure 24-14. *A vertical lap joint being welded from bottom to top. The flame is aimed more toward the surface piece and less at the edge. The welding rod is added from above.*

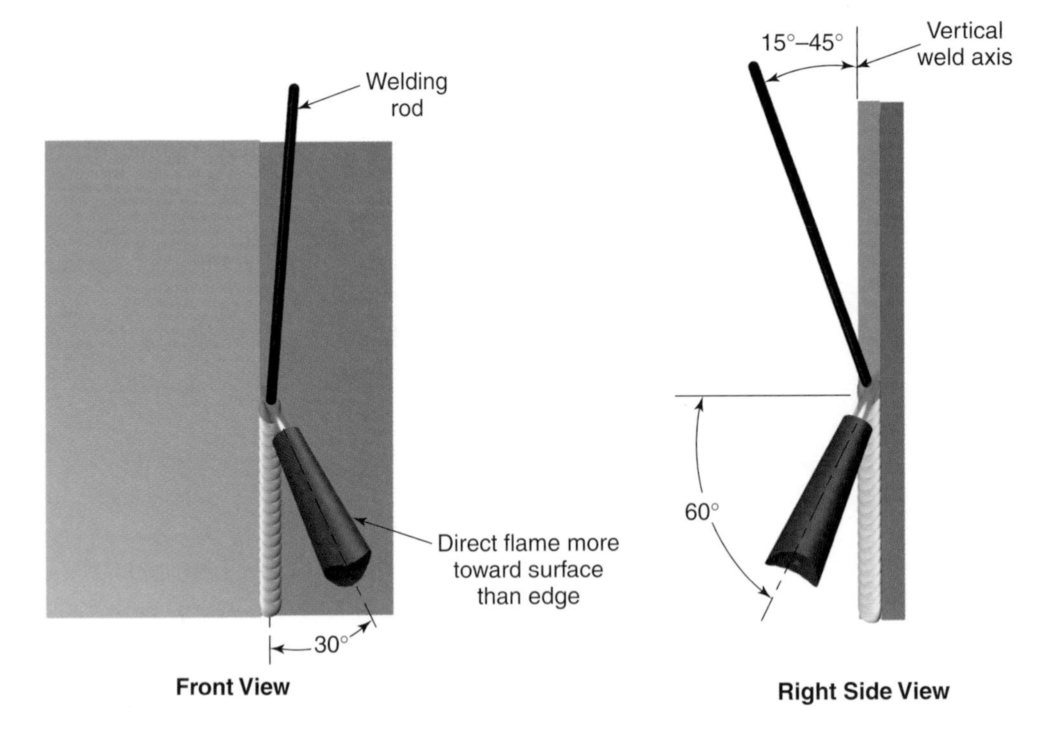

Figure 24-13. *Torch, flame, and welding rod positions for welding a vertical lap joint. Compare the positions with those used for welding a lap joint in the flat welding position.*

Weld Pool

The welding rod is added when a C-shaped weld pool forms. It is added to the upper edge of the weld pool to prevent sagging. The welding rod should be added to the pool often enough to create a convex weld bead.

Corner and T-Joints

A fillet weld is used to weld the inside corner and T-joint on thin metal. A square-groove butt joint may be used for an outside corner joint on metal under 3/16″ (4.8 mm).

The edges of metal over 3/16″ (4.8 mm) thick are cut or ground to provide for better penetration. On thick metal, a fillet weld is commonly added to a bevel-groove weld on the inside corner and T-joint, **Figure 24-15**.

The weld axis for practice welds is usually between 80° and 90° for the vertical position. Practice pieces may be held in a weld positioner, **Figure 24-16**, or supported by firebricks or steel blocks.

Exercise 24-4 Fillet Weld on a Lap Joint in the Vertical Welding Position

1. Obtain two pieces of 1/8″ × 1 1/2″ × 6″ (3.2 mm × 40 mm × 150 mm).
2. Clean the surfaces of both pieces.
3. Place one piece on the other with a 3/4″ (19.1 mm) overlap.
4. Clamp the pieces tightly together.
5. Select the correct torch tip and welding rod size. Refer to Figure 20-36.
6. Start the welding outfit and set the working pressure. Light the torch and adjust to a neutral flame.
7. Tack weld in three places on both sides.
8. Remove the clamps and place the weld axis in the vertical position.
9. Wait for the C-shaped weld pool to form before adding the welding rod. Make a fillet weld on both sides of the metal.

Inspection:

The completed welds should have convex weld beads with an even width and evenly spaced ripples.

Torch and Flame Position

The torch and flame are held at a 45° work angle to the metal surfaces when making a weld on an inside corner or T-joint. The torch and flame are also held at a 45°–60° travel angle, **Figure 24-17**. The weld is usually made from the bottom and is advanced upward using a forehand motion. Heat should be applied evenly, since two surfaces are being joined.

Welding an outside corner joint is similar to welding a vertical butt joint, **Figure 24-18**. However, on the outside corner joint, one piece has an exposed edge and the other piece is a surface. More heat should be aimed toward the surface to avoid melting the edge excessively, **Figure 24-19**. The torch should be held 30°–45° up from the base metal.

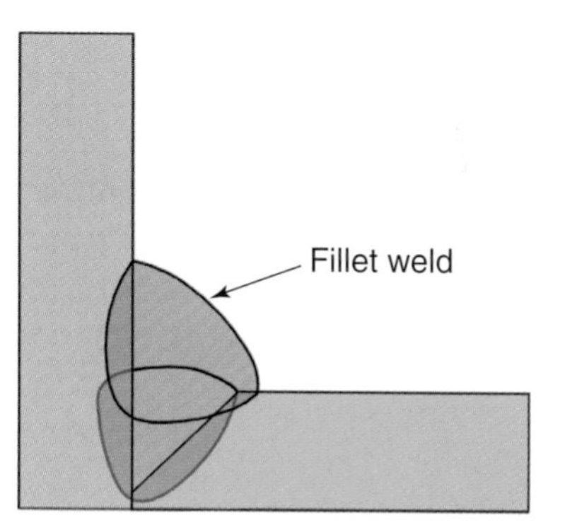

Figure 24-15. *A fillet weld on thick metal. The fillet weld is on top of the weld made on the bevel-groove inside corner joint. This weld requires two weld passes.*

Figure 24-16. *A T-joint held in the vertical position in an adjustable fixture. The heat is applied equally to both surfaces. The welding rod is added from above.*

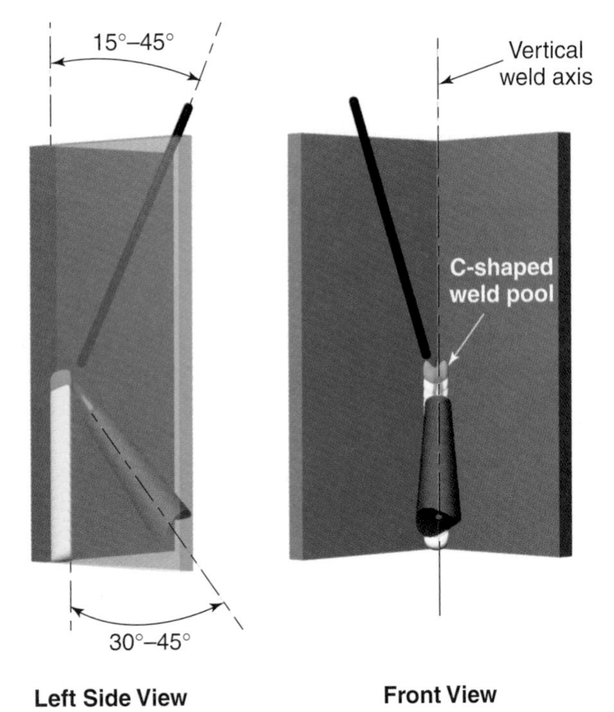

Figure 24-17. A T-joint being welding in the vertical welding position. The torch tip and flame should be aligned with the weld axis and held at a 30°–45° angle from the weld axis.

Figure 24-19. A weld in progress on an outside corner joint in the vertical welding position.

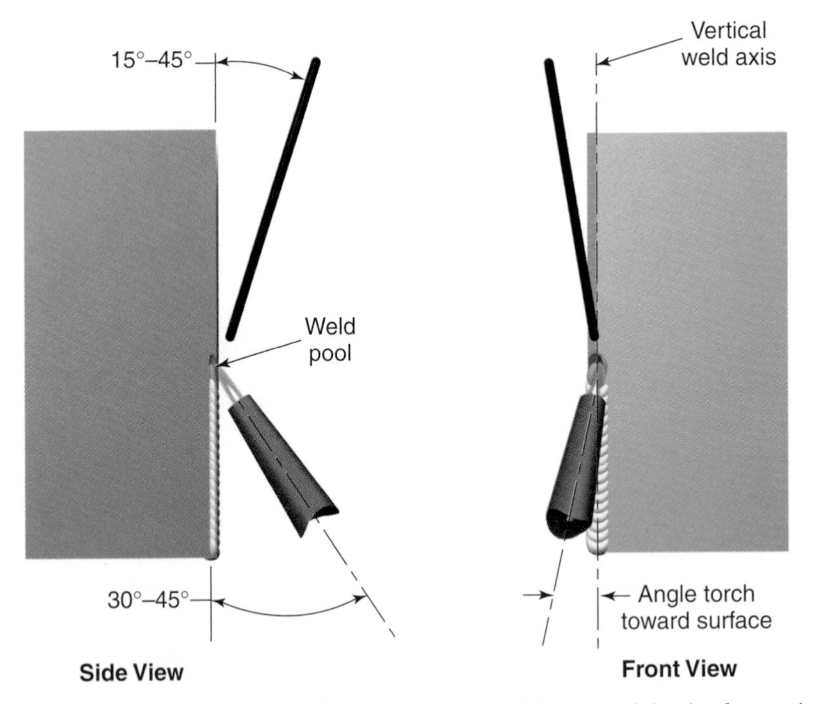

Figure 24-18. Torch, flame, and welding rod positions for welding an outside corner joint in the vertical welding position.

Welding Rod Position

The welding rod should be held between 15° and 45° from the base metal. The end of the welding rod should be held near the weld pool to keep it hot, or preheated. The rod should be added to the upper edge of the weld pool.

Weld Pool

When welding an inside corner or T-joint, a C-shaped weld pool should be formed before adding the welding rod. The C-shaped pool indicates that both pieces have melted. An oval pool is formed when welding a square or bevel-grooved outside corner joint.

Exercise 24-5 Fillet Weld on a T-Joint in the Vertical Welding Position

1. Obtain two pieces of mild steel that measure 1/16″ × 1 1/2″ × 6″ (1.6 mm × 40 mm × 150 mm).
2. Clean the surfaces of both pieces.
3. Select the correct torch tip and welding rod size. Refer to Figure 20-36.
4. Start the welding outfit and set the working pressure. Light the torch and adjust to a neutral flame.
5. Tack weld the pieces in three places to form a T. This may be done in the flat position.
6. Place the weld axis into a vertical position.
7. Hold the torch tip, flame, and welding rod at the suggested angles. See Figure 24-17.
8. Melt the two pieces evenly and watch for a C-shaped weld pool to form.
9. When the weld pool forms, add the welding rod to the upper edge of the pool.
10. Continue this procedure to the end of the joint.

Inspection:

The completed weld should not penetrate through either piece. The weld bead should be convex, even in width, and have evenly spaced ripples.

Butt Joint

When welding a butt joint in the vertical welding position, the weld axis must be 80°–90° from horizontal. The weld face must be 90°–100° from a straight down position, **Figure 24-20**. The top of the plate can be tipped away from the welder but not toward the welder. If the top of the plate is tipped toward the welder, the weld will be made in an overhead position rather than a vertical position.

Both metal surfaces must be heated equally so that the weld pool forms evenly on both sides of the weld axis. On thin metal, very little side-to-side motion is required. On thicker metal, some side-to-side motion is required. A ***crescent-shaped motion*** works well, **Figure 24-21**.

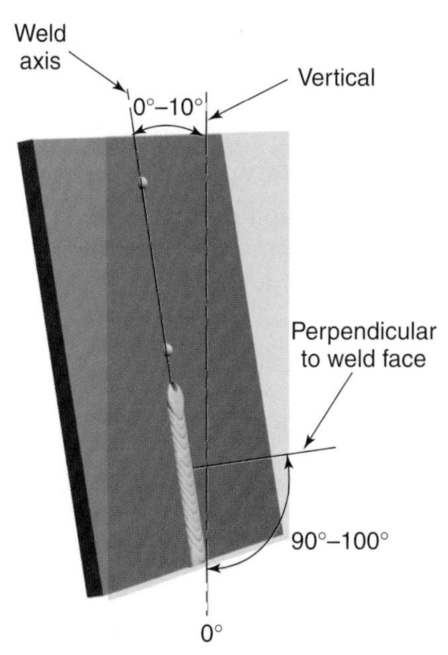

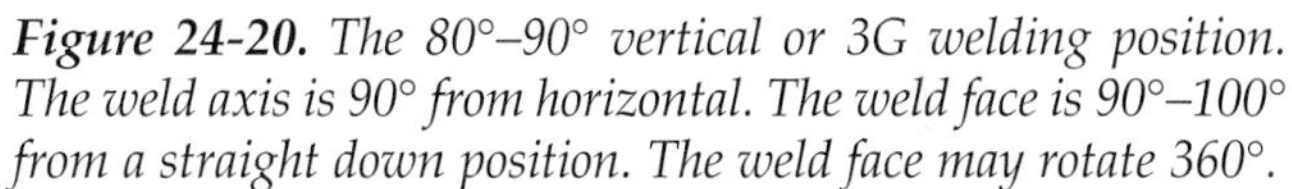

Figure 24-20. *The 80°–90° vertical or 3G welding position. The weld axis is 90° from horizontal. The weld face is 90°–100° from a straight down position. The weld face may rotate 360°.*

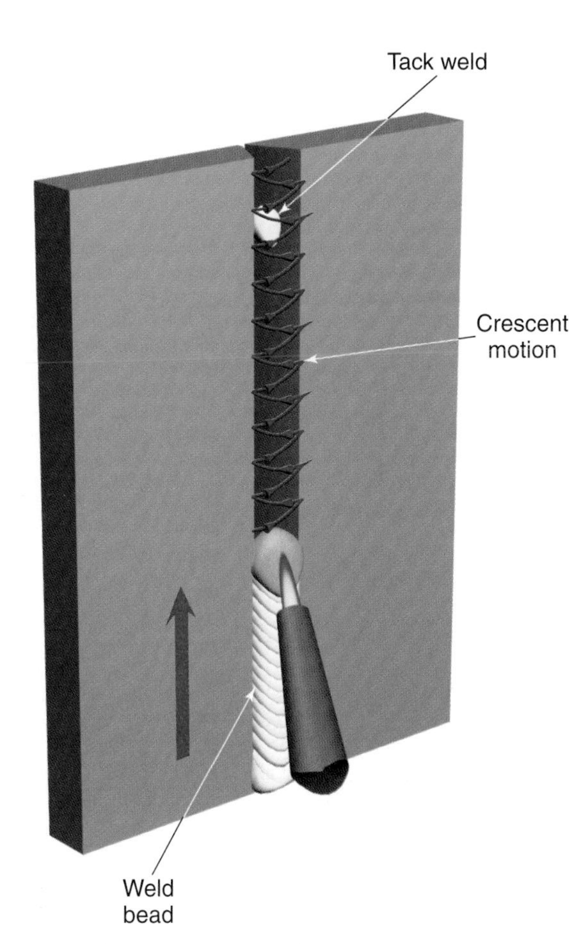

Figure 24-21. *Crescent-shaped weaving motion. This motion is used to make wide weld beads on thick metal joints.*

Torch and Flame Positions

The centerline of the torch and flame should be held at a 0° work angle and a 60° travel angle, **Figure 24-22**. The tip of the flame should be about 1/16″–1/8″ (1.6 mm–3.2 mm) away from the surface of the weld pool for the greatest amount of heat. If the pool becomes too large, remove the flame momentarily, and then return it to its previous height. This flip motion allows for the weld pool to cool and helps prevent sagging of the pool and weld bead.

Welding Rod Position

The welding rod is usually held above the torch and flame, at an angle of 15°–45° from the metal surface. It is preheated by the reflected heat from the welding flame, **Figure 24-23**.

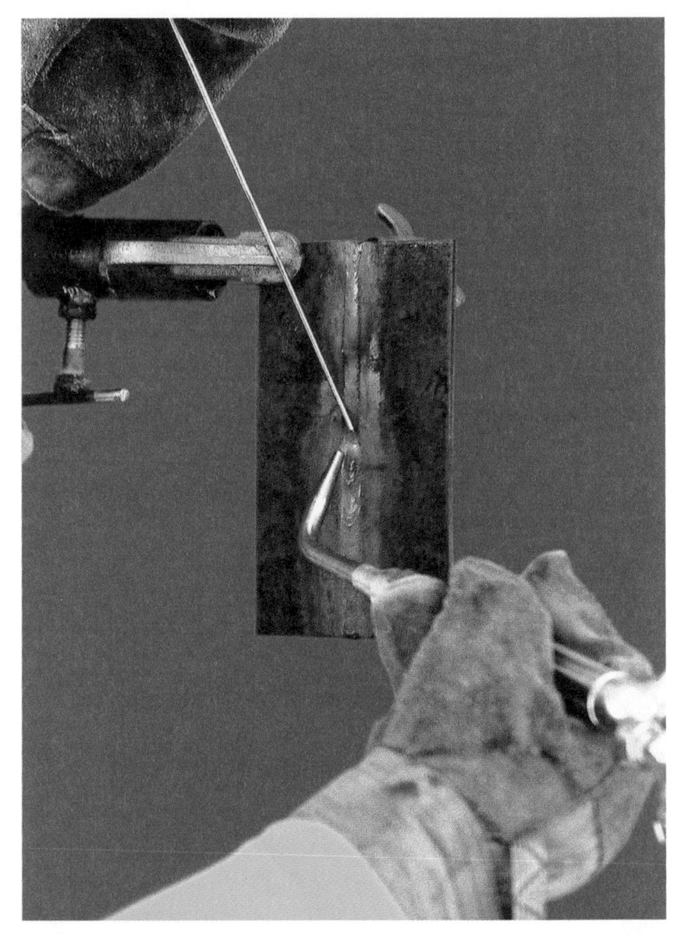

Figure 24-23. *The second weld pass on a square groove butt joint. Note that the welding rod is added from above.*

Welding in the Overhead Welding Position

With care and practice, strong welds and attractive weld beads can be made in the overhead welding position. Welds in the overhead (4G) position must meet the following conditions:

- The weld axis is between 0°–80°.
- The weld face is between 0°–80° or 280°–360°.
- The weld is made from the lower side of the base metal.

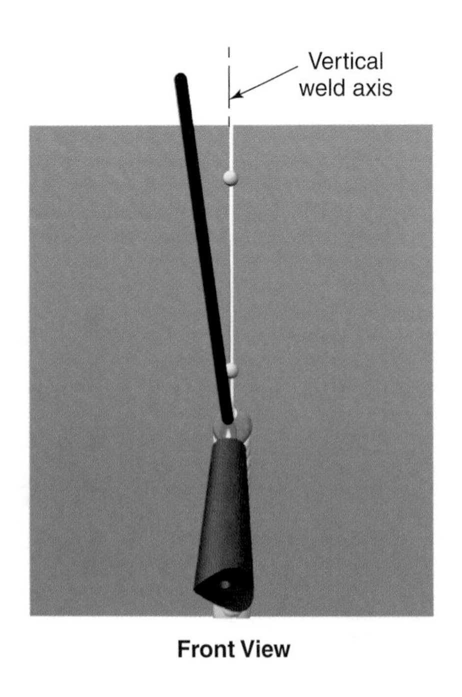

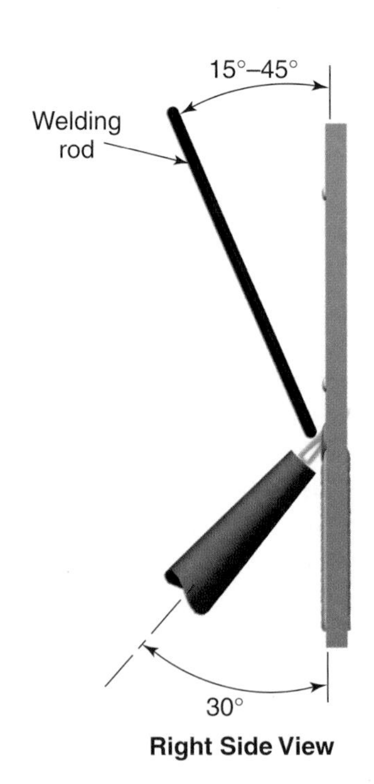

Figure 24-22. *Torch, flame, and welding rod positions for welding a vertical butt joint. Compare the angles to those used in the flat welding position.*

Coveralls or jackets should be buttoned at the collar to prevent hot metal from getting inside clothing. Pockets must be buttoned. Flammable items should not be carried in pockets. A cap should be worn. Flame-resistant leather clothing may also be worn for extra protection.

Practice and qualification welds are generally made with the weld axis horizontal and the weld face at 360° (0°), **Figure 24-24**.

When working in the overhead welding position, a welder usually stands away from the weld axis, similar to flat welding. This allows a good view of the weld in progress. It also decreases the chance of molten metal falling on the welder. Some welders stand in line with the weld axis. This position allows for a good view of the weld and keyhole. However, welding away from the body is not as efficient as welding across the body.

The torch, flame, and welding rod angles are the same for overhead welding and flat welding. However, these angles are rotated 180° into an overhead welding position. See **Figures 24-25** and **24-26**.

Welds made in the overhead welding position are usually made with a forehand motion. Right-handed welders work from right to left, **Figure 24-27**.

The welding techniques learned in previous positions and joints may be used in the overhead welding position. An overheated weld pool is the greatest problem in overhead welding. If the pool overheats, briefly move the flame about 1″ (25 mm) from the pool, and then return it. This torch movement permits the weld pool to cool slightly.

Making a fillet weld in the overhead position is similar to welding in other positions. The welder must wait for a C-shaped pool to form before adding the welding rod.

Exercise 24-6–Square-Groove Butt Joint in the Vertical Welding Position

1. Obtain two pieces of mild steel that measure 1/16″ × 1 1/2″ × 6″ (1.6 mm × 40 mm × 150 mm).
2. Clean the edges of both pieces.
3. Select the correct torch tip and welding rod size. Refer to Figure 20-36.
4. Start the welding outfit and set the working pressure. Light the torch and adjust to a neutral flame.
5. Tack weld the pieces in three places in the flat position.
6. Position the weld axis for vertical welding.
7. Begin welding at the bottom of the joint and progress upward. Use the suggested torch and welding rod angles. See Figure 24-22.

Inspection:

The completed weld should have a convex weld bead with even ripples, an even bead width, and 100% penetration.

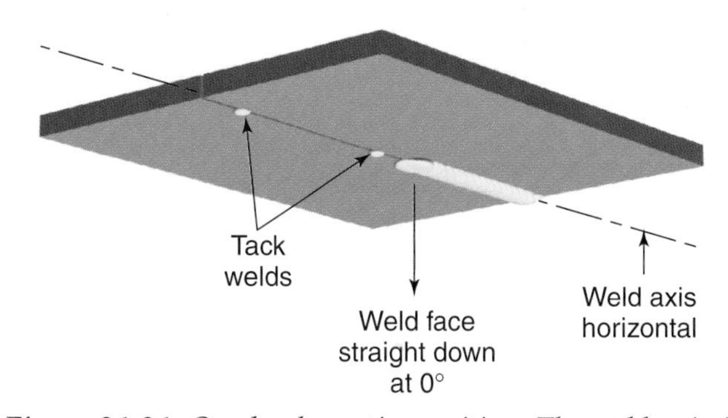

Figure 24-24. Overhead practice position. The weld axis is horizontal and the weld face is at 360°.

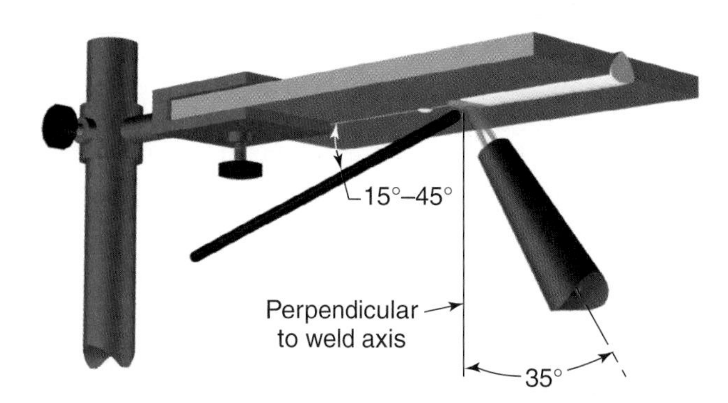

Figure 24-25. Torch, flame, and welding rod positions and angles for welding a butt joint in the overhead welding position. These positions are mirror images of welding in the flat welding position.

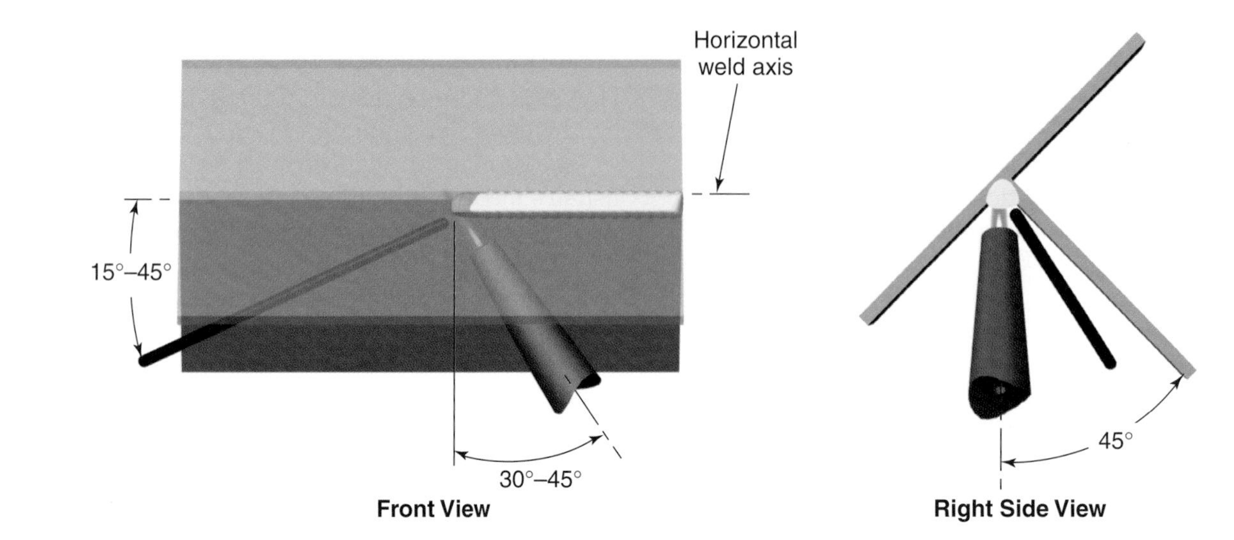

Figure 24-26. Torch, flame, and welding rod positions used for welding a T-joint in the overhead welding position.

Exercise 24-7 Fillet Weld on a T-Joint in the Overhead Welding Position

1. Obtain two pieces of mild steel that measure 1/16″ × 1 1/2″ × 6″ (1.6 mm × 40 mm × 150 mm)
2. Clean both surfaces on each piece.
3. Select the correct torch tip and welding rod size. Refer to Figure 20-36.
4. Start the welding outfit and set the working pressure. Light the torch and adjust to a neutral flame.
5. In three places, tack weld the pieces in a T-joint formation.
6. Using a weld positioner, place the weldment in the overhead welding position as shown in Figure 24-27.
7. Hold the torch tip, flame, and welding rod at the suggested angles. Refer to Figure 24-26.
8. Melt the two pieces evenly and watch for a C-shaped weld pool to form.
9. Add the welding rod to the upper edge of the C-shaped pool when it forms.
10. Continue this procedure to the end of the joint.

Inspection:

The completed weld should not penetrate through either piece. The weld bead should be convex, even in width, and have evenly spaced ripples.

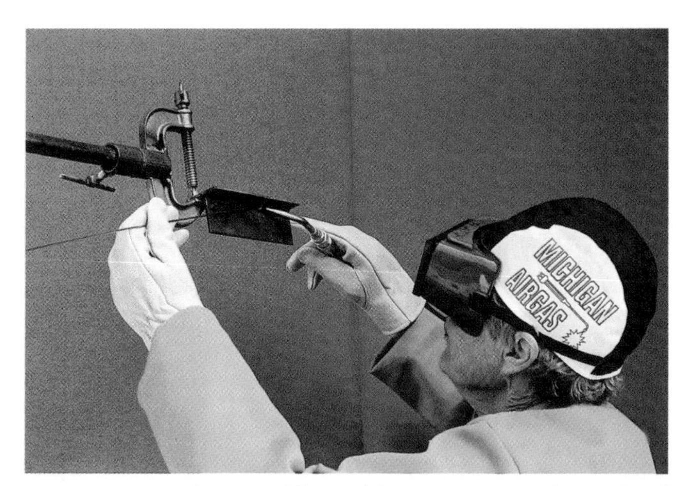

Figure 24-27. Placing a fillet weld on a T-joint in the overhead welding position.

Exercise 24-8 Square-Groove Butt Joint in the Overhead Welding Position

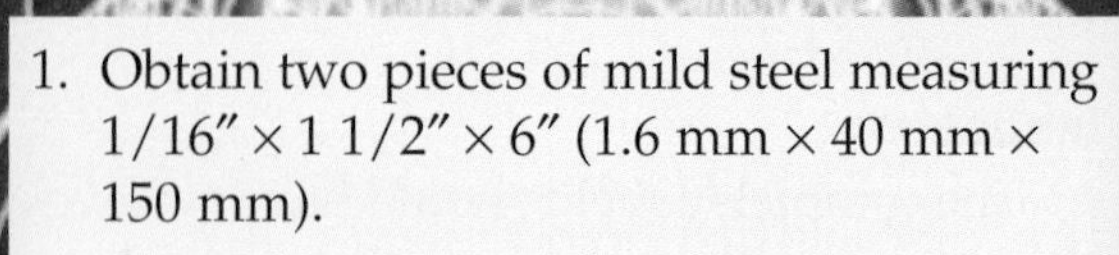

1. Obtain two pieces of mild steel measuring 1/16″ × 1 1/2″ × 6″ (1.6 mm × 40 mm × 150 mm).
2. Clean the edges of the pieces.
3. Select the correct torch tip and welding rod size. Refer to Figure 20-36.
4. Start the welding outfit and set the working pressure. Light the torch and adjust to a neutral flame.
5. Tack weld the pieces in three places in the flat welding position.

6. Position the pieces for overhead welding.
7. Weld the joint using the suggested torch and rod angles, Figure 24-25. Add the welding rod to the upper edge of the molten weld pool.

Inspection:

The completed weld should have a convex weld bead with even ripples and bead width. It should also have 100% penetration.

Summary

- Welding in a position other than the flat position is commonly referred to as out-of-position welding. More training and talent are required to weld out of position. Those who can weld out of position have a chance to earn more money.
- Many out-of-position welds are performed at or above eye level. The protective clothing worn for out-of-position welding is similar to that worn for flat position welding.
- Welds made in the horizontal (2G) welding position must meet the following conditions: the weld axis must be within 15° of horizontal and the weld face must be between 80°–150° or 210°–280°.
- When welding a lap joint in the horizontal position, the centerline of the tip and flame should point more at the surface than at the edge. Hold the torch at a work angle of 30°–45° and a travel angle of 45°. Hold the welding rod at a 15°–45° angle from the metal surface and aligned with the weld axis. A C-shaped weld pool should be formed before adding the welding rod. Insert the welding rod at the upper edge of the pool.
- When welding corner and T-joints in the horizontal position, the centerline of the torch tip and flame should be centered between the surfaces of the two pieces of metal. Hold the torch tip at a 45° work angle and a 35°–45° travel angle. The rod should be held at a 15°–45° angle to the weld axis. After a C-shaped weld pool forms, the welding rod should be placed into the upper edge of the weld pool.
- When welding a butt joint in the horizontal position, the centerline of the torch tip and flame are pointed at an upward 5°–10° work angle and held at an 80°–85° travel angle. The welding rod is held at a 15°–45° angle from the base metal. The welding rod is placed into the upper edge of the molten weld pool.
- Welds in the vertical (3G) welding position must meet either of the following sets of conditions: the weld axis is between 80° and 90° from horizontal and the weld face is between 80° and 90° from horizontal *or* the weld axis is between 15° and 80° from horizontal, the weld face is between 80° and 280° from horizontal, and welding is performed from the upper side of the joint.
- When welding a lap joint in the vertical position, the centerline of the torch tip and flame is held at about a 30° work angle and a 60° travel angle. The flame is pointed in the direction of motion. The welding rod is held about 15°–45° from the base metal or weld axis. When a C-shaped weld pool forms, the welding rod is added to the upper edge of the weld pool.
- When welding corner joints and T-joints in the vertical position, the torch and flame are held at a 45° work angle and a 45°–60° travel angle. The weld is usually made from the bottom and is advanced upward using a forehand motion. The welding rod should be held between 15° and 45° from the base metal. When a C-shaped weld pool forms, the welding rod should be added to the upper edge of the weld pool. (An oval pool is formed when welding a square or bevel-grooved outside corner joint.)

(continued)

- When welding a butt joint in the vertical welding position, the weld axis is 80°–90° from horizontal and the weld face is 90°–100° from a straight down position. The torch is held at a 0° work angle and a 30° travel angle. The welding rod is usually held above the torch and flame, at an angle of 15°–45° from the metal surface.
- Welds in the overhead (4G) position must meet the following conditions: the weld axis is between 0°–80°, the weld face is between 0°–80° or 280°–360°, and the weld is made from the lower side of the base metal.
- The torch, flame, and welding rod angles are the same for overhead welding and flat welding. However, these angles are rotated 180° into an overhead welding position. Welds made in the overhead welding position are usually made with a forehand motion.
- An overheated weld pool is the greatest problem in overhead welding. If the pool overheats, briefly move the flame about 1″ (25 mm) from the pool, and then return it. This torch movement permits the weld pool to cool slightly.

Review Questions

Write your answers on a separate sheet of paper. Please do *not* write in this book.

1. What does the phrase "welding out of position" mean?
2. If the torch tip is held at a 30°–45° travel angle , this angle can also be described as _____°–_____° from the base metal or weld axis. The welding rod should be held between _____° and _____° from the base metal or weld axis for all welding positions.
3. Why should your collar and pockets be buttoned when welding out of position?
4. Name two pieces of protective clothing that are strongly recommended when welding in an overhead welding position.
5. The weld face of a horizontal weld must be within _____° and _____° or _____° and _____°.
6. The weld axis of a horizontal weld must be within _____° of horizontal.
7. List two methods used to overcome a sagging weld pool, undercutting, and the pull of gravity when making a horizontal weld.
8. Describe two techniques used to prevent the edge from melting too rapidly when lap welding.
9. What can the welder do to cool the weld pool when welding out of position?
10. What does a C-shaped weld pool indicate to the welder when fillet welding a lap, inside corner, or T-joint?

Chapter 25

Brazing and Braze Welding

Learning Objectives

After studying this chapter, you will be able to:

- Define brazing and braze welding and state the major difference between these two processes.
- Select the correct torch tip, filler metal, and flux for brazing and braze welding.
- List the special safety precautions for brazing and braze welding.
- Explain the difference between a welded joint and a brazed or braze welded joint.
- Describe the procedure for brazing.
- Braze metal.
- Describe the procedure for braze welding.
- Braze weld metal.

braze welding
brazement
brazing
brazing rod
cadmium (Cd)
capillary action
flux
liquidus
solidus
toxic
zinc (Zn)

Brazing and Braze Welding Principles

The American Welding Society defines ***brazing*** as "a group of welding procedures that cause materials to join by heating them to the brazing temperature in the presence of a filler metal. The filler metal has a liquidus temperature above 840°F (450°C) and below the solidus temperature of the base metal. The filler metal is distributed between the close-fit surfaces of the joint by capillary action." ***Liquidus*** is the temperature at which a metal or alloy becomes completely liquid. ***Solidus*** is the temperature at which a metal or alloy begins to melt.

Braze welding is defined by AWS as "a welding process variation in which a filler metal, having a liquidus above 840°F (450°C) and below the solidus of the base metal, is used. Unlike brazing, in braze welding, the filler metal is not distributed in the joint by capillary action."

The major difference between brazing and braze welding is in the fit of the parts. Parts to be brazed must fit very tightly together. ***Capillary action*** is the process in which a liquid is drawn into the space between two tightly fitted mating surfaces. Capillary action occurs during brazing and depending on joint design may also occur during braze welding. Parts to be braze welded must fit closely together, but they do not need to fit as tightly as parts to be brazed.

The strength of the brazed joint depends upon the molecular attraction and adhesion of the brazed filler metal and the surface of the base metal. Molecular attraction occurs when the particles of the brazed filler metal combine with the surface particles of the base metal and form a bond. When the parts are not closely fitted, a weak brazed joint is created due to poor adhesion of the filler metal.

When fusion welding, the base metal is melted. When brazing or braze welding, the base metal is not melted. The brazing filler metal melts at a temperature above 840°F (450°C) and below the melting point of the base metal. If the filler metal melts below 840°F (450°C), it is a soldering process. Brazing and braze welding on steel is usually done at 840°F–2000°F (450°C–1093°C).

Brazing and braze welding are performed at temperatures lower than those required for fusion welding. Therefore, brazing and braze welding processes may be used to join thin metal. Also, because of the lower temperatures, metals may be joined without changing the characteristics produced by heat treatment.

A ***brazement*** is an assembly of parts to be brazed or braze welded. A brazed or braze welded part is not as strong as a similar part joined by fusion welding. The filler metal used when brazing or braze welding is not as strong as the base metal. In fusion welding, the filler metal is as strong or stronger than the base metal.

Oxyfuel gas welding equipment is commonly used to braze or braze weld. Fuel gases, such as acetylene, MPS (methylacetylene-propadiene), propane, natural gas, hydrogen, and city gas, may be used. Acetylene and MPS are commonly used because they produce the highest temperatures when combined and burned with oxygen.

Brazing can also be done in a furnace. When furnace brazing, a preshaped piece of brazing filler metal is commonly placed in the joint. The part is then heated in a furnace. The filler metal melts when the part reaches the correct temperature. Hundreds of parts can be brazed simultaneously in this manner.

Brazing Filler Metal

Brazing filler metal is available in many shapes and sizes. It is offered as wire, metal strips, rods, powder, and preshaped forms to fit specific joints, **Figure 25-1**. Filler metal ***brazing rods*** are available in various diameters. Rod diameters include: 1/16″ (1.6 mm), 3/32″ (2.4 mm), 1/8″ (3.2 mm), 5/32″ (4.0 mm), 3/16″ (4.8 mm), and 1/4″ (6.4 mm). Rods are usually 36″ (.91 m) long and are sold by the pound (kilogram).

Brazing rods are available as bare wire or with a flux coating. The flux coating is applied during manufacture. Flux must be added to bare brazing rods during the brazing process. Flux-coated brazing rods are more expensive than bare wire rods. Flux-coated rods are more convenient to use but must be handled carefully to prevent the coating from coming off. **Figure 25-2** shows a flux-coated brazing rod being used to braze weld a lap joint.

A

B

Figure 25-1. *A—Brazing filler metal is available in wire or preshaped forms. B—The preshaped forms are placed on or in brazements before they are heated. (Handy & Harmon PMFG/Lucas Milhaupt, Inc.)*

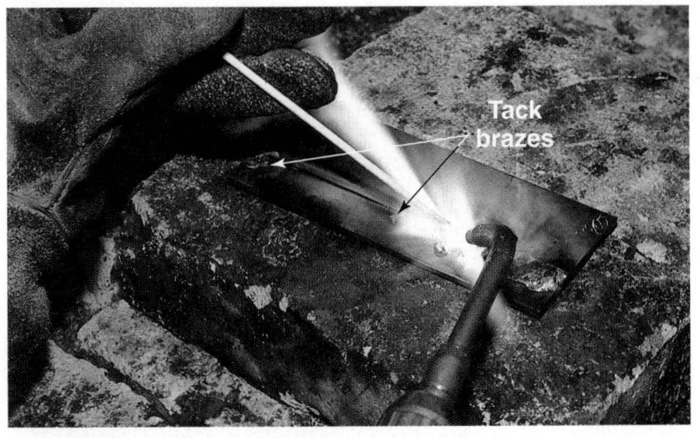

Figure 25-2. Braze welding a lap joint in the flat position. A flux-coated brazing rod is being used. Note the tack brazes.

Brazing filler metal can be any metal that melts between 840°F (450°C) and the melting point of the base metal. Aluminum and alloys with low melting points can be used to braze aluminum alloys with higher melting temperatures. Brass is used to braze or braze weld steel. Brass is an alloy of copper and zinc. Silver alloys are commonly used to braze copper, stainless steel, and a variety of other metals. Silver brazing is often incorrectly referred to as silver soldering. Silver brazing filler metal melts above 840°F (450°C). Therefore, silver brazing is not a soldering process. Gold alloys are also used as a filler metal when brazing. The table in **Figure 25-3** shows the filler metals suggested for use in joining various base metal combinations.

	Al & Al alloys	Mg & Mg alloys	Cu & Cu alloys	Carbon & low-alloy steels	Cast iron	Stainless steels	Ni & Ni alloys
Al & Al alloys	BAlSi	X	X	BAlSi	X	BAlSi	X
Mg & Mg alloys	X	BMg	X	X	X	X	X
Cu & Cu alloys	X	X	BAg, BAu, BCuP, RBCuZn	BAg, BAu, RBCuZn	BAg, BAu, RBCuZn	BAg, BAu,	BAg, BAu, RBCuZn
Carbon & low-alloy steels	BAlSi	X	BAg, BAu, RBCuZn	BAg, BAu, BCu, RBCuZn, BNi	BAg, RBCuZn	BAg, BAu, BCu, BNi	BAg, BAu, BCu, BNi RBCuZn
Cast iron	X	X	BAg, BAu, RBCuZn	BAg, RBCuZn	BAg, RBCuZn, BNi	BAg, BAu, BCu, BNi	BAg, BCu, RBCuZn
Stainless steel	BAlSi	X	BAg, BAu,	BAg, BAu, BCu, BNi	BAg, BAu, BCu, BNi	BAg, BAu, BCu, BNi	BAg, BAu, BCu, BNi
Ni & Ni alloys	X	X	BAg, BAu, RBCuZn	BAg, BAu, BCu, RBCuZn, BNi	BAg, BCu, RBCuZn	BAg, BAu, BCu, BNi	BAg, BAu, BCu, BNi
Ti & Ti alloys	BAlSi	X	BAg	BAg	BAg	BAg	X
Be, Zr & alloys (reactive metals)	X BAlSi (Be)	X	BAg	BAg, BNi	BAg, BNi	BAg, BNi	BAg, BNi
W, Mo, Ta, Cb & alloys (refractory metals)	X	X	BAg	BAg, BCu, BNi	BAg, BCu, BNi	BAg, BCu, BNi	BAg, BCu, BNi
Tool steels	X	X	BAg, BAu, RBCuZn, BNi	BAg, BAu, BCu, RBCuZn, BNi	BAg, BAu, RBCuZn, BNi	BAg, BAu, BCu, BNi	BAg, BAu, BCu, RBCuZn, BNi

*Filler metals:
BAlSi - Aluminum-silicon alloy
BAg - Silver-based alloy
BAu - Gold-based alloy
BCu - Copper-based alloy
BCuP - Copper-phosphorous alloy
RBCuZn - Copper-zinc alloy
BMg - Magnesium-based alloy
BNi - Nickel-based alloy

***Figure 25-3.** Brazeable metal combinations and suggested filler metals. This table is used when you are brazing two different metals. Locate one metal in the left column and the other metal in the top row of the table. The suggested filler metal is found at the intersection of the vertical column and horizontal row. An X indicates that brazing this combination is not recommended. RB indicates that the metal is suitable for resistance brazing.*

Brazing filler metal is generally an alloy of two or more metals. **Figure 25-4** lists the metals (with their chemical symbols) that are combined to create various AWS brazing filler metals. The chemical composition of a filler metal determines its solidus and liquidus temperatures.

Generally, when an alloy is formed of two metals, the liquidus of the alloy will be lower than the melting temperature of either of the metals. For example, the liquidus for the alloy BAu-2 in **Figure 25-5** is 1635°F (891°C). The liquidus for gold (Au) is 1945°F (1064°C); for copper (Cu), it is 1981°F (1083°C). When an alloy is formed of three or more metals, the resulting liquidus is not always below the melting temperature of each metal.

Good ventilation should be provided wherever brazing and braze welding are done. Some filler metals contain *zinc (Zn)* or *cadmium (Cd)*. Both zinc and cadmium are *toxic* (poisonous). Also, fumes from some fluxes and metal coatings may be toxic.

Chemical Symbols			
Ag	Silver	Mn	Manganese
Al	Aluminum	Mo	Molybdenum
Au	Gold	Ni	Nickel
B	Boron	P	Phosphorous
Be	Beryllium	Pd	Palladium
C	Carbon	Si	Silicon
Cb	Columbium	Sn	Tin
Cd	Cadmuim	Ta	Tantalum
Cr	Chromium	Ti	Titanium
Cy	Copper	W	Tungsten
Fe	Iron	Zn	Zinc
Li	Lithium	Zr	Zirconium
Mg	Magnesium		

Figure 25-4. *A list of the chemical abbreviations for base metals and filler metals indicated in Figures 25-3, 25-5, and 25-6.*

Preparing to Braze or Braze Weld

When brazing or braze welding, wear clothing similar to that worn when oxyfuel gas welding. Coveralls or a cotton shirt and pants are suggested. A cap, gloves, and hard-toed shoes should also be worn. All pockets should be covered and buttoned, if possible. Button the top button on coveralls or a shirt to prevent hot metal or sparks from entering. Leathers (jacket, cape, apron, and pants) are recommended when brazing or braze welding in the overhead position.

Cleanliness is one of the most important factors in creating a strong brazed or braze welded joint. The metal in the joint area must be clean. Before beginning to braze or braze weld, all rust, oil, dirt, paint, and metallic dust must be removed from the surfaces to be joined.

Mechanical or chemical cleaning can be used for the joint area. Mechanical cleaning is done with some type of tool or abrasive material. Grinding, filing, grit blasting, wire brushing, and emery cloth are commonly used. The base metal may also be rinsed and dried after mechanical cleaning. Chemical cleaning includes any process that uses solutions of various chemicals. Chemical cleaning may be done with approved solvents.

Brazing and Braze Welding Fluxes

A ***flux*** can be defined as "a material used to prevent, dissolve, or facilitate removal of oxides and other undesirable surface substances." As previously noted, the most important step in preparing to braze or braze weld is the cleaning of the base metal surfaces by both mechanical and chemical means. The base metal is again chemically cleaned by the flux during the brazing or braze welding process. As the heat causes the flux to become liquid, the flux acts on both the filler metal and the base metals being joined to keep them clean. The flux also prevents additional oxides from forming after the process is completed.

The AWS classifies fluxes into six categories. These flux categories are recommended for use when brazing or braze welding various base metals, including aluminum, magnesium, iron, steel, and nickel. Some metals require a flux that contains chlorides or fluorides. Other metals require a flux that contains boric acid, borates, fluoborates, or wetting agents. **Figure 25-6** lists the flux types, their suggested uses, and the ingredients of each type.

Fluxes are available in bottles or cans and come in powder, paste, or liquid form. Flux may be applied to the surface of the base metal or to the brazing rod. A small, clean brush may be used to apply flux to the base metal surfaces. For large areas, flux may be sprayed onto the base metal. A syringe-like applicator can be used to apply flux to a joint (see Figure 26-5).

Flux may also be applied to the brazing rod. To apply flux to bare rod, heat the last 3″–4″ (76 mm–102 mm) of the brazing rod to a few hundred degrees. A color change should not be seen when heating the rod. Dip the heated end of the rod into the flux container. Flux will stick to the heated section of the rod. When the applied flux is used up, reheat the end of the rod and dip it into the flux container again. During brazing or braze welding operations, this process is continually repeated as required. When using manufactured brazing filler metal with a flux coating, this procedure is unnecessary.

AWS filler metal classification	Nominal chemical composition (percentage)																				Temperatures			
																					Solidus[1]		Liquidus[2]	
	Ag	Al	Au	Cd	Cr	Cu	Ni	Si	Zn	B	C	Fe	Mg	Mn	Sn	Ti	P	Pb	Pd	Zr	Others	°F	°C	°F	°C
BAlSi-2		91.0				.25		7.5	.20			.8		.10							.15	1070	577	1135	613
BAlSi-3		84.4			.15	4.0		10.0	.20			.8	.15	.15								970	521	1085	585
BAlSi-4		86.3				.30		12.0	.20			.8	.10	.15								1070	577	1080	582
BAlSi-5		88.4				.30		10.0	.10			.8	.05	.05		.20						1070	577	1095	591
BAlSi-7		87.0				.25		10.0	.20			.8	1.5	.10								1038	559	1105	596
BCuP-2						92.6											7.3					1310	710	1460	793
BCuP-3	5.0					88.9											6.0					1190	643	1495	813
BCuP-4	6.0					86.6											7.3					1190	643	1325	718
BCuP-5	15.0					79.9											5.0					1190	643	1475	802
BCuP-6	2.0					90.9											7.0					1190	643	1450	788
BCuP-7	5.0					88.1											6.8					1190	643	1420	771
BAu-1			37.5			62.4															.15	1815	991	1860	1016
BAu-2			80.0			19.9																1635	891	1635	891
BAu-3			35.0			64.9																1785	974	1885	1029
BAu-4			82.0			17.9																1740	949	1740	949
BAu-5			30.0				36.0												34.0			2075	1135	2130	1166
BMg-1		9.0				.05	.005	.05	2.0			.005	88.4	.15							.30	830	443	1110	599
BCu-1						99.9*															.1	1980	1082	1980	1082
BCu-1a						99.0*															.3	1980	1082	1980	1082
BCu-2						86.5*															.5	1980	1082	1980	1082
BAg-1	45.0			24.0		15.0			16.0												.15	1125	607	1145	618
BAg-1a	50.0			18.0		15.5			16.5													1160	627	1175	635
BAg-2	35.0			18.0		26.0			21.0													1125	607	1295	702
BAg-2a	30.0			20.0		27.0			23.0													1125	607	1310	710
BAg-3	50.0			16.0		15.5	3.0		15.5													1170	632	1270	688
BAg-4	40.0					30.0	2.0		28.0													1240	671	1435	779
BAg-7	56.0					22.0			17.0						5.0							1145	618	1205	652
BAg-13	54.0					40.0	1.0		5.0													1325	718	1575	857
BAg-13a	56.0					42.0	2.0															1420	771	1640	893
BAg-20	30.0					38.0			32.0													1250	677	1410	766
BAg-22	49.0					16.0	4.5		23.0					7.5								1260	680	1290	700
BAg-24	50.0					20.0	2.0		28.0													1220	660	1305	705
BAg-27	25.0			13.5		35.0			26.5													1125	605	1375	745
BAg-28	40.0					30.0			28.0						2.0							1200	650	1310	710
BAg-33	25.0			17.5		30.0			27.5													1180	640	1320	715
BAg-34	38.0					32.0			28.0						2.0							1200	650	1330	720
BNi-1		.05			14.0		72.4	4.5		3.2	.75	4.5				.05	.02			.05	.50	1790	977	1900	1038
BNi-1a					14.0		73.1	4.5		3.2	.06	4.5					.02					1790	977	1970	1077
BNi-2					7.0		81.3	4.5		3.2	.06	3.0					.02					1780	971	1830	999
BNi-3							91.1	4.5		3.2	.06	.5					.02					1800	982	1900	1038
BNi-4							92.5	3.5		1.8	.06	1.5					.02					1800	982	1950	1066
BNi-5					19.0		70.0	10.2		.03	.10						.02					1975	1079	2075	1135
BNi-6							88.3				.10						11.0					1610	877	1610	877
BNi-7					14.0		74.9	.10		.01	.08	.2					10.1					1630	888	1630	888
BNi-8						4.5	64.7	7.0			.10			23.0			.02					1800	982	1850	1010
RBCuZn-A		.01				59.0			39.815						.625			.05				1615	880	1635	890
RBCuZn-B		.01				58.0	.05	.12	39.34			.725		.255	.95			.05				1590	866	1620	882
RBCuZn-C		.01				58.0		.095	39.415			.725		.255	.95			.05				1590	866	1630	888
RBCuZn-D		.01				48.0	10.0	.145	41.045								.25	.05				1635	890	1685	920

1-Solidus = melting temperature 2-Liquidus = flow temperature * - Minimum

Figure 25-5. Brazing filler metals. Solidus and liquidus temperatures are given. Note: Cadmium (Cd) alloys are listed in red. Cadmium fumes are extremely poisonous and can be lethal. These brazing filler metal element values are the average of a range as specified by the AWS. (Adapted from AWS A5.8:2004, Tables 1–5, Table of Brazing Filler Metal Alloys, reproduced with permission from the American Welding Society, Miami, FL)

Flux type	Brazable metal	Recommended filler metal	Brazing temperature range °F	Brazing temperature range °C	Flux ingredients
FB1-A FB1-B FB1-C	Aluminum alloys	BAlSi BAlSi BAlSi	1000-1140	538-615	Chlorides, fluorides
FB2-A	Magnesium alloys	BMg	900-1150	482-621	Chlorides, fluorides
FB3-A FB3-C	Aluminum bronze, aluminum brass, iron or nickel containing Al or Ti	BAg & BCuP	1050-1700	565-927	Chlorides, fluorides, borates, wetting agents
FB3-D FB3-H FB3-I FB3-J FB3-K	All metals not listed above	BCu, BAg, BNi, BAu, RBCuZn	1350-2200	732-1203	Boric acid, fluoborates, borates, wetting agents
FB3-E FB3-F FB3-G FB4-A	All metals not listed above	BCuP & BAg	1050-1600	565-886	Boric acid, fluoborates, borates, wetting agents
Note: These fluxes may come in either a liquid, powder, paste, or slurry form.					

Figure 25-6. American Welding Society (AWS) flux classifications for brazing. Consult suppliers of commercial fluxes for specific metals and applications. RB indicates the metal is suitable for resistance brazing. (ANSI/AWS A5.31:1992, various tables, Specifications for Fluxes for Brazing and Braze Welding, reproduced with permission from the American Welding Society, Miami, FL)

Selecting the Torch Tip and Filler Rod

The recommended tip orifice and rod diameter for braze welding is shown in **Figure 25-7**. For brazing, use a tip orifice at least two to three sizes larger than recommended for braze welding. Remember that a smaller tip size number indicates a larger tip orifice.

Setting Working Pressures

The welding outfit is turned on and lit in the same manner as for oxyfuel gas welding. Chapter 21 provides details for turning on an oxyfuel gas welding outfit. The oxygen and acetylene working pressures are set to the pressures recommended for the torch tip size, Figure 25-7.

Metal thickness	Braze welding tip orifice size*	Filler rod diameter	Oxygen		Acetylene		Speed ft/hr
			Pressure (psig)	cfh	Pressure (psig)	cfh	
1/32″	74	1/16″	1	1.1	1	1	
1/16″	69	1/16″	1	2.2	1	2	
3/32″	64	1/16″ or 3/32″	2	5.5	2	5	20
1/8″	57	3/32″ or 1/8″	3	9.9	3	9	16
3/16″	55	1/8″	4	17.6	4	16	14
1/4″	52	1/8″ or 3/16″	5	27.5	5	25	12
5/16″	49	1/8″ or 3/16″	6	33	6	30	10
3/8″	45	3/16″	7	44	7	40	9
1/2″	42	3/16″	7	66	7	60	8

*Note the tip orifice size as shown is the number drill size. These recommendations are approximate. The torch manufacturers' recommendations should be carefully followed.

Figure 25-7. Torch tip and rod diameter recommendations. For brazing, use a tip with an orifice that is two to three sizes larger than the one recommended for braze welding. The larger the tip orifice size, the smaller the tip size number.

Adjusting the Flame

A neutral or slightly carburizing flame is generally used when brazing or braze welding. This is done to prevent the braze joint from oxidizing. The carburizing flame usually produces a braze weld bead with a better appearance.

Brazing and Braze Welding Safety Precautions

To ensure the welder's safety during brazing and braze welding operations, standard precautions for the safe use of oxyfuel gas welding equipment must be followed. Chapters 20 and 21 include detailed information about the safe use of oxyfuel gas welding equipment. Special precautions that should be followed when brazing or braze welding include:

1. **Excellent ventilation is required. Toxic metal and flux fumes are often present when brazing or braze welding.**
2. **Do not allow brazing fluxes to contact your skin. Many fluxes are harmful to the skin, so care should be taken in handling them. If the fluxes come into contact with your skin, wash the area thoroughly with soap and water.**
3. **Only trained personnel should handle acids used for chemical cleaning.**
4. Some brazing filler metals contain cadmium. When molten, and especially if overheated, they emit cadmium oxide fumes to the atmosphere. **Cadmium oxide fumes are very dangerous if inhaled. Inhaling these fumes can result in injury to the respiratory passages. The exposure limit for cadmium oxide fumes is 0.1 milligrams per cubic meter of air for a daily eight-hour period. This value represents the maximum exposure that workers may experience without adverse effects.** If you have concerns regarding cadmium exposure, contact the local industrial hygiene department. **Cadmium fumes have no odor, and a lethal dose will not always cause immediate discomfort. When the worker has absorbed an excessive quantity of cadmium, his or her life can be in immediate danger. However, the symptoms—headache, fever, irritation of the throat, vomiting, nausea, chills, weakness, and diarrhea—may not appear until some hours after exposure.**

Brazing

A brazing joint is made between two close-fitting surfaces, **Figure 25-8**. It may be a pipe or tube joint. When brazing is used for joining T-joints, corner joints, or butt joints, a flanged joint is used. A flanged joint is used to increase surface area and provide for capillary action.

When brazing, the parts are fitted together, the base metal is heated, and filler metal is added to the joint. The filler metal melts at a temperature above 840°F (450°C).

The parts of a brazement must be fitted tightly together, **Figure 25-9**. Only .001″–.010″ (.025 mm–.254 mm) clearance should be allowed between parts of a brazement.

When brazing, the area around the joint is heated; when fusion or braze welding, only a small spot is heated. A larger tip orifice is used in brazing in order to broadly distribute the heat throughout a larger area of the base metal. The brazing torch flame should be held 1″–3″ (25 mm–75 mm) from the metal, **Figure 25-10**. Do not melt the base metal when brazing. The base metal's melting point must be above the melting point of the filler metal.

The base metal is heated enough to melt the filler metal. The brazing rod is touched to the base metal. A brazing rod may be held at practically any angle when it is touched to the base metal. The rod should be held close to the reflected flame to preheat it. As the filler metal melts, it is drawn into the joint by capillary action.

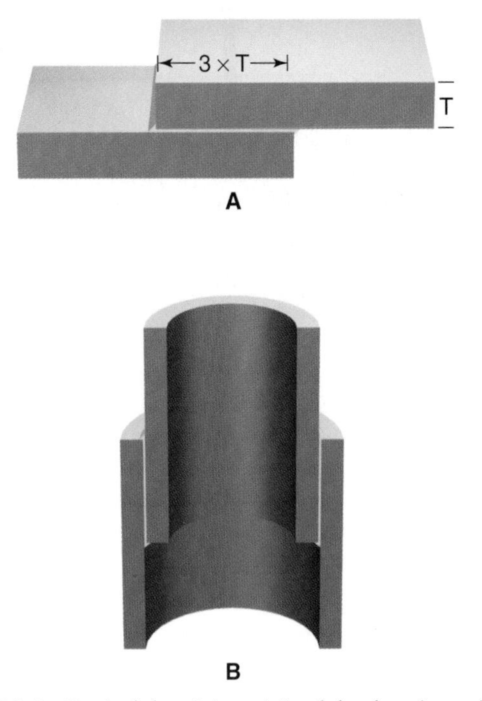

Figure 25-8. Typical lap joints joined by brazing. A—Lap joint. B—Pipe fitting lap joint. When forming a lap joint, use as much surface contact as possible. The strength of the joint increases as the area of surface contact increases.

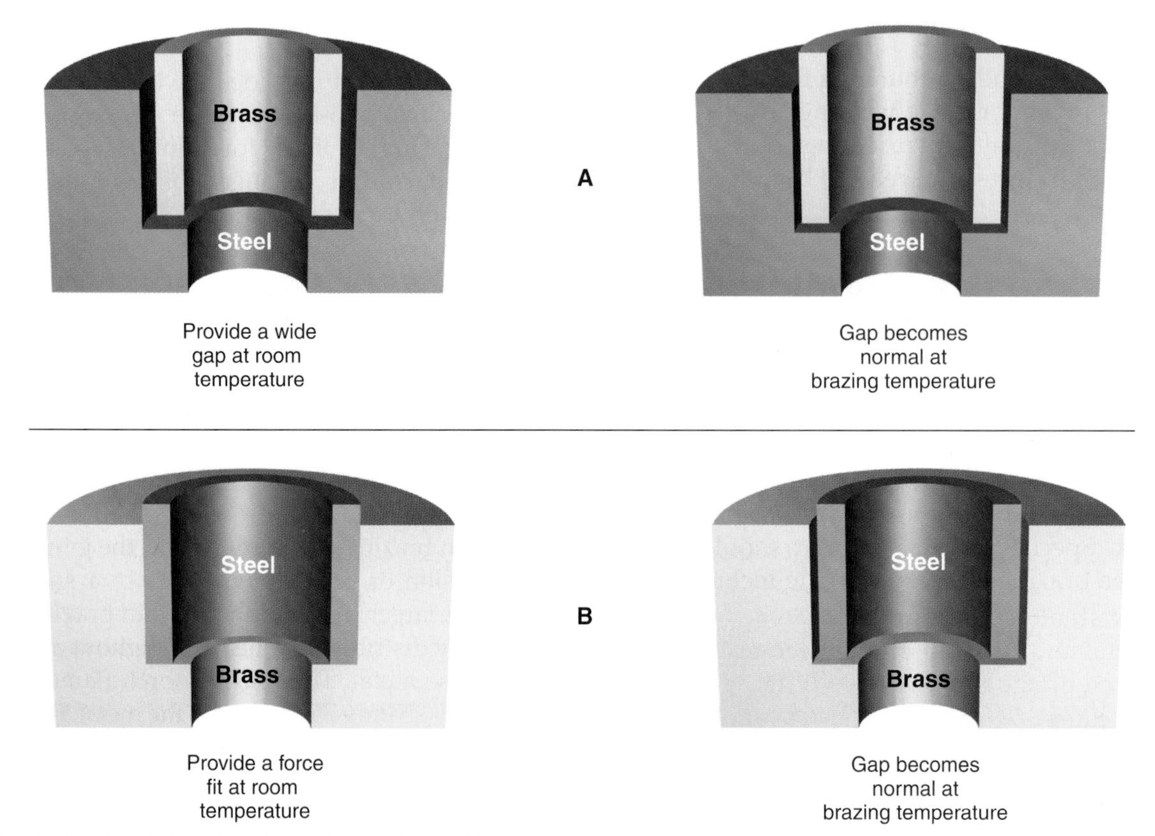

Figure 25-9. *Brazing joints fit with only .001″–.010″(.025 mm–.254 mm) clearance. The fitting of dissimilar metals must allow for different rates of expansion that occur as the metals are heated. Brass expands at a faster rate than steel. A—Loosely fitted brass in steel fits properly at brazing temperatures. B—Steel force-fitted into brass at room temperature has the correct clearance at brazing temperature.*

Figure 25-10. *Torch, flame, and rod positions for brazing. The flame is held 1″–3″ (25 mm–75 mm) from the joint. The flame is moved along the joint while touching the rod to the metal. Continue adding the brazing rod until the filler metal is seen throughout the joint. Note: Joint clearances are enlarged in this figure for clarity.*

Braze Welding–Flat Position

Braze welding differs from brazing in several ways. Braze welded parts are not fitted together as tightly as brazed parts. Therefore, the filler metal is not drawn into the joint by capillary action. Unlike brazing, braze welding is performed on joints similar to those used in fusion welding. See **Figure 25-11**.

The torch tip and filler rod positions for braze welding are generally the same as for oxyfuel gas welding. The torch tip is held 1/16″–1/8″ (1.6 mm–3.2 mm) from the base metal and has a smaller diameter tip orifice than one used for brazing. For a butt joint, the torch tip is aligned with the weld axis and held at a 0° work angle. The torch tip should be held at a 30°–45° travel angle. For a fillet weld on a lap or T-joint, the tip is held at about a 45° work angle and a 30°–45° travel angle, **Figure 25-12**. On a lap joint, the tip should point more toward the surface than toward the edge of the metal. This reduces the tendency for the edge to melt.

When braze welding, the brazing rod is generally held at a 15°–45° angle from the base metal surface. The end of the rod should be held close to, but not in, the molten pool to preheat it. See Figure 25-12.

A thick layer of brazing filler metal is added to form a braze weld bead on the joint. If the base metal overheats while braze welding, withdraw the flame occasionally to control the braze weld bead width. After the base metal cools slightly, return the flame to 1/16″–1/8″ (1.6 mm–3.2 mm) from the surface.

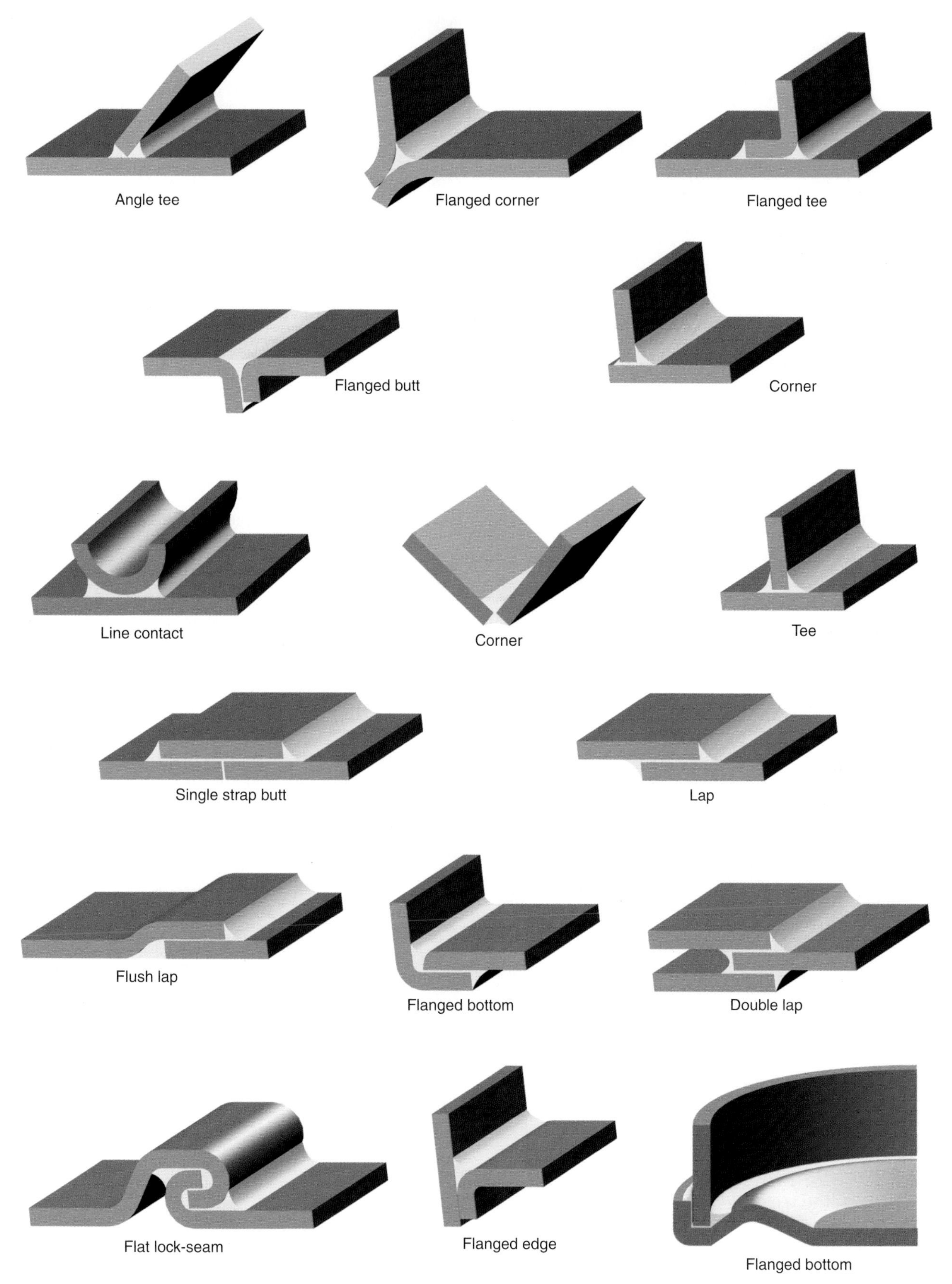

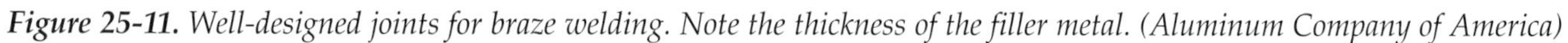

Figure 25-11. *Well-designed joints for braze welding. Note the thickness of the filler metal. (Aluminum Company of America)*

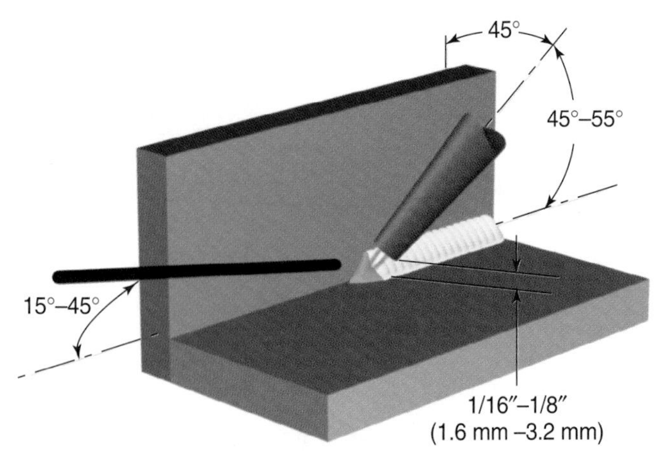

Figure 25-12. Torch, flame, and brazing rod positions for braze welding. Note the thickness of the braze weld bead.

Exercise 25-1 Brazing a Lap Joint in the Flat Position

1. Obtain the following:
   - Two pieces of mild steel measuring 1/8″ × 3″ × 6″ (3.2 mm × 75 mm × 150 mm).
   - 1/16″ (1.6 mm) BCu or RBCuZn uncoated brazing rod.
   - A container of liquid or paste flux and a container of powdered flux. Refer to Figure 25-6. Pour some of the flux into a smaller container for use in this exercise.
2. Clean the overlapping surfaces of the mild steel with a stainless steel brush or emery cloth and wipe away the dust. Be sure the adjoining surfaces have good contact.
3. Apply liquid or paste flux to both surfaces. Completely cover the 3/8″ (9.6 mm) area to be brazed.
4. Arrange the pieces so they overlap 3/8″ (9.6 mm) as shown in the figure at the end of the exercise.
5. Select the correct torch tip size for the metal thickness. See Figure 25-7 for the suggested tip orifice.
6. Turn on the welding outfit, adjust the working pressures, and light the torch. Adjust for a neutral or slightly carburizing flame.
7. Heat the end of the filler metal and dip it into the powdered flux. The flux will stick to the filler metal. Repeat as often as necessary to keep the rod covered with flux while brazing.
8. Tack braze the two pieces at each end of the joint. See Figure 25-2.
9. Direct the flame to heat the center of the overlapped area as shown in the figure at the end of the exercise. Distribute the heat by moving the flame in a circular motion. Keep the flame at least 1″ (25.4 mm) from the top piece. Keep the flame away from the edge of the top piece to avoid melting the base metal. The flux will heat and start to boil, which indicates the metal is close to the correct brazing temperature.
10. Touch the brazing rod to the edge of the joint. When the correct temperature is reached, the filler metal will melt and be pulled into the joint by capillary action.
11. Continue moving the flame down the length of the lap joint in a circular motion and continue adding filler metal along the entire length of the lap joint. Be careful not to add excessive brazing rod. A small fillet will be created at the edge of the joint.
12. Allow the completed joint to cool.
13. Wash the joint in warm water to remove excess flux from the brazement.

Inspection:

Examine the edges on both sides. If the joint was properly cleaned, fluxed, and heated, a thin line of filler metal should be visible along the weld axis between the two parts.

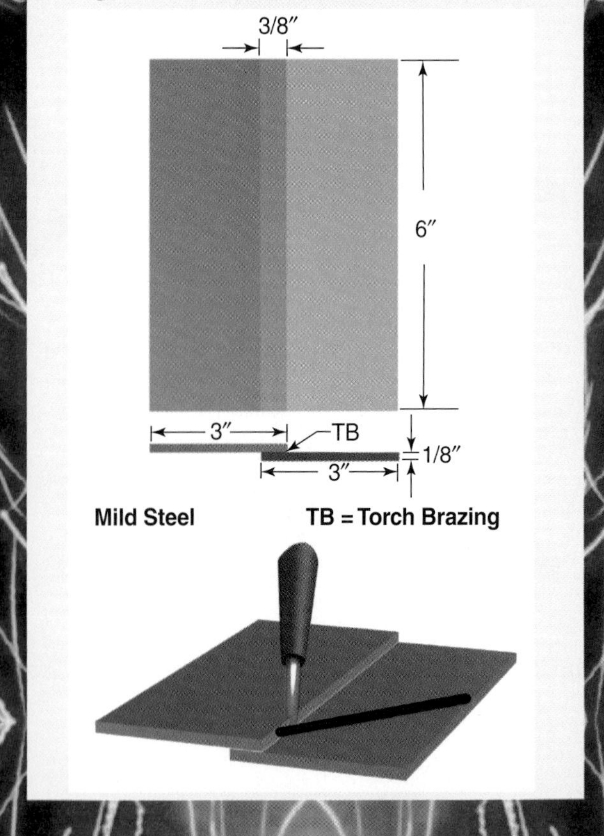

Exercise 25-2 Laying a Braze Weld Bead

1. Obtain the following:
   - A piece of mild steel measuring 1/16″ × 3″ × 6″ (1.6 mm × 75 mm × 150 mm).
   - One 36″ (.91 m) long brass uncoated brazing rod of the correct diameter. See Figure 25-7 for the suggested brazing rod diameter.
   - A container of suitable flux. See Figure 25-6. Pour some of the flux into a smaller container for use in this exercise.
2. Clean the surface with emery cloth and wipe away the dust.
3. Mark two lines on the surface using a soapstone or chalk. The lines should be 1″ (25.4 mm) apart and 1″ (25.4 mm) from the edges.
4. Select the correct torch tip size. See Figure 25-7 for the suggested tip orifice.
5. Turn on the welding outfit, adjust the working pressures, and light the torch. Adjust for a neutral or slightly carburizing flame.
6. Heat the end of the brazing rod and place it into the small flux container. This coats the end of the rod with flux. Repeat this procedure as often as necessary to keep the rod covered with flux.
7. Begin laying a braze weld bead by heating the metal about 1/8″ (3.2 mm) on each side of the line. Do not melt the base metal. Also, apply heat to only the base metal, not the brazing rod.
8. Touch the brazing rod to the metal occasionally. The base metal is hot enough when the rod melts.
9. Move the torch and brazing rod ahead at a constant speed, while heating the base metal and adding the rod. Continue this procedure to the end of the line.
10. If the braze weld bead is too fluid and wide, allow the base metal to cool. Withdraw the flame and allow the metal to cool slightly. Return the flame to about 1/16″–1/8″ (1.6 mm–3.2 mm) above the metal and continue. Withdraw the flame as often as needed to maintain the desired braze weld bead width.

Recommendations:

- Use a larger tip size if the base metal heats slowly.
- Use a smaller tip size if the braze weld bead is hard to control or spreads too far from the line.

Inspection:

The braze weld bead should be about 1/4″ (6.4 mm) wide. It should have a constant width and evenly spaced ripples.

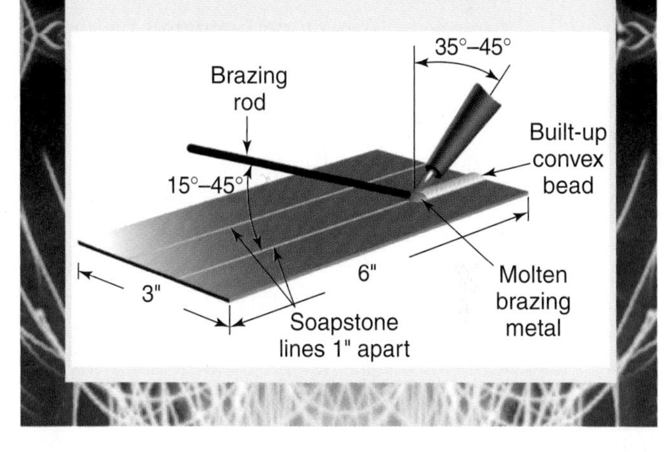

Brazing and Braze Welding out of Position

Brazing and braze welding are most easily done in the flat position. However, these processes can also be done out of position. Torch and brazing rod angles are the same as for fusion welding out of position. Torch and rod angles for fusion welding out of position are covered in Chapter 24.

Overheating causes the braze weld bead to spread and possibly sag downward due to gravity. The brazing filler metal must be kept from overheating. Overheating is controlled as follows:

1. Heat the base metal until it melts the filler metal.
2. Add the filler metal.
3. Remove the flame and allow the braze weld bead and base metal to cool slightly.
4. Reheat the base metal and add filler metal again.
5. Continue the heating and cooling process, as required, to the end of the joint.

Exercise 25-3 Braze Welding a Fillet on an Inside Corner in the Flat Position

1. Obtain the following:
   - Two pieces of mild steel measuring 1/16″ × 1 1/2″ × 6″ (1.6 mm × 40 mm × 150 mm).
   - One 36″ (.91 m) long uncoated brass brazing rod of the correct diameter. See Figure 25-7 for suggested brazing rod diameter.
   - A container of the correct flux. See Figure 25-6. Pour some of the flux into a smaller container for use in this exercise.
2. Clean one surface of each piece with emery cloth and wipe away dust.
3. Arrange the pieces as shown in the figure at the end of this exercise.
4. Select the correct torch tip for the thickness of metal that you will be braze welding.
5. Turn on the welding outfit and light the torch. Adjust for a neutral or slightly carburizing flame.
6. Heat the end of the brazing rod. Apply flux to the heated end.
7. Tack braze the two pieces at each end of the joint.
8. Begin braze welding by heating an area about 1/8″ (3.2 mm) on each side of the joint. Do not melt the base metal. Apply heat to the base metal only, not to the brazing rod.
9. Touch the brazing rod to the base metal occasionally. When the rod melts, the base metal is hot enough. Add the brazing rod to the same area until a flat or slightly concave braze weld bead is formed. See the figure at the end of the exercise.
10. Move the torch and rod ahead at a constant rate, heating the metal and adding the filler metal. Continue this procedure to the end of the joint.
11. If the braze weld bead is too fluid and wide, allow the base metal to cool. Withdraw the flame and let the metal cool slightly. Return the flame to about 1/16″–1/8″ (1.6 mm–3.2 mm) above the metal and continue. Withdraw the flame as often as necessary to keep the desired braze weld bead width.

Recommendations:

- Use a larger tip size if the base metal heats slowly.
- Use a smaller tip size if the braze weld bead spreads too far.

Inspection:

The braze weld bead should be about 1/4″ (6.4 mm) wide. It should have a constant width and evenly spaced ripples.

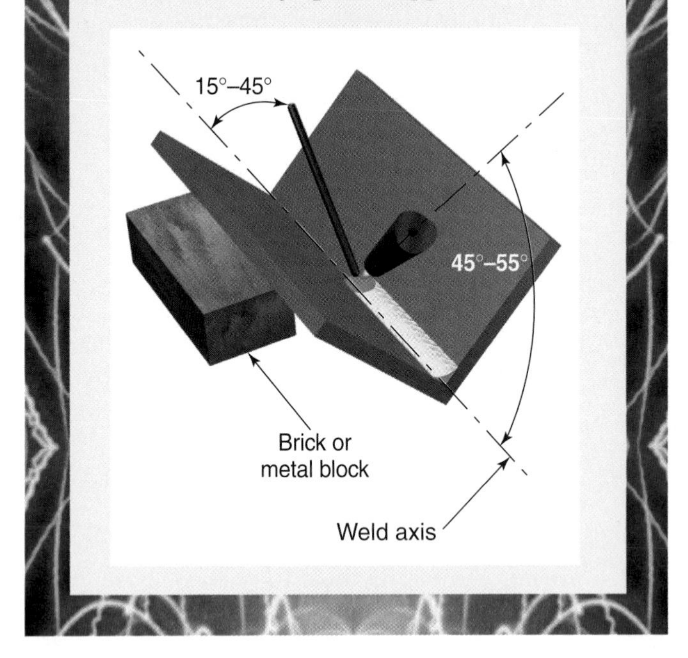

Summary

- The American Welding Society defines brazing as "a group of welding procedures that cause materials to join by heating them to the brazing temperature in the presence of a filler metal. The filler metal has a liquidus temperature above 840°F (450°C) and below the solidus temperature of the base metal. The filler metal is distributed between the close-fit surfaces of the joint by capillary action."
- Braze welding is defined by AWS as "a welding process variation in which a filler metal, having a liquidus above 840°F (450°C) and below the solidus of the base metal, is used. Unlike brazing, in braze welding, the filler metal is not distributed in the joint by capillary action."
- Parts to be brazed must fit very tightly together. Capillary action is the process by which filler metal is drawn into the space between the two tightly fitted mating surfaces. Parts to be braze welded must fit closely together, but they do not need to fit as tightly as parts to be brazed since the process does not rely on capillary action.
- Brazing filler metal can be any metal that melts between 840°F (450°C) and the melting point of the base metal. Aluminum and alloys with low melting points can be used to braze aluminum alloys with higher melting temperatures. Brass is used to braze or braze weld steel. Silver alloys are commonly used to braze copper, stainless steel, and a variety of other metals.
- Cleanliness is one of the most important factors in creating a strong brazed or braze welded joint. The metal in the joint area must be clean. Before beginning to braze or braze weld, all rust, oil, dirt, paint, and metallic dust must be removed from the surfaces to be joined.
- A flux can be defined as "a material used to prevent, dissolve, or facilitate removal of oxides and other undesirable surface substances." The flux provides additional chemical cleaning of the base metal during the brazing or braze welding process.
- The recommended tip orifice and rod diameter are determined by the thickness of the base metal. For brazing, use a tip orifice at least two to three sizes larger than recommended for braze welding. The oxygen and acetylene working pressures are set to the pressures recommended for the torch tip size.
- A neutral or slightly carburizing flame is generally used when brazing or braze welding. This is done to prevent the braze joint from oxidizing.
- To ensure the welder's safety during brazing and braze welding operations, standard precautions for the safe use of oxyfuel gas welding equipment must be followed.
- When brazing, the base metal is heated enough to melt the filler metal. The brazing rod is touched to the base metal. A brazing rod may be held at practically any angle when it is touched to the base metal. The rod should be held close to the reflected flame to preheat it. As the filler metal melts, it is drawn into the joint by capillary action.
- Braze welding is performed on joints similar to those used in fusion welding. The torch tip and filler rod positions for braze welding are generally the same as for oxyfuel gas welding. The end of the brazing rod should be held close to, but not in, the molten pool to preheat it. A thick layer of brazing filler metal is added to form a braze weld bead on the joint.
- Brazing and braze welding are most easily done in the flat position. However, these processes can also be done out of position. Torch and brazing rod angles are the same as for fusion welding out of position.

Review Questions

Write your answers on a separate sheet of paper. Please do *not* write in this book.

1. Brazing and braze welding filler metal melts at a temperature above _____°F (_____°C) but below the melting temperature of the base metal.
2. List two reasons why brazing or braze welding is used instead of fusion welding.
3. A(n) _____ is an assembly of parts joined by brazing or braze welding.
4. *True or False?* A brazed or braze welded joint is as strong as a fusion welded joint.
5. Is *silver soldering* a soldering process, a brazing process, or neither? Explain your reasoning.
6. Refer to Figure 25-3. What alloy of brazing filler metal is used when brazing carbon steel to aluminum?
7. List four things that can be used to clean the base metal prior to brazing or braze welding.
8. Refer to Figure 25-6. What types of AWS brazing flux and filler metal are used to braze aluminum alloys?
9. The brazing process draws filler metal into the joint through _____ action.
10. List two things that can be done if the braze weld bead created during braze welding is too wide and flat.

Chapter 26

Soldering

Learning Objectives

After studying this chapter, you will be able to:

- Describe the principles of soldering.
- Select the appropriate filler metal and flux for soldering.
- Identify acceptable solders for drinking water systems.
- Describe the soldering process.
- Solder a lap joint.
- Solder a pipe joint.

Technical Terms

air-fuel gas torch
applicator
bridge
inorganic fluxes
liquidus
organic fluxes
rosin fluxes
solder
soldering
soldering alloy
solidus

Soldering Principles

Soldering is similar to brazing in many respects. Like brazing, soldering forms a bond between the filler metal and the surface of the base metals where two parts are fitted tightly together. The base metals are not melted. Only the filler metal melts. Capillary action draws the filler metal into the tight-fitting joint. The filler metal bonds to the base metal by molecular action to give the joint its strength. Properly made soldered joints are airtight and watertight. Because soldering makes a leakproof joint, it is the most commonly used process in joining copper tubes that carry air and water.

One of the main differences between soldering and brazing is the temperature at which the operation is performed. ***Soldering*** is defined as "a group of welding processes that join materials by heating them to the soldering temperature and by using a filler metal having a liquidus *below* 840°F (450°C) and below the solidus of the base metals. The filler metal is distributed between closely fitting surfaces of the joint by capillary action." Brazing is done at temperatures *above* 840°F (450°C).

Parts to be soldered must fit tightly together. The clearance between parts in a soldered joint is commonly .001″–.003″ (.025 mm–.076 mm). The standard clearance between copper tubing and fittings is .002″–.006″ (.050 mm–.152 mm). When the tube is centered in the fitting, a clearance of .001″–.003″ (.025 mm–.076 mm) exists around the tube. **Figure 26-1** shows several soldered joints.

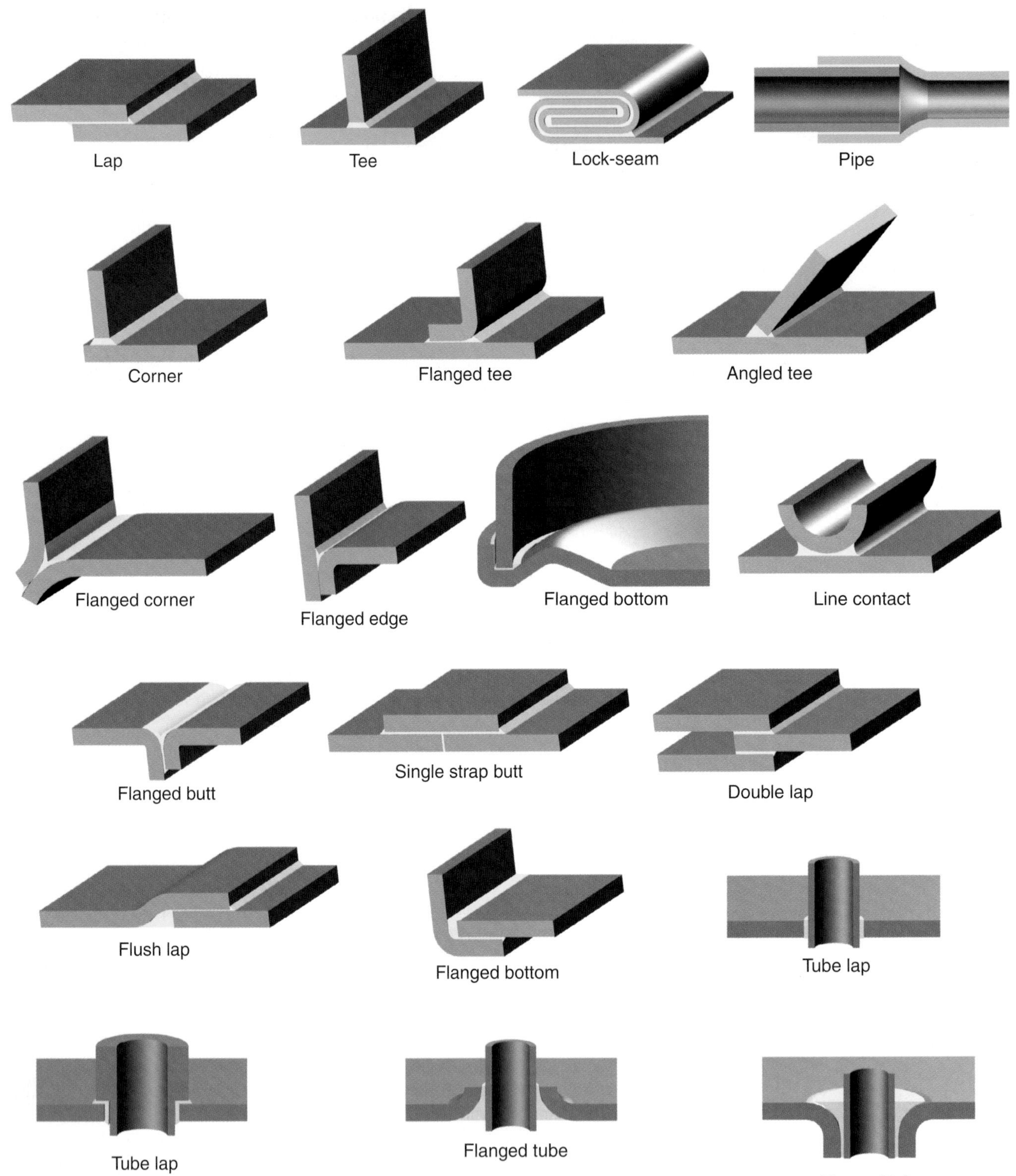

Figure 26-1. *Suggested joint designs for soldered and brazed joints. The spacing in this figure is exaggerated for illustration purposes. Normal clearance spacing is .001″–.003″ (.025 mm–.076 mm). A thin layer of solder is stronger than a thick layer. A larger surface area produces a stronger soldered joint. (The Aluminum Association)*

Before soldering can be done, the base metal near the joint area must be mechanically and/or chemically cleaned. Generally, a wire brush or emery cloth is used. After mechanical cleaning is completed, a suitable flux is applied around the joint.

The base metal is heated until it is hot enough to melt the filler metal. Parts to be soldered may be heated using a soldering iron (soldering copper), oxyfuel gas torch, air-fuel gas torch, furnace, or electrical resistance coils. The filler metal used in soldering is called ***solder***.

The solder flows when it is touched to the hot base metal near the joint. The liquid solder is drawn into the tight-fitting joint by capillary action. Liquid solder flows over all surfaces that are heated to the melting point of the solder.

Soldering filler metals, such as lead, tin, and zinc, are not as strong as filler metals used in welding, brazing, or braze welding. However, thin layers of solder in contact with a large enough surface in a joint will make a joint strong enough for its intended purpose. Soldering should not be used when extremely strong joints are required. Thick and wide beads of solder are seldom used, since soldering filler metals are weak. However, thick layers of solder may be used to fill a depression (such as in auto body repair). Thick layers of solder are applied for appearance, not for strength.

Advantages of Soldering

Soldered joints have excellent electrical and thermal (heat) conductivity. For this reason, soldering is commonly used in making electrical connections. Soldered joints also have a very clean appearance. Since soldering is a low-temperature process, it is effectively used for the following purposes:

- Connecting thin metals that would otherwise melt if higher temperatures were used.
- Connecting electronic parts to circuit boards and electrical parts to wires.
- Connecting metals without affecting their heat treatment characteristics.
- Jewelry repair and manufacturing.
- Finishing metal surfaces.
- Connecting plumbing pipes and fittings.

Soldering Filler Metals

Solder, like brazing and braze welding filler metals, is available as wire, paste, and a variety of standard and special preshaped forms. Most solders are available in 1-pound (.45 kg) spools of wire. Common solder wire diameters are 1/16″ (1.6 mm), 3/32″ (2.4 mm), and 1/8″ (3.2 mm).

Solders contain such metals as aluminum, antimony, cadmium, copper, indium, lead, nickel, silver, tin, or zinc. These metals are often mixed to form ***soldering alloys***. The physical properties of these solders differ widely. **Figure 26-2** lists a number of soldering alloys

Solder Alloy Compositions and Melting Temperatures

Alloy	Composition (% by Weight)								Solidus Temperature		Liquidus Temperature	
	Tin	Lead	Silver	Antimony	Cadmium	Zinc	Aluminum	Indium	°F	°C	°F	°C
Tin-Antimony	95			5					450	232	464	240
Lead-Tin-Silver	96		4						430	221	430	221
	62	36	2						354	180	372	190
	5	94.5	0.5						561	294	574	301
	2.5	97	0.5						577	303	590	310
	1.0	97.5	1.5						588	309	588	309
Tin-Zinc	91					9			390	199	390	199
	80					20			390	199	518	269
	70					30			390	199	592	311
	60					40			390	199	645	340
	30					70			390	199	708	375
Silver-Cadmium			5		95				640	338	740	393
Cadmium-Zinc					82.5	17.5			509	265	509	265
					40	60			509	265	635	335
					10	90			509	265	750	399
Zinc-Aluminum						95	5		720	382	720	382
Tin-Lead-Indium	50							50	243	117	257	125
	37.5	37.5						25	230	138	230	138
		50						50	356	180	408	209

Figure 26-2. *The chemical composition of various soldering alloys. Solder begins to melt at its solidus temperature. Solder is completely liquid at its liquidus temperature.*

and the composition of each one. The solidus and liquidus temperatures for each solder are also shown. The *solidus* is the temperature at which the solder begins to melt. The *liquidus* is the temperature at which the solder is totally liquid. The solidus and liquidus temperatures are important to know when selecting a solder for a particular job. Solder flows easily into small spaces when the solidus and liquidus are close together. Solder fills wide spaces in joints and depressions in base metal more easily when the solidus and liquidus are farther apart. Compare the temperatures of several alloys in Figure 26-2.

Soldering Fluxes

A soldering flux has two purposes:

- To clean the metal surface by removing oxides and other undesirable substances.
- To prevent surface oxidation while soldering.

Fluxes are available as liquids or pastes. They are sold in leakproof containers intended to keep the flux fresh and clean, **Figure 26-3**. Remove from the container only enough flux to do the job. The container should be tightly resealed after use.

Classifications

There are three basic classifications of flux: organic, inorganic, and rosin fluxes. *Organic fluxes* contain carbon. They have a medium level of cleaning ability. Organic fluxes are corrosive during the soldering operation, but generally become noncorrosive when the soldering is completed. They are water soluble and can be washed away with water.

Inorganic fluxes do not contain carbon. Inorganic fluxes clean better than organic or rosin fluxes. They do not char or burn easily. They can be used for torch, oven, and other soldering methods. Inorganic fluxes are very corrosive. Parts soldered with inorganic flux must be cleaned after soldering. Inorganic fluxes are not used on electrical or electronic parts because they are so corrosive.

Rosin fluxes are the least effective cleaners. They are recommended for electrical and electronic parts because they are noncorrosive.

Fluxes are used for soldering on a variety of metals. Some types of fluxes are used for soldering on all metals. **Figure 26-4** shows the recommended fluxes for soldering various metals.

Fluxes may be applied in a variety of ways. The most common means of application is with a clean brush. Flux is brushed onto the surface of the joint area. The flux flows around the surfaces of the joint as the part is heated. It is also drawn into tightly fitted joints by capillary action.

Fluxes may also be applied by means of a large syringe-like *applicator*. In high-volume production, the flux is added in exact amounts. A pressurized flux paste applicator, **Figure 26-5**, is used to apply the flux. The flux paste will then flow around the joint when the part is heated.

Figure 26-3. Several types of fluxes and flux containers. (Handy & Harmon PMFG/ Lucas-Milhaupt, Inc.)

Recommended Fluxes for Various Metals			
Base metal, alloy or applied finish	Flux recommendations		
	Corrosive (organic and inorganic)	Non-corrosive (rosin)	Special flux and/ or solder
Aluminum			X
Aluminum-bronze			X
Beryllium copper	X		
Brass	X	X	
Cadmium	X	X	
Cast iron			X
Copper	X	X	
Copper-chromium	X		
Copper-nickel	X		
Copper-silicon	X		
Gold		X	
Inconel			X
Lead	X	X	
Magnesium			X
Monel	X		
Nickel	X		
Nichrome			X
Palladium		X	
Platinum		X	
Rhodium	X		
Silver	X	X	
Stainless steel			X
Steel	X		
Tin-zinc	X	X	
Tin-bronze	X	X	
Tin-lead	X	X	
Tin-nickel	X	X	
Tin-zinc	X	X	
Zinc	X		
Zinc die castings			X

Figure 26-4. Recommended fluxes for soldering various metals.

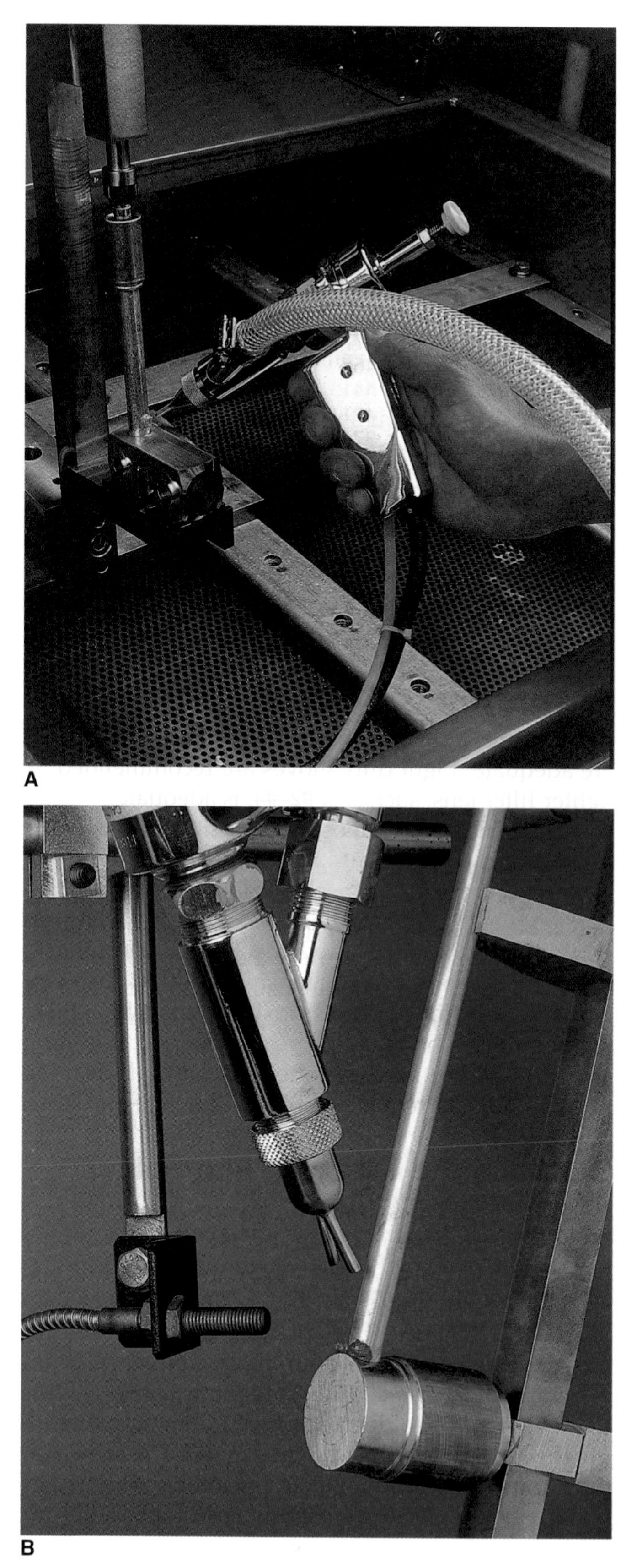

Figure 26-5. A—A close-up view of a manual solder flux applicator. B—Automatic solder flux paste applicator. The amount of flux paste applied is controlled by the paste pressure and by a timer that is accurate to within 1/20 of a second. For high-volume production soldering, the applicator can be mounted on a robot. (Fusion, Inc.)

Hazards of Solders and Fluxes Containing Lead

According to the Environmental Protection Agency (EPA) studies, excess lead in drinking water supplies can cause a variety of health problems. Public Law 99-339, *The Safe Drinking Water Act Amendment of 1986*, banned the use of lead solder, flux, and pipe from use in drinking water installations. Since 1988, all states have been required to enforce this ban.

Solders containing lead in excess of 0.2% may still be used in water systems for drainage, heating, fire sprinklers, air conditioning, and machine cooling. Tin and lead alloys are commonly used on parts and systems that do not carry drinking water. These tin-lead alloy solders may contain from 5% tin and 95% lead to 70% tin and 30% lead. The compositions of various tin-lead alloys are shown in **Figure 26-6**.

Tin-lead alloy solder is available in rolls of wire. The solder is available in a wide range of tin to lead ratios. The 50% tin and 50% lead (50/50) solder is most common.

Acceptable Solders for Drinking Water Systems

Several companies have developed new lead-free solders. These solders produce joints that are as strong or stronger than those made with tin-lead solders. The new lead-free solders have a 40°F–60°F (22°C–33°C) difference between their solidus and liquidus. This wide range of temperatures makes it possible to apply the new lead-free solders using the same techniques as 50/50 tin-lead solder. The range of temperatures also makes it possible for these solders to ***bridge*** (fill) wide gaps in poorly fitted assemblies. **Figure 26-7** shows the solidus and liquidus of several lead-free solders. Similar equipment and flux is used as when applying 50/50 solder. Lead-free alloys that have been recently developed include the following:

- 95.5% tin, 4% copper, 0.5% silver.
- 88% tin, less than 2% silver, 4% antimony, less than 2% copper, and 4% zinc.
- 89% tin, less than 2% silver, 5% antimony, 3% copper, and less than 1% nickel.

Several other lead-free soldering alloys are harder to apply because their solidus and liquidus temperatures are closer together. The following is a list of some of these solders:

- 95% tin, 5% antimony.
- 96% tin, 4% silver.
- 95% tin, 5% silver.
- 94% tin, 6% silver.
- 97% tin, 3% copper.

Section 7

Resistance Welding

Miller Electric Mfg. Co.

Review Questions

Write your answers on a separate sheet of paper. Please do *not* write in this book.

1. Soldering is done below _____°F (_____°C).
2. *True or False?* Strong soldered joints are made with convex beads on loosely fitted joints.
3. *True or False?* The base metal melts the filler metal when soldering is performed properly.
4. *True or False?* Soldered joints are strongest when a thin layer of filler metal is applied to a large surface.
5. List three applications for which soldering is the best process to use.
6. The temperature at which solder begins to melt is the _____ temperature.
7. *True or False?* For a tightly fitted joint, the solidus and liquidus temperatures should be close together.
8. Refer to Figure 26-7. What is the solidus and liquidus temperature of a 95/5 tin-silver solder?
9. Refer to Figure 26-4. What type of flux is recommended for soldering steel?
10. The flame should be held _____″–_____″ (_____ mm–_____ mm) away from the base metal when soldering.

Section 7

Resistance Welding

- A soldering flux cleans the metal surface by removing oxides and other undesirable substances and prevents surface oxidation while soldering.
- Organic fluxes contain carbon, have a medium level of cleaning ability, are water soluble, and are corrosive during the soldering operation, but generally become noncorrosive when the soldering is completed.
- Inorganic fluxes do not contain carbon. Inorganic fluxes clean better than organic or rosin fluxes, do not char or burn easily, and are very corrosive. Parts soldered with inorganic flux must be cleaned after soldering. Inorganic fluxes are not used on electrical or electronic parts because they are so corrosive.
- Rosin fluxes are the least effective cleaners. They are recommended for electrical and electronic parts because they are noncorrosive.
- Public Law 99-339, *The Safe Drinking Water Act Amendment of 1986*, banned the use of lead solder, flux, and pipe from use in drinking water installations. Since then, new lead-free solders have been developed These solders are acceptable for drinking water installations.
- An oxyfuel gas torch or an air-fuel gas torch (acetylene, propane, butane, or MPS) is commonly used to heat the base metal when soldering. Generally, soldering is done on metal that is less than 1/8″ (3.2 mm) thick. The soldering torch tip used for air-acetylene, air-propane, air-butane, or air-MPS gas generally produces a wide, spreading flame. A tip orifice between #74 and #55 is generally used for oxyfuel gas soldering.
- The diameter of the solder wire used for an operation depends on the base metal thickness. If the diameter is too small, the solder wire will have to be held on the heated joint longer to completely fill it. Solder wire with a 1/8″ (3.2 mm) diameter is generally used on copper pipe joints.
- An air-propane, air-butane, or air-MPS gas torch is started by opening the torch valve slightly and lighting the flame with a spark lighter. The torch valve is adjusted until a neutral flame is obtained.
- Air-acetylene and oxyfuel gas cylinders are opened and the pressures set according to the tip size. The flame is adjusted on the air-acetylene torch using only the acetylene torch valve. Adjust the flame until the correct amount of acetylene is flowing to provide a neutral or slightly carburizing flame.
- Observe all standard safety precautions for welding with oxyfuel gas welding equipment.
- When torch soldering, properly clean the base metals, check the fit of the parts to be joined, make sure the metals are firmly supported, apply the appropriate flux to the surfaces to be joined, light the torch, and apply heat to the joint. The solder should be melted only by the heat of the base metals, not by the flame. The solder should be added to the joint as quickly as possible to reduce oxidation. Most fluxes should be removed from the finished joint as soon as possible.

Exercise 26-2 Soldering a Pipe Joint

1. Obtain the following:
   - Two pieces of 3/4″ (20 mm) copper pipe approximately 6″ (150 mm) long.
   - Two 3/4″ × 90° elbows.
   - A coil of a suitable solder with a 1/8″ (3.2 mm) diameter.
   - A container of suitable soldering flux and a small clean brush for application of the flux, Figure 26-4.
   - An oxyfuel gas welding outfit or other fuel gas torch.
2. Clean the ends of the pipe and the inside of the fittings using emery cloth. The pipe should be cleaned for a distance of at least 1″ (25 mm) from the end.
3. Apply flux to the pipe and fittings using a flux brush.
4. Fit the parts together and support them. A fire brick or metal block may be used.
5. For an oxyacetylene torch, use a tip that is two or three number sizes larger than one used for welding the same thickness of metal. Refer to Figure 20-36 for the appropriate tip sizes for welding. If an air-fuel gas (propane, butane, acetylene, or MPS) torch is used, select a large heating tip.
6. Turn on the fuel gas slightly. Light the gas using a spark lighter. Adjust for a neutral or slightly carburizing flame.
7. Hold the flame 1″–3″ (25 mm–76 mm) from the joint. Heat the area around the joint. Touch the solder wire to the base metal to check the temperature.
8. When the solder melts, continue to heat the base metal and add solder.
9. Control the heat to the joint by withdrawing the torch. This allows the joint to cool. Apply the flame again when the joint cools.
10. Withdraw the flame and solder when the joint appears to be full of solder.
11. Quickly and carefully wipe all excess solder from the joint area with a clean cloth. Do this before the solder becomes firm. Solder the second joint using the same procedure.

Inspection:

The joints must be filled with solder. They should have a smooth fillet completely around the diameter. Low spots or holes should not be seen in the solder joint. The joint should be wiped clean.

Summary

- Soldering is defined as "a group of welding processes that join materials by heating them to the soldering temperature and by using a filler metal having a liquidus *below* 840°F (450°C) and below the solidus of the base metals. The filler metal is distributed between closely fitting surfaces of the joint by capillary action."
- Before soldering can be done, the base metal near the joint area must be mechanically and/or chemically cleaned. Generally, a wire brush or emery cloth is used. After mechanical cleaning is completed, a suitable flux is applied around the joint.
- In the soldering process, the base metal is heated until it is hot enough to melt the filler metal. The filler metal (solder) flows when it is touched to the hot base metal near the joint. The liquid solder is drawn into the tight-fitting joint by capillary action.
- Solders contain such metals as aluminum, antimony, cadmium, copper, indium, lead, nickel, silver, tin, or zinc. These metals are often mixed to form soldering alloys. The physical properties of these solders differ widely.

(continued)

2. Check the fit of the parts to be joined. Be sure they have close contact with each other.
3. Be sure the metals are firmly supported during and after the soldering operation. They should be supported as they cool. As the joint cools, it increases in strength.
4. Apply the appropriate flux to the surfaces to be joined. Flux must be fresh and as chemically pure as possible.
5. Light the torch and apply heat to the joint. Metals to be soldered must be heated to the melting temperature of the solder. Be careful not to overheat the base metal.
6. Solder should be melted only by the heat of the metals to be soldered and not by the flame. Add only enough solder to complete the job. Excess solder on the joint is unsightly and does not add strength as might be supposed. Add the solder as quickly as possible to reduce oxidation.
7. Most fluxes should be removed from the joint as soon as possible after the soldering operation is completed. This keeps corrosion from continuing.

Hold the flame about 1″–3″ (25 mm–76 mm) above the solder joint at a 45° angle from the base metal. The end of the solder wire does not have to be preheated, as in welding or brazing. The solder wire may be held at any convenient angle.

The flame is moved back and forth to heat the entire joint to the soldering temperature at the same time. When soldering pipes or tubes, the flame is moved completely around the joint. As the joint is heated, the solder is touched to it to test the temperature. To prevent overheating the joint as the solder melts, the flame is withdrawn to about 3″–5″ (76 mm–127 mm) from the joint. The solder wire is placed on the joint and continually added until solder can be seen along the entire length of the joint. If the solder cools before the joint is full, reheat the joint and add more solder. In a well-made joint, solder should not extend more than about 3/8″–1/2″ (9.6 mm–12.7 mm) to either side of the joint centerline. If the solder is wider than 1/2″, too large an area of the joint has been heated.

Exercise 26-1 Soldering a Lap Joint

1. Obtain the following:
   - Six pieces of mild steel that are less than 1/16″ (1.6 mm) thick and measure 1 1/2″ × 6″ (40 mm × 150 mm).
   - A coil of a suitable solder with a 1/8″ (3.2 mm) diameter.
   - A container of the appropriate flux and a small clean brush to apply the flux. See Figure 26-4 to determine the appropriate type of flux.
   - An oxyfuel gas welding outfit or air-fuel gas torch.
2. Clean the edges and surfaces of all pieces. For the surfaces, begin at the edges and clean inward at least 3/4″ (19.1 mm).
3. Use the small brush to apply flux to the areas of the two pieces that were cleaned in step 2.
4. Form a lap joint with the two pieces and clamp the pieces together.
5. Turn on the fuel gas and light the torch using a spark lighter. Adjust the gas flow to produce a neutral or slightly carburizing flame.
6. Hold the flame about 1″–3″ (25 mm–76 mm) above the centerline of the joint. Move the flame back and forth along the entire length of the joint. Touch the end of the solder wire to the joint occasionally to test the joint temperature. When the solder melts, withdraw the flame to 3″–5″ (76 mm–127 mm). Continue adding solder until it is visible along the entire length of the joint. If the solder cools before the joint is full, reheat the joint and add the solder again. Use the other pieces of metal to complete the exercise by soldering two more lap joints.

Inspection:

Solder must be seen between the pieces and on each side of the solder joint. The width of the solder on each side of the joint should be uniform and not wider than about 3/8″ (9.5 mm).

Solder wire is commonly available in the following diameters: 1/16″ (1.6 mm), 3/32″ (2.4 mm), and 1/8″ (3.2 mm). The size used for an operation depends on the base metal thickness. If the diameter is too small, the solder wire will have to be held on the heated joint longer to completely fill it. Solder with a 1/8″ (3.2 mm) diameter is generally used on copper pipe joints.

Adjusting the Flame

An air-propane, air-butane, or air-MPS gas torch is started by opening the torch valve slightly and lighting the flame with a spark lighter. The torch valve is adjusted until a neutral flame is obtained.

Air-acetylene and oxyfuel gas cylinders are opened and the pressures set according to the tip size. The flame is adjusted on the air-acetylene torch using only the acetylene torch valve. Adjust the flame until the correct amount of acetylene is flowing to provide a neutral or slightly carburizing flame. On an oxyfuel gas torch, open the torch acetylene valve until the smoke stops. Then, open the oxygen torch valve until the flame is neutral or slightly carburizing. Flame adjustment is detailed in Chapter 21.

Soldering Safety Precautions

Observe all standard safety precautions for welding with oxyfuel gas welding equipment, regardless of the type of soldering equipment used. A number of general safety precautions are presented in Chapter 20. The following special precautions for soldering safety must be carefully observed:

- **Excellent ventilation is required.**
- **Care must be taken when handling fluxes that are harmful to the skin.**
- **Only trained personnel should use acids for chemically cleaning metal.**
- **Some solders contain cadmium and emit cadmium oxide fumes when molten or overheated. These fumes are extremely toxic. See Chapter 25 for more information regarding cadmium hazards.**
- **The proper protective clothing for oxyfuel gas welding should be worn. A #2–#4 welding goggle filter lens is adequate for most soldering jobs.**

Figure 26-8. A MAPP (MPS) gas cylinder, torch, and tip combination. This type of torch is used for torch soldering in a variety of applications.

Procedures for Torch Soldering

Nonproduction and repair soldering can be done using a variety of hand-held torches. **Figure 26-8** shows a typical fuel gas cylinder, torch, and tip combination. An oxyfuel gas flame is generally not needed for soldering. Air-fuel gas torches, which produce lower temperatures, work very well for soldering. Propane, butane, acetylene, or MPS gas can be used to obtain the low-temperature flame.

When soldering, pinpoint heating is not required or recommended. A large tip that produces a wide flame should be used to spread the heat over a broad area.

These standard soldering procedures must be followed to produce an acceptable soldered joint:

1. Properly clean the metals mechanically and/or chemically. All oxides, grease, paint, and dirt must be removed.

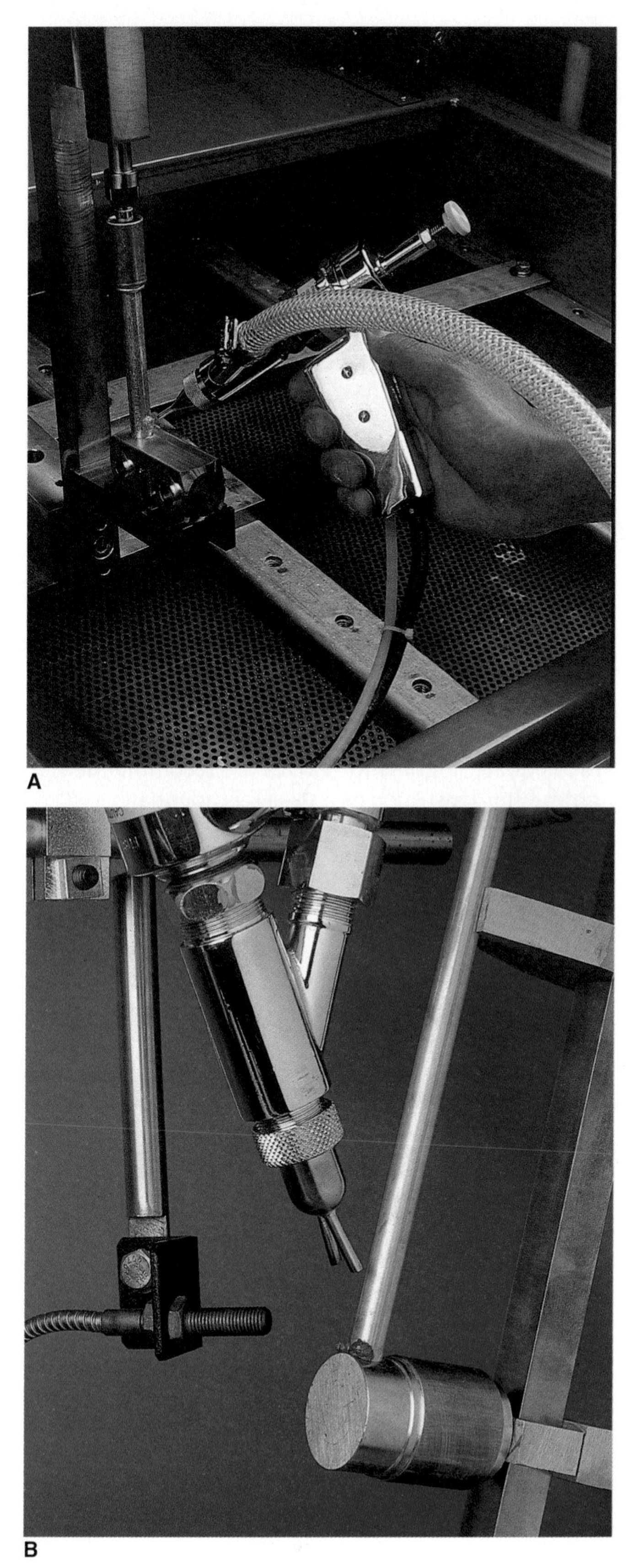

Figure 26-5. A—A close-up view of a manual solder flux applicator. B—Automatic solder flux paste applicator. The amount of flux paste applied is controlled by the paste pressure and by a timer that is accurate to within 1/20 of a second. For high-volume production soldering, the applicator can be mounted on a robot. (Fusion, Inc.)

Hazards of Solders and Fluxes Containing Lead

According to the Environmental Protection Agency (EPA) studies, excess lead in drinking water supplies can cause a variety of health problems. Public Law 99-339, *The Safe Drinking Water Act Amendment of 1986*, banned the use of lead solder, flux, and pipe from use in drinking water installations. Since 1988, all states have been required to enforce this ban.

Solders containing lead in excess of 0.2% may still be used in water systems for drainage, heating, fire sprinklers, air conditioning, and machine cooling. Tin and lead alloys are commonly used on parts and systems that do not carry drinking water. These tin-lead alloy solders may contain from 5% tin and 95% lead to 70% tin and 30% lead. The compositions of various tin-lead alloys are shown in **Figure 26-6**.

Tin-lead alloy solder is available in rolls of wire. The solder is available in a wide range of tin to lead ratios. The 50% tin and 50% lead (50/50) solder is most common.

Acceptable Solders for Drinking Water Systems

Several companies have developed new lead-free solders. These solders produce joints that are as strong or stronger than those made with tin-lead solders. The new lead-free solders have a 40°F–60°F (22°C–33°C) difference between their solidus and liquidus. This wide range of temperatures makes it possible to apply the new lead-free solders using the same techniques as 50/50 tin-lead solder. The range of temperatures also makes it possible for these solders to ***bridge*** (fill) wide gaps in poorly fitted assemblies. **Figure 26-7** shows the solidus and liquidus of several lead-free solders. Similar equipment and flux is used as when applying 50/50 solder. Lead-free alloys that have been recently developed include the following:

- 95.5% tin, 4% copper, 0.5% silver.
- 88% tin, less than 2% silver, 4% antimony, less than 2% copper, and 4% zinc.
- 89% tin, less than 2% silver, 5% antimony, 3% copper, and less than 1% nickel.

Several other lead-free soldering alloys are harder to apply because their solidus and liquidus temperatures are closer together. The following is a list of some of these solders:

- 95% tin, 5% antimony.
- 96% tin, 4% silver.
- 95% tin, 5% silver.
- 94% tin, 6% silver.
- 97% tin, 3% copper.

ASTM Solder Classification	Composition (% by Weight)		Solidus Temperature		Liquidus Temperature	
	Tin	Lead	°F	°C	°F	°C
5	5	95	572	300	596	314
10	10	90	514	268	573	301
15	15	85	437	225	553	290
20	20	80	361	183	535	280
25	25	75	361	183	511	267
30	30	70	361	183	491	255
35	35	65	361	183	477	247
40	40	60	361	183	455	235
45	45	55	361	183	441	228
50	50	50	361	183	421	217
60	60	40	361	183	374	190
*	62	38	361	183	361	183
70	70	30	361	183	378	192

*Figure 26-6. Tin-lead solder compositions. ASTM 60 solder is commonly used for electrical connections. The * represents the Eutectic alloy. Its solidus and liquidus temperatures are the same.*

Lead-free Soldering Alloy	Solidus		Liquidus	
	°F	°C	°F	°C
95% tin, 5% antimony	450	232	464	240
96% tin, 4% silver	430	221	460	238
95% tin, 5% silver	430	221	473	245
94% tin, 6% silver	430	221	535	279
97% tin, 3% copper	450[1]	232	500[1]	260
Silvabrite 100™ 95.5% tin, 4% copper, 0.5% silver	440	227	500	260
Stay-Safe 50™ 88% tin, <2% silver, 4% antimony, <2% copper, 4% zinc	400	204	440	227
Stay-Safe Bridgit™ 89% tin, <2% silver, 5% antimony, 3% copper, <1% nickel	460	238	630	332
*50% tin, 50% lead	361	183	421	217

*Figure 26-7. Chemical composition, solidus, and liquidus for several lead-free alloys. Notes: [1]These temperatures are approximate. *50/50 tin-lead information is given for comparison only.*

Lead-free solders are more expensive per pound than tin-lead solders. However, since lead-free solders are lighter, more wire length is available in a one-pound roll. The cost per soldered joint is approximately the same for lead-free solders and tin-lead solders.

Preparing to Solder

For most light soldering jobs, regular work clothes are adequate. Light duty gloves are recommended. A lighter filter lens, such as a #2–#4, is adequate.

Cleanliness is the most important factor in creating a good solder joint. Metals to be soldered are cleaned chemically or mechanically. Chemical cleaning, on most soldering jobs, is done by using an appropriate soldering flux. Mechanical cleaning is done with a wire brush, abrasive cloth or paper, or abrasive blasting. Wire brushes must be clean. Stainless steel brushes are recommended. Washing and drying the surfaces after mechanical cleaning must be done to remove all particles on the surfaces.

Selecting the Torch Tip and Solder

An oxyfuel gas torch or an ***air-fuel gas torch*** (acetylene, propane, butane, or MPS) is commonly used to heat the base metal when soldering. Generally, soldering is done on metal that is less than 1/8″ (3.2 mm) thick. The soldering torch tip used for air-acetylene, air-propane, air-butane, or air-MPS gas generally produces a wide, spreading flame. A wide flame is used in soldering in order to evenly distribute the heat throughout the base metal. The same size tip is commonly used for all applications. A tip orifice between #74 and #55 is generally used for oxyfuel gas soldering. However, the correct size is a matter of personal preference.

Chapter 27

Resistance Welding: Equipment and Supplies

Learning Objectives

After studying this chapter, you will be able to:

- Explain the principle of electrical resistance and how it is used in resistance welding.
- List the four most common resistance welding machine designs.
- Explain how a step-down transformer affects voltage and current.
- List the properties of a material suitable for use as an electrode in resistance welding.
- Describe the regular checks to be made for safe operation of a resistance spot welding machine.

electrode face diameter
expulsion weld
hold time
horn spacing
off time
percent heat control
resistance
resistance spot welding (RSW)
seam weld
spot weld
squeeze time
tap settings
throat depth
weld nugget
weld time
wheel electrode

Principles of Resistance Spot Welding

When electricity flows through metal, the metal heats up. This occurs because there is ***resistance*** (opposition) to the flow of electricity through the metal. For example, when electricity flows through the metal coil in an oven, the coil gets red-hot. The heat from the glowing coil is used to cook foods.

If two metals are touching and an electrical current is passed through them, heat will be generated. The largest amount of heat is created at the surfaces where the two metals are touching.

In ***resistance spot welding (RSW)***, two or more metal pieces are stacked on top of one another. Pressure is applied to the pieces. Then electrical current flows through the metal.

As the current flows, heat develops at the surfaces where the metal pieces touch. When enough heat is generated, the metal will melt in the area where the pieces are touching. After the current stops, the molten metal becomes solid again. The parts are welded together. The metal that is welded together is called a ***spot weld*** or a ***weld nugget***. **Figure 27-1** illustrates the steps that take place when a spot weld is made.

Welding Machines

A resistance welding machine must accurately apply the desired force to the metal to be welded. Different methods are used to apply force to the metal. The most common method is pneumatic pressure. Hydraulic pressure and mechanical leverage are also used.

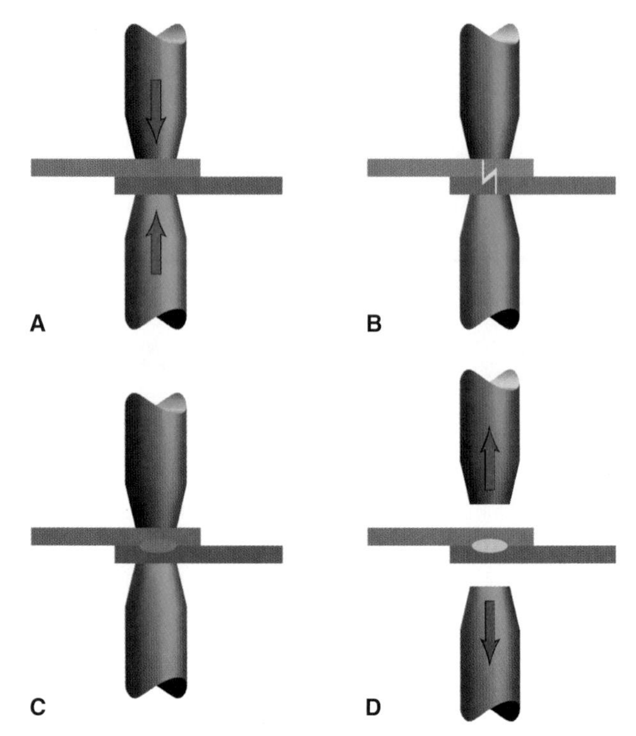

Figure 27-1. *Steps in making a spot weld. A—The metal is clamped between the electrodes under pressure. B—Current flows. C—Spot weld nugget is formed. D—Electrodes are removed after the weld becomes solid.*

Exercise 27-1 Learning the Parts of a Spot Welding Machine

1. Your instructor will assign a specific piece of spot welding equipment for you to study.
2. Make sure the equipment is off. You will not need to operate the equipment for this exercise.
3. Sketch the equipment. Show each of the items listed, if present on your machine:
   - Electrical connections.
   - Water connection and drain.
   - Pneumatic or hydraulic connections.
   - Regulators for controlling the pneumatic or hydraulic pressure.
   - Pneumatic or hydraulic cylinder or force spring.
   - Electrodes.
   - Tap switch or switches.
   - Control panel.

Various types of resistance welding machines use these force systems. The most common machine designs are:

- Rocker arm, see **Figure 27-2**.
- Press, see **Figure 27-3A**.
- Portable gun, see **Figure 27-3B**.

Rocker arm, press, and gun welding machines may use either pneumatic or hydraulic pressure.

Two important dimensions must be determined when a welding machine is selected. They are the throat depth and the horn spacing. ***Throat depth*** is the distance from the electrodes to the frame of the machine. It limits how far the electrodes can reach over a piece to be welded. ***Horn spacing*** is the distance between the arms of the welding machine, measured when the electrodes are touching. This dimension determines how tall of a part can be welded. Throat depth and horn spacing are illustrated in **Figure 27-4**.

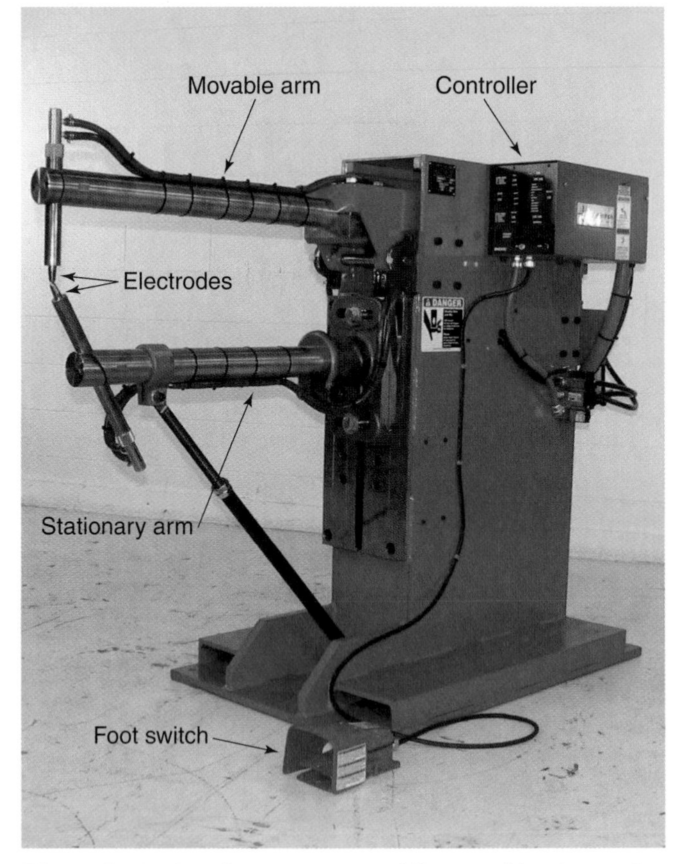

Figure 27-2. *A rocker arm spot welding machine is starting using a foot switch. (Janda Company, Inc.)*

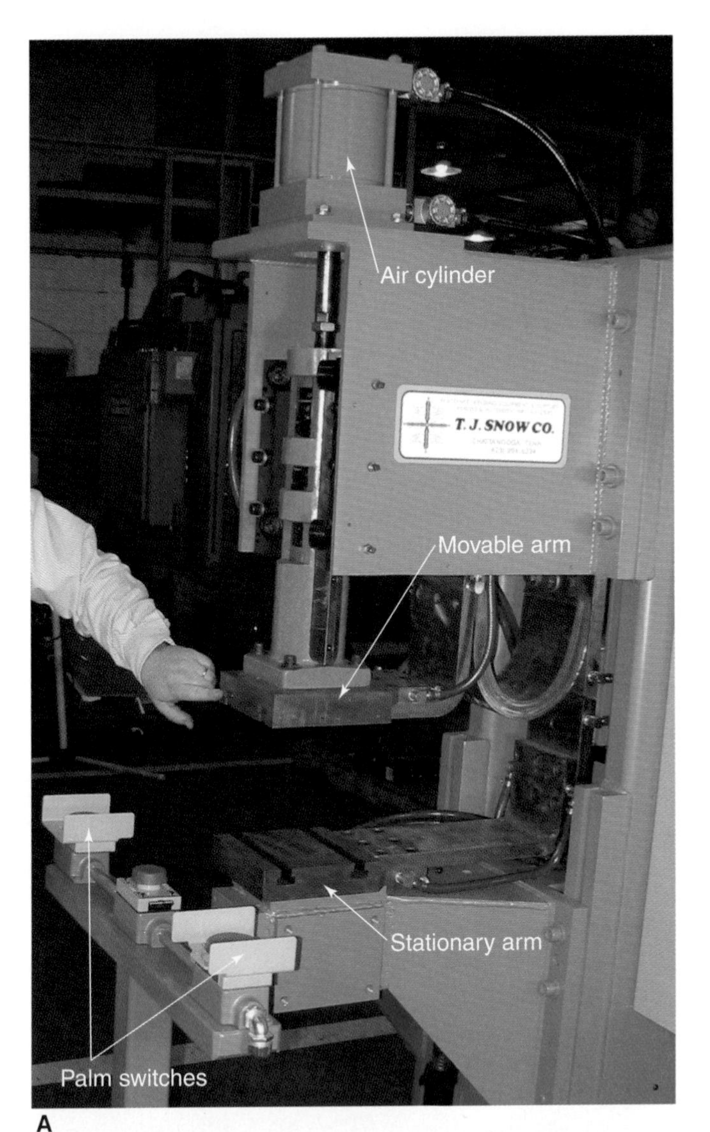

A

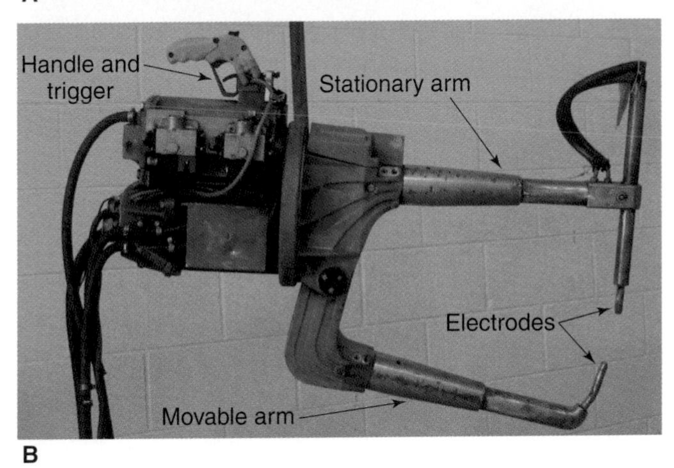

B

Figure 27-3. A—On a press spot welding machine, the air cylinder is located above the parts being welded. Tooling or electrodes are attached to the platens. Welding is started by pushing both palm switches. (T.J. Snow Company, Inc.) B—This portable spot welding gun is suspended from above. The gun is moved to the parts being welded. Welding is started by pulling the trigger. (T.J. Snow Company, Inc.)

Transformers

Most resistance welding uses alternating current (ac). Power supplied to a welding machine can vary from 115V to 460V with current varying from a few amps to over 1000A.

A transformer is used to reduce the supplied high voltage to a low voltage. The same transformer also increases the supplied low current to a very high current.

Transformers have a primary winding and a secondary winding, as shown in **Figure 27-5**. The primary winding is connected to the supplied power, usually 460V or 230V. The secondary winding is connected to the electrodes.

Transformers that reduce the supplied voltage are called step-down transformers. In a spot welding machine, a step-down transformer may reduce the voltage supplied to less than 1V or as high as 40V. The transformer increases the current to thousands or tens of thousands of amps.

Changing the *voltage* of the secondary windings results in large changes in welding *current*. The voltage is changed by using ***tap settings***. Thus, when you change the tap setting on a machine, you are changing the secondary voltage and current. Some machines have a high and a low position; others have multiple settings. Some machines have both. **Figure 27-6** shows a tap switch on a resistance welding machine.

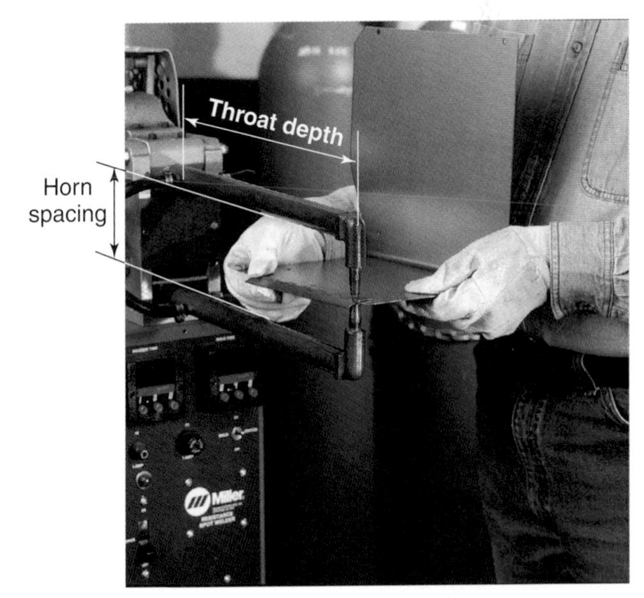

Figure 27-4. A floor-mounted spot welding machine with its throat depth and horn spacing indicated. (Miller Electric Mfg. Co.)

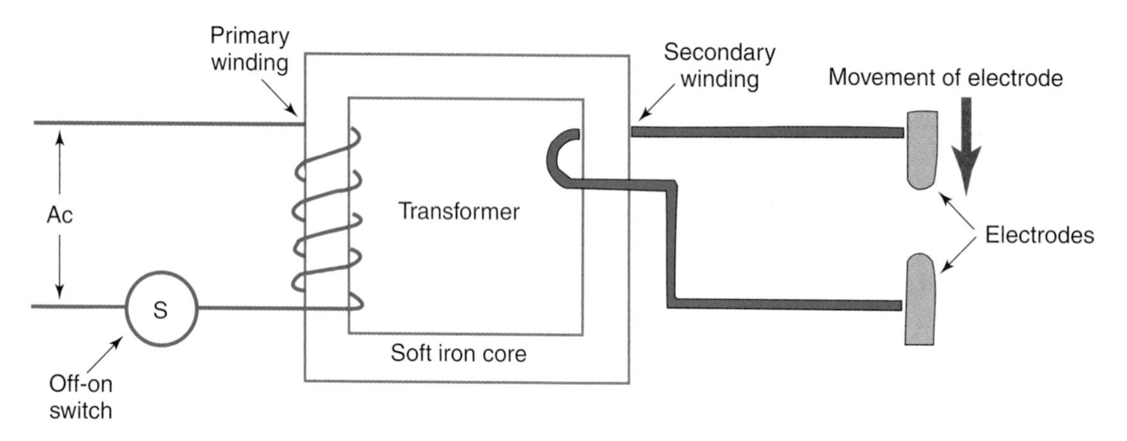

Figure 27-5. *A schematic drawing of the primary and secondary circuits of a spot welding machine transformer. This is a step-down transformer. High-voltage, low-amperage current runs through the primary windings. This induces a low-voltage, high-amperage current in the secondary windings. The amperage is increased by the same factor that the voltage is reduced. For example, when the voltage is reduced from 230V to 2.3V (a factor of 100), the amperage increases from 100A to 10,000A (also a factor of 100).*

Process Controls

There are five variables that must be taken into account when spot welding. These process controls or variables are:

- Time.
- Current.
- Electrode force.
- Electrode face diameter.
- Type of welding machine.

Resistance spot welding is controlled by three different time intervals. These are:

- Squeeze time.
- Weld time.
- Hold time.

Squeeze time is the time needed for the welding electrodes to close and apply the correct force to the metal to be welded. ***Weld time*** is the time during which the electrical current flows. ***Hold time*** occurs after the weld is made. It allows the molten metal to solidify (become solid). At the end of the hold time, the electrodes lift off the metal, and the spot weld is completed.

Off time is an additional time interval that is rarely used. It is the time between one weld sequence and the automatic start of the next sequence. Off time is used when making a resistance seam weld.

Force is required to press together the parts to be welded. This force creates a path for the electricity to flow. It also keeps the molten metal within the weld area. Molten metal will squirt out of the weld if there is not enough force, if the electrode tip diameter is too small, or if the current is too high. This is called an ***expulsion weld***, which is not a desirable weld.

Electrode face diameter is the size of the electrode that contacts the weldments. This needs to be maintained to a fairly consistent size to make good quality welds every time. If the electrode face diameter increases in size, the welding current needs to be increased also.

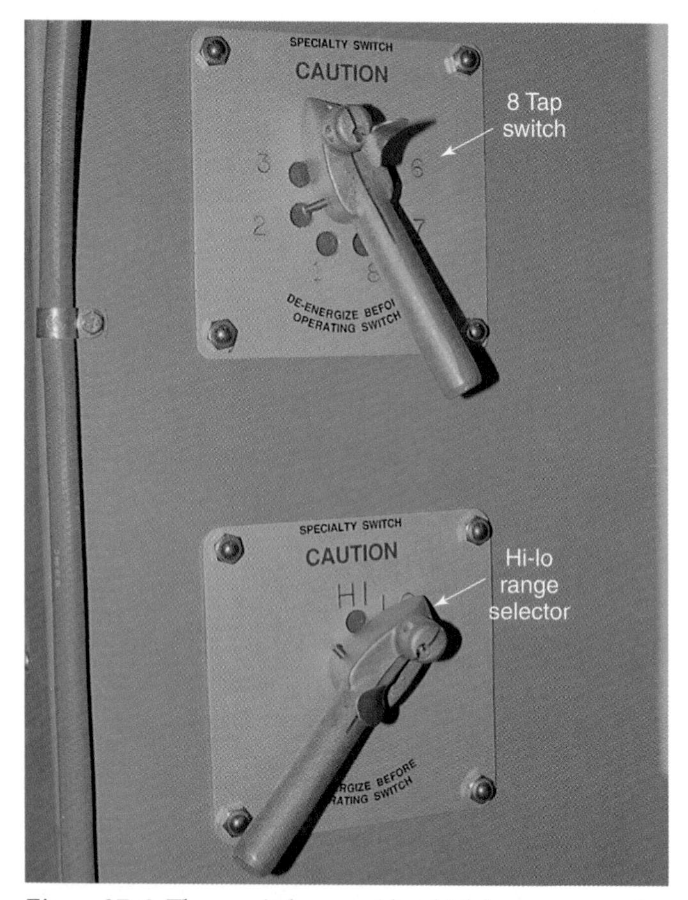

Figure 27-6. *These switches provide a high/low amperage tap selection and a choice of eight transformer tap settings. (Taylor-Winfield Corp.)*

Controllers

Spot welding is a fast welding process. A spot weld is often started and completed in less than one second. As discussed in the preceding section, there are different times (squeeze, weld, and hold) that must be controlled.

In spot welding, a controller is used to correctly set the length of each of these. Times used in spot welding are set in cycles (one cycle = 1/60 second). Squeeze time must be long enough for the electrodes to close on the work and apply the correct force. Squeeze time is usually 10–40 cycles. Weld time can vary from 1–99 cycles; however, 5–30 cycles is typical. Hold time allows the molten metal to cool. It is often 10–30 cycles. Off time is rarely used. When it is not used, the time is set to 0 cycles. Off time may be used to create a continuous or intermittent seam. The parts or electrode tips are automatically moved a small distance during the off time. After the short off time, the welding sequence is repeated. An example of a complete weld schedule is shown in **Figure 27-7**.

In addition to controlling times, a controller regulates the current within the tap setting. Large changes in current are made by changing the tap setting. However, within a given tap setting, the controller changes the current by using the ***percent heat control***. Percent heat is the amount of the available current that will be used. For example, a 100% heat setting will use all the available current. A 50% heat setting will use only half the available current. Note: Do not use a percent heat setting of less than 20%. Instead, lower the tap setting and use a higher percent heat.

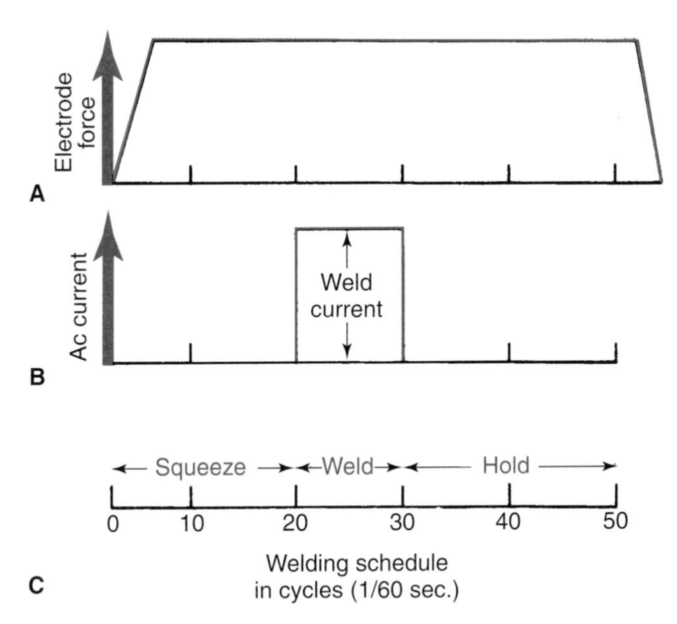

Figure 27-7. A complete welding schedule. A—The electrode force plotted against time. B—Welding current versus time. C—The desired times are set in cycles (1 cycle = 1/60 second).

Many controllers used in industry have additional features that improve the quality of the weld. These features include upslope and downslope. Upslope gradually increases the current at the beginning of the weld. This can be used when welding galvanized or galvannealed mild steel. The upslope can be used to melt off the coating between sheets. Downslope gradually decreases the current at the end of the weld. This can be used to slow the cooling rate, which is important on some types of alloy steel. Another feature is multiple pulses. This allows more than one pulse of weld current while creating a single spot weld. The short intervals between pulses allow the weld to cool.

Manual Control

Mechanical leverage-type spot welding machines often use a limited electric controller. This controller controls only the weld time. Using a foot pedal, the welding operator controls all other times. The pedal has two positions and two switches. When the pedal is pressed to the first position, the first switch closes. This closes the electrodes and applies force to the work. This is the squeeze time. When the foot pedal is pressed to the second position, the weld time begins. The electric controller controls the length of the weld time. After the current stops, the force is still applied. This is the hold time. When the operator releases the foot pedal, the electrodes open, and the weld is complete.

Electrodes

Electrodes used in resistance welding do *not* become part of the weld. Instead, they are used to squeeze the metal together and carry the current into the metal. A material used as an electrode should have the following properties:

- Good electrical conductivity.
- Good thermal conductivity.
- Good mechanical strength and hardness.

Often, a material that conducts electricity and heat well does not have good strength and hardness. An example is pure copper. A material that is very strong and hard often does not conduct electricity and heat well. Examples are tungsten and molybdenum.

The most common electrode materials are copper alloys. They are used to weld mild steel, stainless steel, nickel, and other metals. There are different types of copper alloy electrodes. As the hardness and strength of a copper alloy increases, its ability to conduct electricity and heat decreases. Other electrodes are made from molybdenum or tungsten. These electrode materials have better strength than copper alloy electrodes, but they do not conduct current as well as copper electrode materials.

Electrode Classification

Electrodes have been grouped into classes by the Resistance Welder Manufacturers' Association (RWMA). The most common electrode materials are listed below.

- *Class 1*. This class of electrodes has the highest thermal and electrical conductivities. Its electrical conductivity is 90% (annealed copper has a conductivity of 100%) and has the highest thermal conductivity. It is also the softest material, having a Rockwell hardness of 70B.
- *Class 2*. This class has good electrical conductivity (85%) and good thermal conductivity. It has a Rockwell hardness of 83B. Electrodes in this class are the most widely used. They are used to weld most steels, stainless steels, nickel, and other metals with low electrical and thermal conductivities.
- *Class 3*. These electrodes have an electrical conductivity of 48% and poorer thermal conductivity than Class 2. They have a Rockwell hardness of 100B. They can withstand high force and high temperatures.
- *Classes 4* and *5*. These electrodes are made from copper alloys that are very hard but have poor electrical and thermal conductivity.
- *Classes 10, 11,* and *12*. These include copper-tungsten alloys. Conductivity is 46%; hardness is Rockwell 90B. These electrodes are used to replace Class 1, 2, or 3. They usually do not stick to the work.
- *Class 13*. Tungsten electrode. This electrode has a conductivity of 32% and is very hard. Tungsten electrodes are used to weld copper, brass, and other very conductive metals.
- *Class 14*. Molybdenum electrode. This electrode has a conductivity of 31% and has a hardness of Rockwell 90B. Like the Class 13 electrodes, these electrodes are used on nonferrous metals (metals that do not contain much iron) with high electrical and thermal conductivities.

Electrode Shapes

Electrodes can be machined, cast, or forged into complex shapes. **Figure 27-8** shows many different shapes of electrodes. Electrodes are also available as caps and adaptors. Caps are much less expensive than large electrodes. When they wear, they can be replaced separately. **Figure 27-9** shows caps and adaptors.

Because of the heat developed in a resistance weld, most electrodes are water-cooled. There is a metal tube in the electrode that carries water very close to the tip. Water flows whenever the machine is on, removing the heat from the electrode, **Figure 27-10**.

Seam Welders

In addition to making individual spot welds, resistance welding is used to make resistance seam welds (RSEW). A ***seam weld*** is a series of spot welds that form a seam. Seam welds can be continuous or a series of individual spots. Seam welding machines use a round electrode that operates like a wheel, rolling along the weld joint. The electrode is often called a ***wheel electrode***, **Figure 27-11**.

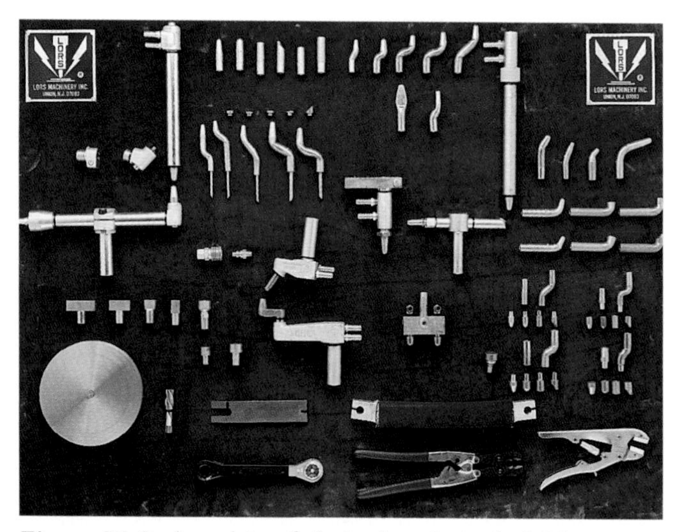

Figure 27-8. *A variety of electrodes, electrode holders, caps, adaptors, and electrode removing and dressing tools. (LORS Machinery, Inc.)*

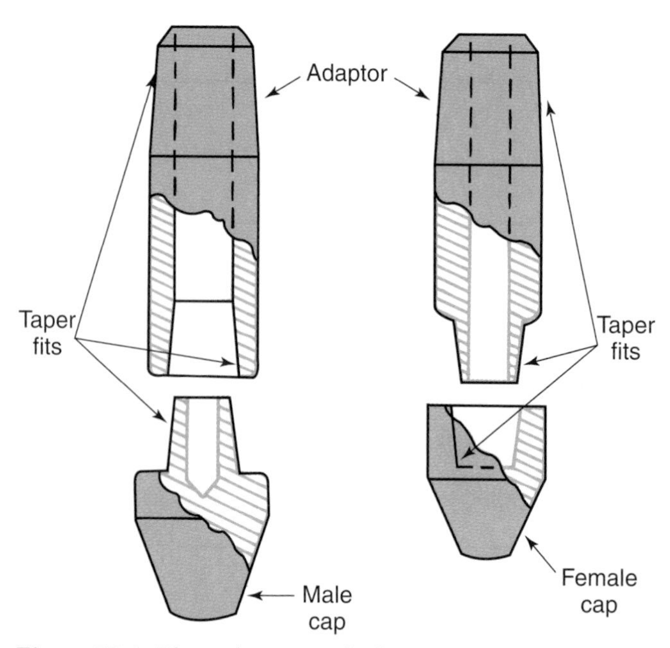

Figure 27-9. *Electrode caps and adaptors. These caps are held in place by a taper fit. (Tuffaloy Products, Inc.)*

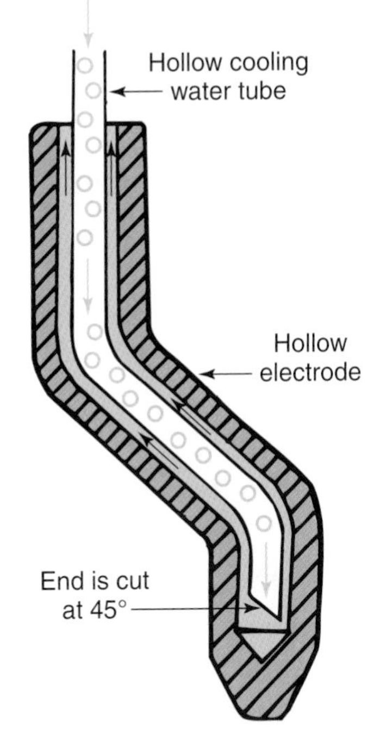

Figure 27-10. *The cooling water tube carries water to the tip of the electrode.*

Figure 27-11. *A seam welding machine makes continuous or intermittent spot welds. Shafts from the frame of the machine drive two small friction wheels, which drive the large wheel electrodes. (T.J. Snow Company, Inc.)*

The current used to make a seam weld can flow constantly as a seam weld is made, or it can repeatedly turn on and off at an adjustable interval. When the current turns on and off, it makes individual spot welds. Depending on the duration of the off time, these spot welds can overlap to form a continuous seam, or can be a series of individual spots that do not overlap. Spot welds that do not overlap form an *intermittent* seam weld. See **Figure 27-12**. Overlapping or intermittent seam welds are applications in which off time in the welding sequence is used.

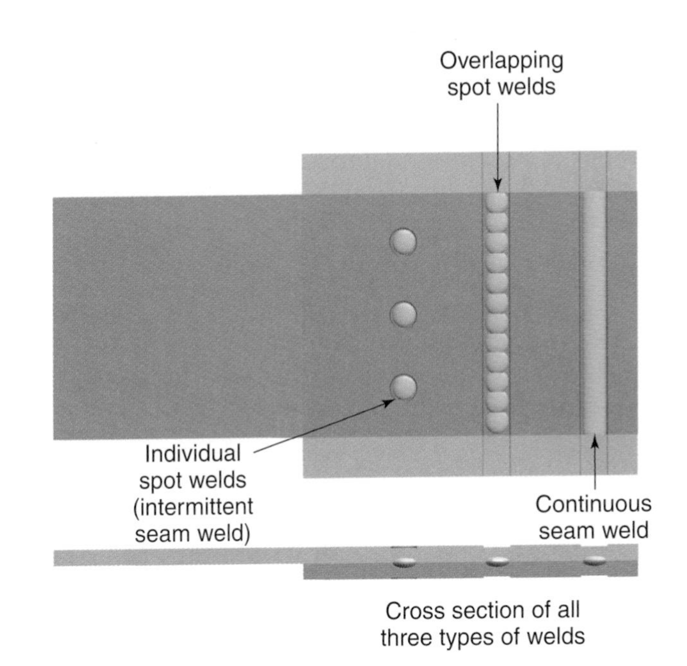

Figure 27-12. *Welds that are possible with a seam welding machine: individual spots (intermittent seam weld), overlapping spots, or a continuous seam.*

Safety

Resistance welding machines can be much more complex than arc welding machines. When using a resistance welding machine, you should perform the following tasks regularly:

- **Check all electrical connections.**
- **Check the electrodes and other parts that carry electricity to be sure they are installed correctly.**
- **Check all pneumatic and hydraulic connections and hoses. Look and listen for any leaks.**
- **Check all water connections and hoses. Make sure water is flowing to cool the welding machine.**

The voltages and currents used in resistance welding can be high enough to shock you and possibly kill you. The pressures used in resistance welding can crush any part of your body that may get in between the electrodes. If a problem develops with a machine you are using, turn it off first, and then notify your instructor or supervisor.

When resistance welding, molten metal can squirt out of a spot weld and be thrown into the air. Always wear safety glasses when spot welding or when working near spot welding equipment.

You should also wear gloves when spot welding. This will prevent cuts from sharp metal edges and will protect your hands from molten metal that may squirt out during a weld. You should wear long pants without cuffs, a long-sleeved shirt, and steel-toed work shoes.

Summary

- When electricity flows through metal, the metal heats up. This occurs because there is resistance (opposition) to the flow of electricity through the metal. If two metals are touching and an electrical current is passed through them, the largest amount of heat is created at the surfaces where the two metals are touching.
- In resistance spot welding (RSW), two or more metal pieces are stacked on top of one another. Pressure is applied to the pieces and electrical current flows through the metal. As the current flows, heat develops at the surfaces where the pieces touch, melting the base metal at that spot. After the current stops, the molten metal becomes solid again, welding the parts together.
- A resistance welding machine must accurately apply the desired force to the metal to be welded. Different methods are used to apply force to the metal. They are pneumatic pressure, hydraulic pressure, and mechanical leverage.
- The most common resistance welding machine designs are rocker arm, press, and gun.
- Throat depth and the horn spacing are important dimensions that must be considered when selecting a welding machine. They determine the size of a part that can be welded.
- Most resistance welding uses alternating current (ac). Power supplied to a welding machine can vary from 115V to 460V with current varying from a few amps to over 1000A.
- Transformers that reduce the supplied voltage are called step-down transformers. In a spot welding machine, a step-down transformer may reduce the voltage supplied to less than 1V or as high as 40V. The current increases to thousands or tens of thousands of amps.
- Resistance spot welding is described in terms of three different time intervals: squeeze time, weld time, and hold time. Squeeze time is the time needed for the welding electrodes to close and apply the correct force to the metal to be welded. Weld time is the time during which the electrical current flows. Hold time is the time during which the electrodes continue to apply pressure after the weld is made. A fourth time interval called off time may be used during resistance seam welding. Off time is the time between one weld sequence and the automatic start of the next sequence.
- Electrodes used in resistance welding do not become part of the weld. They are used to squeeze the metal together and carry the current into the metal. A material used as an electrode should have good electrical conductivity, good thermal conductivity, and good mechanical strength and hardness. The most common electrode materials are copper alloys.
- Electrodes can be machined, cast, or forged into complex shapes. Electrodes are also available as caps and adaptors. Caps are much less expensive than large electrodes. When they wear, they can be replaced separately.
- In addition to making individual spot welds, resistance welding is used to make resistance seam welds (RSEW). A seam weld is a series of spot welds that forms a seam. Seam welding machines use a round electrode that operates like a wheel, rolling along the weld joint.
- When using a resistance welding machine, you should perform the following tasks regularly: **check all electrical connections, check the electrodes and other parts to be sure they are installed correctly, check all pneumatic and hydraulic connections and hoses for leaks, check all water connections and hoses for proper flow and leaks.**
- **When resistance welding, always wear safety glasses, leather gloves, long cuffless pants, a long sleeved shirt, and steel-toed safety shoes.**

Review Questions

Write your answers on a separate sheet of paper. Please do *not* write in this book.

1. What causes the heat used to make a resistance spot weld?
2. Where is most of the heat developed when electricity flows through two or more pieces of metal?
3. List the five variables, or process controls, involved in spot welding.
4. Explain the difference between hold time and squeeze time.
5. In a step-down transformer, does current increase or decrease? Does voltage increase or decrease?
6. List two ways to change the current used for a spot weld.
7. Which RWMA electrode class is the one most commonly used? What metals are those electrodes used to weld?
8. What method is used to remove the heat developed in electrodes during spot welding?
9. What is the electrode used in seam welding often called?
10. If a problem develops with a resistance welding machine, what is the *first* thing you should do?

This specialized multi-head resistance welder is designed for welding electric motor housings. Note the radial arrangement of the electrodes. (LORS Machinery, Inc.)

Chapter 28

Resistance Welding: Procedures

Learning Objectives

After studying this chapter, you will be able to:

- Set up and adjust a spot welding machine.
- Describe the methods used to determine the correct force for spot welding.
- Determine the weld time needed for resistance welding mild steel.
- Describe the means used to test for a good spot weld and the signs that indicate a weld is of the desired quality.

Technical Terms

current density
electrode tip diameter
expulsion weld
force gauge
newtons (N)
projection welding (PW)

Selecting the Welding Machine

There are several types of resistance spot welding machines, as discussed in Chapter 27. Some are very large and can handle large pieces of material. Some are small and can weld only small parts. Some machines have wheel electrodes and are used to make seams. Refer to Chapter 27 for help in selecting a machine that is the right size for the welding you will be doing.

The type and the thickness of the metal you will weld dictates how much force, time, and current the machine must apply. Pneumatic force systems are the most common machines and can weld all common resistance welding applications. They are widely used in automotive, electronic, appliance, and general sheet metal industries. Current can range from a few hundred amps for welding very small electronic assemblies to tens of thousand of amps for thin aluminum sheet metal. Select a welding machine that will supply enough force and current for the job you are welding.

Selecting the Welding Variables

Once the welding machine is selected, there are four additional variables that must be set. The following is a list of these variables, in order of importance:

- Current.
- Time.
- Electrode force.
- Electrode selection.

The Resistance Welding Manufacturing Alliance (RWMA) publishes information on suggested welding conditions. **Figure 28-1** is a table of recommended

Recommended Practices for Single-Pulse Spot Welds in Low Carbon Steel

		DATA COMMON TO ALL CLASSES OF SPOT WELDS				WELDING SETUP FOR BEST QUALITY CLASS "A" WELDS				
Thickness of Thinness Outside Piece		Electrode Major Diameter and Shape		Minimum Weld Spacing[1]	Minumum Contacting Overlap	Weld Time (Single Pulse)	Net Electrode Force	Welding Current*	Diameter of Fused Zone	Minimum Tension-Shear Strength
MFG GAUGE	"T" THICKNESS Inch (mm)	"D" MIN. Inch	"d" MAX Inch	"C" INCH	"L" INCH	CYCLES (60 HZ)	POUNDS	AMPERES (approx.)	"Dw" INCH (approx.)	POUNDS
32	.010 (0.25)	1/2	1/8	1/4	3/8	4	200	4,000	.13	235
25	.021 (0.53)	1/2	3/16	3/8	7/16	6	300	6,100	.17	530
22	.030 (0.76)	1/2	3/16	1/2	7/16	8	400	8,000	.21	980
20	.036 (0.91)	1/2	1/4	3/4	1/2	10	500	9,200	.23	1,350
18	.048 (1.22)	1/2	1/4	7/8	9/16	12	650	10,300	.25	1,820
16	.060 (1.52)	1/2	1/4	1-1/16	5/8	14	800	11,600	.27	2,350
14	.075 (1.91)	5/8	5/16	1-3/8	11/16	21	1,100	13,300	.31	3,225
13	.090 (2.29)	5/8	5/16	1-5/8	3/4	25	1,300	14,700	.34	4,100
12	.105 (2.67)	5/8	3/8	1-13/16	13/16	29	1,600	16,100	.37	5,300
11	.120 (3.05)	5/8	3/8	2	7/8	30	1,800	17,500	.40	6,900

		WELDING SETUP FOR BEST QUALITY CLASS "B" WELDS					WELDING SETUP FOR BEST QUALITY CLASS "C" WELDS				
Thickness of Thinness Outside Piece		Weld Time (Single Pulse)	Net Electrode Force	Welding Current*	Diameter of Fused Zone	Minimum Tension-Shear Strength	Weld Time (Single Pulse)	Net Electrode Force	Welding Current*	Diameter of Fused Zone	Minimum Tension-Shear Strength
MFG GAUGE	"T" THICKNESS Inch (mm)	CYCLES (60HZ)	POUNDS	AMPERES (approx.)	"Dw" INCH (approx.)	POUNDS	CYCLES (60 HZ)	POUNDS	AMPERES (approx)	"Dw" INCH (approx.)	POUNDS
32	.010 (0.25)	5	130	3,700	.12	200	15	65	3,000	.11	160
25	.021 (0.53)	10	200	5,100	.16	460	22	100	3,800	.14	390
22	.030 (0.76)	15	275	6,300	.20	850	29	135	4,700	.18	790
20	.036 (0.91)	21	360	7,500	.22	1,230	38	180	5,600	.21	1,180
18	.048 (1.22)	24	410	8,000	.23	1,700	42	205	6,100	.22	1,600
16	.060 (1.52)	29	500	9,000	.26	2,150	48	250	6,800	.25	2,050
14	.075 (1.91)	36	650	10,400	.30	3,025	58	325	7,900	.28	2,900
13	.090 (2.29)	44	790	11,400	.33	3,900	66	390	8,800	.31	3,750
12	.105 (2.67)	50	960	12,200	.36	5,050	72	480	9,500	.35	4,850
11	.120 (3.05)	60	1,140	12,900	.39	6,500	78	570	10,000	.37	6,150

* **Starting values** shown are based on experience of member companies
1. Minimum spacing shown is for welding of two pieces. Increase spacing by 30% when welding three pieces. Smaller minimum spacing requires higher current.

- Type of steel: SAE 1008-1010
- Material should be free from scale oxides, paint grease, and heavy oil.
- Table is for a 3:1 maximum ratio of thickest to thinnest piece, and a maximum stackup thickness of 4"T"
- Electrode material: RWMA CLASS 2

Figure 28-1. *Starting values used to create quality spot welds in low-carbon steel. Class A welds are the highest quality and have the greatest shear strength. If the welding machine being used is not capable of providing the electrode force or the welding current required, use alternate settings from Class B or Class C welding conditions. (Resistance Welding Manufacturing Alliance)*

welding conditions for low-carbon steel. This is a good table to use to establish welding conditions.

Figure 28-1 lists three classes of welds. Class A welds have the largest spot weld diameters and are the strongest welds. Class B welds are slightly smaller than Class A welds. Class C welds have the smallest weld diameters and the lowest strength. Some weld applications require Class A welds. Class A welds use high electrode force, high weld current, and a relatively short weld time. Class B and C welds use lower electrode force, lower weld currents, and longer weld times. Not all welding machines can achieve a Class A weld. A Class A weld should be made whenever possible, as it is superior to the other classes of welds.

There are some approximation formulas that have been developed for use in setting up resistance welding conditions. **Figure 28-2** lists these formulas. Many companies develop a set of welding conditions they use for welding in their shop. These welding conditions are often set up for a specific job or a specific piece of equipment. Welding conditions are dependent on a few variables. These include the following:

- Type and alloy of metal being welded.
- Thickness of metal being welded.
- Type of equipment being used.

The purpose of using established welding conditions is to produce quality welds and maintain that quality.

Selecting and Preparing the Electrodes

Two important areas must be considered when selecting the electrode. The first consideration is the type of electrode material to be used. Chapter 27 discusses electrode materials. Select the correct material for the metal to be welded. A very common electrode material is the class 2 copper alloys for welding mild and coated steels.

Second select the proper ***electrode tip diameter***. The required diameter of a spot weld increases as the base metal thickness increases. A larger diameter spot weld makes a stronger weld. Figure 28-1 lists the recommended spot weld diameter and recommended electrode tip diameters for various steel thicknesses. To use Figure 28-1, determine the thickness of steel to be welded. If the thicknesses are not the same, determine the thickness of the thinner sheet. Find this thickness in the first column. Find the recommended electrode tip diameter in the second column.

If the electrodes are not parallel or are not in line, only *part* of the tip will contact the metal, and poor welds will result. See **Figure 28-3**. After installing the electrodes into the welding machine, adjust them so they meet correctly.

After making a number of welds, the tips of the electrodes need to be cleaned. Cleaning may be needed after as few as ten welds or as many as 5000 welds. Cleaning removes dirt, oxides, and metal particles from the end of the electrodes. Electrodes that are used with too much current, too much pressure, or for too long a period will flare out (mushroom) at the end. See **Figure 28-4**.

As the tips become larger in diameter, the current used to make the spot weld is spread over a larger area, and the ***current density***, which is the amount of current divided by the electrode tip area, decreases. This lower current density causes a smaller and weaker weld. If electrodes mushroom, replace them or return them to the correct diameter and shape by dressing, filing, or machining.

Remember, when choosing an electrode, you must select the correct tip diameter with the tip design you need. You must also select the correct electrode material for the job you are welding.

Calculating Resistance Spot Welding Variables				
Variable	Equation (Conventional)	Units of Measure	Equation (Metric)	Units of Measure
Contact Tip Dia.	.18 + (total sheet thickness)	inches	4.54 + (total sheet thickness)	mm
Weld Time	120 × (total sheet thickness)	cycles	4.73 × (total sheet thickness)	cycles
Current	125,000 × (total sheet thickness	amps	4935 × (total sheet thickness)	amps
Electrode Force	8500 × (total sheet thickness)	lbs.	1500 × (total sheet thickness)	newtons
Weld Size	The weld size is the same size or slightly smaller than the contact tip diameter.			

Figure 28-2. *This table of equations approximates the values of the variables used when resistance spot welding low-carbon steel. The total sheet thickness is the combination of the two thicknesses being welded.*

Selecting and Setting the Force

After selecting the welding machine and the electrodes, the remaining variables to set are force, time, and current.

The force applied by the electrodes presses together the metals being welded. As noted in Chapter 27, the applied force creates a path for the electrical current to flow through the metal. Refer to Figure 28-1 and find the recommended force for the material thickness being welded. Find the thickness of the thinner sheet to be welded in the first column. Follow that row across to the column titled Net Electrode Force for either a Class A, Class B, or Class C weld. The number listed in the selected column is the recommended force. The force must not be too great or it will damage the metal. The proper force will indent the metal slightly, but will not crush the area under the electrodes. **Figure 28-5** shows two welds: one made with the correct force and one made with excessive force.

On machines that use a pneumatic cylinder, force is increased or decreased by adjusting the air pressure. The higher the pressure going into the pneumatic cylinder, the higher the force at the electrodes. Machines that use hydraulic pressure are adjusted similarly.

Force is measured by a ***force gauge***. To use a force gauge, first turn the weld current off. If the machine has a switch that has a *No Weld* position, place the switch in the *No Weld* position. Place the gauge between the electrodes. Make the welding machine close the electrodes onto the force gauge. The gauge will measure the force applied by the electrodes, **Figure 28-6**. Force is measured in pounds or ***newtons (N)***.

On machines that use mechanical leverage, a spring is usually adjusted to obtain the desired force. By compressing the spring more, the force increases. Adjust the compression of the spring to obtain the desired force.

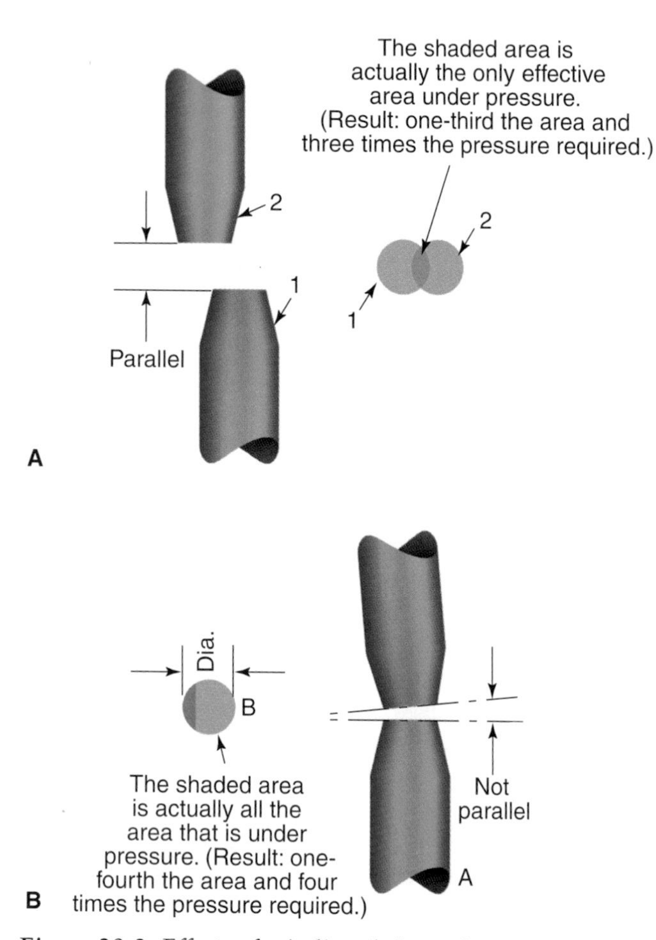

Figure 28-3. *Effects of misaligned electrodes. A—These electrodes are off-center. B—These electrodes are not parallel. In either case, pressure on the reduced contact area is too great.*

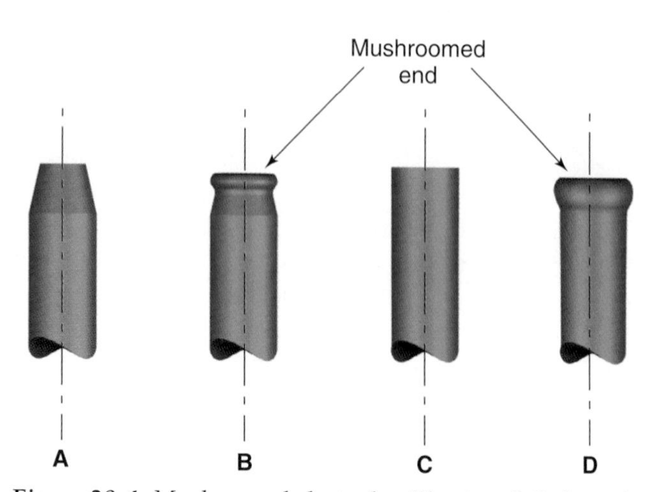

Figure 28-4. *Mushroomed electrodes. The A and C electrodes are in good condition. The B and D electrodes are flared or mushroomed on the end from too much heat and pressure.*

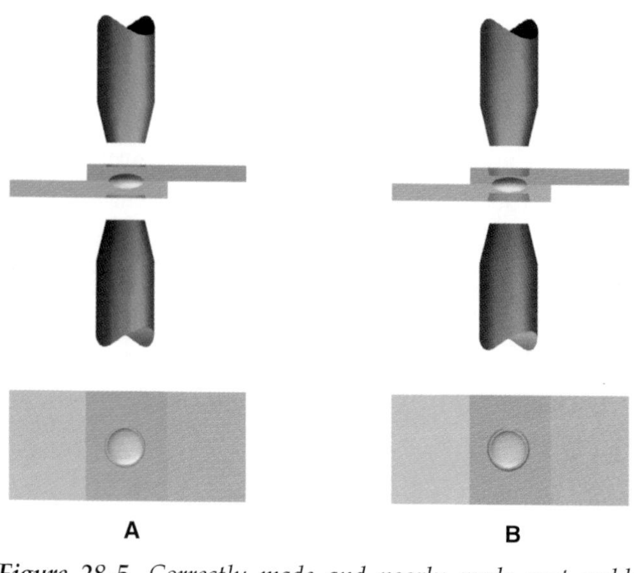

Figure 28-5. *Correctly made and poorly made spot welds shown in cross section. A—This weld was made with the correct heat and force. B–This weld is deeply indented because too much force was used.*

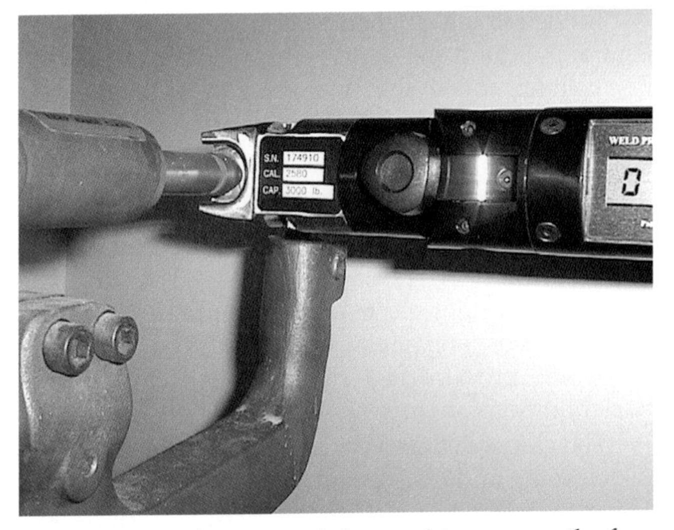

Figure 28-6. *A force gauge being used to measure the force applied by the electrodes. Changing the pressure to the pneumatic or hydraulic force cylinders varies the amount of electrode force. (Sensor Development, Inc.)*

Another way to approximate the correct force on mild steel is to use the welding machine. Turn the weld current off. If the machine has a switch that has a *No Weld* position, place the switch in that position. Another way to prevent the machine from welding is to set the weld time to zero.

Place two pieces of metal between the electrodes. When you press the foot pedal, the electrodes will close on the metal but no welding will occur. Release the pedal, so the electrodes lift off. Look at the metal. If there is no indentation from the electrodes, increase the force until a slight dent is made. Then, reduce the force about 10% so no dent is made. This provides a starting point. When welding, you may need to readjust the force to obtain a good weld. This method of determining force is only applicable to mild steel.

Selecting and Setting the Time and Current

Time and current are the two variables that have the greatest effect on the quality of a weld. A high-quality spot weld requires that all the variables be set correctly, but time and current are the most important. If they are not set correctly, the weld may be too small and too weak, or it may be too large and squirt out liquid metal. When a weld squirts out liquid metal, it is called an ***expulsion weld***.

Making Time Settings

As noted in Chapter 27, there are three different times that make up a complete weld schedule. They are the squeeze, weld, and hold times. A fourth time, off time, is normally set to zero.

Squeeze time must be long enough to allow the electrodes to close and apply the correct pressure on the metal being welded. Typical squeeze time settings are 10–40 cycles.

Weld time (time during which the current flows) must be long enough to allow the metal to melt and the weld to grow to the correct size. The weld time varies, depending on the type and thickness of the metal being welded. Typical values for welding sheet metal are 5–30 cycles.

To find the correct weld time from Figure 28-1, begin by locating the thickness of the thinner sheet to be welded in the first column. Follow that row across to the column titled Weld Time (Single pulse) for either Class A, Class B, or Class C welds, depending on the quality of weld you wish to make. The number in the selected column indicates the recommended weld time in number of cycles.

Hold time must be long enough for the molten metal to become solid. This time typically varies from 5 to 30 cycles.

Off time is rarely used. When it is not used, set the off time to zero. One application of off time is during a seam weld. The weld current can be set to make a series of intermittent or overlapping spot welds.

When operating a mechanical lever system, you may control the squeeze and hold times. If these times are not controlled by a controller, the person doing the welding controls the times. Press the foot pedal down so the electrodes close on the work. Then, press down farther so the weld current will flow for the set time. Keep the electrodes closed on the work for one second or longer. Finally, release the pressure on the foot pedal, and the electrodes will open, completing the weld cycle.

Making Current Settings

Current is adjusted by changing the tap setting, changing the percent heat setting, or by setting the desired current value. Tap changes are for large increases in current; percent heat is for fine adjustments. Some weld controllers allow the desired weld current to be set digitally. The controller uses set values and feedback to deliver the desired weld current.

Several methods are used to properly set the current. If you have operated the equipment before, you know the approximate settings that are required. You can also use a trial-and-error method to determine the current. A third way is to calculate the required current, and then use special equipment to set and monitor it.

After using a resistance welding machine for some time, the controls and settings become familiar. You will learn what tap and percent heat settings are needed to weld a particular combination of metals.

When you are first learning about spot welding, the trial-and-error method of setting the current is useful. First, set the tap switch. A higher setting is used for thicker metals and a lower setting for thin metals. Set the percent heat to 70%. Make two spot welds, as shown in **Figure 28-7**. If metal squirts out, turn down the tap setting. If the metal pieces do not weld together, turn up the tap setting. If the pieces do weld together, tear them apart and look at the spot welds. The weld size should be slightly smaller than the diameter of electrode being used. A good weld will tear some metal away from one of the pieces as they are pulled apart. **Figure 28-8** shows pieces of metal pulled apart to check welds for quality.

Adjust the tap setting and percent heat up or down. This increases or decreases the current and affects the size of the weld. Make two more spot welds. Tear them apart and look at the weld quality. Adjust the current up or down. Continue this process until a good-quality weld is made.

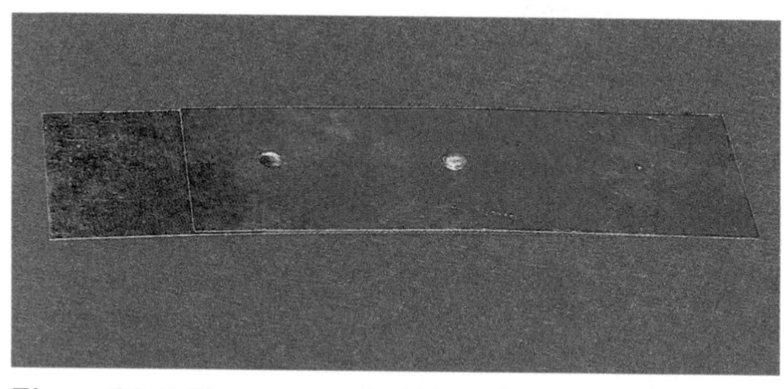

Figure 28-7. *Two pieces of mild steel overlapped about 4″ (100 mm) and spot welded.*

Figure 28-8. *A spot weld nugget torn out of one piece during a destructive peel test. Test welds are made to determine the correct machine settings to get the proper spot weld size and strength.*

To determine a good starting current, refer to Figure 28-1. Find the thinner material thickness in the first column. Follow that row across to find the recommended welding current for the class of weld to be made.

Welding Mild Steel

Properly setting up the equipment is the most important part of resistance welding. This involves choosing the right electrodes and setting the correct current, times, and pressure. Once the equipment is set, it will produce high-quality spot welds if these two conditions are met:

- The electrodes are kept in good condition by cleaning and resurfacing as needed.
- The surface of the metal to be welded is thoroughly cleaned and free of oil and rust.

Each weld schedule is set for one combination of metals. If a spot welding machine is set up to weld two pieces of 1/16″ (1.6 mm) mild steel, the same setting cannot be used to weld 1/8″ (3.2 mm) mild steel. Nor can that setting be used to weld aluminum. Each type and thickness of metal requires a different setting or weld schedule.

The resistance spot welding values shown in Figure 28-1 are good starting conditions when welding mild steel. The formulas shown in Figure 28-2 can also be used to determine starting values for spot welding mild steel. Welding current is the most important variable that affects the weld nugget diameter and strength.

Welding Aluminum

Aluminum is softer than steel. It is also more electrically and thermally conductive than steel. This means that electricity and heat will flow through aluminum more easily than they do through steel. For these reasons, the welding setup for aluminum is different from the setup for steel. A 2″ or 3″ (50 mm or 75 mm) radiused electrode tip is used. Class 1 or Class 2 electrode material is often used.

Less force is used, and much more current is required. The current is about three times greater when welding aluminum than it is for welding similar thicknesses of steel. Currents are often in 20,000 to 30,000 amp range for thin aluminum sheet.

Different types of machines can be used when welding aluminum to obtain these high currents. These include the following:

- A single-phase transformer.
- A three-phase transformer.
- A three-phase rectified transformer. This produces dc current.
- A three-phase mid-frequency rectified transformer. This produces dc current from a smaller sized transformer.

- A stored energy system that stores primary power (often in capacitors) and rapidly discharges to make the spot weld.

Welding time is shorter for welding aluminum than it is for welding steel. When using a single-phase machine, the weld times are about 20% less for welding aluminum than they are for welding mild steel. When using a three-phase machine, the weld times are about one-half the times used to weld mild steel.

To obtain the highest-quality welds, aluminum must be cleaned to remove oxides from the surfaces before it is welded. Good-quality welds can be made without cleaning, but cleaning improves the repeatability of the welding process. A stainless steel brush or a cleaning solution can be used.

Exercise 28-1 Spot Welding Mild Steel

1. Obtain ten or more pieces of mild steel measuring approximately 1 1/2″ × 8″ × 1/16″ (40 mm × 200 mm × 1.6 mm).
2. Clean both surfaces of the pieces. This can be done with steel wool, a wire brush, or a degreasing solution supplied by your instructor.
3. Use Figure 28-1 and determine what values should be used to create a Class A weld. Write your answers on a separate sheet of paper.
   Electrode tip diameter: _____
   Electrode force: _____
   Weld time: _____
   Weld current: _____
4. Select the proper electrodes and install them in the machine.
5. Set the force by adjusting the pneumatic pressure, the hydraulic pressure, or the force spring.
6. If you are using a machine with a controller, set the squeeze time to 30 cycles. Set the weld time to the time determined in Step 3. Set the hold time to 30 cycles. Set the off time to zero.
7. Set the starting weld current as described in this chapter.
8. If the machine is water-cooled, turn on the water.
9. Overlap two pieces of mild steel about 7″ (180 mm), as shown in Figure 28-7. Make a spot weld. Adjust the current as required. Make additional welds, adjusting the force, time, and current as needed, until you obtain a spot weld that is about 1/4″ (6.4 mm) in diameter. You may make more than one weld on each sample while setting the machine.
10. Once you have set the machine correctly, weld two pieces together. Make welds, one every inch (25 mm) along the overlapped portion. Do not weld closer than .5″ (13 mm) to the end of each piece.
11. On a separate sheet of paper, record the settings you used to make these welds:
    Electrode tip diameter: _____
    Electrode force: _____
    Weld time: _____
    Tap setting: _____
    Percent heat setting: _____
12. Tear or peel apart at least five spot welds. See Figure 28-8. Measure and record the size of five different spot welds on a separate sheet of paper.

Projection Welding

In ***projection welding (PW)***, welds similar to spot welds are made at locations where projections or bumps have been formed in one of the pieces, **Figures 28-9** and **28-10**. The pieces to be welded are forced together by the electrodes. When the current flows, a weld is formed at each point where there is a projection or bump. The projections accurately locate the weld points.

A welding schedule for projection welding is similar to spot welding the same material and thickness without a projection. More than one projection weld can be made at a time. This requires more force and more current than is needed to make a single weld.

Resistance Seam Welding

Resistance seam welding (RSEW) is done with two wheel electrodes. These wheels roll along opposite surfaces of the metal to be welded. A seam weld can be continuous or can be a series of spot welds that are evenly spaced. Such a line of evenly spaced spot welds is called an intermittent seam weld.

Exercise 28-2–Spot Welding Aluminum

Discuss with your instructor the type of welding machine that will be used for this exercise. Your instructor will provide data on the proper force, weld time, and weld current settings for the type machine that will be used.

1. Obtain ten or more pieces of aluminum measuring approximately 1 1/2″ × 8″ × 1/32″ (40 mm × 200 mm × 0.8 mm).
2. Clean both surfaces of the pieces. This can be done with a stainless steel or a solution supplied by your instructor.
3. Install 2″ or 3″ (50 mm or 75 mm) radiused electrodes.
4. Set the electrode force by adjusting the pneumatic pressure. The force should be less than is used for welding steel of the same thickness.
5. Set the required times in the controller. Set the squeeze time to 30 cycles. Set the proper weld time, which should be only a few cycles. Set the hold time to 30 cycles. Set the off time to zero.
6. Set the required weld current. Set the tap setting to the high position. Set the percent heat or actual current value.
7. Turn the water cooling on.
8. Overlap two pieces of aluminum about 7″ (175 mm), as shown in Figure 28-7. Make one or two spot welds. Follow the trial-and-error approach described earlier in this chapter, to determine the weld time and weld current. Make additional welds, adjusting the force, time, and current, until you have spot welds that are about 1/8″ (3.2 mm) in diameter.
9. Once you have set the machine correctly, weld two pieces together. Make welds, every 1″ (25 mm), along the overlapped portion. Do not weld closer than .5″ (13 mm) to the end of each piece.
10. On a separate sheet of paper, record the settings you used to make welds.
    Electrode tip diameter: ____
    Electrode force: ____
    Weld time: ____
    Tap setting: ____
    Percent heat or current setting: ____

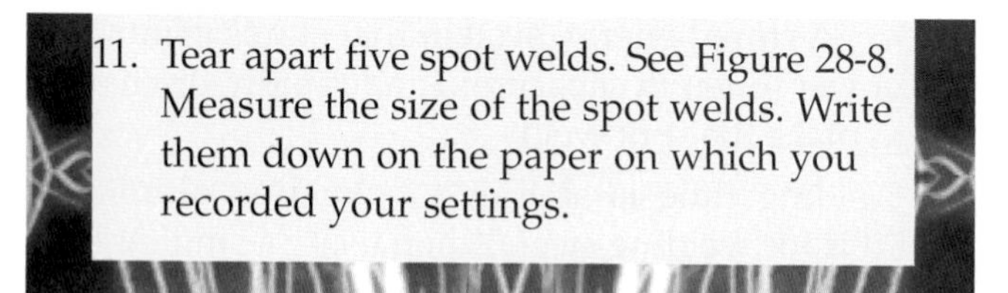

11. Tear apart five spot welds. See Figure 28-8. Measure the size of the spot welds. Write them down on the paper on which you recorded your settings.

The pressure needed to make a seam weld is the same or slightly higher than that used to make individual spot welds. The current required to make a continuous seam weld is much higher than is necessary to make individual spot welds. An intermittent seam weld requires more current than individual spots but less than a continuous seam weld.

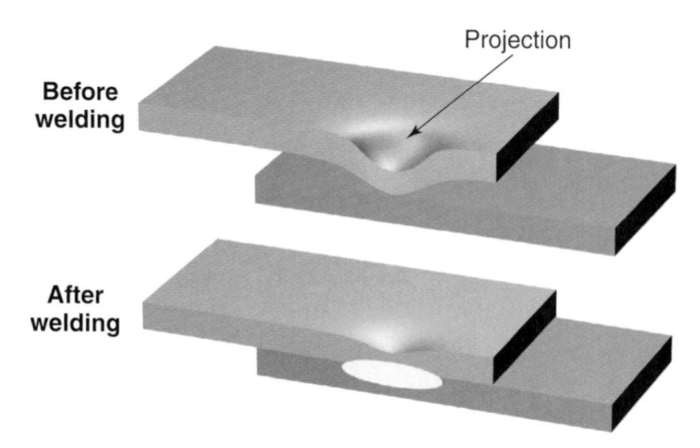

Figure 28-9. A drawing of metal before and after projection welding. The metal is stamped to form a projection. Part of the depression from stamping remains after the weld is completed.

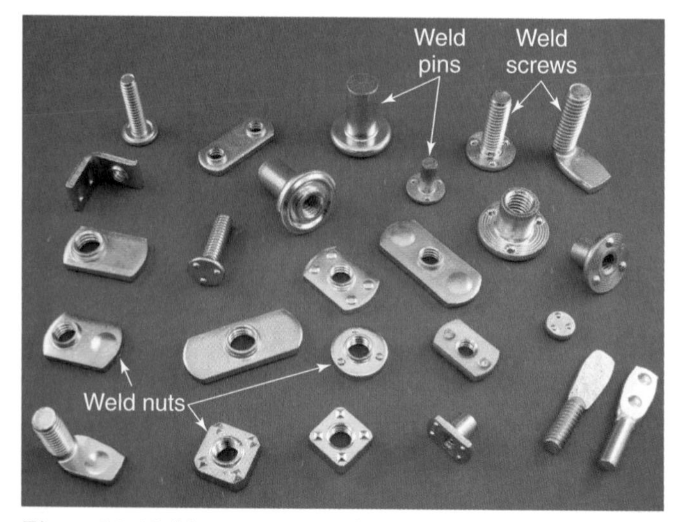

Figure 28-10. Nuts, screws, and pins with projections formed on them for projection welding. Notice the different projection types and sizes. (The Ohio Nut and Bolt Co.)

Summary

- Most resistance welding applications use a pneumatic (air pressure) force system. Welding currents can range from a few hundred amps for welding very small electronic assemblies to tens of thousands of amps for welding thin aluminum sheet metal. Select a welding machine that will supply enough force and current for the job you are welding.
- When selecting electrodes, determine type of electrode material and electrode tip diameter you need. Not every size tip is available, so use one that is close to the size you need. After installing the electrodes into the welding machine, adjust them so they meet correctly. After making a number of welds, the tips of the electrodes need to be cleaned. If electrodes mushroom, replace them or return them to the correct diameter and shape by dressing, filing, or machining.
- The force applied by the electrodes presses together the metals being welded. The force must not be too great or it will damage the metal. The proper force will indent the metal slightly, but will not crush the area under the electrodes.
- Squeeze time must be long enough to allow the electrodes to close and apply the correct pressure on the metal being welded. Weld time must be long enough to allow the metal to melt and the weld to grow to the correct size. Hold time must be long enough for the molten metal to become solid.
- Current is adjusted by changing the tap setting or the percent heat setting. Tap changes are for large increases in current; percent heat is for fine adjustments.
- The most important part of resistance welding is the equipment setup. Once the equipment is set, it will produce high-quality spot welds if the surface of the metal is clean and the electrodes are in good condition.
- The welding setup for aluminum is different from the setup for steel. Less force is used. Much more current is required, but the weld time is 20% to 50% less than is needed for steel. Class 1 or 2 electrodes are often used to weld aluminum.
- In projection welding (PW), the pieces to be welded have small projections (bumps) formed in them at the desired locations for the welds. The pieces are forced together by the electrodes. When the current flows, a weld is formed at each projection, or bump.
- Resistance seam welding (RSEW) is done with two wheel electrodes. These wheels roll along opposite surfaces of the metal to be welded. A seam weld can be continuous or can be a series of spot welds that are evenly spaced.

Review Questions

Write your answers on a separate sheet of paper. Please do *not* write in this book.

1. What are the five variables in resistance welding?
2. Which two variables have the greatest effect on weld quality?
3. To produce a good weld, what must be done to the electrodes after they are installed into the welding machine?
4. Why must the electrode tips be cleaned?
5. If the pneumatic or hydraulic pressure is increased, does the electrode force increase or decrease?
6. Use Figure 28-1 to determine the following values for creating a Class A weld between two pieces of .032″ (0.8 mm) mild steel.

   Electrode tip diameter: ________
   Electrode force: ________
7. Use Figure 28-1 to determine the following values for welding (Class A) one piece of .048″ (1.2 mm) mild steel to a piece of .060″ (1.5 mm) mild steel.

   Weld time: ________
   Weld current: ________
8. After a welding machine is set up properly, what two conditions must be met to continue making high-quality spot welds?
9. *True or False?* Aluminum conducts heat and electricity more easily than steel.
10. What are the two types of seam welds that can be made?

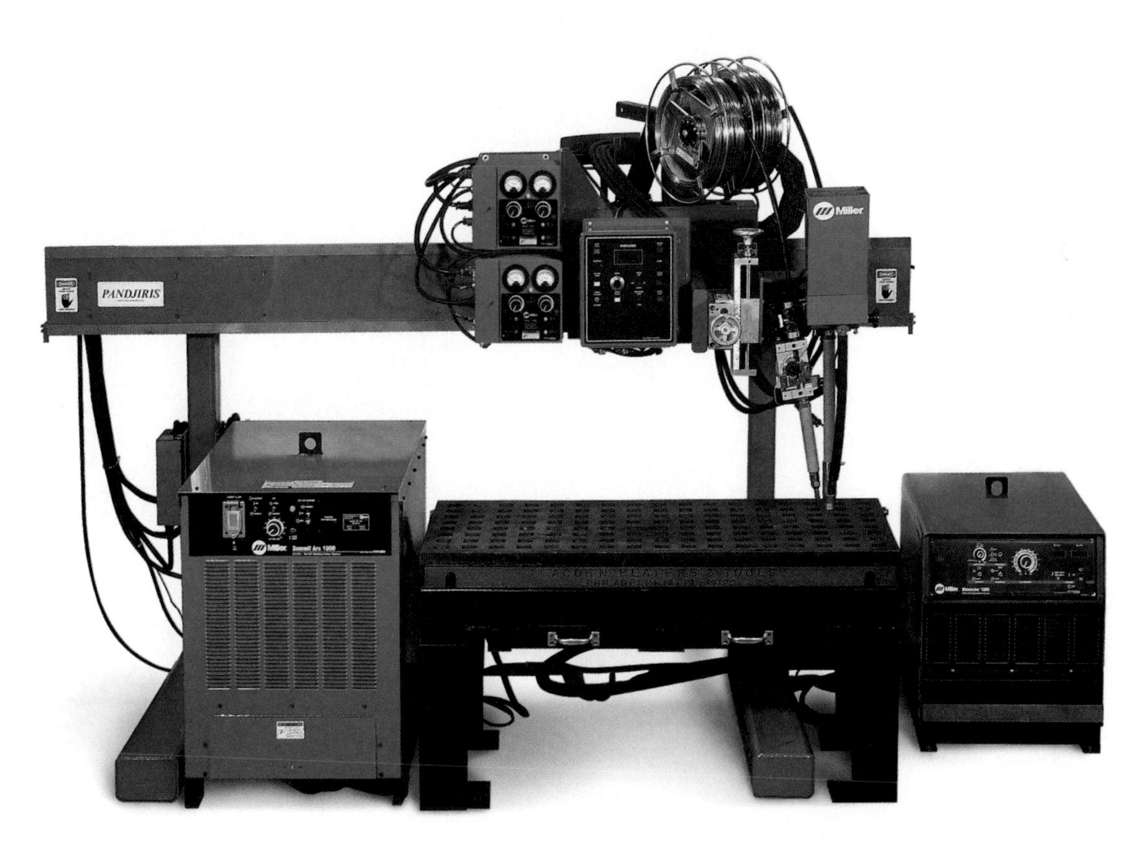

This is one popular submerged arc welding machine. Submerged arc welding is ideal for production jobs that require long, thick weld seams. (Miller Electric Mfg. Co.)

Section 8

Special Processes

FANUC Robot S-12

SULZER

Sulzer Metco

Chapter 29

Special Welding and Cutting Processes

Learning Objectives

After studying this chapter, you will be able to:

- Identify several special welding processes used in industry for unusual metals or unusual positions.
- Identify several special cutting processes used in industry.
- List the advantages of some special welding and cutting processes that are used in industry.

Technical Terms

air carbon arc cutting (CAC-A)
arc stud welding (SW)
burning bars
cutting rods
electrogas welding (EGW)
electron beam welding (EBW)
electroslag welding (ESW)
exothermic cutting
friction stir welding (FSW)
friction welding (FRW)
kinetic energy
laser
laser beam cutting (LBC)
laser beam welding (LBW)
nontransferred arc
orifice
orifice gas
oxyfuel gas cutting (OFC)
oxygen arc cutting (OAC)
photons
plasma
plasma arc welding (PAW)
restricted nozzle
solid-state welding (SSW)
sonotrode
submerged arc welding (SAW)
transferred arc
ultrasonic transducer
ultrasonic welding (USW)
upset
vacuum

Frequently Used Special Processes

The welding, cutting, brazing, soldering, and surfacing processes discussed in earlier chapters of this book are not the only processes available. The American Welding Society lists over 100 different welding and cutting processes. Refer to Chapter 2 (Figure 2-6) for a chart listing all these processes. Among them are special processes used on unusual metals, or in locations that are unusual. Extremely thin or thick metals also require special welding and cutting processes. Many of the special processes discussed in the following sections are not found in small production shops, body shops, home or farm shops, or school shops. These special welding and cutting processes are, however, being found in ever-increasing numbers in large manufacturing and repair businesses. The terms and abbreviations used in this chapter to describe these processes are those approved by the American Welding Society in AWS A3.0, *Standard Welding Terms and Definitions.*

Special Arc Welding Processes

There are many welding situations that require development of a special form of welding. Welding on metal over 1′ (300 mm) thick is very difficult to do using SMAW, GMAW, or GTAW. Electrogas arc welding and electroslag welding were developed to meet this need.

When scientists created plasmas experimentally, they discovered that extremely high temperatures and heat outputs were also created. Welding engineers developed the plasma arc welding and cutting processes to take advantage of this new heat source. Arc stud welding makes it possible to attach fastening devices to steel structures very easily. Specially designed equipment makes it possible to arc weld under water.

Electrogas Welding (EGW)

The ***electrogas welding (EGW)*** process was developed as a means of welding extremely thick metal. Thick metal sections are hard to weld, since the weld area must be must be kept near the melting point at all times. This is difficult with regular welding equipment.

Metal of almost any thickness can be welded with the electrogas process. See **Figure 29-1**. The weld is always made in a vertical position. One or more consumable electrodes are fed into the joint from above. To make the weld, the welding arc is struck, and the consumable electrode melts in the joint to form the weld. Solid electrode wire or flux-cored wire may be used. Two water-cooled copper shoes act as dams. They contain the molten metal in the weld area and cool the completed weld. As the copper shoes and electrode move up, the welding continues to the top of the joint.

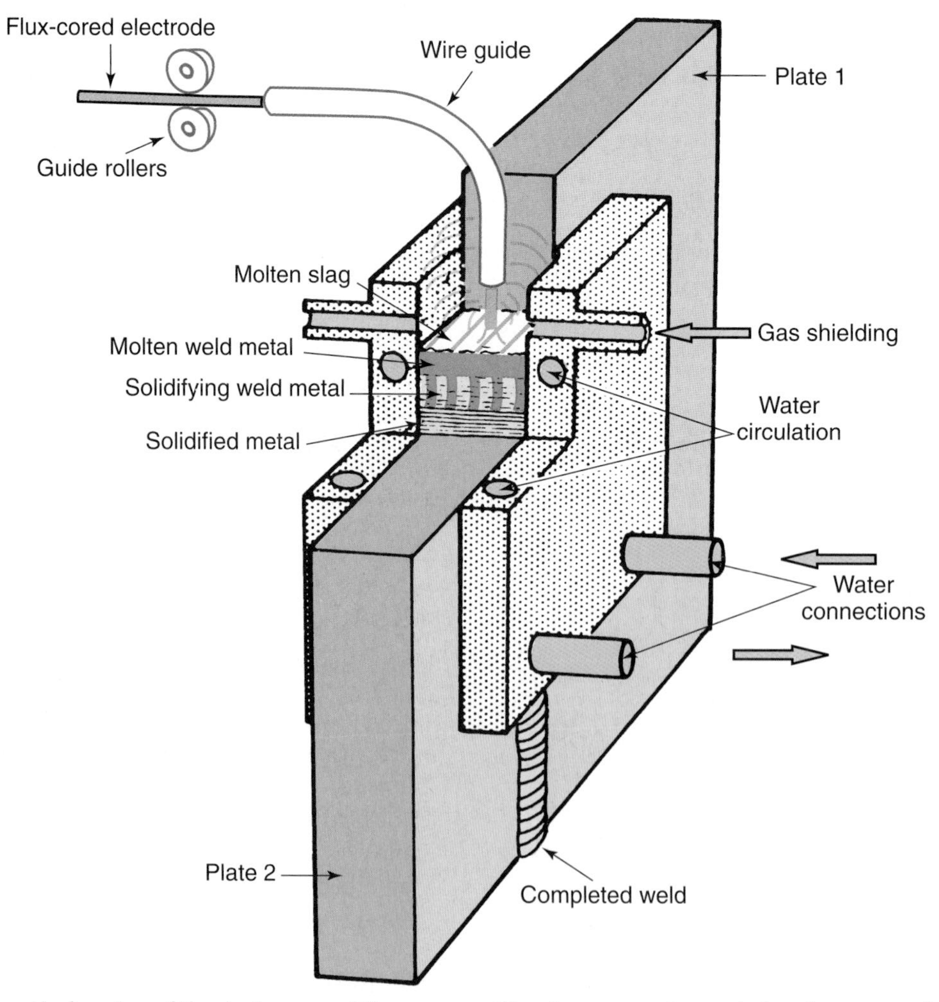

Figure 29-1. *A schematic drawing of the electrogas welding process. The shoes are water-cooled and move up along with the weld. A shielding gas protects the weld and molten metal.*

The weld area is protected from contamination by a shielding gas. The gas may be carbon dioxide, an inert gas, or a combination of two gases. It is supplied from above through openings in the shoes and covers the molten metal.

Electroslag Welding (ESW)

The ***electroslag welding (ESW)*** process is similar to electrogas welding. A thick layer of powdered flux is placed in the joint before welding begins. Once welding starts, the powdered flux melts. It forms a protective, floating slag above the weld area as shown in **Figure 29-2**. Impurities in the molten metal float to the top and mix with the flux. The molten flux protects the weld metal from impurities in the air above the weld.

Submerged Arc Welding (SAW)

Submerged arc welding (SAW) has several advantages over other welding processes. The larger electrodes, higher currents, and narrower groove angles used in SAW permit faster welding. The following is a list of the advantages of SAW:

- Faster than manual arc welding.
- No visible arc.
- No spatter.
- High-quality weld.
- The welding groove angle may be smaller.

In SAW, as in GMAW, a consumable electrode is fed into the weld joint. A thick layer of a powdered flux is deposited ahead of the electrode, as shown in **Figure 29-3**. The arc between the electrode and base metal occurs beneath this thick flux layer. The flux covers the arc and prevents any spatter. This is normally a fully automated machine welding process, **Figure 29-4**. It may be done semiautomatically, as well. In that case, the welder guides the wire and flux-feeding mechanism.

Plasma Arc Welding (PAW)

Scientists say that there are four states (conditions) of matter: solid, liquid, gas, and plasma. ***Plasma*** is an ionized gas (a gas made up of atoms that have lost or gained electrons). Ionized gas or plasma is extremely hot—reaching temperatures as high as 43,000°F (24,000°C). The plasma arc is an excellent heat source for welding or cutting. Cross sections of the two types of arc plasma torches, transferred and nontransferred, are shown in **Figure 29-5**.

An inert gas, such as helium, argon, or nitrogen, is used to create the plasma. The gas, which flows through the ***restricted nozzle***, is called the ***orifice gas***. Inert shielding gas is passed around the restricted nozzle to shield the weld area. A nonconsumable tungsten electrode is used in the torch. **The noise of the plasma arc process can be harmful to one's hearing; therefore, industrial quality ear protection should be worn when performing plasma arc welding.**

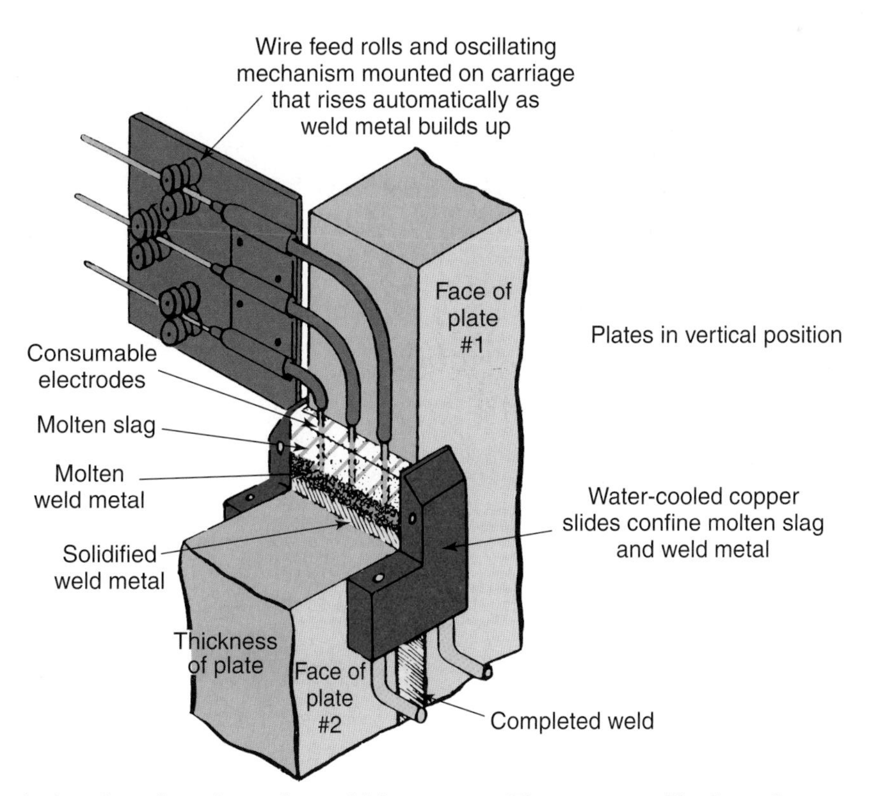

Figure 29-2. A schematic drawing of an electroslag weld in progress. Three consumable electrodes are used in this application. The molten slag floating above the weld prevents oxidation.

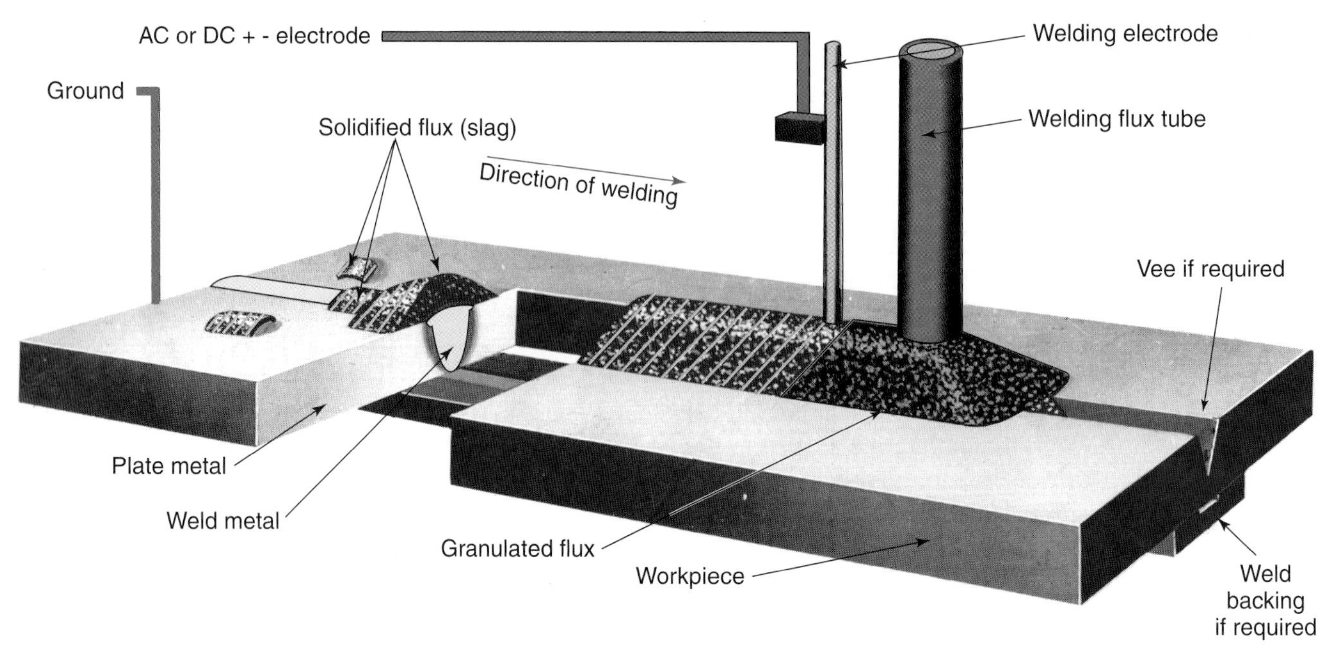

Figure 29-3. A diagram of a submerged arc weld in progress. Note the flux granules are deposited ahead of the consumable electrode. (ESAB Welding and Cutting Products)

Figure 29-4. Two submerged arc welds being made during the assembly of a vertical column exposed in an open atrium. (The Lincoln Electric Co.)

In the ***transferred arc*** process, an arc is struck between the electrode and the work. The inert gas is turned into superheated plasma as it passes through the arc at the ***orifice*** (hole) in the restricted nozzle. It is the plasma that melts the base metal.

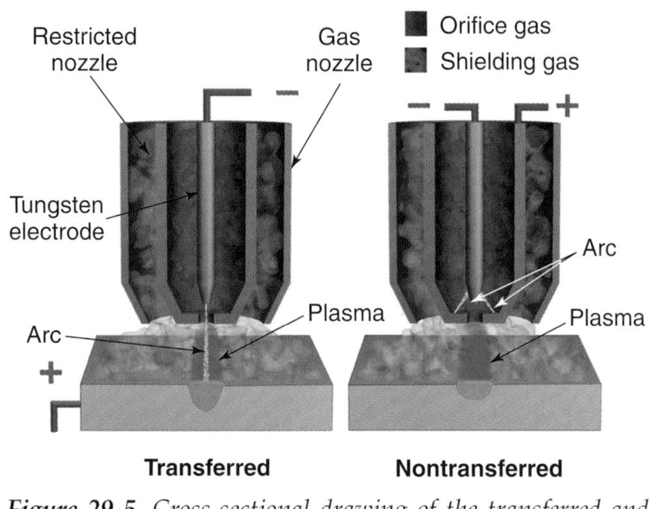

Figure 29-5. Cross-sectional drawing of the transferred and nontransferred plasma arc welding torch.

In a ***nontransferred arc*** plasma torch, the arc occurs between the tungsten electrode and the restricted nozzle. See Figure 29-5. When automatic plasma arc welding (PAW) is done, the filler wire is fed into the weld pool automatically, **Figure 29-6**. Edge and flange joints may be welded without the addition of filler metal.

Arc Stud Welding (SW)

Arc stud welding (SW) was developed to weld threaded studs, location pins, or nails to metal plates. The process is fast and simple. It requires little skill.

To perform the weld, the stud is loaded into a stud welding gun, **Figure 29-7**. The gun and stud are

Figure 29-6. Plasma arc welding equipment used in an automated application welding a lid to a small canister. (Process Welding Systems, Inc.)

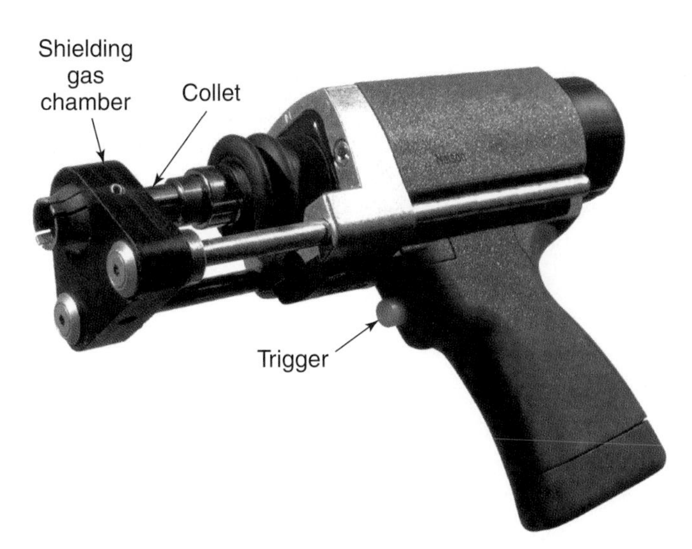

Figure 29-7. An arc stud welding gun. This gun is being used to weld aluminum. Shielding gas is fed into a chamber surrounding the stud. (Nelson Stud Welding, Inc.)

placed against the base metal, as shown in **Figure 29-8.** When the trigger is pulled, electricity flows through the stud to the base metal. The stud is then automatically pulled away from the plate, and an arc is struck. Electricity flows for only a fraction of a second. The gun then forces the molten tip of the stud into the molten base metal. It is held for a second or two and the weld is completed. On nonferrous metals, a shielding gas is used to produce good arc stud welds.

Shielded Metal Arc Welding (SMAW) Underwater

Shielded metal arc welding (SMAW) is done underwater to repair bridges, ships, and oil drilling rigs. The process is similar to regular SMAW. However, the electrode holder for underwater welding has especially good electrical insulation. The coating on the electrode is waterproof. See **Figure 29-9.** **Special training, equipment, and safety procedures are required to weld underwater.**

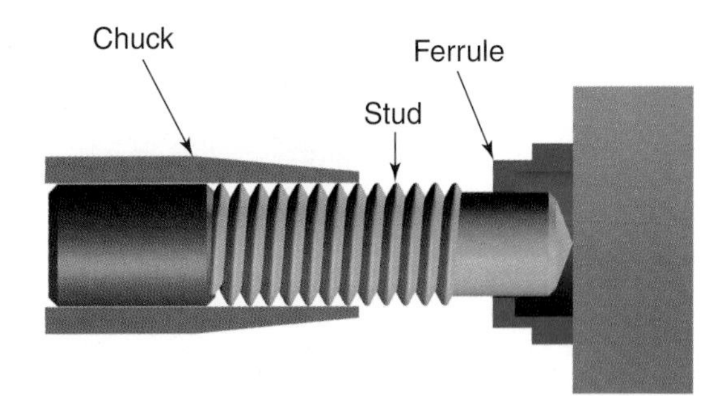

A–Fluxed end of stud is placed in contact with work.

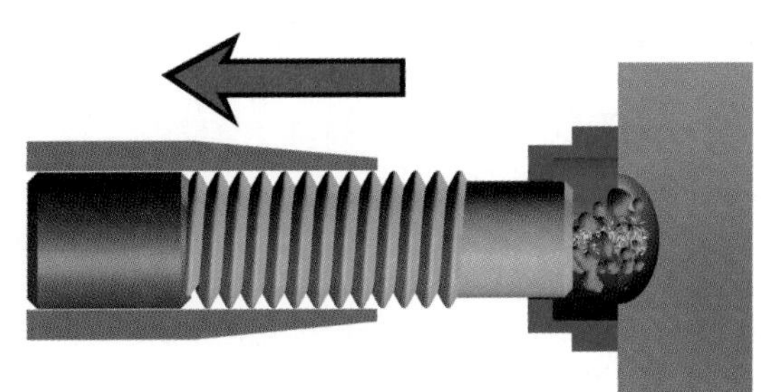

B–Stud is automatically retracted to produce an arc.

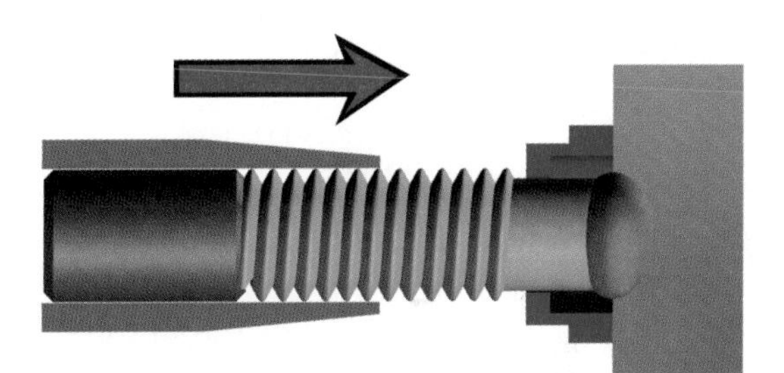

C–Stud is plunged into pool of molten metal.

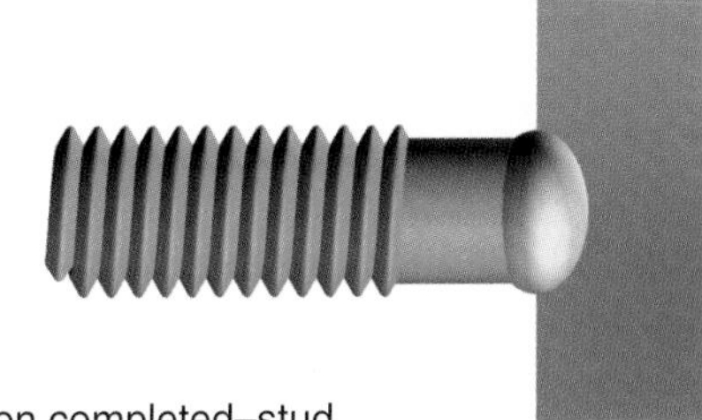

D–Operation completed–stud is welded to work.

Figure 29-8. The steps that take place during an arc stud weld on a steel plate. (Nelson Stud Welding, Inc.)

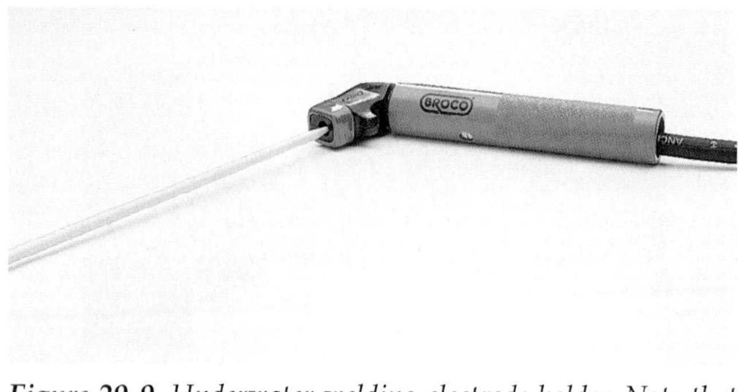

Figure 29-9. *Underwater welding electrode holder. Note that the holder is well insulated, with no bare metal areas exposed. (Broco, Inc.)*

Solid-State Welding Processes

The American Welding Society lists nine processes and four subprocesses in the solid-state welding group. Refer to Figure 2-6. ***Solid-state welding (SSW)*** may be defined as a group of welding processes that produce a fusion weld by application of pressure at a welding temperature below the melting temperature of the base metal and filler metal. Solid-state welding may be done cold, warm, or hot, but never above the melting temperature of the base or filler metal. The following sections describe some of the more common solid-state welding processes.

Friction Welding (FRW)

When two objects are rubbed together, they generate heat. This principle is the basis of ***friction welding (FRW)***. **Figure 29-10** shows the steps in a friction weld. One part is held stationary, while the other part is held in a chuck and rotated rapidly. The parts are pressed tightly together. Friction heats the two parts to their welding temperature. The welding temperature is lower than the solidus of either part, but high enough to allow fusion under pressure. When the welding temperature is reached, the rotation is stopped. The parts are then suddenly forced together under heavy pressure. After the heavy pressure is applied, the parts are held firmly until they cool. The finished weld is strong, with complete fusion over the entire joining surface. As shown in Figure 29-10, the weld is ***upset*** (enlarged) where the parts meet.

Friction welds are often produced in less than 15 seconds. Friction welding has been used successfully to join some dissimilar (unlike) metals that normally cannot be welded by other processes.

Friction Stir Welding (FSW)

Friction stir welding (FSW) is similar to friction welding except a rotating tool, not a rotating part, creates the heat. This welding process was developed at TWI, Cambridge UK in 1991.

A special rotating tool, firmly pressing on the parts to be welded develops heat. The rotating tool has a wide shoulder and a probe or pin that penetrates the parts along the weld joint. Friction created by the rapidly rotating tool softens the base metals and the base metals are mixed together. Full penetration welds can be made. **Figure 29-11** shows a drawing of the process and a completed butt weld.

Ultrasonic Welding (USW)

Sound waves can cause vibrations in objects. For example, a loud stereo can make the walls vibrate. In ***ultrasonic welding (USW)***, a very high-pitched sound is used to vibrate the surfaces of the metals to be welded.

Ultrasonic welding has many advantages. Since there is literally no heat, there is no metal distortion. Fluxes and filler metal are not needed. Very thin metals can be joined easily. Ultrasonic welds are normally small welds like resistance spot welds. Seam welds can also be made.

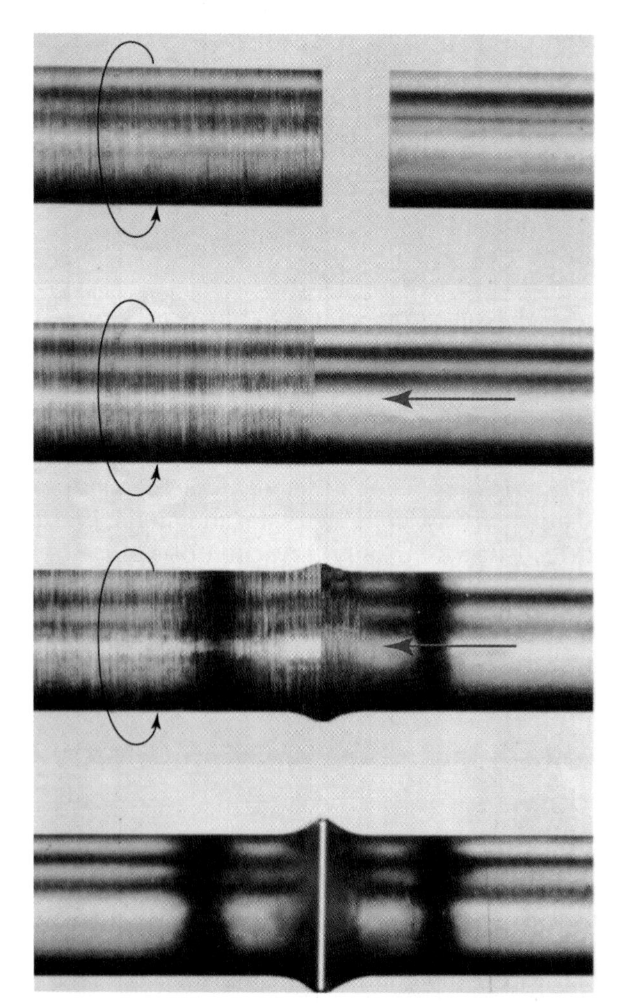

Figure 29-10. *The steps in making a friction weld on two pieces of 1″ (25.4 mm) diameter carbon steel. The rod at the left is spun at a high speed. The rod at the right is then forced against it. Friction creates enough heat to produce a strong weld.*

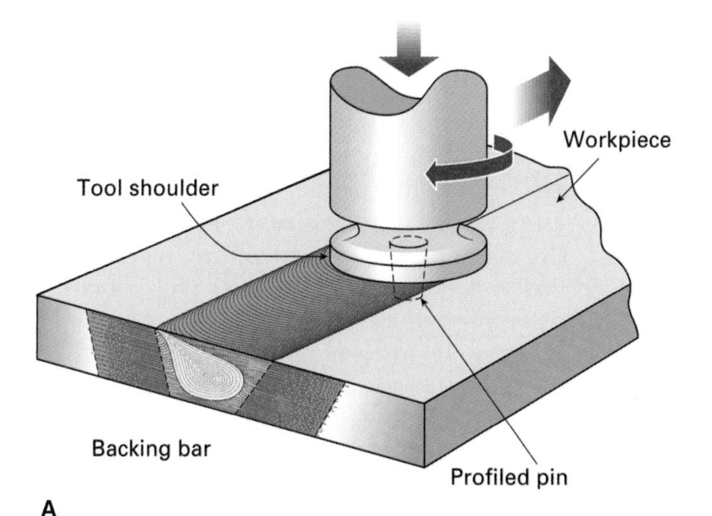

Figure 29-11. *A—A rapidly rotating tool with a large shoulder and a probe to penetrate the parts being welded softens the base metal. The base metals are welded together by stirring and mixing. B—A completed friction stir weld is shown. (TWI)*

Figure 29-12 shows a schematic of an ultrasonic weld in process. The parts to be joined are placed between two tips, as in resistance spot welding. An electronic device called an ***ultrasonic transducer*** causes either a wedge-reed or a lateral drive to vibrate extremely fast. This, in turn, causes a ***sonotrode*** (sound electrode) to vibrate. A slight force is applied to the parts through the sonotrode. The vibrations break up surface films, causing the parts to bond together without heat. An ultrasonic weld is completed faster than a resistance spot weld. Ultrasonic welds can be made on similar or dissimilar metals.

Other Welding Processes

Other welding processes are those that do not meet the definition of arc welding, oxyfuel gas welding, resistance welding, or solid state welding. These other welding processes are listed in Figure 2-6, *Master Chart of Welding, Joining, and Allied Processes*. The most common of these processes will be explained in the following pages.

Laser Beam Welding (LBW)

A ***laser*** creates a beam of light with a high energy density. This light energy is emitted from the ***laser beam welding (LBW)*** machine as energy particles called ***photons***. The photons produce heat by striking, and having their energy absorbed by, the base metal. The most popular lasers include the Nd:YAG (neodymium-doped yttrium aluminum garnet) crystal laser and the CO_2 gas laser. A schematic view of a CO_2 laser is shown in **Figure 29-13**. Electrical energy is

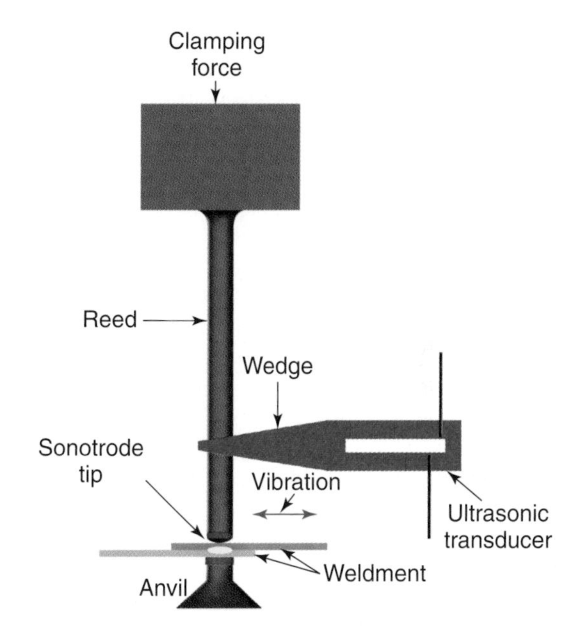

Figure 29-12. *A schematic drawing of the wedge-reed ultrasonic welding principle. A small spot weld is made by vibrating the reed while it is in contact with the base metal. (Sonobond Ultrasonics, Inc.)*

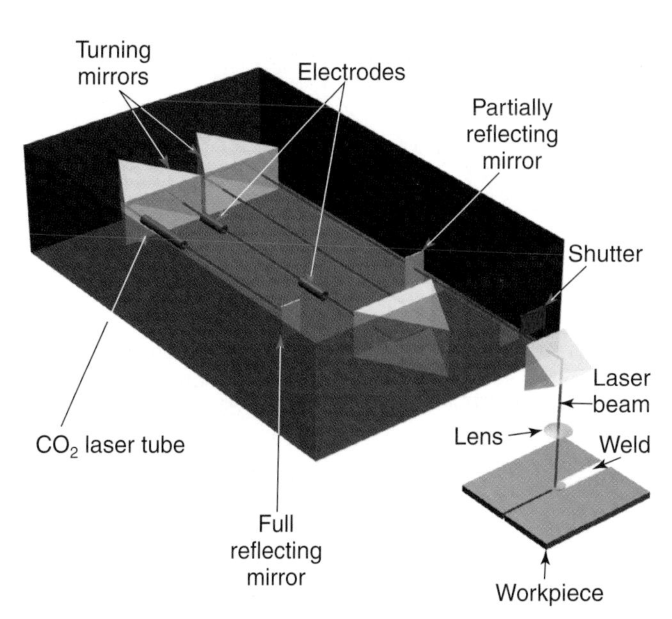

Figure 29-13. *A carbon dioxide (CO_2) laser welding machine. Mirrors reflect and aim the laser beam and a lens focuses it. The beam is released to make the weld when the shutter opens.*

generally used to excite the lasing material in the laser source, releasing photons. Photon energy is allowed to build up in the laser or laser welding machine until the desired level is reached. The energy is then released to fuse the weld joint.

Fusion occurs at the point where the laser beam strikes the weldment. Laser beams may be continuous or pulsed. They can be focused with lenses to a very small and accurate beam. Their direction can be changed through the use of mirrors.

Electron Beam Welding (EBW)

Figure 29-14 shows the basic parts of an ***electron beam welding (EBW)*** machine. Electrons are emitted from the electron gun. They are then focused and directed at the weld joint. The kinetic energy of the electrons creates the heat for welding. ***Kinetic energy*** is the energy of an object in motion. In this case, the objects in motion are the electrons.

Electron beam welding may be done in a high vacuum, a partial vacuum, or under normal atmosphere pressure. A ***vacuum*** is a condition in which atmospheric pressure in a closed vessel has been decreased. This is done by pumping out air. **Figure 29-15** shows an electron beam welder in use in industry. The beam energy is easier to direct under high vacuum conditions. Also, less energy is lost in a vacuum.

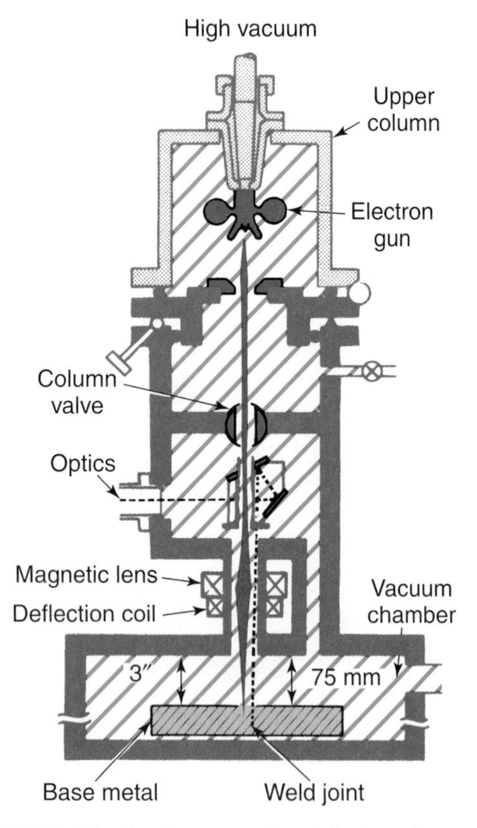

Figure 29-14. The basic parts of an electron beam welding machine shown in schematic form. Note that, in this machine, the weldment is mounted in a vacuum chamber. (PTR-Precision Technologies, Inc.)

Advantages of Laser and Electron Beam Welding

Laser and electron beam welding machines are made in a variety of energy levels. The greater the electrical energy input, the greater the energy output. Nd:YAG lasers are produced in sizes from 100W–4000W (watts). CO_2 lasers are produced in sizes from 500W–25,000W. A 600W Nd:YAG laser will penetrate steel .01″ (2.5 mm) in thickness.

Electron beam welding machines are produced in sizes from 1kW–100kW (kilowatts). A 100kW EB welding machine will produce 100% penetration on steel 10″ (254 mm) thick. See **Figure 29-16**. Laser beam and electron beam welding offer these advantages:

Figure 29-15. This electron beam welding machine comes with an optical viewing system, video, wire feed, vacuum chamber, and CNC technology. (PTR-Precision Technologies, Inc.)

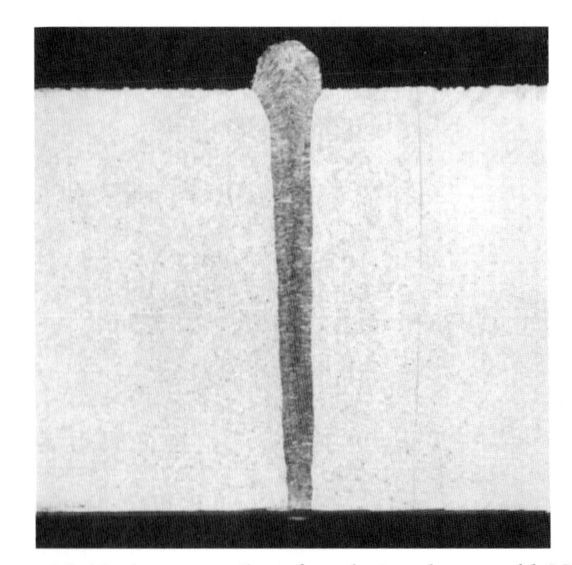

Figure 29-16. A cross section of an electron beam weld. Notice how thin the weld is in relation to the metal thickness. (PTR-Precision Technologies, Inc.)

- Low heat input.
- Controlled welding atmospheres.
- The ability to weld dissimilar metals.
- The ability to weld parts as thin as .001″–.002″ (.025 mm–.050 mm).
- No need for filler metal.
- Very precise aiming of beams to produce accurate and repeatable welds.

Special Cutting Processes

Special cutting processes using oxyfuel gas and oxygen arc cutting equipment have been developed for use in underwater salvaging and repairs. Other special thermal cutting processes were developed to permit the cutting of nonferrous metals and even concrete. These processes make it possible to cut metals that cannot be cut using regular oxyfuel gas cutting.

Oxyfuel Gas Cutting (OFC) Underwater

Underwater cutting is often done to repair underwater structures. It is also used to cut and salvage sunken ships. ***Oxyfuel gas cutting (OFC)*** is a process that can be done underwater, using a special torch. The torch is similar to the regular OFC torch, with two major changes. First, an air jacket is installed around the cutting tip. Second, a tube is added to carry compressed air to the air jacket, **Figure 29-17.**

Oxygen and fuel gas are mixed in the torch and burned at the tip, as in a regular cutting torch. A cutting oxygen valve and lever directs pure oxygen through

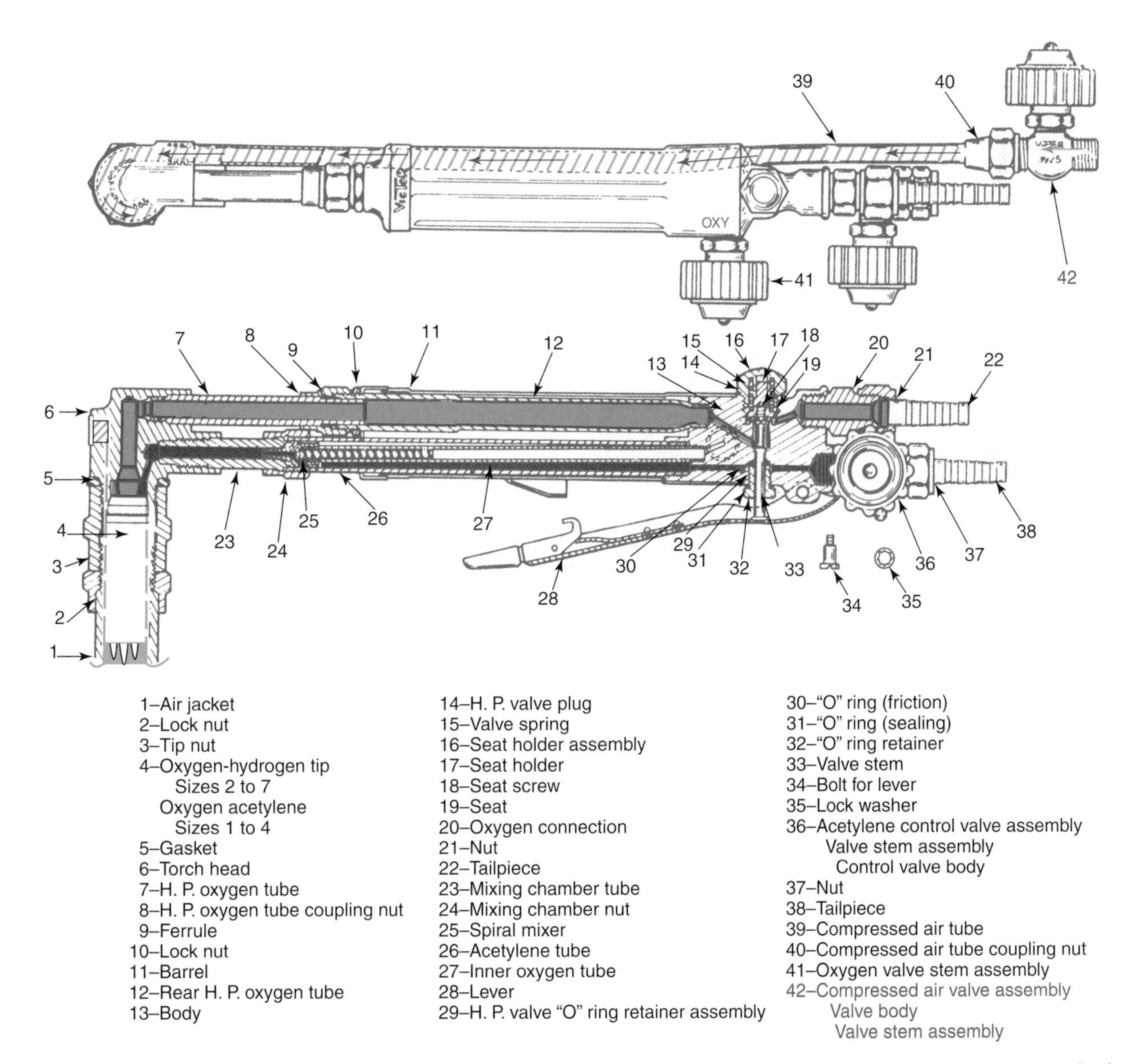

Figure 29-17. *A cross-sectional drawing of an underwater oxyfuel gas cutting torch. The air valve, part #42, controls the airflow to the air jacket, part #1. The air flow is shown in cross-hatched light green with red arrows. This high-pressure air keeps water away from the cutting tip. (Victor Equipment Co.)*

the center orifice for cutting. The compressed air in the air jacket keeps water away from the cutting tip.

Oxygen Arc Cutting (OAC) Underwater

Oxygen arc cutting (OAC) uses an electrode and electrode holder. The electrode holder is equipped with an oxygen passage. The cutting oxygen flow is controlled by a lever, **Figure 29-18**. The oxygen flows through the hollow electrode to the kerf (cut).

Preheating is done by striking an arc between the hollow electrode and the metal to be cut. Once the arc is struck, the cutting oxygen lever is depressed and the cutting begins. The arc and oxygen flow must be maintained as the torch is moved along the cutting line.

Oxygen arc cutting may be done above or below water. Oxygen arc and plasma arc are two of the most effective ways of cutting underwater.

Figure 29-18. An underwater oxygen arc cutting (OAC) torch. The arc furnishes the heat for cutting while oxygen oxidizes the metal and blows it away. A hollow electrode carries the cutting oxygen to the spot that is being cut. (Arcair, a Thermadyne Company)

Air Carbon Arc Cutting and Gouging

Air carbon arc cutting (CAC-A) uses an arc between a carbon electrode and a base metal to heat and melt the base metal. The electrode holder has holes just behind the electrode through which high pressure air is blown. This high pressure air blows the molten metal out of the cutting area. While *cutting* goes through the entire thickness of the base metal, *gouging* only carves out a U-shaped groove in the base metal, **Figure 29-19**.

Exothermic Cutting

Exothermic cutting is a cutting process that once started will continue to cut without any fuel gas or electricity for an arc. Electrodes are filled with small fuel rods that function as the heat source. Once the electrode and rods start to burn, they will continue to burn. Oxygen is used to accelerate the burning process at a temperature of about 7600°F (4200°C). The flow of high pressure oxygen also blows molten metal away from the cutting area. Exothermic cutting can cut any metal and also nonmetals like concrete. This process can be used for underwater cutting.

Fuel rods in the electrode are called exothermic ***cutting rods***, **Figure 29-20**. These are usually 3/16″, 1/4″ or 3/8″ (4.8 mm, 6.4 mm, or 9.5 mm) in diameter. Larger tubes, called ***burning bars***, have a diameter of .540″ up to 1.05″ (13.7 mm up to 26.7 mm), **Figure 29-21**.

Laser Beam Cutting (LBC)

Lasers (particularly the CO_2 laser) can cut mild steel, stainless steel, and even titanium cheaply and quickly. ***Laser beam cutting (LBC)*** produces clean-cut edges. Numerically controlled pulsed laser beams can also be used to drill or pierce clean holes of extremely small diameter in production parts. An oxygen jet is sometimes used with laser beam equipment to drill or pierce. Holes larger than .014″ (.36 mm) are generally made with conventional drill bits.

Figure 29-19. The air carbon arc cutting process is being used to gouge a groove into a base metal. High pressure air blows the molten metal from the gouging area. (Arcair, a division of Thermadyne Industries, Inc.)

Figure 29-20. *An exothermic cutting rod is gouging and removing a 3/4″ (19 mm) fillet weld. A 24″ (600 mm) long weld bead can be removed in 20 seconds. (Oxylance, Inc.)*

Figure 29-21. *This exothermic burning bar is being used during a salvaging operation to cut through 18″ (460 mm) thick steel casting. (Oxylance, Inc.)*

Summary

- The electrogas welding (EGW) process was developed as a means of welding extremely thick metal. To make the weld, the welding arc is struck, and the consumable electrode fed in from the top melts in the joint to form the weld. Two water-cooled copper shoes contain the molten metal in the weld area, supply shielding gas, and cool the completed weld.
- The electroslag welding (ESW) process is similar to electrogas welding. A thick layer of powdered flux is placed in the joint before welding begins. Once welding starts, the powdered flux melts and forms a protective, floating slag above the weld area.
- In submerged arc welding (SAW), a consumable electrode is fed into the weld joint. A thick layer of a powdered flux is deposited ahead of the electrode. The arc between the electrode and base metal occurs beneath this thick flux layer. The flux cover prevents any spatter.
- Plasma arc welding (PAW) uses superheated plasma to melt the base metal. In the transferred arc process, an arc is struck between a nonconsumable tungsten electrode in the torch and the base metal. In the nontransferred arc process, an arc is struck between the electrode and the restricted nozzle. In both processes, an inert gas emitted from the torch turns into superheated plasma as it passes through the arc. It is the plasma that melts the base metal.
- Arc stud welding (SW) was developed to weld threaded studs, location pins, or nails to metal plates. In this process, a stud is loaded into a stud welding gun. The gun and stud are placed against the base metal. When the trigger is pulled, electricity flows through the stud to the base metal. The stud is then automatically pulled away from the plate, and an arc is struck. Electricity flows for only a fraction of a second. The gun then forces the molten tip of the stud into the molten base metal.
- Shielded metal arc welding (SMAW) is done underwater to repair bridges, ships, and oil drilling rigs. The process is similar to regular SMAW. However, the electrode holder for underwater welding has especially good electrical insulation and the electrode coating is waterproof.
- Solid-state welding (SSW) may be defined as a group of welding processes that produce a fusion weld by application of pressure at a welding temperature below the melting temperature of the base metal and filler metal.

(continued)

- In friction welding (FRW), one part is held stationary, while the other part is held in a chuck and rotated rapidly. The parts are pressed tightly together. Friction heats the two parts to their welding temperature. When the welding temperature is reached, the rotation is stopped. The parts are then suddenly forced together under heavy pressure. After the heavy pressure is applied, the parts are held firmly until they cool.
- In friction stir welding, a rotating tool under high pressure develops heat in the base metals. The spinning tool mixes the softened base metals.
- In ultrasonic welding (USW), a very high-pitched sound is used to vibrate the surfaces of the metals to be welded. An ultrasonic transducer causes either a wedge-reed or a lateral drive to vibrate extremely fast. This, in turn, causes a sonotrode (sound electrode) to vibrate. A slight force is applied to the parts through the sonotrode. The vibrations break up surface films, causing the parts to bond together without heat.
- In a laser beam welding (LBW) machine, electrical energy is used to excite a lasing material in the laser source, releasing photons. Photon energy is released as a laser beam. The photons in the laser beam produce heat by striking, and having their energy absorbed by, the base metal, fusing the joint.
- In an electron beam welding (EBW) machine, electrons are emitted from the electron gun. They are then focused and directed at the weld joint. The kinetic energy of the electrons creates the heat for welding.
- Oxyfuel gas cutting (OFC) is a process that can be done underwater, using a special torch. The torch has an air jacket installed around the cutting tip. Also, a tube is added to carry compressed air to the air jacket. The compressed air in the air jacket keeps water away from the cutting tip.
- Oxygen arc cutting (OAC) uses an electrode and electrode holder. The electrode holder is equipped with an oxygen passage. Preheating is done by striking an arc between the hollow electrode and the metal to be cut. Once the arc is struck, the cutting oxygen lever is depressed, oxygen flows through the hollow electrode, and the cutting begins. Oxygen arc cutting may be done above or below water.
- Air carbon arc cutting (CAC-A) and gouging use an arc to melt the base metal. High pressure air blowing through holes in the electrode holder removes molten metal from the cutting area.
- Exothermic cutting uses an electrode filled with small fuel wires and high volume oxygen to burn continually without a fuel gas or an arc. The high pressure oxygen blows the molten metal out of the cutting area. Cutting rods or burning bars are used to cut or gouge any metal or nonmetal.
- Lasers (particularly the CO_2 laser) can cut mild steel, stainless steel, and even titanium cheaply and quickly. Laser beam cutting (LBC) produces clean-cut edges.

Review Questions

Write your answers on a separate sheet of paper. Please do *not* write in this book.

1. How many welding and cutting processes are listed by the AWS?
2. How does EGW differ from ESW?
3. List three gases that may be used to form the plasma for PAW.
4. What process is used to rapidly weld nails, bolts, and location pins to metal plates?
5. List at least four advantages of SAW.
6. List three or more processes used to weld or cut underwater.
7. How is the oxygen delivered to the kerf area in OAC?
8. Which special welding process presses a rotating part against a stationary part?
9. In LBW, the metal is heated by a(n) _____ with high energy density. In EBW, the metal is heated by _____ energy.
10. What temperature does the exothermic cutting process achieve?

Chapter 30

Robotics in Welding

Learning Objectives

After studying this chapter, you will be able to:

- Cite advantages of using robotic welding equipment in manufacturing.
- Identify the main parts of a robot and the components of a robotic welding station.
- Describe the use of a teach pendant in programming a robot to perform its designated tasks.
- Discuss the safety precautions to be taken when working around robots.

Technical Terms

actuator
axis
elbow joint
light curtain
operator controls
parts positioner
program
repeatability
robot
robot cell
robot controller
robotic welding system
robotic welding workstation
servomotor
shoulder joint
teach pendant
waist joint
working volume
wrist joint

The Reason for Robots

Industrial ***robots*** are essentially mechanical devices that can be programmed to perform a task or several tasks under the control of a computerized program. Robots are used because they save the manufacturer money. They save money for these reasons:

- They can weld much faster than a human welder. The actual speed that a robot travels while welding depends on the welding process, metal thickness, joint shape, and other variables.
- They can weld continuously.
- They can maintain the same arc length and travel speed over the entire weld.
- They will produce fewer defects than a human welder.
- They can be programmed to do many welds in a variety of locations. Robotic welding equipment can repeat programmed welds with an accuracy of ±.004″ (.10 mm).

Robots are useful and cost-effective only when large numbers of welds must be made. They are not practical for all welding applications. Robotic welding is not cost-effective for low-volume production and most repair welding.

Types of Industrial Robots

Robots serve many purposes in industry. Some of the more common tasks robots perform include material handling or moving parts, painting, and welding, **Figure 30-1**.

A robot can move in a number of directions. These movements are made possible by the use of mechanical joints and actuators. An ***actuator*** is a device that will cause something to move. The mechanical joints in current robots are moved by electric actuators.

Electric actuators are electric motors called servomotors. ***Servomotors*** are electric motors that rotate very precise distances. These motors are used to move a robot arm or joint to an exact location. Each joint requires its own servomotor, **Figure 30-2**.

Repeatability refers to a robot's ability to accurately perform the same movement again and again. An electric robot can repeat a movement with an accuracy of ±.004″ (.10 mm).

Figure 30-1. *This robot is gas metal arc welding a steel weldment. (Wolf Robotics, Inc.)*

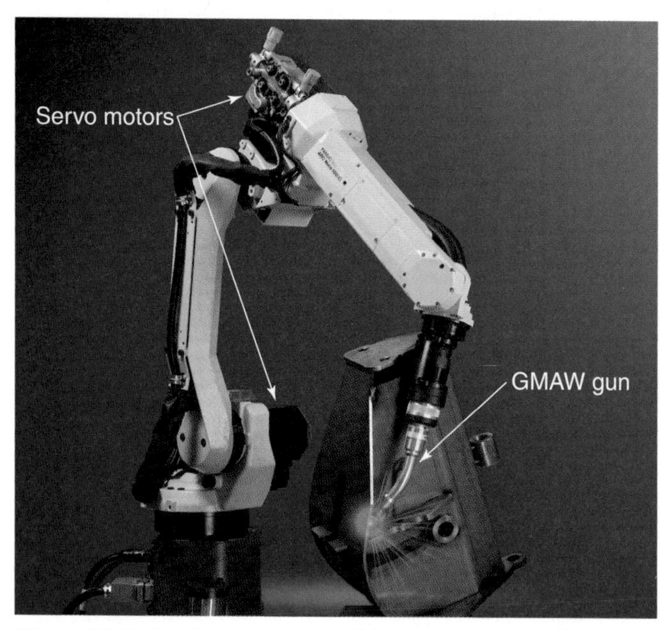

Figure 30-2. *A robot uses a servo drive motor to control each axis of movement. Not all servo motors are openly visible. (Fanuc Robotics America, Inc.)*

Robotic Welding Systems

A robot cannot work by itself. It must function together with other parts to form a ***robotic welding system***. The following are the parts of a robotic welding system:

- The robot.
- Automatic welding or cutting equipment.
- The robot controller.
- A parts positioner.
- Operator controls.
- Safety equipment.

The robotic welding system is also called a ***robotic welding workstation***, **Figure 30-3**.

The Robot

When choosing a robot for any welding application, a manufacturer must consider a number of factors, including the following:

- The weight of the welding equipment to be carried by any part of the robot.
- The location of the welds to be made.
- The accuracy of repeatability required.
- The amount of money that can be spent.
- The future and alternate uses for the robot.

Robot Movements

Each mechanical joint on a robot permits a movement about one ***axis*** (straight line about which rotation takes place). The location of the welds to be made determine the number of axes required. Most industrial robots move about six axes. Each axis is given a name and number. These axes are named after parts of the human body: the waist, shoulder, elbow, and wrist.

The ***waist joint*** allows the robot to rotate about the base. This joint permits the general left and right rotation of the robot about the base.

The ***shoulder joint*** raises or lowers the robot arm in relation to the base or horizontal plane.

The ***elbow joint*** allows the wrist axis to rotate vertically.

The welding gun or torch is mounted on the final segment of the robot arm, called the wrist. Robot

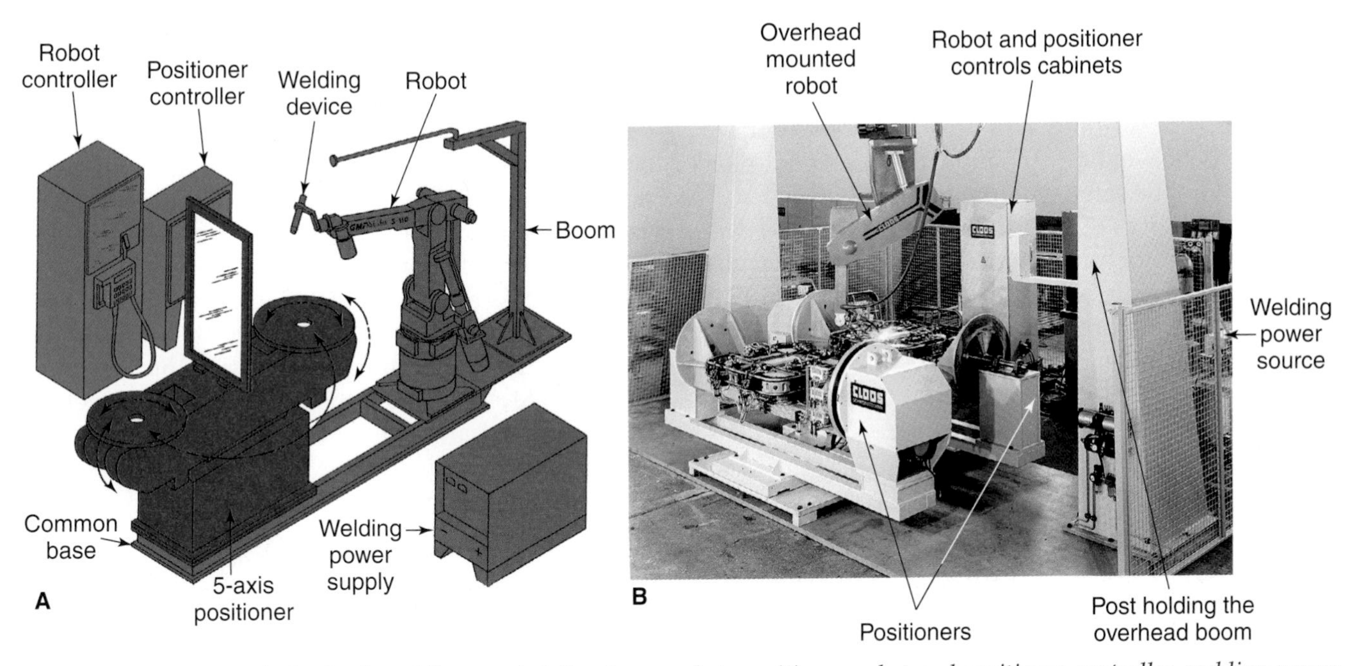

Figure 30-3. *A—A typical robotic welding workstation has a robot, positioner, robot and positioner controller, welding power source, and welding device. (FANUC Robotics North America, Inc.) B—A robotic workstation using two work positioners. Note that the robot is mounted overhead for better access to the parts being welded. The welding power source is located behind the post holding the boom. (Cloos Robotic Welding, Inc.)*

manufacturers often refer to the wrist axes of the robot by letter or number. All axes work together to accurately locate the welding device to the weld joint. The ***wrist joints*** allow the welding gun or torch to rotate vertically in pitch and rotate the wrist joint axes 360° as required. See **Figure 30-4**.

The robot must often reach out toward the welding area. Most of this reaching motion is done by rotating the shoulder and elbow axes.

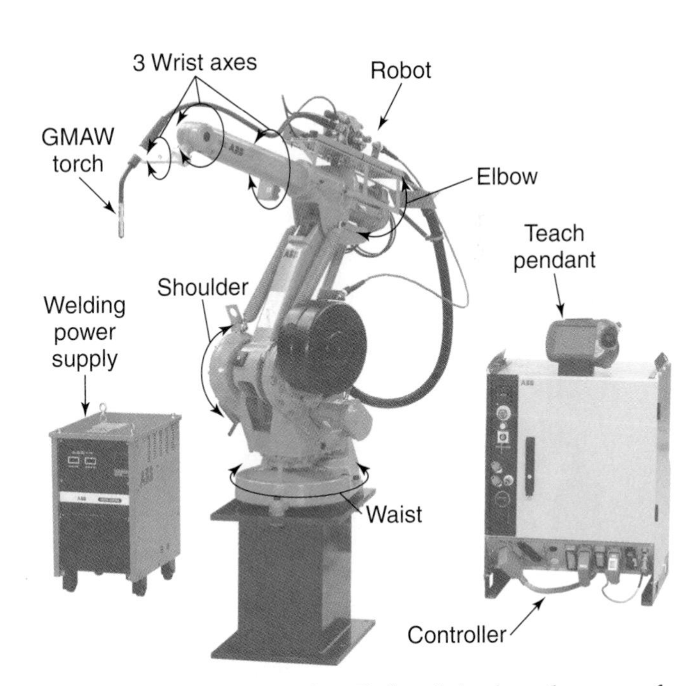

Figure 30-4. *A six-axis robot. Each axis is given the name of a body part: waist, shoulder, elbow, wrist pitch, and wrist roll. These axes can also be referred to by numbers. (Wolf Robotics, Inc.)*

Welding or Cutting Equipment

Any form of welding or cutting equipment can be mounted on a robot. However, cutting is usually done on cutting tables using electronic pattern followers.

The most common welding and cutting processes used with robotic systems are:

- Resistance spot welding.
- Gas metal arc welding.
- Gas tungsten arc welding.
- Plasma arc cutting.
- Laser beam welding and cutting.
- Surfacing.

Robot Controller

The ***robot controller*** is the "brain" of the robot welding system. The robot controller includes a computer that controls the movement of all robot axes. It controls all the variables in the welding equipment and usually controls the parts positioner. Sometimes, however, a separate controller is used for the parts positioner.

The robot controller stores the program of movements and actions that the robot must follow to weld a part. It also checks or monitors all aspects of the robot operation and welding equipment. In addition to monitoring the welding operation, the robot controller also monitors the safety equipment.

If a problem is detected, the robot controller will alert the operator or it may shut down the entire robotic welding system, if programmed to do so. The robot controller is programmed with a teach pendant or with software on a computer.

Teach Pendant

A robot is controlled by the robot controller. The controller must be programmed or taught each step necessary to properly and safely weld parts together. A set of steps is called a ***program***. A ***teach pendant*** is a device used to teach the robot and the controller each step in a program. An electrical cable connects the teach pendant to the controller, **Figure 30-5**.

Before teaching a robot, the parts to be welded are clamped in a jig, fixture, or positioner. An operator moves the robot to a desired location and presses a button to teach the robot that location. The operator moves the robot to each additional location and presses a button. Each of these points is memorized and is part of the program. This program is stored in the controller's memory. When this programmed set of points is run, the robot will repeat the movements each time the operator presses the start button on the operator control panel.

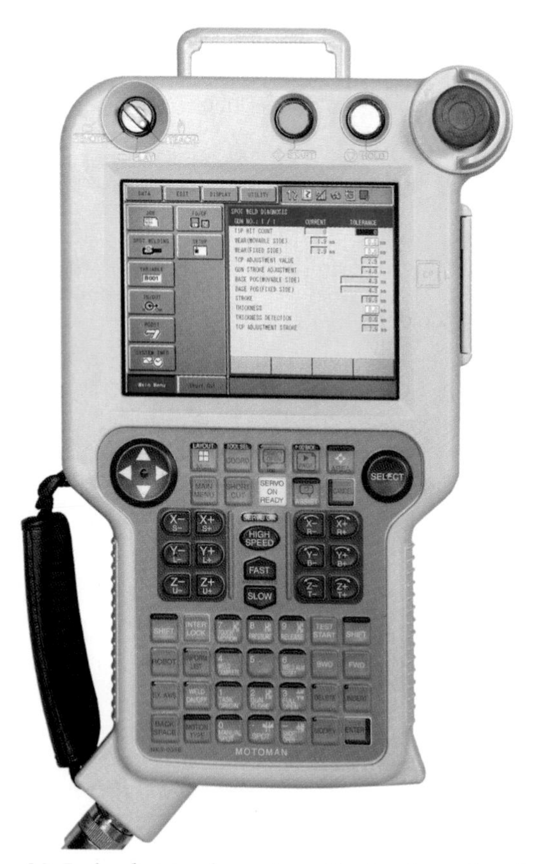

Figure 30-5. *A robot teach pendant. The teach pendant is used to move the robot and weld positioner through all the movements required to make all the welds. Variables, such as amperage, voltage, and wire speed, are set and shown on the visual display. They are then placed into the controller's memory. (Motoman Inc.)*

To teach a robot to arc weld, the robot is moved to the start point of the weld. The nozzle-to-work distance is set by programming the position of the robot. Using the teach pendant, the operator sets the welding conditions or parameters into the computer memory. These parameters include:

- Welding current or wire feed speed.
- Welding voltage.
- Welding travel speed.

The robot is moved to the end point of the weld and the point is programmed. This set of programmed steps determines a start point, the welding conditions, the travel speed, and the end point of the weld. With these, a high quality weld can be made every time, provided identical parts are placed in the same position every time. Using a robot, complex parts can be welded consistently.

A robot can also be taught using software on a computer. A computer model of the positioner, jig, clamps and the parts being welded must be created in the computer. Then the robot program can be created on a computer and downloaded to the robot. Minor touch up work may be required with the actual robot.

Parts Positioner

The ***parts positioner*** is used to bring the part to be welded within reach of the robot and welding torch. The positioner also rotates the part, where possible. The positioner may also be used to move the completed part away from the robot. Once clear of the robot, the completed part is unloaded and a new part is loaded for welding. See **Figure 30-6**. The robot controller usually controls the movements of the positioner.

Operator Controls

The ***operator controls*** are located on a panel and generally include the welding sequence start button, emergency stop button, and system monitoring warning lights. The operator control panel is located at a safe distance from the working robot and parts positioner, **Figure 30-7**. Once the program is completed, welding stops. The controller then directs the positioner to move the part to a safe area away from the robot. Often, a second positioner moves a new set of parts into position for welding. The completed part is unloaded and new parts are loaded on the positioner.

To start the welding program again, the operator pushes the sequence start button on the operator control panel and the sequence repeats. Operator controls can also be used to shut down the system in an emergency.

Figure 30-6. *Blower fan blades are being welded to the fan hub in this robotic welding workstation. Two weld positioners are used. While one fan is being welded a finished part is unloaded and a new tacked assembly is loaded. The positioner is moved in two axes to place the part under the welding gun at the proper welding angle. Note that the robot is mounted from an overhead post for easier access to the workpiece. (Cloos Robotic Welding, Inc.)*

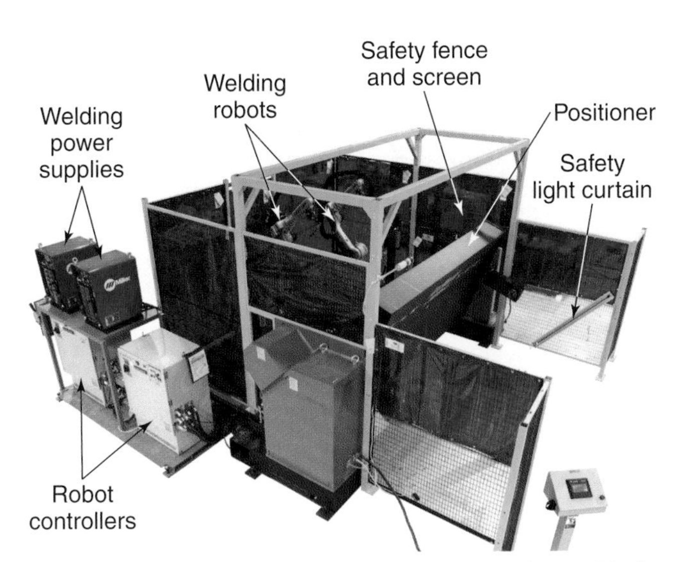

Figure 30-7. *A complete welding workstation is shown. Notice the safety fencing to keep people out of the robot's working area and notice the safety light curtain to make sure no one gets close to the robot while it is operating. The operator controls are a safe distance from the robot and parts positioner. (Motoman Inc.)*

Safety around Robots

A moving robot is extremely dangerous. For this reason, the *robot cell*, which is the floor area containing all the parts of the robotic welding system, is often fenced off. See Figure 30-7. The robot cell is fenced with meshed wire. This is done to keep unauthorized persons from accidentally entering the space.

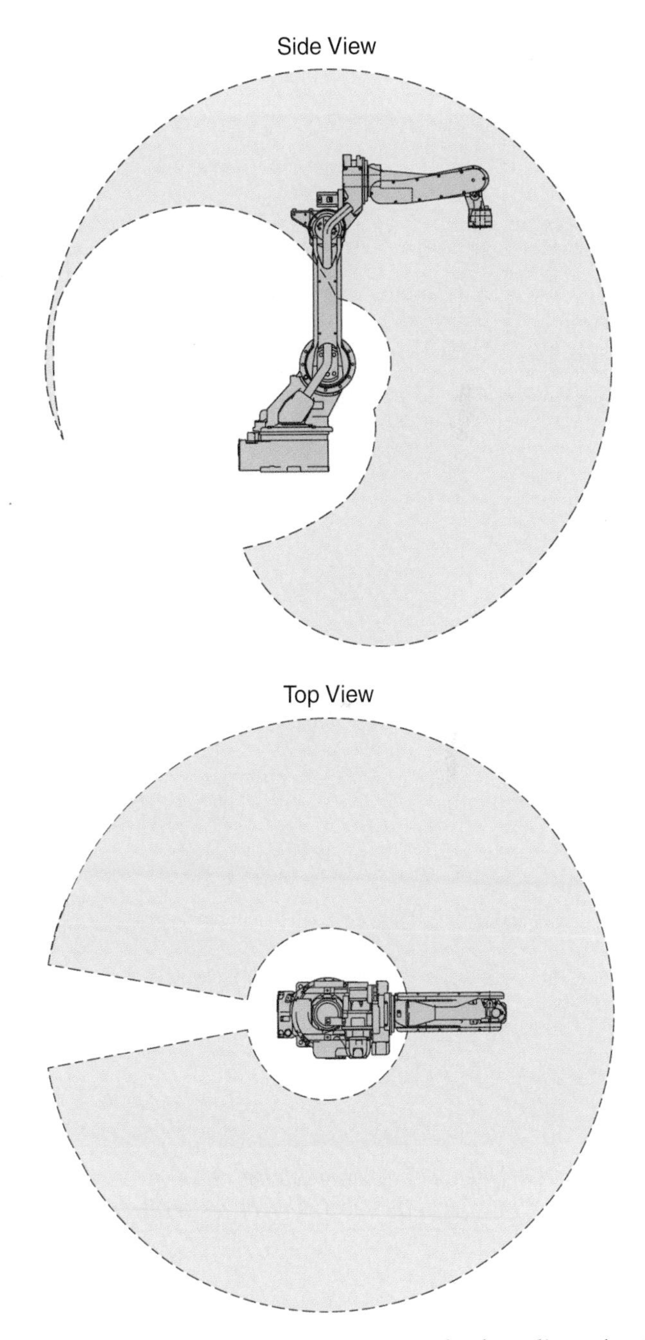

Figure 30-8. *The robot working volume is the three-dimensional space where the robot can move or reach during its movement. Parts to be welded must be inside this work volume. A fence to keep people and equipment from entering the working volume must be present. (Motoman Inc.)*

Robot Working Volume

The robot *working volume* is the three-dimensional space occupied by the robot at the extremes of its movements. **Figure 30-8** shows the working volume of a typical robot.

If something goes wrong, a robot's movements can be erratic and unpredictable. It can move with great power and speed and can cause serious injury if it strikes a person.

The robot working volume should be protected from unauthorized entry. This is usually done with a fence on three sides. An electronic light beam covers the fourth side. The light beam or light barrier is called a *light curtain*. If anything or anyone passes through the light beam, the robot automatically shuts down. See **Figure 30-9**.

Caution: Only trained technicians should enter the robot working volume, and then only under controlled conditions, usually with the robot de-energized.

Figure 30-9. *This robot welding cell has all the necessary equipment, including a light curtain to keep workers from entering the working volume when the robot is moving. (ABB, Inc.)*

Summary

- Industrial robots are essentially mechanical devices that can be programmed to perform a task or several tasks under the control of a computerized program. Robots are used because they save the manufacturer money. Robots can weld faster and with more consistency than a human welder, they can weld continuously, they produce fewer defects than their human counterparts, and they can repeat programmed welds with a high degree of accuracy.
- Servomotors are used to move a robot arm or joint to the exact location where a weld is to be made. Each joint requires its own servomotor.
- A robot cannot work by itself. It must function together with other parts to form a robotic welding system. The parts of a robotic welding system include automatic welding or cutting equipment, the robot controller, an optional parts positioner, and operator controls.
- Each mechanical joint on a robot permits a movement about one axis. Most industrial robots move about six axes. Each axis is given a name and number. These axes are named after parts of the human body.
- The waist joint allows the robot to rotate left and right around its base.
- The shoulder joint raises or lowers the robot arm in relation to the base or horizontal plane.
- The elbow joint allows the wrist axes to rotate vertically.
- The welding gun or torch is mounted on the final segment of the robot arm. Robot manufacturers often refer to the wrist axes of the robot by letters or numbers. The wrist joints allow the welding gun or torch to rotate vertically in pitch and rotate 360° around the wrist joint axes as required.
- The most common welding and cutting processes used with robotic systems are resistance spot welding, gas metal arc welding, gas tungsten arc welding, plasma arc cutting, laser beam welding and cutting, and surfacing.
- The robot controller is the "brain" of the robot welding system. The robot controller includes a computer that controls the movement of the robot in all axes. It controls all the variables in the welding equipment and may also control the parts positioner.
- The robot controller must be programmed before a robotic welding system can be used to make accurate, repeatable welds. A teach pendant is used to program the robot controller to move the robot to different locations.
- The parts positioner brings the part to be welded within reach of the robot and welding torch. It may also rotate the part to a better position. The positioner may also move the completed part away from the robot.
- The operator controls are located on a panel and generally include the welding sequence start button, emergency stop button, and system monitoring warning lights.
- **A moving robot is extremely dangerous. For this reason, the robot cell, which is the floor area containing all the parts of the robotic welding system, is often fenced off.**
- The robot working volume is the three-dimensional space occupied by the robot at the extremes of its movements. **The robot working volume should be protected from unauthorized entry.**

Review Questions

Write your answers on a separate sheet of paper. Please do *not* write in this book.

1. _____ are electric motors that rotate very precise distances and are used to move a robot arm or joint to the exact location where a weld is to be made.
2. List three reasons why robots save money for a manufacturer.
3. Robotic welding equipment can repeat programmed welds with an accuracy of _____″ (_____mm).
4. Name the robot axes shown in the sketch below.

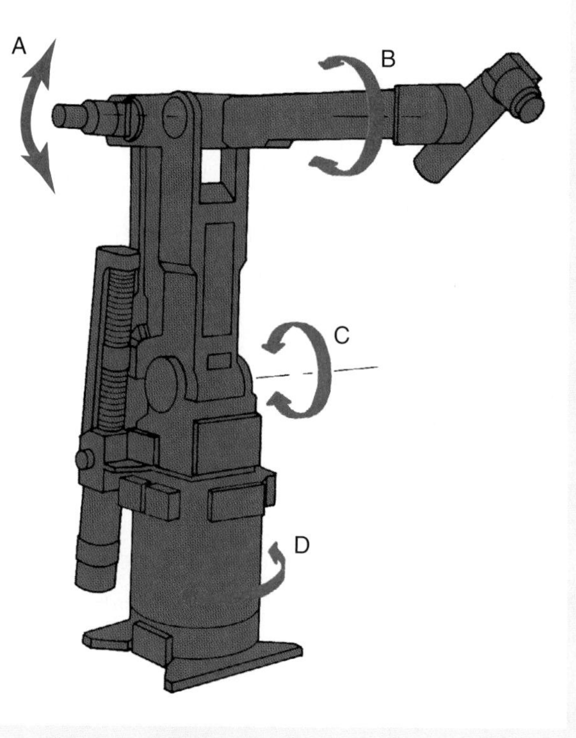

5. What is the name of a set of steps that a robot follows to properly and safely weld parts?
6. The operator starts the robot controller welding program by pressing the sequence start button on the _____.
7. What device can an operator use to program robot movements into the robot controller's computer?
8. Name two devices used to keep unauthorized persons out of the robot working volume.
9. Completed welds are unloaded and new parts loaded by moving the _____ out of the robot working volume.
10. What is the most important safety rule to follow when working around a robot?

Chapter 31

Welding Plastics

Learning Objectives

After studying this chapter, you will be able to:

- Distinguish between the two basic groups of plastics and name representative examples of each.
- Identify various thermoplastics by their flame and odor characteristics.
- Describe the procedure used to join plastics through use of heated gas and a filler rod.

Technical Terms

ABS (acrylonitrile butadiene styrene)
plastics
polyethylene (PE)
polypropylene (PP)
polyurethane (TPUR)
polyvinyl chloride (PVC)
speed welding tip
tack welding tip
thermoplastic
thermosetting

Plastics in Manufacturing and Construction

Plastics are synthetic, nonmetallic, organic materials that can be molded and shaped into a desired form. Often, this molding or shaping is done under heat and pressure.

The use of plastics in manufacturing and construction has increased dramatically in the past twenty years. The use of plastic materials has been attempted in the construction of practically every imaginable object, **Figure 31-1**.

Automobile bodies and bumpers, engines, firearms, aircraft, pipes, ducts, furniture, teeth, houses, luggage, floors, moldings, sinks, skis, and boots are all being made of weldable plastics. The list of products made of plastics is almost endless. See **Figures 31-2** and **31-3**.

Types of Plastics

Many types of plastics have been created to meet specific needs. However, all plastics fall into one of two basic groups. They are either thermosetting or thermoplastic materials.

Thermosetting plastics cannot be welded by heat and fusion. Plastics in this group are initially heated and formed into a desired shape. However, after they have once cooled, they cannot be reheated and reformed.

Figure 31-1. Many consumer products including parts on automobiles are made from weldable plastic. (Subaru of America)

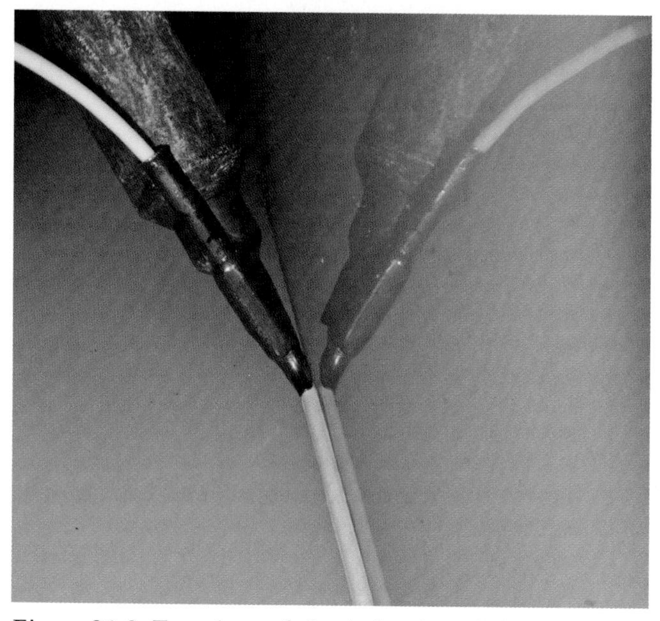

Figure 31-2. Two pieces of plastic forming a T-joint are being welded with a speed welding tip. (JP Plastics, Inc.)

Some plastics in the thermosetting group are: alkyd, allylic, aminoplastic, bakelite, casein, epoxy, phenolic, and silicone.

Thermoplastic materials may be joined by welding in the same manner as metal. Plastics in this group become soft and may be formed when heated. Thermoplastics may be reheated and reformed repeatedly. Important thermoplastics include: ABS, polyethylene, polypropylene, polyurethane, and polyvinyl chloride.

How to Identify Thermoplastics

To make certain that a plastic is a weldable thermoplastic, you must test it. Use the following test to identify thermoplastics:

Figure 31-3. This plastic elbow was welded from multiple pieces. (JP Plastics, Inc.)

1. Remove a small sliver of the plastic to be tested from the back side of the part.
2. Hold the plastic sliver with pliers. Ignite the sliver with a flame.
3. Observe the plastic as it burns.
4. Compare the appearance and odor of the burning sample with those of known thermoplastics. The flame characteristics and odors of the most common thermoplastics are listed below.

- ***ABS (acrylonitrile butadiene styrene)***—Burns with a thick, black, sooty smoke and continues to burn after the flame is removed. It has a sweet odor that cannot be described but is peculiar to ABS plastic.
- ***Polyethylene (PE)***—It melts and drips; drips may continue to burn. It will continue to burn after the ignition flame is removed. Polyethylene smells like burning wax. Drips will float on water.
- ***Polypropylene (PP)***—It burns with no visible flame and has its own peculiar acid-like odor.
- ***Polyurethane (TPUR)***—It burns with a yellow-orange flame and produces black smoke. It will continue to burn with a sputtering flame after the ignition flame is removed.
- ***Polyvinyl chloride (PVC)***—PVC will char and give off a gray smoke; it will self-extinguish when the ignition flame is removed. It has its own unusual odor.

Welding Equipment and Supplies

All the joint designs used for welding metals may be used successfully with thermoplastics. See Chapter 4 for information on joint designs.

The Welding Torch

The fusion welding of thermoplastics is achieved by a pressurized gas heated in the welding torch. The gas may be air or another gas (usually nitrogen). When nitrogen is used, it is supplied under pressure from a cylinder. If air is used, it is pressurized by a motor or engine-driven compressor. The pressurized gas is carried to the torch by a hose. **Figure 31-4** lists the recommended heated gas for welding various thermoplastics.

The gas is heated by electric heating coils enclosed in the special welding torch, **Figure 31-5**. The heated gas is then directed at the surfaces of the plastic joint. A welding outfit used for plastics is shown in **Figure 31-6**. The welding gas temperature is adjusted by regulating the flow of the gas.

Type of Plastic	Type of Welding Gas
ABS (Acrylonitrile Butadiene Styrene)	Compressed Air
Polyethylene (PE)	Nitrogen
Polypropylene (PP)	Nitrogen
Polyurethane (TPUR)	Compressed Air
Polyvinyl Chloride (PVC)	Compressed Air

Figure 31-4. *Welding gases recommended for use with various thermoplastics.*

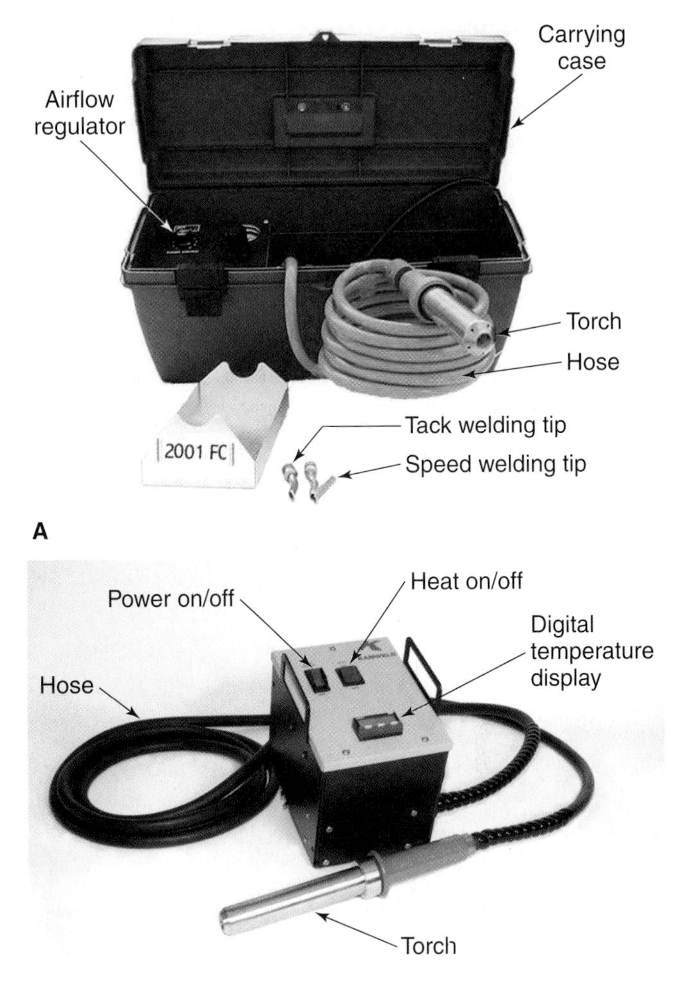

Figure 31-6. *A—A plastic welding outfit consists of a torch, hose, regulator, tips, and carrying case. The regulator and heat adjustment are built into the carrying case. (Seelye, Inc.) B—A self-contained welder has a control knob for adjusting the heat and a digital display of the temperature. The torch is connected to the controller. (Kamweld Technologies, Inc.)*

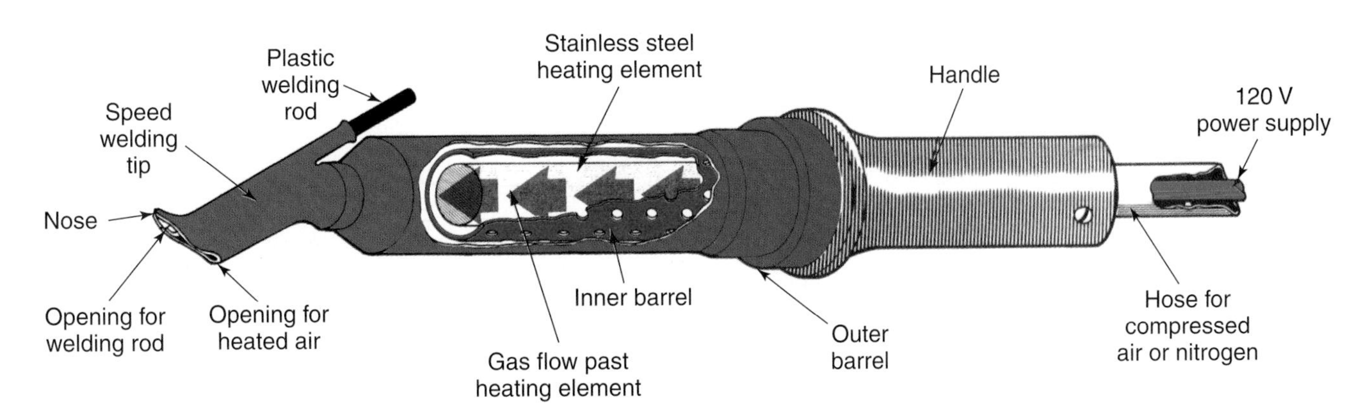

Figure 31-5. *A cutaway diagram of a welding torch for plastics. The electric coils in the torch body heat the gas as it passes. Notice that one passage in the tip heats the base material and the other heats the welding rod. (Kamweld Products Co., Inc.)*

Welding Rod

Most joints require a welding rod. The rod must be made of the same plastic as the part being welded. The cross-sectional shape of a plastic welding rod is usually round, **Figure 31-7**. Round plastic welding rods are generally produced in the following diameters: 1/8″ (3.2 mm), 5/32″ (4.0 mm), and 3/16″ (4.8 mm). **Figure 31-8** is a table of rod diameters recommended for use with various base material thicknesses.

Plastic rods may be oval, triangular, or wide rectangular strips. See **Figures 31-9** and **31-10**. The welder's choice of welding rod is determined by the shape of the joint, the equipment available, and the thickness of the plastic to be welded. **Figures 31-11** and **31-12** show butt and fillet welds.

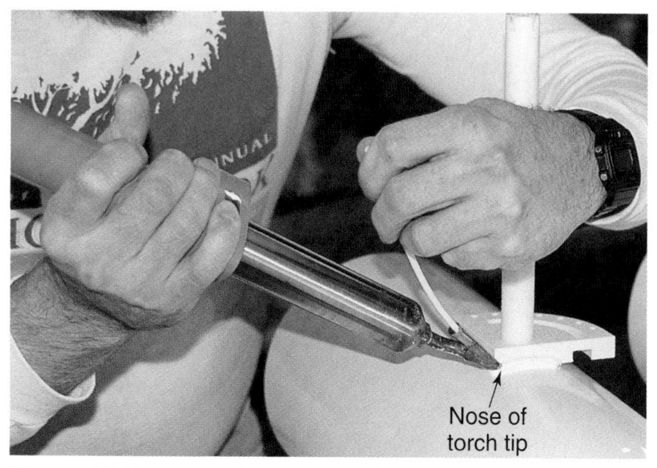

Figure 31-7. *A round plastic welding rod being used with a speed welding tip to weld a control segment onto a plastic pipe fabrication. Note that the tip of the torch pushes the plastic weld rod into the joint. (Laramy Products, LLC)*

Base Material Thickness	Welding Rod Diameter
1/16″ (1.6 mm)	1/8″ (3.2 mm)
1/8″ (3.2 mm)	1/8″ (3.2 mm)
3/16″ (4.8 mm)	3/16″ (4.8 mm)
1/4″ (6.4 mm)	3 passes of 3/16″ (4.8 mm) diameter
For thicknesses larger than 1/4″ (6.4 mm), multiple beads are used.	As many 5/32″ (4.0 mm) or 3/16″ (4.8 mm) diameter rods as required to fill the joint.
Note: Try to select a welding rod diameter close to the thickness size of the base material. For thicknesses larger than 3/16″ (4.8 mm), use several layers of welding rod.	

Figure 31-8. *A table of recommended welding rod sizes for various base material thicknesses.*

Figure 31-9. *A triangular shaped plastic welding rod is being used with a speed welding tip to weld a pipe to a plate. Air is heated to soften both the base plastic and welding rod. (JP Plastics, Inc.)*

Figure 31-10. *Both butt and fillet welds were required to make this plating tank. Both different welding tips and differently shaped welding rods were used to weld the joints. (Seelye, Inc.)*

Plastics-Welding Procedure

Before welding, a joint is often tack welded. A tack welding tip is shown in **Figure 31-13**. The pointed ***tack welding tip*** is pressed into the joint to fuse (weld) the parts together. A tack weld on plastic is generally thin, shallow, and relatively weak. It is, however, strong enough to hold the parts in alignment while they are welded. The tack welded parts may be easily disassembled for realignment.

The welding may be done with a regular tip or speed welding tip. When a regular welding tip is used, the torch is held in one hand. The plastic welding rod is fed into the weld joint with the other hand.

Figure 31-11. This double-V-groove joint was welded with round welding rods. Several welding passes were necessary. (Laramy Products, LLC)

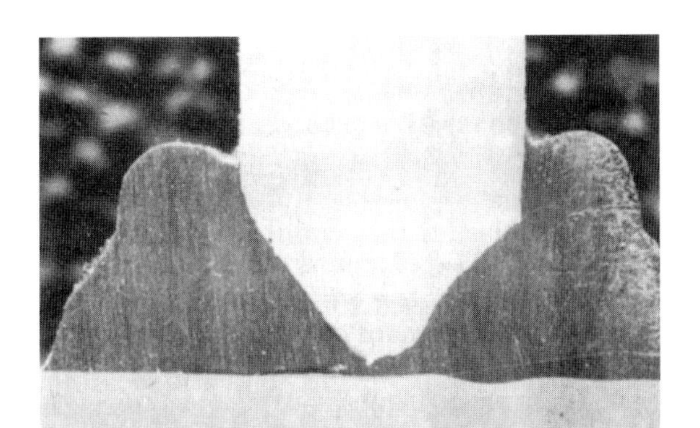

Figure 31-12. The first weld pass on each side of this double-bevel groove corner joint was apparently welded with a triangular welding rod. Additional weld passes were completed with round welding rods. (Laramy Products, LLC)

Figure 31-14 show a regular welding tip in use. The hand holding the welding rod lightly presses the heated rod into the softened plastic joint.

Figures 31-2, 31-7, and 31-9 show a speed welding tip being used to weld a lap joint. In this tip, there are two openings. The heated gas from one opening heats

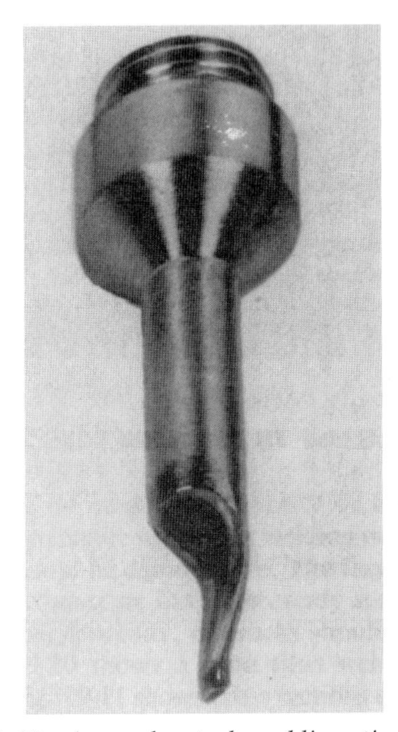

Figure 31-13. A regular tack welding tip. (Kamweld Technologies, Inc.)

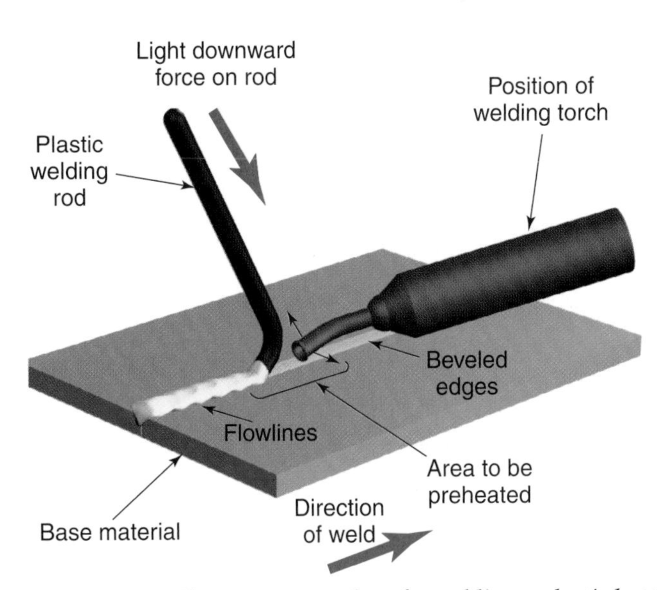

Figure 31-14. The correct procedure for welding a plastic butt joint, using a regular welding tip. Note the positions of the rod and welding tip. (Kamweld Technologies, Inc.)

the plastic to the welding temperature. The welding rod is fed into the joint through the second opening. The ***speed welding tip*** allows one-handed welding and a welding speed 4–6 times faster than is possible with a regular tip.

A speed welding tip is held tightly against the joint while welding. As the tip is moved along the joint, the rod is drawn into the joint through the feed tube, **Figure 31-15**.

The gas flow regulator controls the heat of the welding gas. The less gas per minute that flows over the heating element, the higher the temperature to which it will be heated. If more gas is fed through the torch per minute, it will pick up less heat. Larger heating elements can be used to provide more heat, as well.

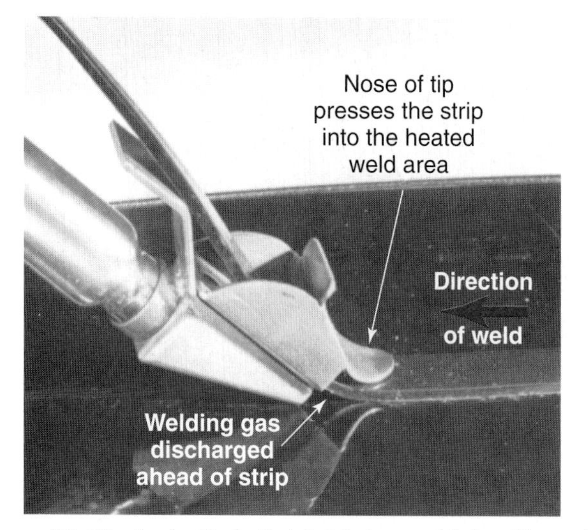

Figure 31-15. A plastic butt joint being welded with a wide flat welding rod. Note the shape of the speed welding tip. (Laramy Products, LLC)

The plastic joint and welding rod are heated to 450°F–800°F (232°C–427°C) for welding. **Figure 31-16** shows welding temperatures for various plastics. At the welding temperature, the plastic surface and welding rod become soft. While in this condition, the rod is forced into the surface with a light pressure, Figures 31-7 and 31-9. When a speed welding tip is used, the nose at the end of the tip applies pressure. When a regular welding tip is used, the welding rod is lightly pressed into the plastic joint by the hand holding the rod. As the parts of the joint and rod are heated and pressed together, a fusion weld occurs.

Type of Plastic	Welding Temperature
ABS (Acrylonitrile Butadiene Styrene)	500°F (260°C)
Polyethylene (PE)	550°F (288°C)
Polypropylene (PP)	575°F (302°C)
Polyurethane (TPUR)	575°F (302°C)
Polyvinyl Chloride (PVC)	525°F (274°C)

Figure 31-16. A welding temperature chart for various plastics.

Summary

- Thermosetting plastics cannot be welded by heat and fusion. Plastics in this group are initially heated and formed into a desired shape. However, after they have cooled, they cannot be reheated and reformed.
- Thermoplastic materials may be joined by welding in the same manner as metal. Plastics in this group become soft and may be formed when heated. Thermoplastics may be reheated and reformed repeatedly. Important thermoplastics include ABS, polyethylene, polypropylene, polyurethane, and polyvinyl chloride.
- To determine whether a plastic is a weldable thermoplastic, begin by removing a small sliver of the plastic from the part. Hold the plastic sliver with pliers and ignite it with a flame. Observe the plastic as it burns and compare its appearance and odor to those of known thermoplastics.
- Thermoplastics are heated to welding temperature, 450°F–800°F (232°C–427°C), by a heated, pressurized gas from a welding torch. The gas may be air or an inert gas. When an inert gas is used, it is supplied under pressure from a cylinder. If air is used, it is pressurized by a motor- or engine-driven compressor. The gas is heated by electric heating coils enclosed in the torch. The heated gas is then directed at the surfaces of the plastic joint. The welding gas temperature is adjusted by regulating the flow of the gas.
- Most joints require a welding rod. The rod must be made of the same plastic as the part being welded.
- The welding may be done with a regular tip or speed welding tip. When a regular welding tip is used, the torch is held in one hand. The plastic welding rod is fed into the weld joint with the other hand. In a speed welding tip, there are two openings. The heated gas from one opening heats the plastic to the welding temperature. The welding rod is fed into the joint through the second opening, allowing one-handed welding.

Review Questions

Write your answers on a separate sheet of paper. Please do *not* write in this book.

1. List five uses for plastics that are not mentioned in this chapter.
2. List four plastics that cannot be welded by heat and fusion.
3. *True or False?* The heated gas used to weld plastics may be air or nitrogen.
4. To increase the temperature of the welding gas, you would (increase/decrease) _____ the flow rate of the gas.
5. _____ in the welding torch are used to heat the gas as it flows through the torch.
6. *True or False?* Welding rods may be round, triangular, or rectangular in shape.
7. The welding gas used for polypropylene is _____.
8. The welding temperature for ABS thermoplastic is _____°F.
9. What is used to press the welding rod into the heated plastic joint when a speed welding tip is used?
10. What is used to push the welding rod into the heated plastic joint when a speed welding tip is not used?

Welding on a cross-country pipeline (Miller Electric Manufacturing Co.)

Chapter 32

Welding Pipe and Tube

Learning Objectives

After studying this chapter, you will be able to:

- Identify the differences between pipes and tubes.
- List the names of the welding passes used in welding pipe with walls more than 3/16″ (4.8 mm) in thickness.
- Describe the procedures used to weld pipes or tubes using SMAW.
- Discuss the differences in technique for uphill and downhill welding.

Technical Terms

backing ring
convex weld bead
cover pass
downhill welding
filler pass
hot pass
inside diameter (ID)
mandrel
outside diameter (OD)
pipe schedule number
plastic state
root pass
socket connection
uphill welding

Pipes and Tubes

Pipes and tubes are hollow cylinders, normally used to carry liquids and gases. They are also sometimes used as structural elements, such as railings or scaffolding. Pipes and tubes are also used in the fuselages (bodies) of small airplanes and in frames for race cars and bicycles.

Pipes are ordered by the ***inside diameter (ID)*** measurement. The wall thickness on a pipe is ordered by a ***pipe schedule number***. Generally, six to eight schedule sizes are available for each diameter of pipe. Some common pipe schedules are 10, 40, 60, and 80. See **Figure 32-1**.

Tubes are ordered by the ***outside diameter (OD)*** size. Tubes can be ordered in virtually any wall thickness. However, the wall thickness of a tube is usually less than the wall thickness of a pipe with the same outside diameter. See **Figure 32-2**.

Pipes may be joined by means of threaded fittings. The thicker walls of pipe allow threads to be used. Tubes are often joined by fittings that are not threaded. These fittings are held together by soft- or hard-soldered joints. Both tubes and pipes can be joined by welding. Tubes and pipes may be made from virtually any metal and some plastics. Low-carbon steel, high-carbon steel, stainless steel, copper, brass, and aluminum are common metals used for pipes and tubes.

Dimensions of Seamless and Welded Steel Pipe

Pipe Size	OD in inches	Pipe Schedules and Wall Thickness													
		5	10	20	30	40	STD.	60	80	E.H.	100	120	140	160	DBLE. E.H.
1/8	.405	.035	.049			.068	.068		.095	.095					
1/4	.540	.049	.065			.088	.088		.119	.119					
3/8	.675	.049	.065			.091	.091		.126	.126					
1/2	.840	.065	.083			.109	.109		.147	.147				.187	.294
3/4	1.050	.065	.083			.113	.113		.154	.154				.218	.308
1	1.315	.065	.109			.133	.133		.179	.179				.250	.358
1 1/4	1.660	.065	.109			.140	.140		.191	.191				.250	.382
1 1/2	1.900	.065	.109			.145	.145		.200	.200				.281	.400
2	2.375	.065	.109			.154	.154		.218	.218				.343	.436
2 1/2	2.875	.083	.120			.203	.203		.276	.276				.375	.552
3	3.500	.083	.120			.216	.216		.300	.300				.437	.600
3 1/2	4.000	.083	.120			.226	.226		.318	.318					.636
4	4.500	.083	.120			.237	.237	.281	.337	.337		.437		.531	.674
5	5.563	.109	.134			.258	.258		.375	.375		.500		.625	.750
6	6.625	.109	.134			.280	.280		.432	.432		.562		.718	.864
8	8.625	.109	.148	.250	.277	.322	.322	.406	.500	.500	.593	.718	.812	.906	.875
10	10.750	.134	.165	.250	.307	.365	.365	.500	.593	.500	.718	.843	1.000	1.125	
12	12.750	.165	.180	.250	.330	.406	.375	.562	.687	.500	.843	1.000	1.125	1.312	
14	14.000		.250	.312	.375	.437	.375	.593	.750	.500	.937	1.093	1.250	1.406	
16	16.000		.250	.312	.375	.500	.375	.656	.843	.500	1.031	1.218	1.437	1.593	
18	18.000		.250	.375	.437	.562	.375	.750	.937	.500	1.156	1.375	1.562	1.781	
20	20.000		.250	.375	.500	.593	.375	.812	1.031	.500	1.280	1.500	1.750	1.968	
24	24.000		.250	.375	.562	.687	.375	.968	1.218	.500	1.531	1.812	2.062	2.343	

Figure 32-1. A table of standard pipe sizes. The inside diameter (ID), outside diameter (OD), and wall thickness is given for each schedule size. (The Gage Co., Taylor Engineering Div.)

Outside Diameter, Inches		Wall Thickness								
		Inch	.028	.035	.049	.065	.083	.095	.109	.120
		Gauge	22	20	18	16	14	13	12	11
1/4	.250		X	X	X					
5/16	.312		X	X	X					
3/8	.375		X	X	X	X				
1/2	.500		X	X	X	X	X	X		
5/8	.625		X	X	X	X	X	X		
3/4	.750			X	X	X	X	X	X	X
7/8	.875			X	X	X	X	X	X	X
1	1.000			X	X	X	X	X	X	X
1 1/8	1.125			X	X	X	X	X	X	X
1 1/4	1.250				X	X	X	X	X	X
1 3/8	1.375					X	X	X	X	X
1 1/2	1.500					X	X	X	X	X
1 3/4	1.75					X	X	X	X	X
2	2.000					X	X	X	X	X

Figure 32-2. A table of common tubing sizes. The outside diameter (OD) is given in fractions and decimals. The wall thickness is given in decimal and gauge sizes. The sizes indicated by an "X" are widely available. (The Gage Co., Taylor Engineering Div.)

Both tubes and pipes can be ordered in either seamless or welded form. Tubes also may be ordered in a variety of cross sections other than round. Three such shapes are square, rectangular, and oval.

Seamless Pipes and Tubes

The wall of a seamless tube or pipe is made from a single piece of metal without a seam or joint. A seamless pipe or tube begins as a solid rod or bar of metal. The metal is heated until it is in the ***plastic state*** (almost molten). A steel ***mandrel***, or forming tool, is then forced through this solid bar. The mandrel forms the inside diameter of the pipe or tube, **Figure 32-3**. The pipe or tube may then be run through forming rolls with a stationary mandrel in place to keep the pipe from crushing as the outside diameter is formed.

Welded Pipes and Tubes

Welded pipes and tubes are formed from flat stock. The flat metal is formed into a round pipe or tube by a series of rolls. As the completed tube shape exits the last roll, the seam is welded. Resistance welding is most common, but GMAW or some other process may be used.

Welding Processes

Tubes and pipes may be joined by virtually any welding process. The processes generally used are shielded metal arc welding (SMAW), gas metal arc welding (GMAW), flux cored arc welding (FCAW), gas tungsten arc welding (GTAW), submerged arc welding (SAW), and oxyfuel gas welding (OFW). See **Figure 32-4**.

Both manual and mechanized welding equipment can be used to join pipe and tubing. **Figure 32-5** shows a GMAW machine used to weld a butt joint in pipe. The welding gun travels around the pipe on a track. This equipment can be made to oscillate, pause, weave, and perform other functions like a manual operation. **Figure 32-6** shows a low profile GTAW machine. The welding torch and wire feed mechanism also move around the pipe or tube on a track. This low profile machine allows welds to be made where clearances around the pipe are small or tight. Both of these machines produce repeatable and very high quality welds. When welding on a pipeline in the field, a shelter or tent is often placed around the welding area. See **Figure 32-7**. This keeps the welders protected from excessive sun, cold, or rain. It also prevents wind from affecting the shielding gas used to protect the weld.

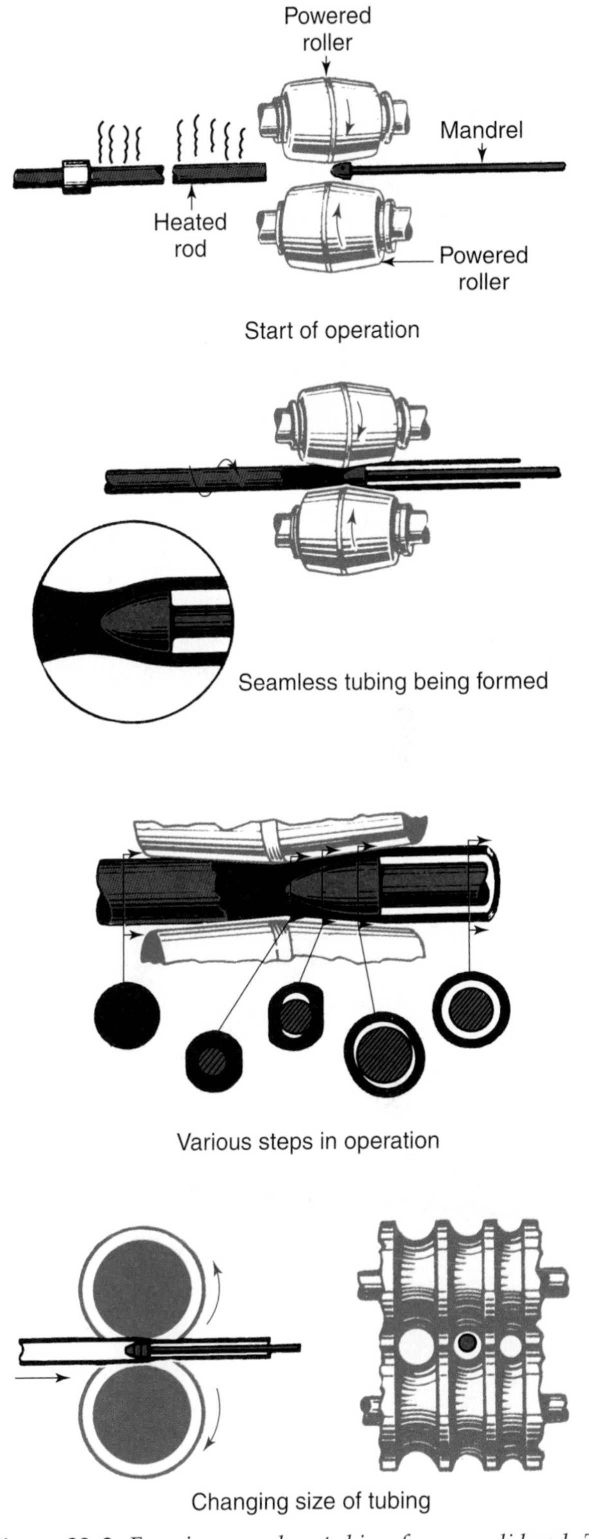

Figure 32-3. Forming seamless tubing from a solid rod. The length of the tubing is determined by the length of the mandrel. (Michigan Seamless Tube Co.)

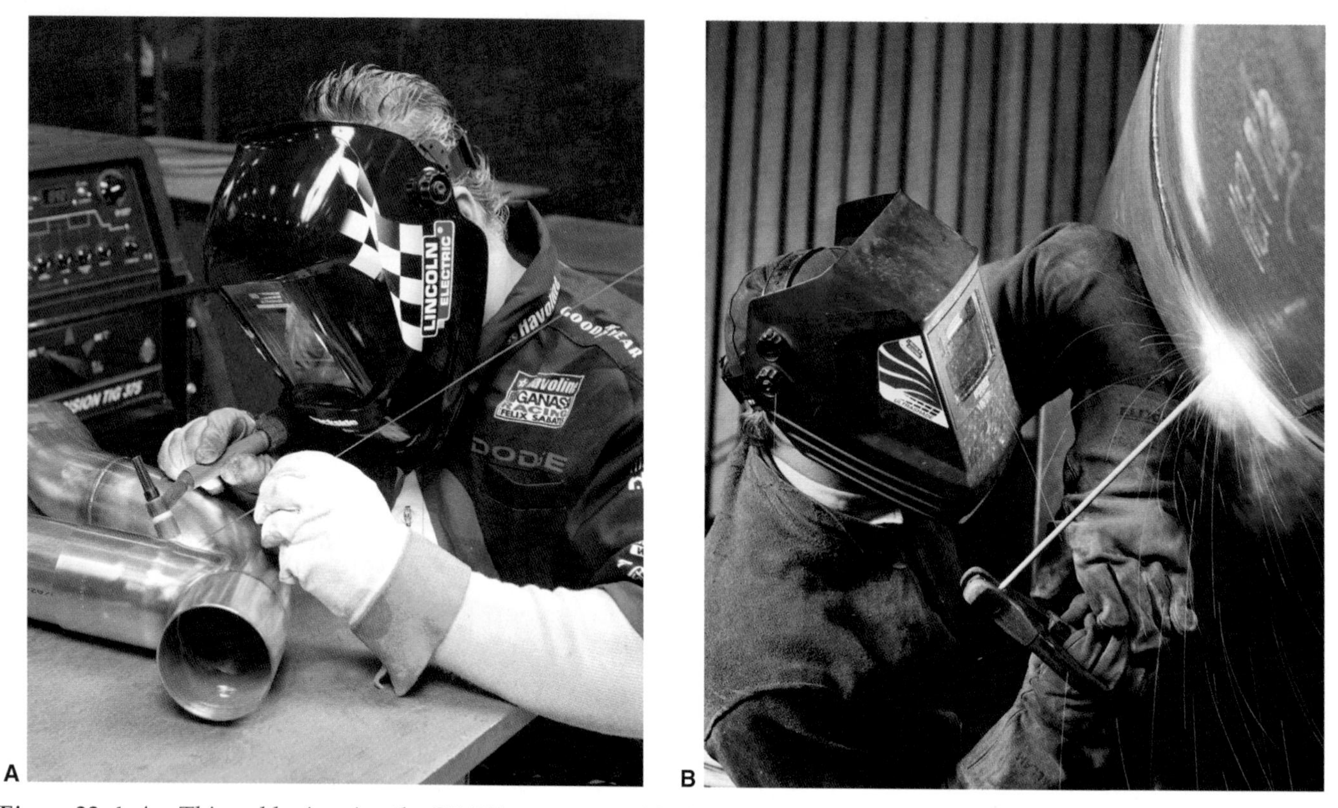

Figure 32-4. A—This welder is using the GTAW process to weld tubing to be used on a race car exhaust. B—This welder is using the SMAW process to butt weld pipe sections. (The Lincoln Electric Company)

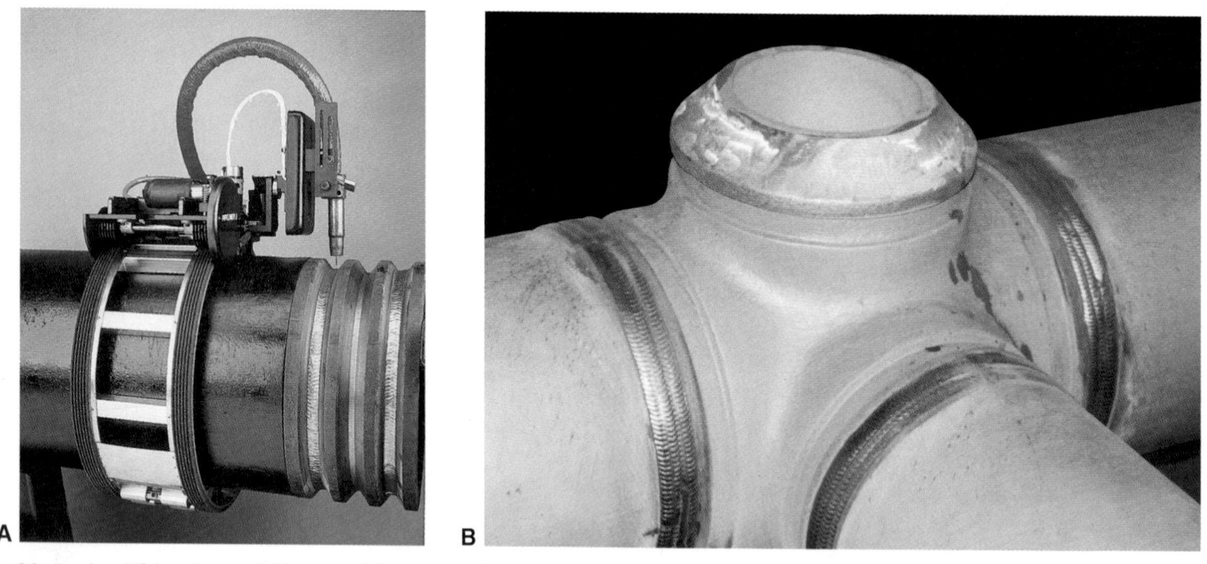

Figure 32-5. A—This pipe welding machine uses the GMAW process. This machine will produce high quality pipe welds in all positions. The weld head travels around the pipe on a track. B—Notice the high quality weld made on this T-fitting. (Magnatech, LLC)

Types of Joints

To weld pipes end to end, a V-groove butt joint is most often used. Connections are often necessary at a number of points along the length of a pipe. These connections are referred to as T, K, and Y connections because of their shapes. See **Figure 32-8**.

To weld T, K, or Y connections, all pieces must be accurately cut. Automatic cutting equipment may be used to cut these shapes. These automatic cutting machines can be programmed to cut special shapes. Metal templates may be used to draw the required shape on a pipe before the metal is cut. Special devices

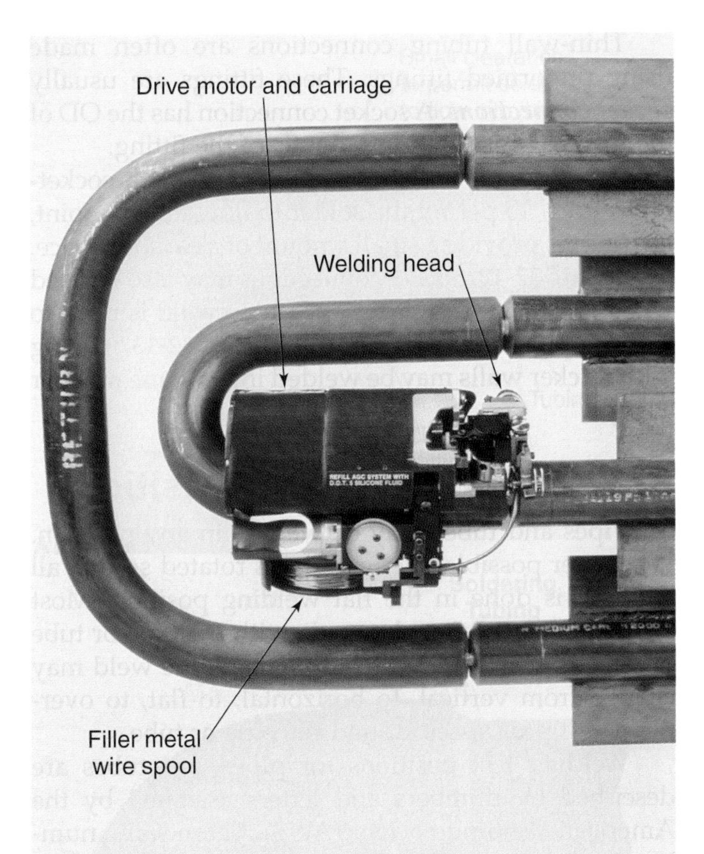

Figure 32-6. *Compact pipe welding machine using the GTAW process. This machine travels around the pipe on a track to complete a high quality weld. The machine design allows multipass welding on small closely spaced pipes. The pipes are tacked and ready for welding. (Magnatech, LLC)*

Figure 32-7. *A shelter is placed by a crane over a weld joint on this pipeline. Shelters protect the welder and the welding area from the elements and wind. (Magnatech, LLC)*

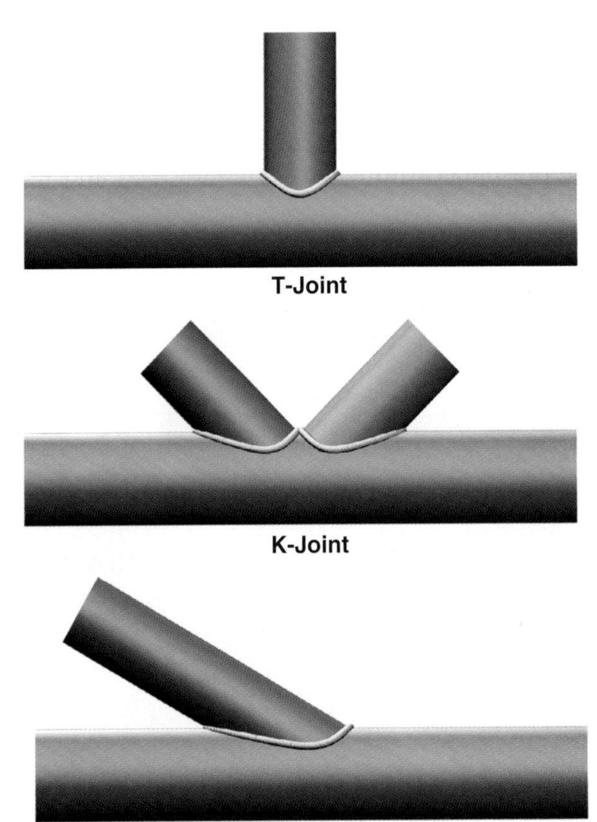

Figure 32-8. *This drawing illustrates T, K, and Y pipe joints. The pipes may be of different diameters.*

Figure 32-9. *This device, made up of many steel rods, is capable of accurately and easily copying the profile of any pipe-to-plate or pipe-to-pipe intersection. The profile copier is manufactured in a variety of diameters. The device is placed in the desired position of the pipe and pushed into contact with the plate or other pipe. The intersection shape is then transferred to the actual pipe using a marker or soapstone. (Profiler, Inc.)*

removed after tack welding is completed. **Figure 32-20** shows a typical backing ring. **Figure 32-21** shows the backing ring in position inside a pipe.

Welding Procedures

Pipe and tube welding using SMAW may be done in any position. If a pipe or tube is capable of being rotated, the welding may be done in the flat welding position.

When a pipe is fixed in a vertical position, the weld must be made in the horizontal position around the entire cylinder. A pipe or tube that is fixed in the horizontal position will be welded in the flat, vertical, and overhead welding positions. These positions change as the weld moves around the outside of the pipe or tube. The vertical welding may be done uphill or downhill.

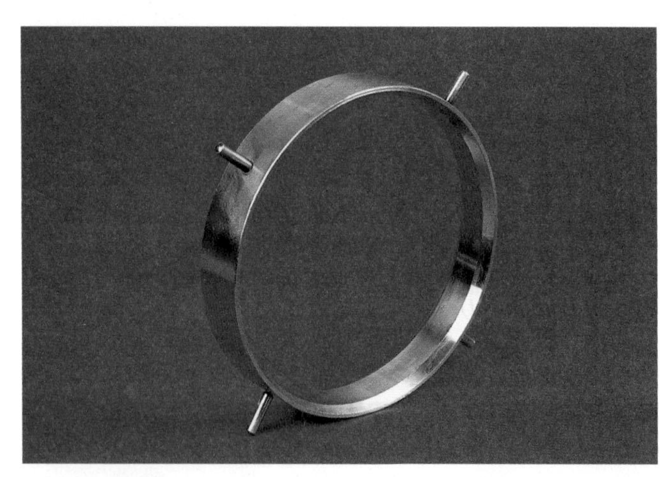

Figure 32-20. A backing ring of the type used in pipe butt welds. This device aligns the pipe, provides a uniform root opening, and controls the amount of penetration. (Tube Turns, Div. Chemetron Corp.)

Figure 32-21. Two sections of pipe being assembled with a backing ring in place. (Imperial Weld Ring Corporation)

Uphill Welding

Uphill welding generally produces the highest quality welds. This welding direction is recommended for pipe with wall thicknesses greater than 1/2″ (12.7 mm). A root opening of 3/32″–1/8″ (2.4 mm–3.2 mm) is recommended. The electrode diameter should be 1/8″ (3.2 mm) or 5/32″ (4.0 mm). The amperage should be on the lower end of the recommended range. Travel speed should be fairly slow. If the weld pool becomes overheated, a whip motion may be used. See **Figure 32-22**. In the whip motion,

Exercise 32-1 Butt Welding Pipe with SMAW

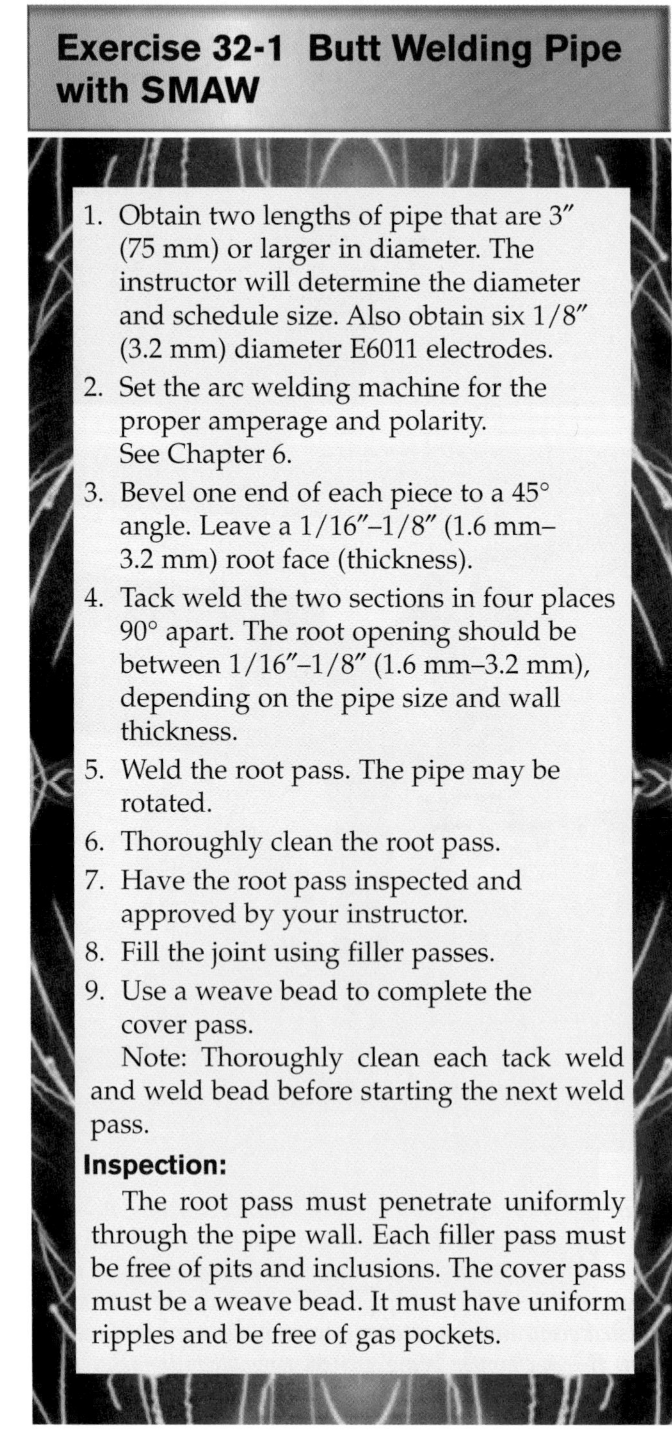

1. Obtain two lengths of pipe that are 3″ (75 mm) or larger in diameter. The instructor will determine the diameter and schedule size. Also obtain six 1/8″ (3.2 mm) diameter E6011 electrodes.
2. Set the arc welding machine for the proper amperage and polarity. See Chapter 6.
3. Bevel one end of each piece to a 45° angle. Leave a 1/16″–1/8″ (1.6 mm–3.2 mm) root face (thickness).
4. Tack weld the two sections in four places 90° apart. The root opening should be between 1/16″–1/8″ (1.6 mm–3.2 mm), depending on the pipe size and wall thickness.
5. Weld the root pass. The pipe may be rotated.
6. Thoroughly clean the root pass.
7. Have the root pass inspected and approved by your instructor.
8. Fill the joint using filler passes.
9. Use a weave bead to complete the cover pass.

Note: Thoroughly clean each tack weld and weld bead before starting the next weld pass.

Inspection:

The root pass must penetrate uniformly through the pipe wall. Each filler pass must be free of pits and inclusions. The cover pass must be a weave bead. It must have uniform ripples and be free of gas pockets.

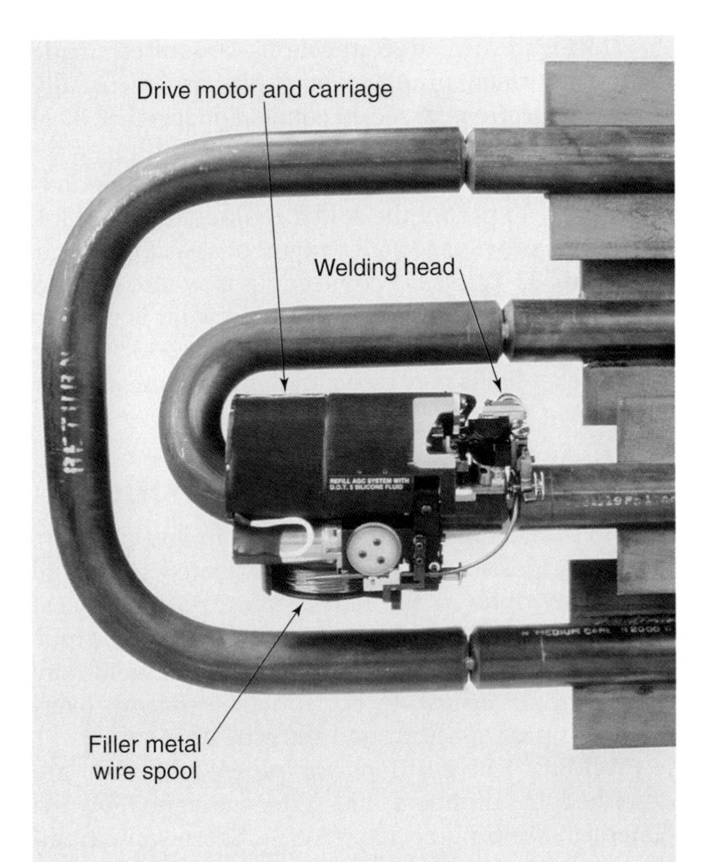

Figure 32-6. Compact pipe welding machine using the GTAW process. This machine travels around the pipe on a track to complete a high quality weld. The machine design allows multipass welding on small closely spaced pipes. The pipes are tacked and ready for welding. (Magnatech, LLC)

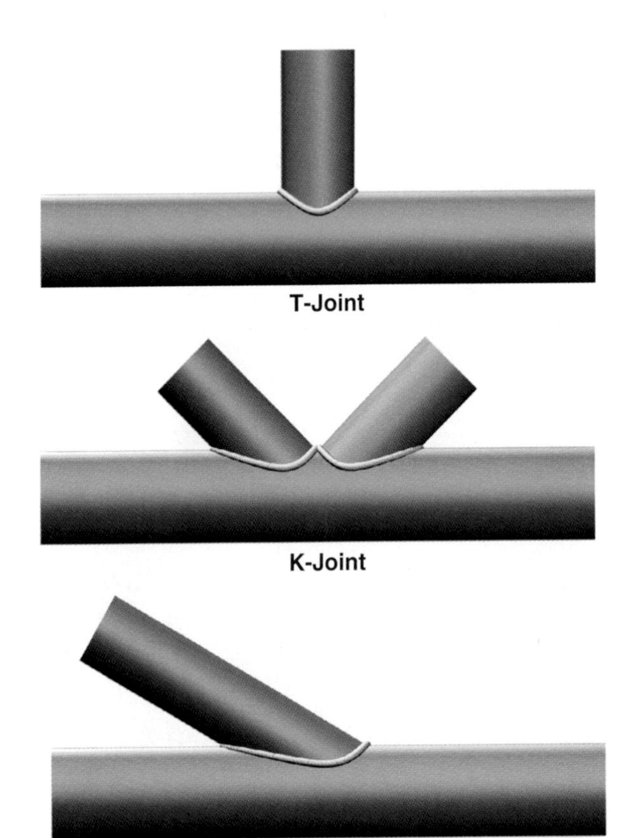

Figure 32-8. This drawing illustrates T, K, and Y pipe joints. The pipes may be of different diameters.

Figure 32-7. A shelter is placed by a crane over a weld joint on this pipeline. Shelters protect the welder and the welding area from the elements and wind. (Magnatech, LLC)

Figure 32-9. This device, made up of many steel rods, is capable of accurately and easily copying the profile of any pipe-to-plate or pipe-to-pipe intersection. The profile copier is manufactured in a variety of diameters. The device is placed in the desired position of the pipe and pushed into contact with the plate or other pipe. The intersection shape is then transferred to the actual pipe using a marker or soapstone. (Profiler, Inc.)

are also available to mark a pipe before cutting a desired shape, as shown in **Figures 32-9** and **32-10**.

Fittings are also available for many of the standard pipe connections. See **Figure 32-11**. The edges of these preshaped fittings are often beveled and ready for welding.

Figure 32-10. A pipe intersection profile can be copied easily and then transferred to the same diameter pipe. Notice that two short Profiler devices are being used with a piece of pipe between them to span the distance desired. (Profiler, Inc.)

Thin-wall tubing connections are often made using preformed fittings. These fittings are usually ***socket connections***. A socket connection has the OD of the tube or pipe fitted into the ID of the fitting.

On thin-wall tubing, solder is used with socket-type joints. To permit the solder to flow into the joint, the fittings provide a small amount of clearance space. See **Figure 32-12**. Socket connections may also be used when joining pipe. In this case, a lap weld is used to join the pipe to the socket. See **Figure 32-13**. Tubing with thicker walls may be welded in the same manner as pipe.

Pipe or Tube Welding Positions

Pipes and tubes may be welded in any position. Whenever possible, the material is rotated so that all welding is done in the flat welding position. Most welds, however, must be made with the pipe or tube in a fixed position. In a fixed position, the weld may change from vertical, to horizontal, to flat, to overhead as it progresses around the pipe or tube.

Welding test positions for pipes and tubes are described by numbers and letters assigned by the American Welding Society (AWS). These same number and letter combinations are often used to describe welding positions on the job. The AWS test positions for groove welds are shown in **Figure 32-14**. The AWS test positions for fillet welds on pipe are shown in **Figure 32-15**.

90° elbows	90° elbows	90° reducing elbows	3 R elbows 45° and 90°	90° elbows	45° elbows	180° returns
180° returns	Tees	Crosses	Concentric reducers	Eccentric reducers	Caps	Lap joint stub ends
Laterals	Shaped nipples	Sleeves	Saddles	Welding rings	Expander flanges	Venturi expander flanges

Figure 32-11. Pipe joint fittings. Such fittings are commercially made in a variety of sizes and thickness. The edges are often beveled and ready for welding.

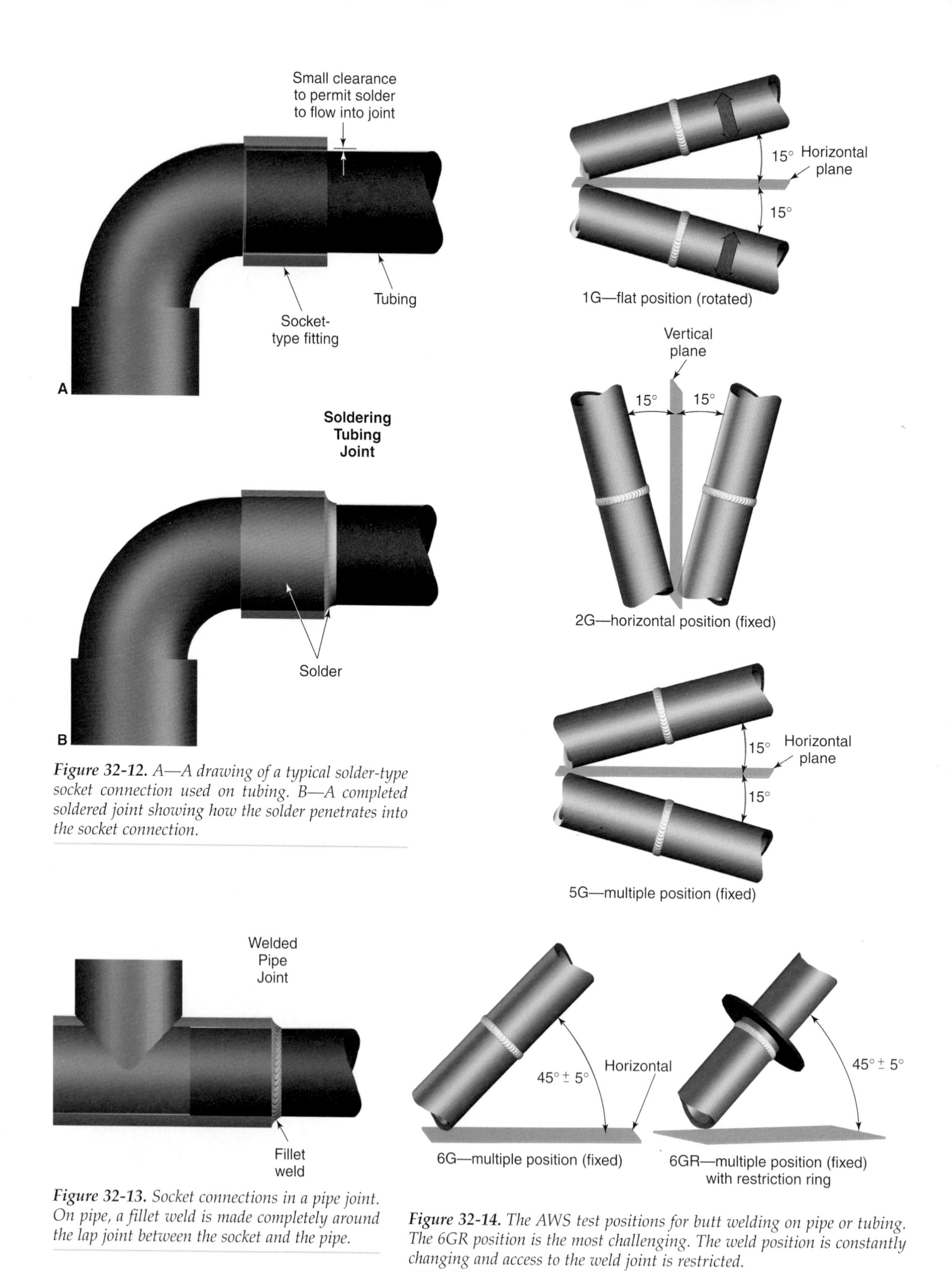

Figure 32-12. *A—A drawing of a typical solder-type socket connection used on tubing. B—A completed soldered joint showing how the solder penetrates into the socket connection.*

Figure 32-13. *Socket connections in a pipe joint. On pipe, a fillet weld is made completely around the lap joint between the socket and the pipe.*

Figure 32-14. *The AWS test positions for butt welding on pipe or tubing. The 6GR position is the most challenging. The weld position is constantly changing and access to the weld joint is restricted.*

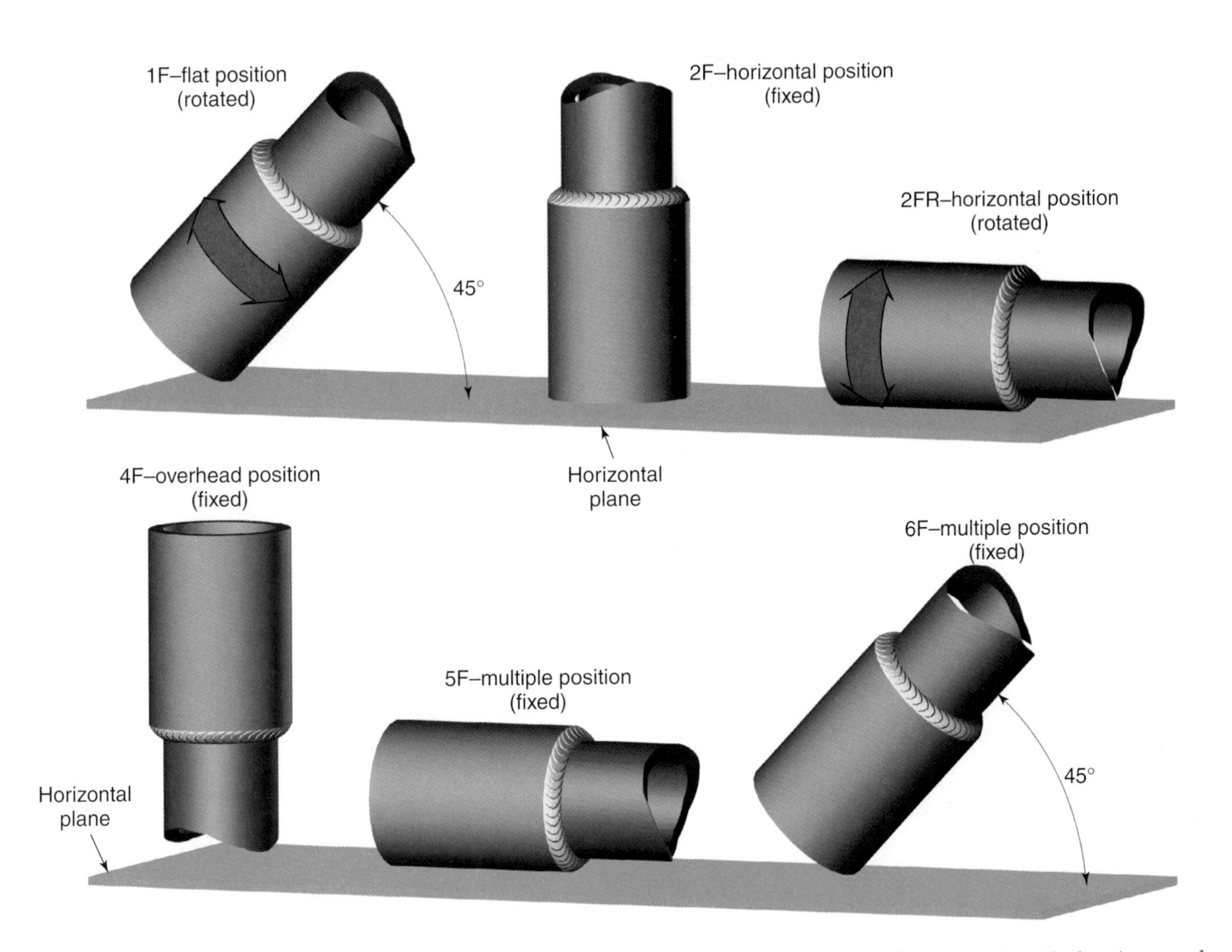

Figure 32-15. The AWS test positions for lap welding pipe or tubing. The 2F position may be requested with the pipe or tube fixed or rotated in the 5F position.

Preparations for Welding

The edges of pipe and tube must be prepared prior to welding. When a butt weld is made on thin-wall tubing, a square-groove butt joint is generally used. Generally, a V-groove butt joint is used on pipe and thick-walled tubing. The angle of the groove is given on the welding symbol. The pipe or tube must be cut to the required angle. The ends can be cut manually, semiautomatically, or automatically.

Figure 32-16 shows a motorized pipe cutter traveling around a pipe on a track. This special tool cuts the pipe to length and cuts the angle at the same time. Two other pipe-beveling cutters are shown in **Figure 32-17**. Pipe ends may also be cut using cutting torches that travel around the pipe or tube. See **Figure 32-18**. The pipe or tube must be aligned and held in position prior to welding. This may be done using special clamps and alignment fixtures. An alignment fixture is shown in **Figure 32-19**.

Figure 32-16. A motorized, traveling pipe cutter. This machine is cutting the pipe to length and chamfering the edge at the same time. (E.H. Wachs Co.)

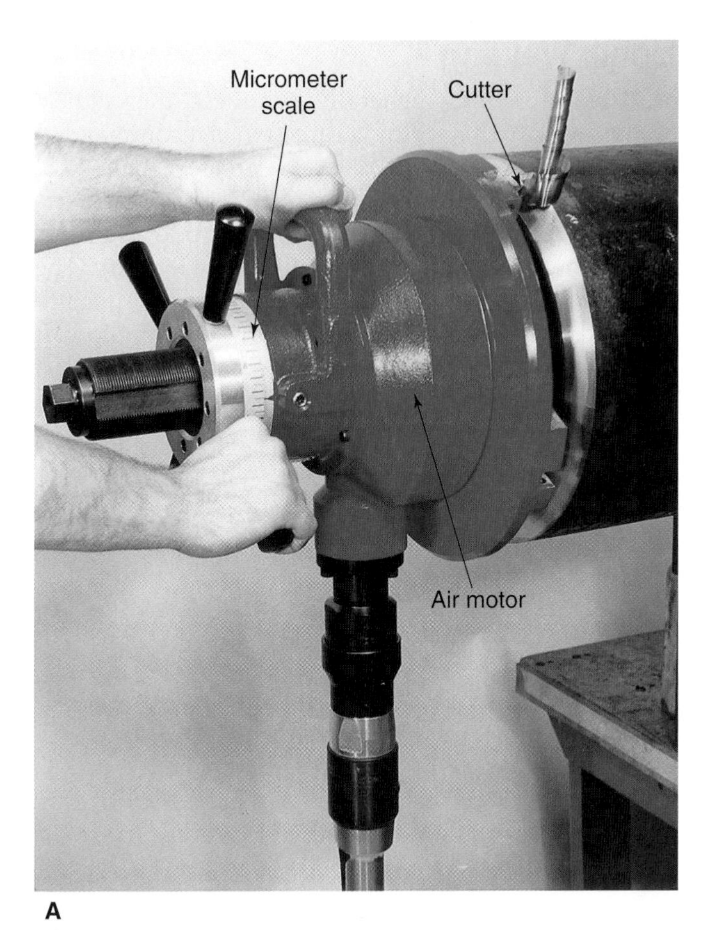

A

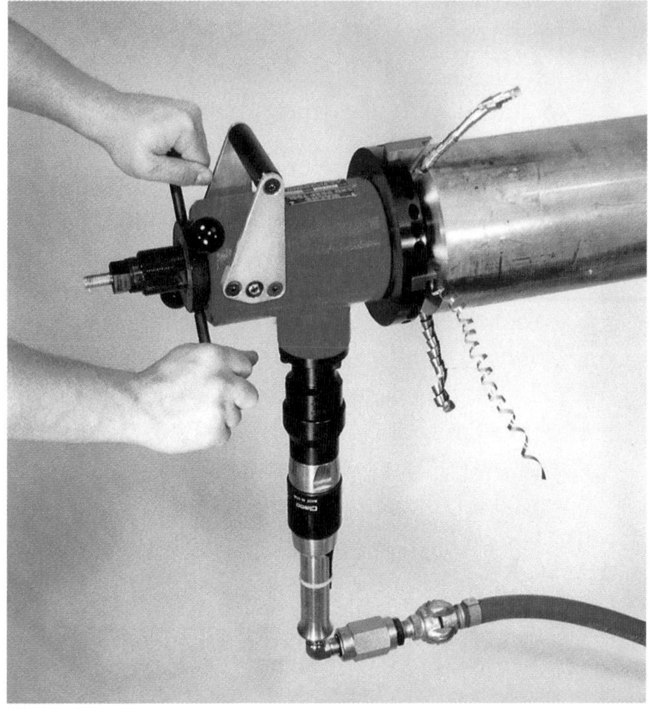

B

Figure 32-17. *Two models of air-powered pipe cutters. A—This pipe cutter is centered and clamped from the inside of the pipe. The cutting tool used determines the shape of the chamfer (bevel) on the end of the pipe. The depth of the cut can be adjusted accurately by setting a micrometer-type feed mechanism. B—A chamfer being cut on a smaller diameter pipe. (E.H. Wachs Co.)*

Figure 32-18. *A section of pipe being cut by a plasma arc cutting (PAC) torch mounted on a motorized cutting torch carriage. The track is flexible and is held in place by strong magnets. (Bug-O System Equipment).*

Figure 32-19. *A large elbow is aligned to the pipe with a chain-type alignment tool. (Dearman, a division of Cogsdill Tool Products, Inc.)*

Backing Rings

Backing rings may be used to align the pipe and control penetration in the butt joint. They may also be used to provide an even root opening while the joint is tack welded. The pins on the backing ring are removed after tack welding is completed. **Figure 32-20** shows a typical backing ring. **Figure 32-21** shows the backing ring in position inside a pipe.

Welding Procedures

Pipe and tube welding using SMAW may be done in any position. If a pipe or tube is capable of being rotated, the welding may be done in the flat welding position.

When a pipe is fixed in a vertical position, the weld must be made in the horizontal position around the entire cylinder. A pipe or tube that is fixed in the horizontal position will be welded in the flat, vertical, and overhead welding positions. These positions change as the weld moves around the outside of the pipe or tube. The vertical welding may be done uphill or downhill.

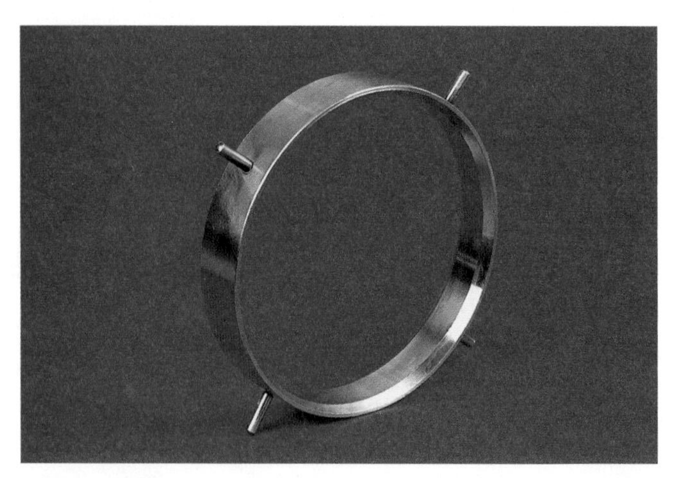

Figure 32-20. *A backing ring of the type used in pipe butt welds. This device aligns the pipe, provides a uniform root opening, and controls the amount of penetration. (Tube Turns, Div. Chemetron Corp.)*

Figure 32-21. *Two sections of pipe being assembled with a backing ring in place. (Imperial Weld Ring Corporation)*

Uphill Welding

Uphill welding generally produces the highest quality welds. This welding direction is recommended for pipe with wall thicknesses greater than 1/2″ (12.7 mm). A root opening of 3/32″–1/8″ (2.4 mm–3.2 mm) is recommended. The electrode diameter should be 1/8″ (3.2 mm) or 5/32″ (4.0 mm). The amperage should be on the lower end of the recommended range. Travel speed should be fairly slow. If the weld pool becomes overheated, a whip motion may be used. See **Figure 32-22**. In the whip motion,

Exercise 32-1 Butt Welding Pipe with SMAW

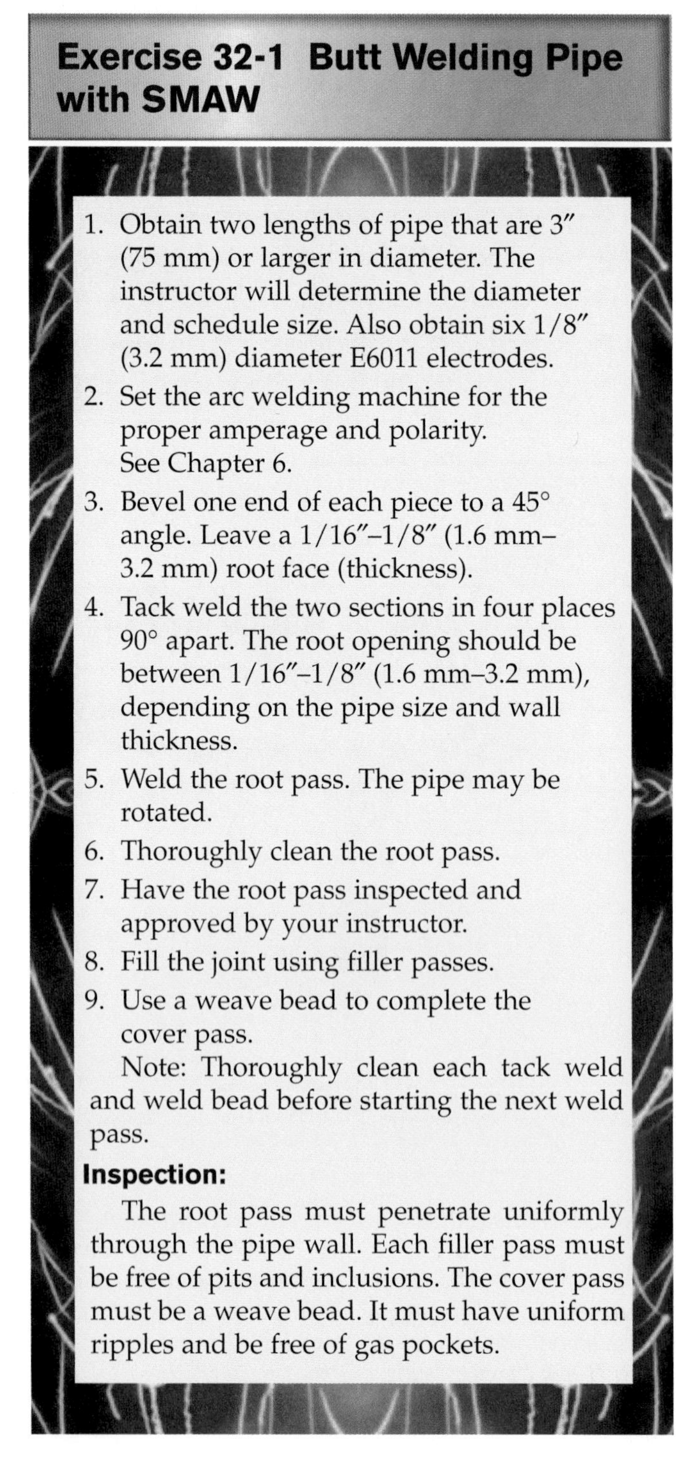

1. Obtain two lengths of pipe that are 3″ (75 mm) or larger in diameter. The instructor will determine the diameter and schedule size. Also obtain six 1/8″ (3.2 mm) diameter E6011 electrodes.
2. Set the arc welding machine for the proper amperage and polarity. See Chapter 6.
3. Bevel one end of each piece to a 45° angle. Leave a 1/16″–1/8″ (1.6 mm–3.2 mm) root face (thickness).
4. Tack weld the two sections in four places 90° apart. The root opening should be between 1/16″–1/8″ (1.6 mm–3.2 mm), depending on the pipe size and wall thickness.
5. Weld the root pass. The pipe may be rotated.
6. Thoroughly clean the root pass.
7. Have the root pass inspected and approved by your instructor.
8. Fill the joint using filler passes.
9. Use a weave bead to complete the cover pass.

Note: Thoroughly clean each tack weld and weld bead before starting the next weld pass.

Inspection:

The root pass must penetrate uniformly through the pipe wall. Each filler pass must be free of pits and inclusions. The cover pass must be a weave bead. It must have uniform ripples and be free of gas pockets.

the electrode and arc are rapidly moved ahead of the weld pool about 1″ (25 mm), then returned to the pool. During this up-and-back motion, the pool has a chance to cool. The weld is made from the 5 o'clock to the 11 o'clock position on each side of the pipe or tube. See **Figure 32-23**. A 10° drag travel angle is used for welding both uphill and downhill. The angle is measured from lines through the center of the cylinder and the center of the electrode. See **Figure 32-24**.

Downhill Welding

On tube or pipe with a wall thickness up to 1/2″ (12.7 mm), ***downhill welding*** is often used. As in all downhill welding, the arc crater must be kept ahead of the slag. The weld will become contaminated if slag runs into the crater. A root opening of 1/16″–1/8″ (1.6 mm–3.2 mm) is used.

A 3/16″ (4.8 mm) or 7/32″ (5.6 mm) electrode may be used. The amperage should be on the high side of the suggested range for the electrode being used. Travel speed must be rapid, so the weld pool stays ahead of the slag. The downhill weld is generally made from the 11 o'clock to 5 o'clock position on each half of the pipe or tube. See **Figure 32-25**. The groove angle used may be less than for an uphill weld. Fewer weld passes are required to complete a downhill weld.

A weaving motion is suggested for the cover pass. The forward motion of the electrode weave should be between 1 and 1 1/2 electrode diameters. See **Figure 32-26**.

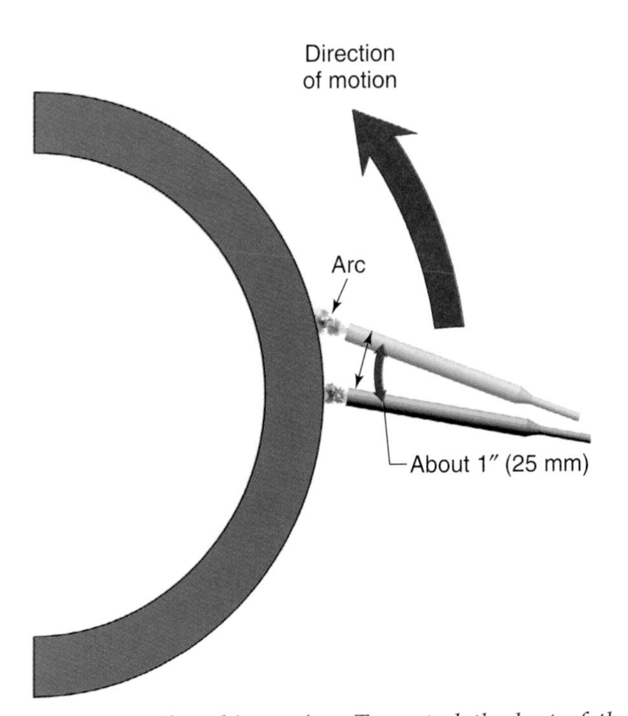

Figure 32-22. *The whip motion. To control the heat of the crater, move the electrode rapidly forward and then back to the crater. The arc should never be broken (stopped) in this motion.*

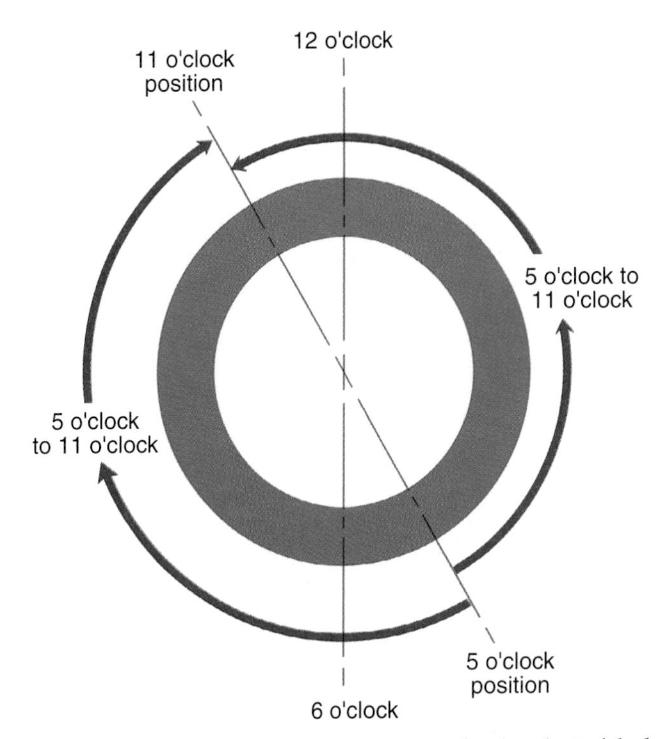

Figure 32-23. *Welding uphill. The weld is started at the 5 o'clock position and is continued to the 11 o'clock position on both halves of the pipe or tube.*

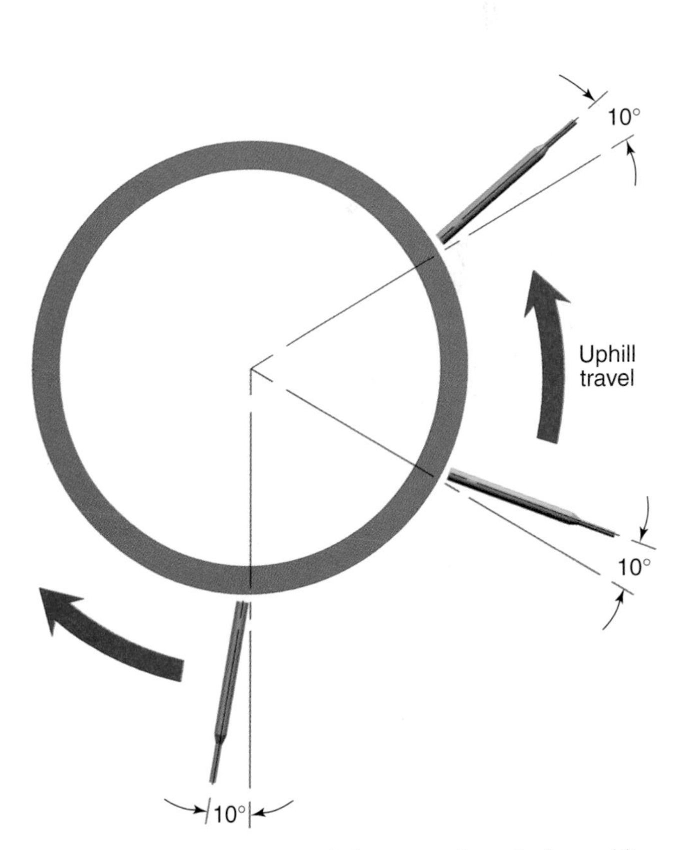

Figure 32-24. *The suggested drag travel angle for welding large diameter pipe is about 10°. The angle is measured between the electrode and a line drawn to the center of the pipe. This 10° angle is continually moving as the weld is made.*

Keyhole

The size of the keyhole is an important indicator of the depth of the penetration. It will also indicate the quality of the penetration.

A keyhole that is too large indicates too much penetration or burn-through. It also indicates possible undercutting of the inside of the pipe. A small keyhole or a lack of a keyhole is a sign of lack of penetration or uneven penetration.

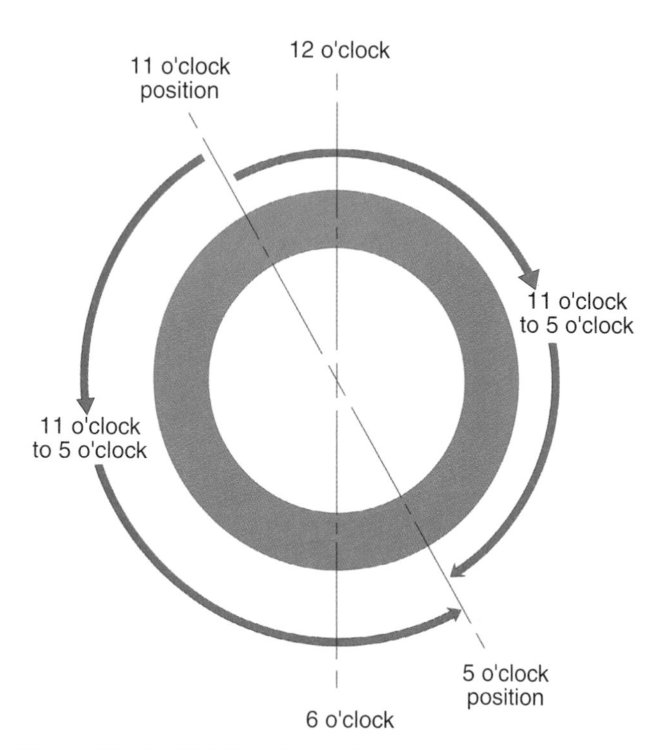

Figure 32-25. *Welding downhill from the 11 o'clock to the 5 o'clock position. The joint should be tack welded in at least three places. It is advisable to weld past the 11 o'clock and 5 o'clock positions to blend the weld beads. Test samples are taken from these positions.*

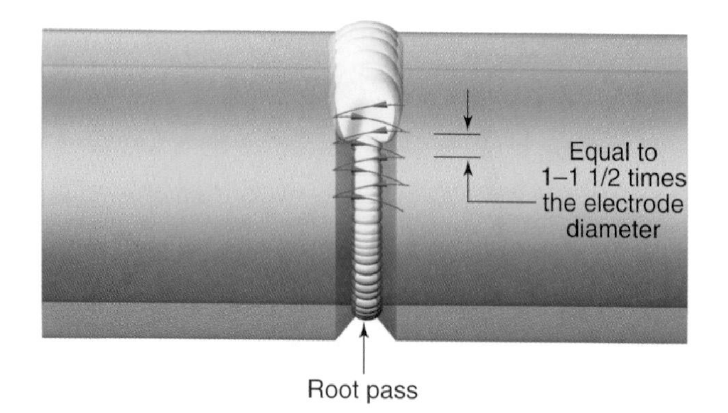

Figure 32-26. *A weave bead is recommended for the cover pass on large groove joints. The electrode should be moved forward a distance equal to 1–1 1/2 electrode diameters during each swing of the weave pattern.*

Multiple Weld Passes

Multiple weld passes are generally required to complete a weld on pipes or tubes with wall thicknesses of 3/16″ (4.8 mm) or greater. **Figure 32-27** shows the names and relationships of the various weld passes. The following is a list of the order of the weld passes and their names:

1st weld pass	Root pass.
2nd weld pass	Filler pass or hot pass.
3rd and succeeding weld passes	Filler passes.
Final weld pass	Cover (cap) pass.

Weld passes should not be thicker (deeper) than 3/16″ (4.8 mm). Thin weld passes allow gases and impurities to reach the surface before the weld cools.

Weld passes on alloy steel pipe or tube should be made smaller in width, preferably about 1/4″ (6.4 mm). Smaller weld passes put less heat into the pipe or tube and reduce the amount of alloy loss due to welding.

Note: When using SMAW to weld pipes or tubes, you must thoroughly clean each weld bead before starting the next weld bead. This reduces the possibility of slag inclusions in the next weld bead.

The ***root pass***, or first weld pass, should be made using an electrode 1/8″ (3.2 mm) in diameter. This is the most important of all the weld passes. The root pass must penetrate through the pipe or tube wall. The bead penetration may be seen from the inside of the pipe or tube. The height of the root reinforcement (weld face for the root pass) should be about 1/16″ (1.6 mm). See Figure 32-27. A higher root reinforcement will interfere with flow through the pipe or tube.

The second weld pass is the first fill pass. It is often called the ***hot pass***. This is not a standard AWS term, but it is often used. This pass is made with the same size electrode as the root pass. However, a higher amperage should be used. This weld pass eliminates any undercutting that may have occurred in the root pass. ***Filler passes*** (the third and subsequent weld passes) should be made with an electrode larger than 1/8″ (3.2 mm).

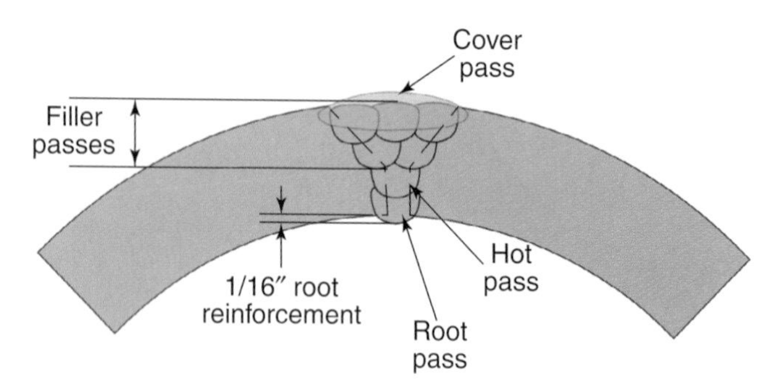

Figure 32-27. *The names given to the various weld passes in a weld joint. Note that there is one hot pass in the second layer and several filler passes and layers.*

The *cover pass*, or final pass, is also made with an electrode larger than 1/8″ (3.2 mm). It must be made with a weaving motion. This weld pass must tie in all the filler passes and form a single, wide ***convex*** (curved outward) ***weld bead***.

Restarting and Ending a Weld

With SMAW, you must stop occasionally to change electrodes. The correct method of restarting is extremely important. If the restart is not done properly, a weakness may be created in the weld bead.

Restarting the Weld

To restart, the arc is struck about 1/2″–1″ (13 mm–25 mm) ahead of the previous crater. The electrode and new weld pool are then moved backward until the ripples of the new weld bead just touch the old weld bead at the rear of the old crater. The electrode and weld bead are then moved forward again at the correct speed.

To restart a root pass when a backup ring is not used, the procedure is slightly different. Strike the arc, as explained above. When the electrode reaches the keyhole, push the tip of the electrode through the keyhole at a 45° angle. See **Figure 32-28**. Hold the electrode there momentarily, then withdraw it to the normal distance and angle. Inserting the electrode through the keyhole ensures that the penetration inside the pipe is continuous.

Ending the Weld

Run the electrode and weld pool to the end of the weld. Then, reverse the weld bead for a short distance as you slowly increase the electrode gap until the arc stops. This reversal of the electrode causes the weld pool at the end of the weld to fill. The end of the weld bead will be relatively smooth, without a depression.

GTAW and GMAW of Pipes and Tubes

Both GTAW and GMAW may be used to weld pipes and tubes. GTAW is often used to make the very important root pass on both tubes and pipes. Generally, GTAW produces the strongest possible weld in such applications, **Figure 32-29**. It is unnecessary to clean the weld beads between weld passes. GTAW is, however, a slow and expensive process.

GMAW is widely used as a pipe and tube welding process. This process is fast and clean. It is usually not necessary to clean the weld beads between weld passes. GMAW out of position may be done using either the short circuiting or pulsed spray transfer methods. Completed welds that are properly made are usually free of inclusions and other flaws.

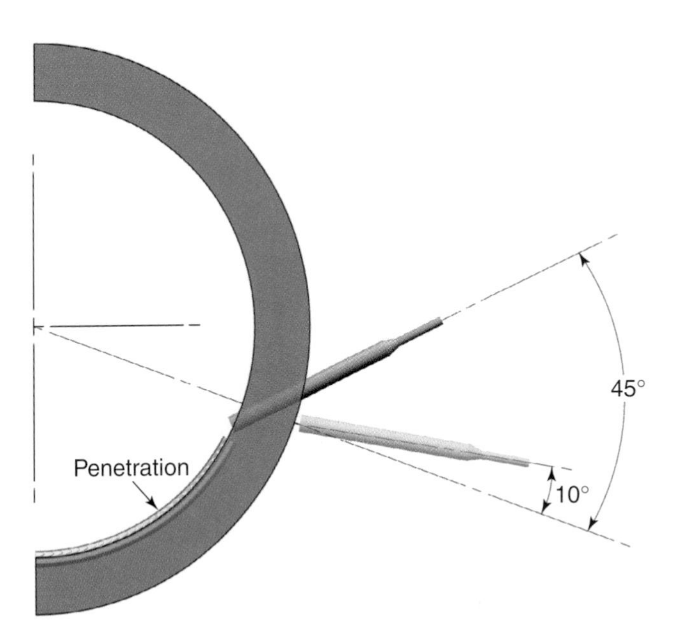

Figure 32-28. *Restarting the root pass without a backing ring. The electrode is momentarily inserted through the keyhole at a 45° to ensure that the penetration is continuous. It is withdrawn and the weld continued.*

Figure 32-29. *GTAW being used to produce welds on stainless steel pipe. (Miller Electric Mfg. Co.)*

Summary

- Pipes are ordered by the inside diameter (ID) measurement. The wall thickness on a pipe is ordered by a pipe schedule number.
- Tubes are ordered by the outside diameter (OD) size. Tubes can be ordered in virtually any wall thickness. However, the wall thickness of a tube is usually less than the wall thickness of a pipe with the same outside diameter.
- To weld pipes end to end, a V-groove butt joint is most often used. Connections are often necessary at a number of points along the length of a pipe. These connections are referred to as T, K, and Y connections because of their shapes.
- The edges of pipe and tube must be prepared prior to welding. When a butt weld is made on thin-wall tubing, a square-groove butt joint is generally used. Generally, a V-groove butt joint is used on pipe and thick-walled tubing. The angle of the groove is given on the welding symbol.
- Backing rings may be used to align the pipe and control penetration in the butt joint. They may also be used to provide an even root opening while the joint is tack welded. The pins on the backing ring are removed after tack welding is completed.
- If a pipe or tube is capable of being rotated, the welding may be done in the flat welding position. When a pipe is fixed in a vertical position, the weld must be made in the horizontal position around the entire cylinder. A pipe or tube that is fixed in the horizontal position will be welded in the flat, vertical, and overhead welding positions.
- Vertical welding may be done uphill or downhill. Uphill welding generally produces the soundest welds. This welding direction is recommended for pipe with wall thicknesses greater than 1/2″ (12.7 mm). Downhill welding is often used on tube or pipe with a wall thickness up to 1/2″ (12.7 mm). As in all downhill welding, the arc crater must be kept ahead of the slag.
- Multiple weld passes are generally required to complete a weld on pipes or tubes with wall thicknesses of 3/16″ (4.8 mm) or greater. The first weld pass is called the root pass. The second weld pass is the first fill pass. It is often called the hot pass. The third and succeeding weld passes are called filler passes. The final weld pass is called the cover pass.
- The root pass should be made using an electrode 1/8″ (3.2 mm) in diameter. This is the most important of all the weld passes. The root pass must penetrate through the pipe or tube wall. The bead penetration may be seen from the inside of the pipe or tube. The height of the root reinforcement (weld face for the root pass) should be about 1/16″ (1.6 mm).
- The first fill pass or hot pass is made with the same size electrode. However, a higher amperage should be used. This weld pass eliminates any undercutting that may have occurred in the root pass. Filler passes should be made with an electrode larger than 1/8″ (3.2 mm). The cover pass is also made with an electrode larger than 1/8″ (3.2 mm). It must be made with a weaving motion. The cover pass must tie in all the filler passes and form a single, wide convex weld bead.
- With SMAW, you must stop occasionally to change electrodes. The correct method of restarting is extremely important. If the restart is not done properly, a weakness may be created in the weld bead. The procedure for restarting a weld varies depending on whether a backup ring is being used.
- To end a weld in the SMAW process, run the electrode and weld pool to the end of the weld. Then, reverse the weld bead for a short distance as you slowly increase the electrode gap until the arc stops.
- Both GTAW and GMAW may be used to weld pipes and tubes. GTAW is often used to make the very important root pass on both tubes and pipes. Generally, GTAW produces the strongest possible weld in such applications and eliminates the need to clean the weld beads between weld passes.

Review Questions

Write your answers on a separate sheet of paper. Please do *not* write in this book.

1. Pipe is ordered by ID size. The wall thicknesses must be chosen from a list of _____ numbers.
2. *True or False?* A mandrel and rolls are used to form the hole in seamless pipes and tubes.
3. When butt welding pipe, the horizontal fixed position is called the _____ position.
4. List two reasons for using a backing ring.
5. Name three different types of weld passes used when pipe or tube welding.
6. Uphill welding is done from the _____ o'clock to the _____ o'clock position on both sides of the pipe or tube.
7. When being moved with a(n) _____ motion to complete a downhill cover pass, the electrode should move forward in steps of 1 to 1 1/2 electrode diameters.
8. A(n) _____ keyhole indicates a lack of penetration.
   a. small
   b. large
   c. elongated
   d. distorted
9. When restarting a weld bead, the arc should be struck 1/2″–1″ (13 mm–25 mm) ahead of the old weld _____.
10. When pipe is welded with the _____ process, each weld bead must be cleaned before the next weld bead is started.
    a. SMAW
    b. GTAW
    c. GMAW
    d. OFW

Section 9

Technical Information

Chapter 33

Welding Symbols

Learning Objectives

After studying this chapter, you will be able to:

- Describe the method for making a mechanical drawing of a three-dimensional object, using the orthographic projection process.
- List the names of the views used in an orthographic projection.
- Identify the basic types of welds indicated on the ANSI/AWS welding symbol.
- Locate information on the weld symbol to determine the size of the root opening, the groove angle, and the desired size, contour, and finish of the weld.

Technical Terms

arrow side
ANSI/AWS welding symbol
backing weld symbol
basic weld symbol
chain intermittent welding
contour symbol
depth of preparation
dimensions
drafting machine
effective throat
field weld symbol
fillet weld leg size
fillet weld size
finish symbol
groove weld size
intermittent welding
legs
mechanical drawings
melt-through symbol
orthographic projection
other side
plug weld
reference line
root opening
sketches
slot weld
spot weld symbol
staggered intermittent welding
tail
weld-all-around symbol
weld contour symbol
weld size

The Three-View Drawing

Parts and assemblies used in industry are produced from mechanical drawings or sketches. These drawings are called ***mechanical drawings*** because they are made by mechanical means, using drafting tools or computers.

For high-production parts, drawings have traditionally been made with compasses, triangles, and a T-square, or with a ***drafting machine*** (a device attached to a drawing board that combines the functions of T-square and triangle). In recent years, computers running CADD (computer-aided drafting and design) programs have become widely used for making drawings. Low-production or one-of-a-kind parts are often made from ***sketches***—simple drawings made without drafting tools.

These sketches or mechanical drawings must show the shape and size of the part or assembly. All the size information needed to make the part must be given on the drawing or sketch. These sizes are called ***dimensions***. Dimensions may be given in US conventional or SI metric sizes.

Orthographic Projection

To completely describe the shape of a part, it may be necessary to look at it from a number of viewpoints. Almost all industrial drawings are created using a process called orthographic projection.

In ***orthographic projection***, the object to be drawn is treated as if it were inside a clear plastic cube, **Figure 33-1**. All points and lines are "projected" onto the surface of the cube by the viewer's eyes, as shown in **Figure 33-2**. The object is drawn on the cube as it is seen from the front, back, top, bottom, left, or right sides. Thus, the object may be seen or viewed from these six different sides. Normally, no more than three views are required to fully describe the shape of an object. The imaginary cube is then unfolded so it lies flat like a sheet of paper. See **Figure 33-3**.

In reality, of course, industrial drawings are not made on a cube, but rather on flat sheets of paper or plastic. However, the top, front, and side views are in the same positions that they would be on the unfolded cube.

To obtain information from a drawing, you must often follow a line or a point from one view to another. Straight lines have two ends. The ends of these lines may be identified by letters or numbers. When viewed from the end, a line will look like a point or a corner. See **Figure 33-4**.

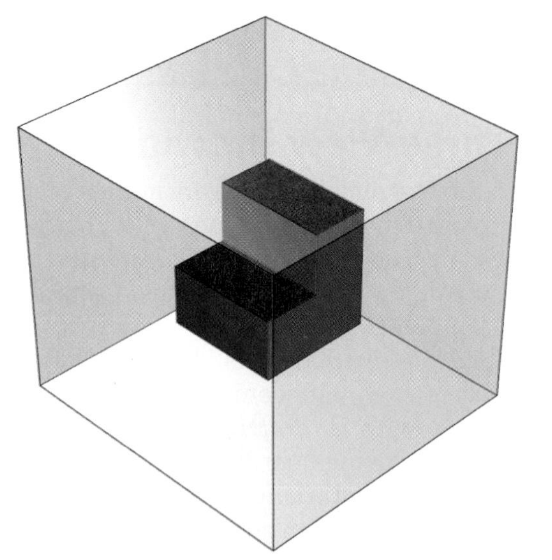

Figure 33-1. *The theory of orthographic projection. A part, such as this step block, is viewed from all sides as if it were in a plastic box. Six different views are possible.*

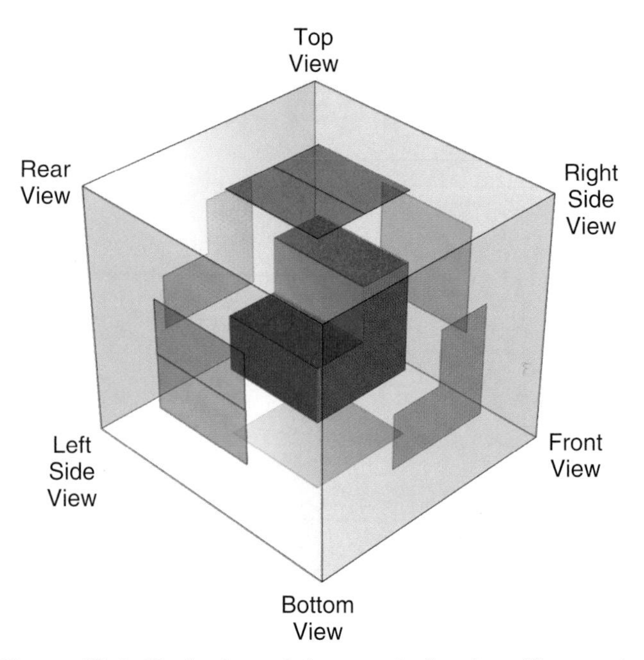

Figure 33-2. *Each view of the part is "projected" onto the cube's surfaces by the person viewing it. The front view is the one that shows the most about the shape of the part. All other views are named in relation to the front view. The right side view is always to the right of the front view.*

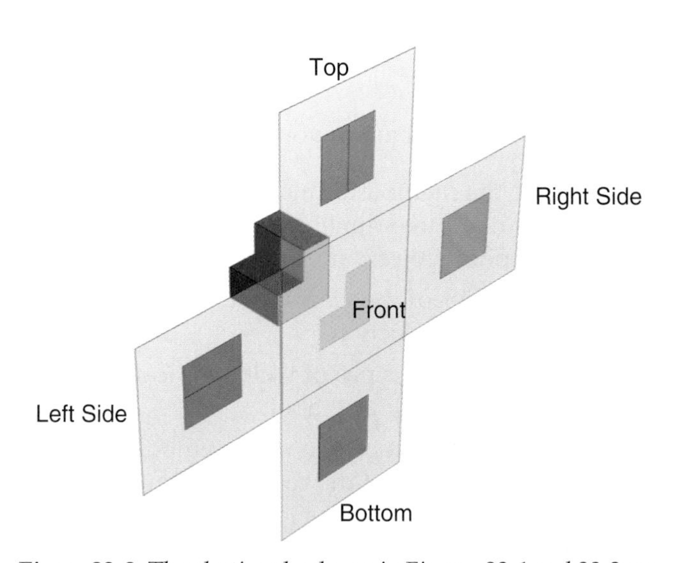

Figure 33-3. *The plastic cube shown in Figures 33-1 and 33-2 can be unfolded so that it lies flat. This is the way the various views of a part will appear on a sheet of paper. The rear view is seldom used.*

Top View

Left Side View

Front View

Figure 33-4. *A three-view orthographic drawing of a wire. This drawing illustrates that a point in one view can appear as a line in two other views.*

Exercise 33-1 Identifying Lines and Points on a Three-View Drawing

On the three-view drawing in this exercise, the lines in the top and front views have been identified with numbers or letters. All points on the upper surfaces have been identified with numbers. All points on the lower surface have been identified with letters.

Note: A point or corner in one view is usually a line in the other two views. For example, line 1,6 in the top view (the line that runs between point 1,a and point 6,d) appears as point 1,6 in the front view. Line 2,5 in the top view (the line between point 2 and point 5) appears as point 2,5 in the front view, and line c,4 in the front view (the line between point b,c and point 3,4) appears as point 4,c in the top view.

On a separate sheet of paper, sketch the right side view as shown. Next, identify all points and lines in this view. The number or letter of the point closest to you must be given first (2,3 has been identified as an example). Show your completed sketch to your instructor for checking.

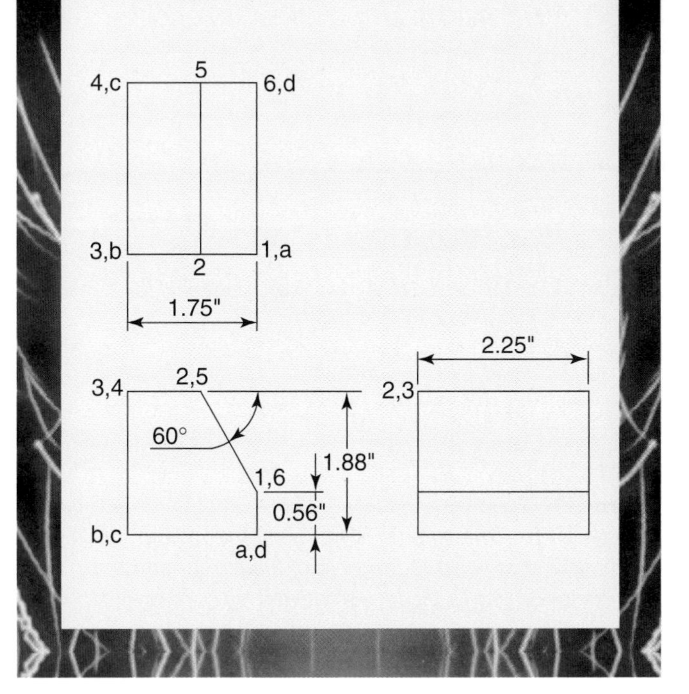

The ANSI/AWS Welding Symbol

Each weld joint must be described in detail. This is done so the welder knows precisely how to make each weld. The American National Standards Institute (ANSI) and the American Welding Society (AWS) have developed a welding symbol to convey all the information needed to properly make the weld on a joint. The ***ANSI/AWS welding symbol*** is used on mechanical drawings or sketches of welded parts. The basic weld joint to be used is shown by the position of the metal parts on the mechanical drawing or sketch. There are five basic weld joints: butt, lap, edge, T-, and corner.

Information given on the welding symbol must always be shown in the designated location on the symbol, as shown in **Figure 33-5**. Important rules for using the ANSI/AWS welding symbol will be shown in *italic* type in this chapter. For example: *Each time the weld joint changes direction, a new welding symbol must be drawn. There are a few exceptions to this rule. They will be described later.*

The Reference Line and Arrow

*The **reference line** is always drawn horizontally.* See **Figure 33-6**. On the drawing, it is placed as close as possible to the weld joint it describes. All other information is placed above or below this horizontal reference line, as shown in Figure 33-5.

An arrow may be drawn from either end of the reference line. The arrowhead must touch the line to be welded, Figure 33-6. The arrowhead may be shown touching the weld line in any of the views. However, the arrowhead usually touches the weld line in a view where the weld line appears as a point or corner.

Some joints, such as the bevel, J, and flare-bevel groove, have only one edge prepared for welding. For these welds, the arrow is bent to point at the edge that is cut, ground, or bent in preparation for welding. See **Figure 33-7**. When the same preparation is used for both edges, the arrow is not bent. The ANSI/AWS welding symbol may appear in any view, but must appear once for each weld joint.

The Tail

The ***tail*** is added to the symbol only when special notes are required. See Figure 33-6. A number or letter code inside the tail directs the welder to special notes located elsewhere on the drawing. These notes may specify the heat treatment, welding process used, or other information not given on the welding symbol.

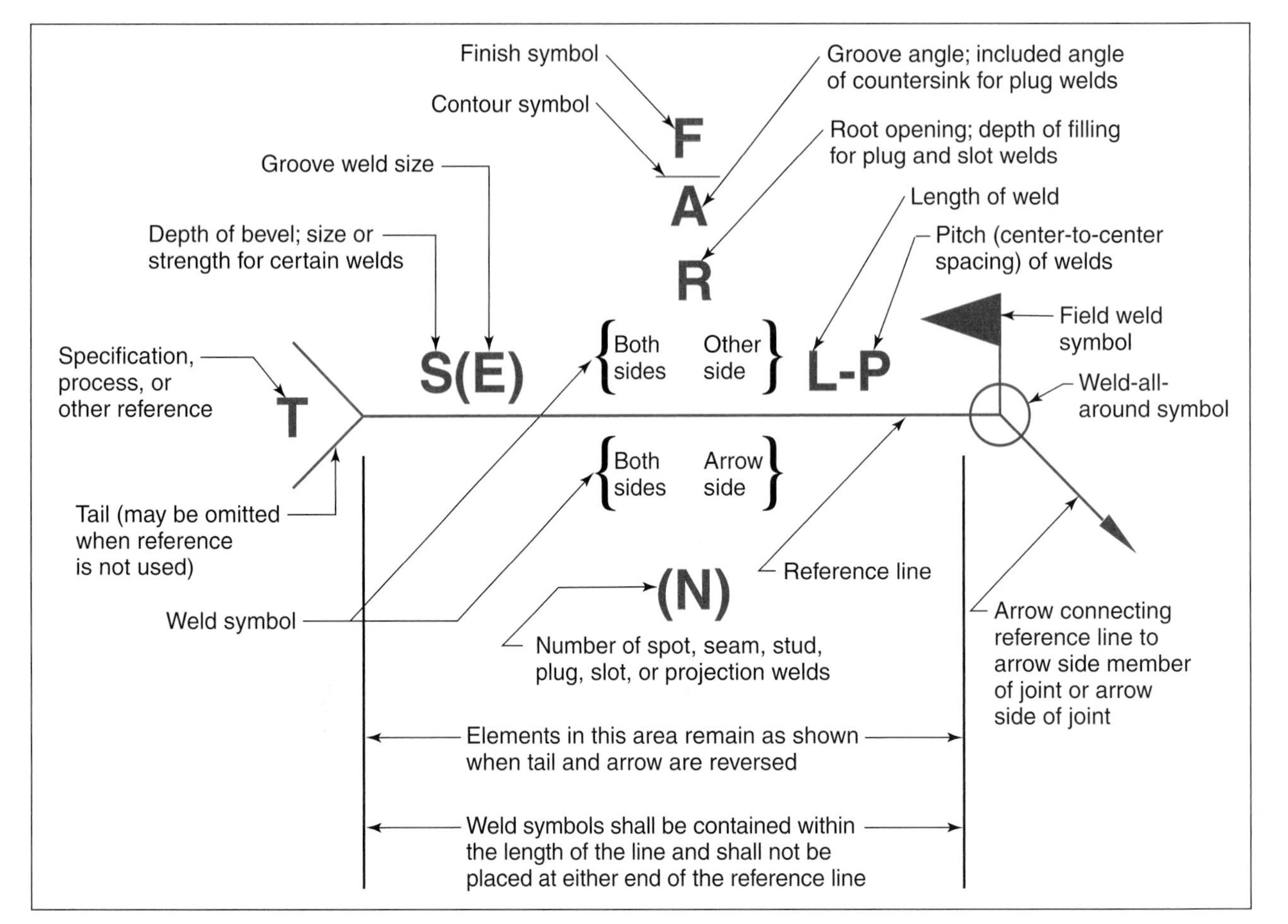

Figure 33-5. *The ANSI/AWS welding symbol. The standard locations for information on the welding symbol are marked and explained. (AWS A2.4:2007, Figure 3, Standard Location of the Elements of a Welding Symbol, reproduced with permission from the American Welding Society, Miami, FL)*

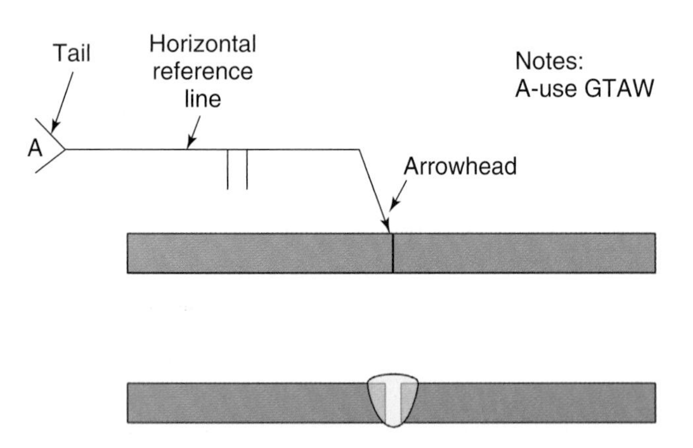

Figure 33-6. *The arrow, reference line, and tail. The arrowhead must touch the line to be welded, usually in the view where it appears as a point. The reference line is always drawn horizontally. The tail may be used to hold a symbol for a note elsewhere on the drawing.*

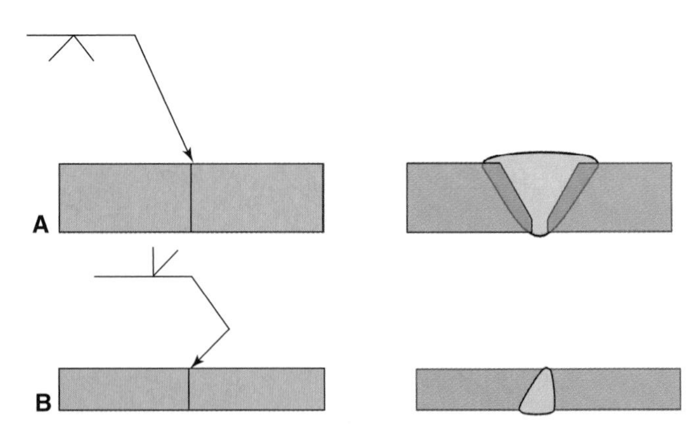

Figure 33-7. *Groove welds. A—The symbol is for a V-groove butt weld. B—The symbol is for bevel-groove butt weld. The arrow is bent to point to the piece of metal that will be beveled. The arrow is bent to the left, so the left piece is beveled.*

Basic Weld Symbols

The ***basic weld symbol*** describes the type of weld to be made. This symbol is a miniature drawing of the metal's edge preparation prior to welding. See **Figure 33-8**. The basic weld symbol is only part of the entire ANSI/AWS welding symbol. **Figure 33-9** illustrates the basic weld symbols used by the AWS.

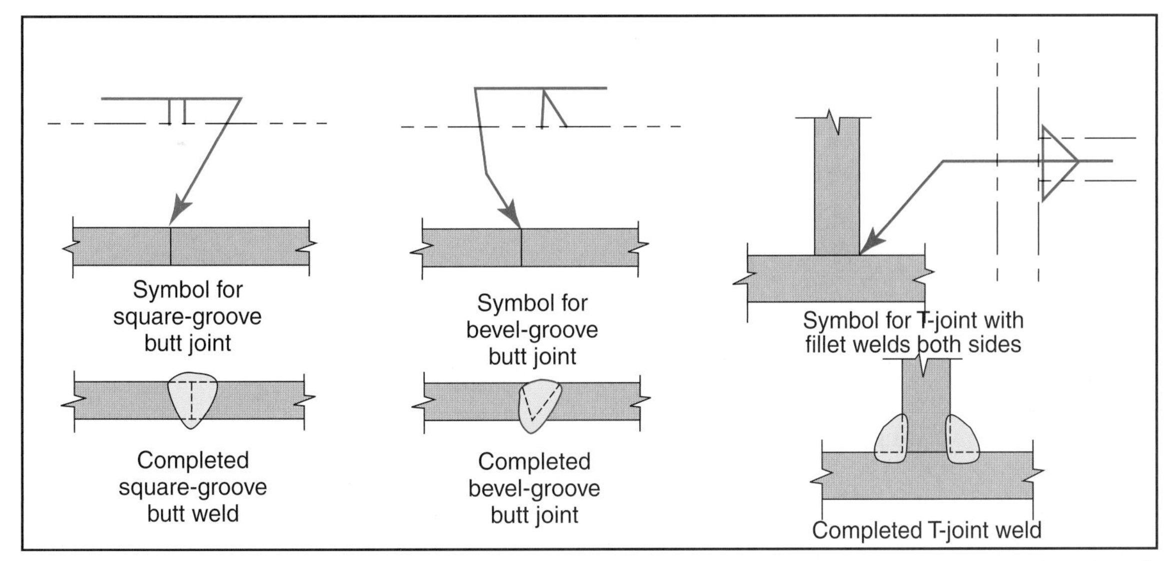

Figure 33-8. *Weld symbols. The weld symbol is a miniature drawing of the joint prior to welding. Phantom lines have been added to these drawings to show the joint better. They are not used on a real welding symbol. The vertical line in the bevel-groove and fillet weld symbols is always drawn at the left-hand side of the weld symbol.*

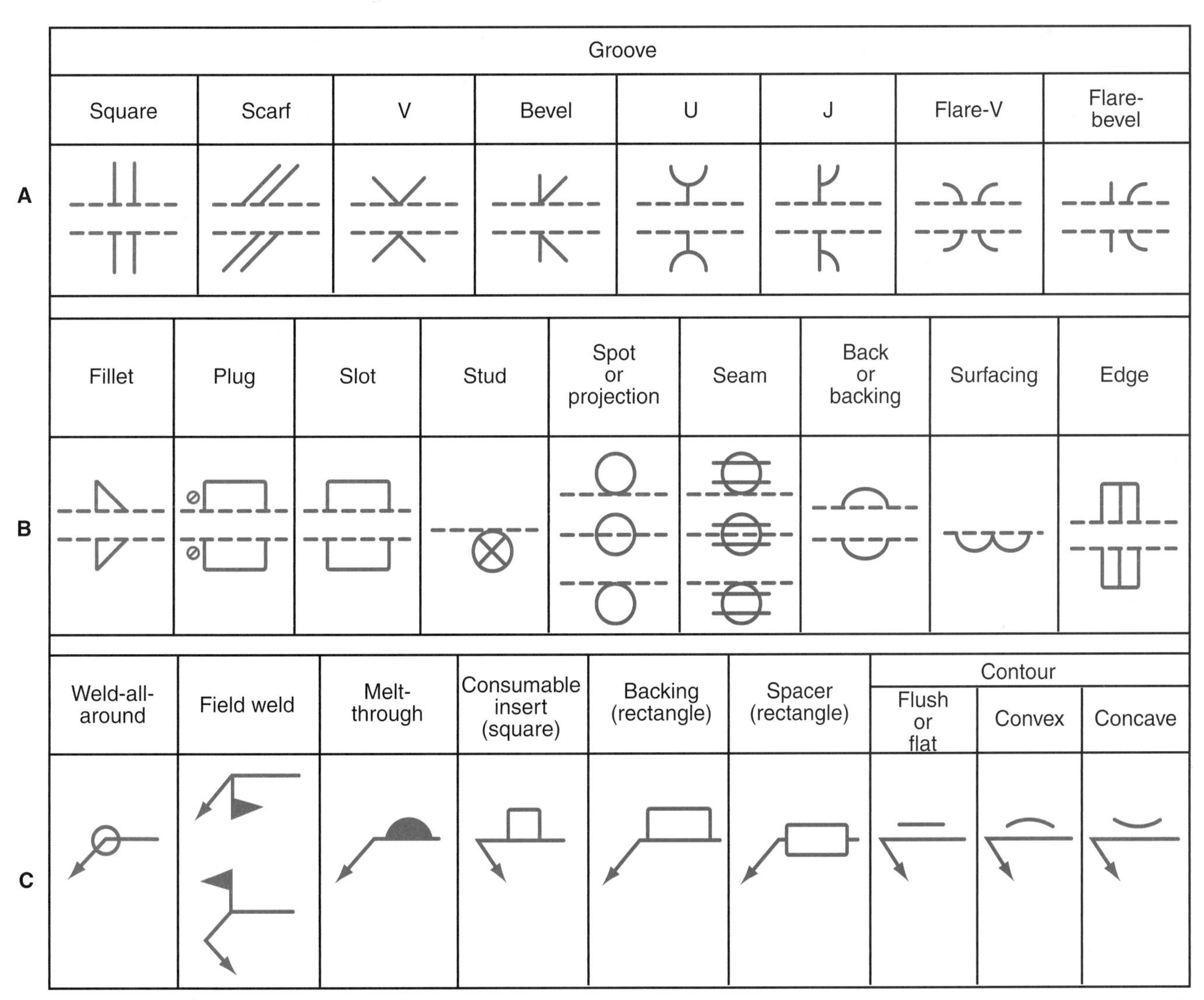

Figure 33-9. *A & B—Basic weld symbols. The basic weld symbol is only a part of the welding symbol. C—Supplemental symbols used on the ANSI/AWS welding symbol. (AWS A2.4:2007, Figures 1 and 2, Weld Symbols and Supplementary Symbols, reproduced with permission from the American Welding Society, Miami, FL)*

The Arrow Side and Other Side

On the drawing, the arrowhead of the welding symbol always touches the line or joint to be welded. The joint, however, has two sides. A weld may be needed on one side or the other. In some cases, both sides of a joint are welded.

*The side of the joint that the arrow touches is always called the **arrow side**. The opposite side of the joint is called the **other side**.*

On the welding symbol, weld information for the arrow side is shown below the horizontal reference line. Weld information for the other side is shown above the reference line. See **Figure 33-10** for examples of arrow side and other side information.

Root Opening and Groove Angle

The ***root opening*** is the space between the base metal pieces at the root (bottom) of the joint. The root opening is held constant in a jig or fixture, or by tack welding. On the welding symbol, the size of the root opening may be given in fractions of an inch, metric units, or decimals. Root opening size is given inside the basic weld symbol. See **Figure 33-11**.

The edge of the metal for a groove-type weld is flame or arc cut, or machined to a specific angle before welding. The angle at which the metal is cut is shown on the welding symbol. If a V-groove weld is being made, the angle cut on each piece is half that shown on the welding symbol. The total angle of the groove-type welded joint is shown just beyond the outer edge of the basic weld symbol. See Figure 33-11.

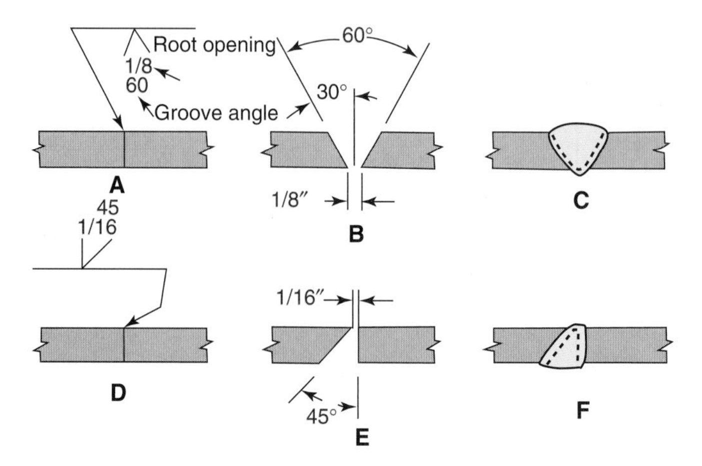

Figure 33-11. *Groove weld preparation. The root opening size is given inside the weld symbol. The groove angle is shown just outside the weld symbol. A—V-groove welding symbol. B—The joint preparation. C—The completed V-groove weld. D—Bevel-groove joint symbol. E—Joint preparation. F—The finished weld.*

Weld Contour Symbol

To indicate the desired contour for the completed weld, a curved or straight line is used on the welding symbol. The contour (shape) of the face of the completed weld may be convex (curved outward), flat, or concave (curved inward).

The contour line is drawn between the basic weld symbol and the finish symbol, as shown in **Figure 33-12**. If no ***contour symbol*** is shown, the weld face should have the normal convex shape.

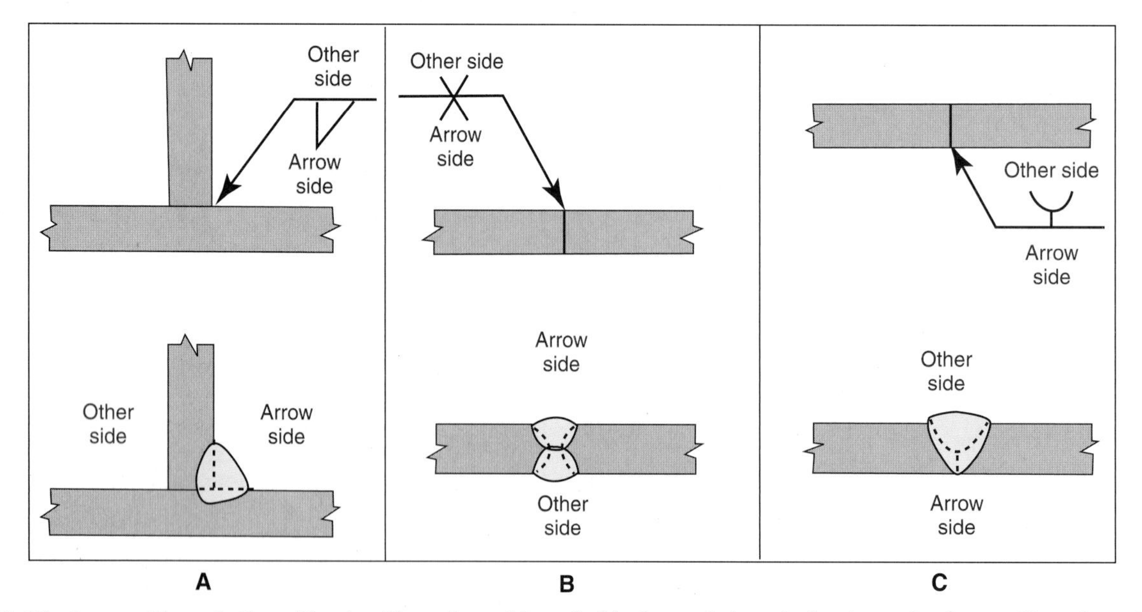

Figure 33-10. *Arrow side and other side. A—Since the weld symbol is drawn below the horizontal reference line, the weld is made on the arrow side. B—Welds are made on both sides of the joint. C—The weld symbol is above the reference line, so the weld is made on the side opposite the arrow.*

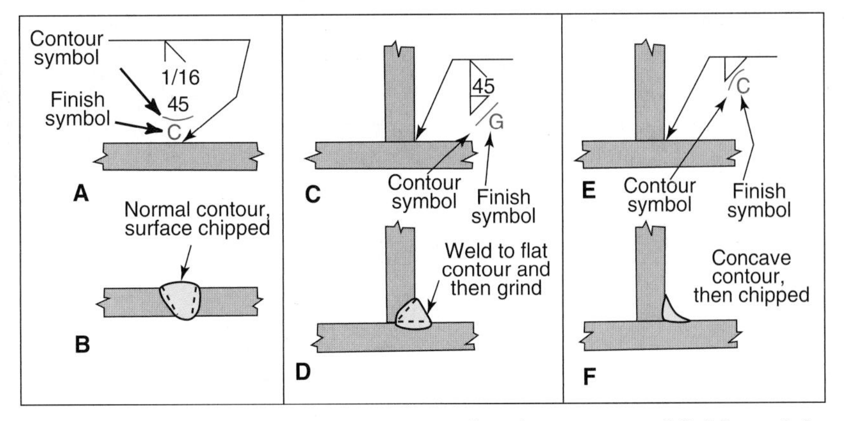

Figure 33-12. Weld contour symbols and finish symbols. In A, C, and E, the contour and finish symbols are shown on the welding symbol. B, D, and F show the shape and finish of the completed weld faces.

Finish Symbol

If the weld face is not to be left "as welded," a symbol designating the required finish is shown on the welding symbol. The ANSI/AWS A2.4 lists the following methods of finishing a completed weld:

- C (chipping).
- G (grinding).
- M (machining).
- R (rolling).
- H (hammering).

The ***finish symbol*** is a symbol placed beyond the contour symbol, designating the finish of the weld, **Figure 33-12**. Users of finish symbols may create their own symbols.

Sometimes, all welds are to be finished in the same manner. In this case, a general note may be placed on the drawing. This avoids the need for a finish symbol on each welding symbol.

Depth of Preparation and Groove Weld Size

The ***depth of preparation*** is given in the "S" position on the welding symbol for groove-type welds. See **Figure 33-13**. Weld strength, or weld size for welds other than a groove weld, may also be given in this same location. Edges of metal over 1/4" (6.4 mm) thick are always prepared in some way before making a groove weld, in order to ensure 100% penetration. The ***groove weld size*** is the depth to which a weld penetrates the joint from the surface of the base metal. See Figure 33-13. The desired depth of preparation and the depth of the weld penetration are generally determined by codes or specifications, or by a welding engineer. Sometimes, experimentation is needed to determine the correct groove weld size and depth of preparation.

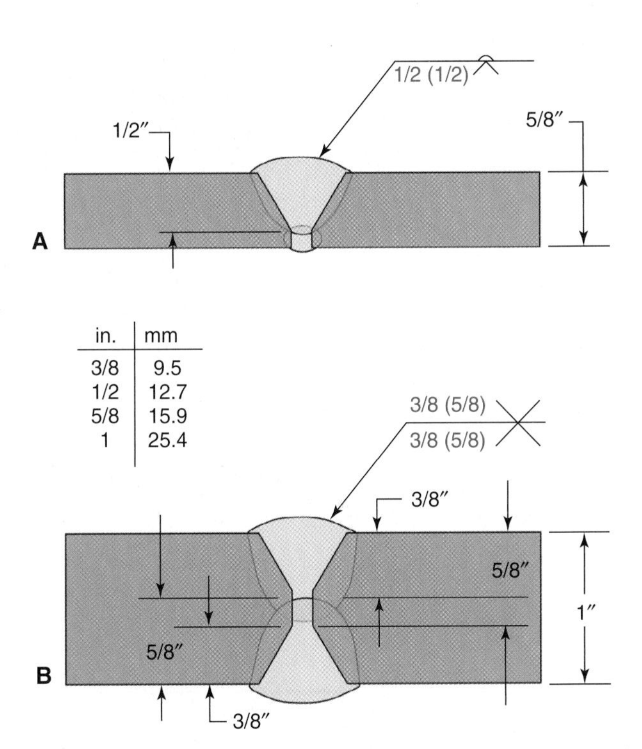

in.	mm
3/8	9.5
1/2	12.7
5/8	15.9
1	25.4

Figure 33-13. Groove weld size and depth of preparation. At A, the groove weld size is the same as the depth of preparation, 1/2" (12.7 mm). At B, the depth of preparation for both welds is 3/8" (9.5 mm), but the groove weld size or depth of penetration is 5/8" (15.9 mm).

Fillet Weld Size

The ***fillet weld size*** is the length of the **legs** (sides) of the triangle that can be drawn inside the cross section of the finished weld. The ***fillet weld leg size*** is the distance from the surface of the base metal to the toe of the fillet weld. On a convex weld bead, the fillet weld size and the fillet weld leg size are the same. However, on a concave weld bead, the fillet weld size is smaller than the fillet weld leg size. See **Figure 33-14**. The length of the fillet weld legs is given in the "S" position of the weld symbol, where the depth of penetration is given for a groove-type weld. See **Figure 33-15**.

A typical fillet weld triangle has legs of equal length. Only one dimension is given for a fillet weld with equal legs. When the legs are unequal in length, the position of the long and short leg are *not* shown by the weld symbol. A standard fillet weld symbol is used. *The positioning of the unequal legs is shown on the drawing of the part.* The size of the two legs of the fillet weld appear to the left of the fillet weld symbol as two dimensions, such as 1/4″ × 1/2″ or 6.4 mm × 12.7 mm.

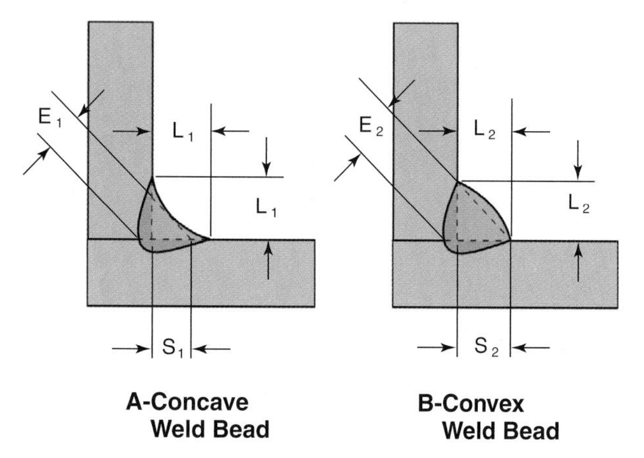

Figure 33-14. *Fillet weld sizes and strengths. The fillet weld leg size (L_1 and L_2) is the distance from the surface of the base metal to the weld toe. The fillet weld size (S_1 and S_2) is the length of the triangle that can be drawn within the completed weld. Note that the triangle in the convex weld bead is larger and therefore stronger than the triangle in the concave weld bead. Note also that the effective throat E_2 is larger than E_1.*

Effective throat is the distance, minus any convexity, between the weld face and the weld root. *The effective throat can never be greater than the metal thickness.* Refer to Figure 33-14. *The effective throat dimension, when used, appears in parentheses to the left of the basic weld symbol and to the right of the fillet weld size.* See **Figure 33-16**. Fillet weld leg size and effective throat size may be shown in fractions of an inch, in decimals, or in SI units.

Intermittent Welding

On many joints, it is not necessary to weld continuously from beginning to end. Where strength is not affected, and to save time and money, the joint may be

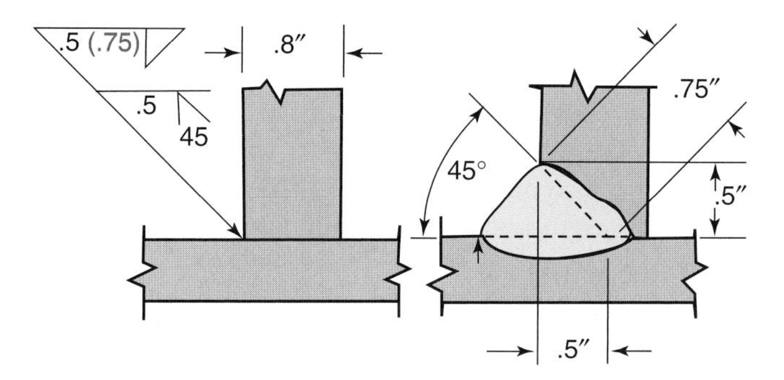

Figure 33-16. *Effective throat size. Effective throat size, or the depth of the weld penetration, is shown in parentheses on the welding symbol. The depth of preparation of the groove weld is .5″ (12.7 mm) and the effective throat of the fillet weld is .75″ (19 mm).*

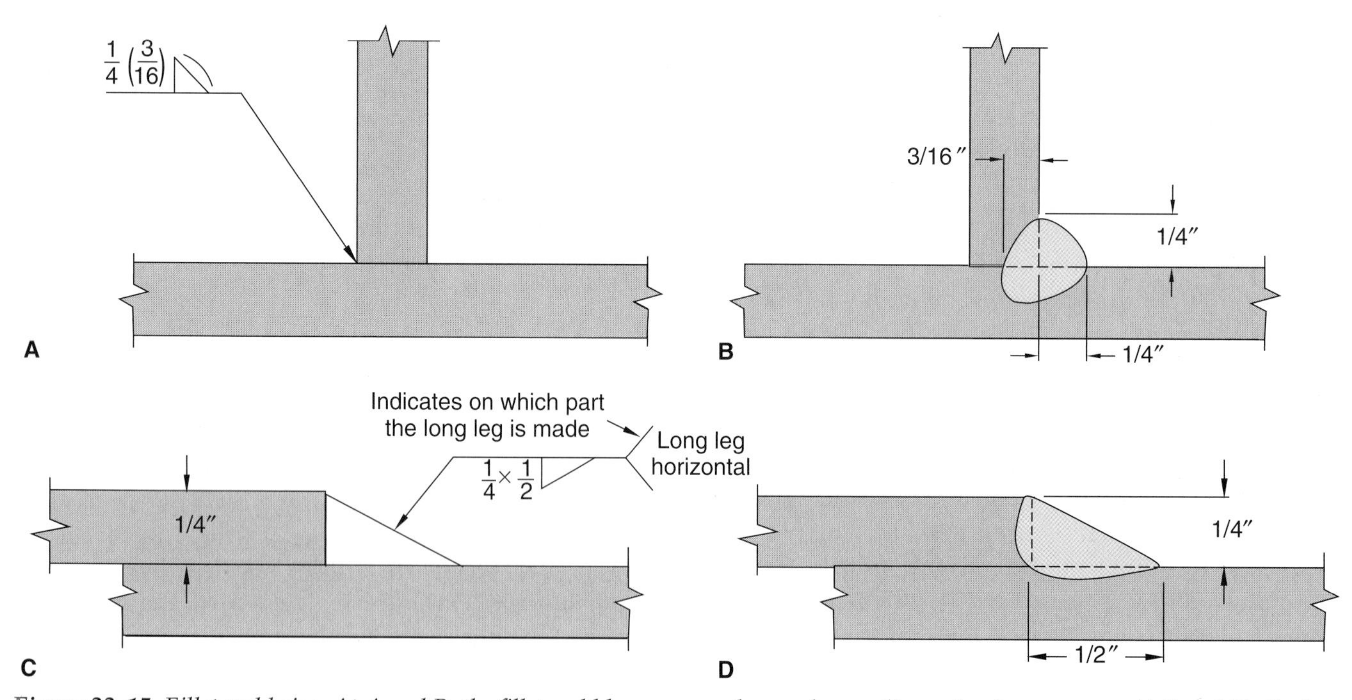

Figure 33-15. *Fillet weld size. At A and B, the fillet weld legs are equal, so only one dimension is necessary. At C and D, the legs are unequal so two dimensions are given. The weld orientation may be shown on the working drawing or as a note in the tail of the weld symbol. Both methods of showing the weld orientation are shown here only as an example.*

made with short sections of weld. See **Figure 33-17**. This type of welding is called ***intermittent welding***.

When intermittent welding is done, two dimensions are given to the right of the basic weld symbol. The first is the length of each weld. The second is the pitch of the weld. This is the distance from the center of one weld to the center of the next weld, as shown in Figure 33-17.

Intermittent welds may be made on both sides of a joint. When ***chain intermittent welding*** is done, the welds on one side of the joint start and stop at the same places as the welds on the other side of the plate. The welds exactly match on the two sides of the joint.

When ***staggered intermittent welding*** is done, the welds do not line up on each side. The welds on one side are staggered, or offset, from the welds on the other side of the joint. See Figure 33-17.

Melt-Through and Backing Weld Symbols

A ***melt-through symbol*** is used when 100% penetration is required. This symbol is used on joints that are welded from one side only. See **Figure 33-18**. The height of the root reinforcement may be placed to the left of the melt-through symbol.

The ***backing weld symbol*** is used when a stringer bead is laid on the root side of the weld to ensure complete penetration. See Figure 33-18.

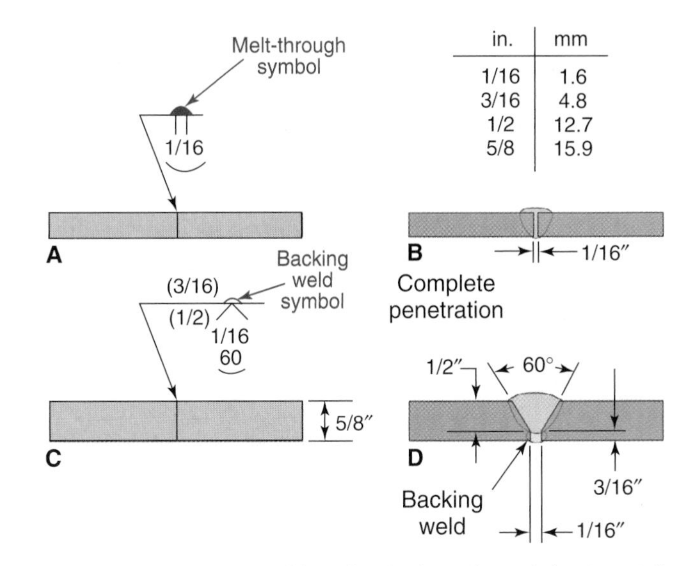

Figure 33-18. Backing welds and melt-through symbols. A,B—The melt-through symbol is used to ensure 100% penetration when welding from one side only. C,D—A backing weld may be used to ensure 100% penetration when welding is possible on both sides of a joint.

Weld-All-Around and Field Weld Symbols

Instructions on a welding symbol are in effect only until the weld joint makes a sharp change in direction. When a joint changes direction at a corner, a new welding symbol generally must be used.

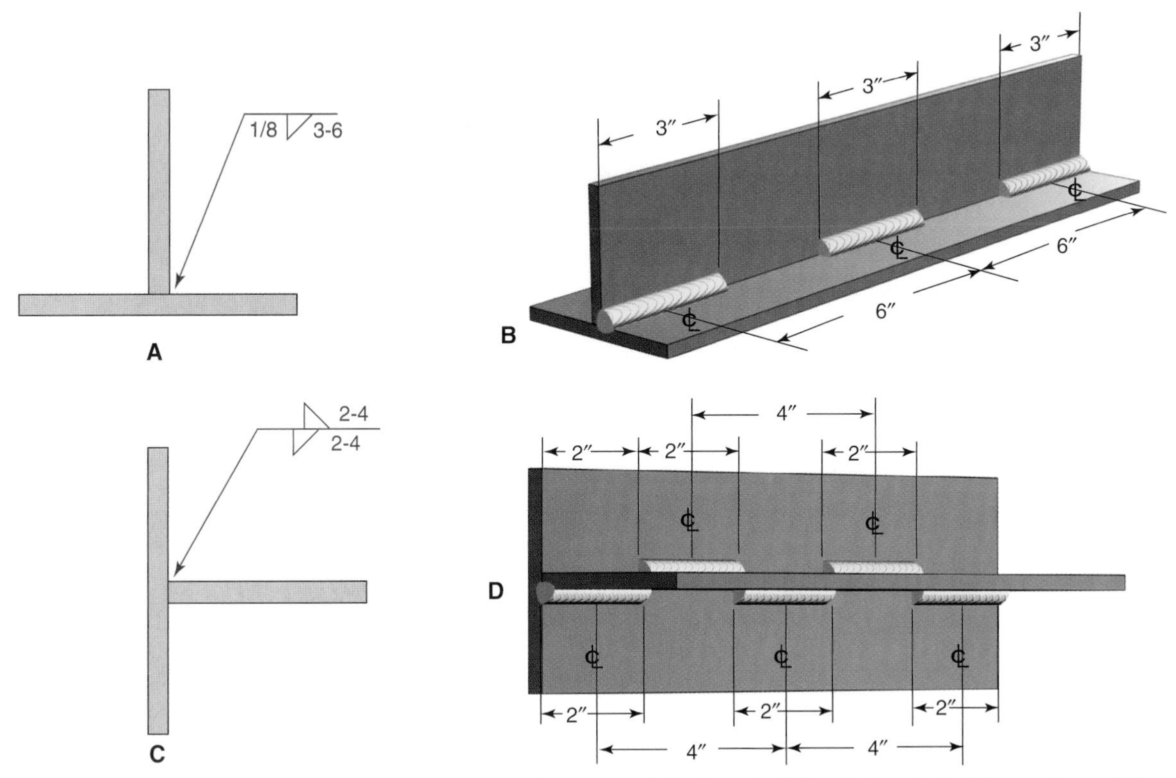

Figure 33-17. Length and pitch of weld. A—Note the placement of the length (3″)and the pitch (6″) on the welding symbol. B—The drawing shows a series of 3″ (76.2 mm) long welds that are 6″ (152.4 mm) apart from center to center. C,D—A staggered weld. Note the staggered fillet symbol at C.

When the same weld is made on all edges of a continuous joint, however, a *weld-all-around symbol* may be used. A continuous joint may occur on a box or cylindrical part. See **Figure 33-19**.

Parts joined by welding are called weldments. Some parts of a weldment may be welded in the manufacturer's shop. On large weldments like bridges and buildings, some welds must be made on the site, or "in the field."

A *field weld symbol* is used when welds are to be made at the construction site. See Figure 33-19.

Multiple Reference Lines

When several operations are to be performed on a welded joint, a corresponding number of reference lines may be used. The reference line nearest the arrowhead indicates the first operation. The last operation on the welded joint is shown on the reference line most distant from the arrowhead. See **Figures 33-20A** and **33-20B**. When only one reference line is used, the operation nearest the reference line is performed first. See **Figures 33-20C** and **33-20D**.

Plug and Slot Welds

Sometimes, two pieces must be welded together at a point away from the edge. This is done by cutting a hole through one piece and welding the pieces together through this hole. See **Figure 33-21**. The hole is often round, but may be any shape. These holes may be drilled, flame-cut, arc-cut, or machined.

A *plug weld* is made through a round hole. A *slot weld* is made through a hole that is not round. The symbol for a plug weld is shown in **Figure 33-22**. The diameter of the plug weld is shown to the left of the basic weld symbol. If the sides of the hole are angled, the total angle is shown below the basic weld symbol. See Figure 33-22. The depth of the weld is shown inside the symbol. If no depth dimension is given for a plug or slot weld, the hole is completely filled.

For a slot weld, the drawing shows the length and width, and the angle of the sides. The drawing will indicate the location and spacing of the holes for both plug and slot welds. If a series of plug or slot welds is to be made, the center-to-center distance of the holes may be shown to the right of the basic weld symbol.

Spot Welds

A spot weld may be accomplished using resistance welding, gas tungsten arc welding, electron beam welding, or ultrasonic welding.

The *spot weld symbol* is a circle 1/4" (6.0 mm) in diameter. The circle may be on either side of the reference line or it may straddle the reference line. If the weld is accomplished from the arrow side, the spot weld symbol should be below the reference line, as in all other welding symbols. If the welding is done on

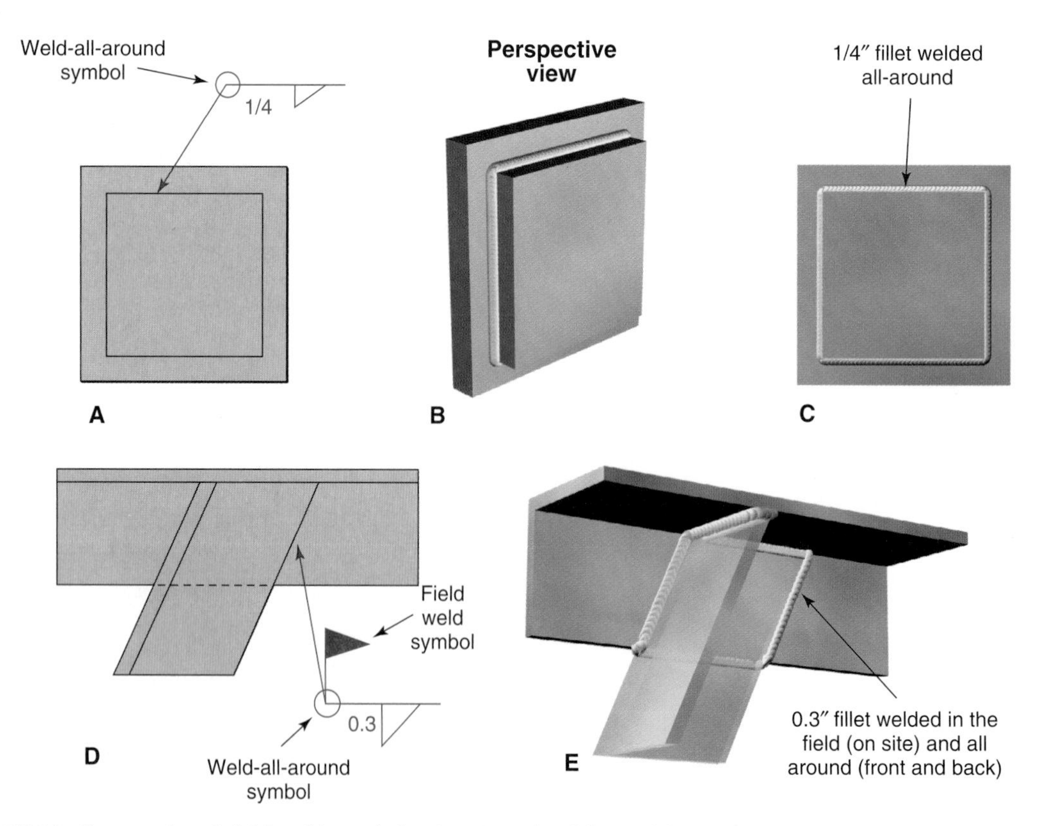

Figure 33-19. Weld-all-around and field weld symbols. A,B,C—This fillet weld is welded all around in the shop. D,E—This 0.3" (8.5 mm) fillet weld is made in the field. It is welded all around the angle iron, both front and back.

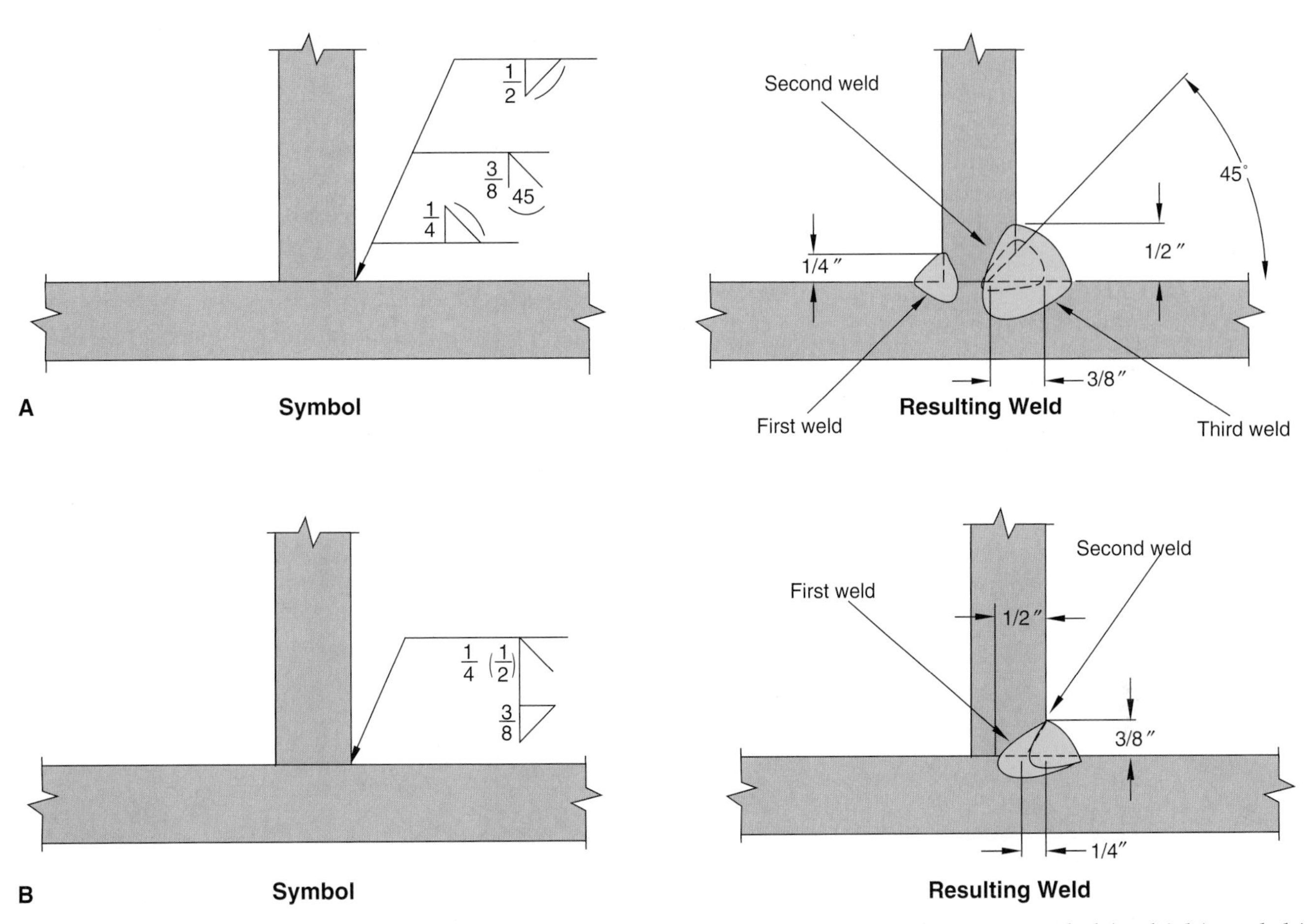

Figure 33-20. A—Multiple reference lines. When several operations are done on the same joint, a stacked (multiple) symbol is used. The weld nearest the arrow is made first. B—When the symbol has only one reference line; the symbol nearest the reference line is used first.

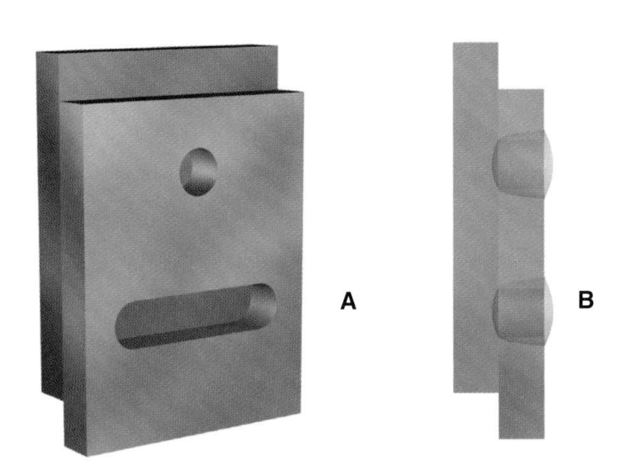

Figure 33-21. Plug and slot welds. The plug weld is made in a round hole. The slot weld is made in a hole that is not round. A—The hole and slot are shown as cut in one piece. B—Cross section of the completed welds.

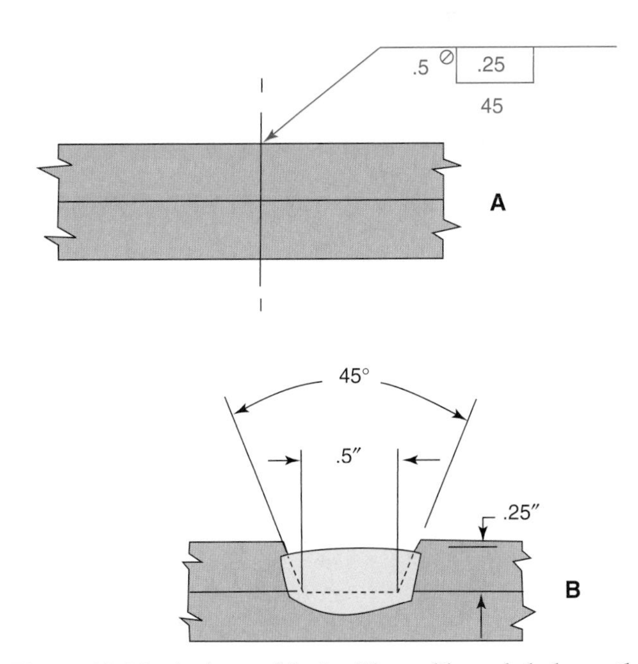

Figure 33-22. A plug weld. A—The weld symbol shows the depth of the weld, the hole diameter, and the angle of the hole (if it is not straight). B—A cross section view of the weld described in A. Note: If no dimension is given inside the weld symbol, the plug or slot weld hole is completely filled with filler metal.

both sides, as in resistance spot welding, the circle straddles the reference line. See **Figure 33-23**.

Projection welding is another form of resistance welding used to produce spot welds. To indicate which piece has the projections on it, the circle is placed above or below the reference line.

Information given for a spot weld includes size or strength, spacing, and number of spot welds. The ***weld size*** or desired strength is given to the left of the weld symbol. The weld size may be given in US customary units or SI metric units. Weld strength is shown in pounds or newtons per spot. The weld spacing is found to the right of welding symbol. Centered above or below the spot welding symbol, in parentheses, is the number of welds desired.

The welding process used is shown in the tail of the welding symbol. See **Figure 33-24** for examples of the welding symbols used for spot welding. The welding symbol may be placed in any view of a drawing.

Seam Welds

Seam welds may be made with a number of processes, such as electron beam, resistance, or gas tungsten arc.

The process used is shown in the tail. The size (width) of the weld, or strength of the weld, are shown to the left of the weld symbol. The strength is given in pounds per linear inch or in newtons per millimeter.

The weld symbol may straddle the reference line if welded from both sides, as in resistance seam welding. For electron beam and gas tungsten arc welding, the symbol is placed above or below the reference line. This indicates from which side of the part the weld is made. The length of the seam may be shown to the right of the weld symbol. See **Figure 33-25**. For more details, see ANSI/AWS publication A2.4.

A periodic review of this chapter as you move through the material covered in this book and the laboratory manual will help build your print-reading ability.

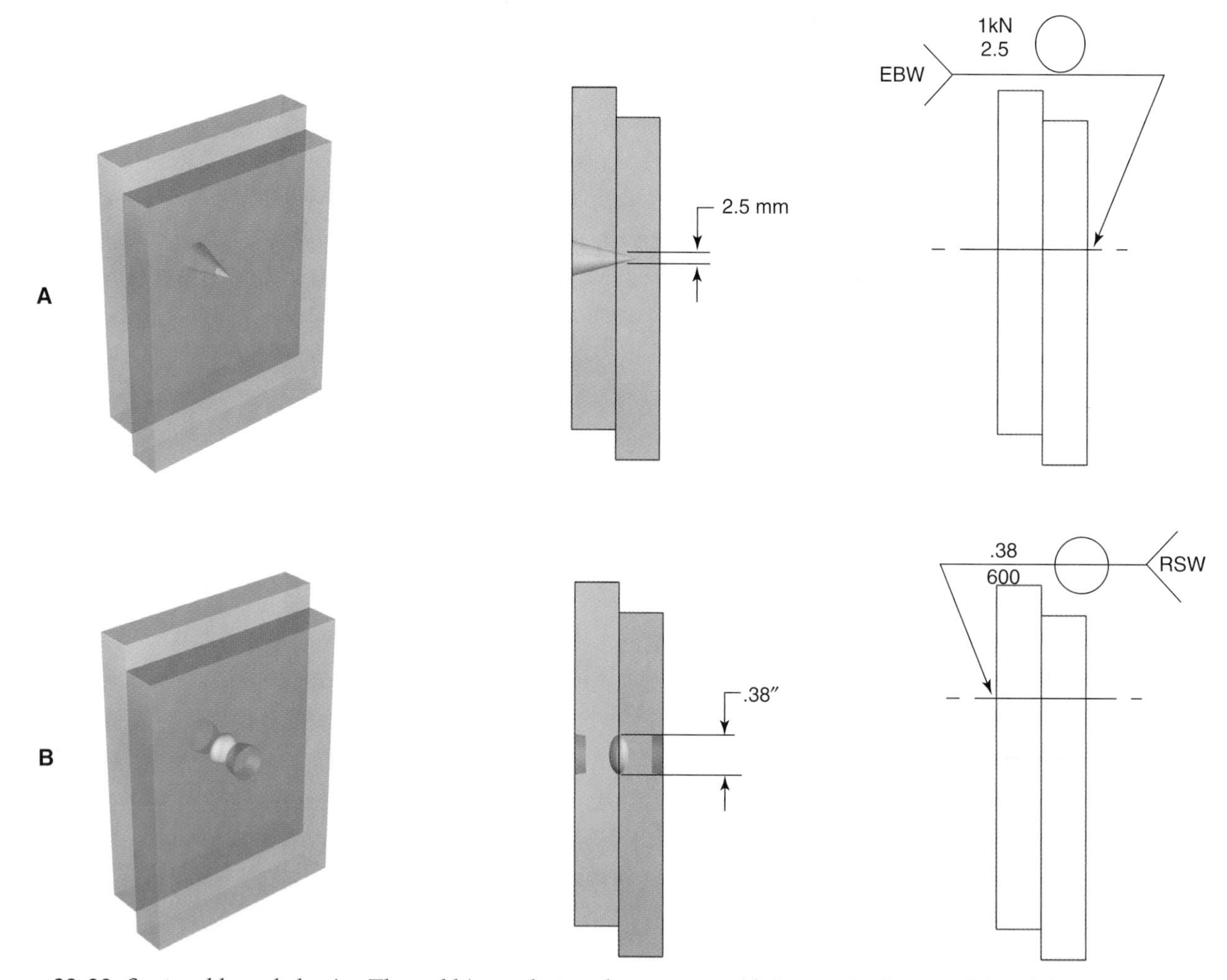

Figure 33-23. Spot weld symbols. A—The weld is an electron beam spot weld. Its required strength is 1 kilonewton. The weld is made from the other side. B—The weld is a resistance spot weld. Its size is .38″ (9.5 mm). Its required strength is 600 pounds/weld. The weld is made from both sides, so the symbol straddles the reference line.

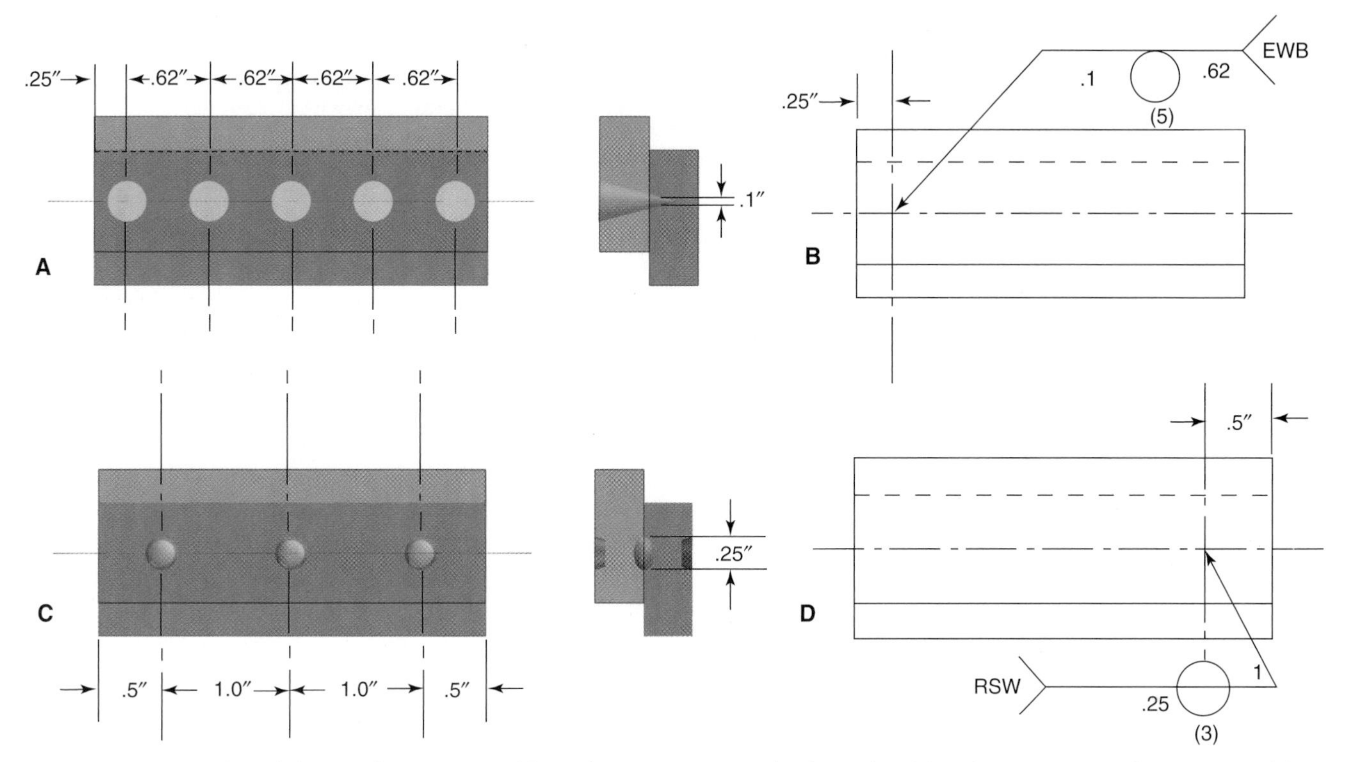

Figure 33-24. A series of electron beam spot welds is shown at A. A—The desired weld and spacing. B—The correct weldment drawing, symbol, and dimensions. C—A resistance spot-welded part. D—The weldment drawing for this part.

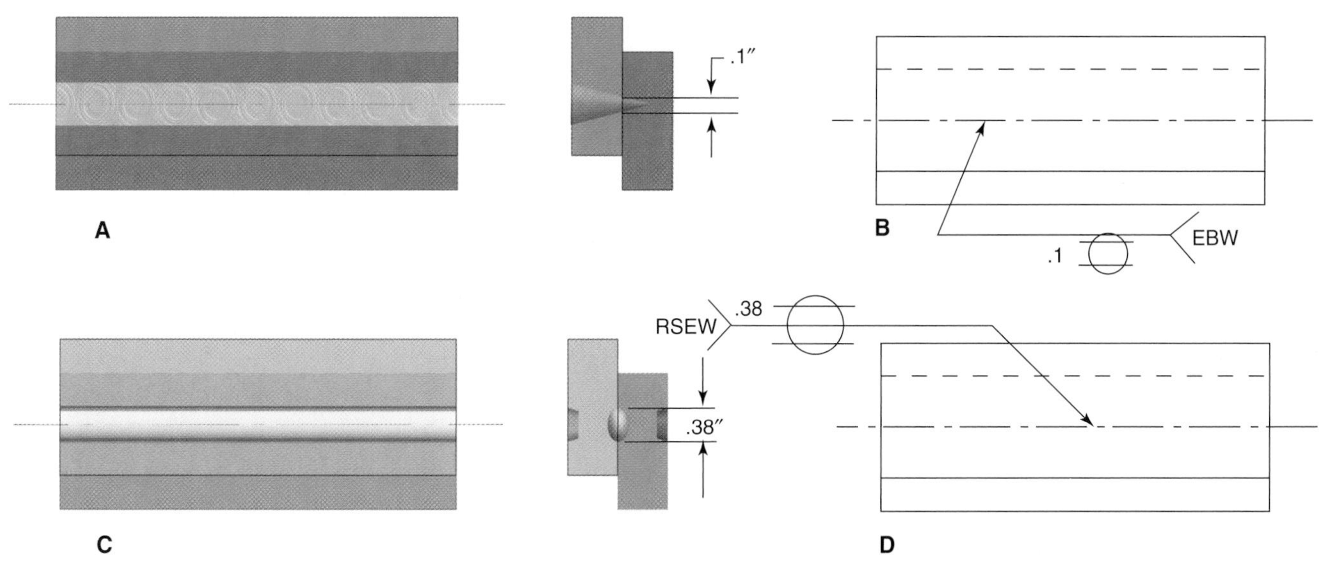

Figure 33-25. Seam welds. A—A seam weld made with the electron beam. Its size at the fusion point is .1″ (2.5 mm). B—The weld symbol and weldment drawing for the seam weld shown at A. C—The resistance welded seam on this part results from the weldment drawing and welding symbol shown at D.

Summary

- Parts and assemblies used in industry are produced from mechanical drawings or sketches. These drawings are called mechanical drawings because they are made by mechanical means, using drafting tools or computers.
- In orthographic projection, the object to be drawn is treated as if it were inside a clear plastic cube. All points and lines are "projected" onto the surface of the cube by the viewer's eyes. The object is drawn as it is seen from the front, back, top, bottom, left, or right sides of the cube. Thus, the object may be seen or viewed from these six different sides. Normally, no more than three views are required to fully describe the shape of an object.
- The American National Standards Institute (ANSI) and the American Welding Society (AWS) have developed a welding symbol to convey all the information needed to properly make the weld on a joint. Information given on the welding symbol must always be shown in the designated location on the symbol. The ANSI/AWS welding symbol is used on mechanical drawings or sketches of welded parts. The basic weld joint to be used is shown by the position of the metal parts on the mechanical drawing or sketch.
- The reference line is always drawn horizontally. On the drawing, it is placed as close as possible to the weld joint it describes. All other information is placed above or below this horizontal reference line. The ANSI/AWS welding symbol may appear in any view, but must appear once for each weld joint.
- An arrow may be drawn from either end of the reference line. The arrowhead must touch the line to be welded. The arrowhead may be shown touching the weld line in any of the views. However, the arrowhead usually touches the weld line in a view where the weld line appears as a point or corner.
- Some joints, such as the bevel, J, and flare-bevel groove, have only one edge prepared for welding. For these welds, the arrow is bent to point at the edge that is cut, ground, or bent in preparation for welding. When the same preparation is used for both edges, the arrow is not bent.
- The tail is added to the symbol only when special notes are required. A number or letter code inside the tail directs the welder to special notes located elsewhere on the drawing. These notes may specify the heat treatment, welding process used, or other information not given on the welding symbol.
- The side of the joint that the arrow touches is always called the arrow side. The opposite side of the joint is called the other side. On the welding symbol, weld information for the arrow side is shown below the horizontal reference line. Weld information for the other side is shown above the reference line.
- On the welding symbol, the size of the root opening may be given in fractions of an inch, metric units, or decimals. The root opening size is given inside the basic weld symbol.
- To indicate the desired contour for the completed weld, a curved or straight line is used on the welding symbol. The contour of the face of the completed weld may be convex, flat, or concave. The contour line is drawn between the basic weld symbol and the finish symbol. If no contour symbol is shown, the weld face should have the normal convex shape.
- If the weld face is not to be left "as welded," a symbol designating the required finish is shown on the welding symbol. The finish symbol is a symbol placed beyond the contour symbol, designating the finish of the weld. Users of finish symbols may create their own symbols.
- The depth of preparation is given in the "S" position on the welding symbol for groove-type welds. Weld strength, or weld size for welds other than a groove weld, may also be given in this same location.
- The length of the fillet weld legs is given in the "S" position of the weld symbol, where the depth of penetration is given for a groove-type weld. Only one dimension is given for a fillet weld with equal legs. When the legs are unequal in length, the position of the long and short leg are *not* usually shown by the weld symbol. A standard fillet weld symbol is used. The positioning of the unequal legs is shown on the drawing of the part. The size of the two legs of the fillet weld appear to the left of the fillet weld symbol as two dimensions, such as 1/4″ × 1/2″ (6.4 mm × 12.7 mm). The long and short legs may be identified in the tail of the welding symbol.

(continued)

- Effective throat is the distance, minus any convexity, between the weld face and the weld root. The effective throat dimension, when used, appears in parentheses to the left of the basic weld symbol and to the right of the fillet weld size. Fillet weld leg size and effective throat size may be shown in fractions of an inch, in decimals, or in SI units.
- When intermittent welding is to be done, two dimensions are given to the right of the basic weld symbol. The first is the length of each weld. The second is the pitch of the weld. Intermittent welds may be made on both sides of a joint. When chain intermittent welding is done, the welds on one side of the joint start and stop at the same places as the welds on the other side of the plate. The welds exactly match on the two sides of the joint. When staggered intermittent welding is done, the welds on one side are staggered, or offset, from the welds on the other side of the joint.
- A melt-through symbol is used when 100% penetration is required. This symbol is used on joints that are welded from one side only. The height of the root reinforcement may be placed to the left of the melt-through symbol. The backing weld symbol is used when a stringer bead is laid on the root side of the weld to ensure complete penetration.
- Instructions on a welding symbol are in effect only until the weld joint makes a sharp change in direction. When a joint changes direction at a corner, a new welding symbol generally must be used. When the same weld is made on all edges of a continuous joint, however, a weld-all-around symbol may be used.
- A field weld symbol is used when welds are to be made at the construction site.
- When several operations are to be performed on a welded joint, a corresponding number of reference lines may be used. The reference line nearest the arrowhead indicates the first operation. The last operation on the welded joint is shown on the reference line most distant from the arrowhead. When only one reference line is used, the operation nearest the reference line is performed first.
- A plug weld is made through a round hole. A slot weld is made through a hole that is not round. The diameter of the plug weld is shown to the left of the basic weld symbol. If the sides of the hole are angled, the total angle is shown below the basic weld symbol. The depth of the weld is shown inside the symbol. If no depth dimension is given for a plug or slot weld, the hole is completely filled. For a slot weld, the drawing shows the length and width, and the angle of the sides. The drawing indicates the location and spacing of the holes for both plug and slot welds. If a series of plug or slot welds is to be made, the center-to-center distance of the holes may be shown to the right of the basic weld symbol.
- The spot weld symbol is a circle 1/4″ (6.0 mm) in diameter. If the weld is accomplished from the arrow side, the spot weld symbol should be below the reference line. If the welding is done on both sides, as in resistance spot welding, the circle straddles the reference line. Projection welding is another form of resistance welding used to produce spot welds. To indicate which piece has the projections on it, the circle is placed above or below the reference line. The weld size or desired strength is given to the left of the weld symbol. The weld size may be given in US customary units or SI metric units. Weld strength is shown in pounds or newtons per spot. The weld spacing is found to the right of welding symbol. Centered above or below the spot welding symbol, in parentheses, is the number of welds desired. The welding process used is shown in the tail of the welding symbol.
- For seam welds, the process used is shown in the tail. The size (width) of the weld, or strength of the weld, are shown to the left of the weld symbol. The strength is given in pounds per linear inch or in newtons per millimeter. The weld symbol may straddle the reference line if welded from both sides. For electron beam and gas tungsten arc welding, the symbol is placed above or below the reference line. This indicates from which side of the part the weld is made. The length of the seam may be shown to the right of the weld symbol.

Review Questions

Write your answers on a separate sheet of paper. Please do *not* write in this book.

1. _____ drawings or _____ may be used to produce welded parts.
2. Most industrial drawings are made using a process called _____ projection.
3. *True or False?* The ANSI/AWS welding symbol shows the type of joint to be welded.
4. List five things that the welding symbol will tell the welder about the weld to be made.
5. *True or False?* A separate weld-all-around symbol will appear on the drawing at each location where the weld changes direction.
6. What types of information may be shown in the tail of the welding symbol? Name two.
7. Sketch the welding symbol for the completed weld shown in the following drawing:

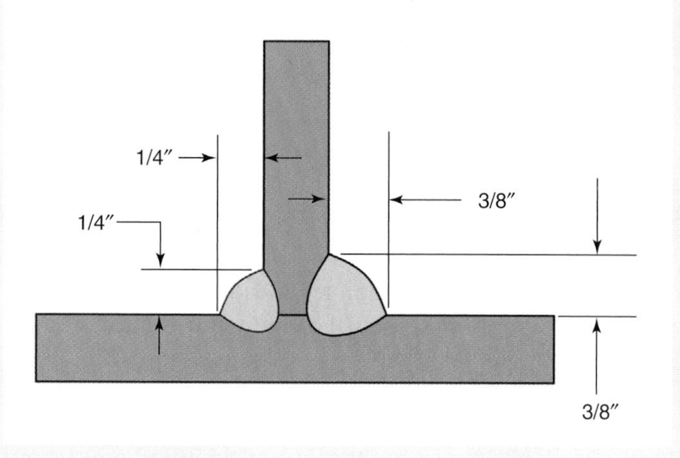

Refer to the following welding symbol when answering question 8–10:

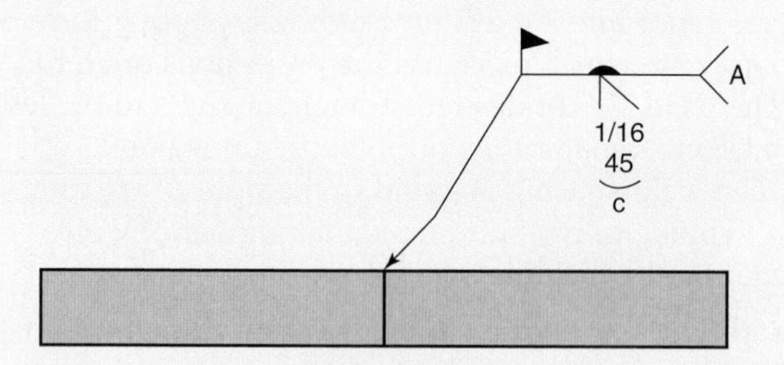

8. Which piece is cut or ground in this bevel-groove weld? What is the angle of the bevel?
9. What is the shape of the face of the weld bead? How is the weld face to be finished?
10. Is this weld made in the shop or in the field (on site)? Is it welded all around?

Chapter 34

Inspecting and Testing Welds

Learning Objectives

After studying this chapter, you will be able to:

- Describe the difference between a welding flaw and a welding defect.
- List the most common types of nondestructive and destructive testing done on welds.
- Perform several basic types of tests on welds to evaluate weld quality.
- Describe the methods used to prepare samples for bend tests.

Technical Terms

air pressure leak test
bend test
code
defect
destructive test
ductility
eddy current inspection
face bend
fillet test
flaw
free bend test
guided bend test
hardness test
impact test
indenter
liquid penetrant inspection
longitudinal sample
magnetic particle inspection
nondestructive examination (NDE)
oscilloscope
peel test
pressure test
radiograph
root bend
side bend
specification
tensile shear test
tensile strength
tensile test
transverse sample
ultrasonic inspection
visual inspection
X-ray inspection

Reasons for Inspecting Welds

Most welds have flaws in them. A ***flaw*** is a part of a weld that is not perfect. Some flaws are so small they can be found only by examining the weld under a microscope. Other flaws are very easily seen. These include large cracks or porosity, as shown in **Figure 34-1**.

A ***defect*** is a flaw that makes a weld unusable for the job it is intended to perform. If you are welding a handle on a garbage can lid at home, and the weld develops a small crack, it can still be used. However, if a small crack develops when welding on a gas pipeline, the crack must be repaired. In both cases, the crack is a flaw, but only in the second case is the crack a defect. The allowable size for a defect varies. Different applications have different requirements.

Welds must be inspected to detect flaws. Then, the flaws must be evaluated to decide if they are acceptable or if they are defects. Most welding is done to

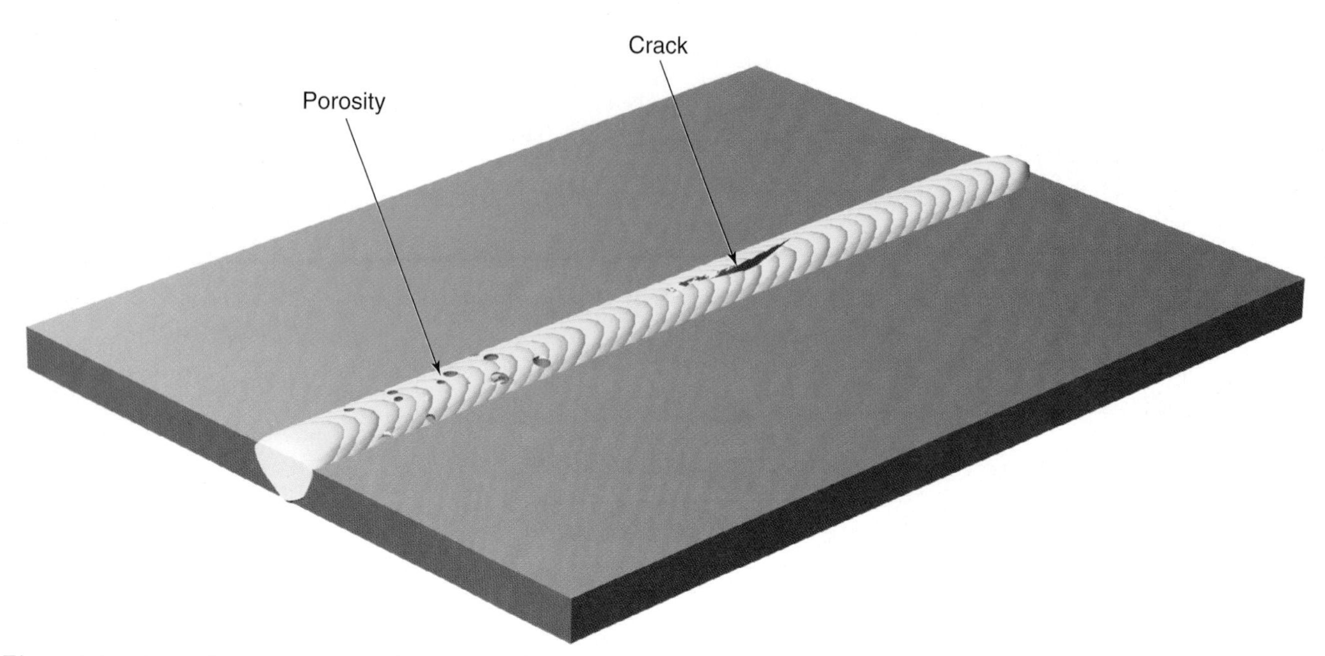

Figure 34-1. Some flaws are seen easily when looking at a completed weld.

requirements of a ***code*** or ***specification***. The code or specification determines how large a flaw can be before it becomes a defect. In some codes, a crack 1/32″ (.8 mm) long may be acceptable. In other codes, a crack of the same length will be a defect.

A weld is inspected to locate all defects. Once located, the defects must be repaired. A weld without defects will perform the job for which it was designed.

Types of Tests

Tests are used to determine the size and location of flaws in a weld. Tests are also used to determine the physical properties of a weld. These tests can be divided into two groups, nondestructive and destructive. A ***nondestructive examination (NDE)*** does not damage the weld or the base metal. A ***destructive test***, however, results in at least some damage to the weld. Destructive tests are normally used to determine the weld's physical properties. They often completely destroy the weld.

Nondestructive Examination (NDE)

Nondestructive examination is often called NDE. It is also called NDT, which stands for nondestructive testing (a nonstandard term). The most typical type of nondestructive examination is a ***visual inspection***. Visual inspection often is sufficient to determine if a weld is a good weld. Other inspections that do not damage a completed weld include the following:

- Liquid penetrant inspection.
- Magnetic particle inspection.
- Ultrasonic inspection.
- X-ray inspection.
- Eddy current inspection.
- Air pressure leak test.

Visual Inspection

Visual inspections are performed on almost all welds. These inspections are quick and do not require much equipment.

Visual inspections are used to locate visible flaws and defects on the surface of the weld. Very small flaws and flaws that are inside the weld are difficult or impossible to find by this method.

A visual inspection is useful to determine the size of a weld. Gauges are available for measuring fillet sizes. See **Figure 34-2**. Visual inspection is also used to check that all dimensions required by a drawing are met. A visual inspection tells a welder or an inspector if there is any undercut, overlap, or other surface flaws.

Whenever you complete a weld, you can look at it to evaluate your welding skills. It also tells you if the joint is filled properly. You can look at the back side of a butt weld to see if there is complete penetration. A visual inspection tells you if the arc is too long, if the travel speed is too fast or too slow, and other information about your welding skills.

Liquid Penetrant Inspection

Liquid penetrant inspection is a good means of locating surface flaws. It cannot locate flaws that are inside a weld, however. This test uses a liquid dye, usually red in color for good visibility.

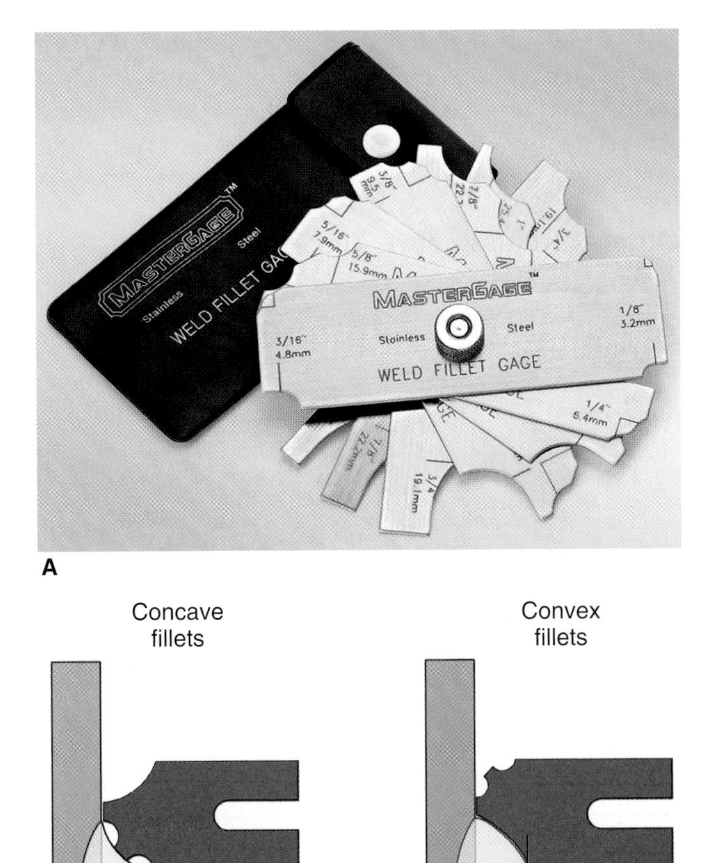

Figure 34-2. *Fillet gages. A—This set of gages is used to check the size of convex and concave fillet welds. (MasterGage Co.) B—This drawing shows how fillet gages are used to measure fillet welds. (Fibre Metal Products Co.)*

First, the surface of the part is cleaned, and then the dye is applied. If there is a flaw on the surface, some of the dye will enter the flaw. The dye is cleaned off the surface after a few minutes. However, any dye that entered a flaw is not removed by the cleaning. A developer is then applied. The developer is usually a white, dry powder. It draws out the dye, showing the location of all surface flaws in the metal or the weld. The steps followed in a liquid penetrant inspection are shown in **Figure 34-3**.

Fluorescent liquids can be used instead of dyes. The surface of the part is cleaned and the fluorescent liquid is applied. After a few minutes, the liquid is cleaned off the surface. Then the developer is applied. After a short while, the part is examined under an ultraviolet lamp, or "black light." The fluorescent liquid glows a light green color and shows the location of all surface flaws.

Magnetic Particle Inspection

Magnetic particle inspection is used to locate flaws that are at the surface or very near the surface of the weld. Flaws that are deeper within the weld cannot be found using this test. Also, magnetic particle inspection can be used only on materials that are magnetic. This eliminates nonferrous metals and many stainless steels.

To examine a part using magnetic particle inspection, the surface must first be cleaned. Then, fine magnetic particles are applied to the surface. The particles can be dry or can be mixed with a liquid. Colored and fluorescent particles are available.

After the particles are applied to the surface, a strong magnetic field is applied. The field can be created by a strong magnet or by passing an electric current through the part.

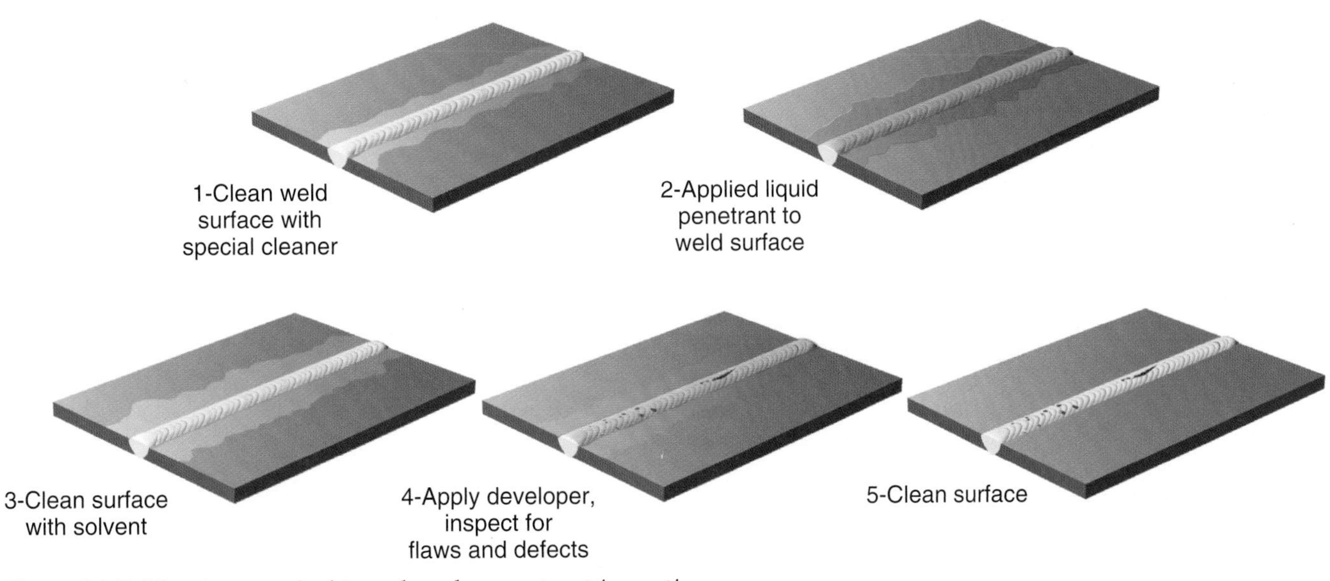

Figure 34-3. *The steps required to make a dye penetrant inspection.*

When the magnetic field is applied to the part, any flaw will act like a small magnet. The fine magnetic particles will be drawn toward the flaw the way a paper clip is drawn toward a strong magnet. These particles outline the flaw. When the magnetic field is removed, the magnetic particles remain in place. The part is then examined to locate any flaws or defects. See **Figure 34-4**.

Ultrasonic Inspection

Ultrasonic inspection can be used on all metals to locate internal flaws. Sound waves are sent into the part being inspected. The sound waves travel until they contact air and are reflected back and create an echo.

The sound waves are displayed on a screen. Each time the sound hits something, a pulse or echo shows up on the screen. Sound will continue to bounce inside the part. With each bounce, sound energy is lost, and each echo appears smaller than the previous one.

When there are no flaws in the metal, each echo that appears represents the thickness of the part. The sound reflects from the back side of the piece. See **Figure 34-5**. If there is a flaw, an additional echo is created when the sound bounces off the flaw. The location of the pulse can indicate where the flaw is. Larger flaws produce larger echoes.

X-Ray Inspection

X rays are used to inspect welds in the same way that a doctor uses X rays to inspect a patient. Film is placed on one side of the part to be inspected. X rays pass through the part and expose the film. After the film is developed, any flaw or defect in the weld can be observed. The film also serves as a permanent record of the quality of the weld. **Figure 34-6** shows an X-ray view, or ***radiograph***, of a weld.

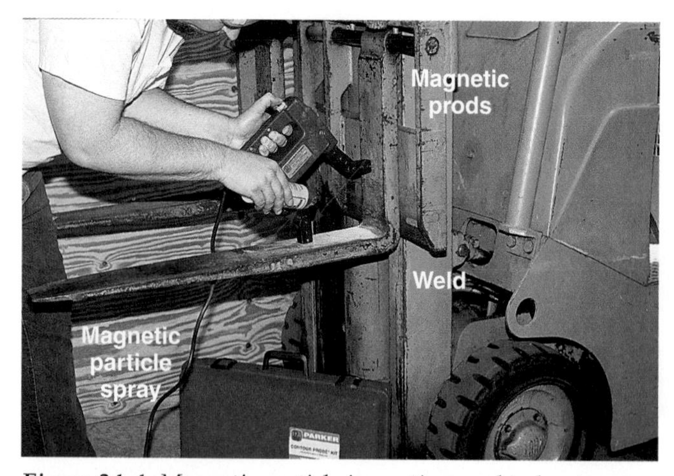

Figure 34-4. Magnetic particle inspection used to locate a suspected crack in a lift-truck fork. Note the white metallic particle material being sprayed onto the fork surface by the inspector. (Parker Research)

Safe working practices are very important when using *X-ray inspection* or working with radioactive materials. Observe all safety procedures so neither you nor anyone working in the area is exposed to the X rays.

Other Nondestructive Tests

Eddy current inspection can detect porosity, cracks, slag inclusions, and lack of fusion at or near

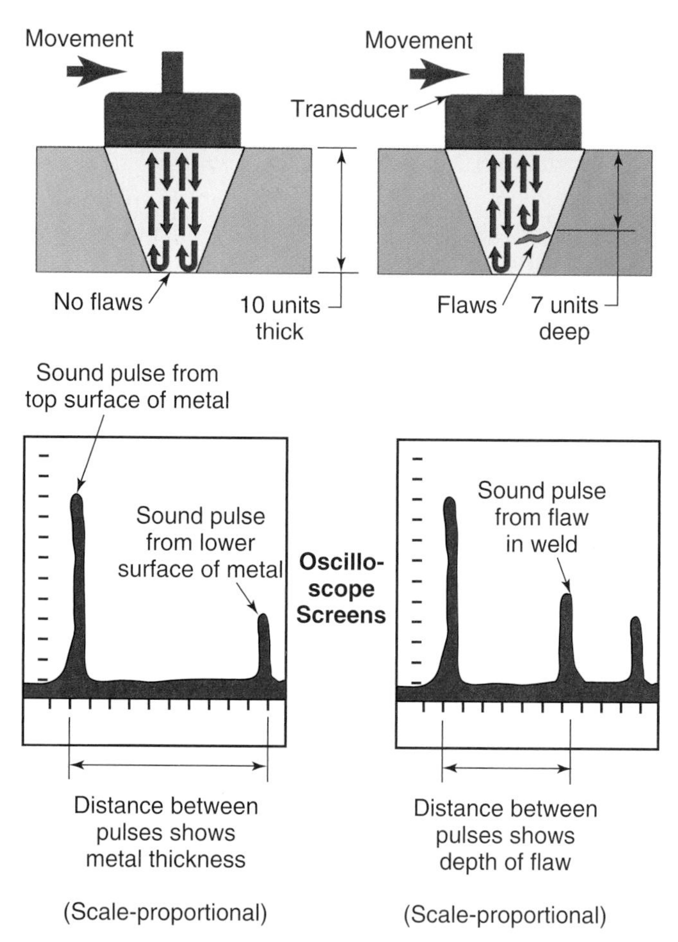

Figure 34-5. Ultrasonic testing. The transducer sends ultrasonic waves through the test piece. The top surface shows up on the oscilloscope screen as a large pulse. Additional pulses on the screen are from the bottom surface and any flaws in the weld. The distance between the top surface, and a flaw or the bottom surface, can be measured on the screen.

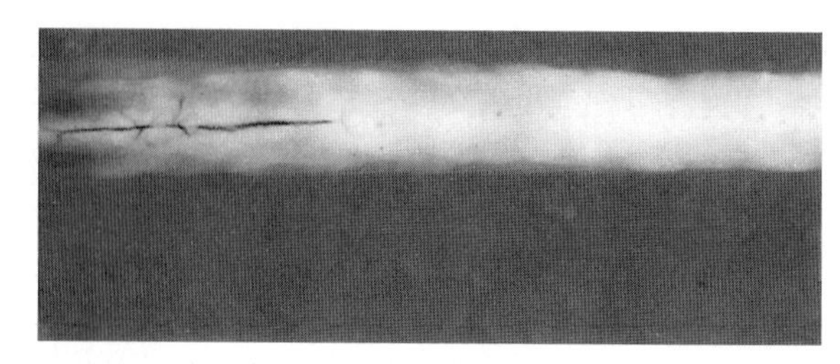

Figure 34-6. This X-ray radiograph shows a large undersurface crack in a weld.

the surface of a weld. A coil carrying high-frequency alternating current is brought close to the part being inspected. This coil creates a current in the part. The current in the part is called an eddy current. Eddy currents flow in a closed circular path.

The coil is moved over the part being inspected. When a flaw is present in the part, the flow of the eddy current is interrupted. This change in current flow is measured on an oscilloscope.

Air-pressure or *water-pressure leak tests* are often used to inspect pipes or tanks that are to contain gases or liquids under pressure. When air is used for the leak test, a soapy solution is used to coat the pipe surface at each weld. When the pipe is under air pressure, a flaw will cause bubbles to form. When water pressure is used, water will spray out from the pipe where there is a leak. The most sensitive inspection for leaks uses helium gas and a spectrometer to detect leaks as small as a few parts per million.

Destructive Tests

Destructive testing is used to determine the physical properties of a weld. To do so, destructive testing requires that a completed weld be damaged in some way. Most such tests totally destroy the weld. The most common destructive tests include the following:

- Tensile test.
- Bend test.
- Fillet test.
- Hardness test.
- Impact test.
- Peel test (for spot welds).
- Tensile shear test (for spot welds).
- Pressure test.

Tensile Test

The ***tensile test*** is a destructive test. This test applies a tensile (stretching or pulling) load to a prepared sample until it breaks.

After a butt weld on a plate or pipe is completed, sections are cut from the weld. These sections are then prepared for a tensile test. See **Figure 34-7**. Note the part of the specimen where the section has a reduced area. **Figure 34-8** shows reduced section tensile specimens after they have been pulled to the breaking point.

A tensile test is used to determine the tensile strength and the ductility of a weld. ***Tensile strength*** is a measurement of the amount of force required to pull the metal apart. ***Ductility*** is a measure of how much the metal stretches before breaking.

Before performing a tensile test, measure the diameter of round parts or the thickness and width of rectangular parts at the reduced section. Then, mark two lines (or use a center punch to mark two points) that are exactly 2″ (50.8 mm) apart. One point should be 1″ (25.4 mm) on one side of the weld, and the other point should be about 1″ (25.4 mm) on the other side of the weld. See **Figure 34-9**. The marks allow you to

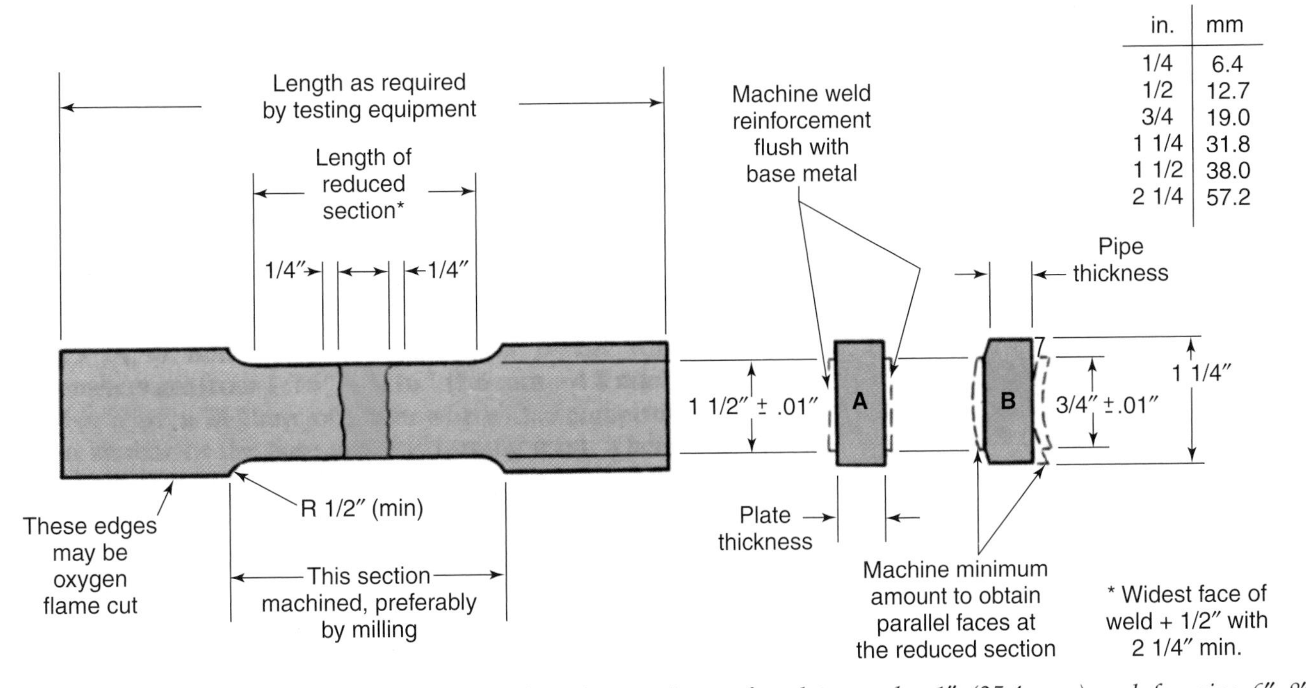

in.	mm
1/4	6.4
1/2	12.7
3/4	19.0
1 1/4	31.8
1 1/2	38.0
2 1/4	57.2

Figure 34-7. *Example of dimensions for reduced-section specimens for plates under 1″ (25.4 mm) and for pipe 6″–8″ (152 mm–203 mm) in diameter. Dimensions for the preparation of various specimens can be found in the* Standard Methods for Mechanical Testing of Welds *publication, ANSI/AWS B4.0.*

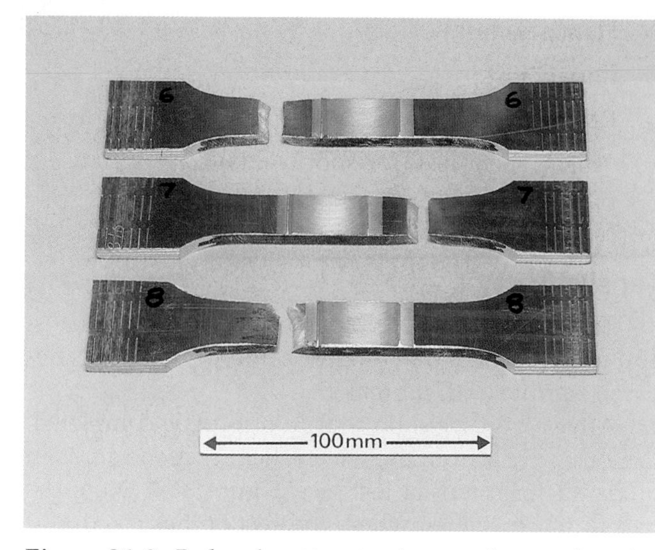

Figure 34-8. Reduced-section tension specimens after they have been tested to failure. Note that the specimens broke outside the weld area, as they should. (TWI)

Figure 34-9. A standard all-weld-metal tensile specimen. Note that the all-weld-metal sample is machined to a cylindrical shape with threads at each end. The threads are used to mount the sample into the tensile test machine. Two small punch marks are made 2″ (50.8 mm) apart before the test begins.

determine how far the metal stretched before breaking. **Figure 34-10** shows an all-weld-metal sample in a tensile test machine.

When performing a tensile test, safety glasses must be worn!

To perform a tensile test, place the prepared section in the machine that will pull the metal apart. See **Figure 34-11**. Clamp each end of the sample into the jaws of the machine. The machine pulls on each end of the sample. During the test, the metal sample stretches and then breaks into two pieces. The machine records the maximum force applied to break the sample. Write down the maximum force applied.

Figure 34-10. An all-weld-metal tension test sample mounted in a tensile test machine. A load is applied to the sample until it fails. This test measures the tensile strength and ductility of the sample.

Remove the pieces from the jaws of the machine. Press the broken edges back together. Measure the distance between the marks that were 2″ (50.8 mm) apart before the test. Record your measurement.

Calculating tensile strength and ductility:

With the data you have recorded, you can find the tensile strength and ductility of a weld. Use the following formulas:

$$\textit{Tensile strength} = \frac{\text{Maximum force applied by machine}}{\text{Cross-sectional area of sample}}$$

$$\textit{Ductility} = \frac{\text{(Distance between points after test–Distance before test)}}{\text{Distance before test}} \times 100$$

Tensile strength is measured in pounds per square inch (psi) or in megapascals (MPa). Ductility is measured as a percentage. For example, low carbon steel has a tensile strength of 55,000 psi (379 MPa) and a ductility of about 25%.

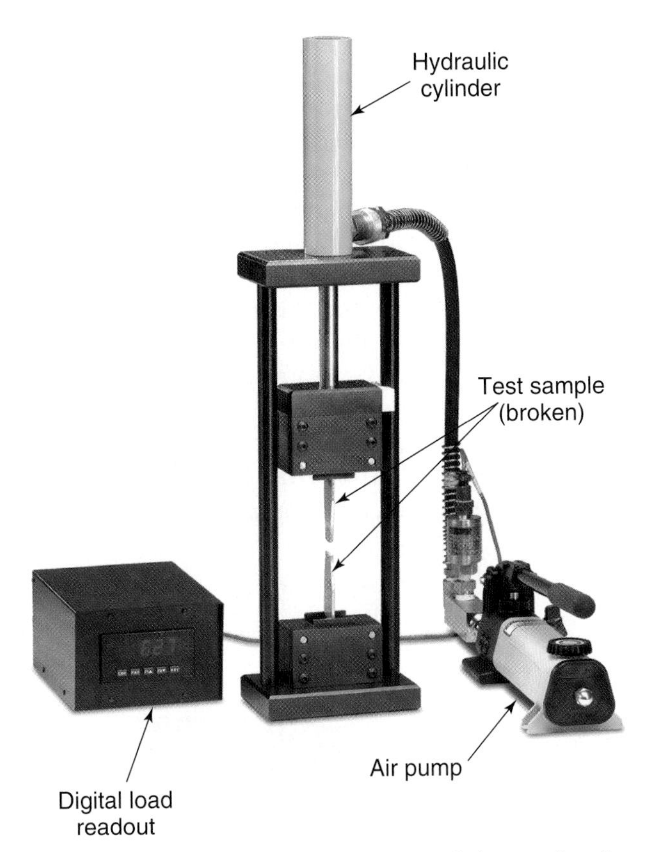

Figure 34-11. An air-operated tensile tester being used to determine the tensile strength of a weld sample. A test sample is pulled until it reaches its tensile limit. A gauge reads the force applied when the test piece breaks. The required force divided by the cross-sectional area of the test sample equals the tensile strength of the metal or weld sample. (Fischer Engineering Company)

Example:

A welded plate is 1/2″ (12.7 mm) thick. The reduced section is 1 1/2″ (38 mm) wide. A maximum force of 48,000 pounds (213,500 newtons) is required to break the sample. The distance between the two points after breaking is 2 1/2″ (63.5 mm).

Calculations:

$$\text{Tensile strength (US customary)} = \frac{48{,}000 \text{ pounds}}{1/2'' \times 1\,1/2''} = 64{,}000 \text{ psi}$$

$$\text{Tensile strength (metric)} = \frac{213{,}500 \text{ N}}{12.7 \text{ mm} \times 38 \text{ mm}} = 442 \text{ MPa}$$

$$\text{Ductility (US customary)} = \frac{2\,1/2'' - 2''}{2''} \times 100 = 25\%$$

$$\text{Ductility (metric)} = \frac{63.5 \text{ mm} - 50.8 \text{ mm}}{50.8 \text{ mm}} \times 100 = 25\%$$

Exercise 34-1 Making a Tensile Test

1. Make a butt weld on plate or pipe, as you did for an exercise earlier in this book.
2. Remove two pieces, each 2″ (50 mm) wide. If the plate or pipe is over 1″ (25 mm) thick, the pieces should be 1 1/2″ (40 mm) wide.
3. Prepare each piece as shown in Figure 34-7.
4. Grind the weld face and weld root flush with the base metal. All grind marks must run lengthwise on the metal samples. Also, be sure to break all sharp edges along the reduced section area.
5. Measure the thickness and width of each piece. Record the information on a separate sheet of paper. Do not write in this book.
   Thickness _____
   Width _____
6. Mark two lines or points exactly 2″ (50 mm) apart. The weld should be centered between the marks.
7. Put the piece to be tested into a tensile testing machine. Apply force until the piece breaks.
8. On your paper, record the maximum force applied.
9. Place the two pieces back together. Measure the distance between the two points. Write down the distance on your paper.
   Distance _____
10. Calculate the tensile strength and ductility of the test piece using the formulas given in this chapter. Show your work.
    Tensile Strength _____
    Ductility _____
11. Did the piece break in the weld, next to the weld, or in the base metal area?
12. Are there any signs of welding flaws along the broken surface?

You should also visually inspect the pieces after they are broken. Note where the sample breaks. The sample should break in the base metal. If it breaks in the weld area, look for signs of porosity, slag inclusions, or other defects.

Bend Test

Bend tests are used to evaluate the quality and ductility of a completed weld. The guided bend test is the most common type. The free bend test is also used. In both tests, a sample is bent into a "U" shape.

In a ***guided bend test***, the size, or radius, of the bend is controlled. The radius selected depends on thickness and type of metal being tested.

The free bend test may also be used on metal up to 1/4" (6.4 mm). In a ***free bend test***, the sample is placed into a large, strong vise and struck with a hammer until it bends to a 90° angle. It is then hammered in the opposite direction to a 90° angle. This back and forth bending continues until the weld sample breaks. A properly made weld will break outside the weld in the base metal. **Because one part of the weld may fly as it breaks, this test must be done in an enclosed area.**

There are three classifications of bend tests: the face bend, the root bend, and the side bend. In a ***face bend***, the weld face is on the outside of the bend. A ***root bend*** has the weld root side of a weld on the outside of the bend, **Figure 34-12**. A ***side bend*** is used on thick material. In each test, the center of the weld must be at the center of the bend. See **Figure 34-13**.

Bend samples can be made with the weld going across the sample or along the length of the sample. The sample is called a ***transverse sample*** when the weld goes across it. A ***longitudinal sample*** has the weld along the length of the sample. See **Figure 34-14**.

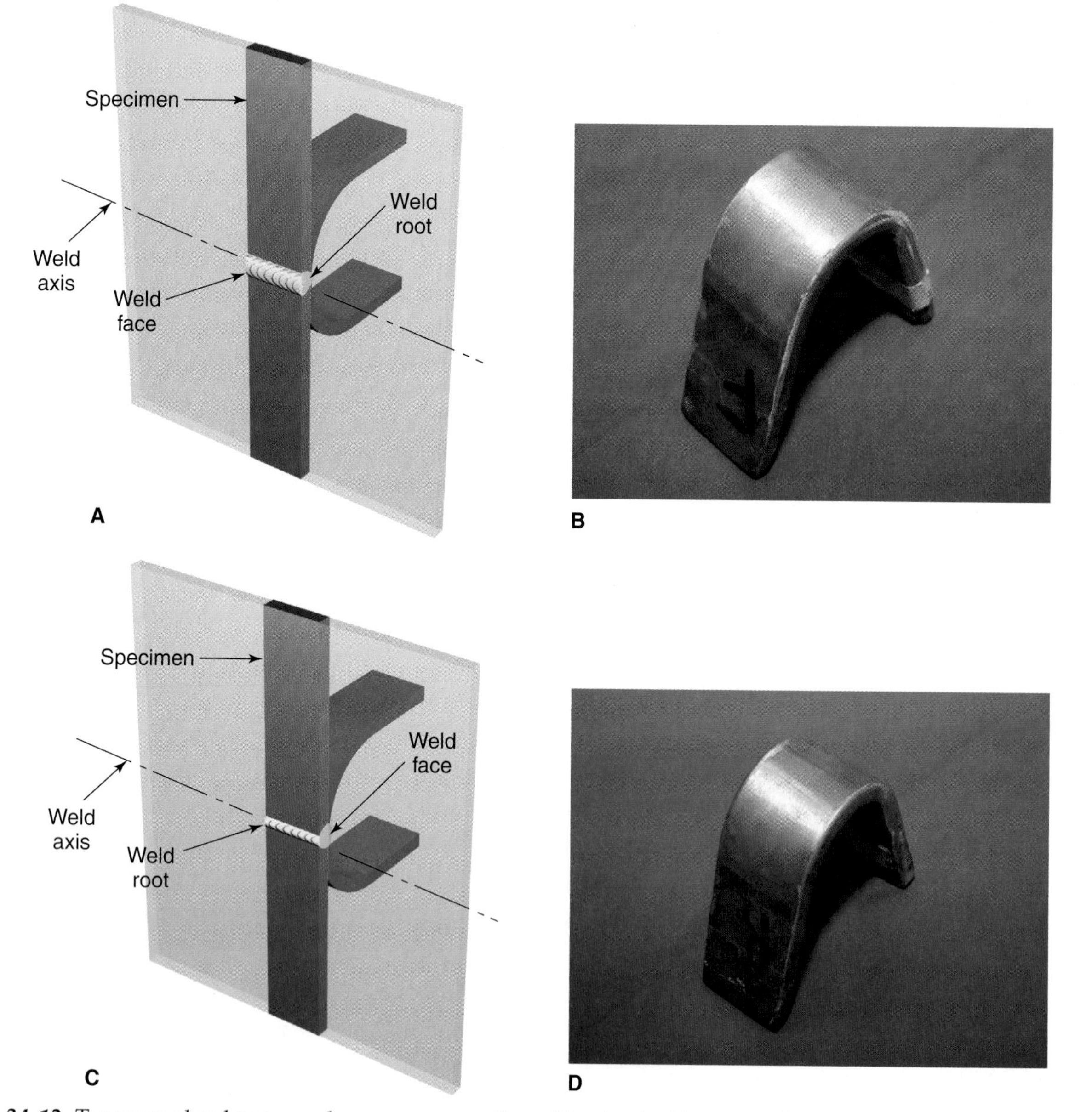

Figure 34-12. Transverse bend test samples are cut across the weld axis. A—Transverse face bend. Note that the weld face is on the outside of the bend. B—A SMAW face bend sample after testing. C—Transverse root bend. Note that the weld root is on the outside of the bend. D—A SMAW root bend sample after testing.

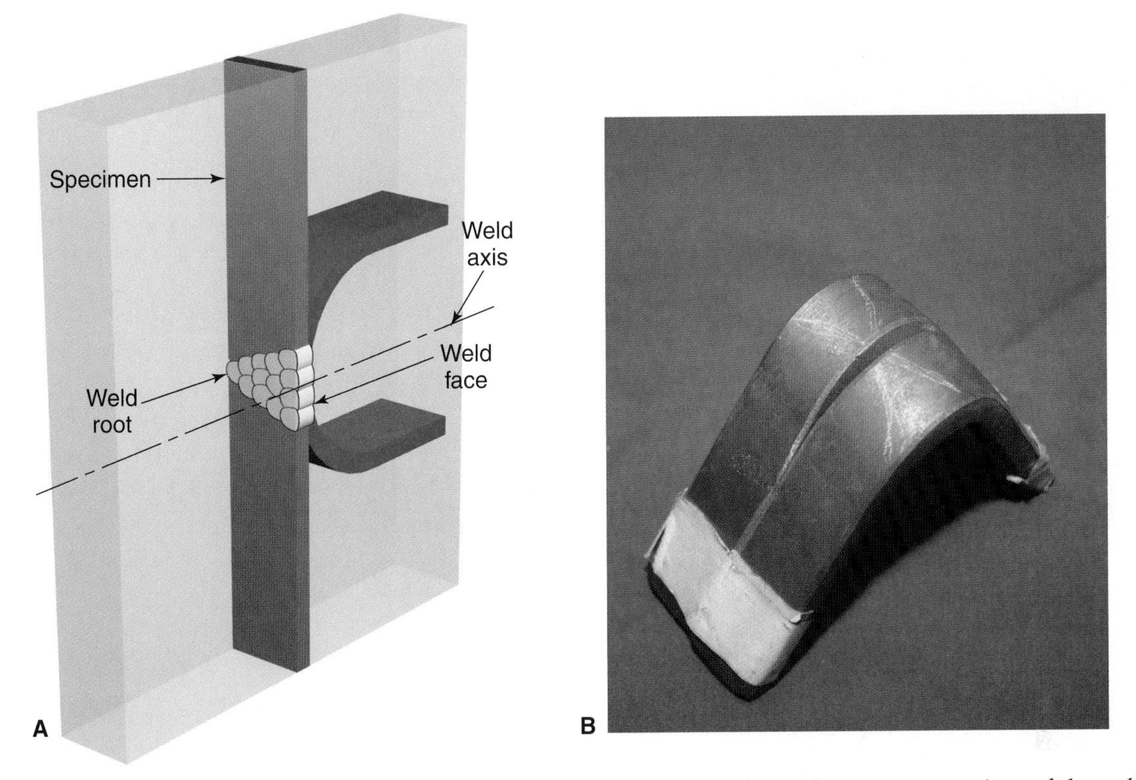

Figure 34-13. *Side bend test samples are cut across the weld axis. A—Side bend samples are cross sections of the welded joint. B—These two side bend samples were taken from the same welded plate and taped together for close examination. Note how the markings mirror each other.*

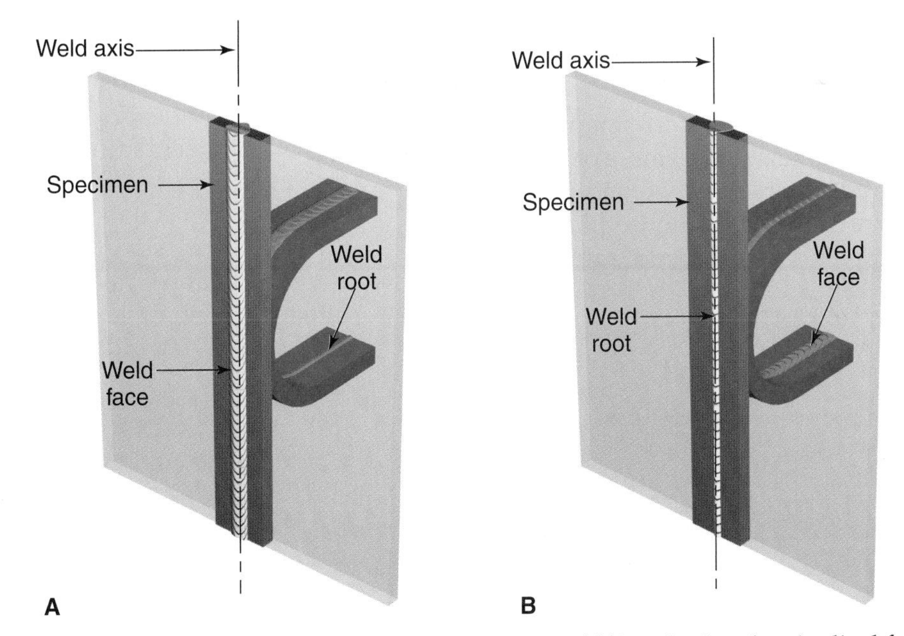

Figure 34-14. *Longitudinal samples are cut in the same direction as the weld line. A—In a longitudinal face bend, the weld face is on the outside of the bend. B—In a longitudinal root bend, the weld root is on the outside of the bend.*

Prior to bending, a section of a completed weld must be prepared. **Figure 34-15** shows the dimensions of a sample to be bent. After preparation, the sample is placed in a jig designed for bending plates. See **Figure 34-16** and **Figure 34-17**.

Force is applied to cause the sample to bend. The sample is bent until it forms a U-shape or until it breaks. If the sample breaks, it has failed the test. If the sample does not break, it is visually examined after bending. There should be no cracks or other

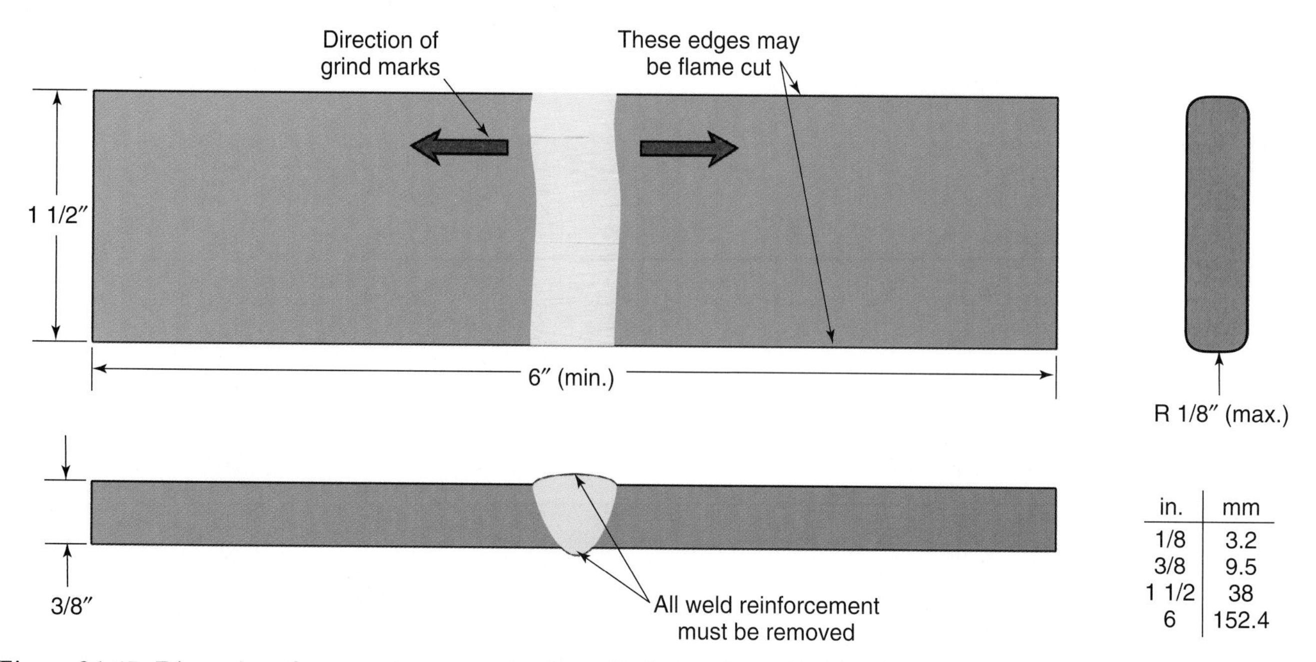

Figure 34-15*. Dimensions for preparing a sample of a welded plate for guided bend test. Remove all weld reinforcement and radius the corners. Notice the direction of the grinding marks.*

defects 1/16″ (1.6 mm) or larger in any direction on the outside bent surface. Some codes allow flaws up to 1/8″ (3.2 mm). Cracks that start at the corner of the sample are not counted, unless there is evidence that the flaw (porosity or lack of fusion, for example) is a result of poor welding skills. If there are no cracks larger than are allowable under code, the sample passes the bend test.

Bend test samples and tensile test samples are often taken from the same weld. This allows the strength, ductility, and quality of a welded joint to be determined. **Figure 34-18** shows how a plate may be cut to obtain samples for both bend and tensile testing.

Exercise 34-2 Making a Bend Test

1. Make a butt weld joint as you did for one of the exercises earlier in this book.
2. Remove two pieces from the joint. Each piece should be 1 1/2″ (40 mm) wide. Using metal stamp or other tool, permanently mark the face side of each piece near the end. This will help you remember which side is the face side.
3. Prepare each piece as shown in Figure 34-15. Grind the weld face and weld root flush with the base metal. Grind marks must run lengthwise on the sample. Do not remove any more metal than is necessary. Be sure to remove all sharp edges from the sample.
4. Place one sample in a guided bend test jig with the weld face up. This will be a root bend. Apply force. If the sample breaks or begins to break, stop applying the force. If the sample does not break, continue to apply force until it is bent into a "U" shape.
5. Bend the second sample with the root side up.
6. If the sample is bent into a "U" shape and there are no flaws larger than 1/16″ (1.6 mm), the sample passes the bend test. Flaws that start from the edge of the sample are not considered unless there is evidence that the flaw is due to poor welding technique.
7. If the sample breaks during the test, examine the broken areas for evidence of poor welding techniques.

Note: Safety glasses must be worn during this exercise.

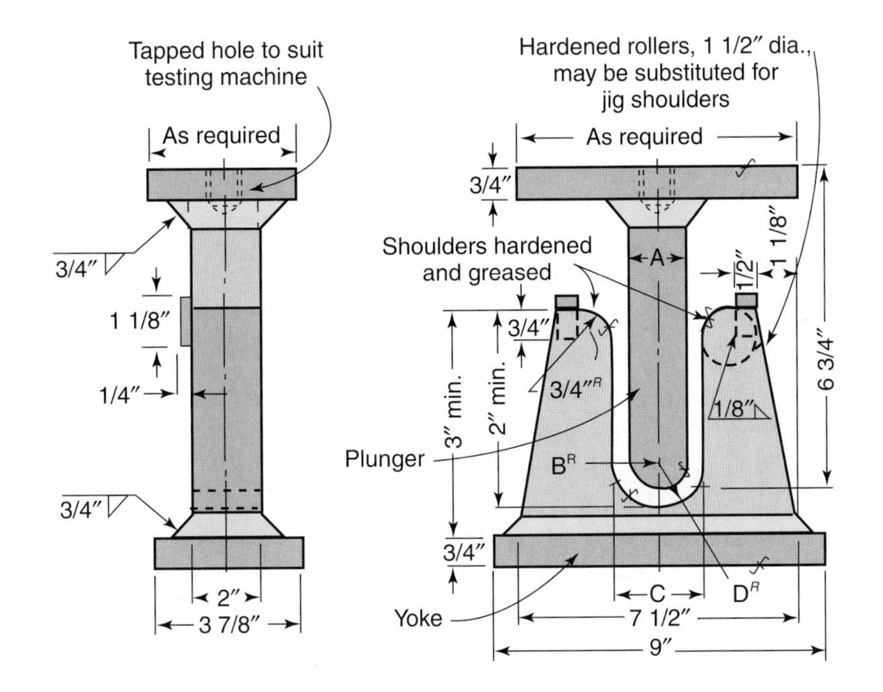

Material	Thickness of Specimen, in.	*A*, in.	*B*, in.	*C*, in.	*D*, in.
P-No. 23 to P-No. 21 through P-No. 25 with F-No. 23; P-No. 35; any P-No. metal with F-No. 33, 36, or 37	1/8 *t* = 1/8 or less	2-1/16 16-1/2*t*	1-1/32 8-1/4*t*	2-3/8 18-1/2*t* + 1/16	1-3/16 9 1/4*t* + 1/32
P-No. 11; P-No. 25 to P-No. 21 or P-No. 22 or P-No. 25	3/8 *t* = 3/8 or less	2-1/2 6-2/3*t*	1-1/4 3-1/3*t*	3-3/8 8-2/3*t* + 1/8	1-11/16 4-1/3*t* + 1/16
P-No. 51	3/8 *t* = 3/8 or less	3 8*t*	1-1/2 4*t*	3-7/8 10*t* + 1/8	1-15/16 5*t* + 1/16
P-No. 52, P-No. 53, P-No. 61, P-No. 62	3/8 *t* = 3/8 or less	3-3/4 10*t*	1-7/8 5*t*	4-5/8 12*t* + 1/8	2-5/16 6*t* + 1/16
All others with greater than or equal to 20% elongation	3/8 *t* = 3/8 or less	1-1/2 4*t*	3/4 2*t*	2-3/8 6*t* + 1/8	1-3/16 3*t* + 1/16
All others with less than 20% elongation	*t* = (see Note b)	32-7/8*t*, max.	16-7/16*t*, max.	34-7/8*t*, + 1/16, max.	17-7/16*t* + 1/32, max.

Figure 34-16*. Specifications for a QW-466 guided bend jig. Note that dimensions A, B, C, and D change as the thickness and type of metal changes. See ASME-Section IX code for an explanation of the "P" and "F" numbers. These dimensions must be adhered to in order to standardize the test results.*

Fillet Test

The ***fillet test*** is used to determine the quality of a fillet weld. To create a sample for the fillet test, two pieces of metal are tack welded to form a T-joint. Next, a fillet weld is made. Welding for this test can be done in any position, using any process. The weld is stopped and restarted at the midway point of the joint. Prior to testing, the ends of the weldment should be cut off as shown in **Figure 34-19**.

Force is applied to the welded joint until it breaks or until the vertical piece is bent flat against the horizontal piece. **Figure 34-20** shows a safe way to apply the necessary force to the sample.

The weld should break along its centerline. This indicates that there is good penetration into the base metal and the legs of the weld are equal. A weld that breaks along one edge may indicate a lack of fusion. Look at the fractured surface for signs of porosity, slag

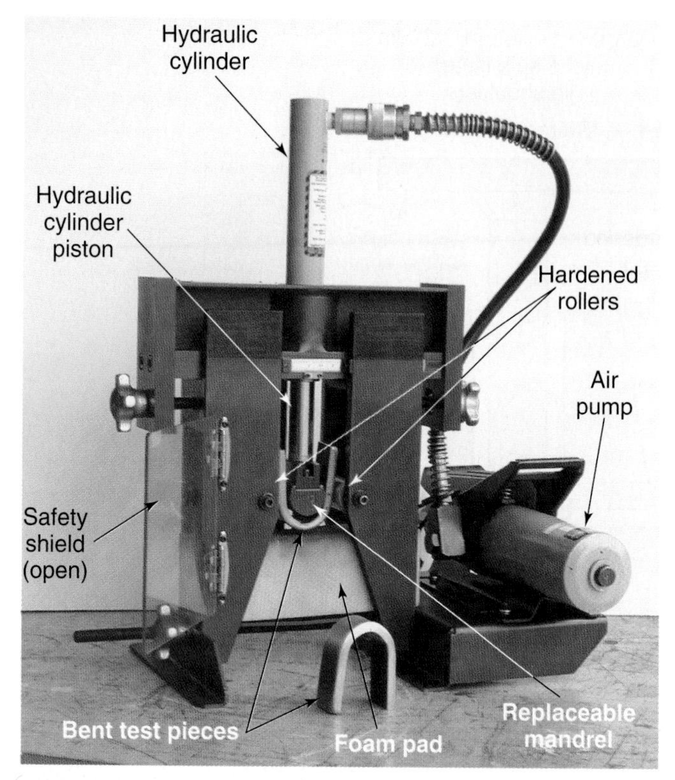

Figure 34-17. A roller-guided bend tester. This unit has a hydraulic cylinder and an air-driven pump. The tester is fully adjustable for the thickness of the test sample. The mandrel size may be changed as required. The prepared test sample is placed on the rollers under the mandrel. The finished bend is ejected downward onto a foam pad. ***It is highly recommended that safety glasses be worn when bend tests are performed.*** *(Fischer Engineering Company)*

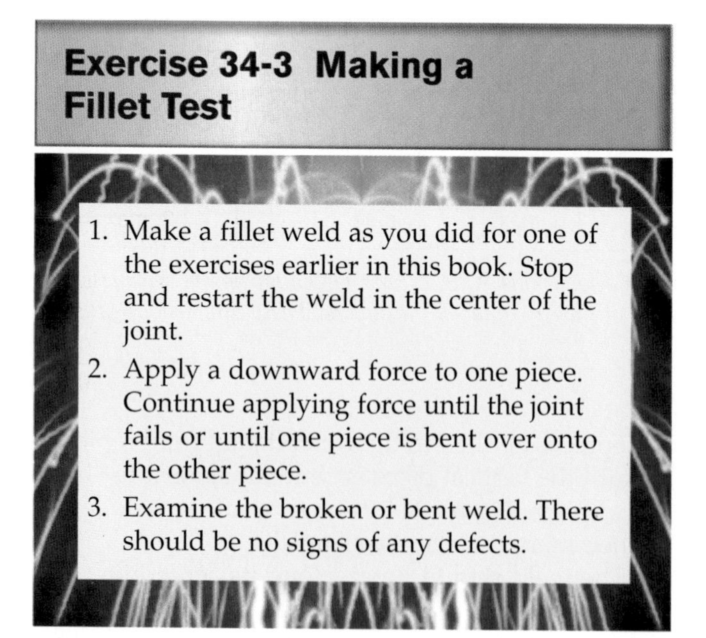

Exercise 34-3 Making a Fillet Test

1. Make a fillet weld as you did for one of the exercises earlier in this book. Stop and restart the weld in the center of the joint.
2. Apply a downward force to one piece. Continue applying force until the joint fails or until one piece is bent over onto the other piece.
3. Examine the broken or bent weld. There should be no signs of any defects.

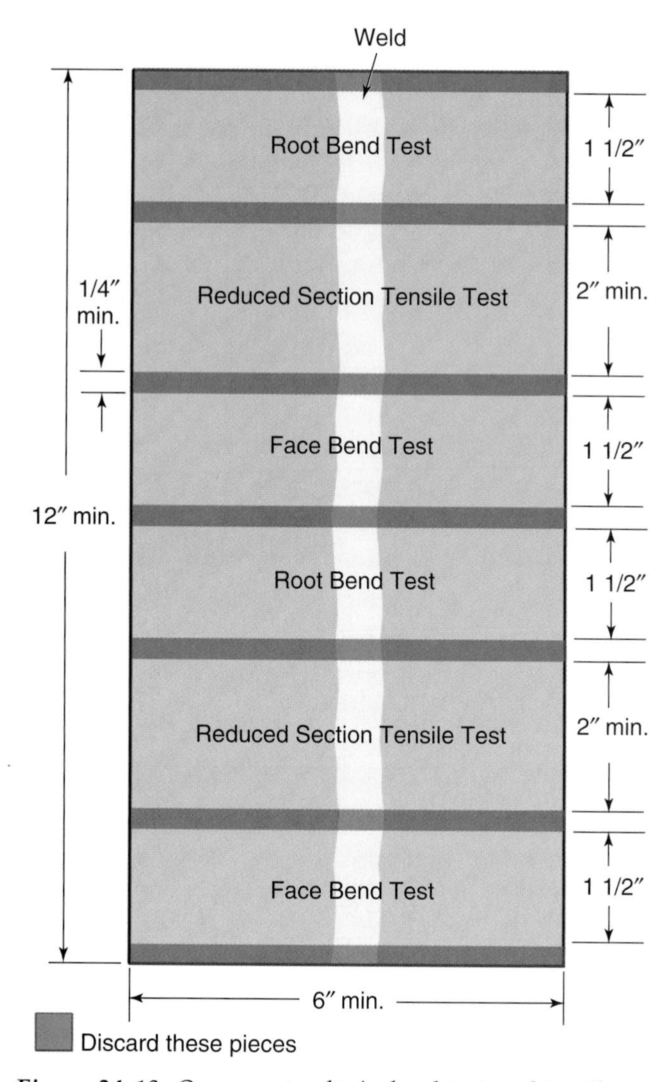

Figure 34-18. One way to obtain bend test and tensile test samples from a test weld. You may weld a 7″ (178 mm) test piece and only perform one root bend, one face bend, and one tensile test. Different codes and specifications require different ways to cut test samples from a test weld.

inclusions, lack of penetration at the root, or lack of fusion. Also look at the size of each leg of the weld. They should be equal. See **Figure 34-21**.

To pass a fillet weld test, your weld should break through the weld metal, or the vertical piece should bend down onto the horizontal piece. The broken surface or bent surface should show no evidence of defects.

Hardness Test

Hardness tests are used to measure a material's resistance to scratching and indentation. Test areas include the weld metal, the heat-affected metal close to a weld, and the base metal. Unlike bend testing and tensile testing, hardness tests do not completely destroy a welded joint. The hardness tests do make impressions or dents in the metal, however. Some parts can be used after hardness testing; other parts cannot.

To determine the hardness of a metal, an ***indenter*** is pressed into the metal by a known force. The indenter is usually a steel ball or a pointed diamond, depending on the type of hardness test. The ball or diamond is pressed into the metal by a specified force. If the metal is soft, the indenter will be pressed into the metal more than if the metal is hard.

Machines used for hardness testing usually have a gauge to show the hardness of the metal being tested. The most common type of tester is a Rockwell hardness testing machine. Other types include the Brinell, Vickers, Scleroscope, Knoop, and microhardness testers. **Figure 34-22** shows two types of hardness testers.

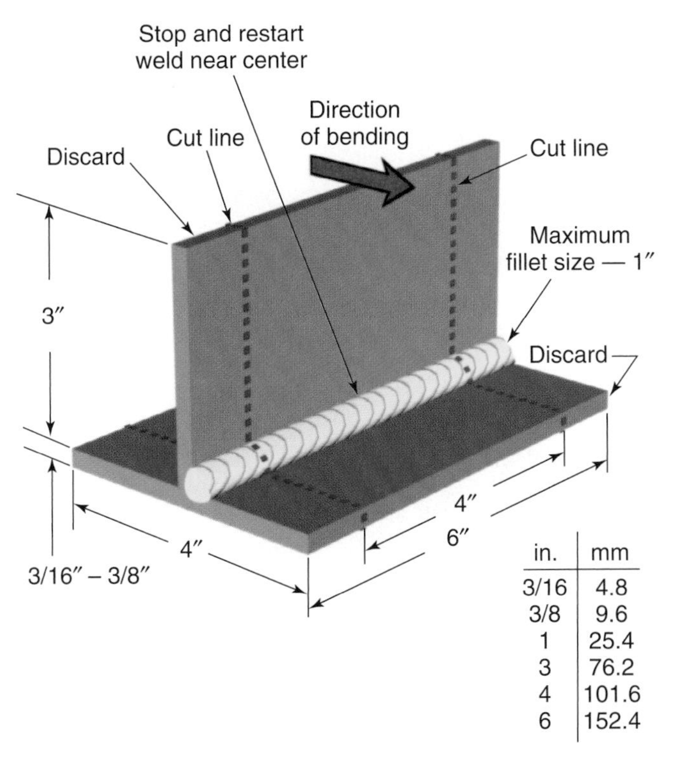

in.	mm
3/16	4.8
3/8	9.6
1	25.4
3	76.2
4	101.6
6	152.4

Figure 34-19*. The dimensions for a fillet weld break test specimen. The welder must stop and restart in the center of this weld. (AWS)*

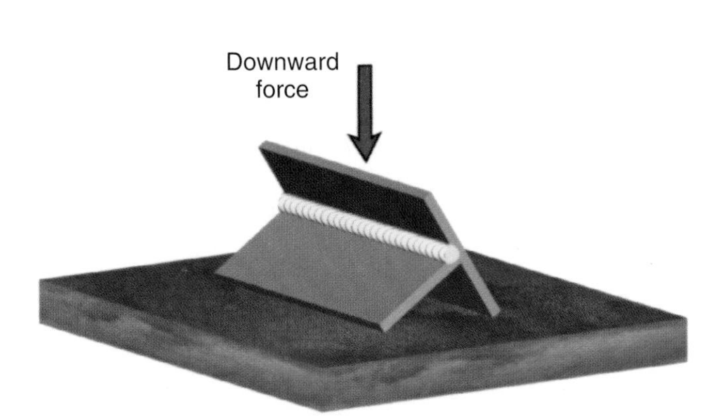

Figure 34-20*. The fillet weld specimen must be bent with a downward force, as shown. The force may be applied with a hammer, hydraulic press, or other suitable means.* ***Safety note: The test should be made in a shielded area to protect others from injury. Safety glasses are highly recommended.***

Figure 34-21*. A fillet weld that has been broken. The fillet weld was well made since it broke in the weld and did not pull away from the weld toes.*

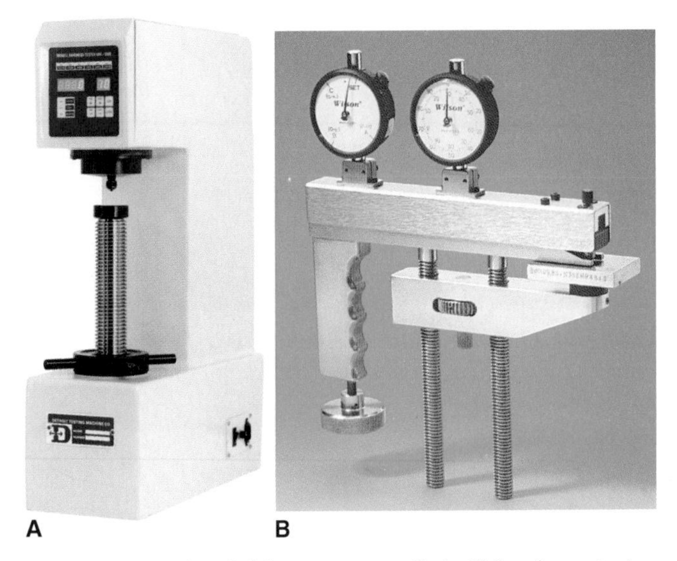

Figure 34-22*. A—A laboratory-type Brinell hardness tester. (Detroit Testing Machine Co.) B—A portable Rockwell hardness tester. A calibrating sample of a known hardness is used to set the tester for accurate readings. This lightweight tester can be used to test metal in the storage racks or on a welding site. (INSTRON Corp.)*

Other Destructive Tests

An *impact test* requires that a test sample be machined to specific dimensions. The sample is held in a clamp and broken by an impact from a large known force. The amount of energy required to break the sample is displayed on a dial.

A *peel test* is used to check the quality of spot welds. One edge of each piece is bent 90°, then force is applied to tear the metal apart. See **Figure 34-23**.

A *tensile shear test* is also used for testing spot welds. The two pieces to be welded are staggered, so that each piece extends about 1″ (25.4 mm) past the other. The pieces of metal are spot welded together. See **Figure 34-24**. Then the pieces are put into a tensile testing machine like the one shown in Figure 34-11. Force is applied and the spot welds are torn apart. This test measures the force required to tear the welds apart, and also allows the spot welds to be examined.

Pressure tests are used to determine how much pressure a pipe or cylinder can withstand or to test for leaks. A cap or valve is used to close each opening in the pipe, piping assembly, or cylinder. The pressure inside the closed system is steadily increased until a leak develops or a specified pressure is reached. See **Figure 34-25**. If pressure testing is done to destruction to determine how much pressure a specific weld can withstand, it is often done under water in an enclosed tank. This is done because the pipe or cylinder can explode from the high pressures applied.

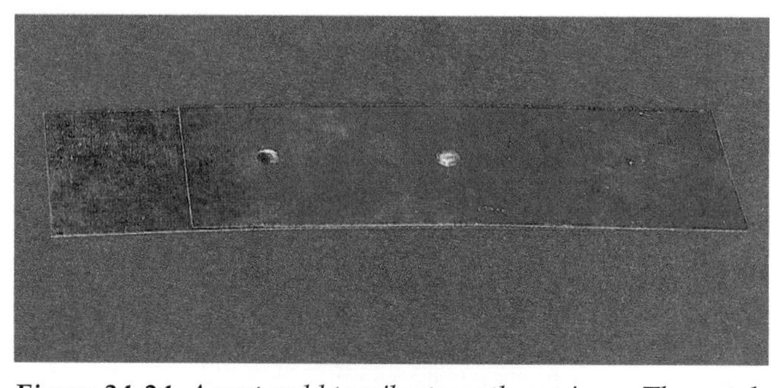

Figure 34-24. A spot weld tensile strength specimen. The metals are overlapped as shown so they can be gripped in the tensile test machine.

Figure 34-23. A destructive peel test. This test is used to determine the size, strength, and proper welding machine settings for resistance spot welds.

Figure 34-25. This long section of gas pipeline has a dome fitting welded at each end. Water is pumped into the pipe to fill it. The pipe is then pressurized to several thousand pounds per square inch to test all of the welds along the pipe for leaks or failure. Note the pressure relief valve, which prevents excessive pressure from developing. After the test, the domes are cut off and the pipe is welded to the next section.

Summary

- Most welds have flaws in them. A flaw is a part of a weld that is not perfect. A defect is a flaw that makes a weld unusable for the job it is intended to perform. The allowable size for a defect varies. Different applications have different requirements.
- Welds must be inspected to detect flaws. Then, the flaws must be evaluated to decide if they are acceptable or if they are defects. Once located, defects must be repaired.
- Weld tests can be divided into two groups, nondestructive and destructive. A nondestructive examination (NDE) does not damage the weld or the base metal. A destructive test results in at least some damage to the weld.
- Nondestructive examinations include liquid penetrant inspections, magnetic particle inspections, ultrasonic inspections, X-ray inspections, eddy current inspections, and air pressure leak tests.
- A visual inspection is useful to determine the size of a weld, to check for undercut, overlap, and other surface flaws, and to check travel speed, arc length, and depth of penetration.
- In a liquid penetrant inspection, a dye is applied to the weld. After a few minutes, the dye is cleaned from the weld and a developer is applied. Any dye trapped in a flaw is developed, revealing the location of the flaw. Liquid penetrant inspections are useful for detecting surface flaws, but cannot detect flaws inside the weld.
- To examine a part using magnetic particle inspection, the surface must first be cleaned. Then, fine magnetic particles are applied to the surface. After the particles are applied to the surface, a strong magnetic field is applied. The fine magnetic particles are drawn toward any flaw at the surface or very near the surface of the part.
- Ultrasonic inspection can be used on all metals and can locate internal flaws. Sound waves are sent into the part being inspected. The waves travel to the far side of the metal and bounce back. Any flaws within the weld are indicated on an oscilloscope.
- In an X-ray inspection, film is placed on one side of the part to be inspected. X rays pass through the part and expose the film. After the film is developed, any flaw or defect in the weld can be observed.
- Eddy current inspection detects porosity, cracks, slag inclusions, and lack of fusion at or near the surface of a weld by inducing eddy currents in the part and monitoring changes in current flow on an oscilloscope.
- Air-pressure or water-pressure leak tests are often used to inspect pipes or tanks that are to contain gases or liquids under pressure. In an air-pressure leak test, a soapy solution is applied to the pipe surface at each weld. When the pipe is under air pressure, a flaw will cause bubbles to form. When water pressure is used, water will spray out from the pipe where there is a leak.
- The most common destructive tests include, tensile tests, bend tests, fillet tests, hardness tests, impact tests, peel test (for spot welds), tensile shear test (for spot welds), and pressure tests.
- A tensile test is a destructive test in which a tensile (stretching or pulling) load is applied to a prepared sample until the sample breaks.
- Bend tests are used to evaluate the quality and ductility of a completed weld. In a guided bend test, the size, or radius, of the bend is controlled. In a free bend test, the sample is placed into a large, strong vise and struck with a hammer until it bends to a 90° angle. It is then hammered in the opposite direction to a 90° angle. This back and forth bending continues until the weld sample breaks.
- In a fillet test, force is applied to the welded joint until it breaks or until the vertical piece is bent flat against the horizontal piece. The weld should break along its centerline. A weld that breaks along the toe may indicate a lack of fusion.
- Hardness tests are used to measure a material's resistance to scratching and indentation. To determine the hardness of a metal, an indenter is pressed into the metal by a known force. The depth of indentation indicates the hardness of the metal.

(continued)

- In an impact test, a sample is held in a clamp and broken by an impact from a large known force. The amount of energy required to break the sample is displayed on a dial.
- A peel test is used to check the quality of spot welds. One edge of each piece is bent 90°, then force is applied to tear the metal apart.
- In a tensile shear test a spot welded sample is put into a tensile testing machine. Force is applied and the spot welds are torn apart.
- Pressure tests are used to determine how much pressure a pipe or cylinder can withstand or to test for leaks. The pressure inside the pipe or cylinder is steadily increased until a leak develops or the desired pressure is reached.

Review Questions

Write your answers on a separate sheet of paper. Please do *not* write in this book.

1. Explain the difference between a flaw and a defect.
2. Why are welds inspected?
3. Which nondestructive examinations cannot detect defects inside a weld?
4. List the five steps involved in performing a magnetic particle inspection.
5. What two values are determined from a tensile test?
6. List two results that will cause a bend test sample to fail the test.
7. Where is the root of the weld positioned on a root bend test sample after bending?
8. Where should a fillet weld test sample fail?
9. What type of indenter is pressed into a metal that is being hardness tested?
10. List the two types of tests used on spot welds.

Chapter 35

Welder Certification

Learning Objectives

After studying this chapter, you will be able to:

- Describe the use of codes and specifications to provide needed information on a required weld.
- Discuss the difference between a welding procedure specification and a welding performance qualification.
- Explain why a welder often must pass a number of welding performance qualifications.
- List the steps that must be followed to conform to most codes.

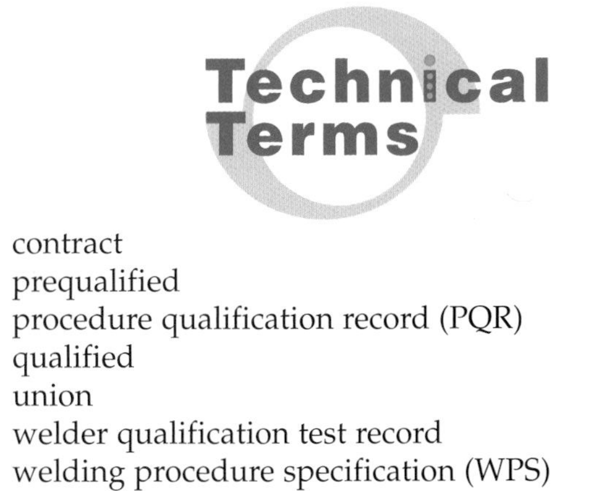

Technical Terms

contract
prequalified
procedure qualification record (PQR)
qualified
union
welder qualification test record
welding procedure specification (WPS)

Codes and Specifications

A ***contract*** is an agreement between two people or organizations. Whenever a welded structure is built, a contract is signed between the organization that is buying the structure and the organization that will build it. Whether the structure is a building, a bridge, a pipeline, an item of military equipment, or something else, a contract is involved. A contract states *what* is to be built, by *when*, for *how much*, and other technical and legal matters.

The contract outlines how all welding is to be performed. Certain government agencies and technical societies and associations have written welding codes and specifications. These codes and specifications cover all the needed information for making a required weld. Thus, the contract need not list all the information for every type of weld. Instead, it can refer to specific codes and specifications that list all required information regarding how a weld is to be made.

Codes and specifications list what can and cannot be done. They cover information on base metals, welding variables, filler metal compositions, and other details. They also cover many safety related topics. These codes and specifications help to standardize welding. Some of the more common codes and specifications are written by the following organizations.

Government Agencies:

- Federal Aviation Administration (FAA).
- Interstate Commerce Commission (ICC).
- Department of Defense—Military Specifications (MIL).
- Occupational Safety and Health Administration (OSHA).

Associations and Societies:

- American Institute of Steel Construction (AISC).
- American National Standards Institute (ANSI).
- American Petroleum Institute (API).
- American Society of Mechanical Engineers (ASME).
- American Welding Society (AWS).

Two widely used codes that deal with welding are the ASME Boiler and Pressure Vessel Code and the AWS Structural Welding Code. There are codes that cover base metal composition, electrode composition, quality of shielding gases, heat treating, and other aspects of welding. See **Figure 35-1**.

Figure 35-1. *Typical AWS and ASME code and standards booklets. These booklets tell how welds are to be made and tested. They also deal with welding symbols, terms, and safety practices.*

Welding Procedure Specification

Before any welding can be done on a structure, the procedure that will be used must first be approved, or qualified. The manufacturer specifically lists how the welding will be done on a ***welding procedure specification (WPS).*** Many welding procedure specifications are required because each major change that affects or could affect the quality of a weld must have its own WPS. See **Appendix A**.

The manufacturer must weld sample plates using the WPS. These samples are then tested to determine the strength, ductility, hardness, and overall ability of this given set of parameters to make a high quality weld. The codes and specifications list how many and what types of tests must be performed. Many of the destructive tests covered in Chapter 34 are used to evaluate the weld samples. The most common are the tensile and bend tests. Results of these tests are listed on a ***procedure qualification record (PQR).***

A procedure qualification record shows the properties of a weld that was made using the procedures in the WPS. When the results of the PQR show that welding meets the code or specification, the WPS is said to be ***qualified.*** Once welding procedure specifications are written and qualified, the WPS can be followed and welding can begin on the project. Both the WPS and the PQR must be kept on file.

A number of welding procedures in AWS structural welding code D1.1 are prequalified. ***Prequalified*** means the welding procedure has been established and tested. The welding company does not need to perform a PQR to prove the procedure. When a contract calls for welds to be made in accordance with D1.1, a prequalified welding procedure can be used.

Welding Performance Qualification

After a procedure has been qualified, as described in the preceding section, it can be followed to produce good quality weld joints. This assumes that a skilled welder is doing the welding.

Every welder who will make welds on a project must also be qualified. The welder's welding ability must be tested. When welders pass the test, they are said to be qualified. Qualified welders can consistently produce high quality welds while following a given set of welding conditions.

Welders must pass a welding performance qualification test before making welds in accordance with a code or specification. The results are recorded on a ***welder qualification test record.*** A welding performance qualification requires the welder to weld test plates following an approved WPS. Testing of the welds is specified by the codes. The employer keeps a record of these tests. See Appendix C. Usually, face and root bends are made, along with tensile tests. On thick plate, side bends are used instead of face and root bends. Welds are sometimes X-rayed to find defects beneath the weld surface.

Welders often must pass a number of welding performance qualifications. A different test is required each time one of the following is changed:

- Welding process.
- Base metal composition.
- Thickness of base metal (large change).
- Welding position.
- Electrode or filler metal type.

Thus, welders would have to pass a welding performance qualification test to weld mild steel using an E6013 SMAW electrode in the horizontal welding position. They would have to pass a different test to weld using an E6013 in the overhead welding position. **Figure 35-2** shows a welder welding a test plate.

If welders do not pass a performance qualification, they are not allowed to weld using that procedure. The company doing the testing can retest welders who do not pass. They may have to receive some training before retaking a test.

In most codes, base metals that have similar properties are grouped into one category. A welder who passes a qualification test on one base metal is qualified to weld on any base metal in that category. For example, a welder who passes a test on one type of stainless steel can also weld on other stainless steels that have similar welding properties.

When a welder passes a qualification test on one thickness of material, that welder becomes qualified for a range of thicknesses. Usually, a welder is qualified for a thickness up to twice the thickness of the test plate. **Figure 35-3** lists the thickness range that a welder is qualified for after completing a test plate of a given thickness.

Figure 35-2. *A welder using GMAW to weld a multi-pass V-groove butt joint in the horizontal (2G) position.*

Certain codes require welders to requalify. This means that welders must be retested to make sure they can still make quality welds. If a welder does not use a given process for three months, or if there is reason to question the welder's ability to follow the welding procedure, retesting is required. Some codes require a welder to requalify once a year.

Positions and Types of Joints

Some codes state that welders who pass a groove weld test are qualified to make groove welds and fillet welds. If welders pass a fillet weld test, they can make only fillet welds, not groove welds. Groove welds are considered more difficult to make.

Flat welding is considered to be the easiest position. Most welder qualifications are done in the horizontal, vertical, or overhead welding positions. Welders who pass a qualification test in one of these positions are qualified for that position and for welding in the flat position. Welders are allowed to weld only in the positions for which they have been qualified. Two separate welding performance qualifications may be required to qualify for uphill and downhill welding. **Figures 35-4** and **35-5** illustrate different positions for plate and pipe welding.

A Typical Welding Performance Qualification

Overhead welding is avoided in most plants, whenever possible. Most welding is done in the flat welding position. Some welding must be done in the horizontal and vertical positions.

Assume a company needs a welder qualified to weld 1/16″–1″ (1.6 mm–25.4 mm) stainless steel in the flat, horizontal and vertical welding positions, using SMAW. One set of welding performance qualification tests that may be given to the welder is listed below:

- Weld a square butt or bevel groove weld on 1/8″ (3.2 mm) stainless steel in the horizontal and vertical positions, using SMAW electrode E308.
- Weld a V-groove weld in 1/2″ (12.7 mm) plate on stainless steel in the horizontal and vertical positions, using SMAW electrode E308.

The welds would be inspected visually for undercut or overlap, and to evaluate the skills of the welder. Two tensile samples, plus one face bend and one root bend sample, would be tested from each weld.

If testing shows that the welds are of high quality, the welder is considered qualified to weld in the flat, horizontal, and vertical welding positions. The welder is qualified only to weld on certain types of stainless steel, using the shielded metal arc welding process. Only certain types of covered electrodes can be used. Welding the 1/8″ (3.2 mm) material qualifies the welder to weld on thicknesses from

Tension Test and Transverse Bend Tests

Thickness *T* of Test Coupon Welded, in. [Note (1)]	Range of Thickness *T* of Base Metal Qualified, in. [Note (2)]		Range of Thickness t of Deposited Weld Metal Qualified, in. [Note (2)]		Type and Number of Tests Required (Tension and Guided Bend Tests – Note (6)			
	Min.	Max.	Min.	Max.	Tension QW-462.1[4]	Side Bend QW-462.2	Face Bend QW-462.3(a)	Root Bend QW-462.3(a)
Less than 1/16	T	$2T$	t	$2t$	2	—	2	2
1/16 to 3/8, incl.	1/16	$2T$	1/16	$2t$	2	—	2 (5)	2 (5)
Over 3/8, but less than 3/4	3/16 (8)	$2T$	3/16 (8)	$2t$	2	Note (3)	2 (5)	2 (5)
3/4 to less than 1 1/2	3/16 (8)	$2T$	3/16 (8)	$2t$ when $t < 3/4$	2	4	—	—
3/4 to less than 1 1/2	3/16 (8)	$2T$	3/16 (8)	$2t$ when $t \geq 3/4$	2	4	—	—
1 1/2 and over	3/16 (8)	8 (7)	3/16 (8)	$2t$ when $t < 3/4$	2	4	—	—
1 1/2 and over	3/16 (8)	8 (7)	3/16 (8)	8 (7) when $t \geq 3/4$	2	4	—	—

Notes:

(1) When the groove weld is filled using two or more welding processes, the thickness *t* of the deposited weld metal for each welding process shall be determined and used in the "Range of thickness *t*" column. The test coupon thickness *T* is applicable for each welding process.
(2) See QW-403 (.2, .3, .6, .7, .9, .10) and QW-407.4 for further limits on range of thickness qualified.
(3) Four side bend tests may be substituted for the required face and root bend tests.
(4) The deposited weld metal of each welding process shall be included in the tension test of QW-462.1(a), (b), (c), or (e); and in the event turned specimens of QW-462.1(d) are used, the deposited weld metal of each welding process shall be included in the reduced section insofar as possible.
(5) Applicable for a combination of welding processes only when the deposited weld metal of each welding process is on the tension side of either the face or root bend.
(6) When toughness testing is a requirement of other Sections, it shall be applied with respect to each welding process.
(7) For the welding processes of QW-403.7 only; otherwise per Note (2) or 2*T*, whichever is applicable.
(8) When the weld metal thickness deposited by a process is 3/8 in. or less, this minimum shall be 1/16 in. for that process.

Tension Tests and Longitudinal Bend Tests

Thickness *T* of Test Coupon Welded, in. [Note (1)]	Range of Thickness *T* of Base Metal Qualified, in. [Note (2)]		Range of Thickness t of Deposited Weld Metal Qualified, in. [Note (2)]		Type and Number of Tests Required (Tension and Guided Bend Tests – Note (5)		
	Min.	Max.	Min.	Max.	Tension QW-462.1[3]	Face Bend QW-462.3(b)	Root Bend QW-462.3(b)
Less than 1/16	T	$2T$	t	$2t$	2	2	2
1/16 to 3/8, incl.	1/16	$2T$	1/16	$2t$	2	2 (4)	2 (4)
Over 3/8	3/16 (6)	$2T$	3/16 (6)	$2t$	2	2 (4)	2 (4)

Notes:

(1) When the groove weld is filled using two or more welding processes, the thickness *t* of the deposited weld metal for each welding process shall be determined and used in the "Range of thickness *t*" column. The test coupon thickness *T* is applicable for each welding process.
(2) See QW-403 (.2, .3, .6, .7, .9, .10) and QW-407.4 for further limits on range of thickness qualified.
(3) The deposited weld metal of each welding process shall be included in the tension test of QW-462.1(a), (b), (c), or (e); and in the event turned specimens of QW-462.1(d) are used, the deposited weld metal of each welding process shall be included in the reduced section insofar as possible.
(4) Applicable for a combination of welding processes only when the deposited weld metal of each welding process is on the tension side of either the face or root bend.
(5) When toughness testing is a requirement of other Sections, it shall be applied with respect to each welding process.
(6) When the weld metal thickness deposited by a process is 3/8 in. or less, this minimum shall be 1/16 in. for that process.

Figure 35-3. *Tables showing required tests for ASME procedure qualification. The tables show that a properly welded test coupon will qualify a welder on a range of thicknesses from* T *to* 2T. T *is the thickness of the metal welded. The QW numbers refer to parts of the ASME code.*

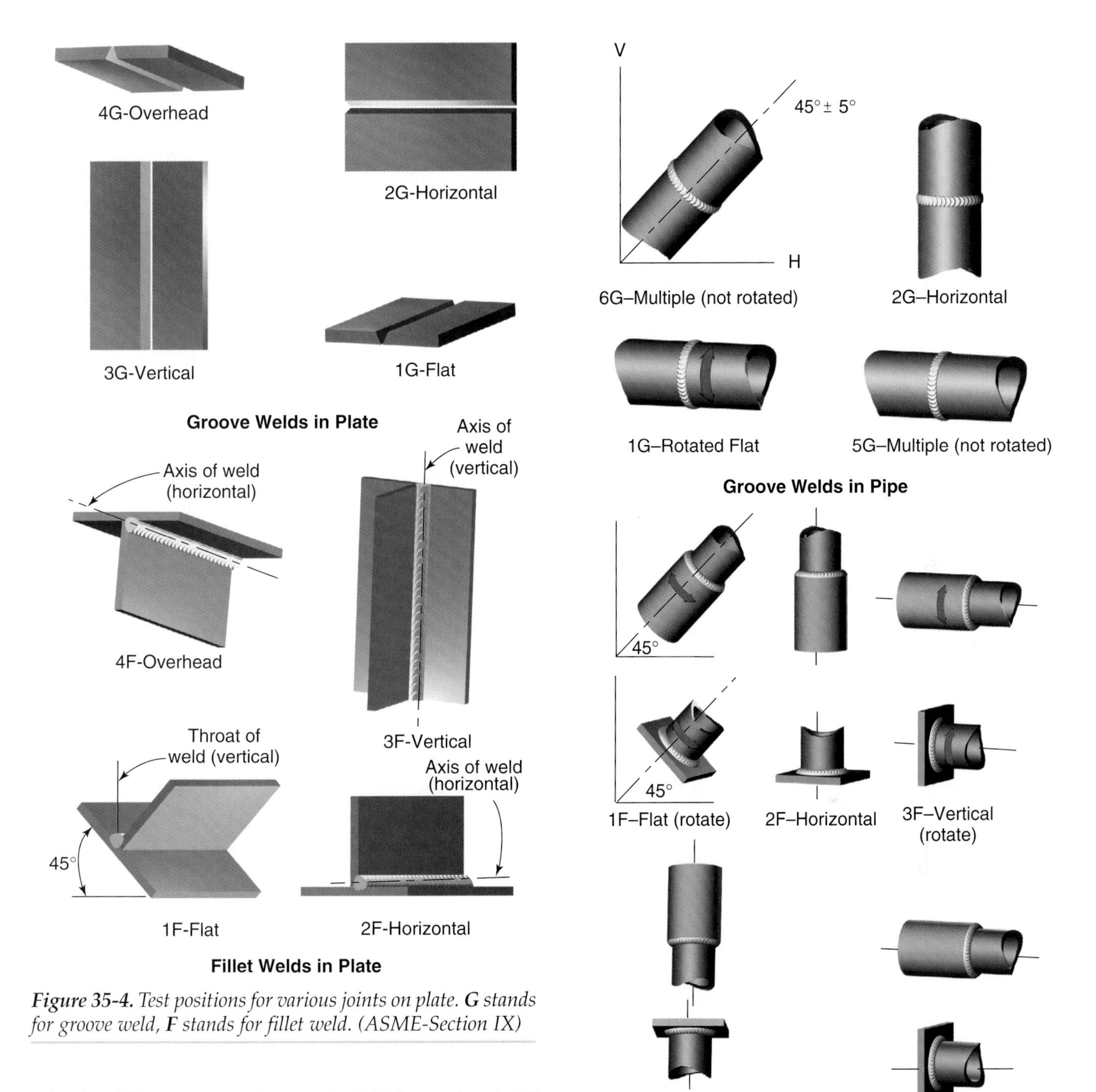

Figure 35-4. *Test positions for various joints on plate.* ***G*** *stands for groove weld,* ***F*** *stands for fillet weld. (ASME-Section IX)*

Figure 35-5. *Test positions for groove and fillet welds on pipe joints. Note that test positions 5F, 5G, and 6G are not rotated. The weld position changes as the weld is made. (ASME-Section IX)*

1/16″–1/4″ (1.6 mm–6.4 mm). Welding the 1/2″ (12.7 mm) plate qualifies the welder for thicknesses from 3/16″–1″ (4.8 mm–25.4 mm). See Figure 35-3.

Summary: Welding per Codes and Specifications

A contract states how the welding on a particular job is to be accomplished. The manufacturer is responsible for making sure that all welding meets the requirements of the codes and specifications listed in the contract. The following are the steps that are followed to conform to most codes.

1. A welding procedure specification (WPS) is written.
2. Test plates are welded, following the WPS.
3. Welded plates are tested. The results are listed on a procedure qualification record (PQR).
4. The PQR must show that the WPS results in good quality welds. Once this is done, the WPS is approved.

5. Welders are tested and the results are recorded on a welder qualification test record. This performance qualification tests the ability of each welder to weld as required by an approved WPS. Welders who pass the welding performance qualification test are allowed to weld.
6. Welders are retested periodically to make sure they are still able to make high quality welds.

Employment Considerations

When reviewing applications for a welding position, an employer often considers three things: how much training, how much education, and how much experience a person has.

Welding training is obtained through industrial education classes in high schools, trade schools, community colleges, and universities. These classes usually teach basic welding techniques. Some classes, especially those offered through a good trade school, can prepare you to take and pass welder performance qualification tests.

Employers often look to see how much education you have completed. A high school diploma is helpful in applying for a welding position.

Experience is very important to an employer. Even though some companies have training programs, they prefer to hire welders that are already experienced. When applying for your first welding position, you can tell the employer how many hours of training you have had.

Labor Unions

On a large construction project, such as a bridge or a pipeline, the company that is building the project may not have any welders who are permanent employees. Instead of interviewing and hiring welders, the company can go to a local union for the welders it needs.

The ***union*** will send a number of qualified welders to the job site. The welders are tested by the company. Those that pass the performance tests are hired by the company to work until the project is completed.

Unions sometimes represent the employees within a company. The workers are members of the union and are also permanent employees of a company.

Summary

- A contract is an agreement between two people or organizations. Whenever a welded structure is built, a contract is signed between the organization that is buying the structure and the organization that will build it. A contract states *what* is to be built, by *when*, for *how much*, and other technical and legal matters.
- Certain government agencies and technical societies and associations have written welding codes and specifications. These codes and specifications cover all the needed information for making a required weld and help to standardize welding.
- Before any welding can be done on a structure, the procedure that will be used must first be approved, or qualified. The manufacturer specifically lists how the welding will be done on a welding procedure specification (WPS).
- The manufacturer must weld sample plates using the welding procedure specification. These samples are then tested to determine the strength, ductility, hardness, and overall ability of this given set of parameters to make a high quality weld. The codes and specifications list how many and what types of tests must be performed. Results of these tests are listed on a procedure qualification record (PQR). When the results of the PQR show that welding meets the code or specification, the WPS is said to be qualified. AWS D1.1 specification has prequalified procedures.
- Every welder who will make welds on a project must be qualified. Welders must pass a welding performance qualification test before making welds in accordance with a code or specification. A welding performance qualification requires the welder to weld test plates following an approved WPS. If welders do not pass a performance qualification test, they are not allowed to weld using that procedure.

(continued)

- Some codes state that welders who pass a groove weld test are qualified to make groove welds and fillet welds. However, if welders pass a fillet weld test, they can make only fillet welds, not groove welds. Most welder qualifications are done in the horizontal, vertical, or overhead welding positions. Welders who pass a qualification test in one of these positions are qualified for that position and for welding in the flat position.
- When reviewing applications for a welding position, an employer often considers how much training, how much education, and how much experience a person has.
- When undertaking a large project, companies may go to labor unions to hire qualified welders on a temporary basis. Permanent employees of a company may also be union members. In these cases, the union represents the employees within the company.

Review Questions

Write your answers on a separate sheet of paper. Please do *not* write in this book.

1. What agreement is signed before a welded structure is built?
2. What groups write codes and specifications?
3. List two widely used codes or specifications that deal with welding.
4. What document must a manufacturer prepare that specifically lists how welding will be performed?
5. On what document are the results from the test plates recorded?
6. On what document are the results of welding performance qualification tests written?
7. Name three destructive tests that are usually performed on a welder's test plates to determine the quality of the weld.
8. If a welder welds a test plate that is 1″ (2.54 mm) thick, what range of thickness is that welder qualified to weld? Refer to Figure 35-3.
9. When qualified to weld on one type of material, what other type of material can the welder weld on?
10. What three things does an employer often consider when a person applies for a job?

Appendix A

WELDING PROCEDURE SPECIFICATION (WPS) Yes ☐
PREQUALIFIED __________ QUALIFIED BY TESTING __________
or PROCEDURE QUALIFICATION RECORDS (PQR) Yes ☐

Identification # ______________________
Revision ________ Date __________ By __________
Authorized by ______________ Date __________
Type—Manual ☐ Semiautomatic ☐
Machine ☐ Automatic ☐

Company Name ______________________
Welding Process(es) ______________________
Supporting PQR No.(s) ______________________

JOINT DESIGN USED
Type:
Single ☐ Double Weld ☐
Backing: Yes ☐ No ☐
Backing Material:
Root Opening ______ Root Face Dimension ________
Groove Angle: __________ Radius (J–U) ________
Back Gouging: Yes ☐ No ☐ Method ________

BASE METALS
Material Spec. ______________________
Type or Grade ______________________
Thickness: Groove __________ Fillet __________
Diameter (Pipe) ______________________

FILLER METALS
AWS Specification ______________________
AWS Classification ______________________

SHIELDING
Flux __________________ Gas __________
Composition __________
Electrode-Flux (Class) ______ Flow Rate __________
__________________ Gas Cup Size __________

PREHEAT
Preheat Temp., Min. ______________________
Interpass Temp., Min. __________ Max. __________

POSITION
Position of Groove: ______________ Fillet: __________
Vertical Progression: Up ☐ Down ☐

ELECTRICAL CHARACTERISTICS

Transfer Mode (GMAW) Short-Circuiting ☐
Globular ☐ Spray ☐
Current: AC ☐ DCEP ☐ DCEN ☐ Pulsed ☐
Power Source: CC ☐ CV ☐
Other ______________________
Tungsten Electrode (GTAW)
Size: __________
Type: __________

TECHNIQUE
Stringer or Weave Bead: __________________
Multi-pass or Single Pass (per side) __________________
Number of Electrodes __________________
Electrode Spacing Longitudinal __________
Lateral __________
Angle __________
Contact Tube to Work Distance __________________
Peening ______________________
Interpass Cleaning: ______________________

POSTWELD HEAT TREATMENT
Temp. ______________________
Time ______________________

WELDING PROCEDURE

Pass or Weld Layer(s)	Process	Filler Metals		Current		Volts	Travel Speed	Joint Details
		Class	Diam.	Type & Polarity	Amps or Wire Feed Speed			

Form N-1 (Front)

Appendix A. *This form is used as both the Welding Procedure Specification (WPS) and as the first page of the Procedure Qualification Record (PQR). (AWS D1.1:2008, Annex N, Welding Procedure Specification (WPS) / Procedure Qualification Records (PQR), reproduced with permission from the American Welding Society, Miami, FL)*

Metric – Inch Equivalents

Inches: Fractions	Inches: Decimals	Millimeters
	.00394	.1
	.00787	.2
	.01181	.3
1/64	.015625	.3969
	.01575	.4
	.01969	.5
	.02362	.6
	.02756	.7
1/32	.03125	.7938
	.0315	.8
	.03543	.9
	.03937	1.00
3/64	.046875	1.1906
1/16	.0625	1.5875
5/64	.078125	1.9844
	.07874	2.00
3/32	.09375	2.3813
7/64	.109375	2.7781
	.11811	3.00
1/8	.125	3.175
9/64	.140625	3.5719
5/32	.15625	3.9688
	.15748	4.00
11/64	.171875	4.3656
3/16	.1875	4.7625
	.19685	5.00
13/64	.203125	5.1594
7/32	.21875	5.5563
15/64	.234375	5.9531
	.23622	6.00
1/4	.2500	6.35
17/64	.265625	6.7469
	.27559	7.00
9/32	.28125	7.1438
19/64	.296875	7.5406
5/16	.3125	7.9375
	.31496	8.00
21/64	.328125	8.3344
11/32	.34375	8.7313
	.35433	9.00
23/64	.359375	9.1281
3/8	.375	9.525
25/64	.390625	9.9219
	.3937	10.00
13/32	.40625	10.3188
27/64	.421875	10.7156
	.43307	11.00
7/16	.4375	11.1125
29/64	.453125	11.5094
15/32	.46875	11.9063
	.47244	12.00
31/64	.484375	12.3031
1/2	.5000	12.70
	.51181	13.00
33/64	.515625	13.0969
17/32	.53125	13.4938
35/64	.546875	13.8907
	.55118	14.00
9/16	.5625	14.2875
37/64	.578125	14.6844
	.59055	15.00
19/32	.59375	15.0813
39/64	.609375	15.4782
5/8	.625	15.875
	.62992	16.00
41/64	.640625	16.2719
21/32	.65625	16.6688
	.66929	17.00
43/64	.671875	17.0657
11/16	.6875	17.4625
45/64	.703125	17.8594
	.70866	18.00
23/32	.71875	18.2563
47/64	.734375	18.6532
	.74803	19.00
3/4	.7500	19.05
49/64	.765625	19.4469
25/32	.78125	19.8438
	.7874	20.00
51/64	.796875	20.2407
13/16	.8125	20.6375
	.82677	21.00
53/64	.828125	21.0344
27/32	.84375	21.4313
55/64	.859375	21.8282
	.86614	22.00
7/8	.875	22.225
57/64	.890625	22.6219
	.90551	23.00
29/32	.90625	23.0188
59/64	.921875	23.4157
15/16	.9375	23.8125
	.94488	24.00
61/64	.953125	24.2094
31/32	.96875	24.6063
	.98425	25.00
63/64	.984375	25.0032
1	1.0000	25.4000

Appendix A

WELDING PROCEDURE SPECIFICATION (WPS) Yes ☐
PREQUALIFIED __________ QUALIFIED BY TESTING __________
or PROCEDURE QUALIFICATION RECORDS (PQR) Yes ☐

Identification # ____________________
Revision _______ Date __________ By __________
Authorized by ______________ Date __________
Type—Manual ☐ Semiautomatic ☐
Machine ☐ Automatic ☐

Company Name ____________________
Welding Process(es) ____________________
Supporting PQR No.(s) ____________________

JOINT DESIGN USED
Type:
Single ☐ Double Weld ☐
Backing: Yes ☐ No ☐
Backing Material:
Root Opening ______ Root Face Dimension ________
Groove Angle: __________ Radius (J–U) ________
Back Gouging: Yes ☐ No ☐ Method _______

BASE METALS
Material Spec. ____________________
Type or Grade ____________________
Thickness: Groove __________ Fillet ________
Diameter (Pipe) ____________________

FILLER METALS
AWS Specification ____________________
AWS Classification ____________________

SHIELDING
Flux ______________ Gas ____________
Composition ________
Electrode-Flux (Class) _____ Flow Rate __________
______________ Gas Cup Size ________

PREHEAT
Preheat Temp., Min. ____________________
Interpass Temp., Min. __________ Max. ________

POSITION
Position of Groove: ______________ Fillet: __________
Vertical Progression: Up ☐ Down ☐

ELECTRICAL CHARACTERISTICS
____
Transfer Mode (GMAW) Short-Circuiting ☐
Globular ☐ Spray ☐
Current: AC ☐ DCEP ☐ DCEN ☐ Pulsed ☐
Power Source: CC ☐ CV ☐
Other ____________________
Tungsten Electrode (GTAW)
Size: __________
Type: __________

TECHNIQUE
Stringer or Weave Bead: ____________________
Multi-pass or Single Pass (per side) ______________
Number of Electrodes ____________________
Electrode Spacing Longitudinal __________
Lateral __________
Angle __________
Contact Tube to Work Distance ______________
Peening ____________________
Interpass Cleaning: ____________________

POSTWELD HEAT TREATMENT
Temp. ____________________
Time ____________________

WELDING PROCEDURE

Pass or Weld Layer(s)	Process	Filler Metals		Current		Volts	Travel Speed	Joint Details
		Class	Diam.	Type & Polarity	Amps or Wire Feed Speed			

Form N-1 (Front)

Appendix A. *This form is used as both the Welding Procedure Specification (WPS) and as the first page of the Procedure Qualification Record (PQR). (AWS D1.1:2008, Annex N, Welding Procedure Specification (WPS) / Procedure Qualification Records (PQR), reproduced with permission from the American Welding Society, Miami, FL)*

Appendix A

Procedure Qualification Record (PQR) # __________
Test Results

TENSILE TEST

Specimen No.	Width	Thickness	Area	Ultimate Tensile Load, lb	Ultimate Unit Stress, psi	Character of Failure and Location

GUIDED BEND TEST

Specimen No.	Type of Bend	Result	Remarks

VISUAL INSPECTION
Appearance__________
Undercut __________
Piping porosity __________
Convexity__________
Test date __________
Witnessed by__________

Radiographic-ultrasonic examination
RT report no.: __________ Result__________
UT report no.:__________ Result__________

FILLET WELD TEST RESULTS

Minimum size multiple pass
Macroetch
1. ______ 3. ______
2. ______

Maximum size single pass
Macroetch
1. ______ 3. ______
2. ______

Other Tests

All-weld-metal tension test

Tensile strength, psi __________
Yield point/strength, psi __________
Elongation in 2 in, % __________
Laboratory test no. __________

Welder's name __________ Clock no. __________ Stamp no.__________

Tests conducted by __________ Laboratory

Test number __________

Per __________

We, the undersigned, certify that the statements in this record are correct and that the test welds were prepared, welded, and tested in conformance with the requirements of Clause 4 of AWS D1.1/D1.1M, (__________) *Structural Welding Code—Steel.*
(year)

Signed __________
Manufacturer or Contractor

By__________

Title __________

Date __________

Form N-1 (Back)

Appendix A. (cont.) *This form is the second page of the Procedure Qualification Record (PQR). (AWS D1.1:2008, Annex N, Procedure Qualification Record Test Results, reproduced with permission from the American Welding Society, Miami, FL)*

Appendix B

WELDER, WELDING OPERATOR, OR TACK WELDER QUALIFICATION TEST RECORD

Type of Welder ____________________

Name ____________________ Identification No.____________________

Welding Procedure Specification No. __________ Rev __________ Date __________

Variables	Record Actual Values Used in Qualification	Qualification Range
Process/Type [Table 4.12, Item (1)]		
Electrode (single or multiple) [Table 4.12, Item (7)]		
Current/Polarity		
Position [Table 4.12, Item (4)]		
Weld Progression [Table 4.12, Item (5)]		
Backing (YES or NO) [Table 4.12, Item (6)]		
Material/Spec.	to	
Base Metal		
Thickness: (Plate)		
Groove		
Fillet		
Thickness: (Pipe/tube)		
Groove		
Fillet		
Diameter: (Pipe)		
Groove		
Fillet		
Filler Metal (Table 4.12)		
Spec. No.		
Class		
F-No. [Table 4.12, Item (2)]		
Gas/Flux Type (Table 4.12)		
Other		

VISUAL INSPECTION (4.8.1)

Acceptable YES or NO _____

Guided Bend Test Results (4.30.5)

Type	Result	Type	Result

Fillet Test Results (4.30.2.3 and 4.30.4.1)

Appearance ____________________ Fillet Size ____________________

Fracture Test Root Penetration ____________________ Macroetch ____________________

(Describe the location, nature, and size of any crack or tearing of the specimen.)

Inspected by ____________________ Test Number ____________________

Organization ____________________ Date ____________________

RADIOGRAPHIC TEST RESULTS (4.30.3.2)

Film Identification Number	Results	Remarks	Film Identification Number	Results	Remarks

Interpreted by ____________________ Test Number ____________________

Organization ____________________ Date ____________________

We, the undersigned, certify that the statements in this record are correct and that the test welds were prepared, welded, and tested in conformance with the requirements of Clause 4 of AWS D1.1/D1.1M, (__________) *Structural Welding Code—Steel.*
(year)

Manufacturer or Contractor ____________________ Authorized By ____________________

Form N-4 Date ____________________

Appendix B. *This form is a copy of the Welder Qualification Test Record. (AWS D1.1:2008, Annex N, Procedure Qualification Record Test Results, reproduced with permission from the American Welding Society, Miami, FL)*

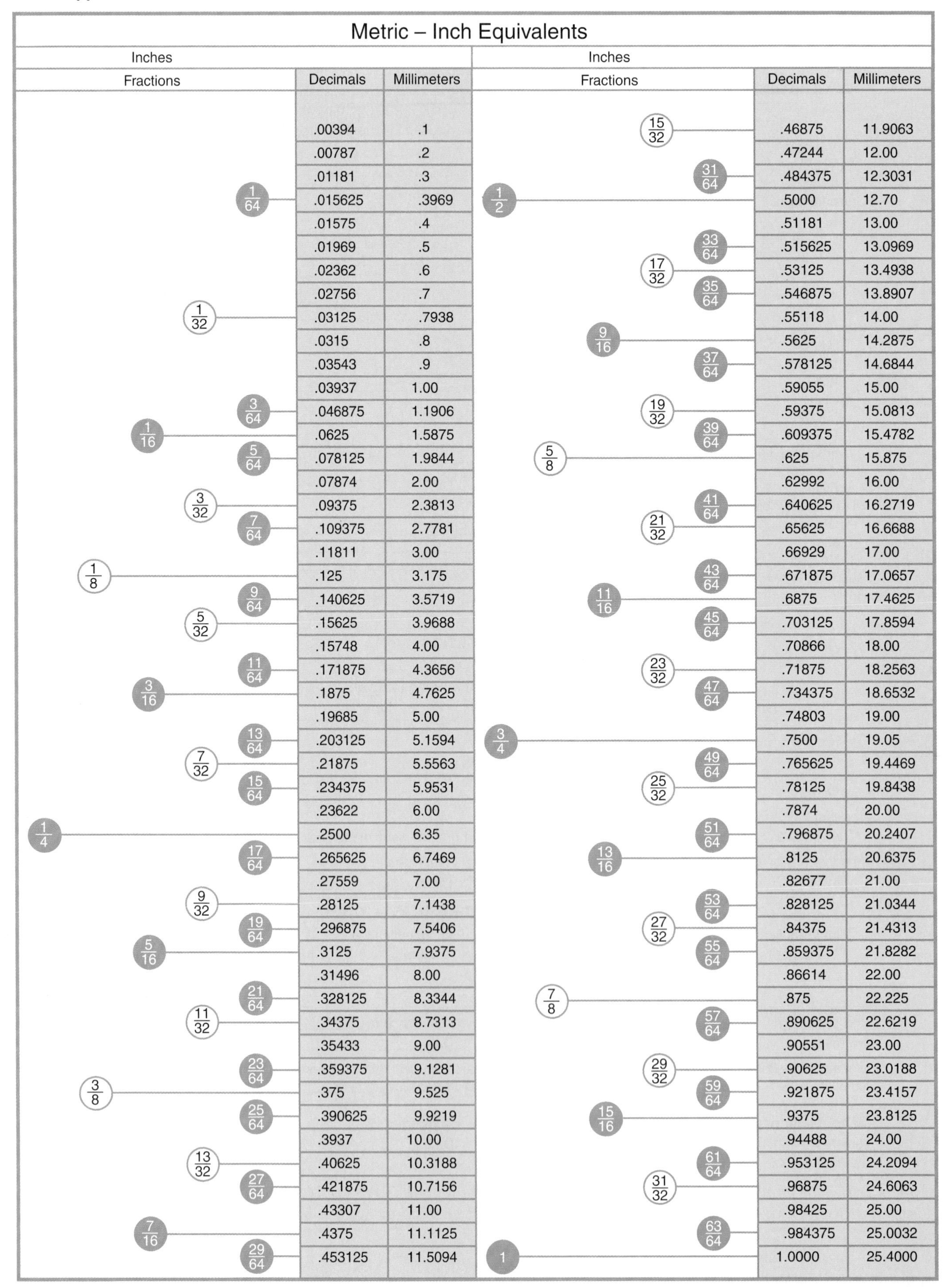

Metric – Inch Equivalents

Inches			Inches		
Fractions	Decimals	Millimeters	Fractions	Decimals	Millimeters
	.00394	.1	15/32	.46875	11.9063
	.00787	.2		.47244	12.00
	.01181	.3	31/64	.484375	12.3031
1/64	.015625	.3969	1/2	.5000	12.70
	.01575	.4		.51181	13.00
	.01969	.5	33/64	.515625	13.0969
	.02362	.6	17/32	.53125	13.4938
	.02756	.7	35/64	.546875	13.8907
1/32	.03125	.7938		.55118	14.00
	.0315	.8	9/16	.5625	14.2875
	.03543	.9	37/64	.578125	14.6844
	.03937	1.00		.59055	15.00
3/64	.046875	1.1906	19/32	.59375	15.0813
1/16	.0625	1.5875	39/64	.609375	15.4782
5/64	.078125	1.9844	5/8	.625	15.875
	.07874	2.00		.62992	16.00
3/32	.09375	2.3813	41/64	.640625	16.2719
7/64	.109375	2.7781	21/32	.65625	16.6688
	.11811	3.00		.66929	17.00
1/8	.125	3.175	43/64	.671875	17.0657
9/64	.140625	3.5719	11/16	.6875	17.4625
5/32	.15625	3.9688	45/64	.703125	17.8594
	.15748	4.00		.70866	18.00
11/64	.171875	4.3656	23/32	.71875	18.2563
3/16	.1875	4.7625	47/64	.734375	18.6532
	.19685	5.00		.74803	19.00
13/64	.203125	5.1594	3/4	.7500	19.05
7/32	.21875	5.5563	49/64	.765625	19.4469
15/64	.234375	5.9531	25/32	.78125	19.8438
	.23622	6.00		.7874	20.00
1/4	.2500	6.35	51/64	.796875	20.2407
17/64	.265625	6.7469	13/16	.8125	20.6375
	.27559	7.00		.82677	21.00
9/32	.28125	7.1438	53/64	.828125	21.0344
19/64	.296875	7.5406	27/32	.84375	21.4313
5/16	.3125	7.9375	55/64	.859375	21.8282
	.31496	8.00		.86614	22.00
21/64	.328125	8.3344	7/8	.875	22.225
11/32	.34375	8.7313	57/64	.890625	22.6219
	.35433	9.00		.90551	23.00
23/64	.359375	9.1281	29/32	.90625	23.0188
3/8	.375	9.525	59/64	.921875	23.4157
25/64	.390625	9.9219	15/16	.9375	23.8125
	.3937	10.00		.94488	24.00
13/32	.40625	10.3188	61/64	.953125	24.2094
27/64	.421875	10.7156	31/32	.96875	24.6063
	.43307	11.00		.98425	25.00
7/16	.4375	11.1125	63/64	.984375	25.0032
29/64	.453125	11.5094	1	1.0000	25.4000

Appendix D

Advantages/Disadvantages of Various Welding Processes Using Consumable Electrodes

Process	Advantages/Disadvantages
SMAW	**Advantages** • Low initial cost. • Flexibility. • Usable in all positions. • Portability. • Many types of filler metal with special characteristics available. **Disadvantage** • Requires slag removal.
GMAW	**Advantages** • Higher deposition rates than SMAW. • Relative flexibility. • Adaptable to mechanized, robotic, or automatic welding method. • Doesn't require slag removal. **Disadvantages** • Needs special power source to be usable in all positions. • Needs external gas supply and wire feeder.
FCAW	**Advantages** • Higher deposition rates than SMAW and GMAW. • Adaptable to mechanized, robotic, or automatic welding methods. • Relative flexibility. • Usable in all positions without special power source. **Disadvantages** • Needs external gas supply for most electrodes. • Requires wire feeder. • Requires slag removal.
SAW	**Advantages** • Very high deposition rate when mechanized. • High-quality, low-cost process when mechanized. **Disadvantages** • Can be used only for flat and horizontal position welding. • Needs large capital investment. • Requires slag and flux removal.

Appendix E

Percent of Welding Filler Metal That Actually Becomes Part of a Completed Weld Joint

Filler Metal Form/Process	Percent Deposited In Weld
Bare solid wire	
Electoslag	95-99
Gas metal arc	90-95
Gas tungsten arc	99
Submerged arc	95-99
Covered electrodes (SMAW)	
14″ length	55-65
18″ length	60-70
28″ length	65-75
Flux-cored electrodes	
Flux-cored arc	80-85

Appendix F

Variables That Affect the Cost of Welding

Type of:
- electrode or filler wire
- joint
- shielding
- weld

Size of:
- electrode or filler wire
- weld

Other variables:
- arc voltage
- electrode deposition efficiency
- flux consumption ratio
- power source efficiency
- shielding gas flow rate
- total time to complete welding
- weld current

Expense of:
- edge preperation
- electric power
- filler metal
- finishing
- flux
- inspection
- labor
- overhead
- set-up
- shielding gas

Glossary of Technical Terms

A

Abrasion: worn condition produced by rubbing.
Abrasion resistance: quality of a material that allows it to resist abrasive wear.
ABS (acrylonitrile butadiene styrene): a plastic widely used for automotive body panels and other components.
Acetone: a colorless and extremely flammable liquid used to fill the air pockets of the porous material inside an acetylene storage cylinder. The acetylene is absorbed into the acetone for safe storage.
Acetylene (C_2H_2): a colorless fuel gas composed of two parts carbon and two parts hydrogen. When burned in the presence of oxygen, acetylene produces one of the highest flame temperatures obtainable.
Acetylene generator: a device used to produce acetylene fuel gas in a controlled environment. Calcium carbide is placed in a hopper and fed into water at a controlled rate to chemically produce acetylene.
Active gas: gas that will combine with the weld metal if given the chance. Carbon dioxide and oxygen are active gases.
Actuator: a device used to move part of a machine. Robots use electric and hydraulic actuators.
Air carbon arc cutting (CAC-A): a cutting process that uses a carbon electrode to carry current for burning the base metal and high-pressure air for blowing away the molten base metal.
Air-fuel gas torch: a torch burning a combination of air and a fuel gas, such as propane, to provide heat for soldering.
Air pressure leak test: inspection method used for pipelines and storage tanks. Welds are coated with a soap solution, then the vessel is filled with gas under pressure. Bubbles show any leaks due to defective welds.
Alloy: a pure metal that has had additional metal or metallic elements added to it while molten. An alloy has mechanical properties that are different (usually improved) from those of the pure metal.
Alternating current (ac): type of electricity in which the direction of electron flow reverses at regular intervals.
Aluminum oxide: compound of aluminum and oxygen with a higher melting temperature than pure aluminum.
Ammeter: instrument used to measure electrical current in amperes.
Ampere: unit of measurement of electrical current. One ampere of current will flow through a conductor that has a resistance of one ohm at a potential (electrical pressure) of one volt.
Annealing: softening metals by heat treatment.
ANSI/AWS welding symbol: a welding symbol developed by the American National Standards Institute (ANSI) and the American Welding Society (AWS) to convey all the information needed to properly make a weld on a joint.
Applicator: a device used to add flux to a joint being soldered.
Arc: flow of electricity across a gaseous space (air gap).
Arc blow: wandering of a welding arc from its normal path because of magnetic forces.
Arc flash: unprotected eye exposure to a welding arc that can result in temporary or permanent eye damage.
Arc length: distance between the electrode and the base metal. In SMAW, this distance should be approximately equal to the electrode diameter.
Arc spraying (ASP): surfacing process using electric arc between two electrodes of surfacing material and a pressurized gas to propel the vaporized material onto the base metal surface.
Arc stud welding (SW): a quick, automatic welding process used to attach connectors, such as threaded studs, location pins, and nails, to metal plates.
Arc welding: a group of welding processes used to melt and weld metal using the heat of an electric arc, with or without filler metal.
Argon (Ar): an inert gas found in the atmosphere and used as a shielding gas in some welding operations.
Arrow: on a welding symbol, a line with a point that touches the line to be welded.
Arrow side: area under the welding symbol's reference line; the side of the weld touched by the arrow.
AWS: American Welding Society.
AWS electrode specifications: strict manufacturing specifications developed by the American Welding Society for electrodes and fluxes. These specifications are published and revised every five years.
AWS welding symbol: see **ANSI/AWS welding symbol.**
Axis: line around which something rotates.

B

Backfire: a short "pop" of the torch flame, followed by extinguishing of the flame or continued burning of the gases.
Background current: a relatively low amperage used to maintain the arc during pulsed spray transfer GMAW.
Backhand welding: a method used when the welding end of an electrode or a gas flame is pointed in a direction opposite the direction of travel.

Backing: material placed beyond the root opening to control penetration and prevent a hot-shortness hole. Backing may be machined to control the shape of the penetration bead.
Backing ring: a metal ring placed inside a pipe before butt welding to ensure complete weld penetration and a smooth inside surface.
Backing weld: a weld placed on the root side of a weld to aid in the control of penetration.
Backing weld symbol: a symbol used when a stringer bead is laid on the root side of the weld to ensure complete penetration.
Backward arc blow: arc blow that occurs at the end of the joint as the magnetic field tries to stay in the metal. This causes the filler metal to blow toward the center of the joint.
Base material: nonmetallic material to be welded, cut, or brazed.
Base metal: metal to be welded, cut, brazed, or soldered.
Basic weld symbol: symbol used to describe the shape of the welding joint. It is used as part of the welding symbol.
Bead: see **Weld bead.**
Bell-mouthed kerf: a kerf that is widened at the bottom as a result of using too much oxygen while cutting.
Bend test: a test, described by code, in which a sample weld plate is bent under specific conditions until it fails.
Bevel angle: the angle formed between the prepared edge of one piece and a plane perpendicular to the surface of that piece.
Bird's nest: a tangle of welding wire in a GMAW wire drive mechanism.
Brazement: assembly joined by brazing.
Braze welding: making an adhesion groove, fillet, or plug connection above 840°F (450°C). The brazing material does not flow in the joint by capillary action.
Brazing: making an adhesion connection with a minimum of alloy that melts above 840°F (450°C). The brazing material flows between the parts by capillary action.
Brazing rod: filler metal rod used in brazing and braze welding.
Bridge: the process of filling wide gaps in poorly fitted assemblies.
Bridge rectifier: a group of diodes or SCRs used to transform ac into dc.
Brinell hardness test: an accurate measurement of a metal's hardness, made with an instrument that presses a hard steel ball into the surface of the metal under standard conditions.
Brittleness: the quality of a material that causes the material to develop cracks when it is deformed (bent) only slightly.
Buildup: the amount that the weld face extends above the surface of the base metal.
Buildup process: applying surfacing materials to a part so that its worn surface is returned to its original dimensions.
Bullet-shaped ripples: proper appearance of the evenly spaced ripples in a completed stringer bead.
Burning: see **Oxyfuel gas cutting.**
Buttering: process used to form a transition layer when welding dissimilar metals. One or more layers of easily welded material are applied to the surface of a part with poor welding characteristics.
Butt joint: assembly in which two pieces joined are in the same plane, with the edge of one piece touching the edge of the other.

C

Cable: see **Lead.**
Cadmium (Cd): a chemical element sometimes used in soldering alloys. It is a toxic material.
Cam lock–style cable connector: a commonly used connector on electrode and workpiece cable leads that function with quick-connect terminals on welding machines.
Capacitor: a component in electrical circuits used to store an electrical charge.
Capillary action: in brazing, the process that occurs when the filler metal is drawn between tightly fitted, mating surfaces.
Carbon dioxide (CO_2): a chemical compound used as a shielding gas. CO_2 is often used when welding carbon steel with GMAW.
Carbon steel: a steel made by adding small, controlled amounts of carbon to pure iron.
Carburizing flame: an oxyfuel gas flame that occurs when there is too much fuel gas in the mixture and not enough oxygen to burn it.
Carrying a weld pool: the process of creating a weld pool of molten base metal and advancing it along a line.
CC: see **Constant current**.
Ceria: a compound (cerium oxide) added to tungsten electrodes.
Chain intermittent welding: intermittent welding on two sides of a joint, arranged so that the welds on the two sides are opposite each other.
Check valve: valve designed to allow a gas or fluid to flow in only one direction, preventing the reverse flow of gases through the torch, hoses, and/or regulators.
Chemical composition: a list of the different metals or elements that are combined to produce a certain metal or metal alloy.
Chemical corrosion: eating away of a metal surface as a result of chemical reaction.
Chemical properties: properties that determine how a material will withstand the effects of the environment.
Chipping: the process of cleaning slag from a weld bead by striking it with a chipping hammer.
Chipping goggles: eye protection worn while chipping slag from a weld bead.
Chipping hammer: cleaning tool with a sharp pointed pick at one end of its head, used to remove slag from weld beads.
Circuit: the path of electron flow from the source through various components and back to the source.
Circuit breaker: a device designed to shut off the flow of current in a circuit if the amperage becomes too high.
Cladding: a process used to apply surfacing materials that improve the corrosion or heat resistance of a part.
Clearance: a gap or space between adjoining or mating surfaces.
Clockwise: rotation to the right when facing the object being rotated; in the same direction as movement of a clock's hands.
Closed arc: an arc enclosed by the heavy covering material on a SMAW electrode.
Coarse amperage range: one of several ranges that provide electrical current output on a welding machine. Coarse amperage ranges usually extend over 50A or more.
Coated electrode: see **Covered electrode**.

Code: a written system of figures, symbols, rules, regulations, and procedures used to regulate the welding industry.
Cold welding (CW): a process that uses high pressure to fuse metal parts. No outside heat is applied.
Collet: a sleeve used in a GMAW gun or GTAW torch to hold the electrode tightly enough to ensure electrical contact with the welding power supply.
Collet body: the part of the torch or gun that accepts the collet.
Combustibles: easily ignited materials such as wood, paper, rags, and flammable liquids.
Compressive strength: greatest stress developed in material under compression.
Concave bead: a weld bead with a surface that curves inward, toward the root of the weld.
Cone: the visible inner flame shape of a neutral or nearly neutral flame.
Constant current (CC): term used for arc welding machines that produce a nearly constant current even though the arc gap voltage may vary.
Constant potential: an alternate term for constant voltage.
Constant voltage (CV): term used for arc welding machines that produce a nearly constant voltage even when the current changes.
Constricting: reducing in size or diameter, as in a constricting orifice.
Consumable electrode: an electrode that melts and becomes part of the weld.
Contactor switch: a switch on a welding machine used for GTAW that allows the welder to choose whether current, shielding gas, and water flow can be controlled remotely with a foot pedal or thumb switch while welding.
Contact tube: a device that transfers current to a continually fed electrode.
Contamination: an undesirable substance in or around a weld.
Continuous weld: completing the weld in one operation, without stopping.
Contour: the shape of the face of the weld.
Contour symbol: see **Weld contour symbol.**
Contract: a legal agreement between two parties.
Contraction: the shortening of metal, usually upon cooling.
Controller: an electronic device (usually a computer) with a memory that can be programmed to signal automatic equipment when and where to move.
Convex bead: a weld bead with a surface that curves outward, away from the root of the weld.
Corner joint: a junction formed by the edges of two pieces of metal meeting at an angle (usually a right angle, or 90°).
Corrosion: chemical and electrochemical reaction of a metal with its surroundings, causing the metal to deteriorate.
Counterclockwise: rotation to the left, when facing the object being rotated; in the direction opposite the movement of a clock's hands.
Corrosion resistance: the ability of a material to withstand corrosive attack.
Covered electrode: in arc welding, a metal rod with a covering of materials that aid in the welding process.
Cover pass: the final weld pass that forms the weld face.
Cover plates: replaceable pieces of clear glass or plastic, used to protect the more costly filter lenses or plates of welding goggles and helmets.
Crack: a break or separation in rigid material, such as weld metal, that runs in more or less one direction.
Crater: a depression in the weld face. Usually found at the termination of the weld.
Creating a continuous weld pool: creating a pool with molten base metal and advancing it along a line, maintaining a consistent width.
Crescent-shaped motion: a curved side-to-side motion of the welding torch or electrode holder while, at the same time, moving forward. This creates a wider weld bead.
C-shaped weld pool: a crescent-shaped weld pool, which indicates that the base metal surfaces have melted enough to create a well-fused weld.
Cup: see **Gas nozzle**.
Current: the flow of electrons through an electrical circuit.
Current density: a measure of the amount of current passing through a given area at a given time.
Current switch: a switch on a GTAW welding machine that allows the welder to choose whether current is adjusted with a foot pedal or thumb switch while welding.
Cutting machine: a motor-driven device with one or more oxyfuel gas cutting torches mounted on it, used to make complex, long, or multiple cuts.
Cutting oxygen: the stream of oxygen that cuts the metal.
Cutting oxygen lever: a control on an oxyfuel gas cutting torch that releases the cutting oxygen.
Cutting oxygen orifice: the center orifice in the cutting tip of an oxyfuel gas cutting torch.
Cutting outfit: the equipment needed to perform cutting operations.
Cutting station: a work area that contains the cutting outfit, booth, ventilation, and all required supplies to perform welds.
Cutting tip: the part of an oxyfuel gas cutting torch from which gases are released for combustion.
Cutting torch: in oxyfuel gas cutting, the device that controls and directs the fuel gases and oxygen needed for cutting and removing metal.
Cutting torch attachment: a device that converts a welding torch body into a cutting torch, eliminating the need for a separate cutting torch.
CV: see **Constant voltage**.
Cycle: a set of events that repeat in a specific order; in alternating current, one complete reversal of current.
Cylinder: a thick-walled container holding, under high pressure, a supply of gas used for welding.
Cylinder pressure gauge: a gauge on a pressure regulator that shows the pressure of the gas in the cylinder.
Cylinder valve: a device that can be opened or closed to control the flow of gas from a cylinder.

D

Dash number: an AWS-assigned number that indicates how an FCAW electrode is to be used.
Defect: an imperfection that, because of its size, shape, location, or makeup, reduces the useful service of a part.
Degree: one unit of a temperature scale. °F or °C are the most common scales.
Density: the weight of a particular metal per unit volume.
Deoxidizer: a substance added to molten metal to remove either free or combined oxygen.

Deoxidizing: the process of removing oxygen from molten metal with a deoxidizer; in metal finishing, removing oxide films with chemical or electrochemical processes.
Deposition rate: the weight of material applied in a unit of time, usually expressed in lb./hr or kg/hr.
Depth of preparation: the distance from the base metal surface to the bottom of the weld preparation.
Destructive test: a testing method that involves applying stress to a sample until it fails. Destructive testing is used to determine how large a discontinuity can be before it is considered a flaw.
Dewar flask: a pressurized container with an insulated double wall, used to store liquid oxygen.
Dimensions: the size information shown on a mechanical drawing of a part or assembly.
Diode: an electrical device that allows current to flow in only one direction.
Direct current (dc): electric current that flows in only one direction.
Direct current electrode negative (DCEN): the arc welding method in which direct current flows from the electrode (cathode) to the workpiece (anode).
Direct current electrode positive (DCEP): the arc welding method in which direct current flows from the workpiece (cathode) to the electrode (anode).
Direct current reverse polarity (DCRP): see **Direct current electrode positive**.
Direct current straight polarity (DCSP): see **Direct current electrode negative**.
Downhand welding: A nonstandard term. See **flat (1G) welding**.
Downhill weld: a vertical weld made from the top of the joint to the bottom.
Downslope: the downward slope of a volt-ampere graph.
Drafting machine: a device used by drafters to simply and accurately draw parallel, perpendicular, and diagonal lines.
Drag: the offset distance between the actual and theoretical exit points of the cutting oxygen stream, measured on the exit side of the material.
Drag angle: the travel angle used during backhand welding when the welding end of an electrode or a gas flame points opposite the direction of travel.
Drag welding: a welding method in which the heavy covering of a covered electrode is dragged across the surface of the workpiece to maintain a constant arc length.
Drive rolls: the rolls in a wire drive unit that are directly driven by the unit's drive motor.
Droopers: a term for constant current welding machines, also called droop curve machines because of the voltage versus amperage curve they produce.
Ductility: the ability of a material to be changed in shape without cracking or breaking.
Duty cycle: a rating that indicates how long a welding machine can be used at its maximum output current without damaging it. Percentage of time in a 10-minute period that a welding machine can be used at its rated output without overloading.

E

Eddy current inspection: an inspection method in which cracks or other weld defects are indicated by changes in the flow of an induced eddy current, which is displayed on an oscilloscope.
Edge joint: a joint formed when two pieces of metal are lapped with at least one edge of each piece at an edge of the other.
Edge preparation: shaping of the edges of the joint prior to welding.
Edge weld: a weld produced on an edge joint.
Effective throat: the minimum distance, minus any convexity, from the weld face to the weld root.
Elbow (joint): one of several axes (straight lines about which rotation can take place) on a typical robot.
Electric arc spraying: see **Arc spraying**.
Electric motor–driven carriage: a type of automatic cutting machine.
Electrode: terminal point to which electricity is brought to produce the arc for welding. In many electric arc welding processes, the electrode is melted and becomes part of the weld.
Electrode covering: flux materials combined into a thick clay-like mixture and then applied to the electrode wire in a very exact thickness.
Electrode dispenser: a device for temporary storage of SMAW electrodes. The dispenser protects the electrodes and keeps them relatively dry.
Electrode drying oven: equipment in which an adequate supply of SMAW electrodes can be kept under ideal conditions.
Electrode extension: the length of unmelted electrode extending beyond the end of the contact tube in GMAW, FCAW, SMAW, and GTAW. Also called an *electrode stick out*.
Electrode face diameter: the size of the electrode that contacts the weldments.
Electrode holder: device in which an electrode is held for welding.
Electrode identification system: a system of identification numbers for all welding electrodes, developed by the American Welding Society.
Electrode lead: the electrical conductor between the welding machine and the electrode holder.
Electrode stick out. see **Electrode extension**.
Electrode tip: in arc welding processes, the extreme end of the electrode. The arc extends from the electrode tip to the base metal. Proper shaping of the tip is especially important in GTAW.
Electrode tip diameter: dimension that determines the size of the spot weld in resistance spot welding.
Electrogas welding (EGW): an arc welding process that uses one or more GMAW or FCAW torches, shielding gas, and a set of water-cooled shoes that move up the joint to contain the molten weld metal as the weld progresses.
Electron: one of the fundamental parts of an atom. It carries a small negative electrical charge.
Electron beam cutting (EBC): a cutting process in which a focused stream of electrons is used to cut metals.
Electron beam welding (EBW): a welding process in which a focused stream of electrons heats and fuses the material being welded.
Electronic pattern tracer: an electronic eye device used to follow a pattern (usually black on white) of a part to be cut. May be used to direct one or more cutting torches.
Electroslag welding (ESW): a welding process in which one or more arcs are established between continuously fed metal electrodes and the base metal. As the weld progresses vertically, moving water-cooled shoes contain the molten metal, flux, and slag.

Element: a chemical substance that cannot be broken down into simpler substances by chemical action.
Elongation: the percentage increase in a specimen's length when that specimen is stressed to its yield strength.
Equal pressure torch: see **Positive pressure torch**.
Exothermic cutting: a continuous cutting process that uses no fuel gas or electricity but rather burning fuel rods and oxygen and can cut through both metals and nonmetals.
Expansion: the increase in the dimensions of a piece of metal as it is heated.
Expulsion weld: a resistance spot weld that squirts (expels) molten metal from the weld area.

F

Face: see **Weld face**.
Face bend: a test performed by bending a weld sample with the weld face on the outside of the bend.
Face reinforcement: the distance from the top of the weld face to the surface of the base metal.
Fatigue: a condition in which repeated stress below the tensile strength of a material (usually a metal) will cause the material to crack.
Feed rate: the length of electrode passing through the welding gun in a unit of time.
Field weld symbol: a symbol added to the basic AWS welding symbol to indicate that a weld is to be made at the job site ("in the field"), rather than in a fabricating shop.
Filler metal: metal or alloy to be added to the base metal to make welded, brazed, or soldered joints.
Filler pass: an intermediate pass used to fill the joint in a multiple pass weld.
Filler rod: see **Welding rod**.
Fillet test: a test used to evaluate fillet welds. The vertical piece of the weldment is bent until it cracks or touches the horizontal piece.
Fillet weld: an inside corner weld made at the intersection of two surfaces that form approximately a right (90°) angle.
Fillet weld size: the length of the legs of the triangle inscribed into the cross section of a completed fillet weld.
Filter lenses or plates: lenses in welding goggles or plates in helmets with optical properties that protect the welder's eyes from infrared, ultraviolet, and visible radiation.
Finish symbol: a letter or number that identifies the method used to finish the weld surface.
Fire watch: a safety practice in which a person is designated to observe welding operations to prevent fires started by stray sparks or heat from welding activities.
Fitting: threaded connectors on the ends of fuel gas hoses, used to join the hoses to regulators and torches.
Fixture: a device used to hold a weldment in proper alignment for welding.
Flame cutting: see **Oxyacetylene gas cutting**.
Flame spraying (FLSP): a thermal metal spraying process in which an oxyfuel gas flame is the heat source for melting the coating material. Compressed gas may be used to propel the coating material onto the base material.
Flange joint: a joint formed when the edge of one or more pieces of the base metal are bent and form a flange(s).
Flare-groove joint: a joint formed when the flanged edges of one or both pieces of base metal are placed together to form a single-flare-bevel or double-flare-V-groove.
Flash: the harmful effect of arc rays on the human eye. Also, the surplus metal thrown out at the seam of a resistance weld.
Flashback: an extremely dangerous occurrence where the torch flame moves into or beyond the mixing chamber of the torch.
Flashback arrestor: a device usually installed between the torch and welding hose to prevent the flow of a burning fuel gas and oxygen mixture from the torch into the hoses, regulators, and cylinders.
Flash goggles: goggles worn under their helmets by arc welders to protect their eyes from flashes from the rear.
Flat bead: a relatively flat bead contour used when the bead is to be ground or machined.
Flat (1G) welding position: welding done with the weld axis and base metal surface nearly horizontal.
Flaw: a discontinuity in a weld that is within the accepted limit.
Flowmeter: a device that controls the amount of gas that goes to the welding torch.
Fluorescent penetrant: a penetrating fluid containing a dye that gives off or reflects light when exposed to short wave radiation.
Flux: material used to prevent, dissolve, or facilitate removal of oxides and other undesirable surface substances.
Flux cored arc welding (FCAW): a welding method in which heat is supplied by an arc between the base metal and a hollow, flux-filled electrode.
Flux cored electrode: a hollow metal electrode filled with a flux material, usually used in FCAW or ESW.
Flux covering: see **Electrode covering**.
Foot pedal: a foot-operated rheostat used for remote control of the output of an arc welding machine.
Force gauge: a gauge used to measure applied force, such as the pressure being exerted by electrodes in resistance spot welding.
Forehand welding: a method used when the welding end of an electrode or a gas flame is pointed in the direction of travel.
Forge welding (FOW): a weld made by heating the parts in a forge, then applying pressure or striking blows with enough force to make them fuse together.
Forward arc blow: arc blow that occurs at the beginning of the joint as the magnetic field tries to stay in the metal. This causes the filler metal to blow toward the center of the joint.
Free bend test: bending a test specimen, usually in a vise, without using a fixture or guide.
Freeze: refers to a welding rod or electrode that becomes stuck in the weld pool.
Frequency: a term used to describe how many cycles per second there are in ac power.
Friction stir welding (FSW): a process similar to friction welding that employs a rotating tool rather than a rotating part to create the heat and mix the base metals into a weld bead.
Friction welding (FRW): a welding process in which frictional heat is created by revolving one part against another under very heavy pressure.
Fuel gases: gases that support combustion, such as acetylene, propane, and butane.
Full anneal: application of the heat-treating (annealing) process to full thickness of the workpiece.
Fusible plug: a steel plug used to prevent acetylene cylinders from exploding in a fire. The plug is filled with a metal that melts at approximately 212°F (100°C).

Fusion: the intimate mixing or combining of molten metals.
Fusion welding: any type of welding that uses fusion as part of the process.

G

Galvanized metal: metal coated with zinc to prevent rust. The zinc coating gives off toxic fumes when heated excessively.
Gas-cooled torch: a GTAW torch used for light-duty work. These torches are not recommended for use with over 200A.
Gas lens: a series of fine stainless steel wire screens that makes the shielding gas exit the nozzle in a column. This allows the electrode to stick out farther, so the welder can see the weld better.
Gas metal arc welding (GMAW): an arc welding process that uses a continuously fed consumable electrode and a shielding gas. Sometimes called *MIG welding*.
Gas nozzle: a ceramic or metal device on a torch or gun used to direct the shielding gas to the weld area.
Gas tungsten arc welding (GTAW): an arc welding process that uses a tungsten electrode and a shielding gas. The filler metal is added using a welding rod.
Gauntlet gloves: gloves with cuffs that extend above the wrist for protection.
Generator: a device that produces electricity; also, a device that produces acetylene.
Globular transfer: in GMAW, the movement of molten metal in large droplets from the consumable electrode across the arc.
Globule: a large droplet of molten metal.
GMAW. see **Gas metal arc welding**.
Gouging: the process of cutting a groove in the surface of a piece of metal, using an oxyfuel gas or arc cutting outfit.
Grain size: an important factor in determining the mechanical properties of a material. Coarse-grained materials are brittle and have low ductility; fine-grained materials are less brittle and are more ductile.
Grit blasting: a mechanical cleaning method in which abrasive particles are propelled at high velocity against the metal surface.
Groove angle: the total angle formed between the groove face on one piece and the groove face on the other piece.
Groove face: the surface formed on the edge of the base metal after it has been machined, bent, or flame cut.
Groove joint: a joint formed when there is a designed space in the form of an angled or shaped groove between the pieces being joined.
Groove weld: a weld made by fusing filler material into a joint that has had base metal removed to form a V, U, or J profile at the edges to be joined.
Groove weld size: Depth to which the weld reaches from the base metal surface.
GTAW. see **Gas tungsten arc welding**.
Guided bend test: a test in which a specimen is bent in a specific way, using a specially designed fixture.

H

Hand truck: a two-wheel cart with a means of securing cylinders to it, such as chains or straps.
Hardfacing: a surfacing process in which hard materials are applied to the surface of a part to reduce wear or loss of materials by impact, abrasion, or corrosion.
Hardness: the ability of a metal to resist plastic deformation; the same term may refer to stiffness or temper, resistance to abrading or scratching.
Hardness test: a measurement of hardness made by forcing a steel ball or pointed diamond under a given force into the surface of a material.
Hard slag: slag that cannot be easily removed.
Heat: molecular energy of motion.
Heat-affected zone (HAZ): the area of the base metal around the joint that has been changed (mechanically or in microstructure) by welding, brazing, or soldering.
Heliarc welding: a welding process that uses two tungsten electrodes to create an arc. Sometimes used incorrectly to refer to GTAW.
Helium (He): an inert, colorless gas used as a shielding gas in welding.
Helmet: a protective hood that fits over the welder's head and includes a filter plate through which the welder can safely observe the electric arc.
Hertz: a unit of frequency equal to one cycle per second.
Hold time: the time that force is maintained on resistance welding electrodes after the current is turned off.
Horizontal: in a plane parallel to the ground.
Horizontal (2G) welding position: a welding position in which the weld axis is nearly horizontal and the surface is nearly vertical.
Horn spacing: the space between the parts of the resistance welding machine that hold the electrodes, measured when the electrodes are touching.
Hose: a flexible rubber tube used to convey gases from a pressure regulator to a welding torch.
Hot hardness: the ability of a metal to retain its strength at high temperatures.
Hot pass: the second pass in a multiple-pass welding joint.
Hot shortness: a weakness of the metal that occurs in the hot forming range (above the recrystallization temperature).
Hydraulic: a term that refers to a device actuated by force transmitted and/or multiplied through a fluid.
Hydrogen (H_2): a chemical element used as a fuel gas in oxyfuel gas welding.

I

Ignition temperature: the temperature at which a material will burn if enough oxygen is present.
Impact strength: the ability of a material to withstand impact or hammering forces without cracking or breaking.
Impact test: a test that carefully measures how materials behave under heavy loading, such as bending, tension, or torsion. Charpy or Izod tests, for example, measure energy absorbed when breaking a specimen.
Impurities: undesirable elements or compounds in a material.
Inch switch: a switch on a welding machine that is used to slowly feed consumable electrode wire through a combination cable to a GTAW torch.
Inclusions: foreign matter introduced to, and remaining in, welds or castings.
Incomplete fusion: failure of weld metal to fuse completely with base metal or the preceding weld bead.
Indentation: a depression left on the surface of base metal after a spot, seam, or projection weld is made.
Indenter: in a hardness test, the ball or diamond that is pressed into the surface being tested.

Inductance: in the presence of a varying current in a circuit, the magnetic field surrounding the conductor generates an electromagnetic force in the circuit itself. If a second circuit is adjacent to the first, the changing magnetic field will cause *(induce)* voltage in the second circuit. An important application of this principle is the step-down transformer used in welding machines.

Inert gas: shielding gases that do not react with the weld, such as argon and helium.

Infrared rays: heat rays produced by either an arc or welding flame.

Injector-type torch: a torch used with a low-pressure acetylene generator. Acetylene is drawn in through the injector and mixed with oxygen in the mixing chamber. The mixed gases flow to the torch tip, where they are burned.

Inorganic: composed of material that was never living; mineral, as compared to plant or animal.

Inorganic fluxes: soldering fluxes that do not contain carbon. They are very corrosive, so they are not used on electrical or electronic parts.

Input power: the electric power required (220V or 440V, for example) to operate a given welding machine.

Inside corner joint: a joint made by welding along the inside of the intersection (usually a 90° angle) of two pieces of base metal.

Inside diameter (ID): the interior size of a pipe or tube, measured at its widest point.

Inspection: the process of examining welds for suitability without damaging or destroying them.

Insulation: a material that will not permit the flow of electricity, used as a covering on wires, cables, and electrode holders.

Intermittent welding: the process of joining two pieces with welds that are not continuous; leaving gaps between welds.

Interpass heating: heating or reheating a joint between the weld passes needed to complete the weld.

Inverter: a type of welding machine used for GMAW.

Iron oxide: a compound of iron and oxygen with a higher melting point than pure iron.

J

Jack: a socket on the front of a welding machine that accepts a connector on the end of the electrode or work lead.

Jig: a fixture or template used to accurately position and hold a part during welding or machining.

Joint geometry: the shape and dimensions of a (weld) joint, in cross section, prior to welding.

Joint penetration: the depth that a weld extends into the joint from the surface.

K

Kerf: the slot or opening produced in the metal when cutting.

Keyhole: an enlarged root opening that looks like an old-fashioned keyhole.

Keyhole welding: a welding technique in which concentrated heat penetrates the workpiece, leaving a hole at the leading edge of the weld. As the heat source moves on, the keyhole is continually formed and filled.

Kinetic energy: the energy in a moving object.

L

Lack of fusion: a weld defect resulting from failure of the weld metal and base metal to mix (fuse) completely.

Lanthana: a metal (lanthanum oxide) used in GTAW electrodes.

Lap joint: a joint formed by overlapping the edges of two pieces of base metal.

Laser: a device that emits a beam of coherent light.

Laser beam cutting (LBC): a process that uses the energy of a laser beam to cut material.

Laser beam welding (LBW): a process that uses the energy of a laser beam to fuse materials.

Layer: one level of a thick welding joint made up of one or more weld passes.

Laying a bead: forming a line of fused weld metal along a line or a joint of the weldment.

Lead: a wire that carries electricity from a power source to the electrode or ground clamp; also, a metallic element (symbol: Pb) that is a major part of some solders.

Leading edge: the forward edge of the weld pool.

Leathers: protective clothing worn by a welder, especially when welding out of position.

Leg: the shortest distance from the weld toe of a fillet to the point where the pieces of base metal touch.

Lens: see **Filter lenses or plates**.

Light curtain: an electronic safety device that will turn off power to machinery when the beam of light that the device creates is broken.

Liner: a flexible tube placed inside the combination cable through which a consumable wire electrode passes on its way to the GMAW torch.

Liquefaction: a process that liquefies air and then separates the gases in it at their various boiling points.

Liquid penetrant inspection: a nondestructive method that uses a special liquid and dye to test welds for flaws.

Liquidus: the temperature at which the filler metal or solder becomes completely liquid.

Low hydrogen electrodes: electrodes used for SMAW that have little or no hydrogen in them.

Lugs: heavy-duty electrical terminals that are cylindrical at one end (to accept the lead wires) and flat on the other end. The flat end has a hole that allows the lug to be fastened to a stud or bolted to a surface.

M

Machinability: the ability of a part to be machined or ground to size.

Magnetic field: a field created around a wire or electrode whenever electricity travels through that wire or electrode.

Magnetic particle inspection: a procedure for using a liquid that contains magnetic particles to check a weld for flaws. The particles are drawn into the flaws when a magnetic field is applied to the weldment.

Mandrel: a solid, pointed shaft that is forced through a nearly molten metal rod to form seamless tubing.

Manifold: an assembly of pipes that delivers gas from several cylinders into a single pipe for distribution to several workstations.

Manual welding: any welding method in which time, distance, speed, and other variables are controlled by the person making the weld.

MAPP: a trade name for stabilized methylacetylene-propadiene fuel gas.
Mechanical drawings: parts or assembly drawings made by mechanical means, either traditional board drafting or CAD (computer-aided drafting).
Mechanical properties: descriptions of a material's behavior when force is applied. The mechanical properties determine that material's suitability for a given use. Examples: tensile strength, hardness, modulus of elasticity, elongation, fatigue limit.
Melt-through symbol: a symbol used on the welding symbol to indicate 100% penetration on a weld made from one side of the base metal.
Metal inert gas (MIG): a term often used in the trade to describe gas metal arc welding.
Metal inert gas welding (MIG): see **Gas metal arc welding (GMAW)**.
Metal-to-metal wear resistance: the ability of a material to resist wear from metal-to-metal contact.
Metal transfer: movement of metal from one surface to another (as in *metal transfer wear).*
Metal transfer wear: wear that occurs when metal leaves one surface and fuses or sticks to another surface.
Metric system: see **SI metric system**.
Microprocessor: a silicon chip that contains computer logic circuits and other components used to control a digital device.
MIG: see **Gas metal arc welding.**
Mill file: a tool used to remove weld metal buildup from an oxyfuel gas welding torch tip.
Mixing chamber: the part of the welding torch in which gases come together and are mixed before combustion.
Motor-driven beam-mounted torch: a type of automatic cutting machine that uses an electronic tracer to follow the edge of the pattern. The tracer may control one or more gas cutting torches mounted on the beam.
Motor-driven magnetic tracer: a relatively inexpensive automatic cutting machine that works best with shapes that are not too complex. The magnetic tracer follows a steel pattern. One torch is generally carried on this type of cutting machine.
MPS: methylacetylene-propadiene (stabilized) fuel gas.
Multiple-pass weld: a welding joint requiring more than one weld pass.

N

Natural gas: a naturally occurring mixture of hydrocarbon gases used as a fuel gas in some applications.
Nd:YAG: neodymium-doped yttrium aluminum garnet; a material used to make one type of laser.
Negative pressure air-purifying respirator: a type of safety equipment worn on the face or over the head of a welder; used to remove harmful fumes and particles by drawing contaminated air through a filter or cartridge in the respirator.
Neutral flame: a flame resulting from combining oxygen and a fuel gas in perfect proportions.
Newton (N): the unit of force in the SI metric system.
Nonconsumable electrode: an electrode that does not melt and become part of the weld.
Nondestructive examination (NDE): a means of testing for defects that does not damage or destroy the weld.
Nonpetroleum-based: made from materials that do not contain petroleum or petroleum products.
Nontransferred arc: in plasma arc welding or spraying, an arc established between the electrode and the restricted nozzle. The workpiece is not part of the circuit.
Normalizing: a heat treating process used to eliminate internal stresses and create uniform grain size. Cooling is more rapid than in the annealing process.
Nozzle: see **Gas nozzle**.

O

Occupation: a person's career or job. The status of an occupation usually depends on the amount of education a person obtains and his or her experience in the chosen field.
Off time: in resistance welding, the time between repeating cycles, when the electrodes are off the work.
Ohm: the unit of measurement for resistance to the flow of electricity through a circuit.
Ohmmeter: an instrument for measuring electrical resistance.
Open arc: a visible arc.
Open circuit voltage (OCV): the voltage in the welding machine circuit when it is on, but the arc is not struck.
Operator controls: in robotics, the controls used by an operator to start a programmed operation.
Organic: having carbon in its chemical composition.
Organic fluxes: fluxes that contain carbon. They are corrosive during the soldering operation, but become noncorrosive when the soldering is completed and can be washed away with water.
Orifice: a precisely bored hole through which gases flow. May be in a regulator, torch, or torch tip.
Orifice gas: the gas that surrounds the electrode in plasma arc welding and cutting. It becomes ionized in the arc to form the plasma.
Orthographic projection: a method of developing a mechanical drawing, based on viewing an object as if it were in a transparent cube.
Oscillating: moving back and forth in an uninterrupted manner.
Oscilloscope: an instrument that shows electrical impulses on a calibrated screen.
OSHA: the Occupational Safety and Health Administration, the federal agency responsible for safety in the workplace.
Other side: the area of the welding symbol above the horizontal reference line. Also, the side of the welding joint opposite the side touched by the welding symbol arrow.
Out-of-position welding: making welds in the vertical, horizontal, or overhead welding positions.
Outside corner joint: a joint made by welding along the outside of the intersection (usually a 90° angle) of two pieces of base metal.
Outside diameter (OD): the exterior width of a pipe or tube, measured at its widest point.
Overhead (4G) welding position: a weld made on the underside of a joint, with the face of the weld in an approximately horizontal welding position.
Overheating: applying excessive heat, and thus damaging the properties of the metal. When the original properties cannot be restored, the damage is referred to as *burning*.

Overlap: a condition in which the weld pool flows onto the base metal surface and is not fused into the base metal.

Oxidation: the process in which oxygen combines with a material to form a chemical compound called an oxide.

Oxidation resistance: the ability of a material to resist the formation of oxides. Metal oxides occur when oxygen is combined with metal.

Oxidized: combined with oxygen.

Oxidizing flame: the flame produced by an excess of oxygen in the torch mixture. The excess oxygen tends to burn the molten metal.

Oxyfuel gas cutting (OFC): a process that uses an oxygen and fuel gas flame for heat and a jet of oxygen to oxidize and cut the molten metal.

Oxyfuel gas cutting (OFC) underwater: a process used for ship repair and similar applications. Oxygen and a fuel gas are burned to provide the necessary heat for cutting.

Oxyfuel gas welding (OFW): a group of welding and cutting processes and methods that use heat produced by a gas flame.

Oxygen (O_2): a colorless, odorless gas making up about one-fifth of the earth's atmosphere. Used in OFW and OFC to support combustion.

Oxygen arc cutting (AOC): a process that uses the arc from a hollow electrode to heat the metal, and pressurized oxygen flowing through the electrode to make the cut.

Oxygen cylinder: a specially built container, manufactured to Interstate Commerce Commission standards, that is used to ship and store quantities of oxygen.

P

Parallel: term describing the relationship of two lines or surfaces that, if extended, will never touch.

Parts positioner: a device that holds the weldment and rotates or otherwise positions it for welding operations. For robotic welding, the positioner may be positioned by a controller and computer program.

Parts positioner controller: an electronic device with a memory capable of executing a program that directs the parts positioner where and when to move.

Pascal: the SI metric unit of pressure; often expressed in terms of kilopascals (kPa), or one thousand pascals.

Pass: see **Weld pass**.

Peak current: during pulsed GMAW, the higher current that is used for welding. The lower *background* current maintains the arc and allows the weld to cool.

Peel test: a destructive test in which a resistance welded lap joint is mechanically separated by peeling one piece away from the other.

Penetrant: a liquid applied to the surface of a weld to locate discontinuities. It enters the cracks and makes them visible.

Penetration: the depth of fusion of the weld below the surface.

Percent heat control: on a resistance welding machine, the control that permits fine adjustment of current within the amperage limits of the tap switch setting.

Petroleum: a natural hydrocarbon compound that is processed into lubricants (grease and oil) and liquid fuels such as gasoline, heating oil, and kerosene.

Petroleum-based: a term used to describe lubricants. Flammable lubricants derived from petroleum must not be used to lubricate any part of a welding or cutting outfit.

Phosgene gas: a highly toxic (poisonous) gas that can be released by the action of heat on cleaning chemicals that contain chlorinated hydrocarbons.

Photons: units of light energy emitted by a laser.

Physical properties: the physical characteristics used to identify or describe a metal, such as color, melting temperature, or density.

Pinch force: in short circuiting GMAW metal transfer, the magnetic force that squeezes off the droplet of the molten electrode metal.

Pipe: a hollow cylinder with a relatively thick wall; pipe size is identified by its inside diameter and schedule number.

Pipe schedule number: a one-, two-, or three-digit number that classifies pipe. Schedule numbers are determined by a combination of the pipe's inside diameter and its wall thickness.

Plasma: a temporary physical state assumed by a gas after it has been exposed to, and reacted to, an extremely high temperature.

Plasma arc cutting (PAC): a process using an electric arc and fast-flowing ionized gases to cut metal.

Plasma arc welding (PAW): a process in which an electric arc ionizes a gas, creating a plasma that generates the heat for welding.

Plasma spraying (PSP): a thermal spraying process that uses a nontransferred plasma arc to melt and propel a coating material onto a base material.

Plastic: soft and easily shaped; a metal that is in a plastic state is almost molten.

Plastics: synthetically produced nonmetallic compounds that can be shaped or molded into desired form.

Plastic state: in welding, the almost molten state of heated metal pieces being joined.

Plastic welding: a process in which heated air is used to soften and fuse plastic materials.

Plug weld: a weld made through a hole in one piece of metal that is lapped over another piece.

Polarity: the direction of the flow of electrons in a closed direct current welding circuit. When the electrons flow from the electrode to the base metal, the polarity is DCEN (direct current electrode negative), or DCSP. When the current flows from the base metal to the electrode, the polarity is DCEP (direct current electrode positive), or DCRP.

Polyethylene: one of the most extensively used thermoplastic resins.

Polypropylene: a hard, tough thermoplastic used for molded articles and fibers (especially rope).

Polyurethane: a plastic resin used to form tough surface coatings; also used as a casting or potting resin to protect electronic components.

Polyvinyl chloride (PVC): a vinyl resin widely used to make pipes and other rigid molded products.

Porosity: the presence of gas pockets or voids in weld metal.

Positioner: see **Parts positioner**.

Positive-pressure respirator: a type of safety equipment that incorporates full-face headgear to deliver a constant flow of clean breathing air to the welder.

Positive pressure torch: a torch in which the working pressure of the oxygen and fuel gas is high enough to force the gases into the mixing chamber.

Post flow: shielding gas flow that continues for a short time after the weld current stops.

Post flow adjustment: a control on a welding machine that allows variation in the length of time that post flow continues.

Pounds per square inch (psi): the unit of measurement for pressure in the US conventional system.

Preheating: a process that heats the metal to a specified temperature prior to a surfacing or welding operation. Also, the process of heating base metal to its kindling temperature before cutting.

Preheating orifices: openings in a cutting torch tip through which oxygen and fuel gas for the preheating flame are supplied.

Prequalified: a term describing a welding procedure that has been tested and established.

Pressure: the force exerted on a given area, expressed in psi or pascals.

Pressure regulator: a device used to reduce the cylinder pressure of a gas to a usable (working) pressure for welding.

Pressure roll: in gas metal arc welding, the drive roll on the wire feeder that applies pressure to electrode wire.

Pressure test: a procedure used to determine the maximum pressure that a cylinder or tank can hold. The test consists of forcing a gas or fluid into the cylinder or tank under increasing pressure until the vessel fails.

Primary circuit: in a transformer, the circuit and windings connected to the input power supplied by the electrical utility.

Primary current: alternating current supplied to the input side of a transformer.

Procedure qualification record: a document containing the actual welding variables used to produce an acceptable weld. The procedure qualification record is used to qualify a welding procedure specification.

Process: an operation used to produce a product.

Prods: a pair of hand-held electrodes used to magnetize a part for magnetic particle inspection.

Program: a series of step-by-step directions and process parameters (times, pressures) set on the automatic controls of a welding machine or placed in the memory of computer/controller.

Projection: a bump stamped into one piece to be resistance welded. A weld nugget forms at the point where the projection touches the surface of the second part.

Projection welding (PW): a resistance spot welding process in which current is concentrated and welds are made at the points where projections on one piece contact the adjacent piece.

Prototype parts: the first models of parts that later may be mass-produced.

psi: see **Pounds per square inch**.

Pull gun: a GMAW welding gun that pulls the electrode wire through the cable.

Pulsed current: similar to direct current except that it has two current levels, a high current period followed by a low current period.

Pulsed spray transfer: a GMAW process in which the current is pulsed to take advantage of the spray mode of metal transfer, but with current values below the spray transition current.

Pulses: regular alternations of level or intensity, as in pulsed current.

Purge switch: a switch on a welding machine that manually controls the flow of shielding gas.

Purging: the process of passing the correct gas through a regulator, torch, and hose to clean out any air or undesirable gas that may be in the system.

Push angle: the travel angle used during forehand welding when the welding end of an electrode or a gas flame points in the direction of travel.

Push-pull system: a system formed when a GMAW pull gun pulls the electrode wire through the cable and the wire feeder pushes the wire.

Q

Qualified: see **Welding performance qualification record**.

Quenching: rapid cooling of metal in a heat-treating process as a means of hardening it.

Quick-connect terminal: a heavy-duty electrical terminal that easily connects and disconnects from a welding machine.

R

Radiation: see **Thermal radiation**.

Radiograph: an image on film produced by X rays.

Range switch: the coarse adjustment lever on a welding machine.

Rated output current: the maximum current flow a welding machine can produce.

Reading the bead: the process of visually inspecting the weld bead to determine whether the weld was made properly.

Rectified current: alternating current that has been made to flow in only one direction through use of a rectifier.

Rectifier: an electronic device, such as a diode, that acts like a one-way valve as current flows through it. It converts ac (alternating current) to dc (direct current).

Reducing flame: an oxyfuel gas flame with a slight excess of fuel gas.

Reducing gas: a gas that removes reactive oxygen from the atmosphere by combining with the oxygen.

Reference line: the horizontal line drawn on a welding symbol. All information about the weld is positioned above or below this line.

Regulator: a device used to control the volume and pressure of a welding or shielding gas as it flows from the cylinder to the torch.

Regulator adjusting screw: a screw that controls the working pressure of gas delivered by the regulator.

Remote contactor control: a receptacle on the welding machine where the wire feed control mechanism is plugged in.

Repeatability: term used to describe a robot's ability to make welds in the same place and of the same quality on part after part in production.

Residual stress: stress that remains in a body, such as a weldment, after the external forces or thermal gradients have been removed.

Resistance: the property of a material that causes the flow of current in a circuit to be retarded.

Resistance seam welding (RSEW): a process that usually uses a rotating wheel electrode to make a continuous seam. A seam can also be formed with overlapping spot welds.

Resistance spot welding (RSW): welding overlapping pieces of metal together in small spots between two electrodes.
Resistance welding: a process that uses the resistance of metals to electrical flow as a source of heat for welding.
Restarting the arc: establishing a new arc after changing electrodes, or after losing the arc due to faulty welding technique.
Restricted nozzle: in plasma arc welding, the nozzle through which orifice gas flows to form the plasma.
Reverse polarity: see **Direct current electrode positive**.
Robot: a device that uses a computer program to direct its movements as it completes a series of welds or other operations.
Robot cell: a three-dimensional space enclosing the robot and its working area, usually fenced off for safety of employees.
Robot controller: the computer and associated devices that direct a robot in carrying out the program developed for it.
Robotic welding system: an industrial production operation using one or more robots, controllers, positioners, and other equipment to make a programmed series of welds on a weldment.
Robotic welding workstation: a location where one or more robotic welders make a specific set of welds.
Rockwell C hardness test: generally used for hard materials, this test uses a special piece of equipment to force a pointed diamond into the metal surface.
Root: See **Weld root**.
Root bend: a test performed by bending a weld sample with the root of the weld on the outside of the bend.
Root face: the distance from the root of the joint to the point where the bevel angle begins.
Root opening: the distance between the two pieces at the weld root.
Root pass: the first weld pass made into the root of the joint.
Root penetration: the depth to which weld metal extends into the root of a welding joint.
Root reinforcement: the distance that the penetration projects from the root side of the joint.
Rosin: a substance used as a welding flux. It is derived from the sap of pine trees.
Rosin fluxes: these fluxes are noncorrosive and are recommended for soldering electrical and electronic parts.
Running a bead: the process of making a weld bead.
RWMA: the Resistance Welding Manufacturing Alliance.

S

Safety cap: a forged steel cap that should be screwed over the cylinder valve to protect it when the cylinder is stored or moved.
Safety valve: a device that prevents a gas cylinder from exploding when exposed to high temperatures. The valve includes a disk that ruptures under increased pressure.
Sagging: the sinking or downward curving of metal (as in a weld pool) due to gravity or pressure.
Seamless tubing: tubing formed on a mandrel so that it has no welded seams.
Seam weld: a series of resistance spot welds on a lap joint that forms a seam. Seam welds can be a series of continuous overlapping spot welds or a series of intermittent spot welds. Other processes can also make a seam weld by welding through one workpiece.
Secondary circuit: in a welding machine transformer, the electrical circuit and winding connected to the electrodes.
Secondary current: the current coming out of a transformer.
Self-shielding electrode: an electrode that produces its own shielding gas and does not require additional shielding gas.
Semiautomatic welding: manual welding with one or more of the welding variables controlled by automatic devices.
Sequence: the order in which operations or events take place.
Servomotor: a small electric motor capable of precisely controlled movement.
Shielded metal arc welding (SMAW): a welding process in which the base metal is heated to fusion temperature by an electric arc created between a covered metal electrode and the base metal.
Shielding gas: a gas, usually inert, that is used to blanket the weld area and prevent contamination from the air.
Short arc: see **Short circuiting transfer**.
Short circuit: in the short circuiting transfer process, the condition that occurs when the electrode touches the base metal, causing metal from the electrode to enter the weld.
Short circuiting transfer: a gas metal arc process that uses relatively low voltage. The arc is constantly interrupted and restarted as the molten electrode shorts out against the base metal.
Shoulder (joint): one of several axes (straight lines about which rotation can take place) on a typical robot.
Side bend: a test performed by bending a weld sample to the side toward the thickness of the metal. The side bend is used on thick material.
Silicon-controlled rectifier: an electrical device that allows current to flow in only one direction.
SI metric system: a measuring system adopted in most countries. It uses such units as the millimeter, kilogram, liter, newton, and pascal.
Single-stage regulator: a regulator that reduces cylinder pressure to working pressure in one stage (step).
Sketches: simple drawings of parts. Sketches are made without drafting tools.
Slag: the hard, brittle metal that covers a finished shielded metal arc, flux cored arc, and submerged arc weld beads; metal oxides and other materials that form on the underside of a flame or arc cut.
Slag inclusions: see **Inclusions**.
Slope: the downward curve of the volt-ampere diagram for an arc welding machine.
Slope adjustment: changing the slope of the volt-amp curve by decreasing the maximum short circuit current. This reduces spatter during short circuiting transfer.
Slot weld: a weld similar to a plug weld, but made through a hole that is not round.
SMAW: see **Shielded metal arc welding**.
Socket connection: a preformed fitting used to connect lengths of tubing or pipe. The ID of the connector fits the OD of the pipe or tube with a small clearance.
Solder: filler metal used in soldering.

Soldering: a group of welding processes that join materials by heating them to the soldering temperature and by using a filler metal having a liquidus below 840°F (450°C) and below the solidus of the base metals.
Soldering alloy: combinations of various metals such as aluminum, antimony, cadmium, copper, indium, lead, nickel, silver, tin, or zinc.
Solidify: to become solid or hard.
Solid-state welding (SSW): a group of welding processes that weld metals at temperatures below the melting point of the base metal, without the addition of filler metal. Examples are friction, explosion, and ultrasonic welding.
Solidus: the temperature at which a metal, alloy, or solder begins to melt.
Sonotrode: the rapidly vibrating "sound electrode" that causes materials to bond together in an ultrasonic weld.
Spark lighter: a device that creates a spark to ignite an oxyfuel gas torch flame.
Spatter: the scattering of molten metal droplets over the surface near an arc weld.
Specification: see **Welding procedure specification**.
Speed welding tip: a tip for a welding torch that allows one-handed welding of plastics.
Spool: the drum, mounted on the wire drive mechanism, that contains the electrode wire for gas metal arc welding.
Spot weld: see **Resistance spot welding**.
Spot weld symbol: a small circle drawn on, above, or below the horizontal reference line of the welding symbol to show from which side of the weldment the spot weld is to be made.
Spray transfer: a gas metal arc process which has an arc voltage high enough to continuously transfer the electrode metal across the arc in small globules.
Squeeze time: in resistance welding, the time, measured in cycles, that the electrodes are under a force to clamp the parts and ensure good electrical flow.
Staggered intermittent welds: intermittent welds, made on both sides of a joint, that are offset from each other.
Stainless steel: an alloy steel containing chromium. It resists corrosion and oxidation (rusting).
Step-down transformer: an electrical device used in welding machines that reduces voltage and increases amperage in its secondary circuit.
Straight polarity: see **Direct current electrode negative**.
Strain: the reaction of an object to stress.
Strength: the ability of a material to withstand applied loads without failure.
Stress: the load imposed on an object.
Stress relieving: a heat treating process that involves heating evenly to a temperature below the material's critical temperature, followed by a slow, even cooling.
Striking an arc: the act of touching the electrode to the workpiece and then withdrawing it a distance sufficient to maintain the electrical flow across the gap.
Stringer bead: a weld bead made by moving a torch or electrode holder along the weld without any side-to-side motion.
Stud arc welding (SW): an arc welding process in which studs, nails, or other fasteners are used as the electrode while they are welded to a surface. Once the base metal melts, the special stud arc gun presses the fastener into the surface.
Submerged arc welding (SAW): process in which the electric arc is submerged under a heavy layer of flux granules.
Surfacing: the application by welding, brazing, or thermal spraying of a layer(s) of material to a surface to obtain desired properties or dimensions, as opposed to making a joint.

T

T-joint: a joint formed by two pieces of base metal placed at approximately a 90° angle to one another to form at "T" shape.
Tack weld: a small weld used to temporarily hold pieces in alignment.
Tack welding tip: a pointed tip for a plastic welding torch, used to tack weld plastic weldments.
Tail: the "V" shape, drawn at one end of a welding symbol, in which special notes are placed.
Tank: a thin-walled container for liquids or gases. Tank walls are thinner than the walls of cylinders used for pressurized gases.
Tap: one of the several electrical contacts available on a constant current arc welding machine or on a resistance welding machine. Each tap provides a different range of amperage for welding.
Tap settings: the various tap positions available on the resistance welding machine.
Teach pendant: an electronic control used to program a robot to perform a series of actions; the sequence of movements is placed in computer memory as a program.
Teflon® liner: a smooth seamless tubing used inside a cable to make electrode wire feed more smoothly to a GMAW welding gun.
Tempering: a heat treating process in which metal is heated to a temperature just below its melting point.
Tensile shear test: a destructive test that measures the amount of force required to pull a spot weld apart in a direction perpendicular to the weld axis or "in shear."
Tensile strength: the ability of a material to resist pulling forces; measurement of the amount of force required to pull metal apart.
Tensile test: a test in which a specially prepared sample is pulled until it fails. The test determines the weld's ability to withstand forces that would pull it apart.
Terminals: the physical connectors on a welding machine for attaching the workpiece lead and the electrode lead.
Thermal radiation: heat rays given off by a welding arc or oxyfuel gas flame, or by the heated base metal.
Thermal spraying: a process in which a material (metallic or nonmetallic) is heated and sprayed onto a surface.
Thermoplastic: a plastic material that can be formed or reformed by applying heat; damaged articles can be repaired by welding.
Thermosetting: a plastic material that, once formed, cannot be reformed by heating. Damaged articles cannot be repaired by welding.
Thoria: thorium dioxide, a metal used in GTAW electrodes.
Throat depth: the distance measured from the center of the electrodes to the frame of the resistance spot welding machine while the electrodes are closed (in contact).
Throat of a fillet weld: the distance from the weld root to the weld face; also known as the *actual throat*.

Thumb switch: a switch mounted on a GTAW torch to control the amperage, usually operated by the welder's thumb.
TIG: see **Gas tungsten arc welding (GTAW)**.
Tip: the end of the torch where the fuel gas mixture burns, producing a high-temperature flame. Also used to refer to the end of the electrode in the spot welding process.
Tip nut: a large threaded nut used to hold a tip in the cutting torch head.
Toe: See **Weld toe**.
Torch: a device used to control and mix the fuel gas and oxygen and to direct the gas flame to the welding, brazing, cutting, or soldering area. Also, the assembly that holds the electrode in gas metal arc welding and gas tungsten arc welding.
Torch angle: see **Travel angle** and **Work angle**.
Torch body: the main portion of the welding torch, to which fuel gas hoses are connected and the torch tip is attached.
Torch tip: the part of the end of the torch where the fuel gas and oxygen are ignited.
Torch valves: the valves that control the flow of oxygen and fuel gas into the torch.
Toughness: the ability of material to withstand all types of stresses without tearing or breaking.
Toxic: poisonous.
Track: a straight or curved guide path for an electric motor-driven carriage that carries a cutting torch.
Transferred arc: an arc that is established and maintained between the electrode and the workpiece.
Transformer: see **Step-down transformer**.
Transformer-rectifier: one type of constant voltage power supply.
Transition current: the amount of current required to convert from globular transfer to spray transfer.
Travel angle: the angle between a line perpendicular (90°) to the weld axis and the axis of the electrode. This angle is measured in a plane determined by the electrode axis and the weld axis.
Tube: a hollow cylinder similar to a pipe, but generally with thinner walls; the size of a tube is specified by its outside diameter.
Tungsten inclusion: in GTAW, a weld defect caused by getting tungsten from the electrode in the weld.
Tungsten inert gas (TIG): see **Gas tungsten arc welding (GTAW)**.
Two-stage regulator: regulator reduces cylinder pressure to working pressure in two stages (steps).

U

Ultimate tensile strength: the greatest tensile stress that a material can withstand.
Ultrasonic: vibrations generated at frequencies above the range of human hearing.
Ultrasonic inspection: a nondestructive examination method in which ultrasonic waves are passed through the material being inspected. Echo patterns will locate any discontinuities.
Ultrasonic transducer: a device used to generate ultrasonic vibrations that, in turn, vibrate the sonotrode to perform ultrasonic welds.
Ultrasonic welding (USW): a process used to weld metal or other material through use of ultrasonic vibrations.
Ultraviolet rays: light rays with very short wavelengths that are given off by electric arcs and welding flames. Filter lenses in welding goggles and helmets screen out such rays, which can be harmful to the eyes.
Undercut: a depression at the weld toe indicating the weld metal is below the level of the base metal.
Underfill: a condition in which the molten metal sags downward and does not fill the top part of the welding joint.
Under load: in arc welding, term used to describe state of the welding machine when current is flowing through the welding circuit, such as while welding or while the electrode holder is touching the table or workpiece.
Unstable arc: an arc that has not established itself in the ionized space between the electrode and the workpiece. An unstable arc may wander or stop.
Uphill weld: a vertical weld made from the bottom of the joint to the top.
Upset: the term used to describe the enlarged weld where the parts meet in a friction weld.
Upslope: the rising line on a current/time program graph.
US conventional system: measuring system used in the United States that employs pounds, feet, and other traditional units.

V

Vacuum: a condition in which atmospheric pressure in a closed vessel has been decreased. This is done by pumping out air.
Vapor degreaser: a device used to clean metal with chemical vapors before applying surfacing materials.
Vertical: perpendicular to the ground.
Vertical down welding: see **Downhill weld**.
Vertical up welding: see **Uphill weld**.
Vertical (3G) welding position: a weld performed with the weld axis and base metal surfaces nearly vertical.
Visual inspection: determining the quality of a weld by looking at it and comparing it to welds of known high quality.
Volt: a unit of measure of electrical pressure.
Voltage: electrical pressure in a circuit; the force that causes current to flow.
Voltage drop: voltage loss that occurs when electricity travels a long distance from the welding machine.
Voltmeter: an instrument for measuring the voltage in a circuit.

W

Waist (joint): one of several axes (straight lines about which rotation can take place) on a typical robot.
Warped: twisted out of shape; in metals, usually a result of improperly applied heat.
Water-cooled torch: a torch that has a continuous flow of water passing through the torch body to remove heat. This type of torch is used to carry currents over 200A.
Weave bead: a weld bead formed by moving a torch or electrode holder from side-to-side as the weld pass progresses along the welding joint.

Weld: the blending or mixing of two or more metals or nonmetals by heating them until they are molten and flow together.
Weld-all-around symbol: a circle drawn on the welding symbol, indicating that the described weld is to be made all around the part.
Weld axis: an imaginary line running through the center of and parallel to a completed weld.
Weld bead: the shape of the finished joint when fusion welding is done. Usually a filler metal is added during welding to form the weld bead.
Weld contour symbol: a symbol shown just outside the weld symbol that indicates the shape of the finished weld face.
Weld face: the outer surface of a weld on the side from which the weld is made.
Weld groove: a cut, ground, or machined surface on the workpiece, designed to provide space for welding.
Weld joint: the point or line at which two pieces come together and are fastened by welding.
Weld metal: the fused portion of the base metal, or fused base metal and filler metal.
Weld nugget: the weld metal in a resistance spot, seam, or projection weld.
Weld pass: one weld bead along a welding joint.
Weld pool: the small molten volume of metal under the torch flame or electrode, prior to its solidification as weld metal.
Weld reinforcement: weld metal extending beyond the upper and lower surfaces of the base metal.
Weld root: the points where the root surface of the weld intersect the base metal.
Weld schedule: see **Welding sequence**.
Weld time: the time, measured in cycles, when current flows to make a resistance weld.
Weld toe: the point where the weld bead contacts the base metal surface.
Welder: a person who performs welding activities. (Use of "welder" to describe a welding machine is incorrect.)
Welder qualification test record: a record of a welder's performance qualification tests.
Welding: the process of making a weld on a joint.
Welding gun: in GMAW, the device used to hold the consumable electrode.
Welding machine: a device that provides and controls the proper voltage and current for a welding task.
Welding outfit: the welding machine and other equipment required to actually create a weld.
Welding position: the orientation of the weld as defined by the weld axis and weld face. The four welding positions are flat, horizontal, vertical, and overhead positions.
Welding procedure: a method by which a weld is to be made, as outlined in a welding procedure specification.
Welding procedure specification (WPS): a document that lists in detail the specifics of the job: the base metal to be welded, the filler metals to be used, the preheat or post-welding treatment to be used, the metal thickness, and all other variables for each welding process. All items in the specification are identified as essential or nonessential.
Welding rod: welding filler metal, usually packaged in straight lengths. Unlike an electrode, a welding rod is not used to conduct electricity. Also called a *filler rod*.
Welding sequence: the order in which components (parts) of a structure are welded.
Welding station: a work area that contains the fuel gas welding or cutting outfit or welding machine, booth, ventilation, and all required supplies to perform welds.
Welding symbol: see **ANSI/AWS welding symbol**.
Welding torch: a device that controls and mixes fuel gas and oxygen. It is also used to direct the gas flame to the welding, brazing, or soldering work area.
Weldment: an assembly of parts joined by welding.
Wheel electrode: the rotating electrode used in resistance seam welding.
Whip motion: a rapid movement of the electrode or flame away from and back to the weld pool or arc crater. The motion allows time for the weld metal to cool.
Wire brush: a brush with bristles of metal wire, used to clean the weld bead between weld passes.
Wire feeder: in GMAW, the device that continuously feeds consumable electrode wire to the welding gun.
Wire tension control knob: the device used to control the pressure applied to the drive rolls of a wire feeder.
Work angle: the angle between a line perpendicular (90°) to the major workpiece surface and a plane determined by the electrode axis and the weld axis. In a T-joint or a corner joint, the work angle is measured from a line perpendicular to the nonbutting surface.
Work booth: an arc welding work area shielded from the view of workers without eye protection by solid walls, canvas curtains, or filtered, transparent plastic curtains.
Working pressure gauge: a gauge on a pressure regulator that shows pressure of the gas being supplied to the torch.
Working volume: the three-dimensional space in which a robot moves while performing programmed tasks.
Workpiece: the object or assembly being welded.
Workpiece lead: the electrical cable that connects the base metal to the welding machine.
Wrist (joint): one of several axes (straight lines about which rotation can take place) on a typical robot.
Wrought iron: an easily welded or forged iron containing about 0.2% carbon.

X

X ray: a stream of high-energy photons; common term for a photographic image made through the use of X rays.
X-ray inspection: the use of X rays to check a weld for flaws or defects.

Y

Yield strength: the point at which, when a metal is being stretched, it takes a permanent set and will not return to its original dimensions when the stretching force is released.

Z

Zinc (Zn): chemical element sometimes used in soldering alloys. It is a toxic material.
Zirconia: zirconium dioxide, a compound used in some GTAW electrodes for ac welding. It permits easier arc starting than pure tungsten.

Acknowledgments

Air Products and Chemicals
American Torch Tip Company
American Welding Society
Arc Machines, Inc.
Arcsmith
Automation International, Inc.
Bernard, A Division of DovaTech, Ltd.
Broco, Inc.
Bug-O Systems Equipment
Century Manufacturing Company
CK Worldwide, Inc.
Cloos Robotic Welding, Inc.
Controls Corporation of America (CONCOA)
Cronatron Welding Systems, Inc.
E.H. Wachs Company
ESAB Welding and Cutting Products
FANUC Robotics North America, Inc.
Ferranti-Sciaky, Inc.
Fischer Engineering Company
Fusion, Inc.
G.A.L. Gauge Company
Genstar Technologies Co., Inc.
Gullco International Limited
Handy & Harmon PMFG/ Lucas-Milhaupt, Inc.
Harris Welco Division of J.W. Harris Co.
Hobart Welding Products
Imperial Weld Ring Corporation
Instron Corporation
Intercon Enterprises, Inc.
Invincible Airflow Systems
Jackson Products, Inc.
James Morton, Inc.
Kamweld Products Company, Inc.
Koike-Aronson, Inc. of Kalamazoo
Laramy Products Co., Inc.
Lincoln Electric Company, The
LORS Machinery, Inc.
Miller Electric Mfg. Co.
Navy Joining Center
Nederman, Inc.
Nelson Stud Welding
Ohio Nut and Bolt Company, The
Profiler, Inc.
PTR - Precision Technologies, Inc.
Rexarc, Inc.
Seelye, Inc.
Stoody - A Thermadyne Company
Sun-Tec Corporation
Taylor-Winfield Corporation
Thermadyne Holdings, Inc.
Tinius Olsen Testing Machine Co., Inc.
Victor, a division of Thermadyne Industries, Inc.
Wall Colmonoy Corporation
Weldcraft, A Division of DovaTech, LTD.
Wolf Robotics, Inc.

The authors wish to thank the following welding instructors and their students for their cooperation in creating many of the required photographs:

Bob Brandell, Welding Instructor, Van Buren Technical Center
Eric Martin, Welding Instructor, Kalamazoo Valley Community College

The authors would also like to thank the management at the following companies for loaning them various pieces of equipment for photographs:

Airgas and Purity Cylinder Gases, Inc. of Kalamazoo

Most of the photographs in this book, unless otherwise credited, were taken by William A. Bowditch.

Index

D

E

H

I

J

K

L

M

N

O

P

Q

R

S

T

U

V

W

X

Z